W0111381

Birkhäuser

Recent Advances in Nuclear Explosion Monitoring Volume II

Edited by
Martin B. Kalinowski
Andreas Becker

Previously published in *Pure and Applied Geophysics* (PAGEOPH), Volume 171, No. 3–5, 2014

Editors
Martin B. Kalinowski
Provisional Technical Secretariat of the
Preparatory Commission for the Comprehensive
Nuclear-Test-Ban Treaty Organization (CTBTO)
Vienna International Centre
P.O. Box 1200
1400 Vienna
Austria

Andreas Becker
Global Precipitation Climatology Centre
c/o Deutscher Wetterdienst
Frankfurter Straße 135
63067 Offenbach am Main
Germany

ISBN 978-3-0348-0818-7 e-ISBN 978-3-0348-0819-4
DOI 10.1007/978-3-0348-0819-4

Library of Congress Control Number: 2014935373

Cover illustration: Based on Fig. 5 from "The Applicability of Incoherent Array Processing to IMS Seismic Arrays" by S.J. Gibbons.

Cover design: deblik, Berlin.

Printed on acid-free paper

Springer Basel is part of Springer Science+Business Media

www.birkhauser-science.com

Contents

Pure Appl. Geophys. 171 (2014), 337–339

DOI 10.1007/s00024-013-0724-4

Pure and Applied Geophysics

Introduction

MARTIN B. KALINOWSKI[1] and ANDREAS BECKER[2]

Abstract—This is the second topical volume on "Recent Advances in Nuclear Explosion Monitoring" that started with Pure and Applied Geophysics Topical Volume 167 (2010), and again reports on the most recent advances in science and technology that have been achieved to monitor compliance with the Comprehensive Nuclear-Test-Ban Treaty (CTBT). This progress in the development and testing of new sensor technologies and analysis methodologies in all relevant scientific disciplines improves the capabilities in detection, location and characterization of CTBT-relevant events. In particular, the latter poses a challenge for smaller events, where natural or manmade but CTBT-irrelevant sources can generate false-positive events. The efficient discrimination of these events pursued at a minimum risk of missing a nuclear explosion is the overall challenge. The 29 papers of this volume can be structured into 16 waveform studies, eight in the field of radionuclide monitoring and related atmospheric backtracking, and five papers related to on-site inspection or overhead detection of relevant events, with many of these originally presented at a special session on "Research and Development in Nuclear Explosion Monitoring" at the most recent annual General Assemblies of the European Geosciences Union.

Key words: CTBT, nuclear explosion monitoring, waveform analysis, seismology, infrasound, atmospheric radionuclides, atmospheric transport modelling, on-site inspection, overhead detection.

According to Article IV, 11 of the Comprehensive Nuclear-Test-Ban Treaty (CTBT), "Each State Party undertakes to cooperate with the Organization and with other State Parties in the improvement of the verification regime, and in the examination of the verification potential of additional monitoring technologies, such as electromagnetic pulse monitoring or satellite monitoring, with a view to developing, when appropriate, specific measures to enhance the efficient and cost-effective verification of this Treaty". The international scientific community is instrumental in achieving the goal of constant improvement of the verification regime. Scientists drive the relevant research and technical innovations that may shape the future of nuclear explosion monitoring.

The International Monitoring System (IMS) has almost been completed and collects on a regular basis seismic, hydroacoustic, infrasound and radionuclide data at what will eventually be 321 stations around the globe. The purpose of these four sensor networks is to be available for verifying compliance with the Comprehensive Nuclear-Test-Ban Treaty. In addition, the raw data and analysis results prepared by the International Data Centre (IDC) can be used for a broad range of civil and scientific applications. Their use by the scientific community brings back additional expertise and research and development that are of benefit for nuclear explosion monitoring.

The Comprehensive Nuclear-Test-Ban Treaty Organization (CTBTO) Preparatory Commission (PrepCom) undertakes special endeavours to ensure that the complete monitoring system, from sensor technology to analysis algorithms, is state-of-the-art. Many technical workshops are organised every year. The Science and Technology conference series at the Hofburg palace in Vienna is an important opportunity for interaction with the broader scientific community. It started in 2006 with a scientific symposium on "CTBT: Synergies with Science 1996–2006 and beyond" (CTBTO 2006), continued in 2009 with the International Scientific Studies (ISS) Conference (CTBTO 2009), and was followed by the CTBT Science and Technology (SnT) Conference in 2011 (CTBTO 2011) and the CTBT SnT Conference in 2013 (CTBTO 2013).

Another platform to instigate a dialogue with the scientific community on research relevant to nuclear explosion monitoring is the annual General Assembly

[1] CTBTO Preparatory Commission, Provisional Technical Secretariat, Vienna International Centre, P.O. Box 1200, 1400 Vienna, Austria. E-mail: martin.kalinowski@ctbto.org

[2] Deutscher Wetterdienst, Department for Hydrometeorology, Frankfurter Straße 135, 63067 Offenbach am Main, Germany. E-mail: andreas.becker@dwd.de

of the European Geosciences Union (EGU) that is convened every April in Vienna. Since 2007, the guest editors of this volume have served as co-conveners for a special session on "Research and Development in Nuclear Explosion Monitoring". The papers presented in the first 2 years formed the basis for the topical volume of Pure and Applied Geophysics, "Recent Advances in Nuclear Explosion Monitoring" (PAGEOPH 2010). The 15 papers cover all relevant fields in treaty monitoring, including seismology, infrasound, hydroacoustics, nuclear physics dealing with aerosol particle-bound radionuclides as well as noble gas isotopes, and atmospheric backtracking. The compilation of this second topical volume on "Recent Advances in Nuclear Explosion Monitoring" results from a call for papers. Again, many of these papers were originally presented at a special session on "Research and Development in Nuclear Explosion Monitoring" at the most recent annual General Assemblies of the European Geosciences Union (EGU).

These two volumes can be considered a follow-up to research published in a series of topical volumes in Pure and Applied Geophysics in the years 2001–2002, with Brian Mitchel acting as the main editor of that series (PAGEOPH 2001a, b, c, d, e, 2002a, b) following the opening for signature of the CTBT in 1996.

In all sensor technologies and relevant scientific disciplines, significant advances in nuclear explosion monitoring were recently achieved. This progress in the development and testing of new methods improves the capabilities in detection, location and characterization of CTBT-relevant events. In particular, the latter poses a challenge for smaller events, where natural or manmade but CTBT-irrelevant sources can generate false positive events. The efficient discrimination of these events pursued at a minimum risk of missing a relevant event is the overall challenge.

The 29 papers of this volume can be structured into 16 waveform studies, eight in the field of radionuclide monitoring and related atmospheric backtracking, and five papers related to on-site inspection or overhead detection of relevant events.

The latter group of papers includes two papers that involve satellite-based sensors as national technical means. One paper combines these with seismic analysis in a multidisciplinary study of the first two announced nuclear tests of the Democratic People's Republic of Korea and one paper investigates multispectral and infrared imaging of underground nuclear explosion sites. Two studies are related to modelling and detection of radioactive noble gases during on-site inspection, and a team at the Semipalatinsk Test Site demonstrates the application of geophysical techniques for identifying signatures of underground nuclear explosions.

Only one paper covers all waveform sensors (seismic, hydroacoustic and infrasound) to investigate global low and high noise models. The largest group of papers is related to seismic data. These papers deal with incoherent array processing, improving detection capabilities by machine learning or cross correlation, estimation of location error by repeating events, attenuation estimates, seismic source characterization and event discrimination. Two of the seismic papers relate to on-site inspections; one deals with super-sonograms for visual event screening with the seismic aftershock monitoring system (SAMS), while the other explores the suppression of periodic disturbances in seismic aftershock recordings for improved localisation of underground explosions. The three papers on infrasound deal with the influence of spatial filters on infrasound array responses, location estimation with a model for error sources and with progress metrics, and ground truth explosion sources for infrasound calibration.

One of the six radionuclide papers describes the modelling of a single-channel beta-gamma coincidence phoswich detector. Two papers closely related to each other present the analysis of the radionuclide releases from the Fukushima Dai-ichi Nuclear Power Plant accident, the first of which is dedicated to the accident features relying on radionuclide detections performed by IMS monitoring stations, while the second applies atmospheric transport analysis (ATM) to retrieve information like the total activity released into the atmosphere for certain radionuclides. Another ATM study investigates the impact of monthly radioxenon source time-resolution on atmospheric concentration predictions. This has implications on the capability to characterise the source of radioxenon detections. A related paper estimates the global xenon-133 emission inventory caused by medical isotope production, as derived from the worldwide technetium-99 m demand. The potential of applying the well-known radioxenon activity ratios for source characterization is studied by one paper, while another one introduces a similar approach for radioiodine isotopes to discriminate nuclear explosions against civilian sources. Another study clearly demonstrates that the source characterisation approach based on isotopic activity ratios of radioiodine and radioxenon is not compromised by any degree of fractionation that might occur for releases from underground nuclear explosions.

The Guest Editors hope that the breadth of the information given in this volume will prove a useful reference for future research in nuclear explosion monitoring.

Acknowledgments

We are grateful for the contributions from all authors and the helpful and constructive advices received from the following 48 scientists during the peer review process: Jürgen Altmann, Robert Annewandter, Rainier Arndt, Steven Biegalski, Istvan Bondar, Jesse Bonner, David Brown, Douglas Christie, Lars Erik De Geer, Russel Detwiler, Läslo Evers, Sean R. Ford, Dirk Gajewski, Steven John Gibbons, Michael Hedlin, Ross Heyburn, Yong Keun Hwang, Bhupendra Jasini, Yan Jia, Manfred Joswig, Martin B. Kalinowski, Gerald Kirchner, Ivan Kitov, Kevin MacKey, Justin Iraqulon McIntyre, Irmgard Niemeyer, Pavel Povinec, Matthew Purss, Paul G. Richards, Anders Ringbom, Jorge Roman-Nieves, Esteban Rougier, David R. Russel, Paul Saey, David Schaff, Jorg Schlittenhardt, Benjamin Sick, Ulrich Stoehlker, Curt Szuberla, Rick Tinker, Harri Toivonen, Matthias Tuma, Josep Vila, Wolfgang Weiss, Rodney Whitaker, Gerhard Wotawa, Matthias Zaehringer, and Cleat Zeiler. We are also grateful for the smooth collaboration with the Editor in Chief, Renata Dmowska. We would also like to express our appreciation to the publisher, Birkhäuser.

References

CTBTO (2006), *CTBT: Synergies with science 1996–2006 and beyond* symposium website on http://www.ctbto.org/the-organization/ctbt-synergies-with-science1996-2006-and-beyond/.

CTBTO (2009), *International Scientific Studies (ISS) Conference* conference website on http://www.ctbto.org/specials/the-international-scientific-studies-project-iss/.

CTBTO (2011), *CTBT Science and Technology Conference 2011* conference website on http://www.ctbto.org/?id=2538.

CTBTO (2013), *CTBT Science and Technology Conference 2013* conference website on http://www.ctbto.org/specials/snt2013/.

PAGEOPH (2001a), *Monitoring the Comprehensive Nuclear-Test-Ban Treaty: Source Location* (Ringdal, F. and Kennett, B.L.N. eds.), Pure Applied Geophys, *158*, No.1/2.

PAGEOPH (2001b), *Monitoring the Comprehensive Nuclear-Test-Ban Treaty: Hydroacoustics* (deGroot–Hedlin, C. and Orcutt, J. eds.), Pure Applied Geophys, *158*, No.3.

PAGEOPH (2001c), *Monitoring the Comprehensive Nuclear-Test-Ban Treaty: Regional Wave Propagation and Crustal Structure*, (Patton, H.J. and Mitchell, B.J. eds.), Pure Applied Geophys, *158*, No.7.

PAGEOPH (2001d), *Monitoring the Comprehensive Nuclear-Test-Ban Treaty: Surface Waves* (Levshin, A.L. and M.H. Ritzwoller eds.), Pure Applied Geophys, *158*, No.8.

PAGEOPH (2001e), *Monitoring the Comprehensive Nuclear-Test-Ban Treaty: Source Processes and Explosion Yield Estimation*, (Ekström, G., Denny, M. and Murphy, J.R., eds.), Pure Applied Geophys, *158*, No.11.

PAGEOPH (2002a), *Monitoring the Comprehensive Nuclear-Test-Ban Treaty: Seismic Event Discrimination and Identification*, (Walter, W.R. and Hartse, H.E., eds.), Pure Applied Geophys, *159*, No.4.

PAGEOPH (2002b), *Monitoring the Comprehensive Nuclear-Test-Ban Treaty: Data Processing and Infrasound*, (Der, Z.A., Shumway, R.H. and Herrin, E.T., eds.), Pure Applied Geophys, *159*, No.5.

PAGEOPH (2010), *Recent Advances in Nuclear Explosion Monitoring*, (Becker, A., Schurr, B., Kalinowski, M.B., Koch, K., and Brown, D., eds.) Pure Applied Geophys, *167*, No. 4–5.

(Received October 7, 2013, revised October 8, 2013, accepted October 8, 2013, Published online October 29, 2013)

Pure Appl. Geophys. 171 (2014), 341–359

DOI 10.1007/s00024-012-0628-8

Pure and Applied Geophysics

A Multidisciplinary Study of the DPRK Nuclear Tests

R. Carluccio,[1] A. Giuntini,[1] V. Materni,[1] S. Chiappini,[1] C. Bignami,[1] F. D'Ajello Caracciolo,[1] A. Pignatelli,[1] S. Stramondo,[1] R. Console,[1] and M. Chiappini[1]

Abstract—The Democratic People Republic of Korea announced two underground nuclear tests carried out in their territory respectively on October 9th, 2006 and May 25th, 2009. The scarce information on the precise location and the size of those explosions has stimulated various kinds of studies, mostly based on seismological observations, by several national agencies concerned with the Nuclear Test Ban Treaty verification. We analysed the available seismological data collected through a global high-quality network for the two tests. After picking up the arrival times at the various stations, a standard location program has been applied to the observed data. If we use all the available data for each single event, due to the different magnitude and different number of available stations, the locations appear quite different. On the contrary, if we use only the common stations, they happen to be only few km apart from each other and within their respective error ellipses. A more accurate relative location has been carried out by the application of algorithms such as double difference joint hypocenter determination (DDJHD) and waveform alignment. The epicentral distance between the two events obtained by these methods is 2 km, with the 2006 event shifted to the ESE with respect to that of 2009. We then used a dataset of VHR TerraSAR-X satellite images to detect possible surface effects of the underground tests. This is the first ever case where these highly performing SAR data have been used to such aim. We applied InSAR processing technique to fully exploit the capabilities of SAR data to measure very short displacements over large areas. Two interferograms have been computed, one co-event and one post-event, to remove possible residual topographic signals. A clear displacement pattern has been highlighted over a mountainous area within the investigated region, measuring a maximum displacement of about 45 mm overall the relief. Hypothesizing that the 2009 nuclear test had been carried out close to the area where the displacement has been observed through the DInSAR technique, its relation with the epicenter location obtained through seismological processing has been discussed as a possible alternative hypothesis with respect to the preferred solutions reported by the nuclear explosion database (NEDB). The distance of about 10 km between the two places can be considered acceptable in light of the possible systematic location shifts commonly observed in the seismological practice over a global scale. The difference between the m_b magnitudes of the two tests could reflect differences in geological conditions of the two test sites, even if the yield of the two explosions had been the same.

[1] Istituto Nazionale di Geofisica e Vulcanologia, Rome, Italy. E-mail: massimo.chiappini@ingv.it

1. Introduction

The most recent nuclear tests in the world were announced by the Democratic People Republic of Korea (DPRK) having been carried out on October 9th, 2006 and May 25th, 2009. In the following, these tests will be called DPRK06 and DPRK09, respectively. The clear seismological signals detected by many stations of the globe clearly characterized these tests as underground explosions. The analysis of the seismic waveforms allowed the location and magnitude estimation for both events. For the 2006 test a body wave magnitude $m_b = 4.0 \pm 0.1$ was obtained, while the relatively high long-period seismic noise covered the surface waves, so that only an upper limit of 3.5 was estimated for M_s. The higher amplitudes of the recorded seismic signals for the 2009 event lead to the estimation of 4.5 ± 0.1 and 3.2 ± 0.2 for m_b and M_s, respectively.

The magnitudes estimated for the two events by the international agencies are fairly well consistent with ours. The m_b magnitude ranged from 4.1 (IDC/REB[1]) to 4.2 (USGS/NEIC[2]) for DPRK06 and from 4.5 (IDC/REB) to 4.7 (USGS/NEIC) for DPRK09. The surface-wave magnitude M_s for the explosion, determined from the IDC REB based on 15 stations, was 3.6 (significantly larger than expected for an explosion with that small m_b). Also Murphy *et al.*, (2010) noted somewhat unusually large amplitudes for surface waves of both events, compared to the historical explosion surface wave measurements, and they justified this peculiarity with the circumstance that the explosions were small and embedded in high-velocity hard rock.

[1] International Data Center Reviewed Event Bulletin.

[2] United States Geological Survey/National earthquake Information Center.

The epicenters of both tests were located in a mountainous area of scarce natural seismicity, geologically belonging to a tectonic region constituted by granitic massifs and not far from volcanic structures, some of which still active at the present time (Fig. 1). For a conversion of the m_b magnitude into yield, numerous relations developed in the past are available. The proper relation to be used in specific cases, like this, depends on an accurate knowledge of the geological framework. Unfortunately, this is not our case, due to the vicinity of granitic intrusions to lava flows and pyroclastic material coming from the volcanoes present in the area. As an example, adopting the m_b values reported above and m_b yield relations obtained for the former Soviet Republic (Ringdal *et al.*, 1992) and the Nevada test site (Murphy 1981), the possible values of yield range between 0.25 and 1.0 kt for the 2006 test and between 1 and 5 kt for the 2009 test.

In the present study, the DPRK09 test has been investigated also by applying DInSAR (Synthetic Aperture Radar Differential Interferometry) technique. Since its first use in 1992 (Massonnet *et al.*, 1993) DInSAR is nowadays considered an effective tool in Earth Sciences studies able to detect centimetric and/or millimetric surface movements over large areas. Recent studies report that although earthquakes of magnitude <4.8 are unlikely to be observable, coseismic surface deformations induced by very shallow events are detectable (Dawson 2007; Dawson *et al.*, 2008). In recent years the capabilities of DInSAR were improved in terms of spatial resolution, accuracies and revisiting time. Indeed, the new generation of high-resolution (HR) and very HR (VHR) SAR sensors is available since 2007, with the Japanese ALOS PALSAR, the Canadian Radarsat-2, and up to the most recent sensors, like the VHR German TerraSAR-X and the Italian COSMO-SkyMed constellation, both achieving 1 m of spatial resolution. The DPRK09 is the first ever nuclear test where VHR DInSAR has been applied.

Previous studies concerning the co-seismic surface deformation signal due to underground nuclear tests have been conducted at the Nevada Test Site in 1992, where SAR data were collected over a 14-month time span to cover three tests (Vincent

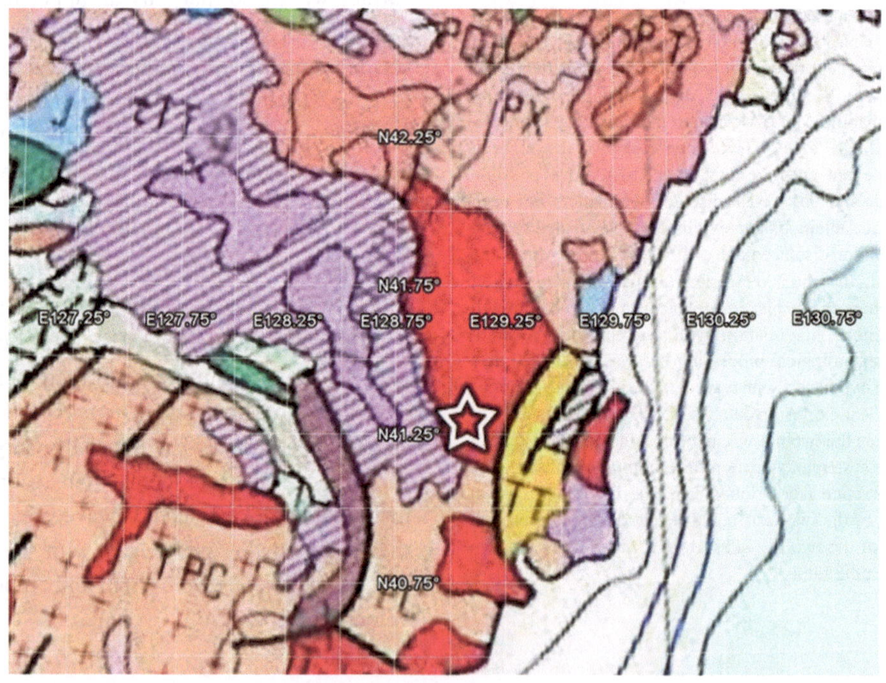

Figure 1
Geological map of the DPRK region where the 2006 and 2009 tests were carried out. *Red* mesozoic granite, *pink* precambrian rock, *purple* quaternary lava. The *white star* shows the location of the test sites

et al., 2003; Vincent 2005). These studies reported a subsidence of a limited area above the detonation point as common co-seismic feature of the underground nuclear explosions.

The present study has exploited the capabilities of DInSAR applied to a dataset consisting of three TerraSAR-X images acquired in Stripmap Mode (at 3 m spatial resolution). The DPRK09 case study has encountered some difficulties which were not present in the former Nevada cases. First, as written above, the location of the nuclear test is known approximately only within the typical uncertainty of seismic location. Second, the topography of the investigated area is far from the flat plate of Nevada test sites. This implies a more complex scenario needing a digital elevation model (DEM) to fully remove the topographic phase. Moreover, the use of three SAR data is expected at least to make some cross-analysis concerning the presence of the atmospheric phase (in particular, the wet troposphere) and possible topographic residual phase.

2. *Single Locations and Systematic Travel Time Residuals*

In the context of verifying compliance of any nuclear test ban treaty, the step immediately following the detection of a waveform event is its location. We collected *P* wave recordings from as many as possible stations belonging to international monitoring networks, and picked the arrival times.

In the framework of the International Scientific Studies (ISS09), an initiative launched by the Comprehensive Nuclear Test-Ban-Treaty Organization (CTBTO) to stimulate the scientific community to improve the Treaty verification, we evaluated some data collected by the International Monitoring System (IMS) established by the Preparatory Commission of CTBTO. Moreover, to evaluate the potential of the Geotool software package developed within the Organization, we picked various phases from the waveforms using Geotool as well. Table 1 reports the arrival times picked up for both the 2006 and 2009 tests. As reported in Fig. 2, it clearly appears that these two events were recorded at mostly, or even only, teleseismic distances.

We located the events using a least-squares location algorithm developed at the INGV. In this algorithm epicentral distances are computed from geographical coordinates through the WGS84 ellipsoid model of the Earth, and travel-times are based on the IASPEI91 tables (Kenneth and Engdahl 1991). Moreover, travel-times are corrected both for the ellipticity of the Earth by the formulation of Dziewonski and Gilbert (1976), and for the station elevation. Our tests showed that these two corrections, even if they are of the order of only few tenths of second, reduce the RMS of the time residuals at the recording stations. We didn't apply any other static station correction nor specific source site correction because such information is unavailable at the present status.

In developing our location algorithm, particular care has been devoted to the computation of the parameters (semiaxis sizes and azimuth) of the epicenter error ellipses. We applied the theoretical framework available in literature for this problem (e.g. Stein and Wysession 2002). Moreover, we carried out numerous tests with a Monte Carlo method, by simulating the location of up to 1,000 events. These locations were obtained starting from synthetic arrival times from a given epicenter, and changing these arrival times in random way with the same standard deviation of the real observation data. The error ellipses statistically obtained from the synthetic epicenters match quite well the theoretical one obtained in connection with the original location.

Our algorithm has been implemented in a FORTRAN 90 code. The locations of the two events using all the available arrival times, and having assigned a zero depth to both of them, are listed in the first two rows of Table 2 and shown in the map of Fig. 3.

Table 2 (lines 1 and 2) and Fig. 3 (orange symbols) show a modest difference between the epicenters of the two events, obtained with the arrival times picked on all the available waveforms. In fact the 2006 event (eight stations used) is located nearly 7 km to the south west of the 2009 event (17 stations used). However, the error ellipses at the 95 % confidence level of the two events partly overlap. We can easily show that this distance is mostly due to the use of different data sets, associated with systematic station-depending discrepancies existing between the

Table 1

Arrival times picked up for the 2006 and 2009 DPRK tests

Station	Phase	Distance (°)	Azimuth (°)	October 9th, 2006	May 25th, 2009
KSRS	Pn	3.95	193.6		00:55:44.697
MJAR	Pn	8.57	120.8		00:56:49.513
SONM	P	17.42	299.8		00:58:48.515
MKAR	P	33.69	295.6	01:42:10.370	01:01:26.220
CMAR	P	34.37	237.5		01:01:31.882
BVAR	P	40.51	307.5		01:02:23.950
ILAR	P	51.13	33.0		01:03:46.929
ARCES	P	56.43	336.0		01:04:26.215
FINES	**P**	**60.34**	**327.7**	**01:45:37.668**	**01:04:52.435**
WRA	**P**	**61.11**	**174.3**	**01:45:43.180**	**01:04:58.114**
YKA	P	64.72	27.0		01:05:21.805
ASAR	**P**	**64.77**	**175.1**	**01:46:07.837**	**01:05:22.722**
AKASG	**P**	**64.84**	**316.5**	**01:46:07.876**	**01:05:22.744**
NOA	P	66.25	332.3		01:05:31.617
GERES	**P**	**73.74**	**322.0**	**01:47:03.385**	**01:06:18.145**
NVAR	**P**	**79.67**	**47.1**	**01:47:38.638**	**01:06:53.646**
PDAR	**P**	**81.03**	**39.1**	**01:47:44.786**	**01:06:59.872**

The common stations for the two events are reported in bold

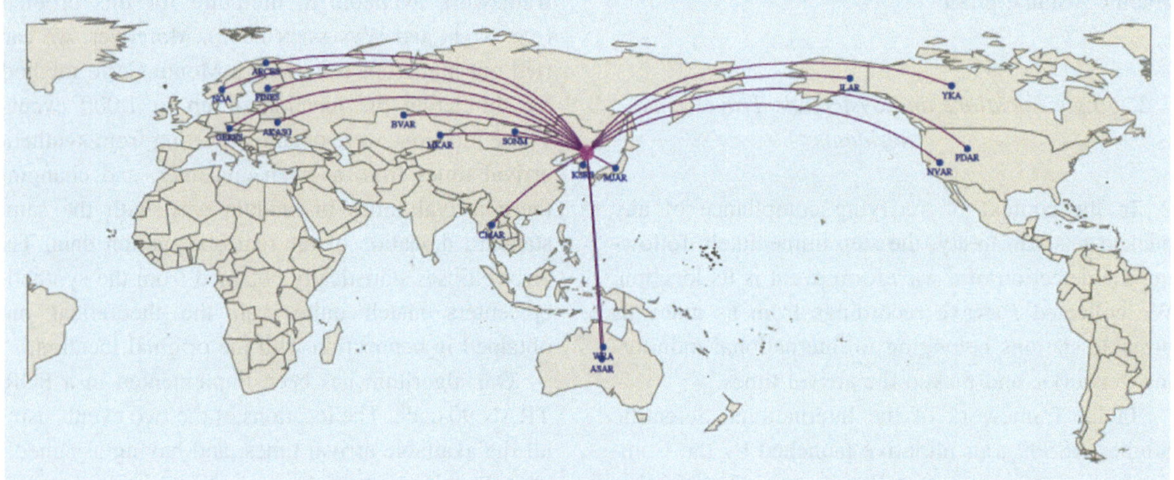

Figure 2
Stations used for event location

real travel-times in the Earth and the theoretical travel-times used in our location algorithm.

By repeating the locations using only seven of the eight common stations for the two events (bold phase in Table 1) the new epicenters, listed in the third and fourth row of Table 2 and showed in Fig. 3 (red symbols), are both located near the epicenter of the 2006 event obtained using all the eight available arrival times, and much closer to each other. In fact, in this case the 2006 event appears located only 3 km to the east of the 2009 event. This distance is contained within the error ellipses at the 95 % confidence level. In this exercise, station MKAR was excluded due to its uncertain time pickings.

We repeated the locations by means of a set of *P* wave arrival times reported in the ISC bulletins. In this case the number of stations used was 27 and 69 for DPRK06 and DPRK09, respectively. The results reported in Table 2 (lines 5 and 6) and Fig. 3 (blue

Table 2

Epicentral coordinates and error ellipse parameters obtained for the 2006 and 2009 DPRK tests using different data sets

ID	Data set	OT	Lat. (°N)	Long. (°E)	RMS (s)	Smax* (km)	Smin* (km)	Azimuth (°)
1	DPRK06 (8 st.)	01:35:27.77	41.250	128.993	0.44	12.3	8.3	70
2	DPRK09 (17 st.)	00:54:43.04	41.292	129.076	0.83	10.8	8.6	72
3	DPRK06 (7 st.)	01:35:27.72	41.247	128.930	0.38	12.8	7.4	77
4	DPRK09 (7 st.)	00:54:42.63	41.244	128.888	0.35	11.8	6.9	77
5	DPRK06 (27 st.)	01:35:27.78	41.303	129.064	0.65	6.4	4.5	64
6	DPRK09 (66 st.)	00:54:42.98	41.312	129.073	0.63	3.5	3.1	91
7	DPRK06 (ISC)	01:35:27.63	41.311	129.055	0.83	5.4	3.8	10
8	DPRK09 (ISC)	00:54:42.85	41.295	129.072	1.17	4.7	3.7	134

* 95 % confidence level

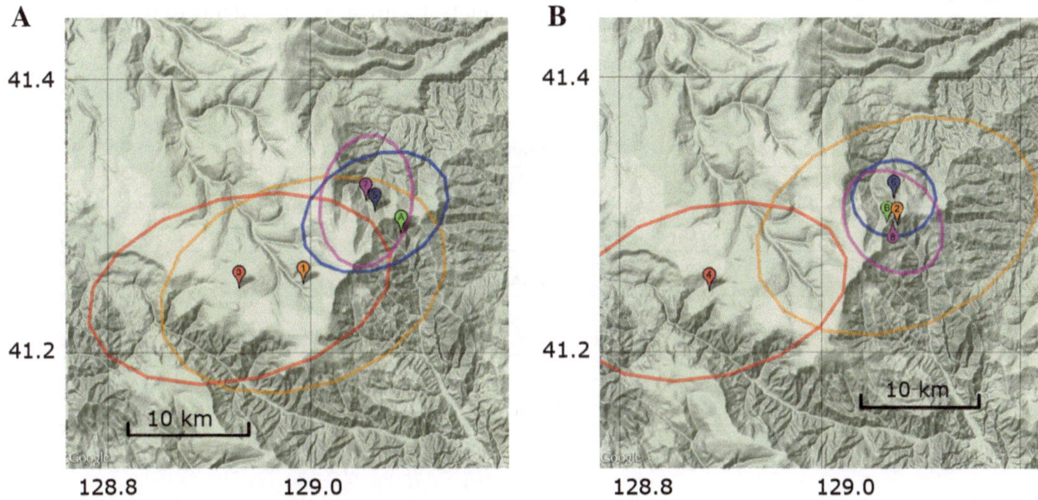

Figure 3

Epicentral locations of the 2006 (**a**) and 2009 (**b**) DPRK tests using the different data sets reported in Table 2. The epicenters are tagged with the same ID numbers as in Table 2. The error ellipses are at a 95 % confidence level. The *green marks* point to the preferred locations reported by NEDB (see Sect. 5)

symbols) show epicenters shifted a few kilometers to the north-east of the previous locations and smaller error ellipses. For sake of completeness, in Table 2 (lines 7 and 8) and in Fig. 3 (purple symbols) we report also the locations obtained by the ISC. These epicenters are very close and their error ellipses are comparable to those reported at lines 5 and 6 of Table 2, which were obtained with a smaller set of arrival times.

Based on the above results, we proceeded to study in more detail the systematic residuals affecting the arrival times at the seven common stations listed in Table 1.

We used a computer code (in the following named *wave-shifter*) specifically developed at the INGV for comparing the relative locations of two or more seismic events. The input for this code consists in the location coordinates, depth and origin time of the events to be compared, besides the digital waveforms recorded at a number of stations from these events, and outputs a plot of such waveforms referred to a time scale the origin of which is, for each single station, the theoretical arrival time computed through the IASPEI[3]91 travel-times tables.

Figure 4 shows the output of *wave-shifter* for the seven common stations used for obtaining the first two epicentral coordinates reported in Table 2. The plots of Fig. 4 show that:

[3] International Association of Seismology and Earth Interior.

(a) The first onsets of the P wavelets at some stations are variably shifted by several tenths of second with respect to the computed arrival times, pointing out the existence of errors in time pickings and/or systematic differences between the IASPEI91 model and the real travel-times in the Earth;
(b) There are also significant shifts between the first onsets of DPRK06 (red line) and those of DPRK09, meaning that at least one or both events are mislocated.

The difference between the two epicenters is greatly reduced, if we use the arrival times of the same stations for both events. As a further test, we applied the *wave-shifter* program by using the epicentral location obtained by the seven common stations as reported in the third and fourth rows of Table 2. The results are plotted in Fig. 5. Now, the first onsets for DPRK09 are much closer to those of DPRK06. Fairly surprisingly, we note a systematic shift between the first onsets for the two events. More precisely, the wavelets from DPRK06 seem to arrive about 0.2–0.3 s earlier than those from DPRK09 at all the stations, the same that we would note delaying the origin time of DPRK06 by that time interval. A possible explanation for this circumstance is that the arrival times picked up for DPRK06 and used for its location (Table 1), due to the lower signal-to-noise level, have been read by the analyst with an average delay of 0.2–0.3 s relatively to those of the other event.

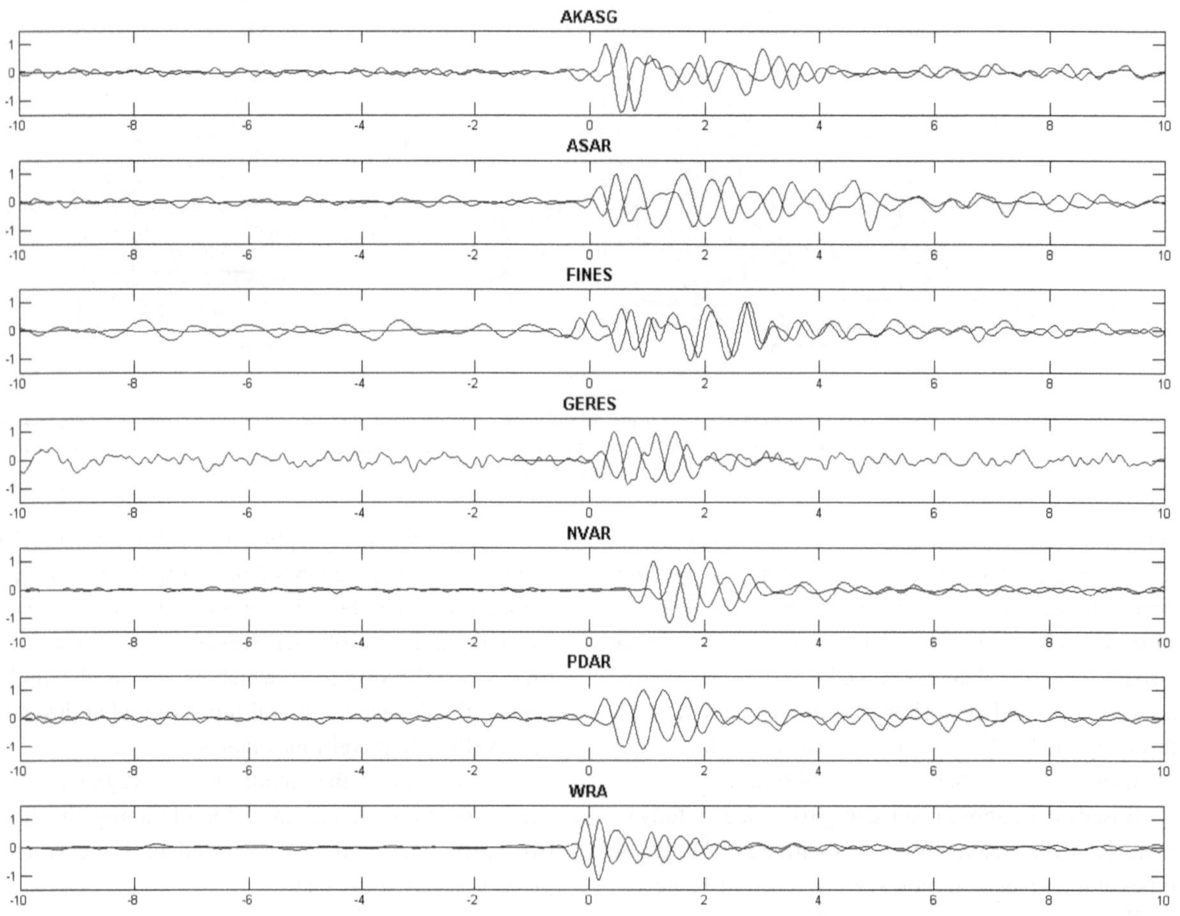

Figure 4
P arrival waveforms plotted by program *wave-shifter* for K06 (*red*) and K09 (*blue*) events. The zero-value on the *x*-axis coincides with the expected arrival time computed for both events at each of the seven common stations, from their respective locations and origin times obtained through all the available data

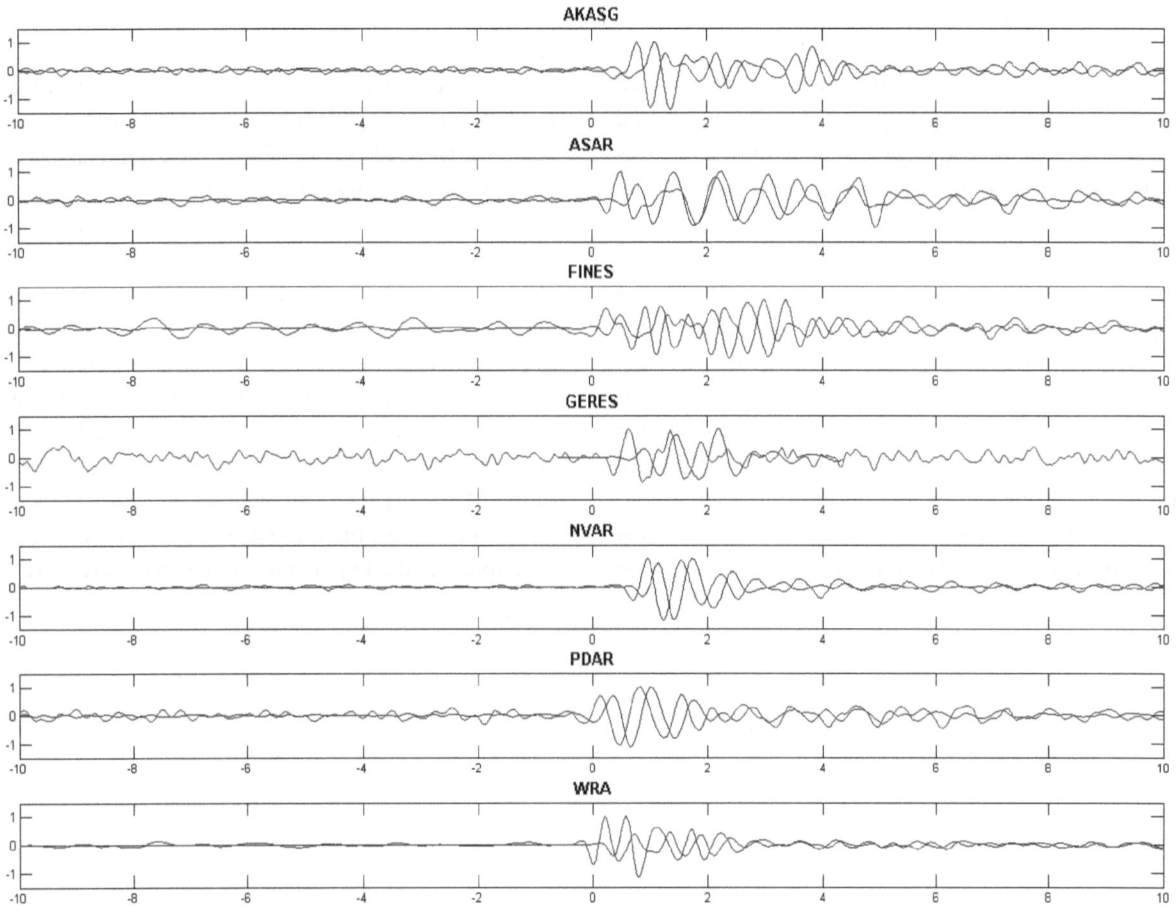

Figure 5
As in Fig. 4, but using the locations and origin times obtained through the data of the seven common stations only

If we applied a correction of this size to the computed origin time of DPRK06, the overlapping of the waveforms of the two events shown in Fig. 5 would be almost perfect. It confirms that the relative location of these epicenters, as reported in the third and fourth rows of Table 2, is very close to the true one.

We may confidently assume that, among our results, the epicentral locations obtained by the largest number of arrival times, reported in the fifth and sixth lines of Table 2, are the most reliable. Nevertheless, based on these data only, and without ground truth information, we ignore the actual size of the systematic mislocation that could still affect the results.

The satellite-based approach introduced earlier and described in more detail later in the following sections is aimed at establishing the missing ground truth.

3. Relative Locations

In the previous section we showed that the plots obtained by our *wave-shifter* program allow considerations about the relative location of DPRK06 and DPRK09. In the following, we shall show that it is possible to obtain a reliable relative location by a trial-and-error procedure, changing the epicentral coordinates and origin time one by one, until a satisfying overlapping of the wavelets at all the stations is obtained. However, this method requires considerable time and skill. At first, we simply tested the hypothesis that the two tests were carried out exactly on the same point; therefore, we applied the *wave-shifter* program, giving the same epicentral coordinates to both of them. For this exercise, we adopted the coordinates and the origin time of the DPRK09 event reported in the fourth line of Table 2. Because the origin time to be associated to the simulated DPRK06 epicenter was unknown, we proceeded tentatively changing the origin time until a very good coincidence of the two wavelets at station PDAR was obtained. The output of the *wave-shifter* program is shown in Fig. 6.

Figure 6 shows that not only stations NVAR and PDAR (in Northern America), but also ASAR and WRA (in Australia) exhibit good correlations. This indicates that the distance between the real epicenters of the two considered events and each of these four stations is respectively about the same. This is not the case for the other three stations AKASG, FINES and GERES, all of them located in the Euro-Asian continent, to the northwest of Northern Korea. The time shift observed for these stations is of the order of 0.2 s, with the signals from DPRK06 (red line) arriving later than those from DPRK09. From this exercise we may conclude that the two tests were not carried out on exactly the same point, and the epicenter of DPRK06 is probably some kilometers to the east-southeast of that of DPRK09.

For relative location of two or more events, the double-difference joint hypocenter determination (DDJHD) method (Waldhauser and Ellsworth 2000) has become very popular in the last decade for high-resolution imaging of clustered seismicity in active areas. At the INGV, we developed our own DDJHD algorithm (Console and Giuntini 2006). We applied it in a global environment, again making use of the IASPEI91 travel-time tables. For both the local and global scales, the DDJHD method is based on the principle that the hypocenters to be located relative to

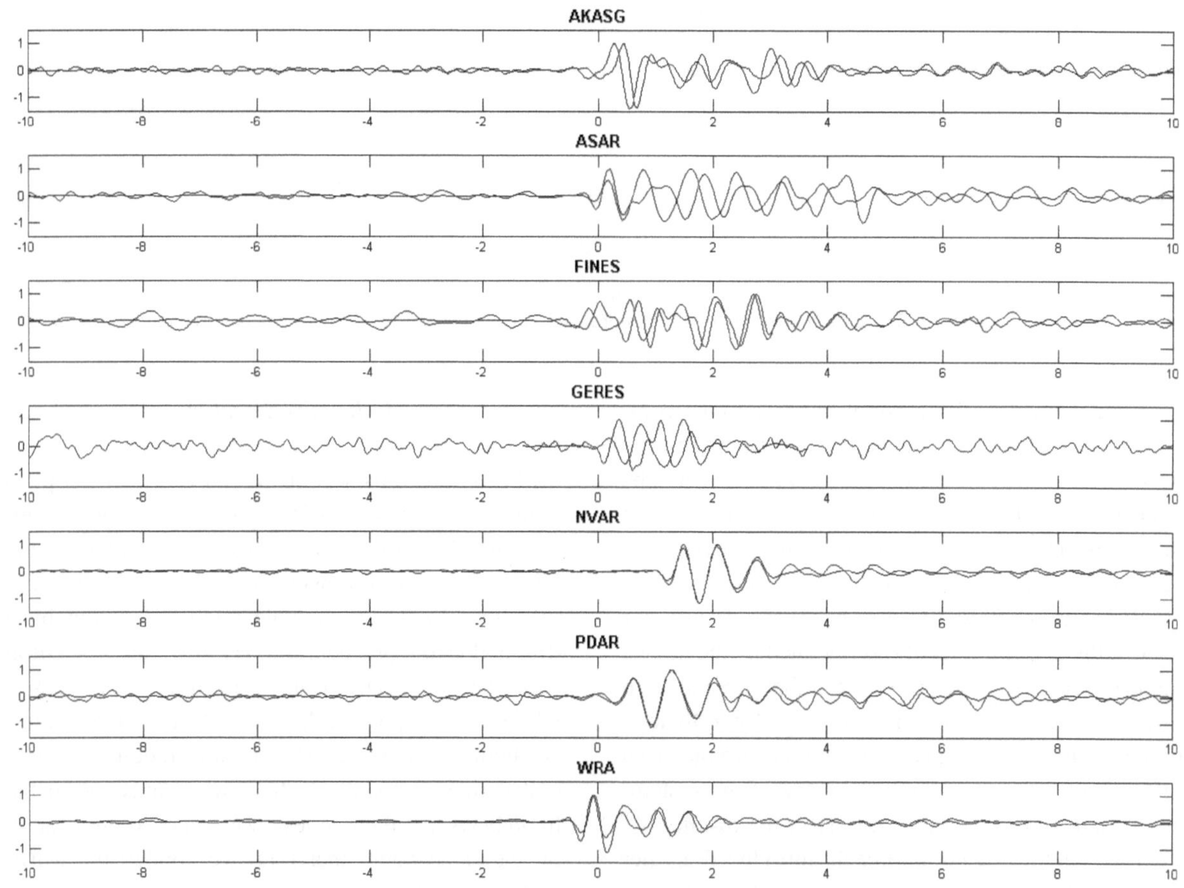

Figure 6
As in Fig. 4, but assuming identical locations for DPRK06 and DPRK09, and assigning the origin time to DPRK06 in order to obtain a perfect coincidence of the two wavelets at station *PDAR*

each other are closely spaced in comparison with the distance between the hypocenters themselves and the recording stations. The advantage of the DDJHD method consists in removing the influence of the systematic travel time residuals from the location process, which, as we have seen before, may be relevant when the different seismic events are located using different sets of stations.

The DDJHD method can be applied to the *P* wave arrival times picked up by the analyst on the seismograms, but its accuracy can be significantly improved by applying a cross-correlation technique to a suitable segment of the *P* waveforms. In such a way, the accuracy of the results is not limited by the skill of the analyst in subjectively reading the first onsets of the *P* arrivals, but is objectively assured by the automatic procedure carried out by the computer in finding out the time by which one segment must be shifted with respect to the other in order to yield the maximum correlation coefficient between these waveform segments. Moreover, the information used by this method comes from a waveform segment of suitable length, and not only from a single sample where the first onset is detected. Nevertheless, according to our experience, the use of the method requires some expertise to assess, for example, the most appropriate filtering band and the length of the waveform segments that must be correlated. Multiple maxima of the correlation function of similar size are a common problem faced when processing the signals recorded from the two considered events. In our algorithm, implemented in a MATLAB code, the solution is obtained through a simple least-squares best fit.

The application of the DDJHD algorithm to the waveforms of the seven common stations has provided the relative location shown in Fig. 7, which is fairly well consistent with what was inferred from the application of the *wave-shifter* program in the test described above.

The RMS of the time shifts residuals obtained by the best fit algorithm is equal to 0.167 s, a small value in comparison with the RMS obtained from all the locations reported in Table 2. This confirms the advantage of using the DDJHD algorithm for relative location of seismic events.

For testing the quality of our DDJHD solution, we applied once again the *wave-shifter* program. In this

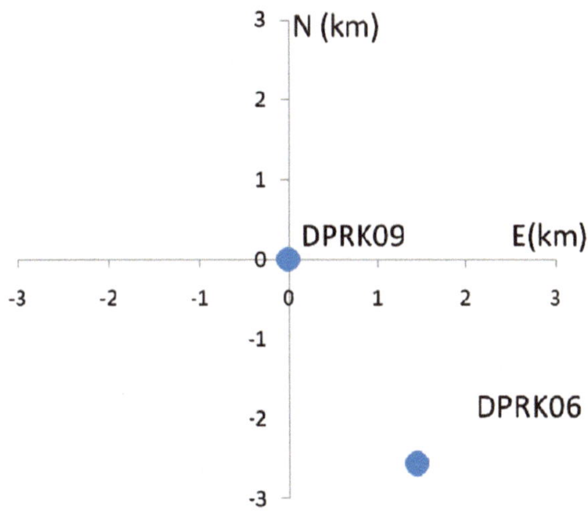

Figure 7
Relative location of the two Northern Korea tests obtained from the DDJHD algorithm. The origin of the coordinates has been arbitrarily put on the epicenter of DPRK09

case we assume that the coordinates of DPRK09 are those reported in the fourth row of Table 2, and the coordinates of DPRK06 are obtained from the former by shifting them by 2.55 km to the south and 1.45 km to the east (see Fig. 7). The origin time of DPRK06 is obtained by the trial-and-error procedure described above in order to obtain a perfect coincidence of the two wavelets at station PDAR. The output of the *wave-shifter* program is shown in Fig. 8.

Figure 8 shows that at all the stations the *P* waves arrive about 0.1 s earlier than expected from the theoretical travel-times, except for station ASAR and WRA, where the actual arrivals are about 0.1 s later. This means that our DDJHD locations are still slightly different from the true ones. As the time differences obtained from the cross-correlation technique are certainly accurate enough, a possible explanation of the misfit observed in Fig. 8 is that the geological structure in the test area is inhomogeneous, representing a violation of the requirements for the application of the DDJHD method.

In order to correct the misfit still existing in Fig. 8 with the DDJHD location, we used the *wave-shifter* program with a procedure of trial-and-error, as explained above. By visual inspection of the plot in Fig. 8, we found out that to improve the relative location of the two events, the epicenter of DPRK06

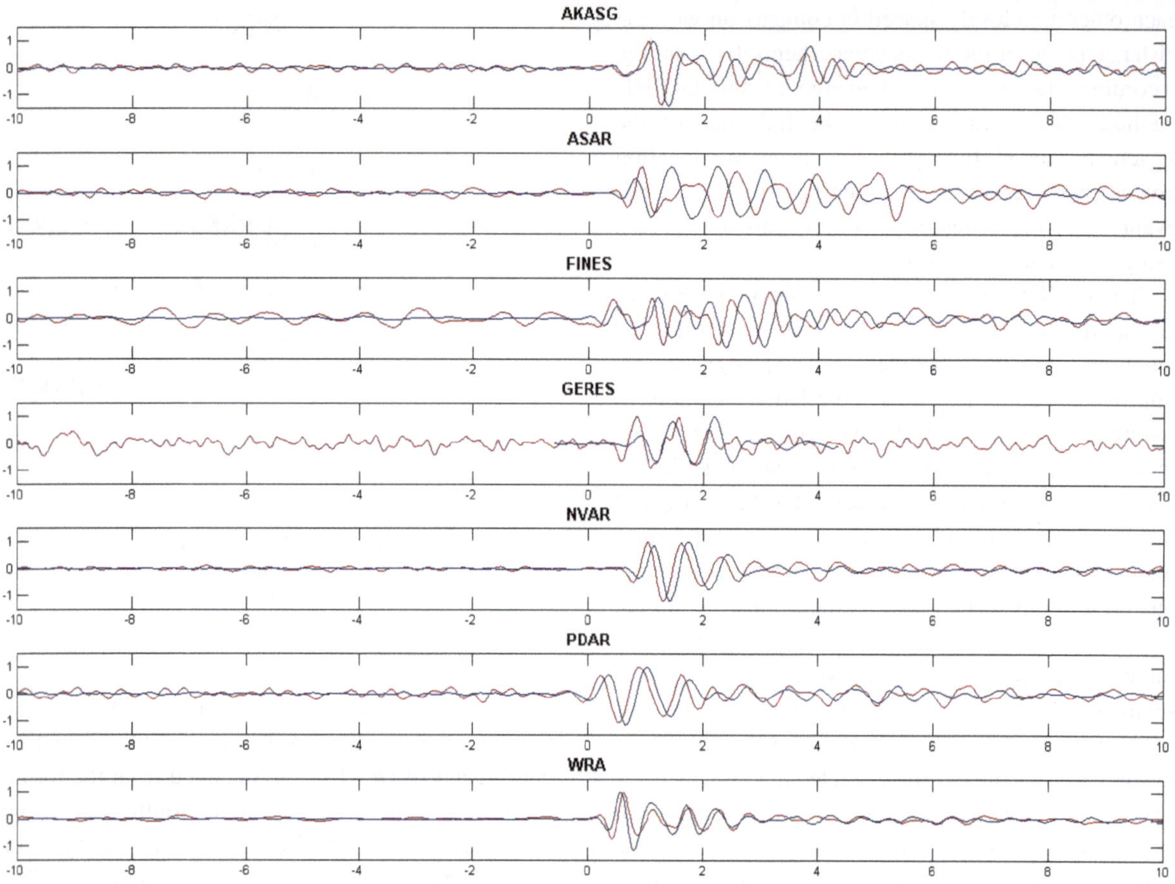

Figure 8
As in Fig. 4, but assuming for DPRK06 the location and origin time obtained through the DDJHD algorithm relatively to DPRK09

has to be shifted to the North. Of course, in addition to the shift of the epicenter, it is also necessary to adjust the origin time of DPRK06 to maintain a perfect overlapping of the two wavelets at station PDAR. After a number of trials, we considered acceptable the solution shown in the map of Fig. 9. The related *wave-shifter* plot is shown in Fig. 10.

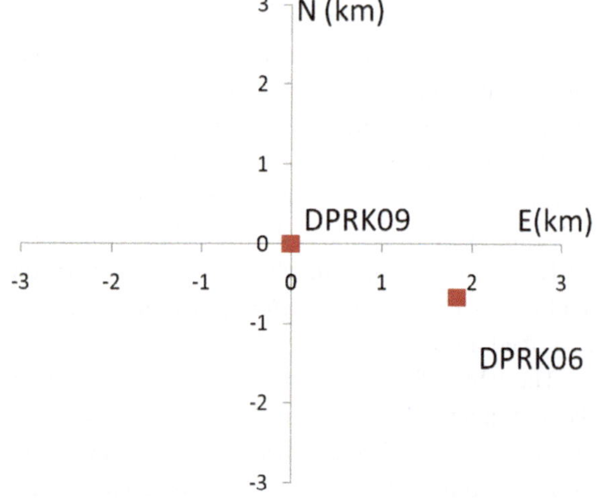

Figure 9
Relative location of the two DPRK tests obtained adjusting the epicentral coordinated of DPRK06 for the maximum correlation of its waveforms with those of DPRK09. The origin of the coordinates has been arbitrarily put on the epicenter of DPRK09

4. *Analysis of Satellite Data*

Remote sensing methods have demonstrated their ability to support the localization of underground nuclear test sites. Both optical and synthetic aperture radar (SAR) data can be used for this purpose (Canty *et al.*, 2009). In particular, in the last decade, optical satellite sensors have increased the resolution of the data imagery, reaching sub-meter per pixel

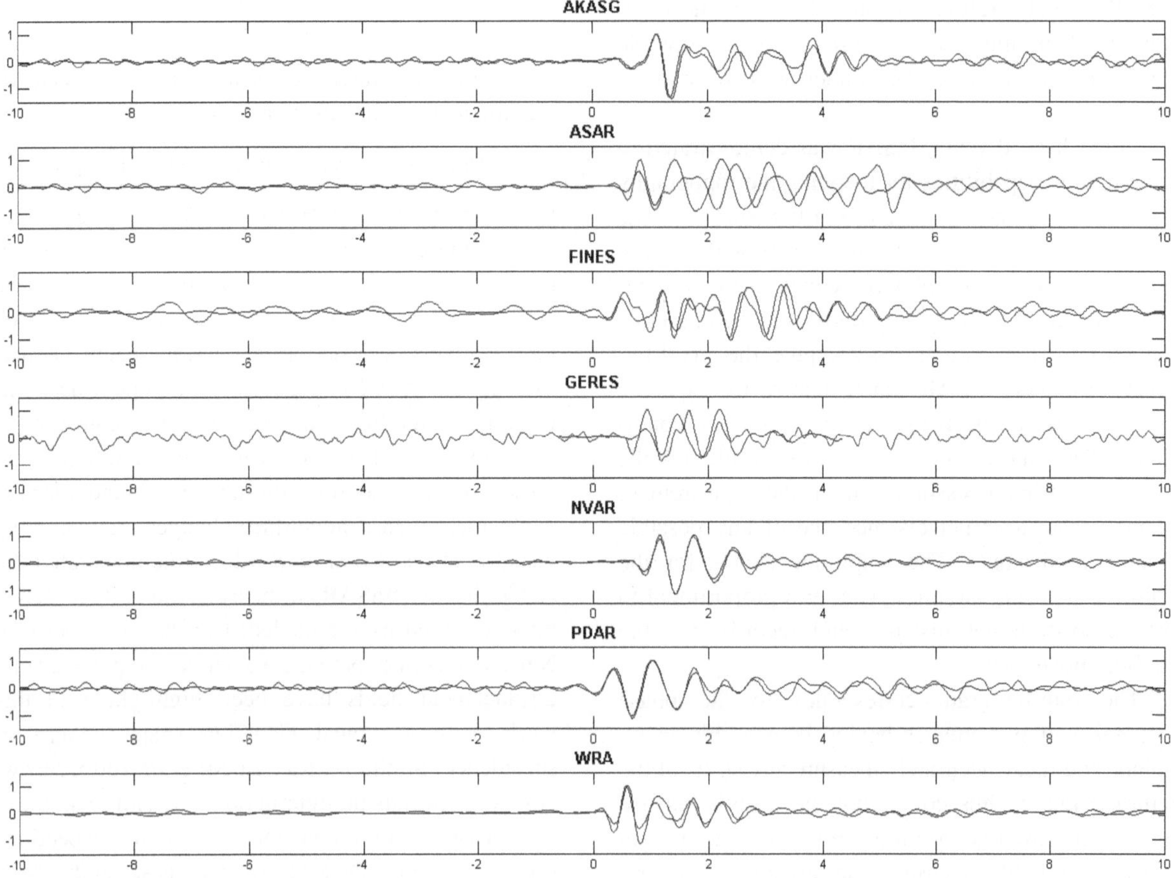

Figure 10
As in Fig. 4, but assuming for DPRK06 the location and origin time obtained adjusting the epicentral coordinated of DPRK06 for the maximum correlation of its waveforms with those of DPRK09

resolution, thus giving a new opportunity for better identify changes directly induced by nuclear test (e.g., damage or strong surface modifications), or indirectly by human activities during site preparation before the nuclear test (e.g., new buildings or roads and spreading of material in the surrounding). Mainly, the change detection approach and visual comparison methods (Schlittenhardt *et al.*, 2010 and references therein) have been applied on optical data to locate area affected by these types of changes in the proximity of nuclear test sites.

Beside optical imagery, SAR data can successfully exploited. SAR are active radar imaging sensors, working on the microwave region of the electromagnetic spectrum and thus they can operate in almost all weather conditions and during day and night time.

4.1. InSAR Rationale

The SAR is a radar imaging sensor exploiting satellite orbit paths to achieve a spatial resolution much better (tens of meters to meters) than standard radar systems. SAR processing significantly improves the resolution of point targets in both the cross-track (range) and along-track (azimuth) directions by focusing the raw radar echoes (Elachi 1988; Curlander and McDonough 1991). In order to exploit SAR data, since 1991 the SAR signal processing technique referred to as SAR interferometry (InSAR) has been developed, and it is now not far from the truth to say that InSAR revolutionized a relevant part of Earth sciences. InSAR is widely used in seismology, volcanology, hydrogeology and landslide studies (Stramondo 2008). Further frameworks wherein

InSAR provides relevant contributions are the monitoring of mining regions, urban areas, strategic infrastructures (bridges, dams, nuclear power plants), etc.

In the last 20 years, InSAR was demonstrated to have unique capabilities for mapping the topography and the deformation of the Earth's surface. The InSAR technique is based on extracting the phase component of the complex SAR data (a two-dimensional record of both the amplitudes and the phases of the returns from targets) to compute the pixel-by-pixel difference of SAR signal relative to a specific area and imaged from two nearby geometric conditions. The interferogram, i.e., the result of the interferometric processing, contains the measurement of the sensor-to-target distance and of any possible change in distance. The amplitude stands for the reflectivity, while the phase is a term proportional to the sensor-to-target distance and records possible surface movements.

The interferogram corresponds to the phase difference of two images having comparable viewpoints, and it can accurately measure any shifts of the returned phase, thus computing the Earth's surface movement towards or away from the satellite. In order to reliably measure the effects of natural disasters generating surface displacements, such as earthquakes, volcanic eruptions, landslides, or man-made activities (e.g., nuclear tests), two images acquired in two different times (one before and one after the event) are needed. This approach is the so called repeat-pass interferometry, characterized by the temporal baseline parameter which corresponds to the time separation between the two SAR scenes.

Describing the technique in deeper detail,, we can state that the interferogram is the combination of the signals S_1 and S_2 received at SAR Sensor 1 and 2, respectively [see Fig. 1 in Stramondo (2008)]. More precisely,

$$S_1 = A_1 e^{-j\frac{4\pi r_1}{\lambda}} S_2 = A_2 e^{-j\frac{4\pi r_2}{\lambda}},$$

where A_1 and A_2 are the amplitudes, and λ is the wavelength. The interferogram is the difference of the phase component and is the product of S_1 versus the complex conjugate of S_2,

$$S_1 \cdot S_2^* = A_1 A_2 e^{-j\frac{4\pi(r_1 - r_2)}{\lambda}}.$$

Therefore, the interferometric phase φ_{int} can be schematically split into five terms:

$$\varphi_{\text{int}} = \varphi_f + \varphi_{\text{topo}} + \varphi_{\text{displ}} + \varphi_{\text{atm}} + \varphi_{\text{err}},$$

where φ_f is the flat Earth component (the *orbital phase*), the topographic phase is φ_{topo}, the displacement phase isφ_{displ}, the atmospheric term φ_{atm} and the error phase φ_{err}. Except for this latter and the φ_f, each term contains information relevant to specific issues. The $\varphi_{\text{displ}} = \frac{4\pi}{\lambda}\Delta R$ is the phase component accounting for the satellite-to-target distance change ΔR.

Conversely from optical data, DInSAR can retrieve terrain modification induced by underground explosions, even if no surface changes are visible by visual analysis. Cong *et al.* (2007) demonstrated the suitability of DInSAR to measure the surface deformation caused by the nuclear test that took place in Nevada Test Site (NTS). Co-seismic and post-seismic displacement fields have been highlighted in this work, with a typical Gaussian shape pattern of subsidence. Similarly, Riechmann *et al.* (http://www.ctbto.org/specials/the-international-scientific-studies-project-iss/scientific-contributions/on-site-inspection-posters/) applied DinSAR for two case study, NST and Lop Nor (China) and, in particular for the second case study, they inferred the position of the nuclear test by observing the pattern of the interferometric coherence loss (i.e., noisy interferogram) and few surrounding fringes.

4.2. InSAR Data and Results

In order to investigate possible surface deformation caused by the registered event, three SAR images were elaborated by means of a differential SAR interferometry technique (DInSAR). In particular, the German TerraSAR-X satellite imaged the investigated area. The available data were acquired in strip-map mode, which is characterized by a swath of about 30 km cross-track and with a ground resolution of 3 × 3 m per pixel (see Table 3). The frames of the SAR images cover an area of about 30 × 60 km (see Fig. 11) which guarantees an optimal coverage of the entire investigation area.

Table 3

Characteristics of the three TerrSAR-X images used for this work

Acquisition date/time (UTC)	Orbit	Mode	Resolution [azimuth × slant range]	Incidence angle	Polarization	Band
18/05/2009, 21:33:15	Descending	Stripmap	3.3 m × 2 m	35.3	HH	9.65 GHz
23/07/2009, 21:33:15	Descending	Stripmap	3.3 m × 2 m	35.3	HH	9.65 GHz
14/08/2009, 21:33:15	Descending	Stripmap	3.3 m × 2 m	35.3	HH	9.65 GHz

The pre-event acquisition is dated May 18, 2009, 1 week prior to the nuclear test, while the post-event roughly span two (July 23) and three (August 14) months after the test. DInSAR has been applied to the overall possible combinations. Both the co-event pairs have very short spatial baselines. Notwithstanding the fact that the looking geometries are very close to each other, and this strongly limits the spatial decorrelation, the high and steep relief, together with the vegetation coverage, affects the interferometric coherence. In order to reduce such effects, a multi-look of 13 × 15 pixels has been applied which does not have any impact, ensuring the minimum spatial resolution (24 m) able to achieve the deformation signal. The topographic phase has been removed using the SRTM (Shuttle Radar Topographic Mission) digital elevation model at 90-m spacing.

In Fig. 12, the entire TSX interferograms of the co-event (upper panel) and of the post-event pairs (bottom panel) are reported. Both interferograms are very noisy, except for some areas: on the central-south section some coherent signals are visible, mainly located inside valleys. The signals of these regions are clearly related to atmospheric artifacts. As far as the co-event interferogram (20090518–20090723) is concerned, we found a significant fringe pattern only in the northern portion of the interferogram (detail in Fig. 13). The result from InSAR shows three fringes of deformation corresponding to 40-45 mm displacement along the satellite LOS (Line of Sight), i.e., 35° from nadir. The fringe pattern is well positioned in the range/azimuth directions, reducing distortion-related phenomena and ensuring the exploitation of the full resolution of SAR sensor.

The analysis of the detected signal after applying a phase unwrapping procedure can be interpreted either as a movement towards the satellite or, in other words, an inflation of the top of the mountain with respect to its southern bottom, or as a lowering of the valley to the south of the mountain with respect to the top (Fig. 14). The very short spatial baselines (64 and 55 m for co- and post-event interferograms, respectively) ensure that the TerraSAR-X pairs are poorly dependant on the topography. However, in order to verify possible residual topography, the post-event pair (20090723–20090814) has been processed too, but no fringe is present. This second interferogram is more coherent than the co-event one, because of the shorter temporal and spatial baselines. Looking closely in the same region where we found some fringes in the co-event interferogram, we detect a small coherent area, but no fringes are present. In any case, the complexity of the area in terms of topography and the dense vegetation coverage have prevented us from obtaining better results. Indeed, the capabilities of X-Band SAR, like TerraSAR-X are hampered by scene properties like those mentioned above ,and the longer temporal baseline, mainly for the co-event data, fosters the loss of interferometric coherence.

5. Discussion

The nuclear explosion database (NEDB, Bennett 2010) maintained by the Research and Development Support Services (RDSS, http://www.rdss.info/) reports, when available, ground truth information for the past nuclear tests. Events DPRK06 and DPRK09 are classified in the NEDB as GT1 (uncertainty of ±1 km on the epicentral coordinates). For these events, the ground truth location estimates are mainly based on the satellite imagery of the entrance to the

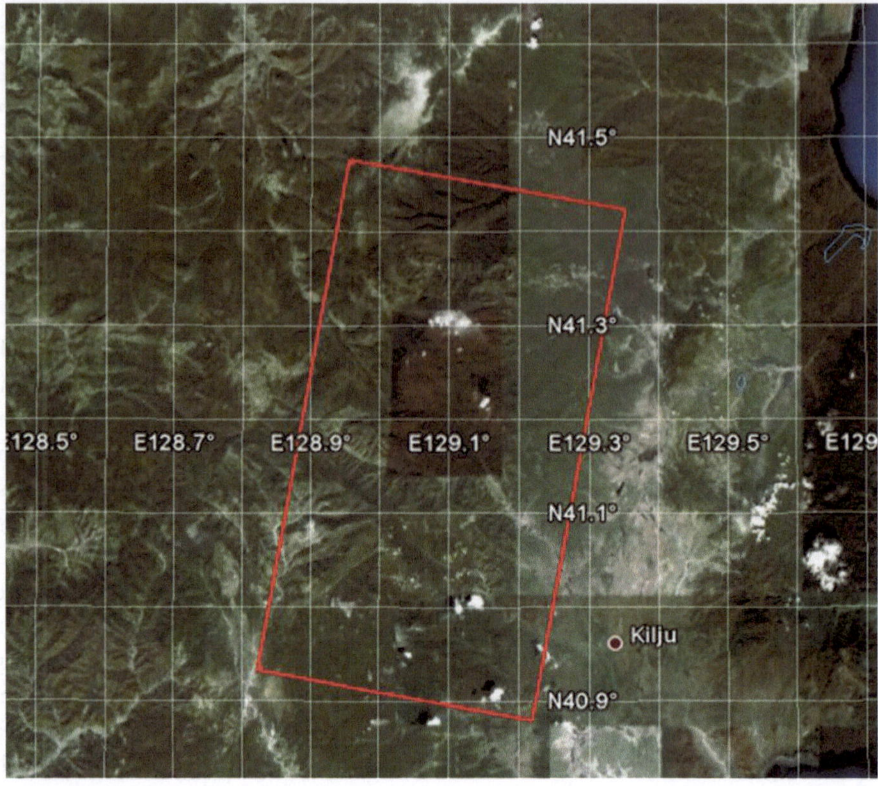

Figure 11
Map of the DPRK region where the 2006 and 2009 tests were carried out. The *red rectangle* shows the area investigated by means of SAR interferometry technique (InSAR)

tunnel that is supposed to have been used for both tests. The evidence that this tunnel was actually used for the nuclear tests is supported by considerations that can not be verified through the methodologies applied in this study (http://www.globalsecurity.org/wmd/world/dprk/kilju.htm). The identification of the tunnel entrance has been combined with analysis of the relative seismic locations of the two events, along with careful assessment of the topography in the vicinity of the tunnel complex and conditions necessary to achieve event containment (assuming normal nuclear testing practice). This analysis produced the so called "preferred" locations for the two events, which are, respectively, 41.287°N, 129.090°E for DPRK06, and 41.293°N, 129.066°E for DPRK09. They are indicated by green marks in Figs. 3 and 15. Both locations are consistent with the epicenters obtained by our seismological analysis, as reported in lines 5 and 6 of Table 2.

Our visual comparison of the pre- and post-event satellite images available from the web for the 2006 test (http://www.globalsecurity.org/wmd/world/dprk/html/kilchu-punggye-yok_comp01.htm) did not provide us with an objective recognition of the NEDB preferred locations as the real test sites. We also applied an accurate analysis on a restricted area around the preferred locations by the interferometric satellite techniques described in the previous section. As a result, we found the following: (1) a majority of the area within the frame does not show any signal coherence, which implies lack of information relevant to potential deformation patterns; (2) the areas characterized by coherence do not show evidence of deformation except for the location focused in Fig. 12.

In the previous section we showed that a few fringes clearly visible on a limited area, covering about 3 km in the east-west direction, denote a

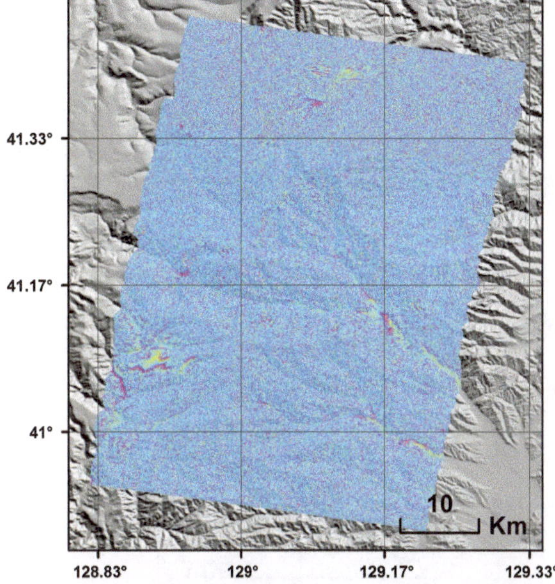

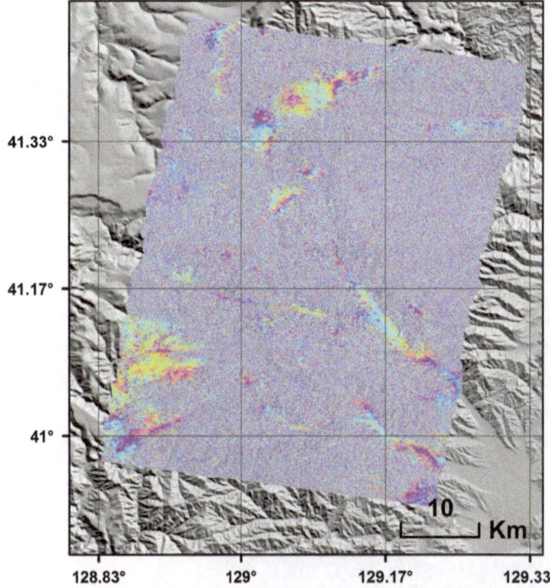

Figure 12
Interferograms of the whole elaborated frames. *Upper panel* refers to the co-event SAR pair, and *bottom panel* refers to the post-event data pair. Both interferograms are noisy except for some coherent areas

displacement of the order of 45 mm occurred between May 18 and July 23, 2009. This time period a little longer than 3 months encompasses the date of the second and larger nuclear test carried out in DPRK on 25 May 2009. The observed displacement is located near the eastern edge of a narrow mountain ridge between two nearly parallel fluvial valleys, at the confluence of two rivers (Fig. 13).

The displacements observed through the DInSAR technique are relative. Therefore, these measurements are usually based on the assumption of absolute stability of a given area with respect to which the displacements of other places are estimated. These displacements are obtained by applying the so-called unwrapping procedure to the detected fringe pattern. In our case, a stable reference zone is not available. Then we assume, as a working hypothesis, the stability of the top of the mountain, which has as a consequence the subsidence of the small red spot shown in Fig. 14 at the bottom of the valley.

As an alternative hypothesis to the preferred solutions reported by the NEDB, we may assume that the observed subsidence was induced by the May 2009 nuclear test and it occurred not far from the place where the nuclear device was detonated. There is no way to prove our assumption objectively, but it appears reasonable at least for the following considerations:

- Subsidence is usually observed as post-seismic effect in the epicentral area of underground nuclear explosions (Houser 1969; Vincent *et al.*, 2003);
- It is reasonable to assume that the epicenter of the explosion is closer to the base of the mountain than to its top;
- The fringes of the interferometric image could close more to the south, but they can not be observed due to the scarce coherence of the InSAR signals in the southern area.

Another possible cause of the measured movement can be the re-activation of a landslide located on the southern flank of the mountain (Moro *et al.*, 2011), triggered by the nuclear test.

If our assumption is true, we must explain the circumstance that the ground truth of the May 2009 test happens to be located about 10 km to the north of the seismological location of such event, out of the 95 % confidence-level error ellipse (Fig. 15). In this respect, it must be recalled that the size of the error ellipse is just a statistical assessment of the consistence among arrival time readings, and it does not represent the absolute reliability of the location. The results of seismological location procedures strongly

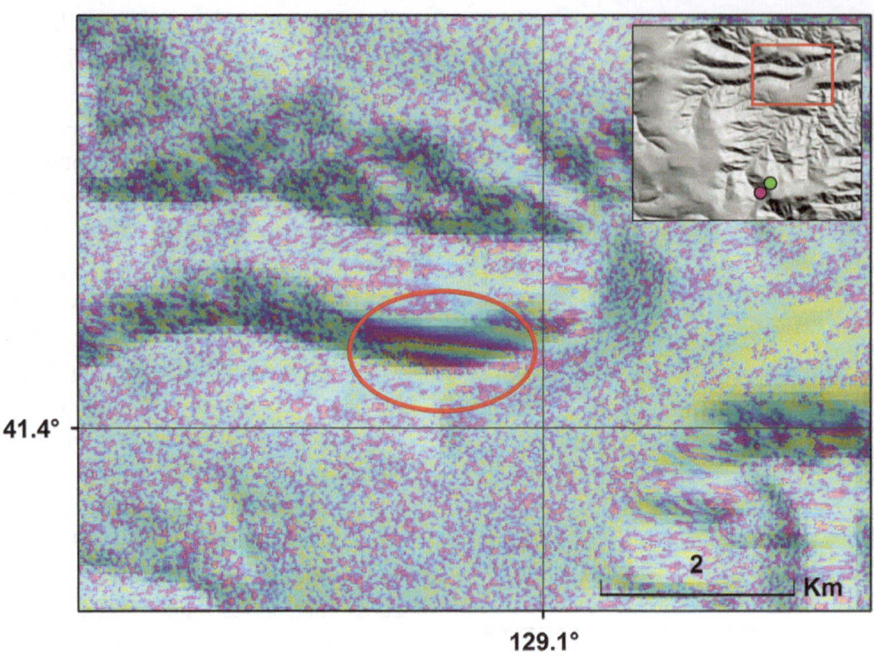

Figure 13
Interferogram showing the detected displacement over the investigated area. A three-fringes pattern is present (*red ellipse*), roughly corresponding to 45-mm LOS movement. The *upper right* sub-panel reports the relative position of the showed interferogram and the *epicenters* locations (rows 5 and 6 in table 2, *magenta* and *green* points, respectively)

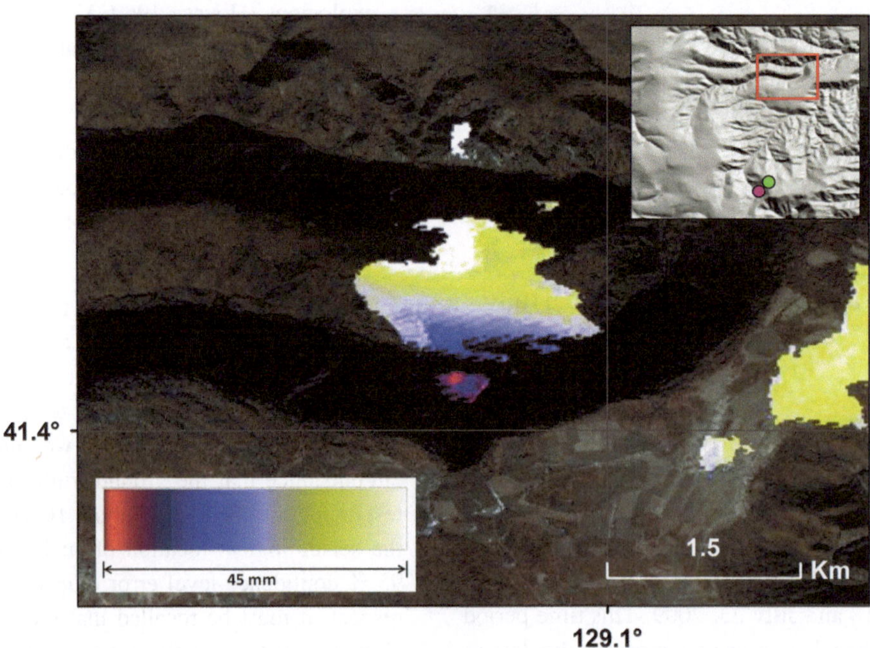

Figure 14
Map showing the results of the unwrapping procedure applied to the fringes of Fig. 12. This map shows a downward movement of the *bottom* of the valley with respect to the *top* of the mountain. The *upper right sub-panel* reports the relative position of the unwrapped interferogram and the *epicenters* locations (rows 5 and 6 in Table 2, *magenta* and *green* points, respectively)

depend on the azimuthal variability of the travel-times due to lateral dishomogeneity in the Earth's structure, and on local time corrections for single stations. These discrepancies between theoretical and real travel-times are not taken into account in our location procedures, and could be responsible of systematic mislocations even larger than the size of the statistical error ellipse, as often noted in the seismological praxis.

To assess the size of the effect of lateral variations of wave velocity on the epicenter locations, we performed tests on seismic events observed in Japan, an area which belongs to the same subduction region as North Korea. We compared the epicenters of two sets of densely clustered earthquakes reported by the ISC with those obtained by the Japan Meteorological Agency (JMA). In consideration of the high density of the JMA national seismological network and the good quality of its locations, these locations were taken as ground truth in our tests. The magnitude of most of the selected events ranged from 4.0 to 5.0 (similar to the m_b magnitudes of DPRK06 and DPRK09), so as to assure a coverage of international seismological stations similar both in azimuth and distance distribution to that used for the nuclear tests. The results of these tests showed that systematic shifts of the epicenters computed by a network composed mostly by teleseismic and regional stations can be as large as 10 km from the ground truth (Giuntini *et al.*, 2012).

The systematic effects due to the dishomogeneity of the Earth structure can be reduced for relative locations, as shown in Sect. 3, by applying such methods as the DDJHD. The application of a relative location method to the waveforms recorded for DPRK06 and DPRK09 showed a shift of the order of 2 km to the east-southeast of the former with respect to the latter. Assuming the ground truth location obtained through the InSAR technique, such shift would locate the DPRK06 detonation point on the same valley and nearly at the same altitude as that of DPRK09. In fact, the trend of the valley bottom in that place is west-northwest to east-southeast.

A distance of 2 km between the detonation points is not very large. However, the strong horizontal variability of the geological conditions in the area (Fig. 1) and a possibly different burial depth of the explosive device could explain the difference between the magnitude of the two tests, even if the yield of the explosions are similar. As we have seen in the introduction, a difference of 0.3 units for the m_b magnitude can be easily obtained for the same yield, assuming regression laws suitable for different lithological environments.

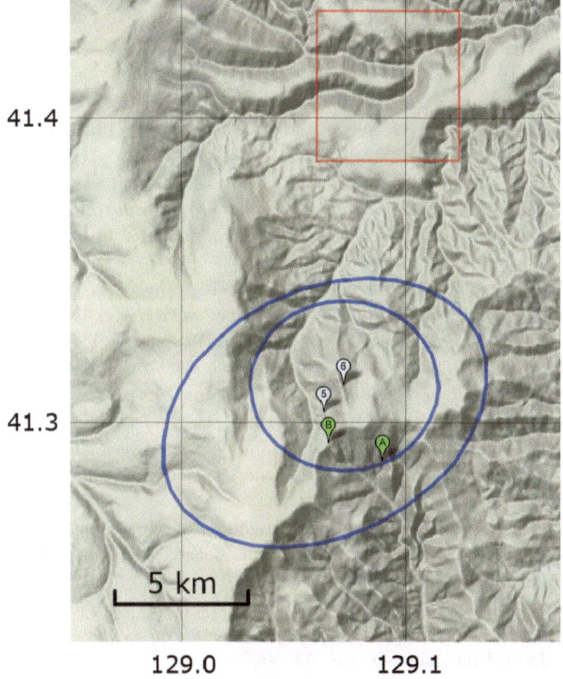

Figure 15
Map of the DPRK region where the 2006 and 2009 tests were carried out. The two *white* tags marked with 5 and 6 point to the epicenters of the DPRK06 and DPRK09 events reported at lines 5 and 6 of Table 2. The respective 95 % confidence level error ellipses are shown in *blue*. The *green tags* point to the preferred locations reported by NEDB. The *red square* shows the area where a significant displacement has been observed through the SAR interferometry technique

6. Conclusions

In Sect. 2 we obtained single locations of the DPRK06 and DPRK09 events by a standard seismological method. In Sect. 3 we compared the locations of the two events by means of the DDJHD and waveform alignment methods, concluding that the two tests were carried out at a distance of 2 km from each other, the DPRK06 epicenter being shifted to the ESE with respect to the DPRK09 epicenter. However, the relative location seen in Sect. 3 doesn't improve the absolute accuracy of any of the two locations.

The NEDB available on the web site (http://www.rdss.info/) reports for DPRK06 and DPRK09 ground truth locations based on the identification of the entrance to the tunnel where the nuclear tests are supposed to be carried out (http://www.globalsecurity.org/wmd/world/dprk/kilju.htm). This identification was supported by activities of intelligence carried out during the years preceding the 2006 test and cannot be verified by means of our purely scientific methods. These ground truth locations are consistent with the epicentral locations obtained by standard seismological methods, within their error ellipses.

The analysis of SAR satellite images does not show any displacement on or close to the NEDB ground truth locations, whereas it puts in evidence an area of subsidence centered about 10 km to the north of the seismological locations. The observed displacement took place during 3 months encompassing the date of the 2009 test. We considered the hypothesis that this subsidence area is strictly related to the nuclear test, possibly as a landslide triggered at a certain distance from the detonation point. This hypothesis can not be rejected just by the separation of 10 km between the epicenter of the event and the subsidence area, by taking into account the systematic shift of the seismological location due to nonhomogeneity of wave velocity affecting the travel times of seismic waves at a global scale.

The present study, together with other recent studies about the use of satellite imageries in underground explosion monitoring, supports the idea of the improvement of the verification regime by additional monitoring technologies such as satellite monitoring, in accordance with Art. IV, Section A, Paragraph 11 of the CTBT.

References

Bennett, T.J., V. Oancea, B. W. Barker, Y.-L. Kung, M. Bahavar, B. C. Kohl, J. R. Murphy, and I. K. Bondár (2010). *The Nuclear Explosion Database (NEDB): A New Database and Web Site for Accessing Nuclear Explosion Source Information and Waveforms*, Seism. Res.Letters, *81*, 1, 12–25, doi:10.1785/gssrl.81.1.12.

Canty, M., Jasani, B., Lingenfelder, I., Nielsen, A. A., Niemeyer, I., Nussbaum, S., Schlittenhardt, J., Shimoni, M., and Skriver, H. (2009). *Treaty Monitoring, in: Remote Sensing from Space—Supporting International Peace and Security* (eds. Jasani B., Pesaresi M., Schneiderbauer, and Zeug, G. *Springer*, p. 167-188.

Cong, X., Schlittenhardt, J., Gutjahr, K., Soergel, U., Canty, M., and Nielsen, A. (2007), *Using differential SAR interferometry for the measurement of surface displacement caused by underground nuclear explosions and comparison with optical change detection results.* In *Global Monitoring for Security and Stability (GMOSS)—Integrated Scientific and Technological Research Supporting Security Aspects of the European Union* (eds. G. Zeug and M. Pesaresi), European Commission—Joint Research Centre, pp. 282–293.

Console, R. and Giuntini, B. (2006). An algorithm for double difference joint hypocenter location: *Application to the 2002 Molise (Central Italy) earthquake sequence*. Annals of Geophysics, *49*, 2/3, 841-852.

Curlander J.C., McDonough R.N.; 1991: *Synthetic Aperture Radar: Systems and Signal Processing*. New York: Wiley-Intersci. 647 pp.

Dawson, J. and Tregoning P. (2007) *Uncertainty analysis of earthquake source parametersdetermined from InSAR: A simulation study*, J. Geophys. Res. *112*, B09406, doi:10.1029/2007JB005209.

Dawson, J. Cummins, P. Tregoning, P., and Leonard, M. (2008). *Shallow intraplate earthquakes in Western Australia observed by Interferometric Synthetic Aperture Radar*. Journal of Geophysical Research *113*, B11408.

Dziewonski, A.M. and Gilbert, F. (1976). *The effect of small, aspherical perturbations on travel times and a re-examination of the correction for ellipticity,*. Geoph. J. R. Astr. Soc., *44*, 7-17.

Elachi C.; 1988: *Spaceborne Radar Remote Sensing: Applications and Techniques*. New York: IEEE. 255 pp.

Giuntini, A., Materni, V., Chiappini, S., Carluccio, R., Console, R., and Chiappini, M. (2012). *Station travel time calibration method improves location accuracy on a global scale,* Seism. Res. Lett. (in press).

Houser, F.N. (1969). *Subsidence related to underground nuclear explosions, Nevada test site*. Bull. Seism. Soc. Am., *56*, 6, 2231–2251.

Kenneth, B.L.N. and Engdahl, E.R. (1991). *Travel times for global earthquake location and phase identification.* Geophysical Journal International, *105*, 429-465.

ISC. On line bulletin of the International Seismological Center, http://www.isc.ac.uk/iscbulletin/search/.

Massonnet, D., Rossi, M., Carmona, C., Adragna, F., Peltzer, G., Feigl, K., and Rabaute, T. (1993). *The displacement field of the Landers earthquake mapped by radar interferometry*, *Nature*, vol. *364*, no. 6433, pp. 138–142, Jul. 1993.

Moro, M., Chini, M., Saroli, M., Atzori, S., Stramondo, S., Salvi, S., (2011). *Analysis of large, seismically induced, gravitational deformations imaged by high resolution COSMO-SkyMed SAR.* Geology, Vol. *39*, No. 6, pp. 527-530, June 2011.

Murphy, J. R. (1981). *P* wave coupling of underground explosions in various geologic media, in Identification of Seismic Sources – Earthquake or Underground Explosion, E. S. Husebye and S. Mykkeltveit (editors), pp. 201–205.

Murphy, J.R., B.C. Kohl, J.L. Stevens, T.J. Bennett, and H.G. Israelsson (2010). *Exploitation of the IMS and other data for a comprehensive, advanced analysis of the North Korean nuclear tests, In: 2010 Monitoring Research Review: Ground-Based Nuclear Explosion Monitoring Technologies*, Science Applications International Corporation, pp. 456–465. https://na22.nnsa.doe.gov/mrr/2010/PAPERS/04-11.PDF.

Ringdal, F., P. D. Marshall, and R. W. Alewine (1992). Seismic yield determination of Soviet underground nuclear explosions at the Shagan River test site, Geophys. J. Int. 109, 65–77.

Schlittenhardt, J., Canty, M., and Grünberg, I. (2010), *Satellite Earth Observation Support CTBT Monitoring: a Case Study of the Nuclear Test in North Korea of Oct. 9, 2006 and comparison with Seismic Results*, Pure and Applied Geophysics, n°*167*, pp 601-618.

Stein, S., and Wysession, M. (2002). *An introduction to seismology, earthquakes, and earthquake structure.* Blackwell Publishing, 498 pp.

Stramondo S., *15 years of SAR Interferometry*, Bollettino di Geofisica Teorica e Applicata, vol. *49*, June 2008.

USGS/NEIC. Online bulletins of the USGS/NEIC Earthquake Hazards Program (http://earthquake.usgs.gov/earthquakes/eqarchives/epic/).

Vincent, P., S. Larsen, D. Gallaway, R.J. Laczniak, B. Foxall, W. Walter, J. Zucca, *New signatures of underground nuclear tests revealed by satellite radar interferometry*, Geophys. Res. Lett., *30*, 22, 2003 (Cover Paper).

Vincent, P., *Detecting Underground Changes from Space*, Science and Technology Review, (Cover Article) LLNL, April, 2005.

Waldhauser, F. and Ellsworth, W.L. (2000). *A Double-Difference Earthquake Location Algorithm: Method and Application to the Northern Hayward Fault*, California. Bull. Seism. Soc. Am., *90*, 1353–1368; doi:10.1785/0120000006.

(Received December 7, 2011, revised September 8, 2012, accepted November 21, 2012, Published online December 29, 2012)

Pure Appl. Geophys. 171 (2014), 361–375

DOI 10.1007/s00024-012-0573-6

Pure and Applied Geophysics

The IDC Seismic, Hydroacoustic and Infrasound Global Low and High Noise Models

David Brown,[1] Lars Ceranna,[2] Mark Prior,[1] Pierrick Mialle,[1] and Ronan J. Le Bras[1]

Abstract—The International Data Centre (IDC) in Vienna, Austria, is determining, as part of automatic processing, sensor noise levels for all seismic, hydroacoustic, and infrasound (SHI) stations in the International Monitoring System (IMS) operated by the Provisional Technical Secretariat of the Comprehensive Nuclear-Test-Ban Treaty Organization (CTBTO). Sensor noise is being determined several times per day as a power spectral density (PSD) using the Welch overlapping method. Based on accumulated PSD statistics a probability density function (PDF) is also determined, from which low and high noise curves for each sensor are extracted. Global low and high noise curves as a function of frequency for each of the SHI technologies are determined as the minimum and maximum of the individual station low and high noise curves, respectively, taken over the entire network of contributing stations. An attempt is made to ensure that only correctly calibrated station data contributes to the global noise models by additionally considering various automatic detection statistics. In this paper global low and high noise curves for 2010 are presented for each of the SHI monitoring technologies. Except for a very slight deviation at the microseism peak, the seismic global low noise model returns identically the Peterson (1993) NLNM low noise curve. The global infrasonic low noise model is found to agree with that of Bowman *et al.* (2005, 2007) but disagrees with the revised results presented in Bowman *et al.* (2009) by a factor of 2 in the calculation of the PSD. The global hydroacoustic low and high noise curves are found to be in quantitative agreement with Urick's oceanic ambient noise curves for light to heavy shipping. Whale noise is found to be a feature of the hydroacoustic high noise curves at around 15 and 25 Hz.

Key words: Seismic, infrasound, hydroacoustic, global noise models.

1. Introduction

The Provisional Technical Secretariat for the Preparatory Commission for the Comprehensive Nuclear-Test-Ban Treaty Organization is tasked with establishing the verification regime for the Comprehensive Nuclear-Test-Ban Treaty (CTBT) that, upon Entry Into Force (EIF), bans the detonation of nuclear devices in any environment. The framework of the verification regime is the global network of 337 seismic, hydroacoustic, infrasound, and radionuclide stations that form the International Monitoring System. The International Data Centre processes in near real time data received from the IMS stations, subsequently generating several event bulletins for the benefit of the States Parties that are signatories to the Treaty.

Along with its regular event processing, the IDC is also mandated to record and monitor station ambient noise with the expectation that knowledge of this sort may be indicative of station state of health (SOH).

Station ambient noise conditions are being represented by the power spectral density (PSD), which provides a measure of the power contained in the signal at each frequency.

Determining both single station and network low and high noise models becomes a straightforward procedure when station ambient noise information is routinely accessible. The purpose of this paper is to present the inferred global low and high noise models for each of the SHI technologies based on data recorded by the IMS network. In doing so, care is taken at all stages to ensure both the integrity of the data and the method used to determine the station noise information.

Section 2 of this paper discusses the method used in determining the station ambient noise, as well as

[1] International Data Centre, Comprehensive Nuclear-Test-Ban Treaty Organization, Preparatory Commission, IDC/CTBTO Vienna International Centre, PO Box 1200, 1400 Vienna, Austria. E-mail: David.Brown@ctbto.org

[2] Bundesanstalt für Geowissenschaften und Rohstoffe (BGR), Geozentrum Hannover Stilleweg 2, 30655 Hannover, Germany.

tests performed to ensure that the chosen PSD method is providing the correct estimation of the PSD.

Section 3 discusses the data used during the analysis, and section four the results for each of the SHI technologies.

2. Numerical Method and Testing

2.1. Numerical Method

The Welch overlapping method (Welch 1967) forms the basis of the procedure used in this paper to determine the PSD. In this method the time interval spanning the data under consideration is divided into a sequence of overlapping sub-intervals and the PSD for each sub-interval is determined. The average of the sub-interval PSD's is assumed to provide an estimate of the required PSD. This averaging procedure reduces the variance on the estimated PSD, which may otherwise be of the order of the contributing sample values (Welch 1967). Strictly speaking, the use of the PSD requires that the waveform under consideration be a Wide-Sense Stationary Process (see, e.g., Schreier and Scharf 2010), implying that both the mean and autocorrelation of the sampled waveform are time independent, in which case the PSD is the Fourier transform of the autocorrelation function as asserted by the Wiener–Khinchin theorem. Here we are assuming that stationarity is assumed to hold as the propagating signals are considered to be short-lived transitory phenomenon and will thus provide an only minor impact on the statistics.

When evaluating the PSD's, careful attention has been given to the spectral windowing process, which is discussed more fully in the Appendix. In this analysis we have chosen to use the nutall4a window of Heinzel *et al.* (2002), which is a good general purpose spectral window with spectral leakage properties superior to that of the Hanning or Hamming windows.

The PSD estimate used in this paper is given by the expression $P_{\mathrm{SD}}(\omega_j) = \frac{2|F(\omega_j)|^2}{\Delta I_W}$ where $\omega_j = \frac{2\pi j}{n\Delta}$ for $j = 0,\ldots,n/2$ is the jth frequency picket, n is the number of samples, Δ is the sample rate, F is the output of a unitary Fourier Transform algorithm and I_w the sum of the squares of the spectral window coefficients. This expression is discussed further in the Appendix and derived more fully in references like Harris (1978), and Heinzel *et al.* (2002).

2.2. Testing

Two levels of testing have been applied to the algorithm described in Sect. 2.1. The first is against an artificial dataset intended to mimic digitizer noise for which the PSD has a theoretically determinable expression. The second is a blind test with a second algorithm on an otherwise random selected data set.

2.2.1 Test 1: Digitizer Noise

The procedure described in Heinzel *et al.* (2002) has been used to test the algorithm. Here, each sample in a synthetic time-series data set has been rounded to multiples of a parameter U_0 defined a priori, which represents the least-significant-bit of a digitizer. In this case, the PSD has a noise-floor that is given by the expression (see, e.g., Lyons 1997) $U^2(\omega) = \frac{U_0^2}{6\Delta}$, where ω is angular frequency.

The following strategy outlined in Heinzel *et al.*, (2002) was used to generate the synthetic time-series data set:

1. The double-sinusoid $u(t) = A_1 \sin(2\pi f_1 t) + A_2 \sin(2\pi f_2 t)$ is used to provide the basic analogue signal. Here, $f_1 = 0.3123456$ Hz, $f_2 = 2.0$ Hz, $A_1 = 2.123456$, $A_2 = 1.0$ is used.
2. $u(t)$ has been sampled at 20 Hz to provide the time-series: $u_j = A_1 \sin\left(\frac{2\pi f_1 j}{\Delta}\right) + A_2 \sin\left(\frac{2\pi f_2 j}{\Delta}\right)$ for $j = 1,\ldots,N$.
3. The new time-series y_j for $j = 1,\ldots,N$ is formed, where $y_j = \mathrm{int}\left[\frac{u_j}{U_0} + 0.5\right] U_0$.
4. With the values for A_1, A_2, f_1, f_2, U_0, and Δ as given above we should expect a white noise background with $\log_{10}\mathrm{PSD} = \log_{10}\left(\frac{U_0}{6\Delta}\right) = -8.079$.

With this procedure a time-series with duration 1-h was generated. After passing the algorithm as described above over the synthetic data set, the PSD as shown in Fig. 1 was obtained.

The desired noise floor is accurately rendered, and further, when multiplying by the Equivalent Noise Bandwidth for the nutall4a window (see the

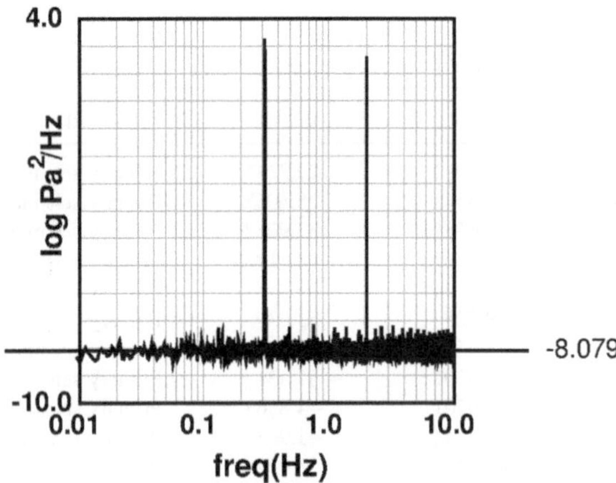

Figure 1
Power spectral density as a function of frequency for the Test 1 data set

Appendix), the amplitude of the spectral peaks is accurately determined, as is shown in Fig. 1.

2.2.2 *Test 2: Blind Test with a Second Independent PSD Algorithm*

One-hour data segments were chosen randomly from the list of infrasound stations provided in Table 1. A completely independent second algorithm that uses the matlab *pwelch* function (MATLAB 2008) to estimate PSD's (denoted as BGR in what follows) was provided by one of the authors and compared with the algorithm described above (denoted CTBTO in what follows). The results are shown in Fig. 2 and indicate excellent agreement.

We conclude from these two tests that the algorithm is functioning, as it should.

3. *The Data*

The IMS seismic, hydroacoustic and infrasound stations that were used in the current analysis are listed in Table 1; the locations of the stations is shown in Fig. 3.

Both primary and auxiliary seismic stations were used in this analysis, and although data from the primary stations is continuously recorded, the auxiliary stations generally send data upon request in order to refine knowledge of a seismic event during the analysis stage. Data from the auxiliary seismic stations may therefore not be present for the entire analysis period.

Data sampled four times per day for the whole of 2010 were used in the current analysis except in situations where better resolution of the low or high noise models was required for a particular station where the data was sampled each hour for the entire year.

Each sampling consisted of 1-h of waveform data divided into 3-min overlapping segments as outlined above, with the seismic data being deconvolved to acceleration and the infrasound and hydroacoustic data deconvolved to displacement, the instrument response has been removed in each case. Note that in the case of the infrasound data only the response of the sensor and digitizer has been removed, not that of the spatial filter system that was likely to be present.

4. *Low and High Noise Models*

Probability Density Functions using the procedure discussed by McNAMARA and BULAND (2004) are determined for each SHI sensor. Data displayed in this format allows a ready estimate of the sensor low and high noise models for the period in which data was contributed. As an example, Fig. 4 shows the PDF obtained for sensor I02AR/I02H1 for the year 2010. The low and high noise curves as a function of frequency are determined by plotting the PDF function at the 5 and 95 % probability levels, respectively. The global low noise curve as a function of frequency is defined to be, for a given frequency, the minimum of the low noise curves from all contributing sensors across all stations at the given frequency. Only stations with sufficient contributing data such that a well-defined PDF is formed with well-behaved low and high noise curves were used in the analysis.

4.1. *Seismic*

A random shift of up to 6 h is applied to the requested processing time for the seismic data in order to reduce the likelihood of contamination by regular cultural noise. Only vertical channels were considered when performing the seismic analysis,

Table 1

IMS seismic, hydroacoustic and infrasound stations that were used in the current analysis

Station	Location	State	Latitude	Longitude	Station type
AKASG	Malin	Ukraine	50.7012	29.2242	Primary seismic
ARCES	Finnmark	Norway	69.5348	25.5057	Primary seismic
ASAR	Alice Springs	Australia	−23.665134	133.90526	Primary seismic
BDFB	Brasilia	Brazil	−15.64178	−48.01485	Primary seismic
BOSA	Boshof	South Africa	−28.61405	25.25542	Primary seismic
BRTR	Belbashi	Turkey	39.725	33.639	Primary seismic
CMAR	Chiang Mai	Thailand	18.4576	98.94315	Primary seismic
CPUP	Villa Florida	Paraguay	−26.3307	−57.331	Primary seismic
DBIC	Dimbroko	Cote d'Ivoire	6.6701	−4.8563	Primary seismic
ESDC	Sonseca	Spain	39.6744	−3.963	Primary seismic
FINES	Lahti	Finland	61.4436	26.0771	Primary seismic
GERES	Freyung	Germany	48.845106	13.701559	Primary seismic
GEYT	Alibeck	Turkmenistan	37.92955	58.11706	Primary seismic
ILAR	Eielson	United States of America	64.771446	−146.88665	Primary seismic
KBZ	Khabaz	Russian Federation	43.726898	42.8996	Primary seismic
KEST	Kesra	Tunisia	35.7317	9.346	Primary seismic
KMBO	Kilimambogo	Kenya	−1.1268	37.2523	Primary seismic
KSRS	Wonju	Republic of Korea	37.4421	127.8844	Primary seismic
LPAZ	La Paz	Bolivia	−16.287927	−68.130706	Primary seismic
MAW	Mawson Antarctica	Australia	−67.6046	62.8713	Primary seismic
MJAR	Matushiro	Japan	36.524717	138.24718	Primary seismic
MKAR	Makanchi	Kazakhstan	46.7937	82.2904	Primary seismic
NOA	Hamar	Norway	61.0397	11.2148	Primary seismic
NVAR	Mina Nevada	United States of America	38.429609	−118.30355	Primary seismic
PDAR	Pinedale Wyoming	United States of America	42.7667	−109.5579	Primary seismic
PETK	Petropavlovsky-Kamchatskiy	Russian Federation	53.108215	157.69885	Primary seismic
PLCA	Paso Flores	Argentina	−40.732733	−70.550835	Primary seismic
PPT	Tahiti	France	−17.5896	−149.5764	Primary seismic
ROSC	El Rosal	Columbia	4.844856	−74.321203	Primary seismic
SCHQ	Schefferville Quebec	Canada	54.832402	−66.833177	Primary seismic
SONM	Songino	Mongolia	47.83469	106.39499	Primary seismic
STKA	Stephens Creek	Australia	−31.8743	141.5964	Primary seismic
TORD	Torodi	Niger	13.14771	1.6947087	Primary seismic
TXAR	Lajitas Texas	United States of America	29.333965	−103.66769	Primary seismic
ULM	Lac Du Bonnet, Manitoba	Canada	50.250261	−95.874956	Primary seismic
USRK	Ussuriysk	Russian Federation	44.1998	131.9888	Primary seismic
VNDA	Vanda, Antarctica	United States of America	−77.5173	161.8528	Primary seismic
WRA	Warramunga NT	Australia	−19.942589	134.33951	Primary seismic
YKA	Yellowknife	Canada	62.4931	−114.6062	Primary seismic
ZALV	Zalesovo	Russian Federation	53.948063	84.818807	Primary seismic
AAK	Ala-Archa	Kyrgyzstan	42.6391	74.4942	Auxiliary seismic
AFI	Afiamalu	Samoa	−13.9093	−171.7773	Auxiliary seismic
AKTO	Aktyubinsk	Kazakhstan	50.4348	58.0164	Auxiliary seismic
ANMO	Albuquerque, New Mexico	United States of America	34.9462	−106.4567	Auxiliary seismic
ASF	Tel Al Asfar	Jordan	32.1723	36.8972	Auxiliary seismic
ATAH	Atahualpa	Peru	−7.13506	−78.39445	Auxiliary seismic
ATD	Arta Tunnel	Djibouti	11.53	42.847	Auxiliary seismic
BATI	Baumata, Nusa Tengarra	Indonesia	−10.206	123.6627	Auxiliary seismic
BBB	Bella Bella	Canada	52.1847	−128.1133	Auxiliary seismic
BBTS	Babate	Senegal	14.6604	−16.5334	Auxiliary seismic
BORG	Borganes	Iceland	64.7474	−21.3268	Auxiliary seismic
BVAR	Borovoye	Kazakhstan	53.0249	70.3885	Auxiliary seismic
CFAA	Coronel Fontana	Argentina	−31.60475	−68.23756	Auxiliary seismic
CMIG	Colonia Cuauhtemoc, Oaxaca	Mexico	17.091	−94.8838	Auxiliary seismic
CTA	Charters Towers	Australia	−20.0876	146.25	Auxiliary seismic

Table 1 *continued*

Station	Location	State	Latitude	Longitude	Station type
DAVOX	Davos	Switzerland	46.7806	9.8797	Auxiliary seismic
DLBC	Dease Lake, British Columbia	Canada	58.43696	−130.03051	Auxiliary seismic
DZM	Mont Dzumad	New Caledonia	−22.068	166.4469	Auxiliary seismic
EIL	Eilath	Israel	29.6725	34.9519	Auxiliary seismic
EKA	Eskdalemuir	United Kingdom	55.3332	−3.1588	Auxiliary seismic
ELK	Elko, Nevada	United States of America	40.7448	−115.2388	Auxiliary seismic
FITZ	Fitzroy Crossing	Australia	−18.09826	125.6403	Auxiliary seismic
FRB	Iqaluit	Canada	63.7467	−68.5467	Auxiliary seismic
GNI	Garni	Armenia	40.1495	44.7414	Auxiliary seismic
GUMO	Guam, Marianas Islands	United States of America	13.5892	144.8684	Auxiliary seismic
HFS	Hagfors	Sweden	60.133474	13.69449	Auxiliary seismic
HNR	Honiara	Solomon Islands	−9.4322	159.9471	Auxiliary seismic
IDI	Anóyia	Greece	35.288	24.89	Auxiliary seismic
INK	Inuvik, Northwest Territory	Canada	68.306516	−133.52543	Auxiliary seismic
JCJ	Chichijima, Ogasawara	Japan	27.095467	142.18463	Auxiliary seismic
JKA	Kamikawa-Asahi, Hokkaido	Japan	44.11895	142.59325	Auxiliary seismic
JMIC	Jan Mayen	Norway	70.9866	−8.50515	Auxiliary seismic
JNU	Ohita, Kyushu	Japan	33.121667	130.87833	Auxiliary seismic
JOW	Kunigami, Okinawa	Japan	26.836	128.2731	Auxiliary seismic
JTS	Las Juntas de Abangares	Costa Rica	10.2908	−84.9525	Auxiliary seismic
KAPI	Kappang, Sulawesi	Indonesia	−5.0142	119.7517	Auxiliary seismic
KDAK	Kodiak Island, Alaska	United States of America	57.7828	−152.5835	Auxiliary seismic
KURK	Kurchatov	Kazakhstan	50.62264	78.53039	Auxiliary seismic
KVAR	Kislovodsk, Stavropol'skiy	Russian Federation	43.9557	42.6952	Auxiliary seismic
LBTB	Lobatse	Botswana	−25.0151	25.5966	Auxiliary seismic
LEM	Lembang, Jawa Barat	Indonesia	−6.82645	107.61748	Auxiliary seismic
LPIG	La Paz, Baja California Sur	Mexico	24.10103	−110.30931	Auxiliary seismic
LSZ	Lusaka	Zambia	−15.2766	28.1882	Auxiliary seismic
MATP	Matopo	Zimbabwe	−20.42583	28.49944	Auxiliary seismic
MBAR	Mbarara	Uganda	−0.6019	30.7382	Auxiliary seismic
MDT	Midelt	Morocco	32.814	−4.607	Auxiliary seismic
MLR	Muntele Rosu	Romania	45.4917	25.9437	Auxiliary seismic
MMAI	Mount Meron	Israel	33.01518	35.40311	Auxiliary seismic
MSKU	MasUnited Kingdomu	Gabon	−1.6557	13.6116	Auxiliary seismic
NEW	Newport, Washington	United States of America	48.26333	−117.12	Auxiliary seismic
NNA	Nana	Peru	−11.9873	−76.8422	Auxiliary seismic
NWAO	Narrogin	Australia	−32.9277	117.239	Auxiliary seismic
OBN	Obninsk	Russian Federation	55.1138	36.5687	Auxiliary seismic
OPO	Ambohidratompo	Madagascar	−18.5706	47.1879	Auxiliary seismic
PALK	Pallekele	Sri Lanka	7.2728	80.7022	Auxiliary seismic
PCRV	Puerto la Cruz	Venezuela	10.1634	−64.58963	Auxiliary seismic
PFO	Pinon Flat, California	United States of America	33.6092	−116.4553	Auxiliary seismic
PMG	Port Moresby	Papua New Guinea	−9.4092	147.1539	Auxiliary seismic
PMSA	Palmer Station, Antarctica	United States of America	−64.7742	−64.049	Auxiliary seismic
PSI	Parapat, Sumatra	Indonesia	2.6952	98.924	Auxiliary seismic
QSPA	South Pole, Antarctica	United States of America	−89.9279	145.0	Auxiliary seismic
RAO	Raoul, Kermadec Islands	New Zealand	−29.2517	−177.9183	Auxiliary seismic
RAR	Raratonga	Cook Islands	−21.2125	−159.7733	Auxiliary seismic
RCBR	Riachuelo	Brazil	−5.82739	−35.90131	Auxiliary seismic
RES	Resolute, Nunavut	Canada	74.689233	−94.896167	Auxiliary seismic
RPN	Easter Island	Chile	−27.1267	−109.3344	Auxiliary seismic
RPZ	Rata Peaks	New Zealand	−43.7146	171.054	Auxiliary seismic
SADO	Sadowa	Canada	44.7694	−79.1417	Auxiliary seismic
SEY	Seymchan	Russian Federation	62.9328	152.3822	Auxiliary seismic
SFJD	Søndre Strømford	Greenland	66.995999	−50.6215	Auxiliary seismic
SIV	San Ignacio	Bolivia	−15.991	−61.072	Auxiliary seismic
SJG	San Juan	Puerto Rico	18.1117	−66.15	Auxiliary seismic

Table 1 *continued*

Station	Location	State	Latitude	Longitude	Station type
SNAA	Sanae Station, Antarctica	Germany/South Africa	−71.6707	−2.8379	Auxiliary seismic
SPITS	Spitsbergen	Norway	78.1777	16.37	Auxiliary seismic
SUR	Sutherland	South Africa	−32.3797	20.8117	Auxiliary seismic
TEIG	Tepich, Yucatan	Mexico	20.2264	−88.2776	Auxiliary seismic
TGY	Tagatay City	Philippines	14.1008	120.93837	Auxiliary seismic
TKL	Tuckaleechee Caverns, Tennessee	United States of America	35.658	−83.774	Auxiliary seismic
TSUM	Tsumeb	Namibia	−19.2022	17.5838	Auxiliary seismic
URZ	Urewera	New Zealand	−38.2592	177.1109	Auxiliary seismic
USHA	Ushuaia	Argentina	−54.83192	−68.43432	Auxiliary seismic
VAE	Valguarnera	Italy	37.469	14.3533	Auxiliary seismic
VRAC	Vranov	Czech Republic	49.30828	16.59351	Auxiliary seismic
YBH	Yreka Blue Horn	United States of America	41.73193	−122.71038	Auxiliary seismic
I02AR	Ushuaia	Argentina	−54.58057	−67.30923	Infrasound
I04AU	Shannon	Australia	−34.59761	116.35669	Infrasound
I05AU	Hobart	Australia	−42.490798	147.68063	Infrasound
I07AU	Warramunga NT	Australia	−19.93482	134.32953	Infrasound
I08BO	Penas-Bolivia	Bolivia	−16.21523	−68.45345	Infrasound
I09BR	Brasilia	Brazil	−15.637967	−48.016422	Infrasound
I10CA	Lac Du Bonnet	Canada	50.201469	−96.026854	Infrasound
I11CV	Maio	Cape Verde	15.25729	−23.18388	Infrasound
I13CL	Easter Island	Chile	−27.12726	−109.36265	Infrasound
I14CL	Robinson Carusoe Island	Chile	−33.65379	−78.79598	Infrasound
I17CI	Dimbokro	Ivory Coast	6.6703566	−4.8569106	Infrasound
I18DK	Qaanaaq	Greenland	77.47556	−69.28776	Infrasound
I21FR	Marquesas Islands	France	−8.86783	−140.15907	Infrasound
I22FR	Port Laguerre New Caldeonoia	FRANCE	−22.18445	166.84592	Infrasound
I23FR	Kerguelen	France	−49.34578	70.24159	Infrasound
I24FR	Tahiti	France	−17.74929	−149.29582	Infrasound
I26DE	Freyung	Germany	48.851617	13.713128	Infrasound
I27DE	Georg von Neumayer Antarctica	Germany	−70.7011	−8.30291	Infrasound
I30JP	Isumi	Japan	35.307756	140.31376	Infrasound
I31KZ	Aktyubinsk	Kazakhstan	50.40697	58.03482	Infrasound
I32KE	Nairobi	Kenya	−1.24216	36.82721	Infrasound
I33MG	Antananarivo	Madagascar	−19.010859	47.305024	Infrasound
I34MN	Songino	Mongolia	47.80172	106.41012	Infrasound
I35NA	Tsumeb	Namibia	−19.19135	17.57678	Infrasound
I36NZ	Chatham Islands	New Zealand	−43.91662	−176.48337	Infrasound
I39PW	Palau	Palau	7.53547	134.54704	Infrasound
I41PY	Villa Florida	Paraguay	−26.3423	−57.31188	Infrasound
I43RU	Dubna	Russian Federation	56.72136	37.21759	Infrasound
I44RU	Petropavlovsk-Kamchatsky	Russian Federation	53.1058	157.7139	Infrasound
I45RU	Ussuriysk	Russian Federation	44.1999	131.9773	Infrasound
I46RU	Zalesovo	Russian Federation	53.94872	84.81891	Infrasound
I47ZA	Boshof	South Africa	−28.621123	25.235228	Infrasound
I48TN	Kesra	Tunisia	35.80523	9.32302	Infrasound
I49 GB	Tristan Da Cunha	United Kingdom	−37.08995	−12.33192	Infrasound
I50 GB	Ascension Island	United Kingdom	−7.93774	−14.37517	Infrasound
I51 GB	Bermuda	United Kingdom	32.36154	−64.69874	Infrasound
I52 GB	Diego Garcia	United Kingdom	−7.37781	72.484161	Infrasound
I53US	Fairbanks Alaska	United States of America	64.875	−147.86114	Infrasound
I55US	Windless Bight Antarctica	United States of America	−77.73149	167.58742	Infrasound
I56US	Newport Washington	United States of America	48.26408	−117.12567	Infrasound
I57US	Pinon Flat California	United States of America	33.605852	−116.45328	Infrasound
I59US	Hawaii	United States of America	19.591532	−155.8936	Infrasound
H01	Cape Leeuwin	Australia	−34.88316	114.13608	Hydroacoustic
H03	Juan Fernandez Island	Chile	−33.825843	−78.909483	Hydroacoustic
H08	Diego Garcia	United Kingdom	−7.6275	72.48383	Hydroacoustic

Table 1 *continued*

Station	Location	State	Latitude	Longitude	Station type
H10	Ascension Island	United Kingdom	−8.95274	−14.6629	Hydroacoustic
H11	Wake Island	United States of America	18.49568	166.68646	Hydroacoustic

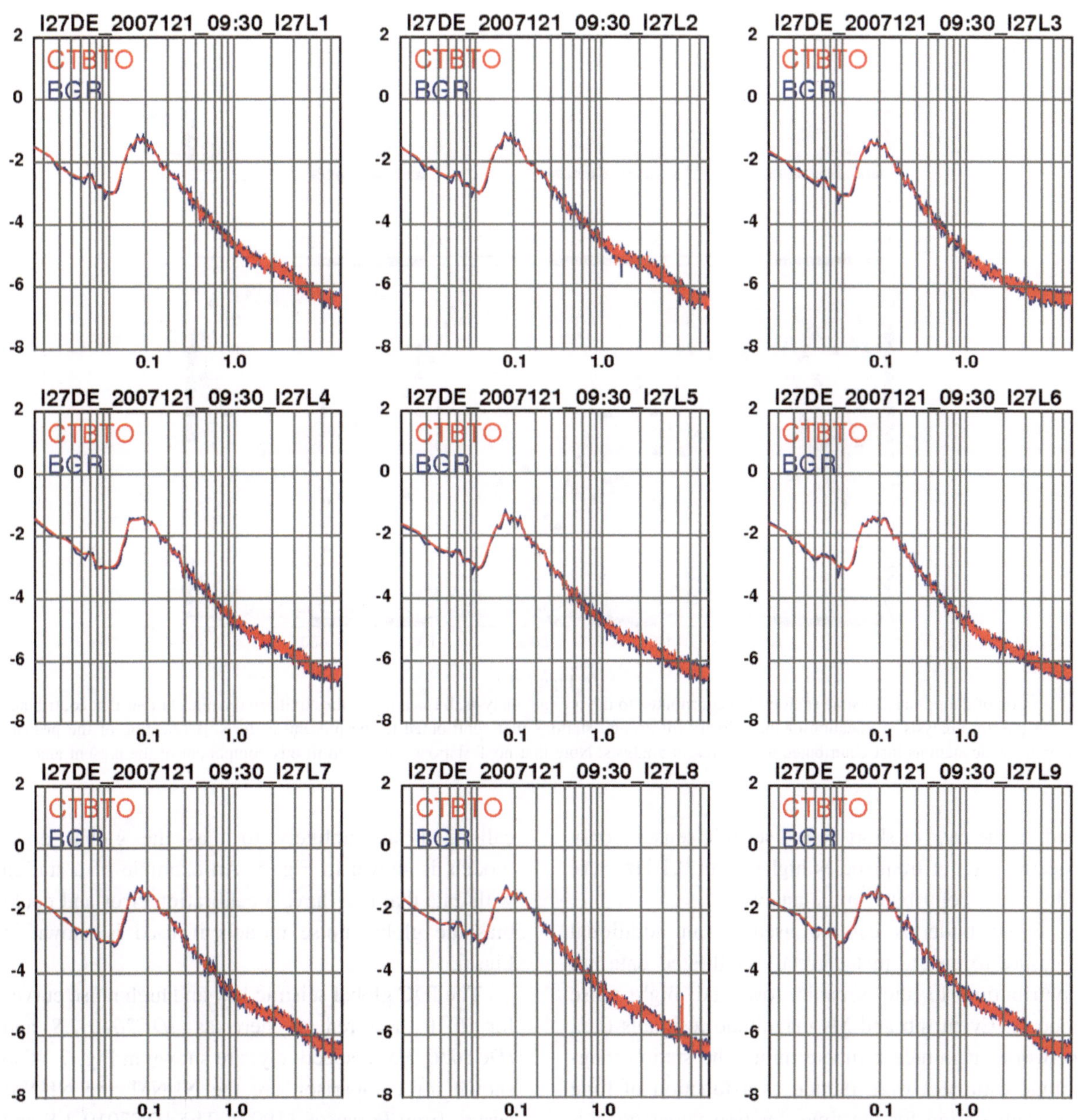

Figure 2
Results of a blind-comparison test between two different PSD algorithms on randomly chosen infrasound data. Abscissa values are frequency in Hz, and ordinate values are logarithm base 10 of the PSD in Pa^2 per Hz. The CTBTO data refer to results generated by the procedure discussed in this Paper. The BGR data refers to a second independent algorithm provided by one of the present authors

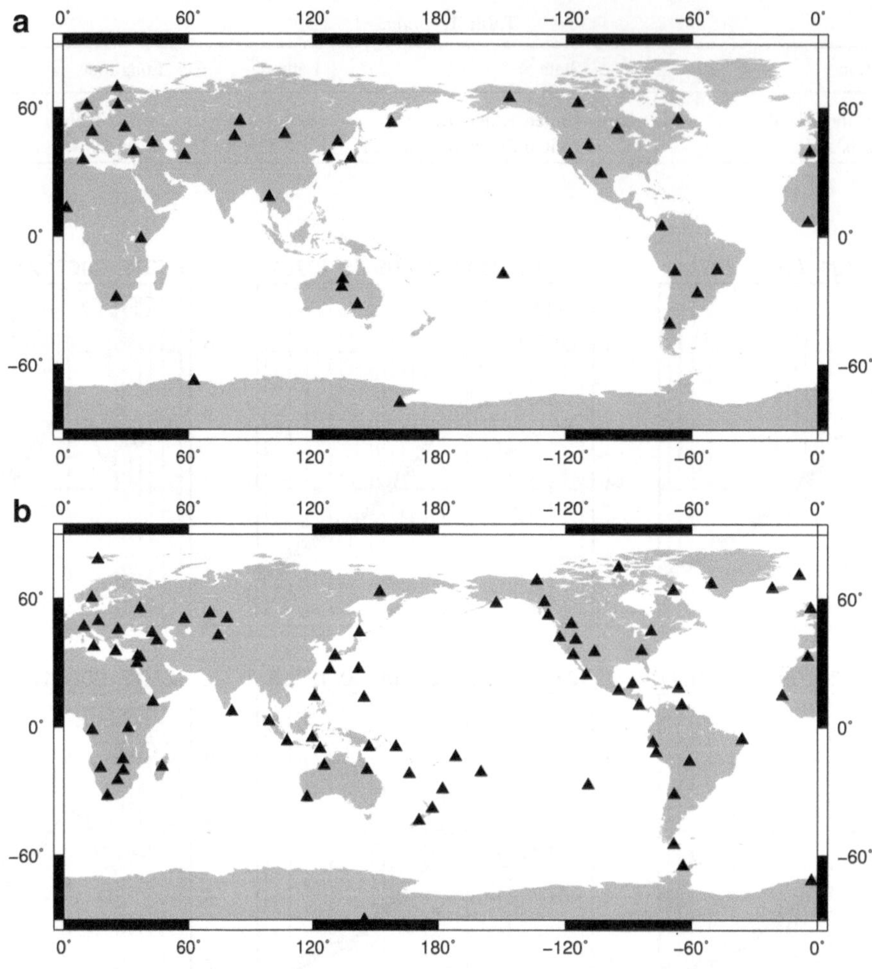

Figure 3

a Location of the primary seismic stations that contributed to the present analysis. **b** Location of the auxiliary seismic stations that contributed to the present analysis. **c** Location of the primary infrasound stations that contributed to the present analysis. **d** Location of the primary hydroacoustic stations that contributed to the present analysis. Note that no T-stations were used in this component of the present analysis

and, in the case of short-period seismic sensors, only spectral data at frequencies higher than 0.1 Hz were allowed to contribute to the analysis.

Event bulletins can be used as an additional measure to ensure that correctly calibrated data are contributing to the seismic low and high noise models. Event mb and Ms magnitude residuals (i.e., network magnitude estimate minus the station magnitude estimate) were plotted as a function of time for each station for the time duration under consideration and a station allowed to contribute to the global noise models if the Ms and mb magnitude residuals were less than 0.3 magnitude units. An example of a station with data that is correctly calibrated and unlikely to bias the global noise models is shown in Fig. 5. An example of a station with data that may have a calibration error and could bias the global noise models if used is shown in Fig. 6.

The IDC global seismic low and high noise curves for 2010, referred to here as *IDC2010_LS* and *IDC2010_HS*, respectively, are shown in Fig. 7. Also shown for comparison are the NLNM and NHNM curves from Peterson (1993). The IDC2010_LS and NLNM curves are in good agreement, particularly in the low noise case. The microseism peak is slightly lower in the case of the IDC2010_LS model; station AAK (Ala Archa, Kyrgyzstan) was found to be

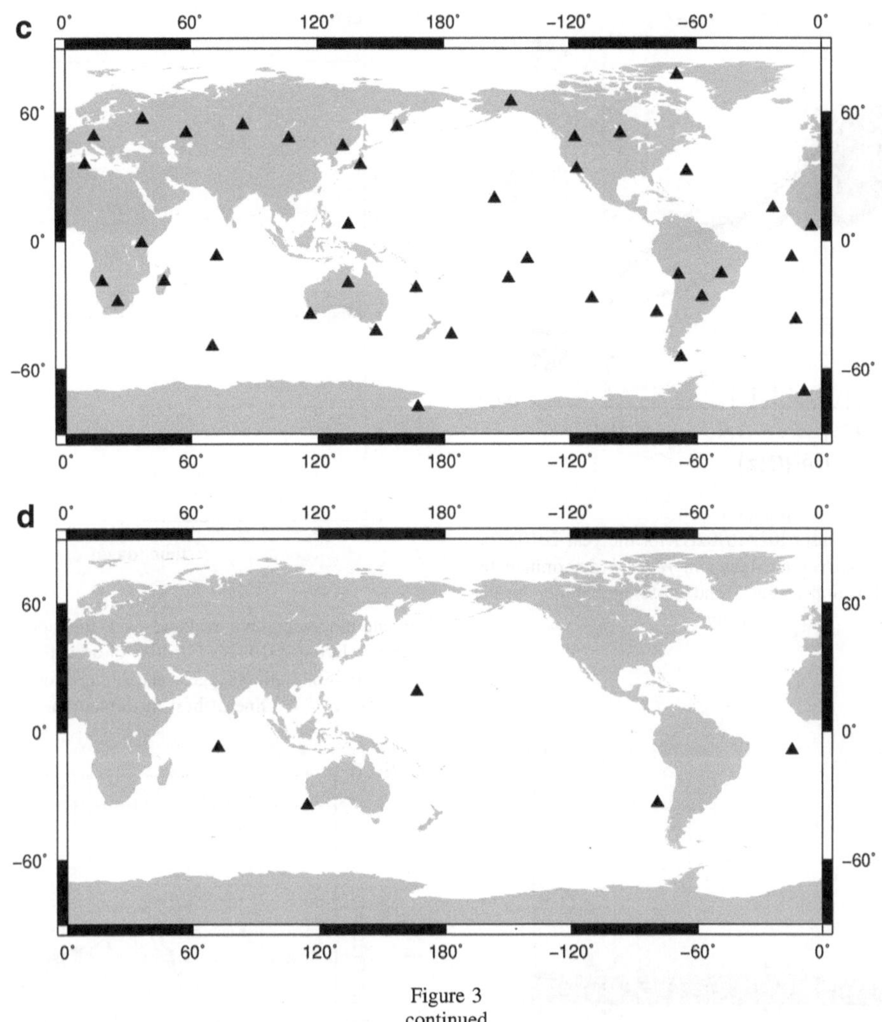

Figure 3
continued

largely responsible for this slight deviation. In order to better resolve the contribution of this station to the low noise curve, data for the entire year, i.e., every hour in the day, for station AAK was processed.

Fairly good agreement exists between IDC2010_HS and the NHNM. The microseismic peak is slightly elevated in the IDC2010_HS model compared to the NHNM model, which is found to be due largely to station BORG (Borgarfjordur, Asbjarnarstadir, Iceland). This station commenced operation in 1994, so it would not have contributed to the PETERSON (1993) analysis. Waveform data for stations BORG was computed each hour for the entire year to better resolve its contribution to the IDC2010_HS model.

4.2. Hydroacoustic

Once again a random time shift of up to 6 h is applied to the requested processing time in order to reduce contamination from regular cultural noise.

The IDC global hydroacoustic low and high noise curves for 2010, referred to here as *IDC2010_LH* and *IDC2010_HH*, respectively, are shown in Fig. 8. Stations that contributed to these curves are shown in Fig. 3d. No T-stations were used during this analysis, so only in-water hydroacoustic signals contributed to the global noise curves. The low and high noise curves differ by a constant 20 dB from 0.01 Hz to around 6 Hz where the low noise curve begins to register shipping noise, making the curve relatively

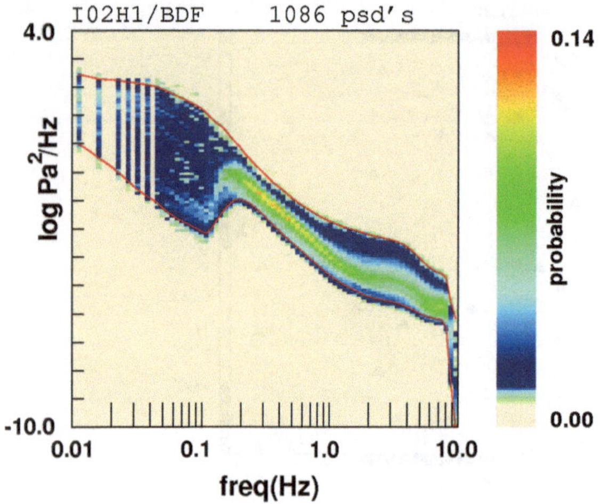

Figure 4
The PDF obtained for sensor I02AR/I02H1 for the year 2010. The *low and high noise curves* as shown in *red* are determined by plotting the PDF function at the 5 and 95 % probability levels, respectively

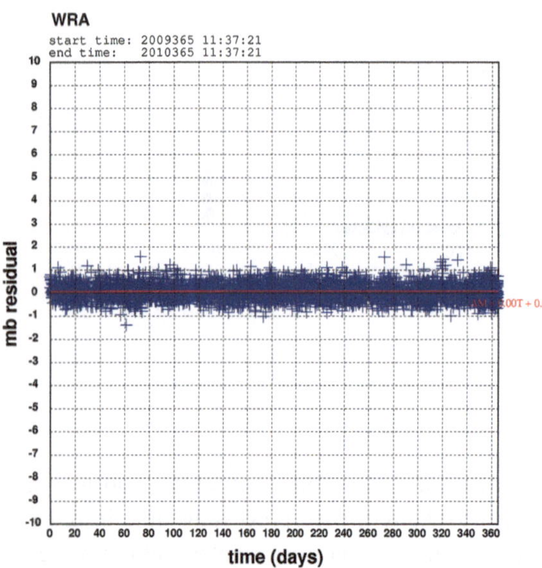

Figure 5
Event mb magnitude residual as a function of time for station WRA during 2010. No bias exists in the magnitude estimate suggesting the calibration for this station is correct

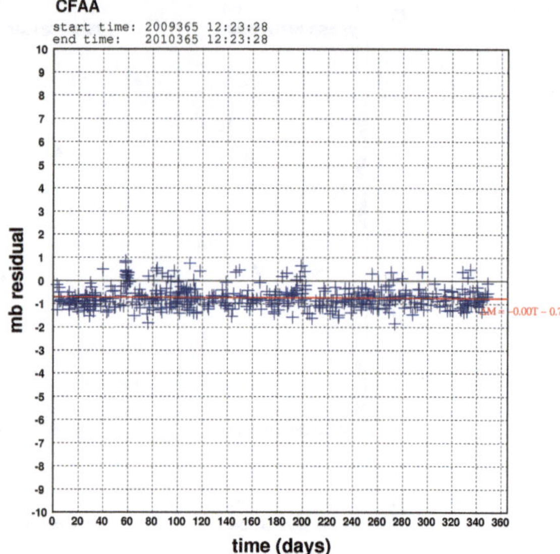

Figure 6
Event mb magnitude residual as a function of time for station CFAA during 2010. A clear bias of around −0.7 magnitude units exists for this station suggests a calibration error. The *red line* is the line of best fit through the data

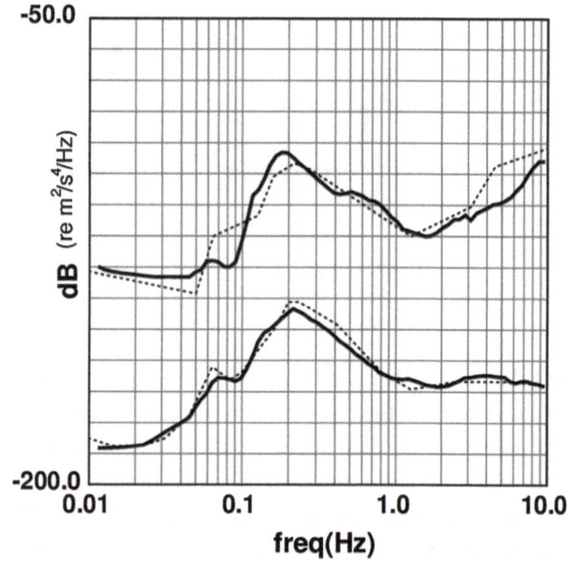

Figure 7
The global seismic low noise model IDC2010_LS and high noise model IDC2010_HS (*solid lines*) compared with the NHNM and NLNM of Peterson (1993) (*dashed lines*)

flat out to around 100 Hz. The microseismic peak at around 5 s period is clearly visible in both curves. It is worth noting that the microseismic energy is being measured directly in the water, which means that it is a more local measurement of surface wave activity than for seismic stations that receive microseism energy from a large area of ocean after propagating through the crust. This explains why the low noise and high noise curves exhibit the same microseismic deviation. Also shown in Fig. 8 for comparison are the oceanic ambient noise curves presented in Urick

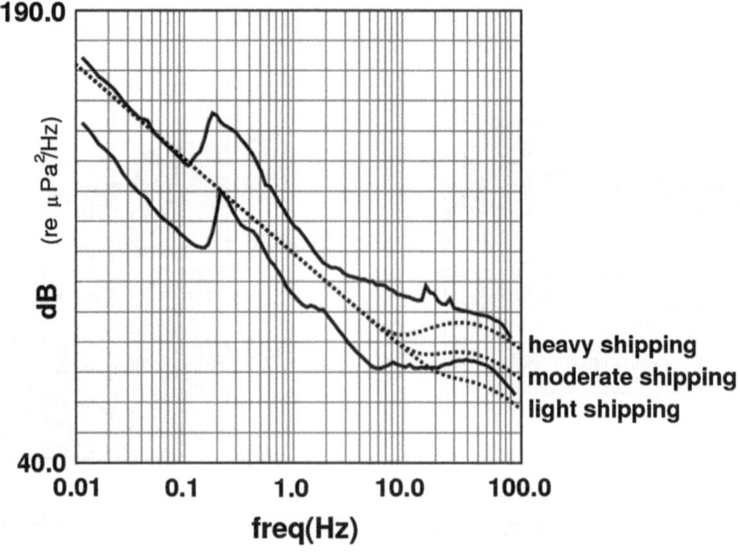

Figure 8
The global hydroacoustic low noise model IDC2010_LH and high noise model IDC2010_HH (*solid curves*) with Urick's noise curves superimposed (Urick 1984) for light, moderate and heavy shipping (*dashed curves*). Note that the wind component in the Urick curves was assumed to be zero, as the contribution due to the wind is not particularly significant below 100 Hz

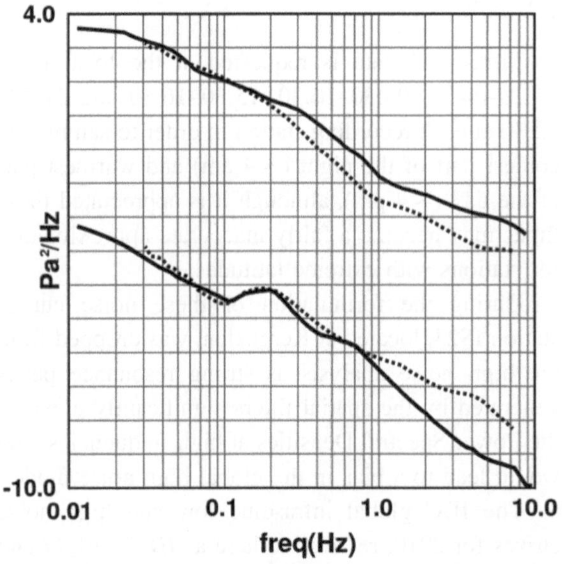

Figure 9
The global infrasonic low noise model IDC2010_LI and high noise model IDC2010_HI (*solid lines*) compared with a two times scaled version of the Bowman (2009) that can be assumed to be representative of the Bowman (2005, 2007) values (*dashed lines*)

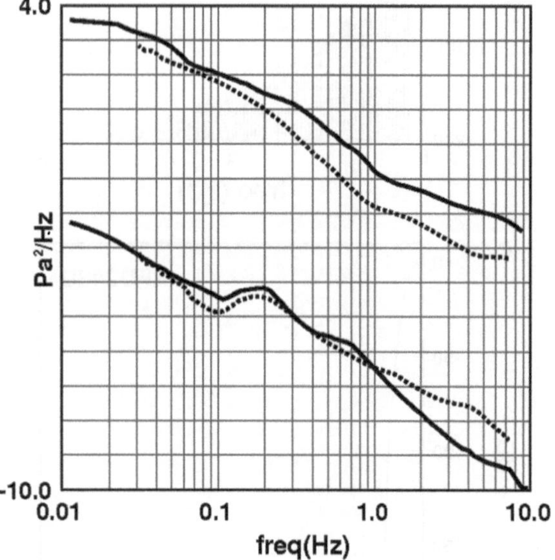

Figure 10
The global infrasonic low noise model IDC2010_LI and high noise model IDC2010_HI (*solid lines*) compared with those of Bowman (2009) (*dashed lines*)

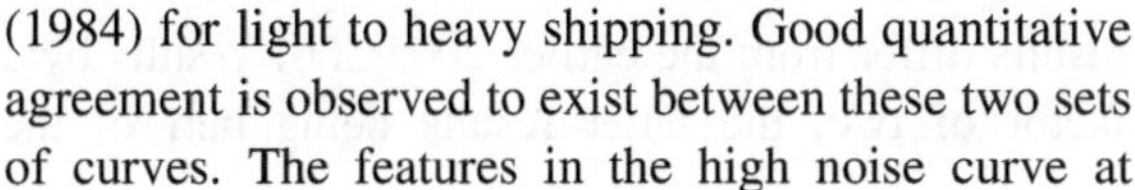
(1984) for light to heavy shipping. Good quantitative agreement is observed to exist between these two sets of curves. The features in the high noise curve at around 15 and 25 Hz are likely to be due to blue and fin whale calling (see, e.g., McCauley *et al.*, 2001; Richardson *et al.* 1995).

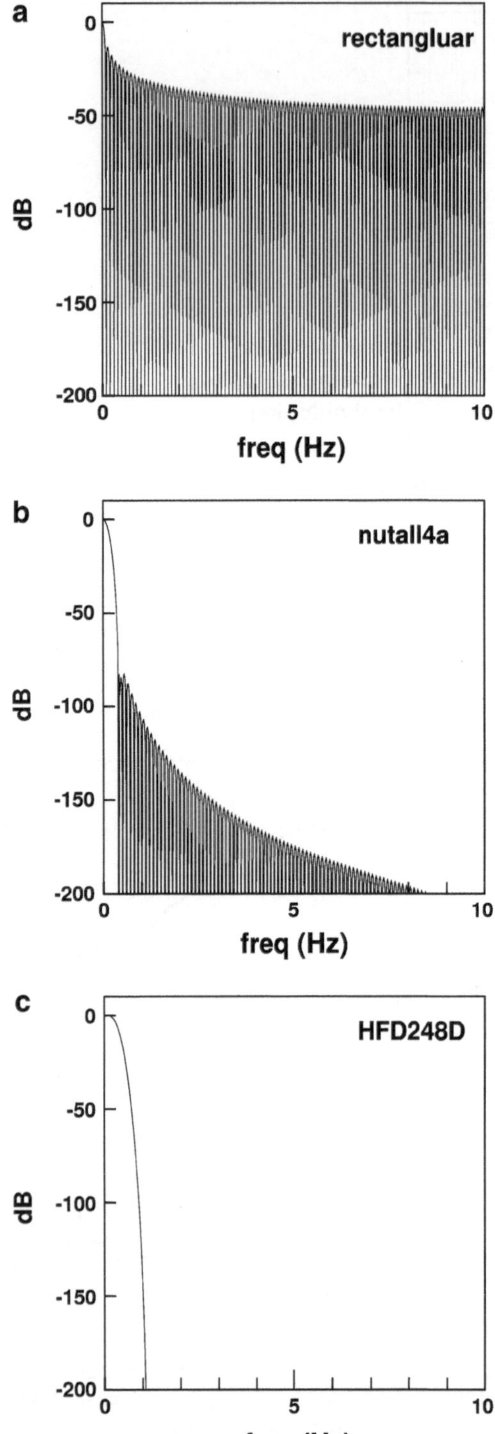

Figure 11
Transform of three spectral windows: **a** rectangular; **b** nutall4a; **c** HFD248D. See HEINZEL *et al.* (2002) for a discussion on these last two windows

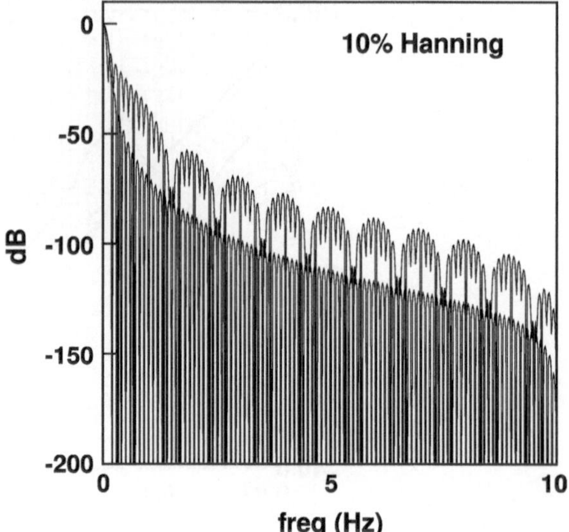

Figure 12
Transform of window function with a Hanning taper applied to the outer 10 % of the window and with the inner 80 % of the window function set to unity (*upper curve*). The transform of the usual Hanning window is also shown for comparison (*lower curve*)

4.3. *Infrasonic*

Infrasound data is requested at the local time: 03:30–04:30, 09:30–10:30, 15:30–16:30 and 21:30–22:30 at each recording station in order to sample the coldest part of the night (~4 am) and warmest part of the day (~4 pm), although it is appreciated these times may become a fairly inaccurate approximation for stations with extreme latitudes.

During the formulation of these noise curves station IS23, located at Kerguelen, was dropped from the high noise analysis as strong resonance peaks generated by the spatial filters significantly elevated the Power Spectral Densities at high frequencies, and would lead to a bias in the global high noise model.

The IDC global infrasonic low and high noise curves for 2010, referred to here as *IDC2010_LI* and *IDC2010_HI*, respectively, are shown in Fig. 9. Also shown in Fig. 9 are the global low and high infrasonic noise curves reported by BOWMAN *et al.* (2009) that have been scaled by two in the PSD's and can be assumed to be representative of the 2005/2007 results, since BOWMAN *et al.* (2009) state their 2009 results differ from the earlier 2005/2007 results by a factor of two, the latter results being half of the earlier results. The revised noise curves presented in Bowman *et al.* (2009) are shown in Fig. 10. Inspection shows that the Bowman (2005, 2007) low noise curves and IDC2010_LI curve are in close agreement around the microbarom peak. The IDC2010_LI curve drops below the Bowman curve at higher frequencies. This is due to the inclusion of station IS55 in the IDC2010_LI curve where it was excluded from the Bowman analysis. Station IS55, located at Windless Bight in the Antarctic, employs sensors manufactured by Chaparral Physics that are known to have very low self-noise at high frequencies. This station was excluded in the Bowman analysis because the authors felt that snow covering the spatial filters may artificially lower the recorded noise levels. However, it was included in this analysis as it is thought that the Chaparral sensors make an important contribution to the high frequency end of the noise curves that would otherwise be ignored. It is felt here that the curve obtained with the IS55 sensors included would more likely be closer to reality than those obtained leaving them out. The revised noise curves presented in Bowman *et al.* (2009) shown in Fig. 10 are consistently out in the value of the PSD's by a factor of 2, suggesting from this analysis that the original analysis of Bowman *et al.* (2005, 2007) is correct and a systematic error was introduced in the more recent work.

5. Conclusions

Data recorded in 2010 from the CTBTO IMS seismic, hydroacoustic and infrasound networks has been used to infer global low and high ambient noise curves.

The determined seismic global low and high noise curves are found to be in excellent agreement with those established by Peterson (1993), whereas the global infrasound low and high noise curves are found to be in general agreement with the Bowman (2005, 2007) models, but disagree with the revised models presented in Bowman (2009), which seem to have a factor of 2 error in the determination of the PSD. The hydroacoustic global low and high noise curves both exhibit contributions from the microseisms at low frequencies and shipping noise at high frequencies. Whale noise is a feature of the high noise curve at 15 and 25 Hz.

Acknowledgments

The authors would like to thank Dr. Roger Bowman for providing copies of his infrasound global noise models.

Disclaimer The views expressed in this paper are those of the authors and do not necessarily reflect those of the Preparatory Commission.

Appendix: Estimation of the Power Spectral Density

The selection of a finite-time interval of digitally sampled data, with supposedly constant sample rate, is equivalent to taking the analogue signal and applying two processes:

1. Sampling using the periodic Dirac Comb function $D(t) = \sum_{k=-\infty}^{\infty} \delta(t - kT)$, where T is the sampling period
2. Windowing using the box-car window function $w(t) = \begin{cases} 1 & \text{when } 0 \le t \le KT \\ 0 & \text{otherwise} \end{cases}$, where it is assumed that time zero is at the beginning of the box-car function, which is of duration KT.

The Fourier transform now becomes:

$$F(f(t)D(t)w(t)) \equiv F(\omega) * D(\omega) * W(\omega) \quad (1)$$

where asterisk (*) indicates the convolution function, and the capitals indicate Fourier Transformation, and $D(\omega) = \sum_{k=-\infty}^{\infty} \delta(\omega - k\frac{2\pi}{T}) = \sum_{k=-\infty}^{\infty} \delta(\omega - \omega_k)$. The Fourier transform of the window function is seen to control the measured frequency content. Examples for several spectral windows are shown in Fig. 11. The phenomenon of 'spectral leakage' is clearly visible in this figure, where energy due to the non-periodic nature of the windowing process is migrated to higher-frequencies. In order to reduce the contribution of spectral leakage to the measured frequency content, it is common practice to taper the window function w, such that the side-lobes are smaller, invariably at the expense of losing some frequency resolution, but generally at the gain in amplitude resolution. Careful comparison of the properties of a large number of window functions is given in HEINZEL *et al.* (2002), where the different windows are classified according to their ability to either resolve frequencies, as in the box-car window (Fig. 11a), resolve amplitudes, such as the flat-top windows (Fig. 11c), or be general purpose with some capability in both areas, such as the nutall windows (Fig. 11b). In this work we use the nutall4a window function of HEINZEL *et al.* (2002) whose transform is displayed in Fig. 11b. This is a general-purpose window that has good amplitude and frequency resolution with side lobes that are around 100 dB below the main lobe. Note that it has become common practice in the literature to apply a fractionally tapered window to waveform data, so that, for example, only the first and last 10 % of the window function differs from unity. This can lead to significantly degraded behaviour and should be avoided. Figure 12 shows the window transform function for a window function that consists of a hanning taper applied to the first and last 10 % of the data.

Several additional concepts are important when considering the use of window functions as applied to the determination of Power Spectral Densities. The first of these is the notion of Incoherent Power Gain. The spatially-extended nature of the main lobe of the window transform function, generally extending across several frequency bins, allows the window to gather energy from those neighbouring bins. HARRIS (1978) shows that if N_0 is the noise-power per bin, then the total power P_w collected by the window function is $P_W = \frac{N_0}{2\pi}\int_{-\pi/\Delta}^{\pi/\Delta}|W(\omega)|^2 d\omega = \frac{N_0}{\Delta}\sum_{j=0}^{n-1} w_j^2 \equiv \frac{N_0}{\Delta} I_W$—where n is the number of samples, Δ is the sample rate, and I_W, which just is the sum of the squares of the window coefficients, is defined to be the Incoherent Power Gain of the window. We are therefore in a position to write an expression for the PSD of the finite-length digitally-sampled analogue signal $f(t)$. It is: $P_{SD}(\omega_j) = \frac{2|F(\omega_j)|^2}{\Delta I_W}$—where $\omega_j = \frac{2\pi j}{n\Delta}$ for $j = 0, \ldots, n/2$, is the jth frequency picket, and F is the output of a unitary Fourier Transform algorithm. Here, we are taking into account the contribution of the negative frequencies with the factor 2. It is important to note that P is the power spectrum contained in each frequency bin, i.e., PSD, and not the power spectrum of an individual spectral component. To determine the power contained in an individual line spectra the concept of the Coherent Power Gain is useful. Application of Eq. (1) to the elemental waveform $f(t) = Ae^{i\omega_k t}$ yields the result.

$F(f(t)D(t)w(t)) = A\sum_{j=-\infty}^{\infty} w_j\delta(\omega - \omega_k) * \delta(\omega - \omega_j) = A\sum_{j=-\infty}^{\infty} w_j$, where w_j, are the values of the window function at times: $t + jT$ for $j = -\infty, \ldots, \infty$. In such a case, the Power Spectrum is given by $P_S(\omega_j) = \frac{2|F(\omega_j)|^2}{C_W}$, where C_W, which is the square of the sum of the window coefficients, is the Coherent Power Gain of the window. One can then estimate the amplitude of the line spectra by taking the square root of $P_S(\omega_j)$. The quantity β is known as the Equivalent Noise Bandwidth of the Window function and through straightforward multiplication allows one to convert *PSD* to *PS* and vice versa.

The Recommended Overlap Value (ROV) for the nutall4a window is 68 % (HEINZEL *et al.*, 2002), implying a total of 63 three-min segments to be evaluated.

REFERENCES

BOWMAN, J.R., BAKER, G.E., and BAHAVAR, M, 2005. *Ambient infrasound noise*. Geophys Res. Lett. *32*. L09803, doi:10.1029/2005GL022486.

BOWMAN, J.R., SHIELDS, G., O'BRIEN, M.S. 2007. *Infrasound Station Ambient Noise Estimates and Models 2003–2006* Infrasound Technology Workshop, Tokyo, 13–16, November 2007.

BOWMAN, J.R., SHIELDS, G., O'BRIEN, M.S. 2009. Infrasound station ambient noise estimates and models 2003–2006 (Erratum), Infrasound Technology Workshop, Brasilia, Brazil, November 2–6, 2009.

HARRIS, F.J.,1978. *On the use of Windows for Harmonic Analysis with the Discrete Fourier Transform*. Proceedings of the IEEE *66* (1): 51–83. doi:10.1109/PROC.1978.10837.

HEINZEL G., RUDIGER, R., and SCHILLING R. 2002. Spectrum and spectral density estimation by the Discrete Fourier Transform (DFT), including a comprehensive list of window functions and some new flat-top windows. http://www.rssd.esa.int/SP/LISA PATHFINDER/docs/Data_Analysis/GH_FFT.pdf.

LYONS, R.G., 1997. "Understanding Digital Signal Processing", Addison-Wesley.

MATLAB, 2008. MATLAB version 7.7.0. Natick, Massachusetts: The MathWorks Inc.

McNamara, D.E. and R.P. Buland, 2004. *Ambient Noise Levels in the Continental United States*, Bull. Seism. Soc. Am., *94*, 4, 1517–1527.

McCauley, R.D., Jenner C., Bannister J.L., Burton C.L.K, Cato, D.H., and Duncan., A. 2001. Blue whale calling in the Rottnest trench—2000, Western Australia. Prepared for Environment Australia, from Centre for Marine Science and Technology, Curtin University, R2001-6, 55 pp.

Peterson, J., 1993. Observation and modeling of seismic background noise, U.S. Geol. Surv. Tech. Rept., 93–322, 1–95.

Richardson, W.J., Greene, C.R., Malme, C.I, and Thomson, D.H. 1995. *Marine Mamals and Noise*, Academic Press, San Diego. pp 576.

Schreier, P.J, and Scharf, L.L, 2010. Statistical signal processing of complex-valued data: the theory of improper and noncircular signals. *Cambridge University Press*. pp. 330.

Urick, R.J. 1984. *Ambient noise in the sea.* Report No. 20070117128. Undersea Warefare Technology Office, Naval Sea Systems Command, Department Of The Navy, Washington DC.

Welch, PD, 1967. The Use of Fast Fourier Transform for the Estimation of Power Spectra: A Method Based on Time Averaging Over Short, Modified Periodograms, IEEE Transactions on Audio Electroacoustics, Volume AU-15 pp. 70–73.

(Received September 30, 2011, revised August 6, 2012, accepted August 12, 2012, Published online September 8, 2012)

Pure Appl. Geophys. 171 (2014), 377–394

DOI 10.1007/s00024-012-0613-2

Pure and Applied Geophysics

The Applicability of Incoherent Array Processing to IMS Seismic Arrays

Steven J. Gibbons[1]

Abstract—The seismic arrays of the International Monitoring System (IMS) for the Comprehensive Nuclear-Test-Ban Treaty (CTBT) are highly diverse in size and configuration, with apertures ranging from under 1 km to over 60 km. Large and medium aperture arrays with large inter-site spacings complicate the detection and estimation of high-frequency phases lacking coherence between sensors. Pipeline detection algorithms often miss such phases, since they only consider frequencies low enough to allow coherent array processing, and phases that are detected are often attributed qualitatively incorrect backazimuth and slowness estimates. This can result in missed events, due to either a lack of contributing phases or by corruption of event hypotheses by spurious detections. It has been demonstrated previously that continuous spectral estimation can both detect and estimate phases on the largest aperture arrays, with arrivals identified as local maxima on beams of transformed spectrograms. The estimation procedure in effect measures group velocity rather than phase velocity, as is the case for classical f–k analysis, and the ability to estimate slowness vectors requires sufficiently large inter-sensor distances to resolve time-delays between pulses with a period of the order 4–5 s. Spectrogram beampacking works well on five IMS arrays with apertures over 20 km (NOA, AKASG, YKA, WRA, and KURK) without additional post-processing. Seven arrays with 10–20 km aperture (MJAR, ESDC, ILAR, KSRS, CMAR, ASAR, and EKA) can provide robust parameter estimates subject to a smoothing of the resulting slowness grids, most effectively achieved by convolving the measured slowness grids with the array response function for a 4 or 5 s period signal. Even for medium aperture arrays which can provide high-quality coherent slowness estimates, a complementary spectrogram beampacking procedure could act as a quality control by providing non-aliased estimates when the coherent slowness grids display significant sidelobes. The detection part of the algorithm is applicable to all IMS arrays, with spectrogram-based processing offering a potential reduction in the false alarm rate for high-frequency signals. Significantly, the local maxima of the scalar functions derived from the transformed spectrogram beams are robust estimates of the signal onset time. High-frequency energy is of greater importance for lower event magnitudes and in the cavity decoupling detection evasion scenario. There is a need to characterize both propagation paths with low attenuation of high-frequency energy and situations in which parameter estimation on array stations fails.

Key words: Seismic arrays, nuclear explosion monitoring, CTBT, signal incoherence, parameter estimation, spectral estimation.

1. Introduction

The International Monitoring System (IMS) for monitoring compliance with the Comprehensive Nuclear-Test-Ban Treaty (CTBT) consists of global networks of seismic, infrasonic, hydroacoustic and radionuclide sensors for detecting potential violations of the treaty (see Dahlman *et al.*, 2009). The seismic network consists of both three-component stations (at single sites) and seismic arrays (deployments of multiple sensors at separate sites over a limited geographical region). The task of any seismic station in the global network is to detect phase arrivals generated by any seismic disturbance and attribute parameters such that signal detections from different stations can be associated to form event hypotheses in an automatic seismic bulletin (e.g., Ringdal and Kværna, 1989). The most important parameter in the phase association process is the arrival time. However, the effectiveness of an association algorithm will also depend upon the reliability of the estimates for two basic parameters: the apparent velocity, v_{app}, from which a phase identification is usually made, and the backazimuth, BAZ, the direction from which the phase arrives. A highly erroneous, or qualitatively incorrect, parameter estimate is likely to prevent the detection from being associated with the correct event hypothesis and may result in the non-detection of a seismic event.

The ability to detect and classify seismic phases is generally superior for array stations than for three-component stations (e.g. Kværna *et al.*, 2011) since the stack-and-delay beamforming can significantly improve the signal-to-noise-ratio (SNR), and the

[1] NORSAR, P.O. Box 53, 2027 Kjeller, Norway. E-mail: steven@norsar.no

time-delay estimation methods for slowness vectors are typically more robust and more widely applicable than methods based on the polarization of particle motion (see e.g. Kværna and Ringdal, 1992). Reviews of array processing techniques for phase detection and parameter estimation are given by, for example, Schweitzer *et al.* (2002) and Rost and Thomas (2002, 2009). However, the classical concept of array processing breaks down when coherence between sensors of an array is lost (e.g., Bungum and Husebye, 1971) and an array for low-frequency seismic waves becomes, in effect, a network for higher frequencies (Levander and Nolet, 2005). In these cases we need to look to incoherent methods, or network techniques, to detect and make useful parameter estimates of seismic arrivals.

The IMS seismic arrays are highly heterogeneous in terms of the geometry and the number of sensors (e.g. Douglas, 2002). The configurations of all primary and auxiliary IMS arrays providing data as of September 2011, with the exception of the so-called medium-period arrays, are displayed in Fig. 1. Over an order of magnitude separates the array apertures between the smallest arrays (aperture typically under 4 km) and the largest (aperture up to 70 km). The reason for the dissimilarity between the different array stations is that they were built at different times to be best suited to differing monitoring priorities. In contrast, the IMS infrasound arrays (Christie and Campus, 2010) have been deployed over a period of approximately one decade with the aim of comprising a unified network, and there is far less variability in the array apertures and sensor configurations.

Many of the larger arrays are so-called legacy stations dating back to the 1960s and 1970s, which were designed to provide optimal detection and estimation for teleseismic P-phases with energy concentrated at around 1 Hz. Figure 2 displays signals from two underground nuclear explosions at the Semipalatinsk test site in the former Soviet Union recorded at the NORSAR array (now labelled NOA) at a distance of 37.8 degrees. One of the events is significantly smaller than the other, with an approximate factor of 10 separating the maximum amplitudes in the frequency bands where the signal exceeds the noise level significantly. It is clear that there is significant energy at frequencies well above 1 Hz, and that it is the SNR at the lower frequencies which is most greatly reduced given an event of smaller size. The SNR above 4 Hz is remarkable both due to the large epicentral distance and the use of an analogue anti-aliasing filter with a cut-off at 4.5 Hz.

Figure 3 illustrates the result of beamforming over the large aperture NOA array for the two events displayed in Fig. 2. The top left panel shows the signals from the larger event, filtered 0.5–1.5 Hz, on three sensors from each of two different subarrays, together with a beam constructed using carefully calibrated time-delays. The signal amplitudes on the NC6 subarray are significantly greater than the amplitudes on the NB2 subarray (c.f. Ringdal and Husebye, 1982), and the amplitude of the signal on the resulting beam is significantly lower than the greatest signal amplitudes recorded on some single sensors. However, the beamforming suppresses the preceding noise sufficiently to ensure an improved SNR on the array beam (this is especially clear for the smaller event in the upper right panel). The SNR in the 0.5–1.5 Hz band essentially does not exceed unity for the single channels, although the signal is clearly visible on the beam. The lower panels in Fig. 3 are the corresponding displays for the 4–8 Hz band. The single-channel SNR is considerably better for all traces, but the waveform dissimilarity between the different sensors is now evident and the beam results in significant signal degradation.

This example illustrates a fundamental problem in array processing: that moving to higher frequencies improves the single-channel SNR at the expense of signal coherency and, consequently, the ability to perform classical array processing. It was pointed out by Ringdal *et al.* (1975) that incoherent (waveform envelope) methods were often necessary to exploit this higher frequency energy on the NORSAR array. The evident limitations of the very large aperture arrays resulted in the development of far smaller arrays capable of processing signals at much higher frequency, in particular for phases from events at regional distances (e.g. Mykkeltveit *et al.*, 1990). The current IMS, including many small and medium aperture arrays, is closer to the global system for monitoring nuclear explosions envisaged in the 1950s (Romney, 1985) than the sparse network of very large aperture arrays developed in the 1960s and 1970s.

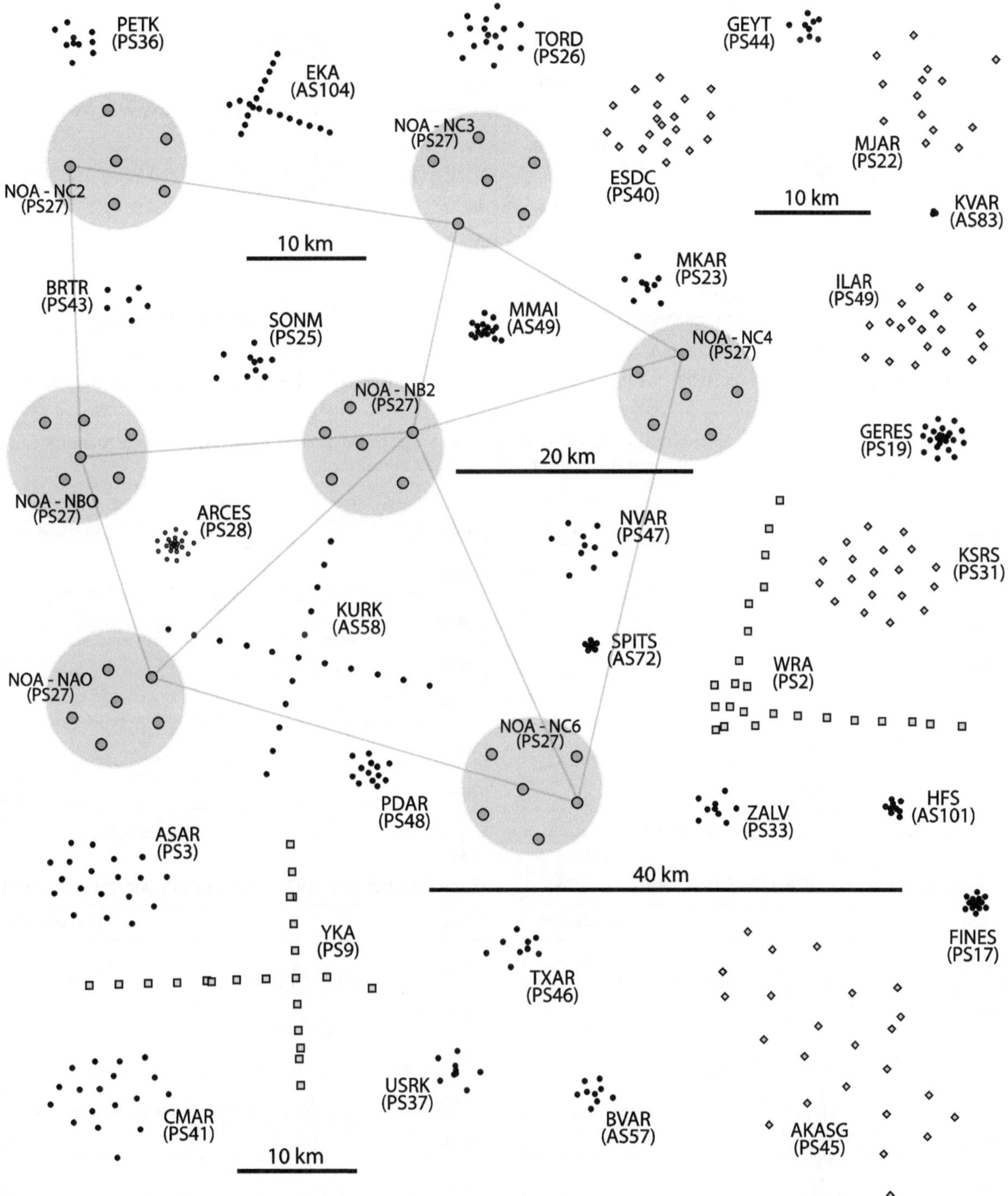

Figure 1
IMS array stations as of September 2011 drawn to a common scale. The seven subarrays of the large aperture NORSAR array (NOA) are *shaded* and labelled in parentheses. The 3-C broadband sensors in each subarray are linked by *grey lines*. The so-called medium period arrays are not displayed

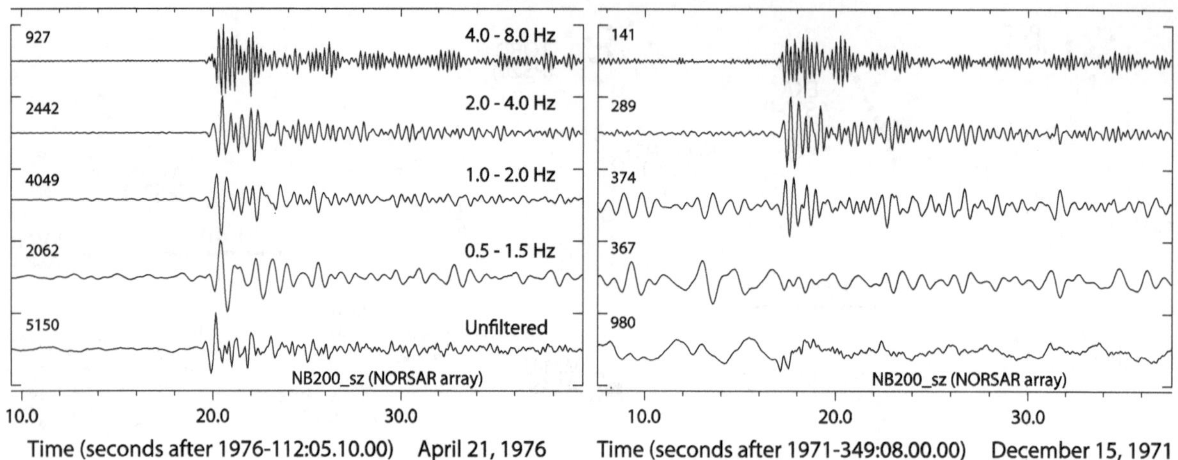

Figure 2
Signals from two explosions at the Semipalatinsk Test Site (STS) in the former Soviet Union recorded on site NB200 of the NORSAR array, raw waveforms and bandpass filtered as indicated

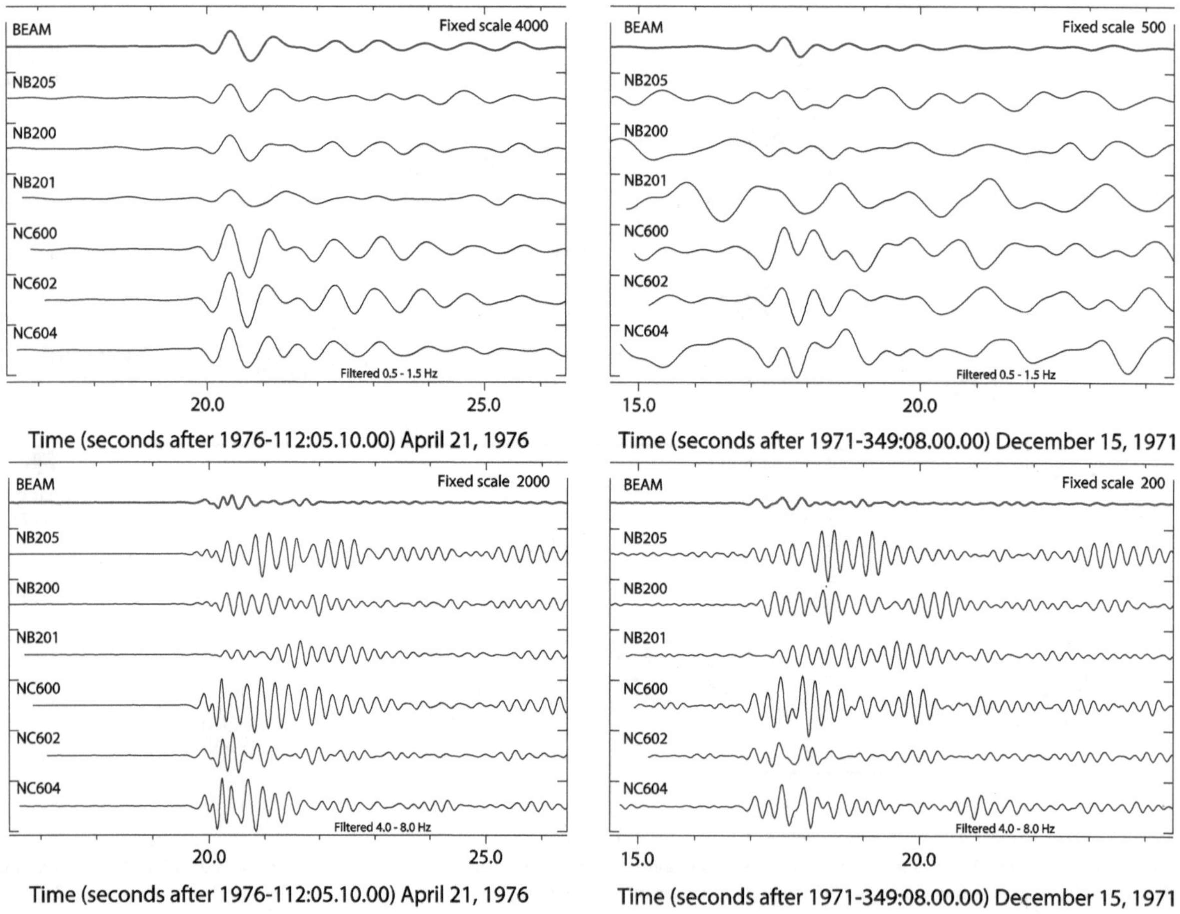

Figure 3
Signals from the two events displayed in Fig. 2 on the channels displayed aligned according to optimal time-shifts. The beams are formed from all 42 sites that comprise the current NORSAR array (PS27). The waveforms in the *upper* and *lower panels* are bandpass filtered 0.5–1.5 Hz and 4.0–8.0 Hz respectively

There is no such thing as a universal seismic array. An array with a sufficiently large aperture, and sufficiently large inter-site spacings, to provide sharp resolution and good noise reduction for teleseismic phases is likely to be unable to process high frequency regional phases using classical beamforming. Similarly, small-aperture arrays with geometries that are optimized for the detection and characterization of high-frequency regional phases are likely to provide poor resolution and low SNR gain for teleseismic phases. Special processing is necessary to improve the detection and estimation of teleseismic phases on small aperture arrays amid strong and coherent noise (SELBY, 2011). Similarly, complementary processing is necessary to detect and resolve phases that are incoherent over very large aperture arrays. The purpose of this paper is to explore the applicability of the spectrogram beamforming method of GIBBONS *et al.* (2008) to different arrays in the IMS network, in particular to arrays where it is demonstrated that classical array processing fails for high-frequency regional phases. We examine the anticipated performance of the method as a function of array aperture, discuss the application of recent innovations in signal processing to improving the performance, and discuss the options available for situations in which the method is likely to fail.

2. *Detection and Parameter Estimation on Seismic Arrays Using Continuous Spectral Estimation*

In the most common framework for seismic phase detection, a single trace is bandpass-filtered in a frequency band deemed to be optimal for the anticipated phase. This single trace may be the output from a single sensor or a beam constructed from the output of many sensors of an array with appropriate time delays applied. Detections are then declared when the signal-to-noise-ratio (SNR), i.e. the ratio between the short-term-average (STA) and the long-term-average (LTA), exceeds a specified threshold. See, for example, WITHERS *et al.* (1998) and references therein for a summary of such detection algorithms. Alternatives to such procedures may operate on 2-dimensional representations of the seismogram, with time on one axis and frequency or scale on the other, to identify amplitude changes characteristic of anticipated phase arrivals or full wavetrains.

Wavelet transform methods (e.g. YOMOGIDA, 1994; OONINCX, 1999; GENDRON *et al.*, 2000; ZHANG *et al.*, 2003; CAPILLA, 2006) provide coefficients for wavelets of different scales from a selected basis function, and phase detection algorithms can trigger on a significant increase in the size of wavelet coefficients over a specified range of scales. Alternatively, the spectrogram or sonogram provides signal amplitude over a uniform discretization of time and frequency. Seismic signal detection on spectrograms has traditionally focused on pattern recognition and identification of characteristic time/frequency signatures of seismic events (e.g. JOSWIG, 1990, 1995; HÄGE and JOSWIG, 2009). Recently, TAYLOR *et al.* (2010) demonstrated a procedure for detecting transient, short-duration, signals on spectrograms using scan statistics. The aim was to detect signals from events at local distances which frequently evaded detection by classical procedures due to a complicated background noise field. They pointed out that a large bandwidth signal would register over a greater number of elements in spectrogram columns than noise transients, such that a 2-D linear filter would emphasize columns of the spectrograms that were relatively rich in energy. The resulting detection procedure was demonstrated to have a lower false alarm rate than a corresponding STA/LTA algorithm.

GIBBONS *et al.* (2008) describe an application of continuous spectral estimation to solving the detection and parameter estimation problem for incoherent signals on large-aperture seismic arrays. Figure 4 displays a typical example of such a signal. The event displayed is included in the Reviewed Event Bulletin (REB) of the International Data Center (IDC), and both Pn and Sn phase arrivals are clearly identifiable on the filtered waveforms on the WRA array (Warramunga, Australia). WRA is an exceptionally quiet station with a superb detection capability for signals from events at teleseismic distances. However, the large-array aperture and inter-site spacings, which are so favourable for teleseismic array processing, make the array almost unusable for events at regional distances. Despite an epicentral distance of around 700 km, high-frequency energy dominates, and both P and S phases are not visible above the background noise below 2 Hz. The velocity and azimuth for the Pn phase reported in the REB are consistent with the phase type and direction, although it is clear from the lower-left f–k plot in Fig. 4 that the relative power for this slowness vector is only marginally greater than that for other very different slowness vectors. The Sn phase in the REB is associated with a velocity and azimuth that are qualitatively incorrect and the lower-right f–k plot in Fig. 4 makes clear that there are no obvious candidate directions to choose from. The incoherence problem is frequently mitigated by removing traces recorded on more distant sensors from the calculation or by selecting a lower frequency band. In this case, it is not possible to use a lower frequency band due to SNR considerations. No frequency band or subset of sensors was found that provided a qualitatively correct slowness estimate for this Sn phase on WRA.

Figure 5 illustrates the steps that the algorithm of Gibbons *et al.* (2008) takes to detect and estimate propagation parameters for such phases. One of very few processing parameters in the procedure is L, the length of the window used for spectral estimation. The window must be long enough to allow a robust estimate of the energy at each of the frequencies of interest (typically 2–15 Hz for regional phases), the multitaper method of Thomson (1982) being deemed ideal for spectral estimation on the short-duration data segments. L must not be so small that the continuous spectral estimates are over-sensitive to rapidly evolving properties of the waveforms (that are likely to vary significantly from sensor to sensor), but must neither be so large that energy is smeared out excessively, diminishing the likelihood of being able to estimate a time-delay between arrivals on neighbouring sensors. Gibbons *et al.* (2008) found empirically that windows between 2.8 and 3.8 s long provided a satisfactory balance between robustness and resolution, and the spectrograms displayed in traces 4–6 of Fig. 5 are calculated using windows of length 3.2 s under the numerical implementation of Prieto *et al.* (2009).

Unlike the transient signals considered by Taylor *et al.* (2010), which appear as vertical lines on the spectrograms, the arrivals in Fig. 5 are followed by coda with amplitude greater than or equal to that of the direct wavefronts, resulting in spectrograms that resemble more closely smoothed step-functions in time than the desired smoothed delta functions. Gibbons *et al.* (2008) appreciated the need to apply a transformation to the spectrograms to make them more comparable to the vertical line spectrograms, with local maxima occurring as close as possible to the phase arrival time. If $A(f)_{t+} = A(f, t, L)$ denotes the amplitude density spectrum measured for the window immediately following a time t, and $A(f)_{t-} = A(f, t - L, L)$ the spectral estimate in the window ending at time t, then the transformed spectrogram denoted

$$S(f,t) = (\log_{10}[A(f)_{t+}] - \log_{10}[A(f)_{t-}])\log_{10}[A(f)_{t+}] \quad (1)$$

provides local maxima at times characterized by both a relatively high amplitude and an increase in amplitude. Such transformed spectrograms are displayed in traces 7–9 of Fig. 5. As pointed out by Taylor *et al.* (2010), the times of the direct phase arrivals are characterized by high values over a broad band of frequencies, whereas noise peaks are localized at rather random frequencies. Local maxima in the transformed spectrograms resulting from noise bursts tend to be suppressed under the spectrogram beamforming operation, whereas the local maxima from the direct phases interfere constructively as the transformed spectrograms are stacked with appropriate time-delays. The beamforming process itself, together with the detection of local maxima in the spectrogram beams, is covered in detail by Gibbons *et al.* (2008).

The two slowness grids in Fig. 5 cover the same parameter space as the f–k spectra in Fig. 4, but display maximum values of spectrogram beams formed in the time domain, averaged over frequencies, rather than relative beam power calculated in the frequency domain. These so-called beampacking grids indicate that a qualitatively correct slowness estimate is obtained for both the Warramunga Pn and Sn phases. Since it is evident from Fig. 4 that a correct classification of these relatively high-frequency phases is either impossible or subject to an unacceptably high likelihood of failure, it is recommended that both forms of estimation are carried out: coherent f–k analysis for high-resolution slowness estimates of lower frequency teleseismic phases, and

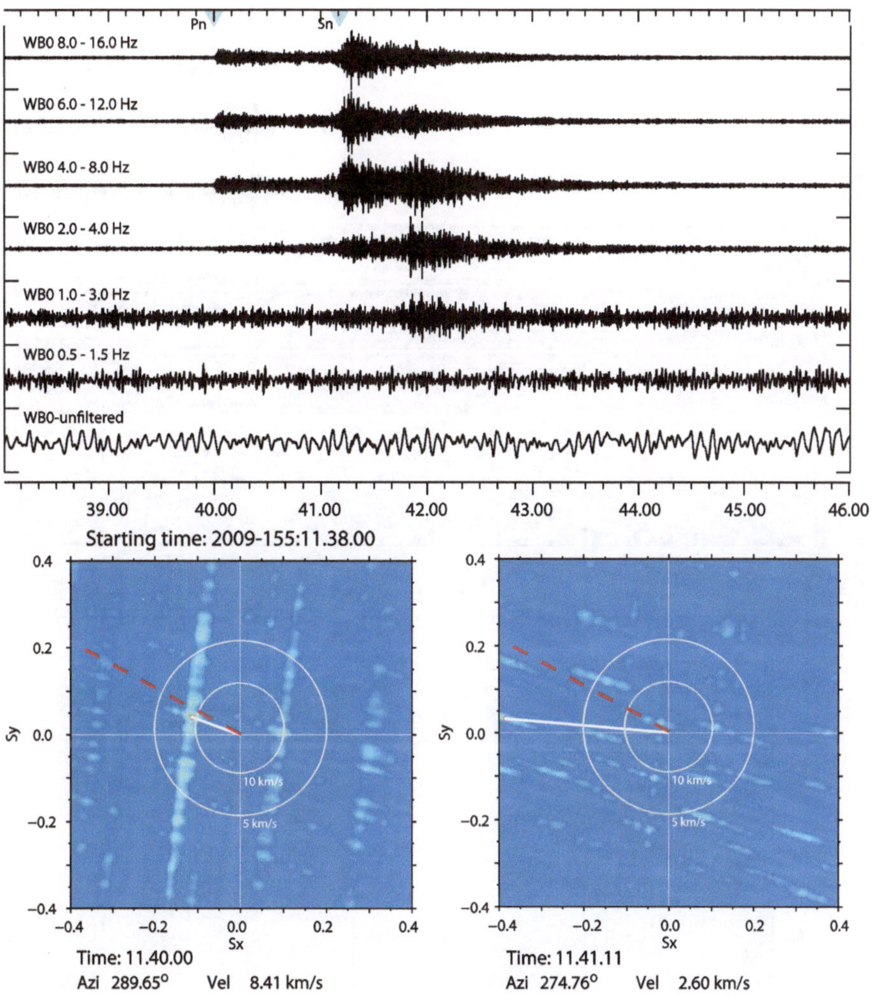

Figure 4

A regional event in the IDC Reviewed Event Bulletin (REB) observed on the Warramunga array (WRA) in Australia June 4, 2009, together with relative-beam-power (P_{rel}) slowness grids for Pn and Sn arrivals using broadband f–k analysis in the 2–4 Hz band. The horizontal slownesses s_x and s_y resulting in the highest P_{rel} are found and then $v_{\text{app}} = (s_x^2 + s_y^2)^{-1/2}$ and BAZ = atan2(s_x, s_y) are calculated. The *dashed red lines* indicate a backazimuth of 298° towards an origin at 16.78°S 128.23°E, 6.6° from WRA, an epicenter determined using regional P and S phases on stations of the AU network operated by Geoscience Australia. The *rings* indicate typical boundaries between teleseismic P-phases ($v_{\text{app}} > 10$ km/s, near the center), regional P-phases (between the rings), and regional S-phases (beyond the outer ring)

incoherent processing for detection and characterization of regional phases.

The regional signals from which these estimates are made contain essentially no energy above the noise level at frequencies below 2 Hz. The transformation applied to the spectrograms of the single channels generates functions which resemble lower frequency arrivals and the estimation problem shifts from one of measuring a phase velocity to one of measuring a group velocity. The period of the scalar functions generated by the process described will depend both upon L, the length of the data window used for spectral estimation, and on how emergent a given signal is. Whether or not the spectrogram beamforming method can be used to estimate a slowness vector on a given seismic array becomes a question of whether or not the distances between sensors are great enough for resolution of these

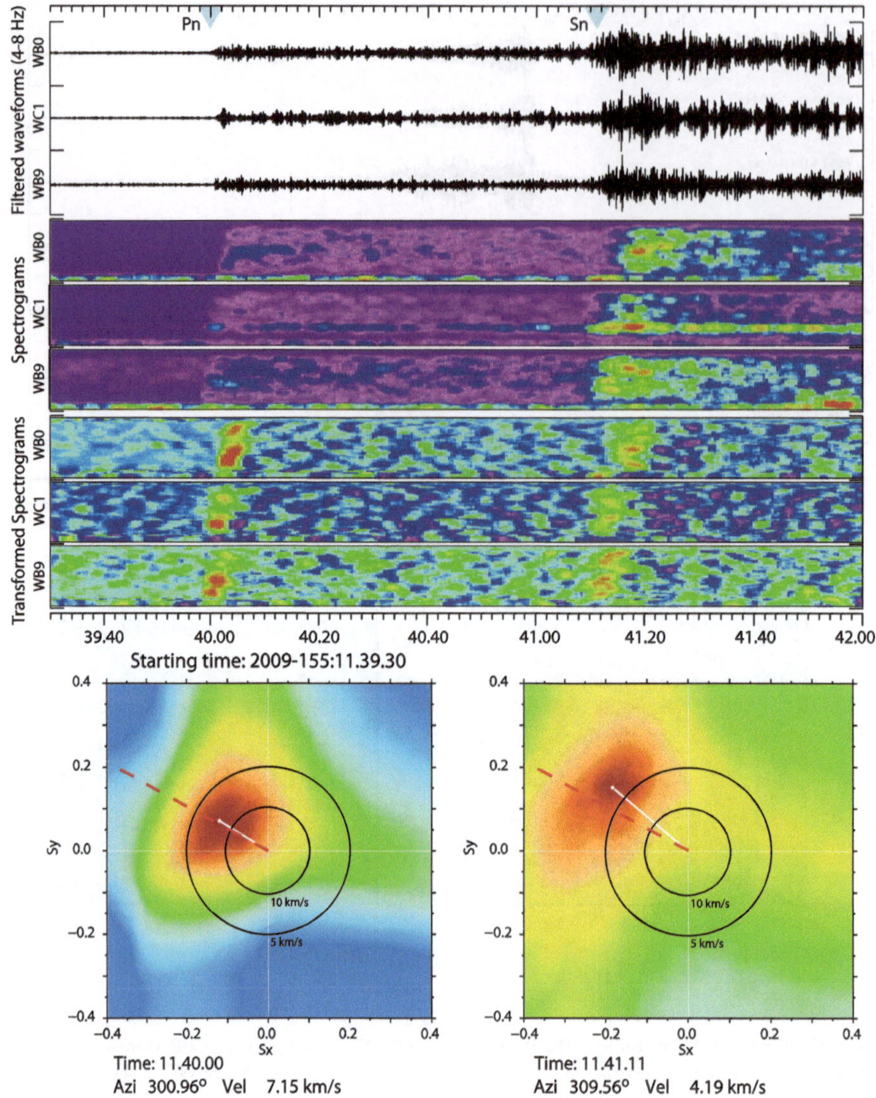

Figure 5

Estimation of slowness vectors using spectrogram beamforming for the Warramunga Pn and Sn arrivals displayed in Fig. 4. The vertical scale for each of the spectrogram panels goes from 0 to 20 Hz, the Nyquist frequency. The slowness grids display the maximum value of the spectrogram beams, averaged over all frequencies, and will be referred to as beampacking grids. The *dashed lines* indicate a backazimuth of 298°

relatively long-period functions. Given L between 3 and 4 s as displayed here, the scalar functions obtained from the transformed spectrograms have a typical frequency of about 0.20–0.25 Hz. Five IMS arrays, NOA, AKASG, YKA, KURK, and WRA, have apertures of the order of 20 km or more and incoherent array processing on each of these arrays is likely to produce results of a similar quality to those displayed in Fig. 5.

3. *Improving Incoherent Slowness Estimates by Smoothing with the Array Response Function*

The incoherent slowness vector estimates displayed in Fig. 5 indicate both apparent velocities which would identify the phase correctly and backazimuth estimates that are sufficiently consistent for any detection association algorithm to use both phases in an event hypothesis. However, an inspection of the slowness grids identifies a cause for concern. The regions of parameter space for which the spectrogram beam maxima are consistent with wavefront hypotheses are very broad. Whereas it appears that the correct quadrant of slowness space is likely to be selected, the risk of choosing a qualitatively incorrect estimate is high. Since the waveforms are incoherent, the transformed spectrograms will also differ from site to site. Although there will be more uniformity in the spectrograms than in the phase structure of the underlying waveforms, local maxima appear in the beampacking grids which are not predicted by the (plane wavefront) propagation model. Inspection of the slowness grids using the human eye identifies credibility contours with a significant degree of symmetry, which may suggest a different slowness vector to that inferred from consideration of the beam maximum alone.

The form of the full f–k spectrum has been considered in numerous previous studies to resolve ambiguity in, or improve the resolution of, slowness vector estimates. Kennett *et al.* (2003) compared the measured f–k spectrum with the corresponding theoretical array response function (ARF) at different frequencies, in order to select the most plausible slowness vector using a neighbourhood algorithm. The primary application of this procedure was to mitigate the effects of spatial aliasing when a high-frequency signal was observed on arrays with relatively sparse sensors. Nishida *et al.* (2008) performed f–k analysis of long-period seismic data and noted that the slowness resolution varied greatly in different directions due to the elongated geometry of the array (in this case a southwest-to-northeast network covering Japan). They noted that this difficulty could be mitigated by performing a deconvolution of the observed f–k spectrum with the ARF. The improvement to parameter estimates resulting from sharpening of f–k spectra by deconvolution with the ARF was demonstrated further by Picozzi *et al.* (2010). The problem we are addressing is fortunately significantly simpler and can be addressed by finding the translation of the ARF which provides the best match with the observed slowness grid. This amounts to a smoothing of the slowness grid with the ARF, which will reduce the influence of the departures from the propagation model.

Figure 6a displays a slowness grid from a spectrogram beamforming procedure for a high-frequency regional phase on the Kurchatov array (KURK) in

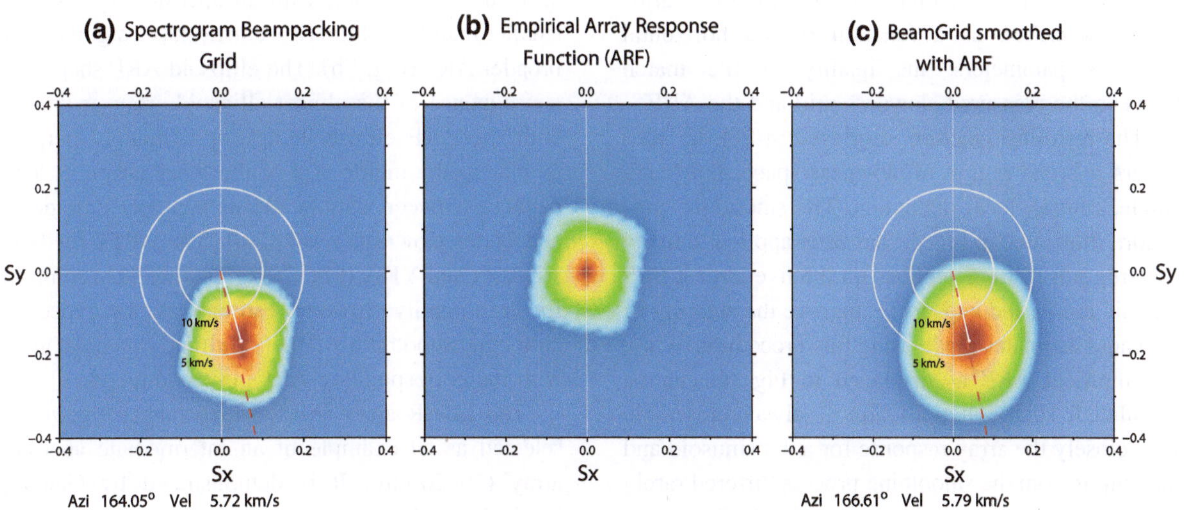

Figure 6
Smoothing (*right*) of a beampacking slowness grid (*left*) with an empirical array response function (ARF, *center*) for the Kurchatov array, KURK, in Kazakhstan. The empirical ARF is calculated by copying the transformed spectrogram from one element to all sensors of the array and then subjecting these identical patterns to the same beamforming procedure used to form the beampacking slowness grid on the *left*. The reference time point for the calculation is 2009-332:07.20.46 and corresponds to a first regional P-wave arrival from a routine quarrying blast at the Kara Zhyra mine at a distance of approximately 0.68°. The coordinates of the blast are 50.02°N, 78.73°E and the receiver-to-source backazimuth of 168° is indicated by the *dashed lines*

Kazakhstan. A qualitatively correct apparent velocity has been measured, i.e., one consistent with a regional *P* type phase, although it is clear that the loci of slowness vectors that are candidates for the optimal array-gain are quite widely spread over slowness space. The ARF for the time-domain spectrogram beampacking procedure is not well defined in the same way that it is for f–k analysis at a given frequency (e.g. SCHWEITZER *et al.*, 2002; KENNETT, 2002). However, in Fig. 6b, we estimate an approximation of an empirical ARF by simply taking the scaled spectrogram from one array element, duplicating it over every sensor of the array, and then measuring the array-gain over the slowness space by stacking the identical channels with the appropriate time-delays. The ARF is approximately straight-edged with the same orientation as the axes of the cross-array geometry (see Fig. 1). A correlation between two time-series, $f(t)$ and $g(t)$, is equivalent to convolving the first time-series with a time-reversed copy of the second time-series, $g(\tau - t)$. This is most effectively computed by taking the inverse Fourier transform of the product of the Fourier transforms of $f(t)$ and $g(\tau - t)$. Similarly, taking Fourier transforms in two-dimensions of the slowness grids in panels (a) and (b) of Fig. 6 allows a rapid frequency domain calculation of the convolution between the two grids which describes, as a function of the horizontal slowness parameters, the quality of the match between the measured slowness grid and the ARF.

The resulting function, displayed in Fig. 6c, also covers a broad region of slowness space, but has a single, central, local maximum. The smoothing procedure illustrated in Fig. 6c has been applied to many spectrogram phase detections on the five largest IMS seismic arrays and appears to improve the stability of slowness estimates made using this procedure. While an empirical ARF as displayed in Fig. 6b can be calculated, such functions almost always resemble very closely the array response for a 4 s sinusoid and the output from the smoothing process differed rarely when such a function was used as the ARF. A very impulsive signal may result in a faster varying transformed spectrogram, and a highly emergent signal may result in a more slowly varying function, but this should affect only the spread and not the shape of the ARF. We propose that the resolution in slowness space of the spectrogram beampacking procedure is well estimated using a 4 s ARF, and that such a function can be used to estimate the applicability of the procedure over arrays of different aperture.

Figure 7 ▶
Array response functions for a 4 s period signal on the seismic arrays as indicated. The geometries of all arrays are provided in Fig. 1 except for BRMAR, a medium-period array with an aperture of approximately 30 km close to the BRTR short period array at Keskin, Turkey (KULELI *et al.*, 2001)

4. *The Applicability of Incoherent Parameter Estimation as a Function of Array Aperture*

Figure 7 displays the ARF for a signal with period 4 s for six IMS arrays with different apertures, all of which have difficulties in processing signals from events at regional distances. As demonstrated by GIBBONS *et al.* (2008), the resolution provided by the NOA array (with aperture of over 60 km) is high enough to provide unambiguous slowness estimates for any regional phases without the need for a smoothing in the slowness grid (Fig. 7a). The AKASG array (Malin, Ukraine) is the second largest array in the IMS but, with an aperture significantly smaller than that of the NOA array, has a significantly broader ARF (Fig. 7b). The ellipsoid ARF shape is a consequence of the almost ellipsoid array geometry, with the ARF ellipse being approximately perpendicular to the major axis of the array shape (a long distance between stations means a greater time-delay and better slowness resolution). The ARFs for both AKASG and YKA (Fig. 7c) are sufficiently compact for high-quality slowness estimates to be expected, although smoothing of the resulting slowness grids with the corresponding ARFs is recommended.

The MJAR array (Matsushiro, Japan, Fig. 7d) is selected as an example of an intermediate aperture array (10–20 km). It is demonstrated by GIBBONS *et al.* (2008) that a plausible slowness vector estimate can be obtained for a regional P phase at MJAR, but it is clear from the beampacking grids that the broad ARF brings a significant risk of selecting a qualitatively incorrect slowness vector. The performance of MJAR has been examined in detail given that it is the

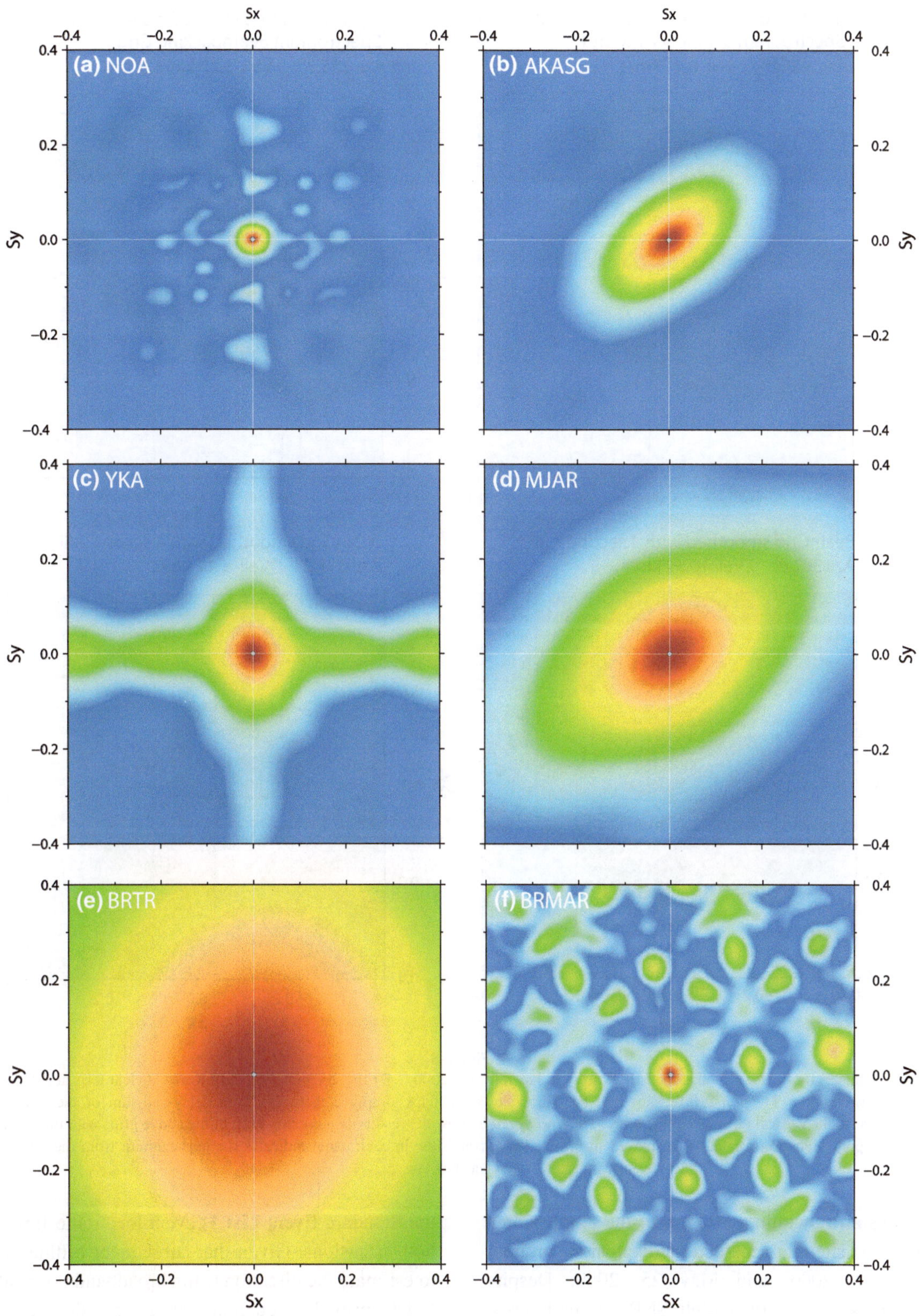
(a) NOA
(b) AKASG
(c) YKA
(d) MJAR
(e) BRTR
(f) BRMAR
Sx
Sy
−0.4
−0.2
0.0
0.2
0.4

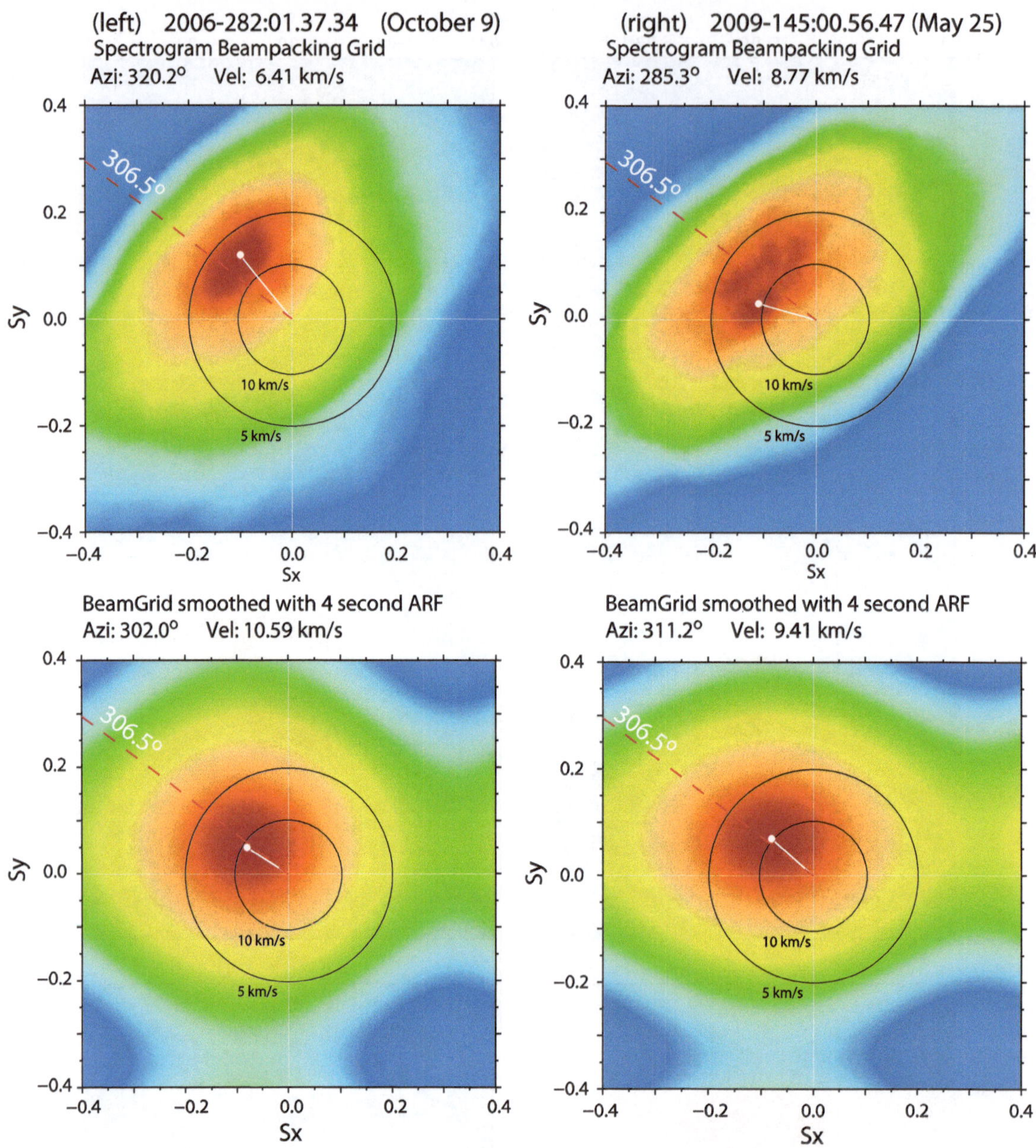

Figure 8

Slowness estimates for the Pn arrival at the MJAR array in Japan for the 2006 (*left*) and 2009 (*right*) North Korea nuclear tests at a distance of approximately 950 km. The *top panels* show the beamgrid slowness maps simply constructed from delay-and-sum of the transformed spectrograms for the vertical channels of the array. The *lower panels* show the result of smoothing each of the above grids with the theoretical 4 s array response function for MJAR displayed in Fig. 7. The *dashed line* in each panel indicates the backazimuth from the array to the nuclear test site

only IMS array within 1,000 km of the North Korea nuclear test site that recorded the explosions on both October 9, 2006, and May 25, 2009. Despite recording signals with a high SNR for both tests, MJAR failed to contribute to the automatic (SEL3—the Standard Event List 3) event location estimate on both occasions. Given that subsequent offline efforts to estimate the direction using broadband f–k analysis fail to provide a qualitatively correct slowness estimate for either of these Pn phases, it is unsurprising

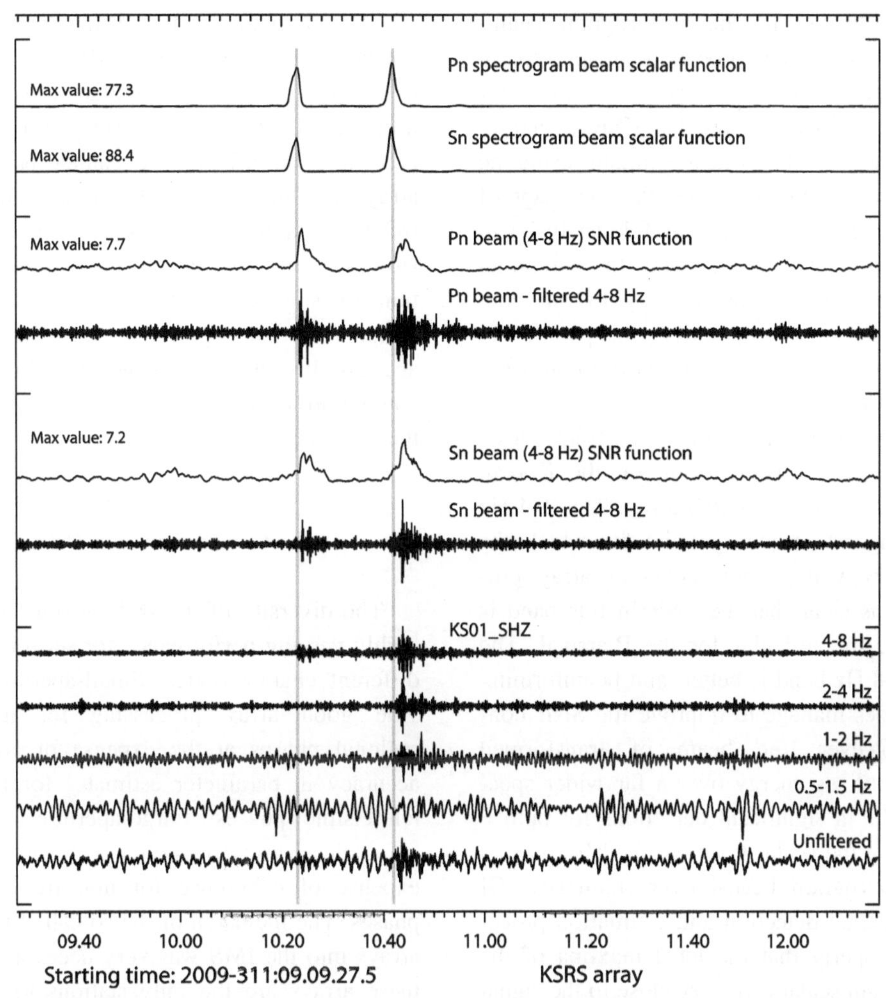

Figure 9
Signals on the KSRS array in the Republic of Korea from a quarry blast within a distance of 2°. It is clear that the SNR is optimal in a frequency band higher than that amenable to coherent processing on this array. The scalar functions derived from the transformed spectrogram beams give the impression of greater significance to the P and S detection peaks than for the classical beam SNR functions. Note also that the local maxima of the SNR functions occur somewhat later than the local maxima of the spectrogram beam scalar functions

that the fully automatic procedures fail. The beam-packing grids in the top row of Fig. 8 clearly show a pattern directed in the correct orientation towards the test site (at 306 degrees backazimuth), although local off-center maxima have led to a significant deviation in the reported backazimuth from the great-circle value. The lower panels, in which the slowness grids have been smoothed with respect to the ARF displayed in Fig. 7d, show a clear improvement with close-to-center local maxima which correspond well with the theoretical backazimuth and apparent velocity. Figure 8 indicates that the smooth functions resulting from the convolution of the observed and ARF slowness grids will provide robust slowness estimates for arrays of this aperture.

Of the 10–20 km aperture arrays (MJAR, ESDC, ILAR, KSRS, CMAR, ASAR, and EKA) there are great differences in the ability to process relatively high-frequency signals. For example, MJAR and KSRS have similar apertures with MJAR having many smaller inter-site spacings than KSRS. However, whereas KSRS can often report stable and accurate slowness estimates of regional phases in the 2–4 Hz band, f–k estimates on MJAR at all but the very lowest frequencies are made impossible by the very complicated geology at that site (see Kato *et al.*, 2005). On MJAR, the spectrogram beampacking with smoothing procedure displayed in Fig. 8 is essentially the only strategy available for high-frequency regional phases. On KSRS, qualitatively correct incoherent slowness estimates using the same procedure are possible, but the accuracy of coherent estimates (using classical f–k analysis) in the 2–4 Hz band is likely to be better. In operational pipelines, it may be a good strategy to run both coherent and incoherent operations in parallel to provide two detection lists which can be compared and cross-validated.

Figure 9 displays filtered waveforms from a low-magnitude industrial seismic event on the Korean peninsula recorded by the KSRS array. The 2–4 Hz band is the highest frequency band for which the KSRS geometry will support coherent array processing, but it is clear that the SNR in this band is extremely low, in particular for the P-arrival. The SNR in the 4–8 Hz band is better, and beamforming in this band does manage to improve the SNR considerably. However, the beam of transformed spectrograms utilizes energy over a far wider spectrum and results in significant detections for both P and S phases on the scalar traces formed from averaging the spectrogram beams over frequency. Of interest for pipeline detection and estimation procedures is the property that the local maxima of the spectrogram beam scalars are very close to the signal onset time (this is confirmed by Song *et al.*, 2010, for signals from microseismic events). The quality of these fully automatic arrival time estimates may reduce the need to apply repicking algorithms.

The 4 s ARF for the BRTR array in Turkey is displayed in Fig. 7e. The fact that such an ARF is unlikely even to be able to identify the quadrant of slowness space to which the true backazimuth and apparent velocity belong suggests that we have now reached the limit in array aperture over which we can apply the spectrogram beampacking procedure for parameter estimation. We note however that spectrogram beamforming is still a valid method for signal detection. The scan statistics algorithm of Taylor *et al.* (2010) is described for a single channel; the beamforming of spectrograms over adjacent sensors is likely only to reduce the variance of the background noise allowing a lower detection threshold.

The final panel of Fig. 7, (*f*), displays a 4 s ARF for the medium period BRMAR array near Ankara; a large aperture (40 km) circular array in close proximity to BRTR (see Kuleli *et al.*, 2001). The procedure described here cannot be used on this array, since the instrumentation does not record high-frequency signals (the sampling interval is 0.25 s). However, this provides an illustration of the resolution in slowness that would be achieved using broadband or short-period sensor deployments at that aperture. It is noted that panel (*f*) of Fig. 7 is the only configuration for which sidelobes appear, also due to the array geometry.

5. *Conclusions and Discussion*

The diversity of the IMS seismic arrays leads to highly varying performance for seismic arrivals with different characteristics. Small-aperture arrays provide good array processing for high-frequency regional phases at the expense of array gain and accuracy in parameter estimates for low-frequency teleseismic phases. Large-aperture arrays provide good teleseismic detection and characterization at the expense of coherence for high-frequency regional phases. The inclusion of the so-called legacy seismic arrays into the IMS was very necessary, as many of these arrays are the only stations to have existing records of many historical nuclear explosions. Waveforms from several of the legacy stations for many nuclear test explosions are openly available on the http://www.rdss.info/database/nedb/nedb_ent.html (see Bennett *et al.*, 2010).

We have demonstrated that the spectrogram beamforming procedure of Gibbons *et al.* (2008) for detecting and characterizing high-frequency regional seismic phases is applicable to the IMS seismic arrays exceeding about 10 km aperture. The quality of the slowness estimates is improved significantly by smoothing the spectrogram beampacking grids using array response functions (ARFs) for signals with a period of 4 or 5 s. The smoothing procedure is deemed to be essential for the arrays with aperture under 20 km. The stability of the procedure (with smoothing applied) is demonstrated in Fig. 8 for the Pn arrivals at MJAR in Japan for the North Korea

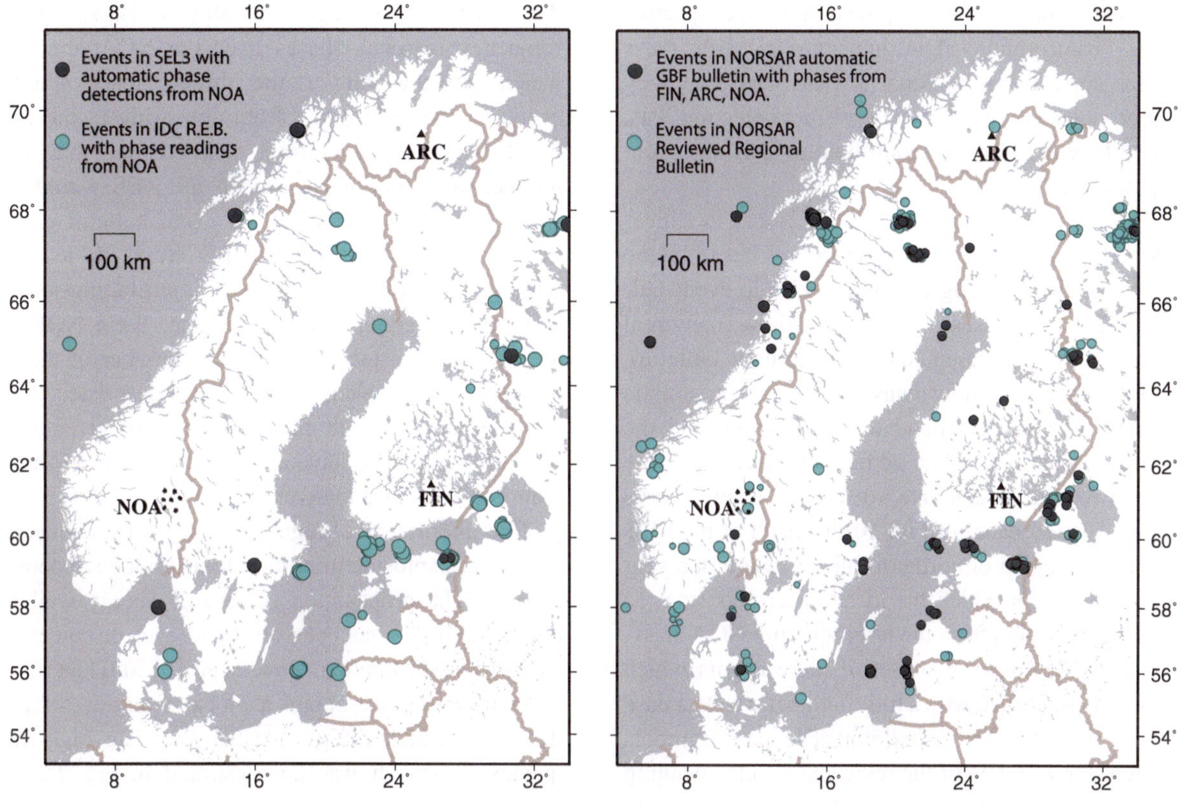

Figure 10
Events between 2007-078 and 2008-078 in the Reviewed Event Bulletins (*light blue symbols*) for the IDC (*left*) and NORSAR (*right*). The IDC REB events (*light blue, left panel*) are restricted to events that include a phase reading from the large aperture NOA array. The *dark symbols* indicate events in the fully automatic bulletins at the same data centers which include at least one NOA phase detection. (*Dark symbols* in the *right hand panel* must also include phases from the primary IMS arrays FINES and ARCES). At NORSAR, two parallel and independent systems process incoming data from the NOA array. A coherent process runs to detect and classify teleseismic phases and a process of spectrogram beamforming described by Gibbons *et al.* (2008) runs to detect and classify regional phases. The fully automatic GBF bulletin (Ringdal and Kværna, 1989) uses phase detections from the second of these processes

nuclear tests in 2006 and 2009. In many cases, coherent processing may be applicable and provides a greater accuracy than the incoherent procedure, but the f–k spectra may be subject to sidelobes due to spatial aliasing. The lower resolution estimates obtained from the smoothed spectrogram beampacking grids may help to identify the correct local maximum in the f–k spectra, in effect increasing the applicability of coherent array processing.

The detection part of the procedure is applicable to all IMS arrays regardless of aperture. There may be improvements to be made in the detection capability by applying algorithms from image processing (see Taylor *et al.*, 2010) to the beams of transformed spectrograms and this may offer complementary detection lists which provide a very low false alarm rate for high frequency signals. The spectrogram beamforming method, in effect, measures a group velocity rather than a phase velocity, and the detection threshold will always be higher for this procedure than for arrays where coherent processing is applied. Experience in operating the algorithm of Gibbons *et al.* (2008) on data from the NOA array (since 2005) indicates that an SNR of about 2.0 on waveforms from individual channels is necessary for the incoherent system to work well. The classification of all detected signals is essential, whether or not the signals in question are of interest to the monitoring aims. For example, many false teleseismic detections at NOA and other large aperture arrays are high-frequency regional signals that have been classified incorrectly in the low-frequency band used for teleseismic processing. Application of the spectrogram beamforming algorithm may classify these phases correctly as regional arrivals such that they can be screened out of the teleseismic detection lists, where they are likely to contribute to spurious event hypotheses.

The measure with which we assess any modification to the pipeline detection algorithms has to be the quality of the resulting automatic seismic event bulletins. This comprises completeness (how many real events are well represented in the automatic bulletin), the false event rate (how many entries in the automatic bulletin do not correspond to real events), and the location accuracy (what additional effort will the analyst require to generate a reviewed location estimate). The left panel of Fig. 10 shows all of the REB events in a region of northern Europe over a one-year time interval for which the location was constrained using a phase pick from the large aperture NOA array. The dark symbols the subset of these events which were also associated with a fully automatic NOA phase detection in the SEL3 bulletin. The dark symbols account for very few of the reviewed events, meaning that significant analyst effort has been required to add phases from NOA. The right panel of Fig. 10 shows events for the same region and time interval from the NORSAR Regional Event Bulletin, the dark symbols indicating events with fully automatic phase associations from each of the three primary seismic IMS arrays NOA, ARCES, and FINES. The spectrogram beamforming procedure operates on NOA array data at NORSAR, but not at the IDC, and the bulletins are not entirely comparable as NORSAR is free to use non-IMS seismic stations that are not available to the IDC. However, there are clearly far more events for which the automatic location estimate is controlled by all three primary seismic arrays when the incoherent processing algorithm is applied to NOA data. Most of the events in southern Norway with pale symbols in the right panel of Fig. 10 are constrained with both regional P and S automatic phase detections at NOA; it is the ARCES array for which detections are missing for most such events.

It is not necessarily the case that, from a resources point of view, it is desirable to increase the number of events as displayed in Fig. 10. However, it is necessary to point out that it is only the absence of an algorithm able to classify high frequency regional phases over the NOA aperture which prevents all of the three-primary station events in the right panel from appearing in the left panel. Hicks *et al.* (2001) pointed out that the number of events in Norway in the REB would be smaller than for the NORSAR event bulletin given that the small aperture NORES array (co-located with part of the NOA array) was contributing to the NORSAR detection lists at the time. Were NORES an IMS array instead of NOA, the number of three-primary station events would be far greater than indicated in the right panel of Fig. 10, given that NORES would also have a lower detection threshold than NOA using incoherent methods.

The importance of high-frequency seismic energy in nuclear explosion monitoring must not be underestimated. The propagation of seismic energy well in excess of 10 Hz has been observed over distances in excess of 1000 km (e.g. Bowers *et al.*, 2001) and it is imperative that we obtain a global overview of the degree to which different paths attenuate high-frequency energy. It was demonstrated in Fig. 2 how weaker events result in a lower signal amplitude over a wide range of frequencies which, due to the spectral level of the background noise, results in better SNR at the higher frequencies. It is precisely the lower frequency energy which is likely to diminish most given a lower explosion yield (compare, for example, the spectra from the 2006 and higher yield 2009 underground nuclear tests at the North Korea test site; Gibbons and Ringdal, 2012). The same is true of the cavity-decoupling evasion scenario (see Stevens *et al.*, 2006), with explosions with lower charge density resulting in regional seismic signals with far less low-frequency energy.

The IMS stations at which high-frequency energy arrivals are observed, but which cannot be processed due to unfavourable array geometries, need to be identified. In situations where events are repeating, or where high-quality calibration information is available, the empirical matched field processing method of Harris and Kvaerna (2010) may be able to exploit phase information associated with high-frequency energy above the limit where classical array processing can be applied. For each instance where signals are being detected by IMS arrays that are not correctly classified, we need to examine the reason. The incorrect classification of a detected phase may have as serious consequences as a missed phase, given that an automatic association algorithm has to process the output from the detection streams at each station. Applying the algorithms described here may significantly reduce the number of phases which are incorrectly characterized, and also may allow the remaining detections to be filtered out for subsequent manual analysis.

Acknowledgements

Data from IMS array stations were obtained via the IDC in Vienna. I am grateful for access to data from additional stations of the Geoscience Australia (AU) network used in this study which were obtained via the IRIS Data Management Center. I am also grateful for the constructive comments of two anonymous referees. Maps were created using GMT software (Wessel and Smith, 1995).

References

Bennett, T.J., Oancea, V., Barker, B.W., Kung, Y.L., Bahavar, M. *et al.*, (2010), *The Nuclear Explosion Database (NEDB): A New Database and Web Site for Accessing Nuclear Explosion Source Information and Waveforms.* Seismological Research Letters *81*, 12–25, doi:10.1785/gssrl.81.1.12.

Bowers, D., Marshall, P.D., and Douglas, A. (2001), *The level of deterrence provided by data from the SPITS seismometer array to possible violations of the Comprehensive Test-Ban in the Novaya Zemlya region.* Geophys. J. Int. *146*, 425–438, doi: 10.1046/j.1365-246x.2001.01462.x.

Bungum, H., and Husebye, E.S. (1971), *Errors in time delay measurements.* Pure appl. geophys. *91*, 56–70.

Capilla, C. (2006), *Application of the Haar wavelet transform to detect microseismic signal arrivals.* Journal of Applied Geophysics *59*, 36–46, doi:10.1016/j.jappgeo.2005.07.005.

Christie, D.R., and Campus, P., *The IMS Infrasound Network: Design and Establishment of Infrasound Stations*, in *Infrasound Monitoring for Atmospheric Studies* (eds. A. Le Pichon, E. Blanc, and A. Hauchecorne) (Springer Science + Business Media B.V., Dordrecht, Netherlands 2010), chap. 2, pp. 29–75. ISBN 978-1-4020-9507-8.

Dahlman, O., Mykkeltveit, S., and Haak, H., *Nuclear Test Ban: Converting Political Visions to Reality* (Springer Media B.V., Dordrecht, The Netherlands 2009). ISBN 978-1-4020-6883-6.

Douglas, A., *Seismometer Arrays - Their Use in Earthquake and Test Ban Seismology*, in *International Handbook of Earthquake and Engineering Seismology* (eds. W.H.K. Lee, H. Kanamori, P.C. Jennings, and C. Kisslinger) (Academic Press 2002), pp. 357–367.

Gendron, P., Ebel, J., and Manolakis, D. (2000), *Rapid Joint Detection and Classification with Wavelet Bases via Bayes Theorem.* Bull Seism Soc Am *90*, 764–774.

Gibbons, S.J., and Ringdal, F. (2012), *Seismic Monitoring of the North Korea Nuclear Test Site Using a Multichannel Correlation Detector.* IEEE Trans. Geosci. Remote Sensing *50*(5), 1897–1909, doi:10.1109/TGRS.2011.2170429

Gibbons, S.J., Ringdal, F., and Kværna, T. (2008), *Detection and characterization of seismic phases using continuous spectral estimation on incoherent and partially coherent arrays.* Geophysical Journal International *172*, 405–421, doi:10.1111/j.1365-246X.2007.03650.x.

Häge, M., and Joswig, M. (2009), *Spatiotemporal characterization of interswarm period seismicity in the focal area Nový Kostel (West Bohemia/Vogtland) by a short-term microseismic study.* Geophysical Journal International *179*, 1071–1079, doi: 10.1111/j.1365-246X.2009.04320.x.

Harris, D.B., and Kvaerna, T. (2010), *Superresolution with seismic arrays using empirical matched field processing.* Geophysical Journal International *182*, 1455–1477, doi:10.1111/j.1365-246X.2010.04684.x.

Hicks, E.C., Bungum, H., and Ringdal, F. (2001), *Earthquake Location Accuracies in Norway Based on a Comparison between Local and Regional Networks.* Pure and Applied Geophysics *158*, 129–141, doi:10.1007/PL00001152.

Joswig, M. (1990), *Pattern recognition for earthquake detection.* Bull. Seism. Soc. Am. *80*, 170–186.

Joswig, M. (1995), *Automated classification of local earthquake data in the BUG small array.* Geophysical Journal International *120*, 262–286, doi:10.1111/j.1365-246X.1995.tb01818.x.

Kato, M., Nakanishi, I., and Takayama, H. (2005), *Variation of teleseismic short-period waveforms at Matsushiro Seismic Array System.* Earth, Planets, Space *57*, 563–570.

Kennett, B.L.N., *The Seismic Wavefield. Volume II: Interpretation of Seismograms on Regional and Global Scales* (Cambridge University Press, Cambridge, United Kingdom 2002). ISBN 0-521-00665-1.

Kennett, B.L.N., Brown, D.J., Sambridge, M., and Tarlowski, C. (2003), *Signal Parameter Estimation for Sparse Arrays.* Bulletin of the Seismological Society of America *93*, 1765–1772.

Kuleli, S., Zor, E., Türkelli, N., Sandvol, E., Seber, D. *et al.*, (2001), *The IMS Belbasi Seismic Array (BRAR) in Central Turkey.* Seismological Research Letters *72*, 60–69.

Kværna, T., and Ringdal, F. (1992), *Integrated array and three-component processing using a seismic microarray.* Bull. Seism. Soc. Am. *82*, 870–882.

Kværna, T., Ringdal, F., and Given, J., *Application of Detection Probabilities in the IDC Global Phase Association Process.* In *Proceedings of the 2011 Monitoring Research Review, Tucson, Arizona, September 13-15, 2011. Ground-based Nuclear Explosion Monitoring Technologies, Report LA-UR-11-04823* (2011), pp. 302–312.

Levander, A., and Nolet, G., *Perspectives on Array Seismology and USArray.* In *Seismic Earth: Array Analysis of Broadband Seismograms* (eds. A. Levander and G. Nolet) (American Geophysical Union 2005), pp. 1–6. Geophysical Monograph 157.

Mykkeltveit, S., Ringdal, F., Kværna, T., and Alewine, R.W. (1990), *Application of regional arrays in seismic verification research.* Bull Seism Soc Am *80*, 1777–1800.

Nishida, K., Kawakatsu, H., Fukao, Y., and Obara, K. (2008), *Background Love and Rayleigh waves simultaneously generated at the Pacific Ocean floors*. Geophysical Research Letters *35*, L16307+, doi:10.1029/2008GL034753.

OONINCX, P.J. (1999), *A wavelet method for detecting S-waves in seismic data*. Computational Geosciences *3*, 111–134, doi: 10.1023/A:1011527009040.

PICOZZI, M., PAROLAI, S., and BINDI, D. (2010), *Deblurring of frequency-wavenumber images from small-scale seismic arrays*. Geophys. J. Int. *181*, 357–368, doi:10.1111/j.1365-246X.2009.04471.x.

PRIETO, G.A., PARKER, R.L., and VERNON, F.L. (2009), *A Fortran 90 library for multitaper spectrum analysis*. Computers and Geosciences *35*, 1701–1710.

RINGDAL, F., and HUSEBYE, E.S. (1982), *Application of Arrays in the Detection, Location, and Identification of Seismic Events*. Bull. Seism. Soc. Am. *72*, S201–S224.

RINGDAL, F., and KVÆRNA, T. (1989), *A multi-channel processing approach to real time network detection, phase association, and threshold monitoring*. Bull. seism. Soc. Am. *79*, 1927–1940.

RINGDAL, F., HUSEBYE, E. S., and DAHLE, A., *P-Wave Envelope Representation in Event Detection Using Array Data*, in *Exploitation of Seismograph Networks* (ed. K. G. Beauchamp), no. 11 in Series E: Applied Sciences (Noordhoff - Leiden 1975), pp. 353–372.

ROMNEY, C. F., *VELA Overview: The Early Years of the Seismic Research Program*, in *The VELA Program: A Twenty-Five Year Review of Basic Research* (ed. A. U. Kerr) (1985), pp. 38–65.

ROST, S., and THOMAS, C. (2002), *Array Seismology: Methods and Applications*. Rev. Geophys. *40*. 1008, doi:10.1029/2000RG000100.

ROST, S., and THOMAS, C. (2009), *Improving Seismic Resolution Through Array Processing Techniques*. Surveys in Geophysics *30*, 271–299, doi:10.1007/s10712-009-9070-6.

SCHWEITZER, J., FYEN, J., MYKKELTVEIT, S., and KVÆRNA, T., *Chapter 9: Seismic Arrays*, in *IASPEI New Manual of Seismological Observatory Practice* (ed. P. Bormann) (GeoForschungsZentrum, Potsdam 2002). 52 pp.

SELBY, N. D. (2011), *Improved Teleseismic Signal Detection at Small-Aperture Arrays*. Bull. Seism. Soc. Am. *101*, 1563–1575, doi:10.1785/0120100253.

SONG, F., KULELI, H. S., TOKSOZ, M. N., AY, E., and ZHANG, H. (2010), *An improved method for hydrofracture-induced microseismic event detection and phase picking*. Geophysics *75*, A47–52, doi:10.1190/1.3484716.

STEVENS, J. L., GIBBONS, S., RIMER, N., XU, H., LINDHOLM, C., RINGDAL, F., KVAERNA, T., and MURPHY, J. R. (2006), *Analysis and simulation of chemical explosions in nonspherical cavities in granite*. Journal of Geophysical Research - Solid Earth *111*, B04306+, doi:10.1029/2005JB003768.

TAYLOR, S. R., ARROWSMITH, S. J., and ANDERSON, D. N. (2010), *Detection of Short Time Transients from Spectrograms Using Scan Statistics*. Bull. Seism. Soc. Am. *100*, 1940–1951, doi: 10.1785/0120100017.

THOMSON, D. J. (1982), *Spectrum estimation and harmonic analysis*. Proc. IEEE *70*, 1055–1096.

WESSEL, P., and SMITH, W. H. F. (1995), *New version of the Generic Mapping Tools*. EOS Trans. Am. Geophys. Union *76*, 329.

WITHERS, M., ASTER, R., YOUNG, C., BEIRIGER, J., HARRIS, M., MOORE, S., and TRUJILLO, J. (1998), *A comparison of select trigger algorithms for automated global seismic phase and event detection*. Bull. Seism. Soc. Am. *88*, 95–106.

YOMOGIDA, K. (1994), *Detection of anomalous seismic phases by the wavelet transform*. Geophysical Journal International *116*, 119–130, doi:10.1111/j.1365-246X.1994.tb02131.x.

ZHANG, H., THURBER, C., and ROWE, C. (2003), *Automatic P-Wave Arrival Detection and Picking with Multiscale Wavelet Analysis for Single-Component Recordings*. Bull. Seism. Soc. Am. *93*, 1904–1912, doi:10.1785/0120020241.

(Received November 30, 2011, revised September 27, 2012, accepted October 1, 2012, Published online October 17, 2012)

Pure Appl. Geophys. 171 (2014), 395–411

DOI 10.1007/s00024-012-0592-3

Pure and Applied Geophysics

A Machine Learning Approach for Improving the Detection Capabilities at 3C Seismic Stations

CARSTEN RIGGELSEN[1] and MATTHIAS OHRNBERGER[1]

Abstract—We apply and evaluate a recent machine learning method for the automatic classification of seismic waveforms. The method relies on Dynamic Bayesian Networks (DBN) and supervised learning to improve the detection capabilities at 3C seismic stations. A time-frequency decomposition provides the basis for the required signal characteristics we need in order to derive the features defining typical "signal" and "noise" patterns. Each pattern class is modeled by a DBN, specifying the interrelationships of the derived features in the time-frequency plane. Subsequently, the models are trained using previously labeled segments of seismic data. The DBN models can now be compared against in order to determine the likelihood of new incoming seismic waveform segments to be either signal or noise. As the noise characteristics of seismic stations varies smoothly in time (seasonal variation as well as anthropogenic influence), we accommodate in our approach for a continuous adaptation of the DBN model that is associated with the noise class. Given the difficulty for obtaining a golden standard for real data (ground truth) the proof of concept and evaluation is shown by conducting experiments based on 3C seismic data from the International Monitoring Stations, BOSA and LPAZ.

1. Introduction

Within the last few decades, significant advances have been achieved in the seismic monitoring task, i.e., detecting and locating sources of seismic wave generation. One of the most influential factors that allowed pushing the detection threshold to low earthquake magnitude values on both global (KVÆRNA *et al.*, 1999; KVAERNA *et al.*, 2002) and regional scales (KVAERNA *et al.*, 2002; HUTTON *et al.*, 1932; NANJO *et al.*, 2010) is the continuously increasing densification of seismic sensors networks in the continental areas of the globe. Thanks to developments in computer technologies, storage and processing of huge waveform data sets acquired at those dense global seismic networks is now feasible and enables real-time scanning the continuous data for relevant information on seismic events in real-time.

One may identify three major steps in a standard processing pipeline that are needed to create accurate seismic bulletins:

1. Discrimination of relevant (usually transient) seismic signal energy at single station level from unwanted waveform portions that are termed "noise". The onset of the detected signal portions are flagged with time marks (picks).
2. Confirmation of single station observations by sensor network considerations, i.e., coincidenting observations consistent with wave propagation principles (travel times, amplitude decay, slowness and azimuth from polarization and/or array processing) leading to an event seed with a draft location.
3. Visual control by human analysts of the set of waveforms related to an event seed created in step 2. Previously determined picks in one may be pruned or adjusted.

The first two steps of the processing chain are usually carried out in an automatic fashion using tuned algorithms to scan for transient signals portions (e.g., WITHERS *et al.*, 1998; NIPPRESS *et al.*, 2010) and associate the resulting time arrival marks (phase picks) to a potential seismic event thus creating an "event seed" with draft location (BACHE *et al.*, 1990; BACHE *et al.*, 1993). It is clear that results obtained in subsequent steps accumulate errors from previous ones, which makes the single station detection task highly relevant to a successful overall outcome. In

[1] University of Potsdam, Potsdam, Germany. E-mail: riggelsen@geo.uni-potsdam.de

particular there is an important trade-off between the number of missed transient detections as well as the false alarm rate at step 1 and the quality of the subsequent pipeline products in steps 2 and 3. If signals are missed on the single station level, the number of events to be seeded will be reduced and probably a significant number of weak events will not be included in an automatic draft bulletin. A visual control of those events will, therefore, never occur and the event remains undetected. In case many false alarms are created at the single station level (spurious detections not related to a true seismic source), the result of the network association procedure will significantly deteriorate. A larger number of "fake events" will be produced that are included in a draft bulletin. All of those events will have to be reviewed in the interactive visual control step and thus increase the manpower workload.

In this work we attempt to improve the accuracy of the initial single station detection problem, i.e., reducing the number of false alarms and decreasing the number of missed events at the same time, by applying a machine learning classification approach developed in RIGGELSEN *et al.* (2007) to 3C-seismic data coming from the International Monitoring Stations (IMS) as operated by CTBTO's International Data Center (IDC); in particular, we focus on seismic stations LPAZ and BOSA, where we had at our disposal in total 5 weeks of data, 1 week of July 2008, October 2008, January 2009, April 2009 and July 2009, plus the associated bulletins: the IDC SEL3 automatic bulletin, the product and the end of the second step of the bulletin pipeline, and the analyst-reviewed IDC LEB bulletin, the product at the end of the third step of the bulletin pipeline. The care with which the LEB bulletin is prepared is rarely seen elsewhere in seismology, and the LEB bulletin comes as close as one may come to a golden standard in seismology. Therefore, the waveform data along with the SEL3 and LEB bulletins of the two IMS 3C-stations considered provide a good opportunity to validate the performance of the classifier. Yet one should be aware that even the LEB bulletin is neither complete nor free of erroneous detections. It is important to note that the approach described (RIGGELSEN *et al.*, 2007) may be applied to 3C waveform data coming from any seismic station. It is merely by the virtue of quality of the bulletins that the two aforementioned IMS stations are chosen.

The paper is organized as follows: in Sect. 2 we describe our approach in general terms and introduce Dynamic Bayesian Networks (DBNs) in the context of supervised learning. We then proceed in Sect. 3 with a discussion on seismic feature selection and extraction, as well as how to use the features in relation to the DBNs. In Sect. 4 we discuss issues relevant to data preparation. In Sect. 5 we evaluate the performance of the proposed DBN classifier on a separate fixed test-set and propose and discuss an adaptive classification approach to deal with the seasonal drift of noise in the context of applying the classifier to continuous data. In Sect. 6 we try to tackle the problem of defining a fair evaluation procedure using a semi-synthetic data set. Finally, in Sect. 7 we draw conclusions.

2. *Approach*

In this section we informally present the approach as described in RIGGELSEN *et al.* (2007), where a machine learning approach based on supervised learning and Dynamic Bayesian Networks (DBN) is introduced for classifying seismic waveform data. In comparison to RIGGELSEN *etal.* (2007) where only the vertical components of seismic signals is considered, we in the present paper take both the vertical and the horizontal components into account.

2.1. *Supervised Learning with DBNs*

The classifier is based on the framework of graphical models, a multivariate statistical framework which has been proven to be fundamental to advanced data analysis, tying together a range of seemingly very different statistical/probabilistic approaches to data analysis. It provides a generic modeling language for many (non-deterministic) real-life problems and processes. DBNs, a particular kind of graphical model, is suitable for analyzing temporal data and allows one to capture, learn and specify the relationship between domain variables of real-life systems and processes where an intrinsic temporal aspect is present. DBNs are attractive as unifying modeling and inference tools for several reasons and have an advantage over many other formalisms such as artificial neural networks, e.g., by the virtue of being "white boxes" rather than "black boxes": the transparent nature of DBNs allows for interpretation and enables for the inclusion of expert knowledge.

Hidden Markov Models (HMMs) are special kinds of DBNs, and have been used extensively with great success in automatic speech recognition (ASR) (Bilmes, 2006). Recently DBNs have started to replace HMMs in ASR (see Zweig, 2003; Zweig and Russell, 1998). Speech and seismic signals share similar characteristics in time-frequency patterns: speech embedded in silence versus transient earthquake signals embedded in ambient wave fields. This similarity has lately been exploited in seismological applications (Hammer *et al.*, 2011; Beyreuther *et al.*, 2008), but until now DBNs in seismology have not been explored.

DBNs are temporal generative models, i.e., statistical models that are capable of modeling intrinsic characteristics of patterns in data streams. Patterns are modeled in a DBN in a non-deterministic way, e.g., a particular pattern may vary in the time dimension (be stretched or compressed along the time axis), yet the DBN formalism is flexible enough to capture all such variations in a single model. This makes them very different from static templates [e.g., cross-correlation approaches as in Gibbons and Ringdal (2006)]; rather, a DBN has the power of acting as a distribution over a particular pattern class. In Sect. 3.1 we will informally describe how DBNs work.

In this work, by employing supervised learning, a classifier is derived from DBNs which enables us to classify seismic signal versus seismic noise. Supervised learning is based on iterative mechanisms that "present" yet untrained DBNs with example patterns, at each iteration tuning/refining the DBNs. When enough patterns have been presented to the DBNs they have essentially picked up the intrinsic characteristics of the patterns. A DBN derived classifier can now be presented with previously unseen seismic patterns, and it will classify the input by returning a measure, that is, a score of how well the seismic waveform belongs to either the signal or noise class; in this classification phase the classifier relies on what the DBNs have learned during the training phase.

In the first phase, the *training phase*, we can roughly distinguish three parts:

1. Stereotypical examples of patterns are picked by domain experts. For each class a number of examples are collected. In our application the number of data examples typically is between 20 and 100; in general these are very costly to obtain.
2. Features extraction from the patterns selected in 1.
3. The DBNs are iteratively trained using generalized expectation maximization (GEM) (see Bilmes and Zweig, 2002 and references therein) to capture the characteristics of the patterns from 2.

For the second phase, the so-called *classification phase*, we can also distinguish three parts:

1. A time segment is buffered of the seismic trace(s) (e.g., 12 s)—this segment is moved in a sliding window fashion as more data arrive.
2. A part that extracts features from the segment in 1.
3. The DBN inference engine based on the Junction Tree approach (see Bilmes and Zweig, 2002 and references therein), which classifies the pattern in 2, given the DBNs as learned in the training phase.

The DBNs are implemented in the GMTK toolkit (Bilmes and Zweig, 2002); this toolkit includes utilities for running GEM and Junction Tree inference on DBNs.

3. *Features Selection and Extraction*

To distinguish signal versus noise, we adhere to signal characteristics which have proven to carry significant information about seismic data streams by decades of research in seismology (see also Köhler *et al.*, 2008). Specifically all relevant features one may think of, may be derived from a coarse time frequency representation of the original waveforms by using a finite impulse response filter (FIR) bank. The FIR filters are designed by computing the complex valued Morlet wavelet $\psi(\tau)$, being a common choice for continuous wavelet transforms (CWT) of time series $s(\tau)$, yielding CWT time/frequency-coefficients $\mathrm{CWT}(t,f) = \int s(\tau) f^{-1/2} \psi^*_{t,f}(\tau)\,\mathrm{d}\tau$, where $*$ denotes complex conjugation for the translated and dilated Morlet mother wavelet $\psi_{t,f}(\tau)$. Henceforth, by

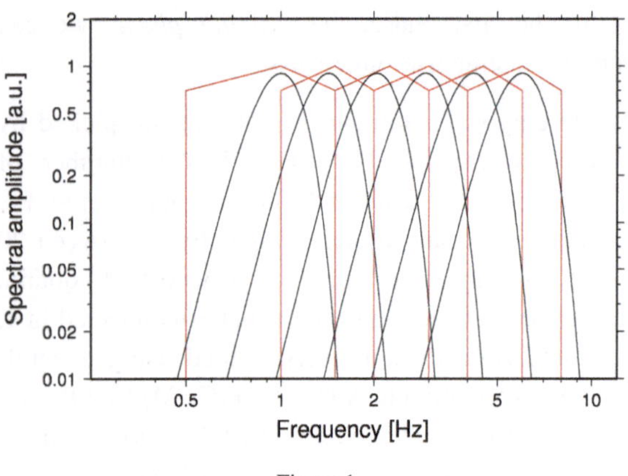

Figure 1
Spectral shape of the FIR filter banks

absolute CWT coefficients we shall mean the absolute value of the complex valued filter output of the FIR filter bank. The CWT coefficients are computed for each seismogram sample, i.e., for each sample in the input channel we obtain one sample in six distinct output channels. From the "raw" CWT coefficients other features (e.g., STA/LTA on CWT-amplitudes) can be derived, however, based on experiments not documented here the CWT features on their own suffice and therefore we will focus on them only.

We apply six FIR filters to each data channel using six distinct frequency bands, each of which are specified by a central frequency and a bandwidth parameter. Figure 1 shows the spectral shape of the six selected FIR filter bands based on the Morlet wavelet (black/smooth) in comparison to the frequency bands (red/edgy) currently in use by the IDC for P and S phase detection; these have been tuned over decades and are based on seismological experience. The center frequencies and the bandwidth parameter of the Morlet wavelet were selected in order to resemble closely the IDC frequency band processing schemes thus making implicit use of this piece of seismological expert knowledge.

3.1. *Features in Dynamic Bayesian Networks*

If we imagine a seismogram of T samples, each sample is a vector of six CWT coefficients, for each band we will have that the coefficients are evolving in time (smoothly); between adjacent frequency bands there is also a dependence: these are the relationships we encode in a DBN. However, for a particular waveform, the degree to which this dependence holds differs; supervised learning is used to determine this. Hence, for each kind of waveform we may define a different DBN. If we now present the DBN with a new waveform, the DBN returns a likelihood of the pattern belonging to the waveform class the DBN has been designed to model. For several DBNs we may then compare likelihoods and classify the new waveform accordingly.

Formally speaking a DBN comprises of *nodes, arcs* and *parameters*; see Fig. 2. Nodes are random variables and are associated with either features (white) or correspond to abstractions of those features which are not observed directly (gray). There are white nodes for each CWT feature: for each frequency band at each sample. The arcs between nodes represent the fact that variables influence each other: the CWTs probabilistically depend on each other (indirectly via the gray nodes) in time (arrows right) and across frequency band (arrows down). The nodes plus the arcs together are referred to as the model structure, and is the qualitative description of how features relate. The degree to which the dependence exists is dictated by the parameters of the DBN, and can be thought of as how the features relate to each other on a quantitative scale; these are learned via supervised learning. As seismograms have to be considered as temporal variable patterns evolving in time, the model structure must be able to cope with this. For a given seismic pattern, this model capability is captured in the shaded part of the DBN (see Fig. 2), which is "unrolled" to accommodate the entire length of the pattern, consisting of T samples; hence, at 40 Hz there would be 40 slices per second times the duration of the pattern in seconds.

The CWT features associated with the white nodes in the DBN are defined as follows (refer to Riggelsen *et al.*, 2007 for the mathematical details). Extract the so-called *average log-amplitude* from the CWT coefficients, and associate O_i^t with the *residuals* instead of the CWT coefficients directly, that is, for the vertical component $O_i^t = \log \mathrm{CWT}(t,i) - \bar{b}^t$ with $\bar{b}^t = \frac{1}{6}\sum_{i=1}^{6} \log \mathrm{CWT}(t,i)$, the average log-amplitude. The ideal earthquake *source spectra* for extended sources provide a scaling relation between the frequency of the spectral maximum and the

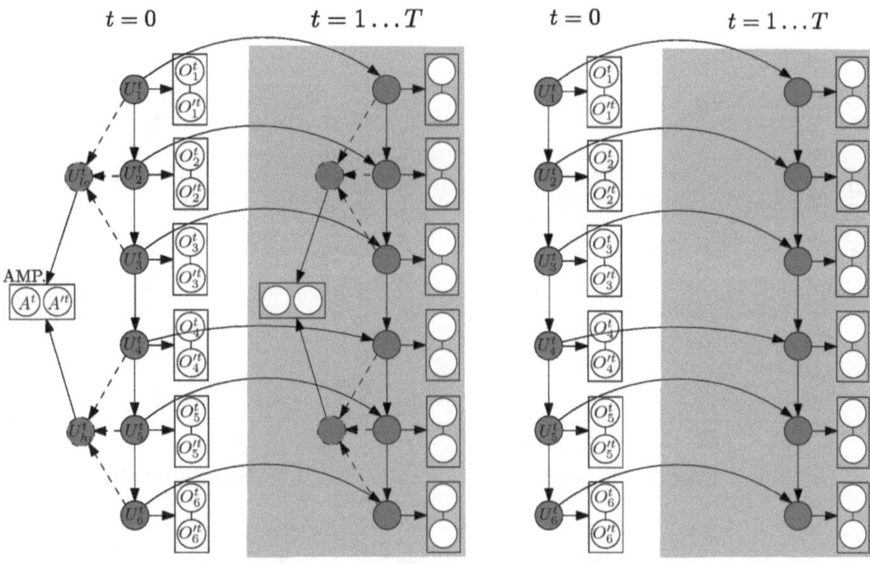

Figure 2
Dynamic Bayesian Networks: *left* with amplitude feature (DBN+A). *Right* without amplitude feature (DBN−A)

average log-amplitude. Figure 3, left, shows the so-called ω^3 model, displayed as proportional to ground velocity (Aki and Richards, 2002, p. 513, eqs. 10.36+10.37), where the signal shape primitives are shown at the bottom (emphasized in the frequency range we are focusing on). Figure 3, right, shows the global "low noise model" and "high noise model" curves (Peterson, 2003) with the noise shape primitives again at the bottom. For each generative DBN model, be it for signal or noise, these "primitives" are essentially learned for each class. The discrimination between signal and noise can therefore be attributed to the differences exhibited by the signal and noise primitives.

The relationship given by the source spectra is explicitly modeled in the DBN by relating the residuals O_i^t to an amplitude node, $\text{AMP}^t = \bar{b}^t$. For O_i^t this is for the vertical component trace only. The geometrical mean of the two horizontal CTWs define the combined horizontal, and the residual is defined analogously to the vertical component and associated with residual $O_i'^t$ and amplitude AMP'^t. The pairs $(O_i^t, O_i'^t)$ are distributed according to bivariate Gaussians, each governed by a mean and a covariance matrix (and together form the parameter set for a Gaussian) and the amplitudes AMP^t and AMP'^t are univariate Gaussians. The hidden nodes are associated with binary (on/off) discrete probability distribution. Moreover, there are also two kinds of arcs: those that correspond to probabilistic dependencies (solid) and those that correspond to deterministic mappings (dashed). For the latter no parameters are associated, but the mapping is given in advance (we refrain from details here). The resulting DBN (referred to as DBN+A) is given in Fig. 2, left.

A too strong dependence on amplitude is not necessarily a desirable property for a seismic signal classifier as it may lead to missing signals with very small amplitudes. The amplitude is modeled in the DBN by a direct observable node; it is, therefore, rather straight forward to alter the model to ignore the amplitude fully by simply removing the corresponding feature node/variable from the DBN structure specification. Figure 2, right, shows the modified DBN (referred to as DBN−A) now without the amplitude feature. The resulting DBN now only captures the spectral shape and relative vertical and horizontal amplitude relations.

4. *Data Preparation*

The classifier needs to be trained using signal and noise "snippets", that is, seismic segments that have

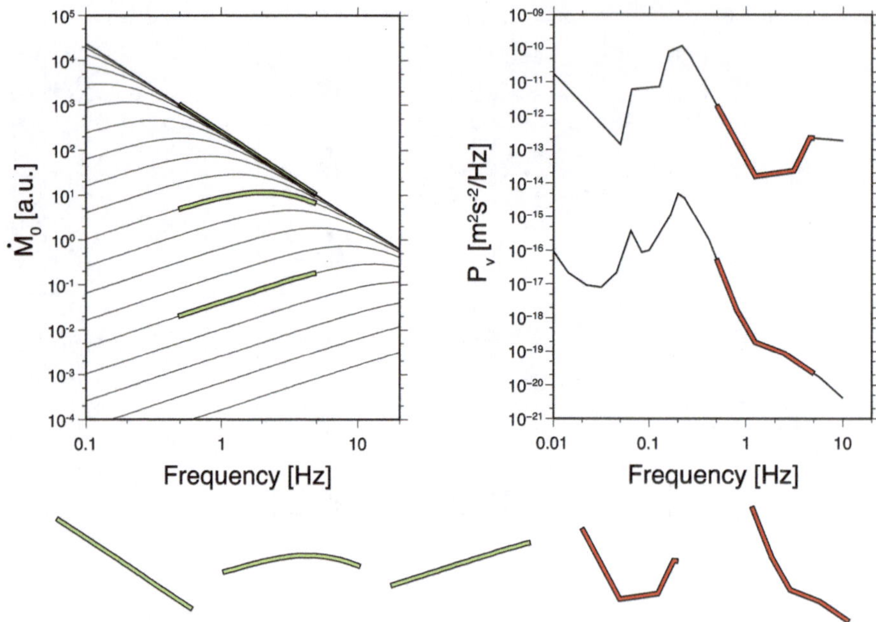

Figure 3
Idealized spectral shape primitives expected for signal and noise spectra at broadband stations. *Top left* source spectra scaling with scalar moment rate (ω^3-model). *Green lines* indicate typical shapes for strong, intermediate and weak sources within the relevant intermediate frequency band 0.5–5.0 Hz. *Top right* Petersen high-noise and low-noise model curves proportional to ground velocity. *Bottom* remaining spectral shape primitives after removing log average spectral power. Signal shape primitives are either linear and inclined or concave shaped whereas noise spectral shape types are convexly formed in the relevant frequency band

been labeled as such by an analysts. However, the analyst only picks or confirms the onset of a signal (and reports these in bulletins), but does not explicitly give the duration, i.e., the end-time is not given. For the noise patterns, no pick nor duration is available at all. Because our approach is segment-based we need to select a duration for both signal and noise ourselves, and we need to select an "onset" for noise. Care should obviously be taken that the noise-segments are outside the range of what one would assign to the signal class. If this is not done with care, the classifier will not perform optimally.

For training the classifier for BOSA and LPAZ, we used 1 week of IMS data from July 2008. The signal training segments are selected according to the associated LEB bulletin and are set to be of 12 s duration: a segment starts 2 s prior to the reported onset and ends 10 s later. Hence, a "signal" pattern consists of a P-wave onset pattern, i.e., the transition from noise to signal. For training we only use first-arrival P-wave onsets as reported in the LEB; S-waves and surface waves are not considered.

Some onsets as reported in the LEB are in our judgment not well discernible, perhaps partly due to a poor SNR. Therefore, in order to get good training examples we have put some effort into visually selecting by hand what we believe are actual onsets/pattern examples that can be discerned by using seismological expert judgment (applying filters, etc.). If we mix up (close to) noise and signal patterns in the signal training set, it might confuse the classifier and poor classification performance will be the result. From visual control we find that a SNR threshold of 5 for 3C-stations BOSA and LPAZ allows a sufficient degree of discernability between signal and noise. In Fig. 4 we show two waveform examples of a very clear (collected as training example) and one quite noisy (not collected as training example) onset for station BOSA.

For the noise training patterns we select 20 s samples long segments, making sure that there is no close signal onset "in the vicinity". Moreover, we did make sure that there is noise present from both day and night in our training set, in order to capture variabilities in the feature space distributions for the noise class.

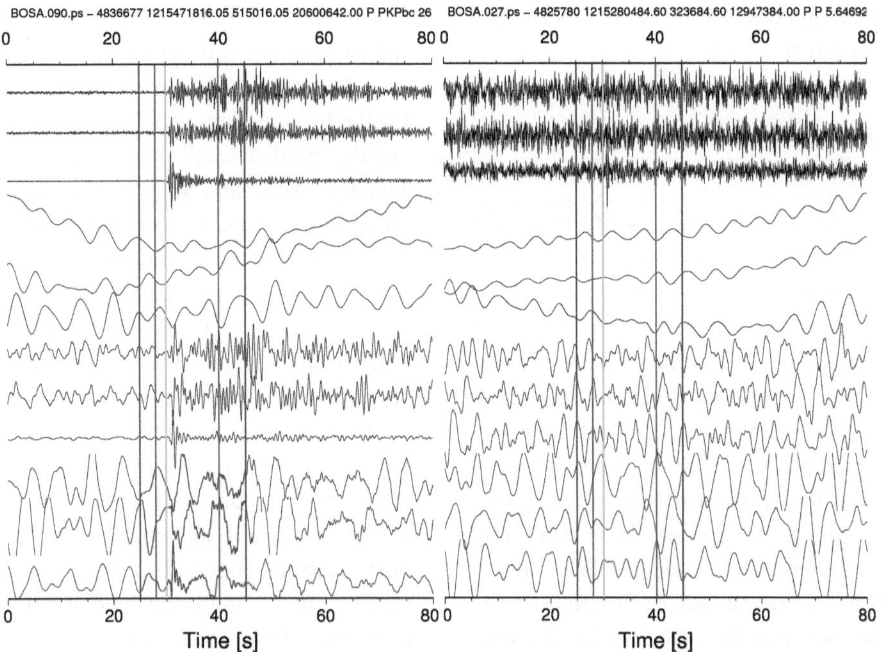

Figure 4

The plots show (from *bottom to top*): unfiltered traces (3C), WWSSN SP filtered waveforms, WWSSN LP filtered waveforms and a high-pass filtered version of the original waveforms (third order Butterworth filter at 3 Hz). All traces are scaled individually to minimum and maximum in the 80 s time window. *Vertical lines* indicate: pick onset as specified in the LEB (*green*), windows for cutting the signal classes (*blue*, *red*). *Blue lines* depict 20 s window with 5 s pre- and 15 s postpick times, *red lines* show the actually 12 s windows used for final training with 2 s pre- and 10 s postpick times

5. Evaluation

The aim is to improve the detection capability (signal vs. noise) using the approach and features described in the previous sections. The classifier should come as close as possible to the analysts classifications. It is improbable that the classifier will outperform the analyst; however, in terms of consistency the classifier might in fact do better. In order to test the performance, a separate test-set is compiled from the data of the same week of July 2008 as was used for training; for a reliable result we make sure that test-set and training-set are disjoint. Table 1 shows the results for DBN−A without the amplitude feature, and for DBN+A with the amplitude feature.

Generally, we get a sensitivity and a specificity in an acceptable range of 0.80–0.97. Although not reported here, we would like to mention that the classifier also exhibits a lower confidence (as defined in RIGGELSEN *et al.*, 2007) in cases where an incorrect classification has been made. This indicates that the patterns that are misclassified are not "typical examples" of the class that they have been assigned to; they are somewhere between signal and noise, but due to the binary nature of the classification approach

Table 1

Performance measures for DBN−A and DBN+A on LPAZ and BOSA

	BOSA DBN−A/DBN+A	LPAZ DBN−A/DBN+A
TP	31/32	72/71
FP	3/3	4/2
FN	6/5	17/18
TN	95/95	21/23
Se	0.84/0.86	0.81/0.80
Sp	0.97/0.97	0.84/0.92

True positives (TP): number of signal patterns also classified as such; false positives (FP): number of noise patterns wrongly classified as signal, also referred to as *false alarms*; false negatives (FN): number of signal patterns wrongly classified as noise also referred to as *missed events*; true negatives (TN): number of noise patterns also classified as such. Sensitivity (Se): a sensitivity of 1 means that the classifier recognizes all signal patterns as such (Se = TP/(TP + FN)); Specificity (Sp): a specificity of 1 means that the classifier recognizes all noise patterns as such (Sp = TN/(FP + TN))

they have to be assigned the most likely label which in this case is incorrect.

5.1. Continuous Classification and Adaptive Noise Modeling

For real-time classification we don't have pre-cut segments aligned neatly around noise and signal patterns. The performance measures given in Table 1 are somewhat artificial because we present the classifier with pre-cut and correctly labeled (LEB-based and confirmed by visual inspection) signal and noise segments. For continuous classification we instead produce and present to the classifier arbitrary segments cut from the waveform data in a moving-window fashion: for signal segments there will rarely be exactly 2 s pre-onset +10 signal falling into a segment which has to be classified. The question is how the classifier performs in such a set-up. We note that in general one may opt for different segment lengths than the 12 s window as we suggest here: too short windows lead to spurious detections as there is too little temporal context for the classifier, and too long windows lead to unsatisfactory timing precision of a signal onset, and may moreover provide the classifier with too much context; the window may contain several adjacent signal and noise patterns. In Sect. 6 this will also be illustrated.

It is well known that the background seismic wave field properties may change in time due to environmental conditions (anthropogenic origin) as well as due to small and large scale atmospheric phenomena (storm, ocean swell, microseisms, see e.g., TANIMOTO *et al.*, 2006). The DBN for noise which was learned based on merely 1 week of data will, therefore, be inadequate to generalize the noise class when considering continuous recordings throughout the year.

To investigate the matter further we have computed power spectral densities for all three components of BOSA and LPAZ. First we divide the 7 day data sets into windows of 30 min length and split these time windows further in 180 s segments for computing individual squared amplitude spectra. Considering the frequency filters that have been used as the primary source of information for the classification procedure, we integrated the power spectral density results within the relevant frequency bands (Morlet shaped narrowband filters) to obtain normalized density plots of the amplitude levels (log scale) for the training data set (7 day time period July 2008) and the test data set (7 day time periods October 2008, January, April and July 2009). Figure 5 shows the comparison of the density functions for the two data sets at LPAZ, and separately for each frequency band and component. For LPAZ, we observe significant shifts in the noise levels for all frequency bands and components (E less than Z and N). For BOSA (not shown) similar observations can be made.

In order to account for the drift in noise characteristics we propose the following algorithm: we continue the supervised learning iterations from the current DBN for the noise class, presenting previously unseen noise training segments based on classification confidence, and interrupt the learning procedure before the learning method converges. Interrupting the convergence effectively avoids fitting the DBN to the new examples only; characteristics of the previous training examples (the "old" ones) will essentially "bias" the new noise DBN model and are implicitly retained in the now updated model. We choose this technique of gradual model adaption over the alternative of keeping a certain percentage of old training material and mix it with the new noise samples to re-train the noise DBN until convergence. However, this procedure would require storage and book-keeping of old and new segments which we wish to avoid for making the adaption procedure as simple as possible. Practically speaking, the training procedure may take place in parallel to the classification procedure as a transparent process "running in the background".

For implementing the proposed adaptation approach, we need to determine how often noise segments should be collected and how many segments are collected in total before spawning the re-training (updating). When doing so it is required to tune the adaption speed towards the expected time scales of changes in noise characteristics. Anthropogenic effects (e.g., switching on/off machines, traffic) are fast processes requiring fast adaption speeds; however, this does not seem to be relevant for the stations LPAZ and BOSA. However, daily, weekly and seasonal variations of the noise characteristics

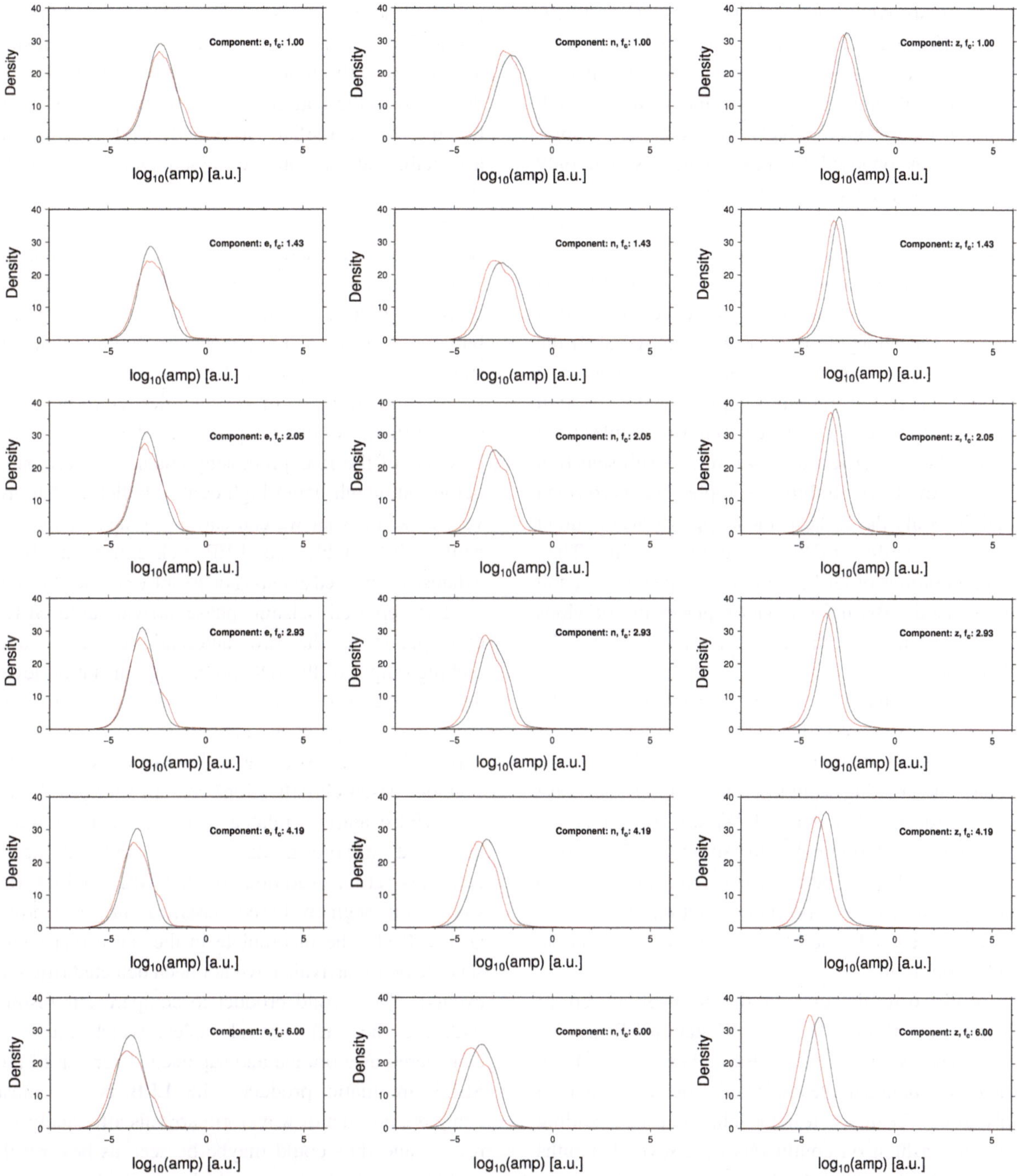

Figure 5
LPAZ: "density" of spectrogram contents. *Top to bottom* 1, 1.43, 2.05, 2.93, 4.19 and 6.00 Hz center frequencies. *Left to right* E, N and Z component. *Red plot* is data used for training (1 week of data) and the *black* is data from the 1 month period (4 weeks of different parts of the year)

are present, and thus we consider a noise adaption scheme able to accommodate as fast as within an hour as an appropriate choice.

Moreover, only a subset of the segments are actually used for training, being those that have the highest confidence of belonging to the noise class (formally speaking, measured in terms of log-likelihood difference between the signal DBN and noise DBN, see Riggelsen *et al.*, 2007). The idea behind this selection strategy is that segments with a high confidence are more likely to be "real" noise segments as opposed to those segments that only have a small confidence. It is exactly those segments we want to use as a basis for adaptive training of the new noise class.

The automated training sample selection policy is as follows: every second minute we check the current segment for belonging to class noise or signal. If the segment belongs to the noise class, it is collected to the set of potential training samples for the adaption of the DBN noise model. Once 20 noise samples have been gathered (earliest after 40 min), a sub-selection process of these training samples is performed retaining only the 15 best (most appropriate) training samples according to their confidence value. Then seven iterations of GEM updating are performed, and the updated DBN noise model replaces the previous DBN noise model; the training pool will start to be filled anew.

Certainly there is strong dependence on the adaptation speed on the actual signal density. If many of the tested time windows are classified as signal, the training sample collection for the noise model update will be delayed. We are aware that this procedure is sub-optimal in the sense that for a noise classifier with poor performance (e.g., due to a too large gap with no adaption) the adaption time will be large or eventually there will be no training sample collection at all.

In Fig. 6 we demonstrate the proof-of-concept of the adaption method for 3C IMS station LPAZ. The noise and signal models have been trained on waveform data samples selected from one week in July 2008. Then we test the classifier in a sliding window fashion on continuous data several months later in October 2008. In the time span from July to October, the noise characteristics apparently have changed so severely that in the beginning nearly all time segments are judged as signal.

With the adaptive mechanism in place we are now ready to classify data in a sliding-window fashion and to slowly accommodate for the change in noise characteristics by re-training the noise DBN using the policy described above. After a small number of re-training rounds, the noise classifier has adjusted to the currently representative noise properties as described in the feature space (Fig. 6, middle). After 6 hours, the classifier has tuned itself towards an acceptable false alarm and missed event rate (Fig. 6, right).

Figure 6 shows qualitatively that the approach works as expected and that signal detections are mostly in agreement to picks made by the automatic CTBTO detection system as listed in the automatic SEL3 bulletin as well as to the entries in the manually supervised bulletin product LEB. A fair evaluation of the absolute accuracy of the classifier seems however quite difficult when comparing to the two bulletin products of the IDC processing chain. As mentioned before, although being high quality bulletins that are very close to what we can call a "golden standard", neither SEL3 tables nor LEB pick entries are fully exhaustive (missed events) nor error free (false alarms).

The reported seismic phase arrival table SEL3 corresponds to the raw automatic detection and picking output of the IDC processing on station level and is, therefore, technically comparable to our approach. Being an automated process, it is to be expected that the SEL3 arrival tables (as well as our segment-based classification) contain spurious detections (false alarms) related to noise bursts or other unwanted waveform disturbances (cultural noise activities, etc.). Additionally, the SEL3 tables (as well as our segment-based classification) most probably will also be incomplete in the sense that some seismic phase arrivals have not been detected (missed events). The second product to compare our results with are the LEB arrival tables, which must be considered as a pruned and superseded version of the SEL3 automatic product. The LEB is the most complete product of waveform arrivals after analyst's review and thus could maybe be seen as best result (golden standard) that should be reached asymptotically by any automatic processing system.

However, the value of comparing the LEB arrival table with the classification result of an automatic processing on station level may be severely hampered by the network philosophy followed by analysts in the interactive review process of individual arrivals. The validity of a particular arrival is reviewed in the

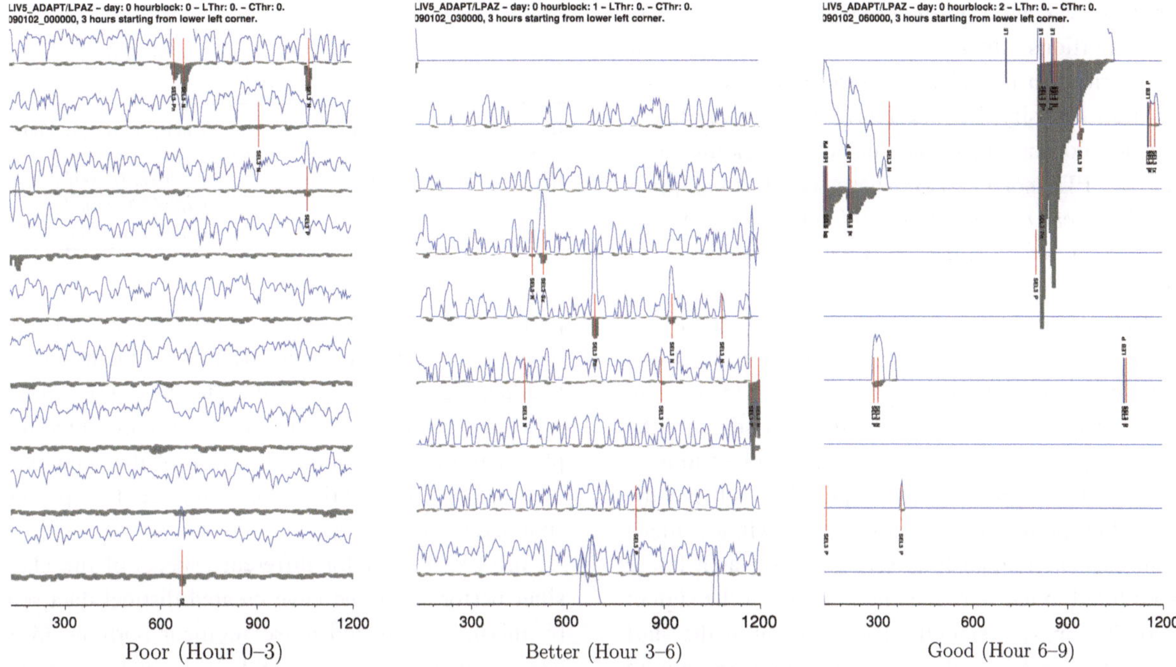

Figure 6
LPAZ: clear improvements when applying an adaptive approach. *The gray bars* indicate the signal confidence of a 12 s window (the taller the *bar* the more confident). The *blue line* provides a smoothed detection measure derived from the *gray bars* also taking into account the immediate temporal "neighborhood confidence". *Red and blue vertical lines* show the automatic and manual picking from the SEL3 and LEB products (with associated phase labels)

context *together* with arrivals at other stations and the coincidence of arrivals among stations that allow associating those arrivals to some event hypothesis. In case of inconsistent arrivals (including clear seismic event arrivals seen only at a single station), the SEL3 arrival may be pruned from the list of automatically detected arrivals.

Unfortunately, given the available information from the SEL3 and LEB bulletins, it is impossible to make a reasonable estimate of the number of false alarms both for the automatically determined SEL3 arrivals as well as for our DBN classifier scheme. A detailed and careful visual inspection by a trained analyst would be required for the full waveform data set in order to derive the true false alarm rate. Again, we think that this procedure does not present an optimal evaluation frame, as decisions made by a single analyst may sometimes be taken on subjective grounds, especially in those cases, where the arrival of a waveform is more difficult to identify. As a consequence we have created a "true" ground-truth data set from real waveform snippets as described in the following section.

6. Performance on Semi-Synthetic Data

We attempt to create a close-to-optimal evaluation framework by generating a "ground-truth semi-synthetic" reference data set. We employ the term "semi-synthetic" in order to indicate that the data set is composed real seismic waveform snippets recorded at 3C stations that are synthetically combined. This procedure allows us to generate exact timings of the occurrence of detectable signals in true noise environments.

In the first step, when creating this reference data set, we choose a number of undebatable transient onsets reported in the LEB and cut a relatively short time window (few tens of seconds) around the signal onset. Relative start and end time of time windows with respect to the reported LEB arrival times is kept for later evaluation. Further, we select longer waveform portions (hundreds of seconds) from the waveform data in which we were unable to identify any transient signal in the relevant frequency bands and that do not contain any detections reported neither in SEL3 nor in LEB. These waveform snippets are then realizations of our reference noise waveforms. For an example of waveform portions see Fig. 7.

Before combining the segments to a semi-synthetic 3C waveform data stream, we treat all waveform segments in the same manner in order to avoid artifacts in the signal characteristics in the merging process. There are two main corrections of the raw time series that are required to achieve the desired properties: firstly, remove time window mean from each individual time series as well as a linear trend fitted to the start and end portions of the signal; secondly, apply a smooth taper function at the start and end of each signal and noise portion for smooth transitions. We then combine the adjusted waveforms by attaching all noise segments in a random order and with random number of repetition until the desired data length (e.g., 12 h continuous waveform) is reached. We define an overlap of the same length as the tapers applied at either side of the waveform portions to enable a smooth transition between segments by merely summing the two segments in the overlapping time range. After the pure noise time series has been created, signal portions are randomly placed in time and summed on top of the semi-synthetic noise time series. The individual steps are summarized in Fig. 7.

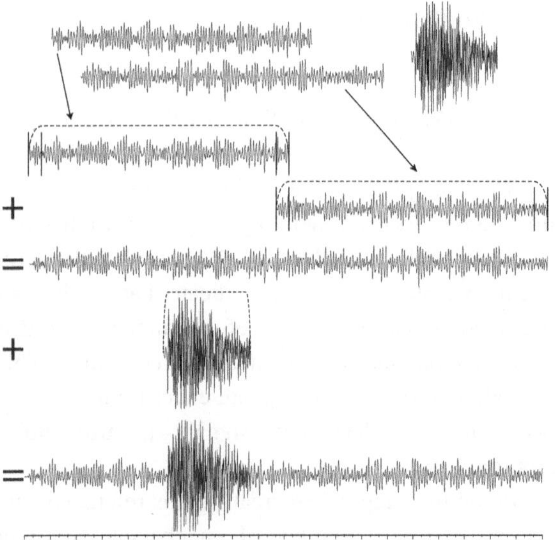

Figure 7
Vertical component recordings from LPAZ. One signal (LEB-reported) and 2 s noise time windows of distinct length are shown to demonstrate the creation of a semi-synthetic evaluation test data set

Table 2

List of data sets: the performance was tested for both 12 and 6 s windows

ID	3C-station	Data-period signal/noise
B1	BOSA	07/2008/07/2008
L1	LPAZ	07/2008/07/2008
L2	LPAZ	07/2008/all years

In order to test for different aspects of the classifier performance we have created distinct data sets by mixing signal and noise segment portions from distinct time periods; see Table 2. In such a way we can address explicitly classifier performance issues related to changes in noise characteristics and amplitude dependencies of the classifier outcome. As before, the DBNs were trained from 1 week of data in July 2008 (see Sect. 4) both DBN+A and DBN−A classifiers (see Fig. 2) were applied to test the sensitivity on absolute amplitude levels.

In the following we display representative examples in Figs. 8, 9 and 10 for the results that have been obtained for Table 2. In the figures, the classification results are presented in the following way: on top the vertical waveform component of the semi-synthetic data set is displayed for visual reference. Below we display the classifier likelihood curves for signal DBN (red line) and noise DBN (black line), indicating further the pieces of overlapping noise windows in green and yellow as well as referencing the additive signal portions with LEB arrival id number by vertical blue lines (start and end times of signal mixed in).

Our most important observation from the overall classification results is the high performance rates we can achieve with this approach. For example for test set L1 we obtain an average hourly missed event rate of 1 (0.25) and an average hourly false alarm rate of 9 (0) using the DBN−A (DBN+A). Although the classifiers have been trained on rather small sets of

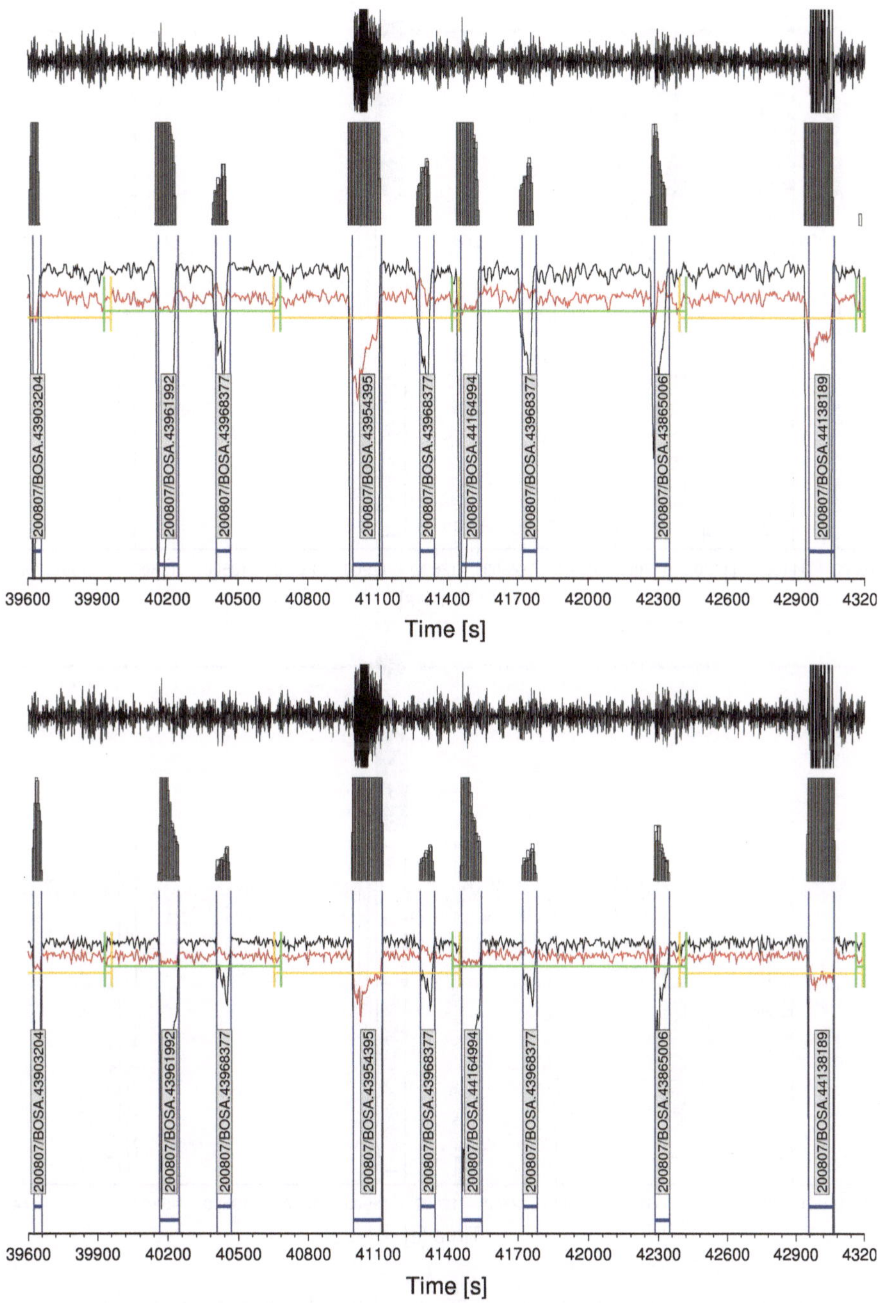

Figure 8
Example figure showing the overall good performance of the classifier approach developed. Here we show an example of 1 h classification result visualized for the data set *B1* and model structure DBN+A. *Upper panel* shows the result for 12 s testing window length; *lower panel* for 6 s testing window length

training data, we obtain close to optimal detection rates for time windows of additive signal portions within continuous noise. As expected the performance decreases when using smaller testing window lengths for the sliding window (B1 test on 6 and 12 s testing length) as the information content presented to the classifier is reduced. On the positive side we find that the timing accuracy of the detection windows

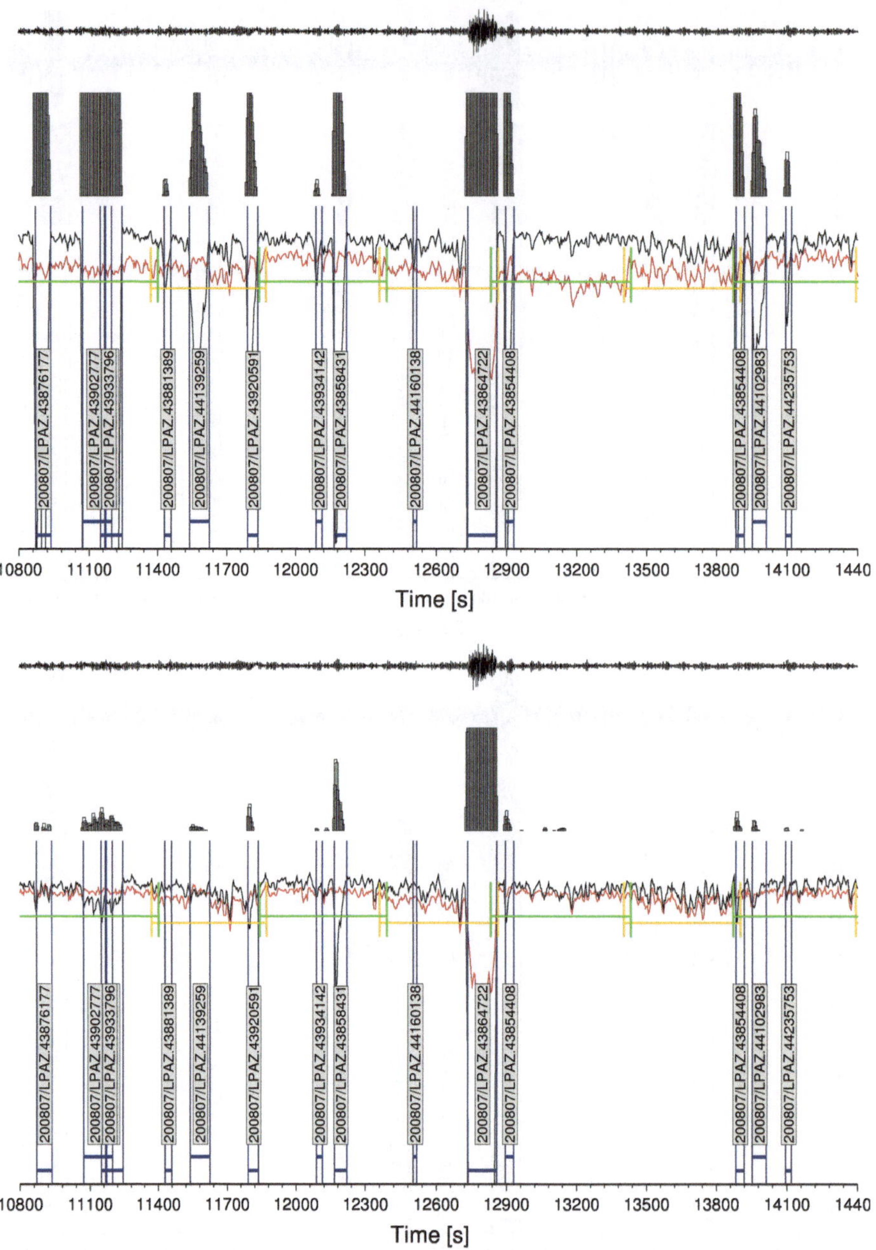

Figure 9
Example figure showing the shift in confidence level between classifiers DBN+A and DBN−A. The example chose presents 1 hour classification result visualized for the data set *L1*. *Upper panel* shows the result for classifier DBN+A; *lower panel* for classifier DBN−A

improves. Those characteristics are shown in Fig. 8. We further note that the overall good performance is mainly a result of the capability of the signal DBN model to explain all kinds of wave field characteristics. Although the probability curves increase slightly for some (true) signal portions, it provides a nearly constant base level for any other waveform portions. On the contrary, the noise DBN reacts rather sensitive to anything that is atypical for its own kind. We conclude, therefore, that the strength of the method relies on the precise modeling of the noise characteristics that must possess a rather confined joint distribution in the modeled feature space. It must be noted that in this case we did not use the adaptive

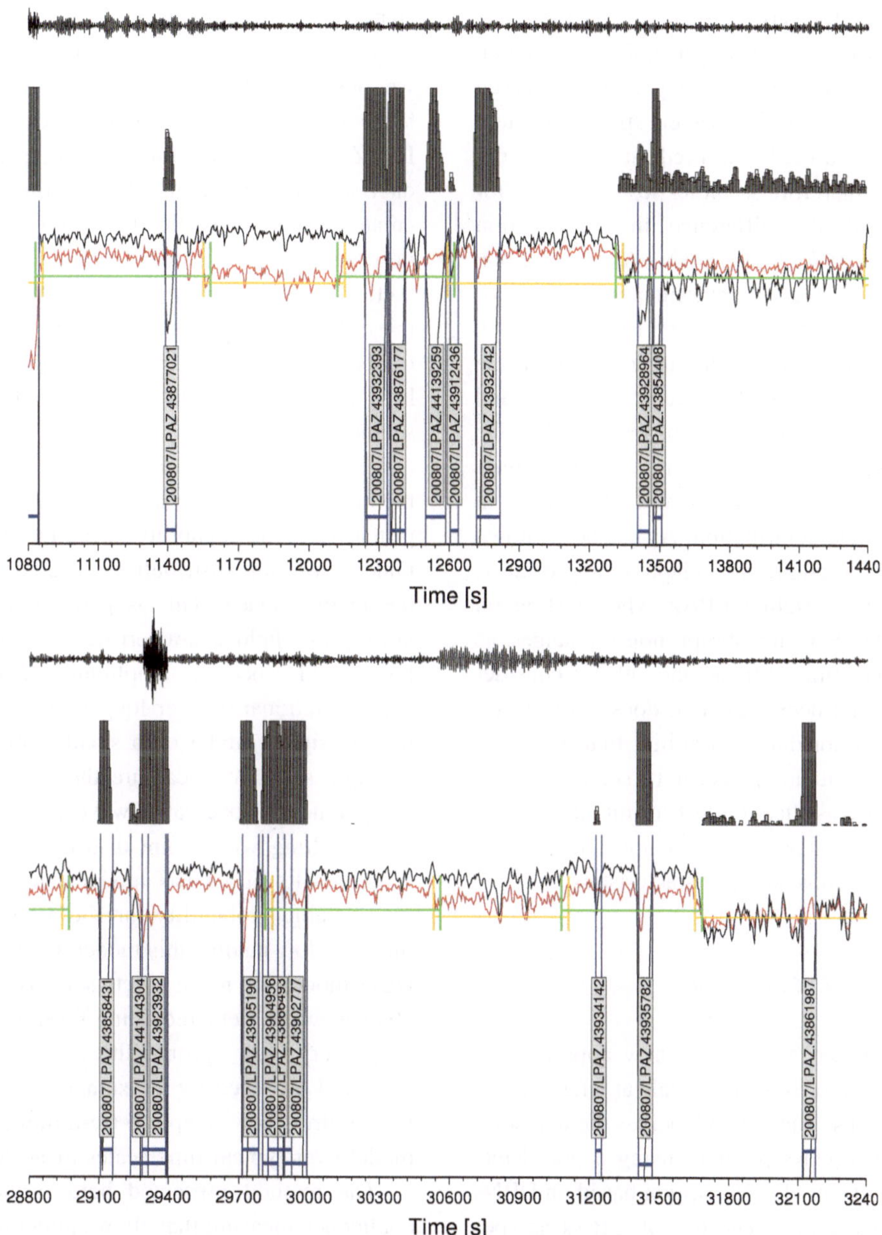

Figure 10
Example figure showing the sensitivity of the noise model classifier to changes in the pattern characteristics of noise. The example gives 1 h classification result visualized for the data set *L2* mixing noise segments from different times of the year. *Upper and lower panels* show the result for two different hours of the data set using model structure DBN+A

approach as the change of noise characteristics in test set L2 is much shorter than the implemented adaption time.

When testing the different DBN+A against DBN−A we find a slightly better classification result when including the information on the absolute amplitude node (DBN+A structure). In particular we observe a much higher average confidence measure for classifier based on DBN+A. This behavior is not surprising being the amplitude level one of the most important discriminating features when visually analyzing waveforms. The importance of absolute amplitude apparently varies depending on the overlapping spectral energies between noise and signal. The high number of high-frequency events observed at LPAZ seem quite similar in their spectral content to the noise wave-field observed at this station. The confidence therefore is rather low, whereas for 3C-station BOSA the difference in classification results between DBN+A and DBN−A is negligible. As an example we show the classification results for semi-synthetic data set *L1* for station LPAZ in Fig. 9.

The clear indication of the importance of the absolute amplitude level observed at a station and the narrowness of observations to be explained by the noise DBN suggests the need for noise model adaption within characteristic time scales of changes in the noise wave field contribution. As has been shown in the previous section, the adaptive approach is indeed required for station LPAZ when testing on continuous data. For the abrupt noise changes as introduced in the semi-synthetic data sets we consider this exercise as not-necessary as it does not reflect a realistic seasonal nor daily variability that would be expected in a real situation. As can be seen in Fig. 10 this leads in our classification test to immediate long term misclassification as long as the noise model does not fit to the particular noise characteristics.

7. Conclusions

We demonstrated a relatively new approach for real-time detection of seismic signals at single three-component stations. The method tackles the detection problem as a two-class problem using a non-deterministic generative model description based on DBNs and supervised learning. The use of DBNs as specialized graphical model allows for modeling the temporal structure of the spectral evolution of characteristic ground velocity recordings in three orthogonal directions. The DBN structure that connects horizontal with vertical wave energy levels in different frequency bands is based on seismological observation standards whereas the parameters of the DBN are estimated from small sets of training examples.

The precise performance of the classifier is difficult to assess due to the lack of a ground truth (golden standard). However, testing on a fixed segment test data with known labels (subset of continuous data) we obtain performance measures between 85 and 97 % for three-component IMS stations BOSA and LPAZ. Beyond that, for mimicking a ground truth data set for continuous classification we choose to create a semi-synthetic data set consisting of real observed waveform portions stitched together synthetically. Applying the classification approach to these data sets enables to illustrate particular aspects of the classifier functioning and proves further the high performance rates as indicated for the fixed-segment test.

From the experiments performed we believe that the high performance rates obtained can be attributed primarily to the capability of the DBNs to model the underlying joint distributions of signal and noise in the feature space. This is particularly true for the noise wave field characteristics that possess narrow log-normal spectral amplitude distributions and exhibit stationarity over longer time periods. Learning the signal model on a small number of training example suffices to capture the body and range of every kind of observable wave field characteristics. The resulting wider joint distribution in the feature space results in a DBN model that provides a relatively stable probability threshold when testing any time window against this model. On the contrary, the DBN model for noise reacts sensitively to anything atypical wave field recording (non-noise) by exhibiting very low probabilities falling below the threshold provided by the signal model. This behavior requires slow adaption (re-training) of the noise model over typical time scales of noise stationarity.

The method presented here further provides a confidence measure that allows judging the quality of the detection. In future this may provide the basis for a continuous characteristic function acting as input for Bayesian multi-station location and/or association procedures (Myers *et al.*, 2009).

Acknowledgments

We would like to acknowledge CTBTO IDC staff, in particular Dmitry Bobrov, Yan Jia, Jeffrey Given and Ronan Le Bras. Thanks to Conny Hammer and

Andreas Köhler for data preparation, "picking" and programming. Thanks to an anonymous reviewer for comments and many suggestions. Carsten Riggelsen is funded by DFG RI 2037/2-1.

References

K. Aki and P.G. Richards. Quantitative seismology. Geology (University Science Books): Seismology. University Science Books, 2002.

T.C. Bache, S.R. Bratt, H.J. Swanger, G.W. Beall, and F.K. Dashiell. *Knowledge-based interpretation of seismic data in the intelligent monitoring system.* October, *83*(5):1507–1526, 1993.

T.C. Bache, S.R. Bratt, J. Wang, R. M. Fung, C. Kobryn, and J.W. Given. *The intelligent monitoring system.* Bulletin of the Seismological Society of America, *80*(6):1833–1851, 1990.

M. Beyreuther, R. Carniel, and J. Wassermann. *Continuous hidden markov models: Application to automatic earthquake detection and classification at Las Canadas caldera, Tenerife.* Journal of Volcanology and Geothermal Research, *176*:513–518, 2008.

J. Bilmes. What HMMs Can Do. IEICE - Transactions on Information and Systems, 2006.

J. Bilmes and G. Zweig. The graphical models toolkit: An open source software system for speech and time-series processing. In Proceedings of IEEE Int. Conf. Acoust., Speech, and Signal Processing, pages 3916–3919, 2002.

S.J. Gibbons and F. Ringdal. *The detection of low magnitude seismic events using array-based waveform correlation.* Geophysical Journal International, *165*(1):149–166, April 2006.

C. Hammer, M. Beyreuther, and M. Ohrnberger. *A seismic event spotting system for volcano fast response systems.* accepted in Bulletin of the Seismological Society of America, 2011.

K. Hutton, J. Woessner, and E. Hauksson. *Earthquake Monitoring in Southern California for Seventy-Seven Years (1932-2008).* Bulletin of the Seismological Society of America, *100*(2):423–446, March 2010.

A. Köhler, M. Ohrnberger, C. Riggelsen, and F. Scherbaum. Unsupervised feature selection for pattern search in seismic time series. In JMLR: Workshop and Conference Proceedings: New Challenges for Feature Selection in Data Mining and Knowledge Discovery, volume 4, pages 106–121, 2008.

Tormod Kværna and Frode Ringdal. *Seismic Threshold Monitoring for Continuous Assessment of Global Detection Capability.* Bulletin of the Seismological Society of America, (August):946–959, 1999.

T. Kvaerna, F. Ringdal, J. Schweitzer, and L. Taylor. *Optimized Seismic Threshold Monitoring - Part 1: Regional Processing.* Pure and Applied Geophysics, *159*(5):969–987, March 2002.

T. Kvaerna, F. Ringdal, J. Schweitzer, and L. Taylor. *Optimized Seismic Threshold Monitoring - Part 2: Teleseismic Processing.* Pure and Applied Geophysics, *159*(5):989–1004, March 2002.

S.C. Myers, G. Johannesson, and W. Hanley. *Incorporation of probabilistic seismic phase labels into a Bayesian multiple-event seismic locator.* Geophysical Journal International, *177*(1):193–204, 2009.

K. Z. Nanjo, T. Ishibe, H. Tsuruoka, D. Schorlemmer, Y. Ishigaki, and N. Hirata. *Analysis of the Completeness Magnitude and Seismic Network Coverage of Japan.* Bulletin of the Seismological Society of America, *100*(6):3261–3268, December 2010.

S.E. J. Nippress, A. Rietbrock, and A.E. Heath. *Optimized automatic pickers: application to the ANCORP data set.* Geophysical Journal International, pages 911–925, March 2010.

J. Peterson. Observations and Modeling of Seismic Background Noise. DIANE Publishing Company, 2003.

C. Riggelsen, M. Ohrnberger, and F. Scherbaum. Dynamic Bayesian networks for real-time classification of seismic signals. In Lecture Notes in Computer Science, volume 4702, page 565. Springer, 2007.

T. Tanimoto, S. Ishimaru, and C. Alvizuri. *Seasonality in particle motion of microseisms.* Geophysical Journal International, *166*(1):253–266, 2006.

M. Withers, R. Aster, C.J. Young, J. Beiriger, S. Moore, and J. Trujillo. *A Comparison of Select Trigger Algorithms for Automated Global Seismic Phase and Event Detection.* Bulletin of the Seismological Society of America, *88*(1):95–106, 1998.

G. Zweig. *Bayesian network structures and inference techniques for automatic speech recognition.* In Computer Speech & Language, volume *17*, pages 173–193, 2003.

G. Zweig and S. Russell. Speech Recognition with Dynamic Bayesian Networks. AAAI-98: Proceedings of the conference on Artificial intelligence, pages 173–180, 1998.

(Received December 1, 2011, revised August 31, 2012, accepted September 5, 2012, Published online September 20, 2012)

Pure Appl. Geophys. 171 (2014), 413–423

DOI 10.1007/s00024-012-0508-2

Pure and Applied Geophysics

"Repeating Events" as Estimator of Location Precision: The China National Seismograph Network

Changsheng Jiang,[1] Zhongliang Wu,[1,2] Yutong Li,[3] and Tengfei Ma[1]

Abstract—"Repeating earthquakes" identified by waveform cross-correlation, with inter-event separation of no more than 1 km, can be used for assessment of location precision. Assuming that the network-measured apparent inter-epicenter distance X of the "repeating doublets" indicates the location precision, we estimated the regionalized location quality of the China National Seismograph Network by comparing the "repeating events" in and around China by Schaff and Richards (Science 303: 1176–1178, 2004; J Geophys Res 116: B03309, 2011) and the monthly catalogue of the China Earthquake Networks Center. The comparison shows that the average X value of the China National Seismograph Network is approximately 10 km. The mis-location is larger for the Tibetan Plateau, west and north of Xinjiang, and east of Inner Mongolia, as indicated by larger X values. Mis-location is correlated with the completeness magnitude of the earthquake catalogue. Using the data from the Beijing Capital Circle Region, the dependence of the mis-location on the distribution of seismic stations can be further confirmed.

1. Introduction

Assessing the location quality of a seismograph network is important in evaluating the capacity for CTBT monitoring. Here the assessment includes two levels: assessing the location capability, which is potentially accessible by applying the best possible algorithms to a seismograph network, and assessing the existing location quality of the data products produced by a seismograph network on a routine basis. The discussion of this paper focuses on the latter, which is similar to the evaluation of the IDC products.

Location quality can be assessed by several approaches, some of which assess both precision and accuracy whereas others assess precision or accuracy only. The "ground truth event" approach (Lienert, 1997), the comparative approach (Wuester *et al.*, 2000), and the statistical optimization approach (Rabinowitz and Steinberg, 1990; Doufexopoulou and Korakitis, 1992; Bartal *et al.*, 2000) all contributed to the knowledge of such assessment. As one of the feasible approaches, in this paper, we propose use of the method of "repeating earthquakes" (Schaff and Richards, 2004, 2011) to evaluate the location precision of a seismic network. We used the China National Seismograph Network and the Beijing Capital Region Seismograph Network as examples. These produce their monthly earthquake catalogue routinely, with different quality and different cutoff magnitude. This paper is an extension of our previous work (Jiang and Wu, 2006; Jiang *et al.*, 2008) with new data used and with new comparative analysis.

2. "Repeating Events" as Estimator of Location Precision

The concept of "repeating earthquakes" can be traced back to the 1960s (Isacks *et al.*, 1967). Research interest in this concept has been reactivated in recent years with the development of digital seismological observation and the processing of digital seismic waveforms. There is currently no well-accepted unified definition of "repeating

Electronic supplementary material The online version of this article (doi:10.1007/s00024-012-0508-2) contains supplementary material, which is available to authorized users.

[1] Institute of Geophysics, China Earthquake Administration, Beijing 100081, China. E-mail: wuzl@cea-igp.ac.cn

[2] Key Laboratory of Computational Geodynamics, Graduate University of the Chinese Academy of Sciences, Beijing 100049, China.

[3] Earthquake Administration of Liaoning Province, Shenyang 110034, China.

earthquakes". Rubin (2002) characterized "repeating earthquakes" as events with similar locations, magnitudes, and waveforms. The definition of Schaff and Richards (2004) refers to a pair of seismic events with the cross-correlation coefficients (CCC) of their band-pass-filtered waveforms no less than 0.8 for the recordings of at least one station. (In this paper we do not differentiate between cases for which more than one station have CCC $\geq$ 0.8 and those for which only one station has CCC $\geq$ 0.8, although theoretically for the latter case the criterion for "repeating" is apparently weaker). Empirical and theoretical studies indicate that a "repeating pair" in this sense has the inter-event separation distance <1 km (Menke *et al.*, 1990; Menke, 1999). Remarkably, "repeating events" in the sense of waveform similarity occupy a significant portion of seismicity. The results of Schaff and Richards (2004, 2011) showed that "repeating earthquakes" amount to more than 10 % of seismicity in China and its surroundings, indicating the importance of analysis of such seismic events.

Taking the working assumption that the distance between the two events in a "repeating earthquake pair" is <1 km, and their apparent inter-epicenter distance X measured by routine analysis of a seismic network reflects the location precision, comparison between a "repeating event" catalogue and the network-produced catalogue can be conducted for evaluation of location quality. For regionalized analysis, the averaged value of X in a specific region can be used as an indicator of the location quality in that region. In this sense, the role of the "repeating earthquakes" used for comparison corresponds to that of the ground-truth (GT) events (Lienert, 1997), but with a wider geographical coverage, albeit GT events can assess both precision and accuracy.

In the evaluation, one has to consider that, because of the random error in the location of each event, for some cases the X value may cancel out. To overcome this problem, we use the average of several event pairs, rather than a single pair, for the evaluation of the precision of location. This average, sacrificing the resolution as per the case that the variances in the north–south direction and in the east–west direction might be different, avoids the "random co-location problem" as mentioned above.

3. Location Precision Varying with Time

We used the monthly earthquake catalogue[1] provided by the China Earthquake Networks Center (CENC) for the evaluation. This monthly catalogue is produced by collecting and analyzing local catalogues and bulletins, and is widely used for analysis of time-dependent seismic hazard (or earthquake forecasting) in China because of its timing characteristics. The well-known ABCE catalogue is a refining of the monthly catalogue, provided on an annual basis. The data used in this study start from 1990 and end in 2005, overlapping with the "repeating event" database of Schaff and Richards (2011, from 1985 to 2005). In that "repeating event" database including 2,379 events, there are 2,178 pairs recognized as "repeating doublets". To identify the events which are thought to be identical in both the database of Schaff and Richards (2011) and the monthly catalogue of the CENC, we used the threshold of the difference of magnitude Δm, latitude–longitude coordinates $\Delta\Phi$ and $\Delta\theta$, and origin time Δt. We prescribed, somewhat arbitrarily, $-0.6 < \Delta m < 0.6$, $-1.0^\circ < \Delta\Phi < 1.0^\circ$, $-1.0^\circ < \Delta\theta < 1.0^\circ$, and $-66\ \text{s} < \Delta t < 66\ \text{s}$ as the threshold, considering the difference between the CENC monthly catalogue and the catalogue used by Schaff and Richards (2011)—the ABCE. There are 1,810 events identified in the CENC monthly catalogue, which formed 1,678 "repeating pairs" for the analysis, as shown by the black circles in Fig. 1 and the *Supporting Online Materials*. In the "repeating event" database of Schaff and Richards (2011), there are 467 events, with 288 located within the territory of China, missed in the CENC monthly catalogue. These events are also shown in Fig. 1 by red circles. Comparison of the frequency–magnitude distributions of the "associated" and the "missing" events (Fig. 1) indicates that the cause of this "missing" is more likely to be because of the re-location procedure.

Figure 2 shows the histogram of the X values versus time. In each histogram, the horizontal axis is the X value and the vertical axis gives the number of

[1] Partly available at: http://www.csndmc.ac.cn/newweb/data.htm#, in Chinese.

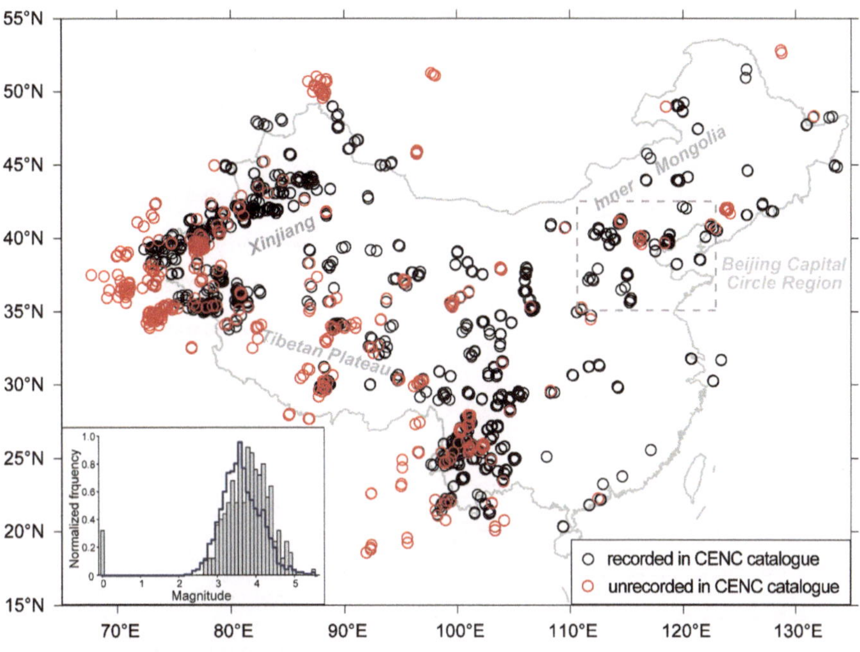

Figure 1

Distribution of "repeating events" in China and its surrounding regions as reported by Schaff and Richards (2011), and comparison with the CENC monthly earthquake catalogue. *The subplot at the bottom left* is the normalized frequency-magnitude distribution of the 288 missing "repeating events" (*gray histogram*) and the 1,709 identified "repeating events" (*blue line*) in the CENC monthly catalogue located within the territory of China. Comparison of these two distributions indicates that the "missing events" are not likely to be caused by the different detection capability but probably by the location procedure

"doublets". Vertical axes are normalized for comparison. The temporal variation chart takes a step of 1 year and a sliding window with a width of 2 years and centered at the first day of the year, as indicated in the figure. Time of the "doublet" is taken as the time of the later event. From Fig. 2 it can be seen that location capability, as represented by the histogram of X, has been stable during the period from 1990 to 2005, with slight improvements since 1995. As a result, all the data can be used to evaluate the regionalized location precision, irrespective of the time of the "doublets". Figure 3 shows, in an alternative way, the average X value for the same sliding step and sliding window as in Fig. 2, with error bars shown on the figure, indicating the fluctuation around the mean value and the number of samples available. In the calculation, the errors are given by one-sigma for the average. Our previous work (Jiang and Wu, 2006) used the "repeating event" database of Schaff and Richards (2004). For comparison, the results of that work are also plotted in Fig. 3, indicating the change of the "repeating event" database of Schaff and Richards (2004, 2011). Figure 3, reaching the same conclusion as Fig. 2, shows more clearly the "slight improvement since 1995".

The dashed line in Fig. 2 has a minimum at $X = 40$ km. Seen from the location procedure, the origins of location errors on the two sides of this value are different. When X is smaller than 40 km, the location error mainly comes from the location calculation and the travel time measurement. Because the X value is an indicator of the location precision, the Earth model is of minor importance, because what is under consideration is the relative location of the two "repeating events". When X is larger than 40 km, problems in phase association and location calculation associated with a sparse network make a major contribution to the error. Considering this property, in the following analysis, we treated these two parts of data separately and focus on the pairs with $X < 40$ km.

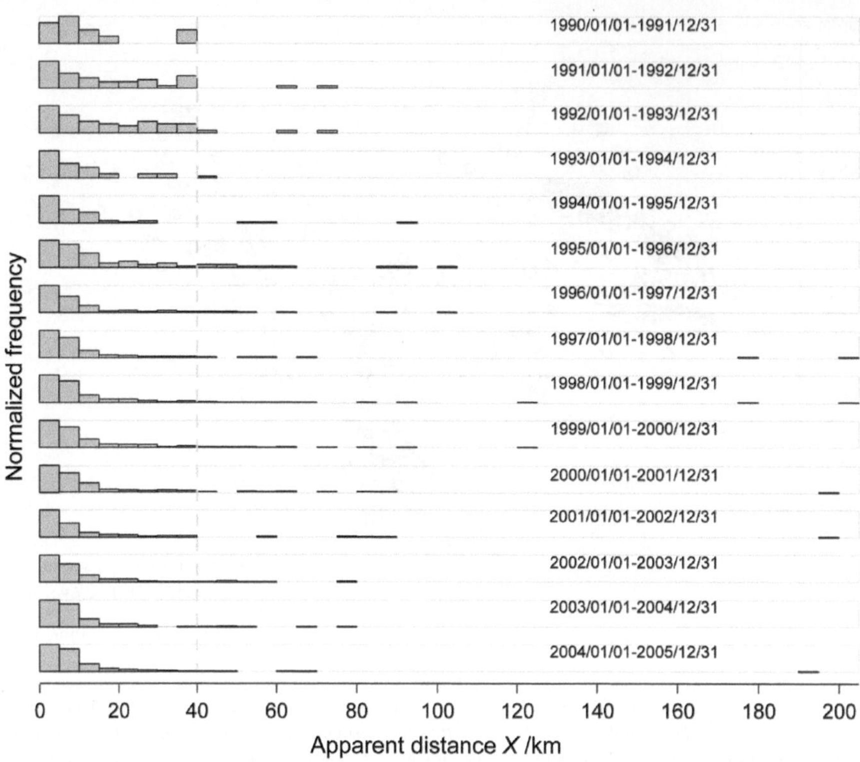

Figure 2
Frequency distribution and temporal variation of the network-measured apparent inter-epicenter distance X between the "repeating events"

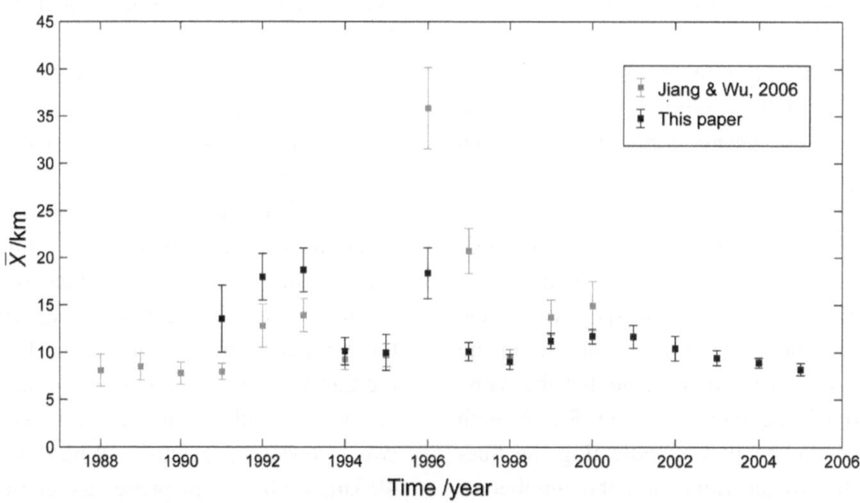

Figure 3
Average value of the network-measured apparent distance X between the "repeating events", and its temporal variation

4. *Regionalized Location Precision*

Figure 4 maps the spatial distribution of the X value, with a grid size of 1° and a circular window of radius 300 km. Each event is associated with its "repeaters", and the distance between this event and its "repeater" is calculated, taken as the apparent inter-epicenter distance X. For more than one association, the largest distance is taken as the X value for the event. For all the events within a spatial window, the average of the X values is taken as the X value of the region covered by this window. In Fig. 4, white circles represent the earthquakes used, and black reverse triangles show the locations of the earthquake pairs with the X value larger than 40 km. For all the pairs, only the later event is plotted in the figure.

Although having obvious unbalanced geographical coverage, Fig. 4 provides a coarse but widely-covered estimate of the regionalized location precision. The regions with warm colors (equivalent to larger X values) are the Tibetan Plateau, west and north of Xinjiang, and east of Inner Mongolia. Events with X values larger than 40 km are further indicators of regions with poor location quality, mainly the Tibetan plateau. From Figs. 2 to 4, it can be seen that the average location precision of the CENC monthly catalogue is approximately 10 km. The Tibetan Plateau, west and north of Xinjiang, and east of Inner Mongolia, plus north of northeastern China, are the regions in need of improvement.

"Repeating events" should also have depth differences of <1 km, and the apparent difference of depths, Y, may reflect the precision of depth determination. In the monthly catalogues there are many events without depth results, or using "fixed depths", for example 5 or 10 km. These events were eliminated from the analysis. After screening of the catalogue on the basis of this consideration, 1,022 events are left, forming 1,008 pairs. Figure 5 shows, in a similar way to that of Fig. 4, the spatial distribution of the average Y value, together with the event pairs whose X is larger than 40 km and are not used in calculation of the Y values. Although the spatial coverage is not as good as that of Fig. 4, Fig. 5 shows a similar pattern to that of Fig. 4. Because of this

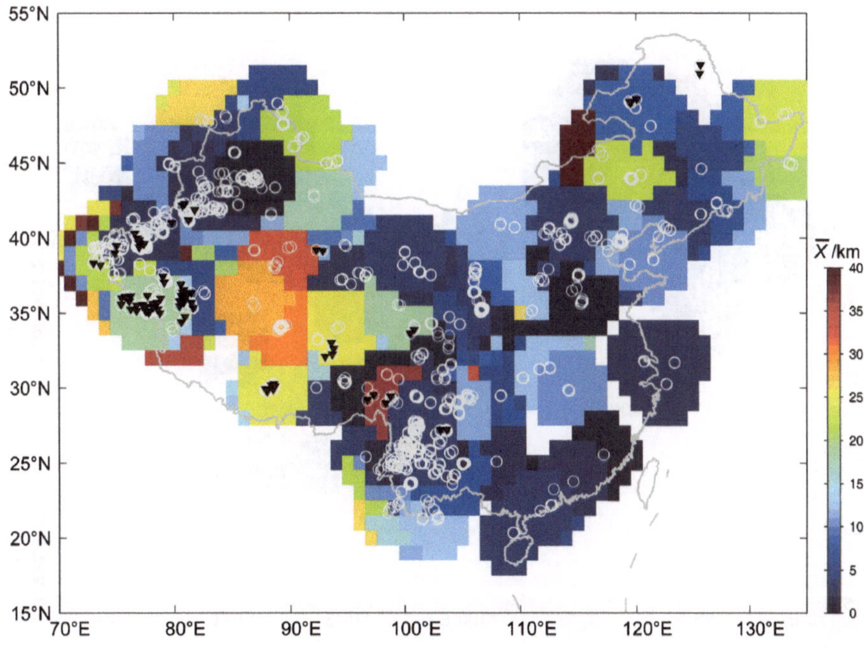

Figure 4

Spatial distribution of the network-measured apparent distance X between the "repeating events". *White circles* indicate the "repeating pairs" used in the analysis. Pairs with apparent distance larger than 40 km are marked by *black reverse triangles*. For all the pairs, only the later event is plotted

similarity, and considering that "fixed depth" is quite common in routine seismological interpretation practice, in evaluation of location quality it is sufficient to use X value only. Typically, if there were no additional efforts, for example use of depth phases, error in depth determination would be larger than those in horizontal coordinate (or epicenter) determination. In Figs. 4 and 5, however, the average X and Y values are apparently of the same order of magnitude. This is partly because X and Y measure the precision of location as the *relative* distance between the "repeaters". But, most importantly, the depth estimate is either unavailable (as shown by the blank pixels in Fig. 5) or of good quality. Theoretically, the precision of horizontal location should be measured by $\left(X/\sqrt{2}\right)$, rather than by X itself, assuming that the mis-location is approximately the same for each event within the "repeating pair". Considering the precision for both horizontal location and for depth determination, however, the X value itself can be used; this reflects the fact that overall precision in cases of (relative) mis-location is approximately the same for the horizontal and vertical directions.

The regionalized inhomogeneous distribution of location precision is related to the distribution of seismic stations, and thus is related to the magnitude of completeness of the catalogue. To demonstrate this point, we applied the Gutenberg–Richter-law-based entire-magnitude-range (EMR) method (Woessner and Wiemer, 2005) to the CENC monthly catalogue to obtain the spatial distribution of the magnitude of completeness (M_c) during the period 1990–2005. For the EMR method, the Gutenberg–Richter's law and normal cumulative distribution function are used to describe the upper and lower parts of the assumed M_c, respectively. Following previous work (Schorlemmer, 2003; Nanjo *et al.*, 2010) on setting values for EMR estimation and using a Monte Carlo approximation of the bootstrap method (Efron, 1979) to calculate the uncertainties δM_c, Fig. 6a and b show, respectively, M_c and δM_c. From the figure it can be seen that the M_c distribution has a similar pattern to the distribution of average X, as seen in Fig. 4.

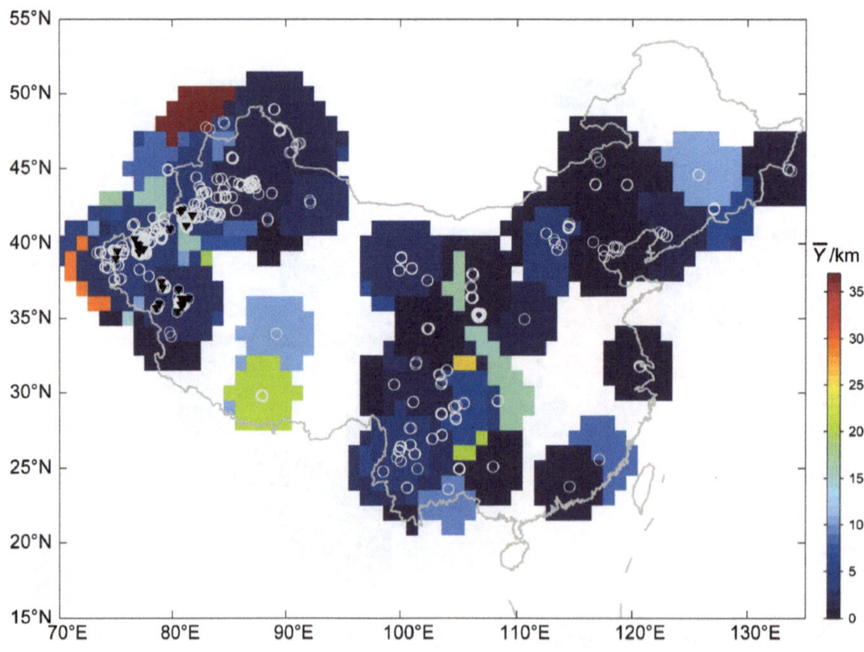

Figure 5
Spatial distribution of the apparent difference of network-measured depths, Y, between "repeating events". *White circles* indicate the "repeating pairs" used in the analysis. *Reverse triangles* show the pairs whose X values are larger than 40 km and are not used in calculating the Y values

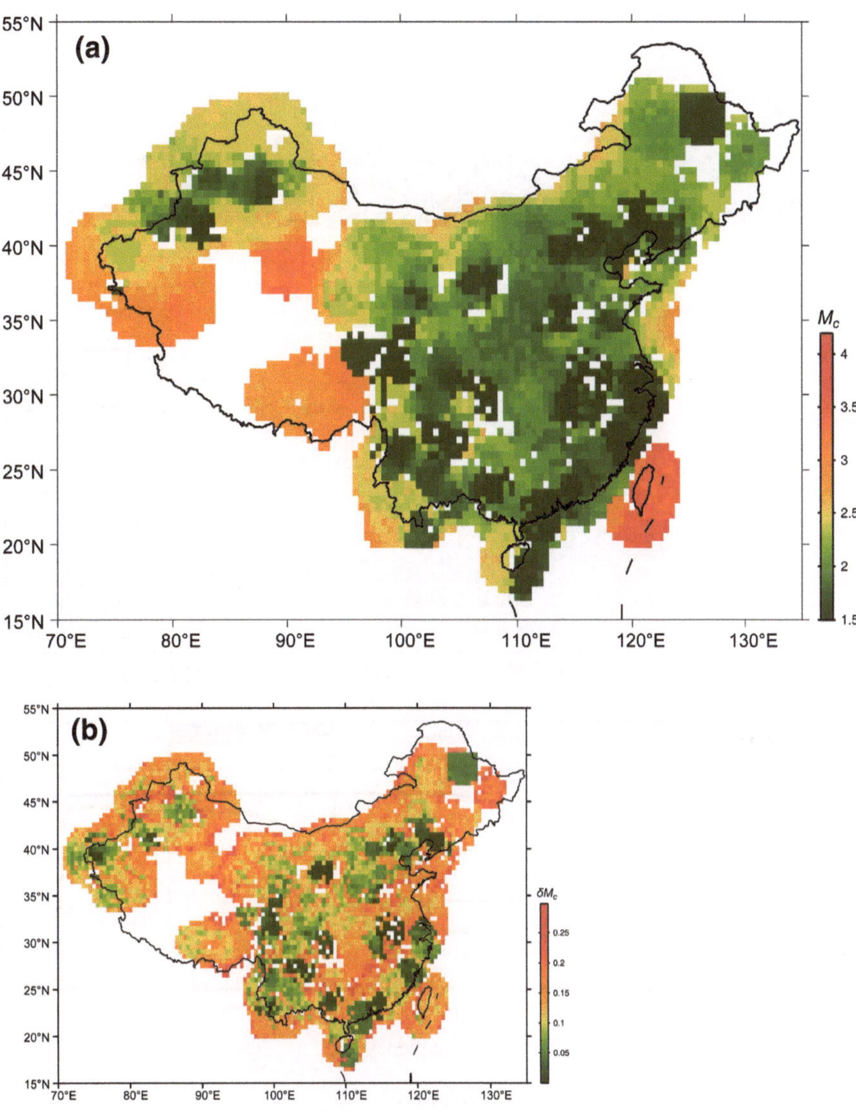

Figure 6
Distribution of completeness magnitude M_c (**a**) and its uncertainty δM_c (**b**) during 1990–2005. M_c is estimated from the nearest 100 earthquakes to nodes of a grid spaced 0.05° apart, with the typical sampling radii 50 km and all smaller than 200 km. *Blank pixels* correspond to the grids with no or lower seismicity under the above criteria

5. *Looking Closely at the Beijing Capital Circle Region*

We had a closer look at the Beijing Capital Circle Region, which has 107 digital seismic stations, including five very broadband (VBB), 44 broadband (BB), and 58 short-period (SP) three-component seismographs, with flat velocity response for frequency bands 0.008–20 Hz, 0.05–20 Hz, and 1–20 Hz, respectively, and sampling frequency 50 Hz. Figure 7 shows the distribution of stations and their frequency response. Digitally-oriented upgrading occurred in 2002. We selected the waveform data of the events with clear P phase pickings at no less than three stations for the period 2002–2006, within the spatial range 35–42°N, 112–122°E. A total of 3,567 events were selected for the study.

Considering the limited coverage of seismic stations, we used both BB/VBB recordings and SP recordings. Before the cross correlation of waveforms,

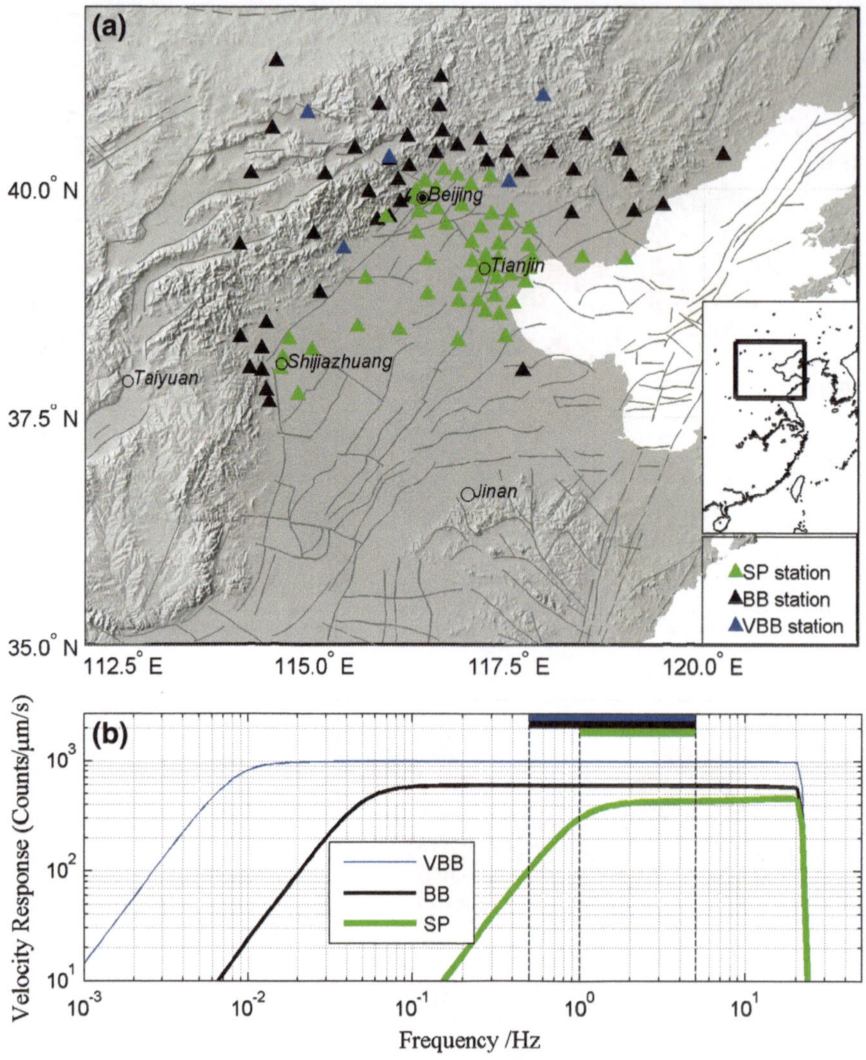

Figure 7
Seismic stations in the Beijing Capital Circle Region, reproduced from Jiang *et al.* (2008). **a** Distribution of stations. *Gray lines* indicate active faults. Indexing figure is to the *lower-right.* Names of stations are provided in Fig. A1. **b** Frequency response for the VBB, BB, and SP seismographs (from the Data Center of the Beijing Digital Seismograph Network), in which the frequency band of the filters in the pre-processing is also marked. See text for details

preprocessing was conducted using a 0.5–5 Hz band-pass filter to BB/VBB recordings and a 1–5 Hz band-pass filter to SP recordings. We selected the whole waveform composed of P, S, and coda phases. Waveform was picked 4 s before the Pg arrival, and the whole waveform length was taken as four times the S–P travel time difference. The sliding window was taken as 4 s less than the length of the whole waveform, sliding from the beginning point (4 s before the Pg arrival), with steps of 1 sample. The maximum in the sliding window is taken as the final value of the cross-correlation coefficient. (For details of the work refer to Jiang *et al.* 2008).

Among the 3,567 events selected, we found that 859 were "repeating" in the sense of waveform cross correlation that at least one station has the CCC no less than 0.8. This constitutes 24 % of the whole dataset, significantly higher than the percentage of "repeating events" in the database of Schaff and Richards (2011) who used fewer stations, often at distances >1,000 km. These 859 "repeaters" comprise 1,568 "earthquake pairs", belonging to 183

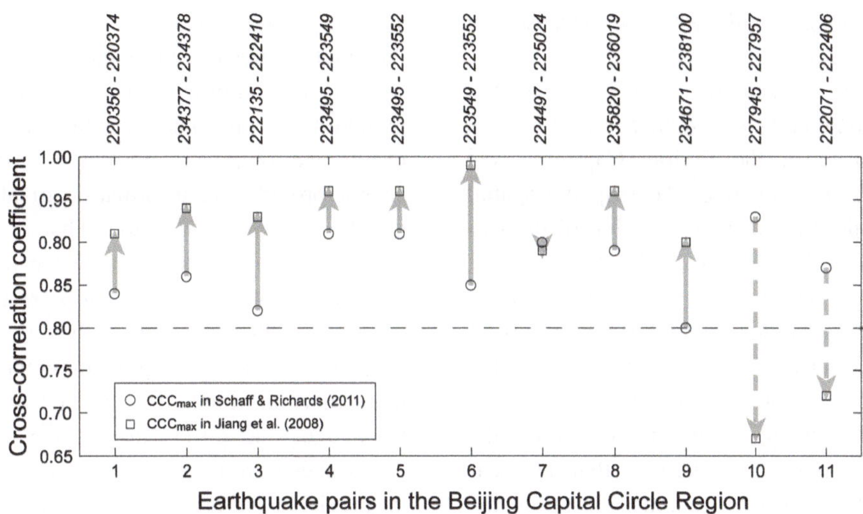

Figure 8

Comparison of the maximum cross-correlation coefficient (CCC) of the "repeating earthquakes" which are registered both in SCHAFF and RICHARDS (2011) and in JIANG *et al.* (2008), in the Beijing Capital Circle Region. In the figure, *circles* are the results of SCHAFF and RICHARDS (2011) from regional or far-regional waveforms and *squares* are the results of JIANG *et al.* (2008) from regional waveforms. *Arrows* indicate the change of the maximum cross-correlation coefficient. *Numbers to the top* marks the codes of the pairs, as used by SCHAFF and RICHARDS (2011)

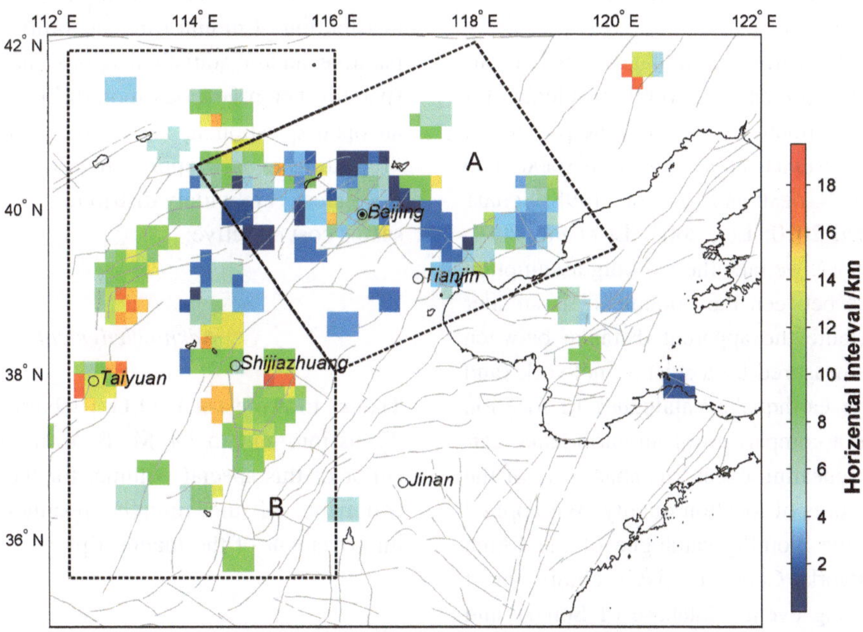

Figure 9

Distribution of *X* for the Beijing Capital Circle Region, reproduced from JIANG *et al.* (2008). *Dashed polygons* indicate regions A and B with distinct differences in location quality. See text for details

"multiplets" in a senso lato sense that one event in a "multiplet" has at least one "partner". Within both databases of "repeaters" (JIANG *et al.*, 2008; SCHAFF and RICHARDS, 2011), 11 pairs are overlapped in the time duration 1 January 2002–31 December 2005 and can be compared with each other. Figure 8 compares the maximum CCC. CCCmax is less for SCHAFF and RICHARDS (2011) than for JIANG *et al.* (2008), probably because of the longer windows used because the stations are further away.

We used these "repeating events" to obtain a picture of location quality with higher resolution. We used the catalogue of the Beijing Capital Circle Region Seismograph Network.[2] To map the spatial distribution we used the average location of the two events to represent the location of the "event pair". We took a grid of size 0.15° × 0.15°. At each node we used a circular window of radius 20 km to pick up the "repeating event pairs". The average for all the "event pairs" falling into the circular window was then mapped on to the grid node. Figure 9 shows the result of such mapping. It can be seen that the location capability has a distinct regional variation: the eastern part is much better than the other parts, mainly because of the distributions (and site conditions) of seismic stations.

6. Conclusions and Discussion

Analysis of "repeating earthquakes", in various senses, has been applied to several problems, for example studies of fracture zones and the physics of seismic sources (Vidale *et al.*, 1994; Nadeau and McEvilly, 1997; Nadeau and Johnson, 1998; Schaff and Waldhauser, 2010; Lin, 2004; Li *et al.*, 2007, 2011). In this paper, we take the working assumption that the distance between the "doublets" is smaller than 1 km, attribute the apparent distance between the "doublets" measured by a seismic network (and registered in the earthquake catalogue) to location error, and make a comparison of an earthquake catalogue with a "repeating event" database one of the assessment measures of location quality. We applied this concept to the monthly catalogue of the China Earthquake Networks Center (CENC), comparing it with the "repeating event" database of Schaff and Richards (2011). We further applied this concept to the waveform and catalogue of the Beijing Digital Seismograph Network.

The working assumption of this paper is that the "repeating events" or "doublets" are separated by no more than 1 km (Schaff and Richards, 2004, 2011), and the network measured apparent distance X of "doublets" indicates the location precision. But this working assumption must still be tested by use of ground truth information, which is a limit of the results we obtained. Besides, "repeating events" cannot provide the information of absolute or systematic bias of the location. That is, the "repeating events" approach can evaluate the precision, rather than the accuracy, of the location result.

It is worth emphasizing that assessment of the monitoring capability of a seismograph network needs more than one method, each of which has its own advantages and disadvantages. Up to now, various approaches have been proposed and used to evaluate the precision and/or accuracy of network locations, for example the ground-truth "calibration" using known seismic events (either natural or artificial), comparison of results obtained by different networks, and statistical approaches to evaluation and optimization, among others. Because of the limited location of "ground truth" events, and the difficulty of obtaining the spatial distribution of monitoring capabilities by using comparative and/or statistical approaches, evaluation of spatially heterogeneous monitoring capability is still an open question in seismological monitoring. The "repeating event" approach has the advantage that "repeaters" are widely distributed, and the evaluation seems cost effective.

Acknowledgments

Thanks to Professor Paul G. Richards for stimulating discussion, and to Dr M. B. Kalinowski, the guest editor of this special volume, for the invitation. The comments of the referees resulted in substantial improvement of the manuscript.

Appendix: The Catalogues and Related Data Used in this Study

The monthly earthquake catalogue is partly available through the web (http://www.csndmc.ac.cn/newweb/data.htm#, in Chinese) provided by the China Earthquake Networks Center (CENC). The earthquake catalogue for the Beijing Capital Circle

[2] Partly available at: http://data.earthquake.cn/data/, in Chinese.

Region is partly available through the web (http://data.earthquake.cn/data/, in Chinese). "Partly" means that in this study some earlier data, which are not included on the web, were also used. Supporting Online Materials provide the following data:

1. Catalogue of "repeating events" in the monthly earthquake catalogue of CENC for evaluation of location precision;
2. "Repeating pairs" in the monthly earthquake catalogue of CENC for evaluation of location precision;
3. Catalogue of "repeating events" in the Beijing Capital Circle Region, identified by waveform cross-correlation; and
4. "Repeating pairs" in the Beijing Capital Circle Region, together with the cross-correlation coefficient, with information about seismic stations shown in Fig. A1.

References

BARTAL, Y., SOMER, Z., LEONARD, G., STEINBERG, D. M. and HORIN, Y. B.: *Optimal seismic networks in Israel in the context of the Comprehensive Test Ban Treaty*, Bull. Seismol. Soc. Amer., *90*, 151–165, 2000.

DOUFEXOPOULOU, M. and KORAKITIS, R.: *Resolution analysis of seismic networks*, Phys. Earth Planet. Interi., *75*, 121–129, 1992.

EFRON, B.: *Bootstrap methods: another look at the Jackknife*, The Annals of Statistics, *7*, 1–26, 1979.

ISACKS, B. L., SYKES, L. R. and OLIVER, J.: *Spatial and temporal clustering of deep and shallow earthquakes in the Fiji-Tonga-Kermadec region*, Bull. Seismol. Soc. Amer., *57*, 935–958, 1967.

JIANG, C. S. and WU, Z. L.: *Location accuracy of the China National Seismograph Network estimated by repeating events*, Earthquake Research in China, *20*, No.1, 67–74, 2006.

JIANG, C. S., WU, Z. L. and LI, Y. T.: *Estimating the location capability of the Beijing Capital Digital Seismograph Network using repeating events*, Chin. J. Geophys., *51*, 817–827, 2008, in Chinese with English abstract.

LI, L., CHEN, Q. F., CHENG, X., and NIU, F. L.: *Spatial clustering and repeating of seismic events observed along the 1976 Tangshan fault, north China*, Geophys. Res. Lett., *34*, L23309. doi: 10.1029/2007GL031594, 2007.

LI, Y. T., WU, Z. L., PENG, H. P., JIANG, C. S. and LI, G. P.: *Time-lapse slip variation associated with a medium-size earthquake revealed by "repeating" micro-earthquakes: the 1999 Xiuyan, Liaoning, M_S = 5.4 earthquake*, Nat. Hazards Earth Syst. Sci., *11*, 1969–1981, doi:10.5194/nhess-11-1969-2011, 2011.

LIENERT, B. R.: *Assessment of earthquake location accuracy and confidence region estimates using known nuclear tests*, Bull. Seismol. Soc. Amer., *87*, 1150–1157, 1997.

LIN, C.-H.: *Repeated foreshock sequences in the thrust faulting environment of eastern Taiwan*, Geophys. Res. Lett., *31*, L13601, doi:10.1029/2004GL019833, 2004.

MENKE, W.: *Using waveform similarity to constrain earthquake locations*, Bull. Seismol. Soc. Amer., *89*, 1143–1146, 1999.

MENKE, W., LERNER-LAM, A. L., DUBENDORFF, B. and PACHECO, J.: *Polarization and coherence of 5 to 30 Hz seismic wave fields at a hard-rock site and their relevance to velocity heterogeneities in the crust*, Bull. Seismol. Soc. Amer., *80*, 430–449, 1990.

NADEAU, R. M. and JOHNSON, L. R.: *Seismological studies at Parkfield VI: Moment release rates and estimates of source parameters for small repeating earthquakes*, Bull. Seismol. Soc. Amer., *88*, 790–814, 1998.

NADEAU, R. M. and MCEVILLY, T. V.: *Seismological studies at Parkfield V: Characteristic microearthquake sequences as fault-zone drilling targets*, Bull. Seismol. Soc. Amer., *87*, 1463–1472, 1997.

NANJO, K. Z., SCHORLEMMER, D., WOESSNER, J., WIEMER, S. and GIARDINI, D.: *Earthquake detection capability of the Swiss Seismic Network*, Geophys. J. Int., *181*, 1713–1724, 2010.

RABINOWITZ, N. and STEINBERG, D. M.: *Optimal configuration of a seismographic network: a statistical approach*, Bull. Seismol. Soc. Amer., *80*, 187–196, 1990.

RUBIN, A. M.: *Using repeating earthquakes to correct high-precision earthquake catalogs for time-dependent station delays*, Bull. Seismol. Soc. Amer., *92*, 1647–1659, 2002.

SCHAFF, D. P. and RICHARDS, P. G.: *Repeating seismic events in China*, Science, *303*, 1176–1178, 2004.

SCHAFF, D. P. and RICHARDS, P. G.: *On finding and using repeating seismic events in and near China*, J. Geophys. Res., *116*, B03309. doi:10.1029/2010JB007895, 2011.

SCHAFF, D. P. and WALDHAUSER, F.: *One magnitude unit reduction in detection threshold by cross correlation applied to Parkfield (California) and China seismicity*, Bull. Seismol. Soc. Amer., *100*, 3224–3238, doi:10.1785/0120100042, 2010.

SCHORLEMMER, D., NERI, G., WIEMER, S. and MOSTACCIO, A.: *Stability and significance tests for b-value anomalies: example from the Tyrrhenian Sea*, Geophys. Res. Letts., *30*, 1835, doi: 10.1029/2003GL017335, 2003.

VIDALE, J. E., ELLSWORTH, W. L., COLE, A., and MARONE, C.: *Variations in rupture process with recurrence interval in a repeated small earthquake*, Nature, *36*, 8624–8626, 1994.

WOESSNER, J. and WIEMER, S.: *Assessing the quality of earthquake catalogs: Estimating the magnitude of completeness and its uncertainties*, Bull. Seismol. Soc. Amer., *95*, 684–698, 2005.

WUESTER, J., RIVIERE, F., CRUSEM, R., PLANTET, J.-L., MASSINON, B. and CARISTAN, Y.: *GSETT-3: Evaluation of the detection and location capabilities of an experimental global seismic monitoring system*, Bull. Seismol. Soc. Amer., *90*, 166–186, 2000.

(Received September 30, 2011, revised March 8, 2012, accepted May 24, 2012, Published online July 7, 2012)

Pure Appl. Geophys. 171 (2014), 425–437

DOI 10.1007/s00024-012-0515-3

Pure and Applied Geophysics

Cross-correlation Coefficients for the Study of Repeating Earthquakes: An Investigation of Two Empirical Assumptions/Conventions in Seismological Interpretation Practice

LIBO HAN,[1] ZHONGLIANG WU,[1,2] YUTONG LI,[3] and CHANGSHENG JIANG[1]

Abstract—For the identification and analysis of 'repeating earthquakes,' there are two empirical concepts. The first is the assumption that the cross-correlation coefficient of the filtered seismograms of closely spaced 'repeaters' depends exponentially on the inter-event separation distance. The second is the convention that in processing regional seismograms, a 0.5–5.0-Hz band pass filter is used. In this article, using a simple layered structure model, we investigated the cross-correlation coefficient of the filtered synthetic seismograms of two closely located events, that is, a 'doublet.' We investigated the relation between the cross-correlation coefficient and the inter-event separation distance. Simulation shows that in the 0.5–5.0-Hz frequency band, even if for simple synthetic seismograms without considering lateral heterogeneity or scattering, the exponential dependence is only a first order approximation concept. To check the frequency dependence of the cross-correlation coefficient, we analyzed a group of seismograms of a 'multiplet' in Xiuyan, Liaoning, northeast China, recorded by the Regional Seismographic Network of Liaoning Province. The cross-correlation coefficients were observed to be relatively stable against frequency for the 0.5–5.0-Hz frequency band.

1. Introduction

'Repeating earthquakes' identified by waveform cross correlation have significant potential for enhancing the detection capability for small events and the precision of earthquake location (VIDALE *et al.* 1994; NADEAU and JOHNSON 1998; NADEAU and MCEVILLY 1999; MENKE 1999; IGARASHI *et al.* 2003; SCHAFF and RICHARDS 2011; RICHARDS *et al.* 2006; CHENG *et al.* 2007; LI *et al.* 2007; RAU *et al.* 2007), which is potentially useful for CTBT monitoring (WU and RICHARDS 2011). This potential is more significantly noticeable because of the generality of 'repeating events.' Three decades ago, it was not clear whether 'repeating' was a typical or an anomalous characteristic of natural seismic activity (GELLER and MUELLER 1980). Today, the answer is clearly the former. For example, for China and its surrounding regions, over 10 % of earthquakes are 'repeaters' (SCHAFF and RICHARDS 2004), providing an 'alternative' estimator of the location precision (JIANG and WU 2006; JIANG *et al.* 2008, 2012). Using waveform similarities, the detection threshold for some regions (such as California and China) has been reported to be reduced by one magnitude unit (SCHAFF and WALDHAUSER 2010). The theoretical and empirical basis for the study of 'repeating earthquakes,' therefore, is apparently worth further investigation in the context of seismology for CTBT monitoring.

The identification and analysis of 'repeating events' are generally based on some empirical/phenomenological concepts that are practical in operation but need further discussion using theoretical and/or numerical tools, and need further tests using experimental and/or observational data. For doing this, several works have explained the phenomenology of 'repeating events.' BAISCH *et al.* (2008) used wave-field simulations to investigate how the waveform similarity of closely spaced hypocenters changes with inter-event separation and discussed the 'quarter lambda criterion,' which is often used in the analysis of seismograms. SCHAFF (2008) conducted a statistical investigation of the performance of 'correlation detectors,' providing practical guidelines for the data analysis of seismographic networks. In this article, we try to investigate

[1] Institute of Geophysics, China Earthquake Administration, Beijing 100081, China. E-mail: wuzl@cea-igp.ac.cn

[2] Laboratory of Computational Geodynamics, Graduate University of the Chinese Academy of Sciences, Beijing 100049, China.

[3] Earthquake Administration of Liaoning Province, Shenyang 110034, China.

two empirical/phenomenological concepts that are often used in the study of 'repeating events.' One is the assumption that the cross-correlation coefficient is exponentially dependent on the separation distance of the two 'repeaters' (MENKE *et al.* 1990; MENKE 1999), and the other is the convention that in identifying 'doublets' and/or 'multiplets,' a 0.5–5.0-Hz band pass filter is often used (SCHAFF and RICHARDS 2004, 2011).

2. *Dependence of the Cross-correlation Coefficient on the Inter-event Separation Distance: Simple Analytical Considerations*

The assumption that the cross-correlation coefficient of the filtered seismograms of 'repeaters' depends exponentially on the separation distance is empirically based on observations (MENKE 1999). Summarizing the results of MENKE *et al.* (1990) and NADEAU *et al.* (1995), MENKE (1999) proposed that the cross-correlation coefficient (*CCC*) of a closely located 'doublet' depends on the inter-event separation distance, r_{ij}, via $CCC = \exp(-r_{ij}/s)$, in which s is a frequency-dependent correlation length. MENKE *et al.* (1990) pointed out that 'We do not claim that the true coherence is necessarily an exponential, just that this formula satisfactorily describes the observed pattern of variation (which, after all, has considerable scatter).' However, in later work, MENKE (1999) tried to explore the possibility of using this relation to determine the 'internal structure' of a 'multiplet,' making the discussion of this empirical concept interesting for seismological interpretation.

Qualitatively, the understanding of this dependence is straightforward. Ground motion along the i-th component at position $\mathbf{r}$, u_i, generated by a seismic source at position $\mathbf{r}^0$ with seismic moment tensor M_{ij}, can be represented (in the frequency domain; by AKI and RICHARDS 1980) as:

$$u_i(\mathbf{r},\omega) = \frac{\partial G_{ij}(\mathbf{r},\omega;\ \mathbf{r}^0)}{\partial x_k^0} M_{jk}(\mathbf{r}^0,\omega) \qquad (1.1)$$

where G_{ij}, Green's function, is the displacement generated by an impulse point source (described by a Dirac δ function). If another form of Green's function G_{ij}^H, representing the displacement generated by a step point source (described by a Heaviside unit step function), is used, then:

$$u_i(\mathbf{r},\omega) = \frac{\partial G_{ij}^H(\mathbf{r},\omega;\ \mathbf{r}^0)}{\partial x_k^0} \dot{M}_{jk}(\mathbf{r}^0,\omega) \qquad (1.2)$$

which is more familiar to earthquake seismologists. Considering the symmetry of a seismic moment tensor, Eq. (1.2) can be simplified as:

$$u_i(\mathbf{r},\omega) = g_{ij}(\mathbf{r},\omega;\ \mathbf{r}^0)\dot{M}_j(\mathbf{r}^0,\omega) \qquad (2)$$

in which $\{M_1, M_2, M_3, M_4, M_5, M_6\} = \{M_{11}, M_{12}, M_{13}, M_{22}, M_{23}, M_{33}\}$ and g_{ij}, a linear combination of the components of the spatial derivative of Green's function, represents the 'coefficients' of $\dot{M}_j$. Consider the cross correlation of two ground motion seismograms, u_i^1 and u_i^2, with:

$$u_i^1(\mathbf{r},\omega;\ \mathbf{r}^0) = g_{ij}(\mathbf{r},\omega;\ \mathbf{r}^0)\dot{M}_j(\mathbf{r}^0,\omega) \qquad (3.1)$$

$$u_i^2(\mathbf{r},\omega;\ \mathbf{r}^0+\delta\mathbf{r}) = g_{ij}(\mathbf{r},\omega;\ \mathbf{r}^0+\delta\mathbf{r})\dot{M}_j(\mathbf{r}^0+\delta\mathbf{r},\omega) \qquad (3.2)$$

which are the ground motion generated by a closely spaced 'doublet' with inter-event separation vector $\delta\mathbf{r}$. Cross correlation of the two seismograms can be represented by (BÅTH 1974):

$$CCC(\omega;\ \delta\mathbf{r}) = \tilde{u}_i^1(\mathbf{r},\omega;\ \mathbf{r}^0)u_i^2(\mathbf{r},\omega;\ \mathbf{r}^0+\delta\mathbf{r}) \qquad (4)$$

where $\tilde{u}$ stands for the complex conjugate. Expanding the function *CCC* with respect to $\delta\mathbf{r}$ gives:

$$CCC(\omega;\ \delta\mathbf{r}) = CCC(\omega;0) \times \left[1+\frac{1}{CCC(\omega;0)}\nabla CCC(\omega;0)\delta\mathbf{r}+\cdots\right] \qquad (5)$$

On the other hand, an exponential function approximates $\exp(-x) \approx 1 - x + \ldots$. Therefore, in the sense of the first order approximation, and only in the sense of the first order approximation, the dependence of the cross-correlation coefficient can be approximated by an exponential function. How good the description by an exponential function is depends on the heterogeneity of the structure and the seismic waves propagating through it, that is, on the higher order terms in Eq. (5).

3. Dependence of the Cross-correlation Coefficient on the Inter-event Separation Distance: Synthetic Seismogram Simulations

To investigate the variation of the cross-correlation coefficient with the separation distance between the two 'repeaters,' we calculated a series of seismic waveforms corresponding to regional seismograms used in the waveform cross correlation (e.g., Li *et al.* 2011). The synthetic seismograms were generated with a frequency-wavenumber (F–K) synthetic seismogram package (Zhu and Rivera 2002), which has been successfully used in several works, such as the inversion of seismic source parameters by the cut-and-paste (CAP) method (Zhao and Helmberger 1994; Zhu and Helmberger 1996). Compared with real data (see below in the next section), the station-source configuration is taken as: source depth: 10 km; station-epicenter distance: 50 km; station azimuth: N60°E. The velocity structure model used in the calculation is taken as the regional structure model of southern Liaoning Province, northeast China (shown in Fig. 1).

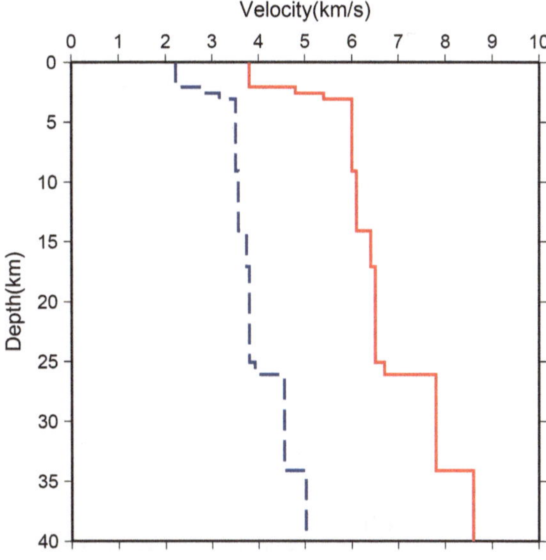

Figure 1
Layered structure model used to calculate the synthetic seismograms. *Red solid lines* denote the P wave and *blue broken lines* the S wave. Data are from the Earthquake Administration of Liaoning Province (Lu *et al.* 2002). Qs and Qp are taken as 500 and 1,000, respectively

To cope with the general case of an arbitrary seismic source, we used the 'elementary seismic moment tensors' of Kikuchi and Kanamori (1991), such as:

$$M_1 = \begin{bmatrix} 0 & 1 & 0 \\ 1 & 0 & 0 \\ 0 & 0 & 0 \end{bmatrix} \tag{6.1}$$

$$M_2 = \begin{bmatrix} 1 & 0 & 0 \\ 0 & -1 & 0 \\ 0 & 0 & 0 \end{bmatrix} \tag{6.2}$$

$$M_3 = \begin{bmatrix} 0 & 0 & 0 \\ 0 & 0 & 1 \\ 0 & 1 & 0 \end{bmatrix} \tag{6.3}$$

$$M_4 = \begin{bmatrix} 0 & 0 & 1 \\ 0 & 0 & 0 \\ 1 & 0 & 0 \end{bmatrix} \tag{6.4}$$

$$M_5 = \begin{bmatrix} -1 & 0 & 0 \\ 1 & 0 & 0 \\ 0 & 0 & 1 \end{bmatrix} \tag{6.5}$$

in which the coordinates (x, y, z) for M_j correspond to north, east and downward directions, and any arbitrary deviatoric moment tensor (corresponding to natural shallow earthquakes) can be represented by a linear combination of M_j. Figures 2, 3, 4, 5 and 6 show the results for these five 'elementary seismic moment tensors,' respectively.

Separation between the two 'repeaters' can have different geometries. Without losing generality, we consider only some simple ('elementary') cases. Figures 2, 3, 4, 5 and 6 show the waveform cross correlation of synthetic seismograms of closely spaced hypocenters, changing with inter-event separation, in which the first event is located with 10 km depth and 50 km station-epicenter distance, and its 'partner' in the 'doublet' is separated by *Z*-direction vertical perturbations as well as *R*-direction and *T*-direction horizontal perturbations, respectively, up to 3.0 km by a step of 0.1 km. Shown in Figs. 2, 3, 4, 5 and 6 are only the cases of the increase of depth and horizontal *R*-distance, together with the counter-clockwise change along the *T*-direction. To mimic the real situation in seismological interpretation, the synthetic seismograms are purposely 'contaminated' by 'seismic' (but not 'random/white') noise—the noise is taken from real seismic recordings, obeying Peterson's law in the frequency domain. The starting

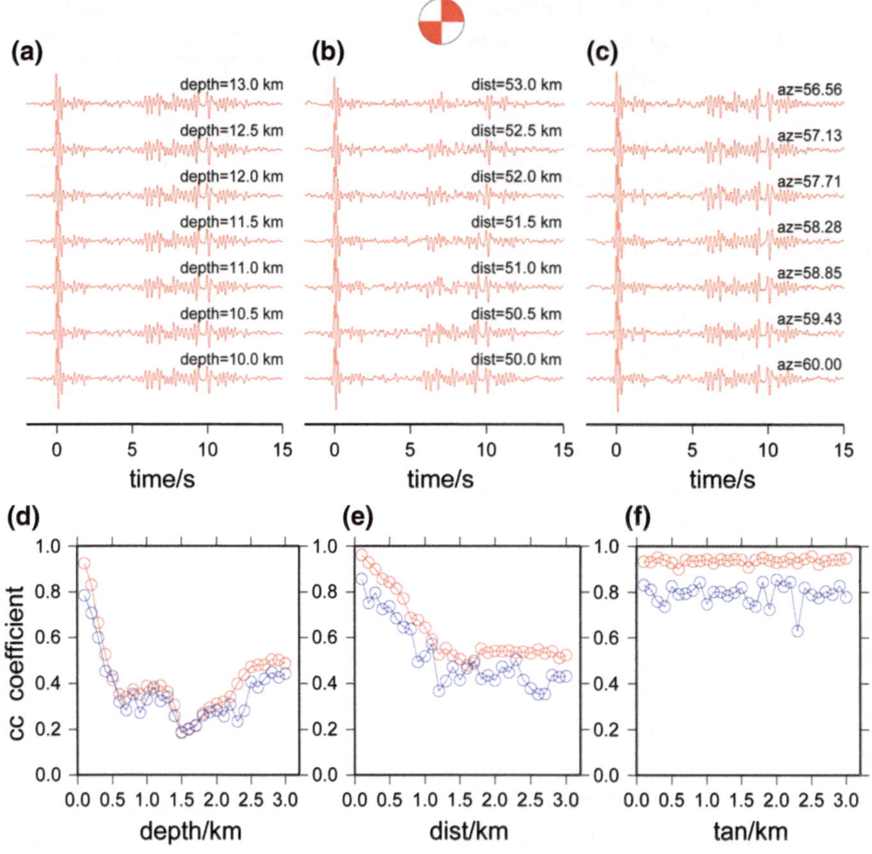

Figure 2

Waveform cross correlation of synthetic seismograms of closely spaced hypocenters, changing with inter-event separation. Moment tensor of the source is M_1. The first event is located with 10 km depth and 50 km epicenter distance, with the azimuth of the station N60°E. Its 'repeater' is separated with vertical perturbations as well as *R*-direction and *T*-direction horizontal perturbations, respectively, up to 3.0 km by a step of 0.1 km. Panels (**a**) to (**c**) show the cases with *Z*, *R* and *T*-direction perturbations, respectively. Panels (**d**) to (**f**) show, respectively, the cross-correlation coefficient as a function of inter-event separation, corresponding to the traces in (**a**) to (**c**), respectively. *Red* and *blue dots* correspond to the cases in which the synthetic seismograms are mixed with 10 % and 20 % seismic noise, respectively. Seismograms in panels (**a**) to (**c**) show the case with 10 % seismic noise. See text for details

point of the noise to be mixed into the seismograms is determined by a random number generator. The noise level is taken, somewhat arbitrarily, so that the peak of the noise signal is 10 and 20 %, respectively, of the peak of the synthetic seismograms. To compare with the real situation of 'repeating event' analysis, these seismograms are 0.5–5.0-Hz band pass filtered before calculating the cross-correlation coefficient. Calculation of the cross-correlation coefficient follows a similar procedure as that used in real seismological analysis, using the whole seismogram (in this case, starting 1 s before the P-arrival, with a total length of 20 s) and sliding the seismogram of one event from 1 s before the synchronized P-arrival to 1 s after, with the step of one sample. The peak absolute value of the cross-correlation coefficient during the sliding is selected as the 'final' cross-correlation coefficient.

Shown in Figs. 2, 3, 4, 5 and 6, panels (a) to (c), are the seismograms of the 'location-perturbed' event in a 'doublet,' with *Z*, *R* and *T*-direction perturbations, respectively. Panels (d) to (e) show, respectively, the cross-correlation coefficient as a function of the inter-event separation distance along different directions. It can be seen that the overall shape of the curve of the cross-correlation coefficient versus the inter-event separation distance can be approximated, to some extent, by an exponential-like

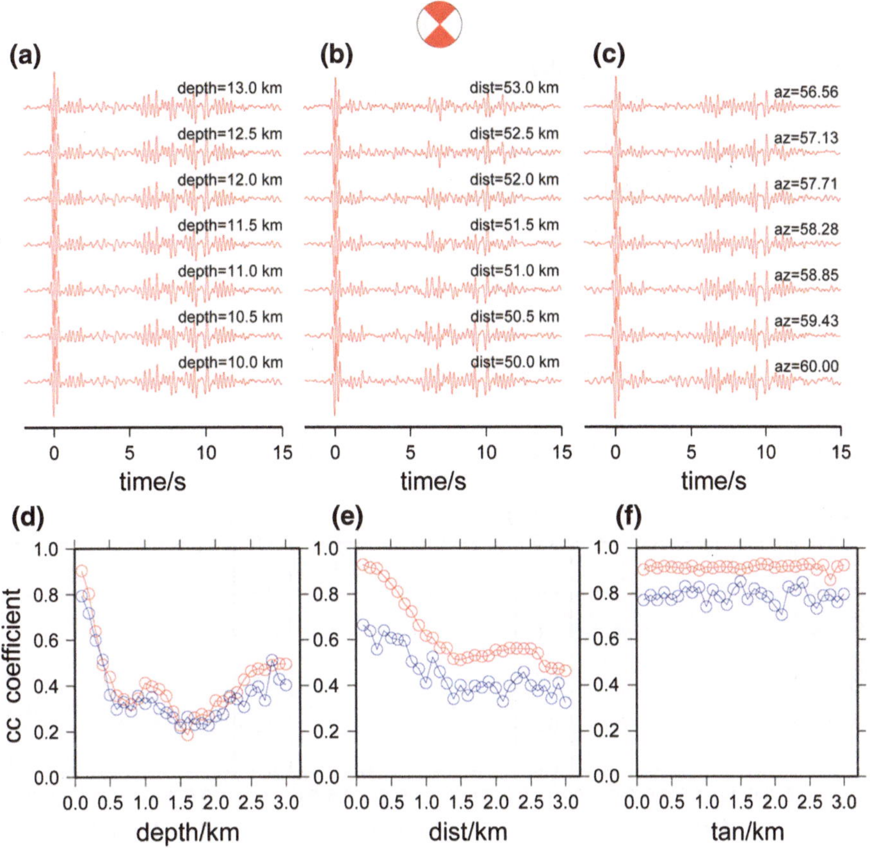

Figure 3

Waveform cross correlation of synthetic seismograms of closely spaced hypocenters, changing with inter-event separation, for 'elementary moment tensor' M_2. For captions, see Fig. 2

dependence. Despite the simplicity of the model, it can be seen that within the 0.5–5.0-Hz frequency band, even if for the simple case without considering lateral heterogeneity and stochastic scattering (seismic coda), the exponential dependence is only an approximation. Looking into the 'internal structure' of a 'repeating multiplet' (e.g., Menke 1999), this feature has to be taken into consideration.

For the *T*-direction perturbation, the variation of the cross-correlation coefficient is much less, and, for some cases, there is almost no significant change in the perturbation up to 3.0 km. Based on this, it might be an important reminder for the identification of 'repeating pairs' that for some special cases, such as when the station-event configuration makes the inter-event separation mainly tangential, the inter-event distance cannot be detected by the cross-correlation coefficient for this 'special' station. Considering the 3D heterogeneity of the earth's structure, this 1D-model-based conclusion may be weakened to some extent. But this result suggests that in the identification of 'repeating events,' generally the criterion based on more than one station (with different azimuths) will be better.

In the calculation, the source time function (STF) of the 'repeating pairs' is taken as triangular with a duration of 0.05 s, which means that the seismograms are close to Green's function (impulse response) or the seismograms generated by a very small earthquake. Considering real earthquakes, on the one hand, the convolution of Green's function with the STF may make the cross-correlation coefficient increase because such a convolution corresponds to

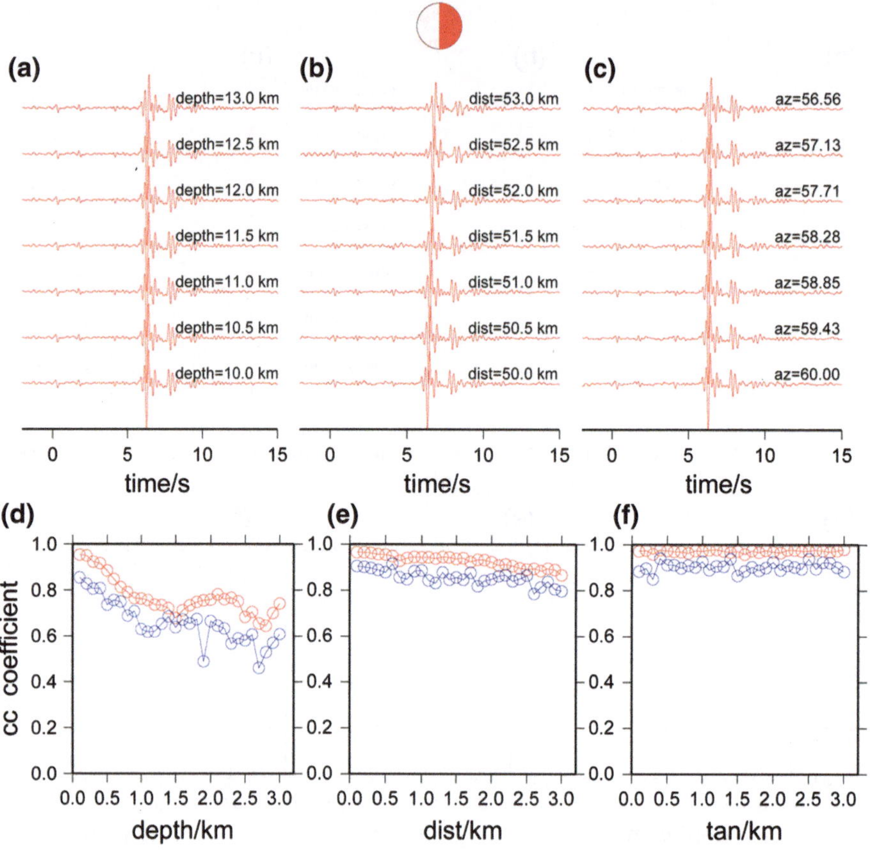

Figure 4
Waveform cross correlation of synthetic seismograms of closely spaced hypocenters, changing with inter-event separation, for 'elementary moment tensor' M_3. For captions, see Fig. 2

some kinds of low-pass filtering; on the other hand, differences in the STFs of the 'repeating pair' (either in shape or the duration) may decrease the cross-correlation coefficient.

4. Frequency-band Dependence of the Cross-correlation Coefficient: Analysis of the Observational Data

To analyze the frequency-band dependence of the cross-correlation coefficients, we used the observational data from a '57-plet' located in Xiuyan, Liaoning, northeast China, recorded by the Liaoning Regional Seismographic Network. Li *et al.* (2011) provided details for the data and the analysis. 'Repeating events' related to the Xiuyan region were also analyzed by Schaff (2010). Figure 7a, b, reproducing Fig. 2 and Fig. A1 of Li *et al.* (2011), shows the distribution of the seismic stations and earthquakes under study, together with part of the filtered seismograms of this '57-plet' recorded by the VBB seismic station in Yingkou. In Fig. 7b, seismograms are aligned by their P-arrivals picked from the seismic waveforms. From the figure it can be seen that the observed seismograms have similar envelopes as the synthetic ones, as shown in Figs. 2, 3, 4, 5 and 6. Note that different event pairs have different station combinations, and the 'multiplet' is identified by the criterion that no fewer than three stations have cross-correlation coefficients above 0.8. In this '57-plet,' there are only 17 event pairs all having the same six stations recording them, as shown in Table 1.

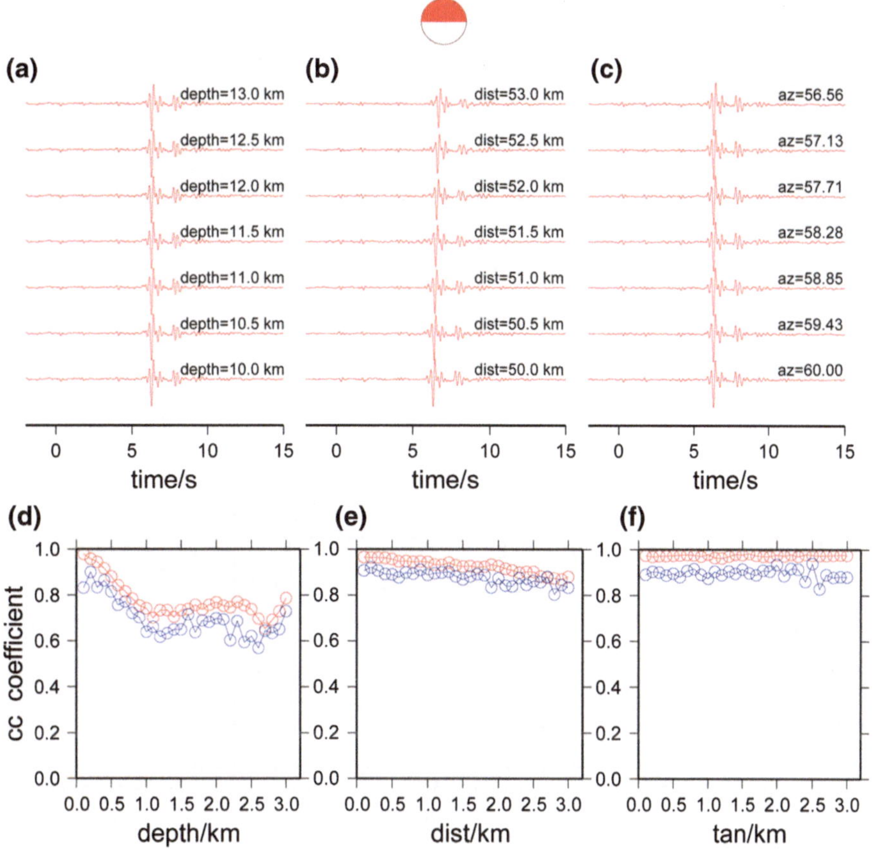

Figure 5
Waveform cross correlation of synthetic seismograms of closely spaced hypocenters, changing with inter-event separation, for 'elementary moment tensor' M_4. For captions, see Fig. 2

Cross-correlation coefficients are calculated using a similar and common procedure: pre-processing using a 0.5–5-Hz band pass filter for BB recordings and a 1–5-Hz band pass filter for SP recordings. The whole waveform composed of P, S and coda phases is selected, starting the waveform 4 s before the P arrival and ending the waveform at 4 times the S–P travel time difference. For one event in a 'pair,' sliding is done for the seismogram of its 'repeater,' with its waveform length 4 s less than the length of the whole waveform, and sliding from the beginning point (4 s before the P arrival) with a step of one sample, taking the peak absolute correlation coefficient as the 'final' cross-correlation coefficient. Jiang *et al.* (2008) systematically discussed the effect of such a mixed use of BB and SP seismic recordings. In many regions such as China, this mixed use is somehow inevitable because of the configuration of seismographic networks according to site conditions. Figure 8 plots the cross-correlation coefficients for different event pairs and different stations. From the figure it can be seen that, for the same 'pair,' different stations have different correlation coefficients. Meanwhile, for different stations, nearly synchronized relative variation with different 'event pairs' can be seen. The explanation for this could be that the 'repeating pairs' are not exactly co-located, but separated by different amounts, while the other reason, which might be more important, is the 'site condition,' which is well known in seismological observations and interpretation. Generally, the (band-pass-filtered) VBB/BB recordings (YKo and Don)

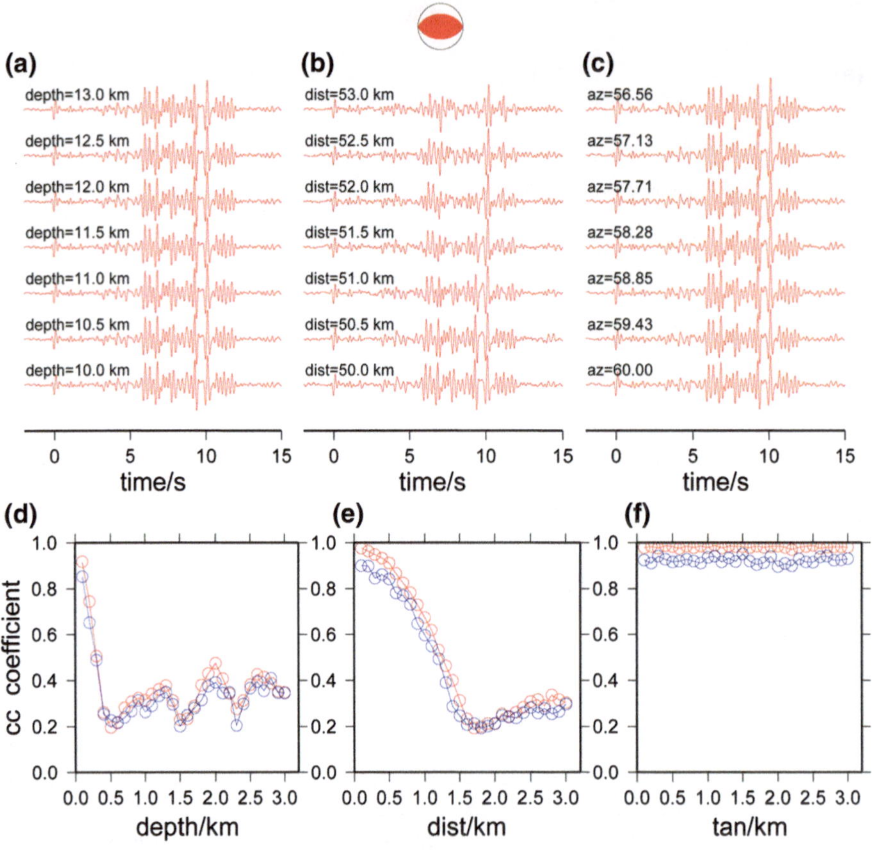

Figure 6

Waveform cross correlation of synthetic seismograms of closely spaced hypocenters, changing with inter-event separation, for 'elementary moment tensor' M_5. For captions, see Fig. 2

seem to have a higher CCC than the (band-pass-filtered) SP recordings, but this conclusion is not definite because of the limited number of stations.

To investigate the influence of the frequency band on the cross-correlation coefficients, examples of one-event pair/different stations and one station/different event pairs, respectively, were taken. Figures 9 and 10 show the cross-correlation coefficients changing with the frequency band. In pre-processing, the seismograms are band pass filtered, with the central frequency shown on the horizontal coordinates in Figs. 9 and 10 and a band width of 1 Hz. From the figure it can be seen that, above 5 Hz, the variation is complicated, while below 5 Hz, the cross-correlation coefficients are relatively stable. From this perspective two conclusions can be reached. Firstly, the frequency term in the relation of the cross-correlation coefficient versus frequency, as mentioned in Menke *et al.* (1990), seems valid for the high frequency part, but with significant complexity. Secondly, below 5 Hz the cross-correlation coefficients are 'stable' against the changing frequency. This could be one of the phenomenological bases for the empirical convention that in processing the waveforms for cross-correlation, generally a band pass filter of 0.5–5-Hz is taken, which is probably the result of a series of trials in practice. Note that due to the logarithm coordinates used in the figures, similar to the exponential-like CCC-distance relation, such a concept of 'stable,' even if 'relatively stable,' is only approximate.

5. Discussion and Conclusions

In this article, using simulation and observation results, we investigated two empirical concepts in the

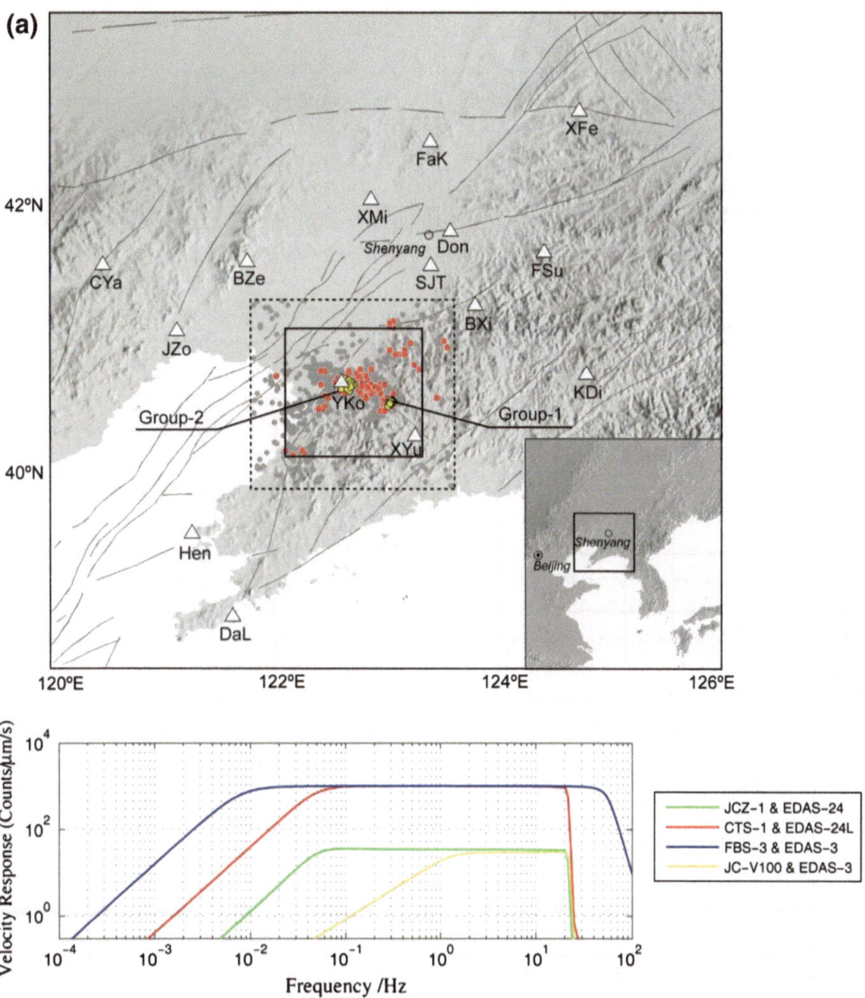

Figure 7

a *Top* distribution of earthquakes and seismic stations used for the analysis of 'repeating events,' reproduced from Li *et al.* (2011). Earthquakes are shown by *gray dots*. Shown in *red* are the 'repeating earthquakes' identified by waveform cross correlation. *Yellow dots* show the two 'multiplets' in the Xiuyan-Yingkou region, with the one to the east (group 1) being the 'Xiuyan 57-plet' discussed in this paper. *Gray lines* show tectonic faults. The *box* in the indexing figure indicates the whole region of Liaoning and its surroundings. In the region, the *box* with *solid lines* highlights the Xiuyan-Yingkou region and its surroundings (40.1 ~ 41.1°N, 122 ~ 123.4°E), encompassed by a *larger box* with *dashed lines* (39.9 ~ 41.3°N, 121.8 ~ 123.6°E) to ensure that there are no missing 'repeaters' due to events located near the margin of the region. Seismic stations are shown by *triangles*, with *letters* nearby indicating their codes. *Bottom* instrumental response of the seismographs (from the Data Center of the Liaoning Digital Seismographic Network). The sensors are very broadband (FBS-3 at JZo and YKo), broadband (JCZ-1 at Don and CTS-1 at DaL), or short-period (JC-V100-3D at Hen, KDi, BZe, XYu, CYa, BXi, FSu, FaK, XFe and XMi) with the EDAS data acquisition system. **b** Filtered seismograms of the Xiuyan 'multiplet' recorded at the Yingkou (YKo) station, reproduced from Li *et al.* (2011). The figure shows part of the seismograms recorded by the Yingkou station (YKo, VBB seismograph), in which 'part of' means that for the '57-plet,' only the seismograms of the first 31 events before and immediately after the 1999 Xiuyan earthquake are shown. In the figure, each trace has been normalized and pre-processed by a 0.5–5-Hz band pass filter. The *bottom trace* is the stacking of all the above traces. The clearness of the *stacked trace* indicates the similarity of the waveforms. Note that the 'repeaters' are identified by the criteria that filtered waveforms of at least three stations (not necessarily including the YKo station) have cross-correlation coefficients ≥0.8. For some of the events, there were no waveform recordings at the YKo station

identification and analysis of 'repeating earthquakes.' One is the semi-quantitative exponential-like relation between the cross-correlation coefficient of seismograms recorded at the same station and the separation distance between the two 'doublets;' the other is the convention that in the pre-processing of

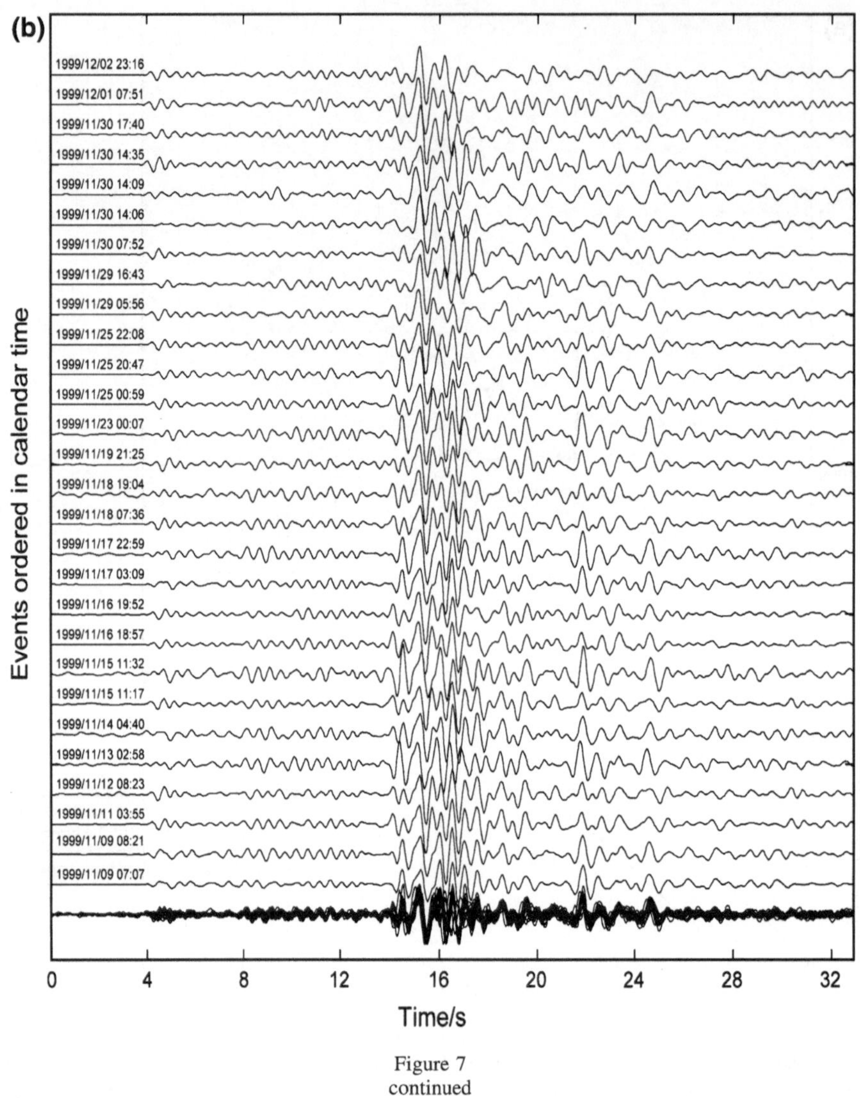

Figure 7
continued

the waveform, a 0.5–5.0-Hz band pass filter is used. Our approach is similar to the one used by Baisch *et al.* (2008), but focuses on the 0.5–5.0-Hz frequency band, which is often considered in the analysis of 'repeating events' using regional seismographic networks. What has been obtained lends support to the assumptions that 'repeating events' defined solely based on cross-correlation coefficients are separated by small amounts for certain criteria. The empirical assumption of the exponential dependence of the cross-correlation coefficient on the inter-event separation distance holds only as the first order approximation for the 0.5–5.0-Hz frequency band, which cautions us against using the difference of the cross-correlation coefficient for mapping of the 'internal structure' of a 'multiplet.' Real observational data also confirm the empirical choice of 0.5–5-Hz as a relatively stable filter band. Analyzing the perturbations along the *T*-direction suggests that in the identification of 'repeaters,' using more than one station with different azimuths will be better.

Acknowledgments

Thanks to Prof. Lupei Zhu for providing the codes and for help with calculating the synthetic seismograms. Observational data are from the Liaoning

Table 1

Event pairs in the Xiuyan '57-plet,' which all have the same six stations at which cross-correlation coefficients are available

No.	Event1	Event2	YKo	XYu	BXi	Don	FaK	XMi
1	199911090707.00	199911090821.00	0.938947	0.895247	0.885643	0.945240	0.900656	0.831180
2	199911090707.00	199911170309.00	0.939424	0.879921	0.897043	0.934387	0.888890	0.799464
3	199911090707.00	199911172259.00	0.914170	0.861791	0.828952	0.926417	0.830891	0.733525
4	199911090821.00	199911170309.00	0.965254	0.913462	0.933924	0.938353	0.878912	0.843797
5	199911090821.00	199911172259.00	0.937119	0.957071	0.929018	0.968114	0.900160	0.820116
6	199911090821.00	199911230007.00	0.912216	0.890484	0.861233	0.904145	0.798338	0.708260
7	199911110355.0	199911250059.00	0.976104	0.972905	0.912480	0.955242	0.877011	0.935391
8	199911161857.00	199911161952.00	0.946800	0.935689	0.817810	0.888831	0.883679	0.905917
9	199911161857.00	199911252208.00	0.962623	0.908552	0.829178	0.918048	0.929806	0.913866
10	199911161952.00	199911252208.00	0.976753	0.946811	0.909382	0.942642	0.909618	0.952618
11	199911161952.00	199911290556.00	0.938970	0.935087	0.852276	0.899135	0.906539	0.914315
12	199911170309.00	199911172259.00	0.900530	0.872583	0.872850	0.942177	0.855462	0.830320
13	199911170309.00	199911230007.00	0.942041	0.918670	0.842221	0.938113	0.808584	0.853293
14	199911172259.00	199911230007.00	0.921564	0.902131	0.849853	0.919352	0.814384	0.833455
15	199911230007.00	199911252047.00	0.893287	0.856404	0.823825	0.911391	0.835229	0.855694
16	199911252208.00	199911290556.00	0.918352	0.907182	0.828014	0.868627	0.925717	0.881175
17	200207020700.00	200401020038.00	0.928429	0.922094	0.935192	0.757500	0.869325	0.950917

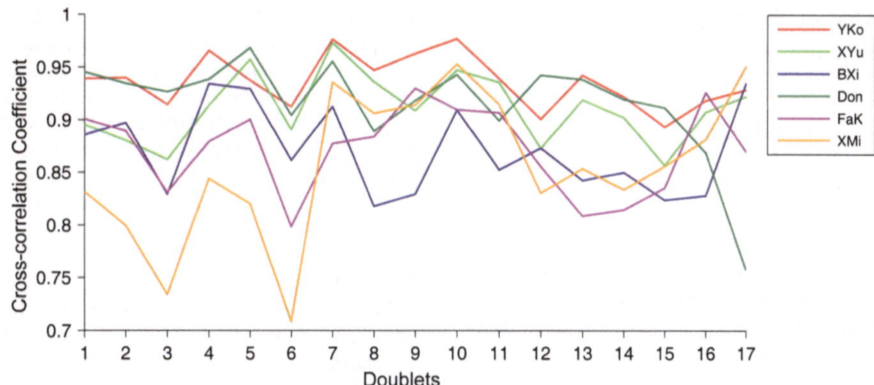

Figure 8
Cross-correlation coefficients at six different stations, YKo, XYu, BXi, Don, FaK and XMi, for 17 event pairs in the '57-plet.' See Table 1

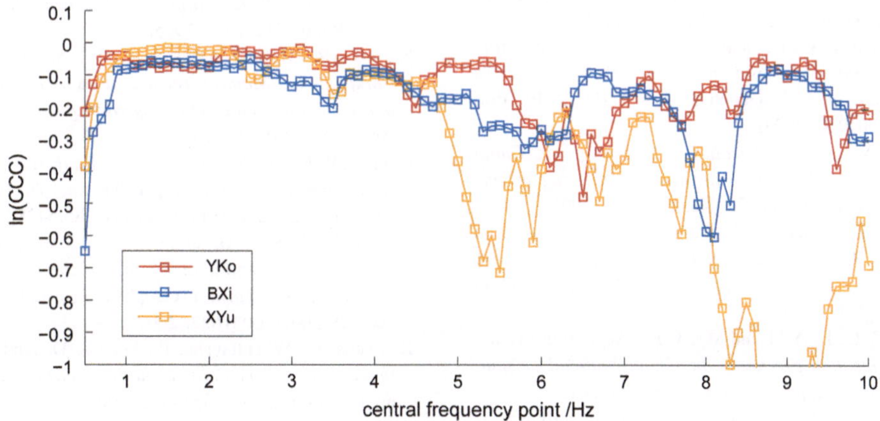

Figure 9
Cross-correlation coefficients for different frequency bands for the first event pair in Table 1 recorded at three stations, YKo, BXi and XYu

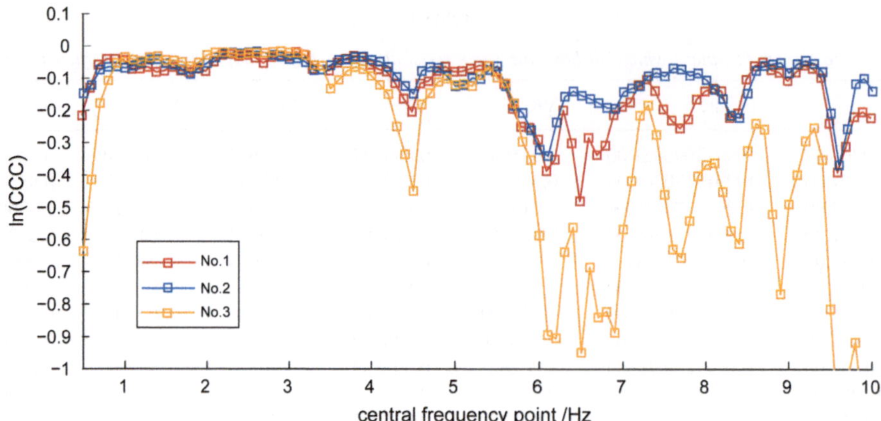

Figure 10
Cross-correlation coefficients for different frequency bands for the first three event pairs recorded at station YKo

Regional Seismographic Network. We thank the guest editors of the present special issue, Martin B. Kalinowski, Andreas Becker and Lars Ceranna, for the invitation. One of the referees suggested investigating the *T*-direction perturbations.

References

Aki, K. and Richards, P. G.: *Quantitative Seismology: Theory and Methods*, W. H. Freeman, San Francisco, 1980.

Baisch, S., Ceranna, L., Harjes, H.-P.: *Earthquake cluster: what can we learn from waveform similarity*? Bull. Seismol. Soc. Amer., *98*, 2806–2814, doi:10.1785/0120080018, 2008.

Båth, M.: *Spectral Analysis in Geophysics*, Elsevier, Amsterdam, 1974.

Cheng, X., Niu, F., Silver, P. G., Horiuchi, S., Takai, K., Iio, Y., and Ito, H.: *Similar microearthquakes observed in western Nagano, Japan, and implications for rupture mechanics*, J. Geophys. Res., *112*, B04306, doi:10.1029/2006JB004416, 2007.

Geller, R. J., Mueller, C. S.: *Four similar earthquake in central California*, Geophys. Res. Lett., *10*, 821–824, 1980.

Igarashi, T. Matsuzawa, T., and Hasegawa, A.: *Repeating earthquakes and interplate aseismic slip in the northeastern Japan subduction zone*, J. Geophys. Res., *108*, 2249, doi: 10.1029/2002JB001920, 2003.

Jiang, C. S. and Wu, Z. L.: *Location accuracy of the China National Seismograph Network estimated by repeating events*, Earthquake Research in China, *20*, No. 1, 67–74, 2006.

Jiang, C. S., Wu, Z. L. and Li, Y. T.: *Estimating the location capability of the Beijing Capital Digital Seismograph Network using repeating events, Chin.* J. Geophys., *51*, 817–827, 2008, in Chinese with English abstract.

Jiang, C. S., Wu, Z. L., Li, Y. T and Ma, T. F.: *'Repeating Events' as estimator of location precision: the China National Seismograph Network*, Pure Appl. Geophys., Accepted, 2012.

Kikuchi, M. and Kanamori, H.: *Inversion of complex body waves-III*, Bull. Seismol. Soc. Amer., *81*, 2335–2350, 1991.

Li, L., Chen, Q. F., Cheng, X., and Niu, F. L.: *Spatial clustering and repeating of seismic events observed along the 1976 Tangshan fault, north China*, Geophys. Res. Lett., *34*, L23309, doi: 10.1029/2007GL031594, 2007.

Li, Y. T., Wu, Z. L., Peng, H. P., Jiang, C. S. and Li, G. P.: *Time-lapse slip variation associated with a medium-size earthquake revealed by "repeating" micro-earthquakes: the 1999 Xiuyan, Liaoning, M_S = 5.4 earthquake*, Nat. Hazards Earth Syst. Sci., *11*, 1969–1981, doi:10.5194/nhess-11-1969-2011, 2011.

Lu, Z.-X., Jiang, X.-Q., Pan, K., Bai, Y., Jiang, D.-L., Xiao, L.-P., Liu, J.-H., Liu, F.-T., Chen, H. and He, J.-K.: *Seismic tomography in the northeast margin area of Sino-Korean platform*, Chin. J. Geophys., *45*, 338–351, 2002, in Chinese with English abstract.

Menke, W.: *Using waveform similarity to constrain earthquake locations*, Bull. Seismol. Soc. Amer., *89*, 1143–1146, 1999.

Menke, W., Lerner-Lam, A. L., Dubendorff, B. and Pacheco, J.: *Polarization and coherence of 5 to 30 Hz seismic wave fields at a hard-rock site and their relevance to velocity heterogeneities in the crust*, Bull. Seismol. Soc. Amer., *80*, 430–449, 1990.

Nadeau, R. M., Foxall, W. and McEvilly, T. V.: *Clustering and periodic recurrence of microearthquakes on the San Andreas fault at Parkfield, California*, Science, *267*, 503–507, doi: 10.1126/267.5197.503, 1995.

Nadeau, R. M. and Johnson, L. R.: *Seismological studies at Parkfield VI: Moment release rates and estimates of source parameters for small repeating earthquakes*, Bull. Seismol. Soc. Amer., *88*, 790–814, 1998.

Nadeau, R. M. and McEvilly, T. V.: *Fault slip rates at depth from recurrence intervals of repeating microearthquakes*, Science, *285*, 718–721, doi:10.1126/science.285.5428.718, 1999.

Rau, R. J., Chen, K. H., and Ching, K. E.: *Repeating earthquakes and seismic potential along the northern Longitudinal Valley fault of eastern Taiwan*. Geophys. Res. Lett., *34*, L24301, doi: 10.1029/2007GL031622, 2007.

Richards, P., Waldhauser, F., Schaff, D. And Kim, W.-Y.: *The applicability of modern methods of earthquake location*, Pure appl. Geophys., *163*, 351–372, 2006.

Schaff, D.: *Semiempirical statistics of correlation-detector performance*, Bull. Seismol. Soc. Amer., *98*, 1495–1507, 2008.

SCHAFF, D.: *Improvements to detection capability by cross-correlating for similar events: a case study of the 1999 Xiuyan, China, sequence and synthetic sensitivity tests*, Geophys. J. Int., *180*, 829–846, 2010.

SCHAFF, D. P. and RICHARDS, P. G.: *Repeating seismic events in China*, Science, *303*, 1176–1178, 2004.

SCHAFF, D. P. and RICHARDS, P. G.: *On finding and using repeating seismic events in and near China*, J. Geophys. Res., *116*, B03309, doi:10.1029/2010JB007895, 2011.

SCHAFF, D. P. and WALDHAUSER, F.: *One magnitude unit reduction in detection threshold by cross correlation applied to Parkfield (California) and China seismicity*, Bull. Seismol. Soc. Amer., *100*, 3224–3238, doi:10.1785/0120100042, 2010.

VIDALE, J. E., ELLSWORTH, W. L., COLE, A., and MARONE, C.: *Variations in rupture process with recurrence interval in a repeated small earthquake*, Nature, *36*, 8624–8626, 1994.

WU, Z. L. and RICHARDS, P. G.: Seismology, Monitoring of CTBT—scientific and technical advances in seismology and their relevance, In: GUPTA, H. K. (eds.) *Encyclopedia of Solid Earth Geophysics*, 2nd Edition, Springer, Amsterdam, 1340–1343, 2011.

ZHAO, L. and HELMBERGER, D. V.: *Source estimation from broadband regional seismograms*, Bull. Seismol. Soc. Amer., *84*, 91–104, 1994.

ZHU, L. and HELMBERGER, D. V.: *Advancement in source estimation techniques using broadband regional seismograms*, Bull. Seismol. Soc. Amer., *86*, 1634–1641, 1996.

ZHU, L. and RIVERA L. A.: *A note on the dynamic and static displacements from a point source in multilayered media*, Geophys. J. Int., *148*, 619–627, 2002.

(Received September 30, 2011, revised May 8, 2012, accepted May 24, 2012, Published online July 7, 2012)

Pure Appl. Geophys. 171 (2014), 439–468

DOI 10.1007/s00024-012-0626-x

Pure and Applied Geophysics

Perspectives of Cross-Correlation in Seismic Monitoring at the International Data Centre

Dmitry Bobrov,[1] Ivan Kitov,[1] and Lassina Zerbo[1]

Abstract—We demonstrate that several techniques based on waveform cross-correlation are able to significantly reduce the detection threshold of seismic sources worldwide and to improve the reliability of arrivals by a more accurate estimation of their defining parameters. A master event and the events it can find using waveform cross-correlation at array stations of the International Monitoring System (IMS) have to be close. For the purposes of the International Data Centre (IDC), one can use the spatial closeness of the master and slave events in order to construct a new automatic processing pipeline: all qualified arrivals detected using cross-correlation are associated with events matching the current IDC event definition criteria (EDC) in a local association procedure. Considering the repeating character of global seismicity, more than 90 % of events in the reviewed event bulletin (REB) can be built in this automatic processing. Due to the reduced detection threshold, waveform cross-correlation may increase the number of valid REB events by a factor of 1.5–2.0. Therefore, the new pipeline may produce a more comprehensive bulletin than the current pipeline—the goal of seismic monitoring. The analysts' experience with the cross correlation event list (XSEL) shows that the workload of interactive processing might be reduced by a factor of two or even more. Since cross-correlation produces a comprehensive list of detections for a given master event, no additional arrivals from primary stations are expected to be associated with the XSEL events. The number of false alarms, relative to the number of events rejected from the standard event list 3 (SEL3) in the current interactive processing—can also be reduced by the use of several powerful filters. The principal filter is the difference between the arrival times of the master and newly built events at three or more primary stations, which should lie in a narrow range of a few seconds. In this study, one event at a distance of about 2,000 km from the main shock was formed by three stations, with the stations and both events on the same great circle. Such spurious events are rejected by checking consistency between detections at stations at different back azimuths from the source region. Two additional effective pre-filters are *f*–*k* analysis and F_{prob} based on correlation traces instead of original waveforms. Overall, waveform cross-correlation is able to improve the REB completeness, to reduce the workload related to IDC interactive analysis, and to provide a precise tool for quality check for both arrivals and events. Some major improvements in automatic and interactive processing achieved by cross-correlation are illustrated using an aftershock sequence from a large continental earthquake. Exploring this sequence, we describe schematically the next steps for the development of a processing pipeline parallel to the existing IDC one in order to improve the quality of the REB together with the reduction of the magnitude threshold.

Key words: Waveform cross correlation, Seismic monitoring, Array seismology, IDC, CTBT.

The views expressed in this paper are those of the authors and do not necessarily reflect the views of the CTBTO Preparatory Commission.

[1] Comprehensive Nuclear-Test-Ban Treaty Organization, Vienna, Austria. E-mail: ivan.kitov@ctbto.org

1. Introduction

The comprehensive nuclear-test-ban treaty (CTBT) obligates each state party not "to carry out any weapon nuclear test explosion or any other nuclear explosion". The technical secretariat (TS) of the Comprehensive Nuclear–Test-Ban Treaty Organization will monitor the CTBT. The International Data Centre is an integral part of the (currently provisional) TS. It receives, collects, processes, analyses, reports on and archives data from the international monitoring system. The IDC is responsible for automatic and interactive processing of the IMS data and for standard IDC products. The IDC is also required by the treaty to progressively enhance its technical capabilities (comprehensive nuclear-test-ban treaty (1996), Protocol, part 1, paragraph 18(b)).

The methods based on cross-correlation have recently shown the possibility of significant improvements in many seismological applications such as detection of low-magnitude seismic events (Gibbons and Ringdal, 2006; Harris and Paik, 2006; Schaff, 2008; Schaff and Waldhauser, 2010), location of seismic events (Schaff *et al.*, 2004; Schaff and Waldhauser, 2005; Richards *et al.*, 2006; Schaff and Richards, 2011; Selby, 2010; Waldhauser and Schaff, 2008), event size characterization (Schaff and Richards, 2011) and event clustering (Baisch *et al.*, 2008; Harris and Dodge, 2011). All these improvements are of crucial importance to IDC routine processing, both automatic and interactive, and also for expert technical analysis of specific events as provided for under the Treaty. In this study, we consider building and testing of a new processing pipeline where phase arrivals are detected using cross-correlation with existing waveforms as measured by IMS array stations. From original waveforms, we estimate standard IDC parameters (e.g., slowness, azimuth, amplitude, period, etc.) as they are required for interactive processing. We also estimate RMS amplitudes of all master signals and the signals in the cross-correlation windows associated with the detections. For cross-correlation traces, i.e., the multichannel time series represented by cross-correlation coefficients (CC), F statistics is estimated. Standard frequency-wavenumber (*f–k*) analysis applied to the CC-traces provides (pseudo-) azimuth and slowness estimates which effectively eliminate spurious detections. (The dependence between true [from waveforms] and pseudo [from CC-traces] azimuth/slowness estimates for a slave signal is not linear except in a small area around the master origin beam.) At the stage of event building, we associate as many qualified detections as possible with events matching specific quality criteria. For this, the arrival times at associated stations are reduced by theoretical travel times from the master events. For a given master, any tight (a few seconds) set of the estimated origin times at three or more stations creates an event hypothesis. This stage also includes conflict resolution between event hypotheses built by different masters. Thus, the tentative cross-correlation pipeline consists of detection, phase characterization/identification, and event building in automatic processing. In order to estimate the performance of the new pipeline, all events built using cross-correlation have to be reviewed by analysts in accordance with the IDC rules of interactive analysis.

The cross-correlation technique can be a powerful tool for the detection of similar signals, but depends on the type of seismic station. For 3-C stations, short signals with similar shapes on vertical channels may come from any azimuth and may have any slowness. The use of all three components puts some constraints on the difference between azimuth and slowness for two signals to have a high correlation coefficient, but P-waves are characterized by incident angles close to zero with horizontal components having much smaller amplitudes, often below the microseismic noise level. The horizontal components do not help much for signals with low SNR. Overall, the sensitivity of a 3-C station is lower than that of a standard array station due to higher influence of ambient seismic noise. For a given level of signal, the estimates of azimuth and slowness at a 3-C station are characterized by a lower accuracy and larger uncertainty.

For array stations with many individual vertical and 3-C sensors at distances from few hundred metres to tens of kilometres, a high-correlation coefficient between two multichannel signals in many cases manifests similar vector slowness. Two signals from different azimuths and/or with different apparent velocities across the arrays have different time delays at individual sensors relative to the reference channel. When the difference between time delays at a given channel is large enough (large azimuth and/or slowness difference), the cross-correlation coefficients at this channel are suppressed by destructive interference, even for similar shapes. As a result, the overall cross-correlation coefficient (in this study, this is the coefficient averaged over individual channels) is also close to zero. Therefore, a higher cross-correlation coefficient between a given waveform and some signal from a master event is a good indicator that there is a phase similar to that from the master event and that this phase belongs to a source near the master event.

Reliable detections obtained with the cross-correlation technique together with efficient phase identification and event building applications, have been used to develop an independent automatic processing pipeline and to assess its performance. This technique allows a flexible approach to time windows, frequency bands, cross-correlation thresholds, and other parameters controlling the number and specific characteristics of detections. An optimally filtered set of detections should result in a consistent and reliable list of automatically built events, which we call cross correlation standard event list, XSEL. By design, the XSEL consists of events matching the EDC, i.e., the events ready to be included in the REB, and the portion of bogus events should be as small as possible.

An effective way to test waveform cross-correlation for the purposes of the International Data Centre is to use a set of events which are close in time and space. A natural candidate is the aftershock sequence of a shallow event. After catastrophic earthquakes, aftershocks are distributed over a wide area and their signals are not necessarily well correlated, whereas swarms and aftershock sequences of small and moderate events usually include many small events not recorded at teleseismic distances. It is therefore preferable to choose a sequence of an intermediate size. For an earthquake in China which occurred on 20 March 2008 and its aftershocks (except one), we used the vertical channels of several IMS array stations to calculate continuous time series of cross-correlation coefficients for a five-day interval, including the day of the main shock. Then, all detections obtained by cross correlation were used to build all possible events according to the event definition criteria. Many of the events built by cross-correlation were not found in the REB. These events were reviewed by an experienced analyst according to IDC rules and guidelines. After the review, many of these new events were ready to be added to the existing REB. Since the published REB cannot be modified, these new events are not actually added to the official IDC product.

The remainder of this paper consists of three sections and a Conclusion. Section 2 briefly presents the current scheme of automatic processing at the IDC. Selected products of the IDC are described, and the geographical distribution of seismic events in the reviewed event bulletin is analyzed. It also describes a tentative set of processes and procedures as related to cross-correlation, and evaluates the potential improvements in the performance of automatic processing.

Section 3 introduces several basic procedures and parameters associated with the implementation of the cross correlation technique at the IDC and illustrates these procedures using two announced underground nuclear tests conducted by the Democratic People's Republic of Korea (DPRK) in 2006 and 2009. In Sect. 4, we present a tentative version of the new processing pipeline as applied to the aftershock sequence of the earthquake in China. Since the cross-correlation technique is constrained to distances of tens of kilometers between events and does not depend on detections from remote events, all results obtained from the analysis of this aftershock sequence can be extrapolated to the global level. The globe should be divided into a large number of intersecting cells with an optimal but small number of master events inside each cell.

In the "Conclusion", we touch upon some prospective methods which will be able to optimize calculations and reliability of the events built using cross-correlation and estimate the level of computing capacity necessary for real-time processing.

2. *Automatic Processing at the IDC and the Reviewed Event Bulletin*

According to (Coyne *et al.*, 2012) the objective of seismic processing at the IDC is to produce bulletins that describe the seismic events that occurred in a given period. (We ignore here the infrasound and hydroacoustic components of the IMS which are indispensable parts of the joint seismic/infrasound/hydroacoustic automatic processing and interactive analysis.) At first, a series of standard event lists (SEL) is produced by automatic processing. Then analysts review and modify the generated automated event bulletin as necessary. The quality of the reviewed events is additionally checked, and all qualified events are placed in the reviewed event bulletin which is the final product of the IDC. It contains a list of events with their origin times, coordinates, depths, magnitudes, the associated uncertainties and other characteristics. A large number of these characteristics are calculated in post-location (automatic) processing.

Automatic processing includes station-based processing followed by network processing. Station processing is aimed at detection of appropriate signals and estimation of their parameters, including phase identification. Network processing has to build events from the detections obtained at many stations. Event building encompasses event location: the arrival times and vector slownesses of detections (with their estimated uncertainties) at several stations have to fit the same source position and origin time. In addition, the arrivals associated with the same event have to fit the magnitude criteria as well (Coyne *et al.*, 2012).

At the level of station processing, the following steps are most relevant to this study: signal detection, signal azimuth and slowness estimation, and phase identification. At the IDC, detection is based on the ratio of the short-term energy to the long-term energy (STA/LTA) and the azimuth/slowness estimation uses frequency–wavenumber analysis for array stations or polarization analysis for 3-C stations. In the new pipeline, we would like to replace some (but not all) of the currently implemented detection/characterization procedures and techniques with those based on cross-correlation. For that purpose, a number of REB events will be treated as master events. All seismic phases detected at IMS primary array stations from these master events, which should pass a rigorous quality check, will be used as waveform templates. The length of these templates depends on source-station distance and frequency band of the filters applied to original waveforms. Since the IDC is interested in the smallest possible events worldwide, waveform templates should be mainly represented by teleseismic P-wave signals, which are quite short (say, 2 to 4 s) for small events. When one or a few stations are available at regional distances from the event, Pn-wave and other regional phases can be included in the list of templates depending on the regional wavefield. For regional phases, waveform templates might be from 10 to 60 s in length. For local and regional networks, it was demonstrated in numerous studies (e.g., Gibbons and Ringdal, 2006; Schaff and Waldhauser, 2010) that cross-correlation can reduce the detection threshold by approximately one unit of magnitude, i.e., by a factor of ten in terms of amplitude. Effectively, cross correlation is able to detect signals below the noise level. At teleseismic distances, cross-correlation uses shorter time windows and one cannot expect the same fall in the detection threshold.

Before the new pipeline can partly (or fully) replace the current one, routine detection and azimuth/slowness estimation together with other tasks of station processing (data quality, improving the SNR of the data, determining amplitude and period of the signal, writing detection beams, phase association, determining the type of signal, grouping the signals at each station and identifying the phases, and location of single station events) might remain untouched. The only difference could be that the set of detections obtained by these routine procedures is reduced by those obtained by cross correlation where they have similar arrival time and other defining parameters. The reason for the exclusion of the XSEL arrivals from the IDC routine processing is the superiority of cross-correlation in detection, phase identification, and azimuth/slowness estimation. However, when two signals obtained from routine and cross-correlation procedures have similar arrival times but quite different azimuth and slowness one might use both in the relevant processing type. Since cross-correlation is very sensitive to the position of master events, it will likely miss signals from the areas not yet covered by the master events.

Network processing attempts to associate a number of arrivals with a common event and then locates it. The procedure then seeks to associate additional arrivals with a given event hypothesis before calculating a new location estimate and then testing the phases already associated for consistency with the new location. This process is iterated before some predefined convergence criteria are matched or stops when the number of iterations reaches its limit and the null hypothesis is rejected. The obtained locations allow converting amplitudes and periods of the arrivals in magnitudes of corresponding events, and thus estimating the event sizes. At the IDC, the Global Association (GA) applications perform all of the tasks of network processing simultaneously in several source regions and time intervals; conflicts between time intervals and regions are then resolved (Coyne *et al.*, 2012). For phase association, a grid-based event search procedure is applied, with the grid of cells covering the entire surface of the earth plus depth zones at which events are known to occur based upon historical observations. For each grid cell, a null hypothesis for an event is formulated and all arrivals at the nearest stations are tested against the null. When an arrival at the nearest stations matches the hypothesis for a given cell it is considered as a driver arrival. Using the hypothetical driver event, the GA predicts the arrival times at other seismic stations. All phases within the network and the station specific predefined uncertainty bounds around the predicted arrival times, azimuth, and slowness are associated with the driver event. In automatic processing, there are several event definition criteria to be met for the event hypothesis to be confirmed. These criteria are related to the number of primary stations (one or two depending on the presence of regional phases) with arrival times and azimuth/slowness estimates within the station and cell specific uncertainty bounds. Statistically, the event definition criteria define the probability of event detection at a given rate of false alarms.

Skipping many details of network processing, we would like to highlight some major features of the GA applications which can be extended or enhanced by cross correlation. The grid search covers the areas with known historical seismicity, i.e., mainly the areas with REB events. To some extent, the usage of an optimally selected subset of the REB as master events for cross correlation is equivalent to the grid search.

For a given station, most of detected signals are associated with relatively small events. These signals are weak and have poor estimates of the parameters used by the GA. It should be noted that small events are the major challenge under a comprehensive treaty—the IDC has to build all events which meet the events definition criteria for a given set of IMS data. Larger events are also of concern but they are usually easy to build and locate automatically. Cross-correlation provides a more reliable detection and azimuth/slowness estimation of smaller signals and thus a more reliable association with the driver events. The cross-correlation technique is very efficient in screening out all signals not matching the arrival time (Bobrov and Kitov, 2011) and vector slowness for a given master event (Gibbons and Ringdal, 2012). Therefore, the GA (or its local equivalent described in this paper) will have a much smaller, very reliable, and well-prepared set of detections for network processing which will allow reducing both monitoring threshold and computational requirements. For station processing, however, computation may significantly increase. The number of master events needed for a comprehensive station and network processing can be estimated from the spatial event density in the REB and the cross-correlation dependence on distance between events.

By September 2011, the REB archive contained more than 335,000 events worldwide. Some of these events are characterized by a non-zero depth, which is an effective criterion to screen out the events of natural origin; explosions are likely to be conducted at very shallow depths. There are approximately 250,000 shallow events with depths less than 50 km. Obviously, the level of the cross-correlation coefficient depends on the distance between two events: the larger the distance the lower the expected correlation coefficient. As shown in Sect. 4, for small and medium-size events, one should not expect any significant cross correlation at distances beyond 50 to 100 km. Thus, 50 km is a conservative estimate of the largest distance where cross correlation might be viable. We have calculated distances between all pairs of events from the 250,000 shallow events and constructed two important dependences on distance.

First, we have calculated the number of events which have at least one other event in various distance ranges: less than 10, <20, <30, <40, and <50 km. There is also an open-end bin for those events which have the closest neighbour beyond 50 km. We consider all events in the latter group as "isolated" events, i.e., the events without good cross-correlation with any of the REB events. As mentioned above, the 50-km threshold is a tentative one and should be re-estimated according to the dependence of the correlation coefficient between all REB events on distance and the desired rate of false alarms, as obtained for the detections using cross correlation. The second distribution counts all pairs of events in the same distance ranges. The total number of such pairs may exceed the total number of events. Table 1 lists both distributions. Three figures are most important: the total number of "isolated" events, the number of events having another REB event in the range below 50 km, and the total number of pairs in the same range.

The total number of "isolated" events (3,618 from 250,000 or 1.4 %) is a proxy to the expected rate of events which cannot be built by cross-correlation, since there is no master event within 50 km. Such events have to be built under the current framework of automatic and interactive processing, where detections are associated according to their arrival times, azimuths and slownesses obtained by

Table 1

Total number of events and pairs in various distance ranges

Range (km)	<10 km	<20 km	<30 km	<40 km	<50 km	>50 km
# Events	212,552 (85 %)	236,106 (94 %)	242,412 (97 %)	245,031 (98 %)	246,382 (98.6 %)	3,618 (1.4 %)
# Pairs	2,275,244	7,265,346	13,774,450	21,434,240	30,048,847	–
# Pairs/event	10.7	30.8	56.8	87.5	122.0	–

standard methods. At the same time, approximately 99 % of REB events would have been accurately built using cross correlation with one of the existing REB events. These are only crude estimates extrapolating the past seismicity into the future. However, with any new REB event the coverage of the globe with master events will increase and 3,618 currently isolated events may then find a pair within 50 km. The number of events having at least one event from the REB in the range below 50 km provides a very conservative estimate. Apparently, the number of events having at least one event at distance less than 50 km is 246,382.

Finally, the number of pairs spaced less than 50 km apart is a proxy to the density of events. For 246,382 events we have 30,048,847 pairs, or ~122 pairs per event on average. This value allows crude estimation of the number of templates necessary to cover all REB events. Taking into account all 3,618 isolated events, ~2,100 (≈246,382/122) templates are used to cover all other events. When the correlation radius decreases, the number of necessary templates should increase.

Figure 1 depicts the spatial distribution of all 250,000 REB shallow events (yellow circles) and the isolated events (red circles). The overall pattern is well-known—an overwhelming majority of earthquakes is associated with plate tectonics: subduction, mid-oceanic ridges, and transform faults, i.e. seismically active zones in oceans. Within continents, even seismic events located at shallower depths are mainly associated with deeper portions of subducting plates and several active rifts.

There are two principal types of isolated event. Many events are isolated on the periphery of seismically active zones. This might be a manifestation of seismicity decaying with distance from the most active tectonic movements or the result of inaccurate location. The latter case is of importance for the cross-correlation technique—the level of mislocation can be easily reduced if these events were actually close to other REB events. Therefore, one needs first to test the isolated events for cross-correlation with, say, all events situated within their confidence ellipses. As a result, the number of isolated events around the seismically active zones will be dramatically reduced. This is a crucial methodological procedure to apply before one selects the optimal set of master events with their waveform templates at primary stations.

By definition, the set of master events has to optimally cover all seismically active areas and also include all events which have been proven to be isolated in terms of cross-correlation. Hence, any new (and also past) REB event which is either in these areas or close to the isolated events has to demonstrate a predefined level of cross-correlation with waveform templates for at least one master event. By putting some optimal thresholds for the correlation coefficients at a predefined number of primary stations (currently, three primary stations are necessary to define a valid REB event) we can automatically build a new event. In other words, we assume that the higher correlation coefficient between waveforms from the master event and the event under study at three or more primary stations guarantees the existence of the new event and its closeness to the master event in space at a predefined level of confidence. All involved thresholds and confidence levels have to be estimated from the REB and then are subject to re-estimation in line with new information.

Waveform templates for a master event may include various types of station. The seismic part of the international monitoring system consists of primary and auxiliary stations. The former stations are used to create initial hypotheses on seismic events which then can be corroborated and improved using data from auxiliary stations. The primary seismic

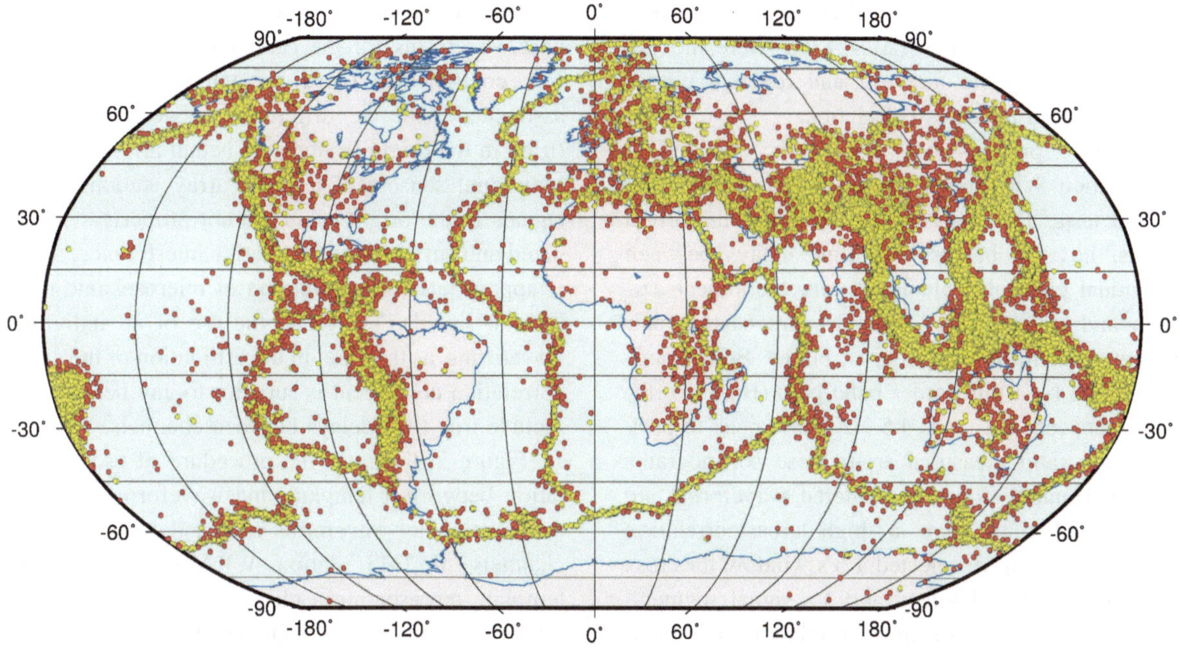

Figure 1
Map of 3,618 isolated REB events (*red circles*) on *top* of 250,000 shallow events (*yellow circles*)

stations are characterized by a continuous data flow and the auxiliary stations deliver data by request. Both primary and auxiliary stations can be either 3-C or array. Most primary stations are arrays which allow for a higher resolution of slowness and azimuth for a given P-wave arrival. (Here we consider only P-waves, which are the only type of waves recorded at teleseismic distances from small seismic events.)

In the next section, we describe the procedures for calculation of cross-correlation coefficient and several important parameters of the signals detected by cross correlation. These procedures and parameters are tested using two announced underground nuclear explosions conducted by the DPRK in 2006 and 2009.

3. *Cross Correlation and Related Techniques*

Following Gibbons and Ringdal (2006), we use a normalized cross-correlation function. Both time series must have the same sample rate. This condition seems to be trivial for a single seismic station but when decimation is used for reduction of the overall computation time one should be careful to use the same rate. The notation $\omega_{N,\Delta t}(t_0)$ is used to denote the discrete vector of N consecutive samples of a continuous time function $\omega(t)$, where t_0 is the time of the first sample and Δt is the spacing between samples:

$$\omega_{N,\Delta t}(t_0) = [\omega(t_0), \omega(t_0 + \Delta t), \ldots, \omega(t_0 + (N-1)\Delta t)]^{\mathrm{T}}.$$

The inner product of $\upsilon_{N,\Delta t}(t_\upsilon)$ and $\omega_{N,\Delta t}(t_\omega)$ is defined by

$$\langle \upsilon(t_\upsilon), \omega(t_\omega) \rangle_{N,\Delta t} = \sum_{i=0}^{N-1} \upsilon(t_\upsilon + i\Delta t)\omega(t_\omega + i\Delta t),$$

and the normalized cross-correlation coefficient by

$$\mathrm{CC}[\upsilon(t_\upsilon), \omega(t_\omega)] = \frac{\langle \upsilon(t_\upsilon), \omega(t_\omega) \rangle_{N,\Delta t}}{\sqrt{\langle \upsilon(t_\upsilon), \upsilon(t_\upsilon) \rangle_{N,\Delta t} \langle \omega(t_\omega), \omega(t_\omega) \rangle_{N,\Delta t}}}.$$

A good example of cross-correlation as applied to seismic monitoring of underground nuclear explosions is the comparison of two tests conducted by the DPRK in 2006 (October 9) and 2009 (May 25). The 2006 event was detected by 22 IMS stations and built as an REB event with m_b (IDC) = 4.1. The 2009 event was detected by 59 IMS stations and had body wave magnitude m_b (IDC) = 4.5.

Unfortunately, the closest IMS stations USRK ($\Delta \sim 3.6°$) and KSRS ($\Delta \sim 4.0°$) were not in operation during the first test and regional phases cannot be used for cross-correlation.

Figure 2 presents two waveforms obtained at IMS station WRA from the 2009 and 2006 underground tests. These waveforms represent the origin beams, i.e., the beams with time delays between individual channels calculated with theoretical azimuth and slowness for a given source/station pair corrected for historically known biases. Both beams are filtered by a third-order band pass (Butterworth) filter between 0.8 and 4.5 Hz. The peak-to-peak amplitude ratio measured from these seismograms is approximately four. The filtered waveforms are similar and thus have a high cross-correlation coefficient. We have selected a 6 s window for cross correlation with a 1-s lead and 5-s signal segment. The estimated value of cross-correlation coefficient is 0.85.

The origin beams, however, do not provide the best waveforms for the estimation of cross-correlation coefficients. There are tangible beam losses associated with the difference between theoretical (used in the origin beams) and actual arrival times at individual sensors of a given array station. Actual signals in the origin beam are not properly synchronized and suffer some destructive interference, which is appropriate for suppression of microseismic noise but not signals. Therefore, the use of all individual waveforms, as they are, in the estimation of the cross-correlation coefficient is superior to any beam—they include true time delays between channels.

Figure 3 illustrates the procedure of cross-correlation between a template and waveform. Here, we use continuous waveforms recorded by 24 vertical channels of IMS station WRA and a 6-s-long template representing a clear signal. One cross-correlation coefficient is calculated over the entire template with all channels aligned (in the order of

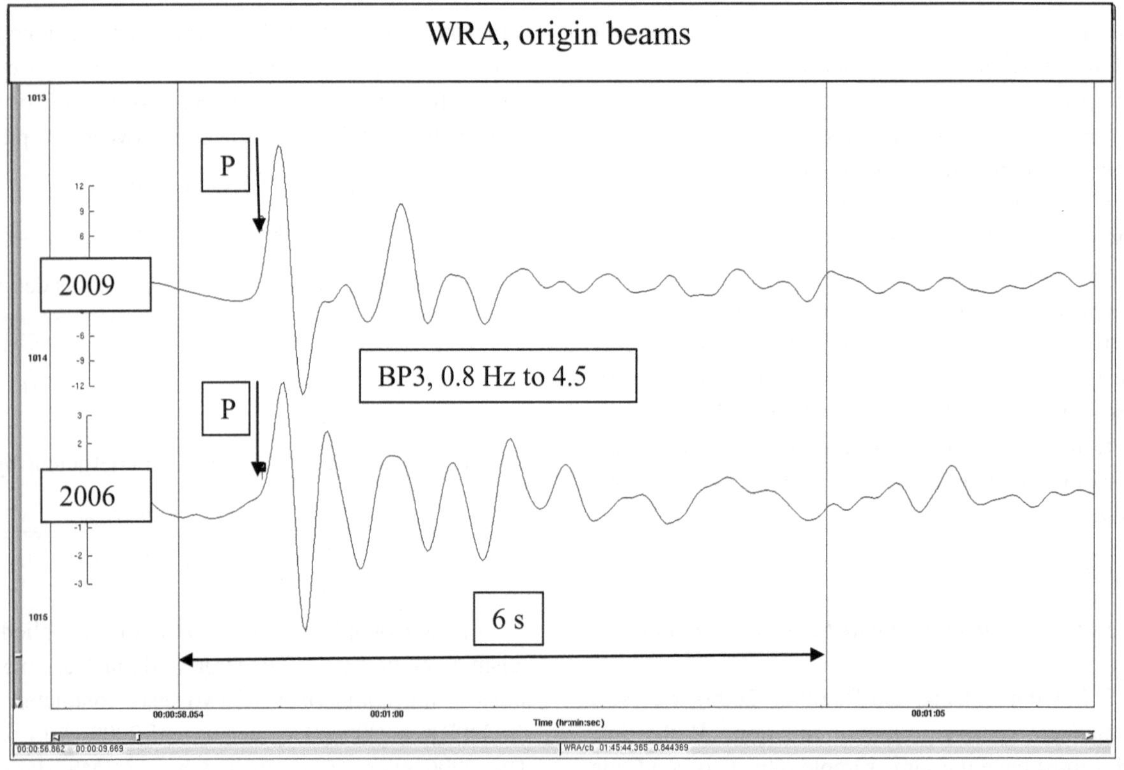

Figure 2
Comparison of two origin beams at station WRA from the 2006 and 2009 tests conducted in the DPRK. Both beams are filtered by a third-order band pass filter between 0.8 and 4.5 Hz. The filtered waveforms are similar. We have selected a 6-s window for cross correlation with a 1-s lead and 5-s signal segments

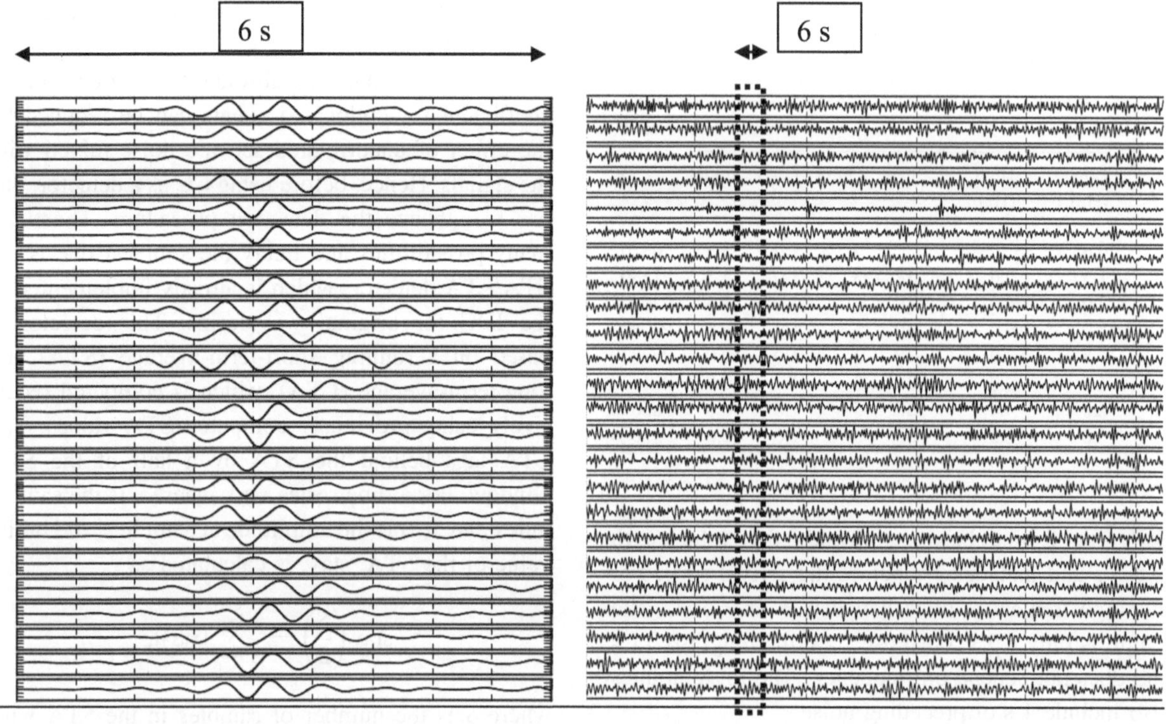

Figure 3
IMS seismic array WRA, a template and continuous waveforms. One correlation coefficient is calculated over the entire template with all channels aligned in one record. In the template, individual channels are shifted in time according to theoretical travel time residuals defined by azimuth and slowness of the origin (source/receiver) beam. All continuous waveforms are also shifted by the same time delays between individual channels

station names) in one time series of 24 × 6 s. This coefficient is associated with the absolute time of the first point in the waveform (usually referred to the reference station of the array under investigation). In the template, individual channels are shifted in time according to theoretical travel time residuals defined by azimuth and slowness of the origin (source/receiver) beam. (As clear from the figure, these theoretical time delays do not guarantee signal synchronization between channels.) In the cross-correlated waveform, all individual channels are also shifted by the same time delays as in the template. Hence, the empirical (true) time delays between the channels are retained when the template is correlated with the waveforms.

The length of seismic signal recorded at teleseismic distances depends on magnitude: smaller events are characterized by shorter visible signals. In addition, smaller earthquakes produce signals enriched by higher frequencies due to higher corner frequencies of their sources. Here, we are looking for the smallest events which are likely to be missed by standard detection algorithms and event building tools used at the IDC. Therefore, the length of template windows should not be large, and should only include valid signals from small and moderate-size events. The difference in frequency content of microseismic noise and signals at IMS stations requires a number of filters covering the whole spectral range of seismic signals from 0.8 to 6 Hz.

Table 2 lists time windows and frequency bands of eight templates used in this study. All templates include several seconds of P- or Pn-wave signal and a short time interval before the signal (lead), which provides additional flexibility in onset time. (At the initial stage of our study, we do not include the PKP phase, which is a primary phase at distances beyond ~115° and might be of importance for specific seismic regions.) The length of a given window depends on its frequency band. For the low-frequency

Table 2

Time windows and frequency bands of the templates for P- and Pn-waves

Phase	Filter				Window (s)		Name
	Low (Hz)	High (Hz)	Type	Order	Lead	Signal	
P	0.8	2.0	BP	3	1.0	5.5	P0820
P	1.5	3.0	BP	3	1.0	4.5	P1530
P	2.0	4.0	BP	3	1.0	3.5	P2040
P	3.0	6.0	BP	3	1.0	3.5	P3060
Pn	0.8	2.0	BP	3	1.0	10.0	Pn0820
Pn	1.5	3.0	BP	3	1.0	10.0	Pn1530
Pn	2.0	4.0	BP	3	1.0	10.0	Pn2040
Pn	3.0	6.0	BP	3	1.0	10.0	Pn3060

(BP, order 3) filter between 0.8 and 2.0 Hz, the length is 6.5 s and includes 1 s before the arrival time. For the high-frequency filter between 3 and 6 Hz, the length is only 4.5 s. For Pn-waves, the length is 11 s and does not depend on frequency. The Pn templates also include 1 s of preceding noise.

When a time series of cross-correlation coefficients is calculated for a given interval, say 2 h, one can apply signal detection algorithms. There is no theoretically justified unique CC threshold, CC_{tr}, used to define a new arrival; rather appropriate thresholds should be determined empirically. These thresholds are likely to be station dependent and vary with geographical coordinates and source depth. For shallow events, free surface reflections may introduce varying interference patterns.

For two neighbouring events, the level of cross-correlation coefficient depends on the distance between them and the similarity of source functions and focal mechanisms, as well as upon signal frequency. As a rule, the larger the distance, the lower the corresponding |CC| as caused by degrading coherency of signals on various channels. The similarity of source functions can also deteriorate with the difference in magnitude, especially for short time windows used in our templates. Shallow earthquakes usually generate emergent signals. For larger events, an early (and different in shape) part of the signal from the same location may emerge from the noise and thus be used in corresponding templates, while it is not seen and used from smaller events. As a result, the level of cross-correlation may decrease even for collocated events.

For weak signals, the absolute level of correlation coefficient for collocated events can be reduced by the effect of uncorrelated seismic noise mixed with the signals. Hence, before using CC as a detector, one has to enhance the detection procedure. There are many possibilities and likely the simplest one is the STA/LTA detector, which is already implemented at the IDC for original waveforms. This detector is based on a running short-term-average (STA) and long-term-average (LTA), which is computed recursively using previously computed STA values. The LTA lags behind the STA by a half of the STA window. For a time series $x(n)$, where n is the sample index and $x(n)$ is the amplitude at sample n, the initial value of the STA, stav, is calculated as

$$\mathrm{stav}\left(\frac{S}{2}\right) = \frac{1}{S}\sum_{s=0}^{S-1}|x(s)|,$$

where S is the number of samples in the STA window. Recursion is used to compute consequent values of the STA:

$$\mathrm{stav}(k) = \mathrm{stav}(k-1) + \frac{1}{S}\left[\left|\left(x(k+\frac{S}{2}\right)\right| - \left|\left(x(k-1-\frac{S}{2}\right)\right|\right],$$

where $(S/2) \le k \le (N-1) - (S/2)$, and N is the number of available samples in the time series. For the end-segment intervals, $k \le S/2$ and $\ge (N-1)-(S/2)$, $\mathrm{stav}(k){=}\mathrm{stav}(S/2)$ and $\mathrm{stav}(k){=}\mathrm{stav}((N-1)-S/2)$, respectively.

The LTA, ltav(k), is computed recursively from the previous STA:

$$\mathrm{ltav}(k) = \left(1-\frac{1}{L}\right)\mathrm{ltav}(k-1) + \frac{1}{L}\mathrm{stav}(k-S),$$

where L is the number of samples in the LTA window. The length of the STA and LTA windows have to be defined empirically as associated with spectral properties of seismic noise and expected signal. We have carried out a brief investigation and determined the following windows: 0.8 s for the STA and 20 s for the LTA.

Figure 4 illustrates the STA/LTA detection process. In the upper panel, a CC time series is shown as

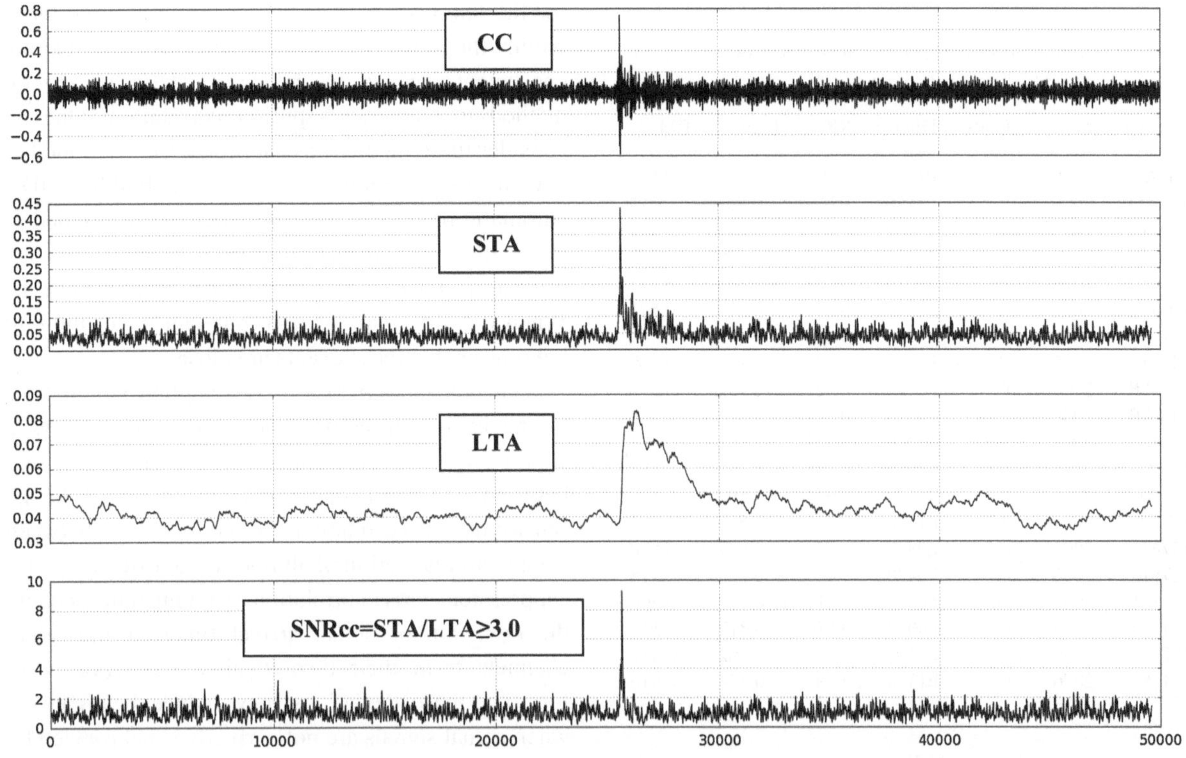

Figure 4
An example of cross-correlation analysis as applied to the 2009 and 2006 DPRK tests

obtained by cross-correlation of the template of the 2009 DPRK event and a 42-min time window centered at the arrival of the 2006 DPRK test at station AKASG. The next two panels demonstrate the relevant STA and LTA. In the lower panel, the STA/LTA ratio is shown which defines the signal-to-noise ratio, SNR_{CC}. A valid signal is detected when the level of SNR_{CC} is above 3.0. This is a tentative but conservative threshold. Before we gather a statistically significant set of arrivals and test them manually it would be premature to reduce the detection threshold from cross-correlation traces. It is worth noting that for original waveforms, a valid signal usually has SNR > 2.0. In this study, we have found that the CC detector can find a valid signal with standard SNR $= 0.7$. For a longer template, Schaff (2008) showed that it is possible for cross-correlation to detect a signal with SNR $= 0.32$. The latter value is put as a preliminary SNR threshold as calculated from the original waveforms. All in all, there is only one valid signal in Fig. 4, which undoubtedly belongs to the 2006 event.

Seismic measurements are not always perfect, and a researcher may face numerous data problems such as spikes, gaps, high noise at a few channels, and wrong polarity in actual waveforms. Among these problems, only wrong polarity is not a challenge for the estimates of cross-correlation coefficient, when the polarity does not change over time. Spikes and short gaps (up to five points) in data can be effectively suppressed and recovered by various interpolation methods used at the IDC (Coyne *et al.*, 2012). Longer data gaps are usually masked and the relevant readings are not used in automatic IDC processing. For cross-correlation, these segments of masked data may introduce artificial steps in the CC time series due to a sudden change in the number of used channels. Such steps might be wrongly interpreted by the STA/LTA detector as signals. In order to reduce the influence of these masked data gaps we have introduced a cross-correlation coefficient averaged over all working channels, which is called beam CC, BCC.

Table 3

Cross-correlation coefficients at IMS primary stations for the 2006 and 2009 tests used as master events

Stations	Master	Phase	Filter	|CC|	logα	RM
AKASG	2009	P	P1530	0.666	−0.740	0.564
AKASG	2006	P	P1530	0.674	0.393	−0.564
ASAR	2009	P	P1530	0.660	−0.652	0.471
ASAR	2006	P	P1530	0.674	0.300	−0.472
GERES	2009	P	P1530	0.571	−0.818	0.575
GERES	2006	P	P1530	0.549	0.311	−0.571
MJAR	2009	Pn	Pn2040	0.677	−0.479	0.309
MJAR	2006	Pn	Pn2040	0.685	0.145	−0.309
MKAR	2009	P	P0820	0.522	−0.952	0.670
MKAR	2006	P	P0820	0.517	0.374	−0.661
NOA	2009	P	P2040	0.757	−0.602	0.481
NOA	2006	P	P2040	0.758	0.365	−0.485
NVAR	2009	P	P1530	0.957	−0.590	0.571
NVAR	2006	P	P1530	0.956	0.553	−0.573
PDAR	2009	P	P0820	0.759	−0.697	0.578
PDAR	2006	P	P0820	0.753	0.455	−0.578
SONM	2009	Pn	Pn0820	0.617	−0.719	0.509
SONM	2006	Pn	Pn0820	0.630	0.317	−0.518
WRA	2009	P	P1530	0.903	−0.462	0.417
WRA	2006	P	P1530	0.907	0.374	−0.416

There are several IMS primary stations which detected both DPRK events. We have calculated cross-correlation coefficients for both events as master ones. There is a slight difference in time delays between individual channels in the relevant waveform templates. These delays are responsible for the difference between cross-correlation coefficients when the master and slave events are swapped. Table 3 lists all peak |CC| values for both tests and proves that the second event could be easily built automatically using cross correlation in line with the distance of several kilometers between the relevant IDC locations. Moreover, the cross-correlation detector has demonstrated its efficiency by finding an additional P-wave arrival at station ARCES from the 2006 events which had been missed in automatic and interactive processing. The 2009 template for ARCES has been used.

Generally, a higher cross-correlation between two waveforms at one station is a very reliable indication of the spatial closeness between their sources. However, there are a few cases when cross-correlation is high for distant events, with the master event having a much smaller magnitude than the slave event obtained by cross correlation: the slave signal can be long enough for the template to find (by chance) a specific place within the slave waveform where cross-correlation coefficient exceeds the threshold. There are several methods to remove such spurious correlations. One can consider these methods as additional filters applied to the flux of detections used for event building. IDC automatic processing uses *f–k* analysis, as applied to the original waveforms, in order to estimate the difference in slowness and azimuth between two events. When this difference is high one can reject the null hypothesis that these events are close.

Gibbons and Ringdal (2006) applied *f–k* analysis to the cross-correlation time series. This allows a significant improvement in the discrimination between valid and inappropriate signals due to the sensitivity of correlation to the distance between events, normalization of all traces, and effective noise suppression: cross-correlation is taking into account the deviation of actual arrival times at individual channels from their theoretical values. (For beam forming, these deviations may induce beam losses when actual signals are not perfectly synchronized by the theoretical time shifts). Thus, we use standard *f–k* analysis to formally estimate (pseudo-) azimuth and slowness using CC time series for all detections obtained by cross correlation. For valid signals, these estimates are very close to the theoretical values associated with the source/receiver path. Other signals are characterized by a higher uncertainty in the azimuth/slowness estimates since their cross correlation traces are just noise. As a result, this procedure allows effectively rejecting the cross-correlation detections from remote events (Gibbons and Ringdal, 2012).

In *f–k* analysis, spectra are computed from the traces of cross-correlation coefficients calculated for individual channels. For each slowness vector, the *f–k* power spectrum, $P(S_n,S_e)$, is calculated as:

$$P(S_n, S_e) = \frac{\sum_{f=f_1}^{f_2} \left| \sum_{i=1}^{J} F_i(f) \cdot e^{2\pi\sqrt{-1}f(S_n \cdot \mathrm{dnorth}_i + S_e \cdot \mathrm{deast}_i)} \right|^2}{J \cdot \sum_{f=f_1}^{f_2} \left\{ \sum_{i=1}^{J} F_i(f)^2 \right\}},$$

where $F_i(f)$ is the Fourier amplitude of the *i*th trace of cross correlation coefficient at frequency *f*, S_e and S_n are the east–west and north–south components of the vector slowness, $deast_i$ and $dnorth_i$ are the east–west and north–south coordinates, respectively, of the ith sensor array element relative to the reference station, f_1 and f_2 are the low and high frequency limits, J is the number of sensor elements in the array. The sum $S_e \cdot deast_i + S_n \cdot dnorth_i$ defines the theoretical time delay between the reference station and the ith sensor element. These delays are used in the templates. The slowness coordinates of the peaks in the f–k power spectrum are used to calculate the azimuth and slowness of the signal's spatially coherent plane wave energy. When the deviation from the master's azimuth and slowness is above some predefined thresholds, the signal under investigation is not considered for event building.

Another method of pre-selection of appropriate arrivals can be based on the ratio of norms used for cross-correlation of a master and slave events. Gibbons and Ringdal (2006) introduced an amplitude scaling factor:

$$\alpha = x \times y / x \times x,$$

where $\boldsymbol{x}$ and $\boldsymbol{y}$ are the vectors of data for the master and slave event, respectively. For two collocated events with the same source time history but different amplitudes, the amplitude scaling factor completely defines the difference in sizes. For close events with similar source functions, the amplitude scaling factor defines the least-squares solution of the equation $\boldsymbol{y} = \boldsymbol{ax} + \boldsymbol{n}$, where $\boldsymbol{n}$ is uncorrelated noise.

Schaff and Richards (2011) noticed that α is equivalent to the unnormalized cross-correlation coefficient divided by the inner product of the master waveform. For close events, they defined a relative magnitude as the logarithm of α and demonstrated a significant ($\sim$5 times) reduction in the variance in this magnitude relative to the conventional magnitude obtained from the catalogue. Hence, for detections obtained by cross-correlation, the relative scaling factor can be a more reliable screening criterion in the automatic event building than the currently used magnitude criterion. It should be noticed that the difference in station magnitudes is a very effective (dynamic) parameter which resolves a substantial proportion of the conflicts between events in the GA when they share the same phase by time, azimuth, and slowness.

For a given event, the relative scaling factors, as based on the same master event at different stations, have to be close. For the current configuration adopted in automatic processing, for a phase to be associated with a given event, the difference between station- and network-averaged magnitudes cannot exceed two units of magnitude. Considering the improvement in standard deviation of magnitude estimates obtained by Schaff and Richards (2011) for the relative scaling factor, this difference should not exceed 10 to 20 times. Since the logarithm of α is equivalent to body wave magnitude, the difference between network averaged and station values of $\log\alpha$ should not exceed 1.3.

The relative scaling factor is not the best measure of relative event sizes when their signals are different but still similar enough to have a cross-correlation coefficient above the predefined threshold. The scaling factor includes the cross-correlation coefficient which depends on the distance between events. For very close events and $|CC| \sim 1.0$, there is almost no bias introduced by CC. For the events with $|CC|$ varying between 0.2 and 1.0 because of distance the bias in $\log\alpha$ may be substantial and the overall dispersion at various stations may grow significantly. In order to reduce the influence of distance, we propose to use the ratio of norms $|x|/|y|$ instead of α. The logarithm of the ratio, $RM = \log(|x|/|y|) = \log|x| - \log|y|$, is essentially the magnitude difference (or relative magnitude, RM) between two events, where the magnitude is based on the RMS amplitude in the template time window instead of the peak-to-peak amplitude. This difference has a clear physical meaning for close events with similar waveforms, i.e., for events with a higher level of cross-correlation. It does not work for remote events because their propagation paths and source functions are quite different, and one has to introduce a standard magnitude scale. Since all variations related to CC are excluded, RM fluctuates less across the stations measuring both events than $\log\alpha$ does, as shown in Table 3 for two tests conducted by the DPRK. Hence, RM is a better dynamic parameter for discrimination between genuine and dynamically inappropriate arrivals for a given event at several stations. The decision line for RM has to be determined empirically from the whole set of REB events matching high quality criteria, say, |CC| > 0.2 and SNR_{CC} > 3.0 for all array stations.

Definitely, logα provides a more precise estimate of the relative size for similar signals with smaller SNR, especially, when the signals are observed in longer time windows. Seismic noise introduces a significant bias in the estimates of RMS amplitude and thus in the RM estimates. As a simple remedy, one can subtract the pre-signal noise RMS amplitude in a similar or longer window from that of the signal and reduce the bias. In any case, this problem needs a further investigation and we estimate both parameters in this study.

The relative magnitude can be extrapolated to the global level, since the seismicity is practically continuous in terms of cross-correlation. From Fig. 1 one can judge that almost any two events, with the exception of those isolated within continents, can be connected through a chain of a few master events. Therefore, one will be able to balance relative magnitudes RM over all chains of neighbouring master events. The extrapolation of the relative magnitude at the global level is of an extraordinary importance for seismic monitoring. When connected by quantitative relationships with the global scale of body wave magnitude, the globalized relative magnitude will allow increasing the accuracy and reducing the uncertainty of the body wave magnitude estimates worldwide. It may significantly improve the performance of the screening criterion based on the difference between m_b and M_s.

Table 3 includes the estimates of logα and RM for all stations where both events were detected. The standard deviation of logα for ten stations is 0.15 and only 0.1 for RM. In this case, the relative magnitude based on the norms of signals in (the same) template windows has a smaller variance and is preferable for discrimination in event building.

Table 4 demonstrates the extremely accurate estimates of arrival time residuals, t_{res}, azimuth residuals, az_{res}, and slowness residuals, slo_{res}. The arrival time residuals are calculated as the difference between the relevant REB arrival times and those estimated by cross correlation. The slowness and azimuth residuals are calculated as the difference between measured values and those corresponding to origin beams (Coyne *et al.*, 2012). When obtained by standard *f–k* analysis applied to cross-correlation time series instead of original waveforms, these residuals (marked as CC_*) are not genuine, i.e., not actually measured in degrees and s/deg because the cross-correlation transformation is not linear and bijective. Therefore, no direct comparison of the residuals estimated from the original and CC-traces is possible. As mentioned before, Gibbons and Ringdal (2012) proposed to use the latter estimates for the elimination of bogus arrivals. It is worth noting that all these estimates were made in automatic processing.

As a complementary study, we have carried out a thorough search for smaller aftershocks of the 2009 event. We have processed seismic data from primary stations 5 days after the event. Both announced tests were used as master events; the Pn-waves at stations USRK and KSRS from the 2009 event have been also used as waveform templates. There was no REB-ready event near the epicenter of the second test. Taking into account the level of correlation between the announced nuclear tests, we would estimate body wave magnitude of the largest possible aftershock as m_b(IDC) $\sim$ 2.5. For a bigger aftershock collocated with the 2009 event, the cross-correlation coefficient has to be above the threshold adopted in our study (|CC| > 0.2) at a few primary stations.

Tables 3 and 4 both evidence that the cross correlation and *f–k* analysis provide a reliable tool for detection of signals from close events and accurate estimates of their arrival time, azimuth and slowness. Due to the shape similarity, one can use the ratio of the RMS amplitudes in a predefined time window (and the same frequency band) as a robust and station independent characteristic of the relative event sizes. Overall, these findings provide a solid basis for automatic detection and event building.

In Sect. 4, we describe principal details of a new processing pipeline as based on cross correlation and present some preliminary results. We have carried out a feasibility study and obtained a tentative XSEL for an aftershock sequence of a large continental earthquake where we presumed no historical seismicity. Therefore, standard automatic and interactive processing is needed to populate the set of master events before any cross-correlation techniques can be applied. This might be a common feature for a few intra-continental earthquakes and likely for underground nuclear explosions. Cross-correlation might be less efficient in the areas without historical

Table 4

Travel time, azimuth and slowness residuals at IMS primary stations for the 2006 and 2009 tests used as master events

Stations	Master	t_{res}	CC_az$_{res}$	az$_{res}$	CC_slo$_{res}$	slo$_{res}$
AKASG	2009	−0.077	−0.07	−3.00	−0.20	0.25
AKASG	2006	0.075	−0.54	−3.10	0.25	0.10
ASAR	2009	−0.154	2.03	2.00	0.15	1.40
ASAR	2006	0.178	−2.65	5.70	−0.04	0.29
GERES	2009	0.210	4.46	−12.50	−0.15	0.12
GERES	2006	−0.212	−0.42	−10.60	1.33	0.12
MJAR	2009	0.147	−0.92	−5.90	−0.12	1.31
MJAR	2006	−0.137	1.93	−4.00	−0.06	−0.47
MKAR	2009	0.716	−0.46	8.60	−2.45	2.23
MKAR	2006	−0.718	3.15	6.50	−0.04	0.82
NOA	2009	−0.402	8.66	−0.10	1.55	0.24
NOA	2006	0.400	4.08	−0.80	−0.25	0.27
NVAR	2009	0.098	−0.26	−2.00	0.47	1.08
NVAR	2006	−0.100	−3.33	−5.30	−0.05	0.49
PDAR	2009	0.048	2.88	18.20	−1.52	−1.80
PDAR	2006	−0.025	−29.79	−2.40	1.25	−1.25
SONM	2009	0.558	3.53	6.20	−0.58	0.24
SONM	2006	−0.556	−1.21	4.80	1.39	−0.24
WRA	2009	0.242	−0.28	1.30	0.05	0.46
WRA	2006	−0.245	−1.03	2.10	0.05	0.37

seismicity and thus is not able to completely replace standard IDC processing.

4. Testing Cross Correlation Procedures

In Sect. 3, we have described general features of cross correlation as a detector and also introduced a few non-standard parameters for the cross-correlation detections together with the relevant estimation procedures. All standard parameters for seismic phases adopted by the IDC are also estimated for all arrivals. Two underground tests conducted by the DPRK have demonstrated the efficiency of the detector and the reliability of such parameters as arrival time, azimuth and slowness as estimated from the traces of cross-correlation coefficient as well as relative magnitude.

The next natural step is to recover a sequence of seismic events from a small area (say, from 50 to 100 km in radius) using master events. Having a flux of detections obtained by cross-correlation from a given master event or a set of master events, one has to then process them in an automatic pipeline to build a cross-correlation standard event list (XSEL). Thus, in this section we delineate the pipeline and also tune the parameters controlling the flux of detections and the quality of XSEL events. This is a feasibility study.

We have chosen a large earthquake in China that occurred at 22:32:56 on 20 March 2008. This earthquake was detected by many primary and auxiliary IMS stations and starting from the automatic location an event was built by IDC analysts with body wave magnitude m_b(IDC) = 5.41. The event had a short but prominent aftershock sequence also recorded by the IMS. There are 142 events (the main shock counted in) in the Reviewed Event Bulletin during 5 days after the earthquake and within 100 km from the main shock. Cross-correlation is a technique which requires extensive computation resources and we have limited our testing to five full days including the whole day of the main shock.

The aftershocks located by IDC provide several opportunities to test cross-correlation as a method for signal detection, phase identification, and event building. First, we must endeavor to reproduce the existing REB using cross-correlation. Secondly, we have to check the REB aftershocks for internal consistency in terms of cross-correlation and populate the list of master events by selecting those events which provide the largest number of good cross-correlations. The aftershocks in the REB without any master event, i.e., those events which do not show appropriate cross-correlation with any other aftershock in the sequence, have to be checked manually for consistency and/or for wrong phase associations. In a sense, this check for consistency is part of a quality check which should be incorporated in interactive processing. Thirdly, we should search for new events (not in the REB) in the same time slot which match the IDC event definition criteria. In short, the EDC require that an REB event has to be detected by at least three IMS primary stations with arrival time, azimuth and slowness within the station and phase-specific uncertainty bounds (Coyne *et al.*, 2012). Fourthly, we must determine those parameters of detection and event building procedures which provide the highest resolution with a predefined false alarm rate. For interactive processing, this rate should be relatively low. The success of these parameters estimations depends on many factors, including time, spatial and magnitude distribution of events. The parameters obtained for the studied aftershock sequence cannot be transported to other areas and from station to station as they are. Any individual region/station pair deserves an independent estimation of all controlling parameters.

The studied aftershock sequence was built by IDC analysts. For the purposes of our study, we have all REB events in order to select the best set of master events in order to reproduce the entire sequence. As a start point, we use all 142 REB events within 100 km of the main shock (Bobrov and Kitov, 2011). These events have magnitudes m_b(IDC) between 2.84 and 5.41 and include from 3 to 14 primary array stations. This exercise should provide a complete set of information on cross-correlation between the events created in an interactive mode in a previously unknown area. Then the best subset has to be retrieved which should be used to process data in an automatic mode to find all valid REB and new events within the correlation radius. It is presumed that all new events obtained in automatic processing have to be reviewed by analysts, before migrated into the XSEL. These new events may also be tested as potential master events.

There is a trade-off between the size of a master event and its efficiency for cross-correlation. For many earthquakes, it takes several seconds for a signal to reach its peak value as related to source mechanism and propagation path. For smaller events, the initial portion of their signals at IMS stations may not exceed the level of microseismic noise, and thus it is likely not included in waveform templates. For larger events (>5.0), the whole signal is usually above the noise level, and thus includes the initial few seconds missed in the small event's templates. The waveform template for a larger event might not find a similar signal from a collocated but much smaller event (which is below the noise level) and cross-correlation fails. At the same time, the waveform template for the smaller event does contain signals repeating some later segments from the large event.

On the other hand, larger events contain more stations and their signals have much higher SNR. Generally, they provide better templates for cross-correlation which are not spoiled by seismic noise. If the difference in shape between the template of a big event and the signal from a collocated but much smaller event is not large then the time delays between individual sensors define the level of cross correlation. In this case, the master event with large magnitudes can be effective in finding of smaller sources. In order to mitigate the risk of missing smallest events one can use collocated events with different magnitudes as master events. The only requirement for a small event to be a master one is clear waveform templates at many stations.

The primary IMS network includes many array stations and few 3-C stations. Theoretically, an array allows signal amplification proportional to the square root of the number of elements, when signals are spatially well-correlated and noise is incoherent. Therefore, array stations are more efficient in detection of weaker signals and smaller events world-wide. However, all 3-C stations are very important for detection of low-amplitude signals when no array stations are available at local and regional distances. Without loss of generality, we use only primary array stations in this study. There is no primary 3-C station at regional distances from the aftershock sequence. Auxiliary IMS stations provide data only by request; their waveforms are not continuous and thus not used.

Figure 5 depicts a map of IMS stations with their roles in the study. For the studied aftershock sequence, thirteen primary array stations at teleseismic distances reported P-wave detections: AKASG, ARCES, CMAR, FINES, GERES, KSRS, MJAR, NOA, PETK, SONM, WRA, YKA, and ZALV. There is one primary station at a distance below 18°—MKAR. It regularly reported detections of the Pn-phase. There are also two auxiliary arrays—BVAR and KURK which could be used for cross-correlation estimates. When continuous waveforms from these stations are available they can be included in templates.

Overall, we have designed the following tentative procedure for event building. As a first step, cross-correlation coefficients have to be calculated for all 142 earthquakes as master events through the entire 5-day-long records. For a given master event, CC are calculated only for time-defining P-waves (Pn-wave for MKAR) with SNR > 2.0. One aftershock with m_b(IDC) = 3.11 has three primary stations but only two P-waves with SNR > 2.0. Low-SNR is

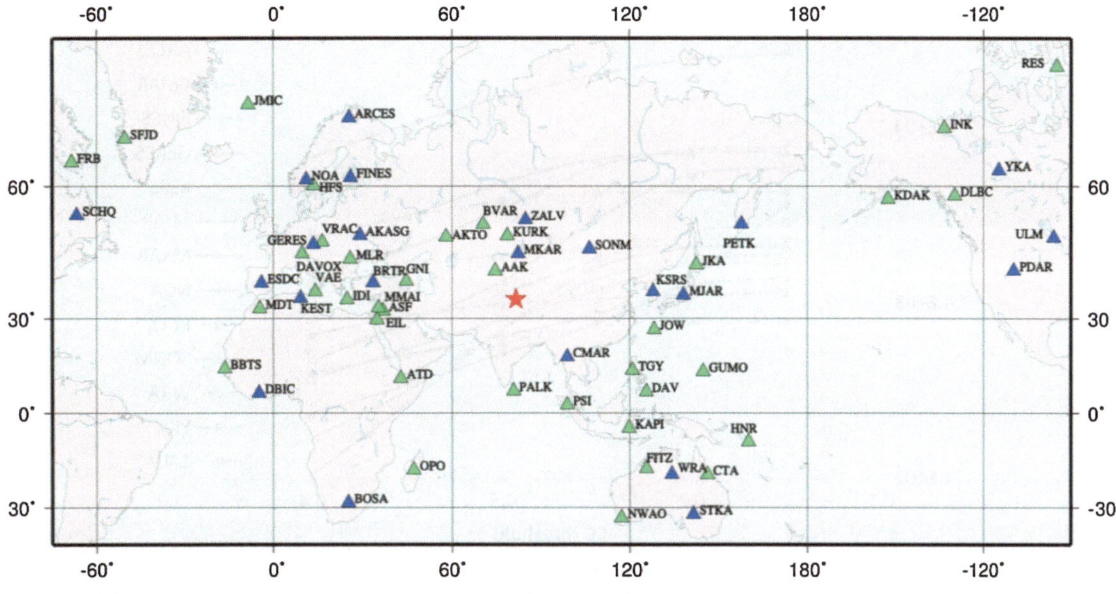

Figure 5
Primary (*blue*) and auxiliary (*green*) IMS stations reported Pn- and P-waves from the main shock (*red star*)

equivalent to poor signals and thus poor templates. Since an event with two primary stations (two templates) cannot find a valid REB event (at least three primary stations are needed) we did not process this aftershock as a master event. As a result, the total number of master events is 141.

For a given multichannel template, the cross-correlation coefficient is a time series of the same length at the original waveform. When the |CC| and its SNR_{CC} exceed some predefined thresholds (these station/region dependent thresholds have to be determined empirically) an arrival is written into a database and several parameters are calculated from the cross-correlation traces (e.g., arrival time, azimuth, and slowness) and original waveforms (e.g., amplitude, period, and RM).

For a station in a master event, we use four templates with different frequency bands and time windows (see Table 2) to generate a single time series of cross-correlation coefficient. It may so happen that there are two or more CC detections from different templates within several seconds and we have to select only one of them. Then we seek for the largest |CC| value among all detections within a four-second interval and attribute its value to the sought detection in the CC series. The onset time and all other parameters (e.g., azimuth, slowness, amplitude, and period) are also estimated and attributed to this detection. All other detections in the studied interval are discarded and the next detection has to be beyond 4 s from the found one.

It is often observed that the P-wave coda may contain signals similar to the direct P-wave (e.g., free surface reflections). Because of this observation we also prohibit any other arrival within 4 s despite the |CC| may exceed the threshold. This interval might be extended but then the risk to miss a valid signal from a different event also rises.

Before the start of event building, all arrivals have to pass a number of quality checks, including the estimation of azimuth and slowness using the *f–k* analysis of the cross correlation traces. However, the possibility cannot be excluded that some higher correlation coefficients may be related to strong signals from sources far away from the master event or associated with coherent noise. To validate the signals detected on the CC traces we use F statistics. Specifically, we calculate maximum F_{prob} for all detections in the time window 4 s from their onset times. For that, F statistics in a running 2-s window is estimated using the approach developed by Douze and Laster (1979). In this study, we tentatively put $F_{prob} > 0.3$ as the threshold for a valid signal.

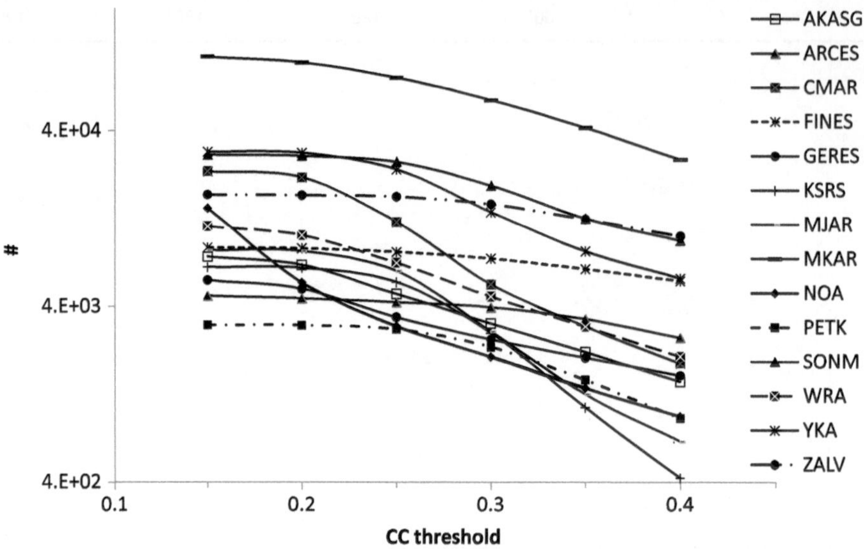

Figure 6
The number of arrivals found by cross correlation at 14 IMS stations from 141 master events as a function of CC threshold

Table 5

Station participation in the master events

Stations	AKASG	ARCES	CMAR	FINES	GERES	KSRS	MJAR	MKAR	NOA	PETK	SONM	WRA	YKA	ZALV
#	54	53	73	107	56	21	27	136	70	11	117	57	113	130

Figure 6 depicts the number of qualified (CC, SNR_{CC}, SNR, F_{prob}, azimuth and slowness residuals) arrivals as a function of CC_{tr} at 14 IMS array stations. These arrivals were found by cross-correlation using all 141 master events. Since the master events have the number of stations from 3 to 14, the total number of arrivals depends on the frequency of participation of a given station in the master events. Table 5 lists the participation frequency for all 14 stations, with the largest input from MKAR—136 master events. ZALV participates in 130 masters and PETK only in 11. Altogether there are 1,025 station (P and Pn) patterns for 141 events, i.e., approximately seven patterns per master on average.

When interpreting the total number of arrivals one has to take into account that different masters predominantly find the same signals as they are characterized by a good cross correlation. The total number of arrivals reflects all valid master/slave pairs with different correlation coefficients and relative positions.

The largest total numbers belong to MKAR, which is the closest station. With the CC_{tr} increasing from 0.15 to 0.40 the number of arrivals at MKAR drops from 105,438 to 27,377. In the lin-log scale in Fig. 6, after some corner CC_{tr}, this fall can be approximated by linear line that manifests an exponential distribution. The transition from a constant level to an exponent fall likely manifests the change from noise detection to true signals from real events and exponential decay of cross-correlation with distance.

Form Fig. 6, the corner CC_{tr} for MKAR is between 0.15 and 0.25. For SONM and YKA, the corner CC_{tr} is slightly above 0.25. For CMAR, one can observe a linear roll-off from 0.2. FINES and ZALV are likely characterized by a corner CC_{tr} of 0.3. NOA demonstrates an unusual behavior—a convex (downward) curve with the quasi-linear portion from $CC_{tr} = 0.3$. This observation supports our earlier assumption on the dependence of CC_{tr} on source region and station. It also gives an accurate tool to separate the flux of true arrivals at a given station from noise detections with a predefined rate of false alarms.

Our first task is to find as many REB events as possible using these cross-correlation detections. The success of this task depends on the definition of a found phase and a found event. We follow the approach adopted at the IDC and consider an REB phase as a found one when there is at least one cross correlation arrival within ± 4 s interval. In IDC interactive processing, an analyst can retime P-wave arrival within this interval which might be considered as the uncertainty of arrival picking. One needs at least three REB arrivals to be found with a given master event to consider the REB event as found by cross correlation.

Figure 7 summarizes principal results of the search for REB events as expressed by the cumulative cross-correlation coefficient, $\Sigma|CCj|$, calculated for all found phases for a given master/slave pair. The results are arranged as a 142 by 142 matrix with the master events ordered in time from top to bottom. The cumulative cross-correlation coefficients for the main shock as a master create the first line in the matrix. Since $|CC|$ exceeds 0.2 at three or more station, the $\Sigma|CCj|$ scale starts from 0.6. When $\Sigma|CCj| > 3.0$, the found REB event is at the level of autocorrelation for a three-station event. Therefore, we limit the cumulative CC scale by 3.0. The m_b scale shows (IDC) magnitudes of the master events in the range between 2.0 and 6.0. The nsta scale shows the number of stations in the master event in the range from 3 to 14. Several masters have magnitude above 4.0 but a few stations. Three events with m_b(IDC) $\sim$ 4.3 include all 14 stations and one event with m_b(IDC) = 3.86 includes 12 stations. These are candidates for the final master event set covering the area around the main shock in routine XSEL building.

Auto-correlation (diagonal cells) finds all REB events except the one with two templates. However we are interested in those REB events which are found by other masters. Figure 8 shows the number of found REB events as a function of magnitude, as obtained from Fig. 7. The largest number of found REB by one master is 113 and belongs to the event with m_b(IDC) = 5.05, i.e. to the largest aftershock. This master also finds the largest number of events overall 165, which includes 113 REB events. The second best master with m_b(IDC) = 4.52 finds 106 REB events (137 in total), and the master with m_b(IDC) = 4.28 finds 152 events in total (97 REB).

The number of REB events found by a given master is almost constant for magnitudes between 4.1 and 5.05. Below 4.0, this number suffers a near-linear decay with magnitude. From five events with magnitude below 3.0 three found only themselves (the auto-correlation). Other three events with auto-correlation only have m_b(IDC) of 3.16, 3.49 and 4.26(!). In total, only the main shock and 134 aftershocks can be used as master events. There are also 5 REB masters which found only one REB event in addition to auto-correlation.

Another view on Fig. 7 reveals the number of REB events which were not found by any master. Nine REB aftershocks are not found using cross-correlation from any master events, with two of them in the set of seven masters which can find only themselves. For various masters, the time shifts between individual channels at a given IMS station are different because of the difference in locations. For two events, the cross-correlation coefficient is not symmetric when master and slave are swapped. As a consequence, an event can find another one but might not be found as a slave for a given CC threshold. Altogether, we have seven events which are worthless as masters and seven not-found events. There are also five aftershocks which were found by only one master event.

There are two principal reasons for an arrival in the REB not to be found by cross correlation with other master events. There is a valid arrival detected by cross-correlation, but it is out of the predefined 4-s window; such an REB arrival is likely a later phase. The missed REB arrival is a misassociated phase from a different event and thus fails to cross-correlate with any of the correct templates. In both cases, the relevant REB events do not contradict IDC rules and guidelines because all defining parameters are within the uncertainty bounds adopted for travel time, azimuth and slowness residuals.

The presence of somehow misassociated phases in the REB does not preclude cross-correlation from finding valid arrivals where they have to be. For some of the poorly built aftershocks, cross-correlation has detected three and more phases when only one or two

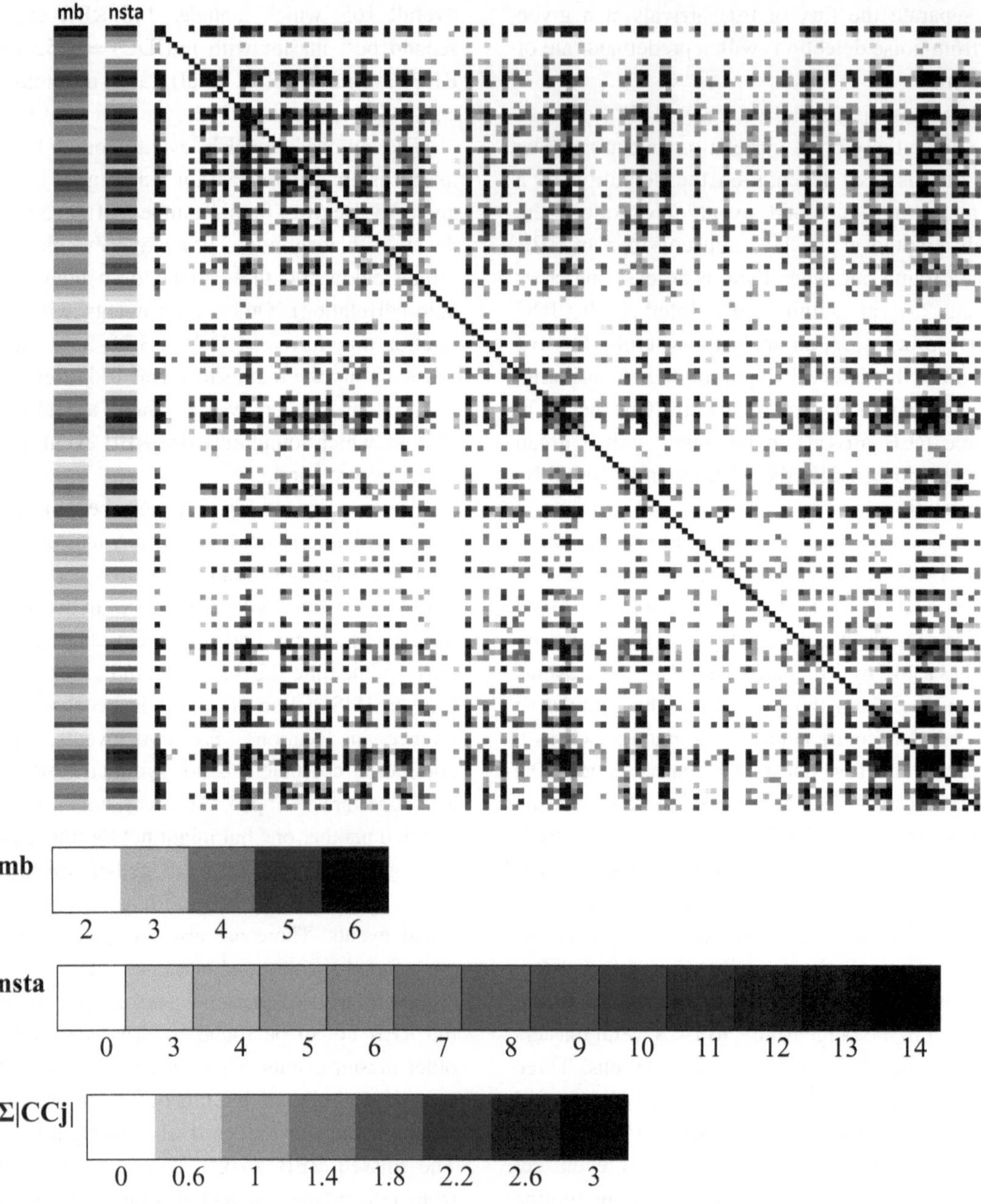

Figure 7
Cumulative cross-correlation coefficients, Σ|CCj|, for all pairs of 142 REB events with at least three stations having cross-correlation coefficient above 0.2. Master events are ordered in time from *top* to *bottom*: cross-correlations with the main shock create the first line in the matrix. The m_b scale shows magnitudes of the master events. The nsta scale shows the number of stations in the master template. Some masters have large magnitude but few stations

REB arrival was matched by time. Even for those aftershocks which are found by many master events, additional arrivals have been found by cross-correlation. These additional arrivals increase the cumulative CCs in Fig. 7.

Figure 9 presents a matrix with the cumulative CC obtained for all cross-correlation arrivals (in and not in the REB). There is a tangible improvement in event finding—eight REB events have one or two phases shared with at least one event built with cross-

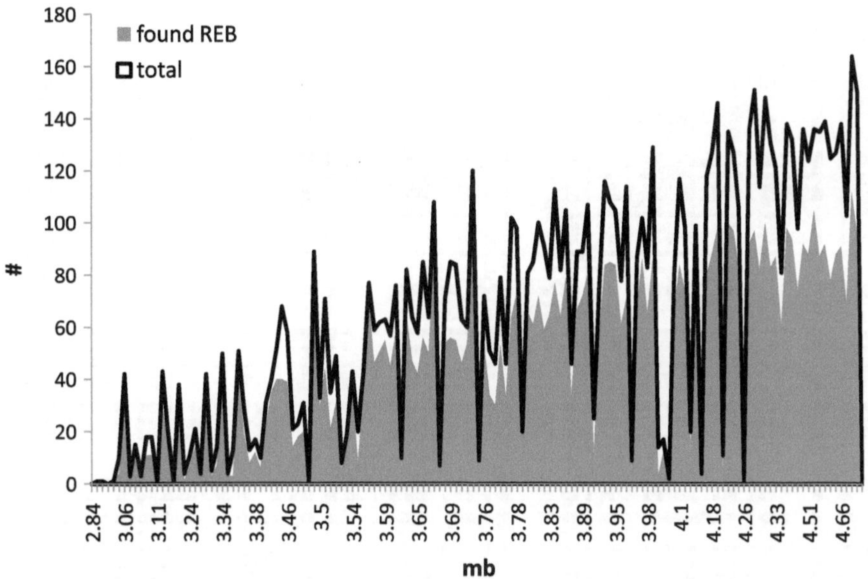

Figure 8
The number of REB events and total number of events found by selected master events, ordered by magnitude. The largest total number of found events (165) and REB events (113) corresponds to the same event with $m_b = 5.05$

correlation by three or more primary stations. In other words, we would build more reliable REB events if cross-correlation detections were used.

There are many questions left on the relations between the REB and cross-correlation arrivals in the studied aftershocks. However, the principal results of the cross-correlation study obtained so far allow addressing the first and second tasks formulated in the beginning of this Section. We have found a larger part of the aftershocks using other aftershocks as master events. 107 aftershocks have been found by at least ten master events and one hundred ten masters have found at least ten aftershocks each. Therefore, approximately 110 from 140 (142 less the main shock and the event with two stations) REB events (79 %) can be considered as mutually consistent ones.

Among the residual 30 REB events, one may distinguish 13 less reliable master events with four to nine found aftershocks and 17 events which, in practice, were able to find only themselves. These 17 REB events have to be considered for internal consistency. Figure 9 gives an obvious hint that many of these 30 events could be built in a more reliable manner when CC is used. In any case, none of these 30 REB events should be used as master ones. The pre-selection of master events is the second part of task two.

Figure 10 shows the number of templates in all aftershocks. Potential master events have to have the largest possible number of templates, i.e., ten and more. This guarantees the best coverage in presence of varying seismic noise at IMS stations. When noise is high at several stations one still has a chance to find enough signals to build an XSEL event. Potential master events should also be able to find as many aftershocks as possible, as shown in Fig. 8. Now we have to introduce the procedure of XSEL events building.

We start with one master event. All CC and SNR_{CC} values are already estimated for all four different frequency band templates for all stations in the master event. After all conflicts between arrivals in various frequency bands are resolved (see Sect. 3), one has a set of detections with their arrival times for each station, t_{ij}, where i is the index of the ith arrival at station j. Apparently, all valid arrivals should belong to some events in proximity to the master event. These are the events we are seeking for. The travel times from these sought events to the relevant stations can be accurately approximated by the master/station travel times, tt_j. Using the approximated travel times (same for all events around the master one) and the measured arrival times one can calculate origin times for all detections:

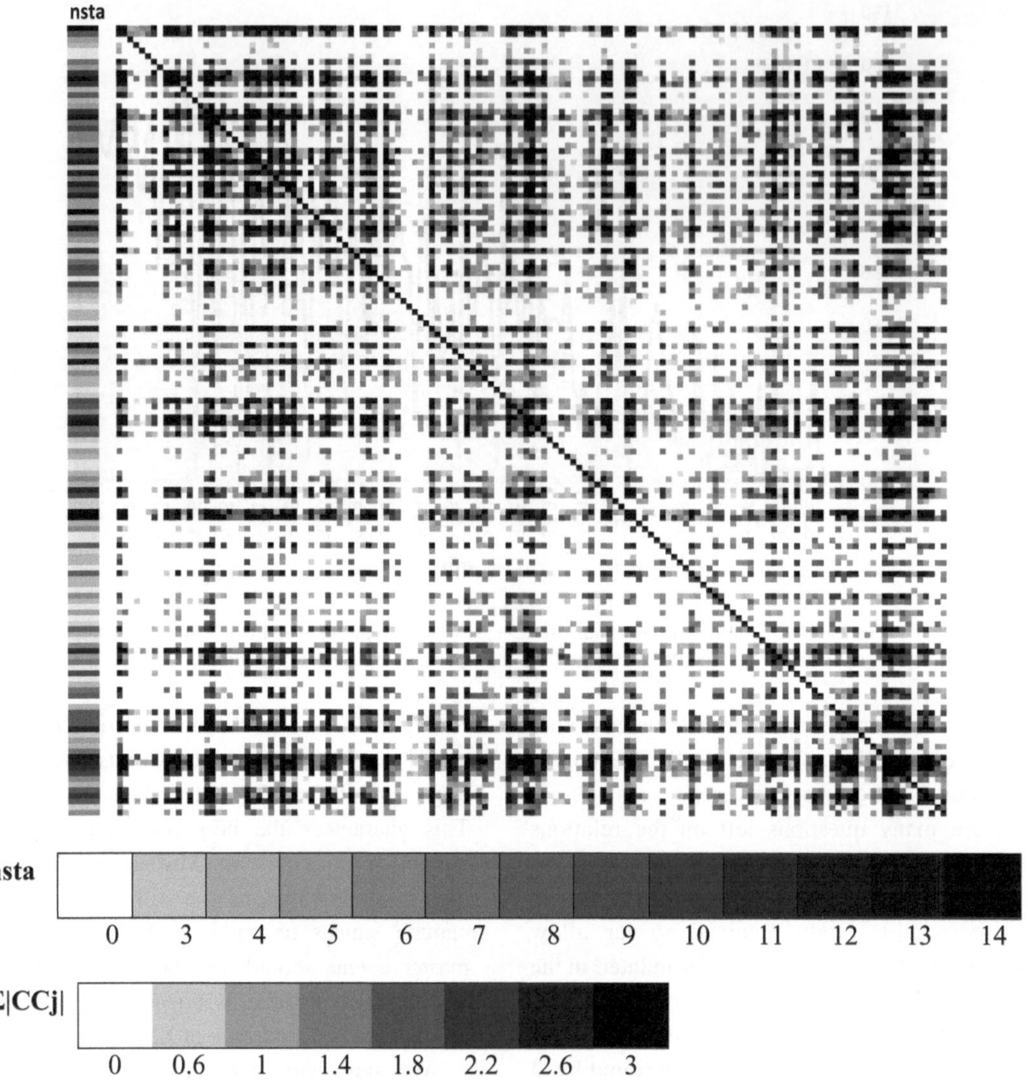

Figure 9
Cumulative cross-correlation coefficients, Σ|CCj|, for all stations with |CC| above 0.2. Master events are ordered in time from *top* to *bottom*

$$ot_{ji} = t_{ij} - tt_j.$$

As a result, one has a set of origin times at several stations instead of arrival times. Origin time is a natural characteristic of source, and when three or more stations detect appropriate signals within a few second one can associate these arrivals with a unique event. (Here we follow up the IDC event definition criteria). At the initial stage, we average all associated origin times and assign the estimated value to the event origin time. One can also use the median value or various weighted sums. There is no need to locate the built event globally since it has to be close to the master one.

Overall, this process can be called "local association", LA, in line with the name of global association, GA. Indeed, only phases from local events should be associated. The LA does not see any events beyond the radius of correlation. That is why two DPRK master events did not find any from hundreds events occurred globally during the 5 days after the second test. The LA had no local arrivals to associate.

We have started with a 6-s time window in the LA for the association into a single event. Considering the travel time uncertainties in the GA (dozens of

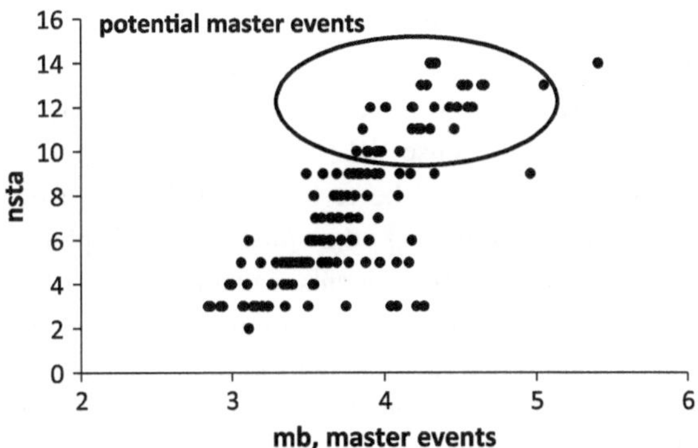

Figure 10
Distribution of the number of (primary array) stations in master events depending on magnitude

seconds), this is a very short interval for origin times. It corresponds to the difference in travel times between the master event and an event on the rim of cross correlation zone (say, 50 km). For a P-wave with 0.05 s/km slowness the travel time difference is of 2.5 s. For two stations in opposite directions, two 2.5 s travel time residuals give a 5 s difference in origin time. This is the worst case scenario. A 1-s uncertainty in onset time may add 2 s to the origin time difference. We have tested longer windows and found 6 s to provide almost as many events as a 9-s window. For the latter window, more noise phases might be wrongly associated and additional events are likely less reliable.

At regional distances, Pn-waves have larger slowness but the cross-correlation coefficient decays much faster with spacing between events. This is the effect of the highly inhomogeneous crustal and upper mantle structure, where the Pn-waves propagate. As a result, the association window of 6 s still works efficiently.

There is an important enhancement of the LA process based on relative magnitudes of associated arrivals. As shown in Sect. 3, all arrivals associated with a given event should have RM estimates within some predefined bounds separating the genuine and bogus signals. We have adopted a tentative value of 0.7, which is much smaller than a similar magnitude difference of 2.0 used in IDC automatic processing. This threshold is then tested on the full set of found events.

The LA is a simplistic process compared to the global association. A big advantage of the LA is a reduced flux of arrivals at a given station from a given master event. At the same time, the total number of arrivals at a station may grow relative to that from the current IDC detector. We have already mentioned that the cross-correlation detector can find valid signals with (standard) SNR < 0.7 which are not seen by the current IDC detector. The increased number of arrivals can be effectively split into a large number of independent sets associated with different masters.

When several master events are close in space, like in the studied aftershock sequence, their templates may correlate with waveforms from the same event and thus create similar new events. To select one from a set of similar events with close origin times (± 15 s) we first count for all individual masters the number of stations used in the LA. When several master events have the largest number of stations we select the one with the highest cumulative cross-correlation coefficient, $\Sigma|CC_j|$. By definition, this event is the most reliable and its parameters are written into the database. All other events in the set are rejected. Thus, for a multiple set of master events the LA provides a unique set of found events.

We have tested the influence of the number of master events on the final catalog of new events. When a found event has origin time within ± 15 s from any of the REB aftershocks it is considered as associated with the REB, not as a new one. We are looking for distinct new events which cannot be confused with the REB events. To populate the set of master events, we progressively included more and more masters and calculate the number of new events. To begin with, we used the best master event which has previously found 113 REB events and 165 events in total. Then we added the second and third best master events from Figs. 8 and 10. Two sets of 10 and 27 (all aftershocks with ten and more stations) events have also been tested for the total number of new events.

Figure 11 depicts the number of new events as a function of the master set size for three different CC thresholds: 0.15, 0.20, 0.25 and 0.35. The CC_{tr} defines the number and quality of used arrivals. At some IMS stations, the threshold can be as low as 0.20. Other stations need 0.3 and even larger threshold in order to reduce the portion of noise arrivals.

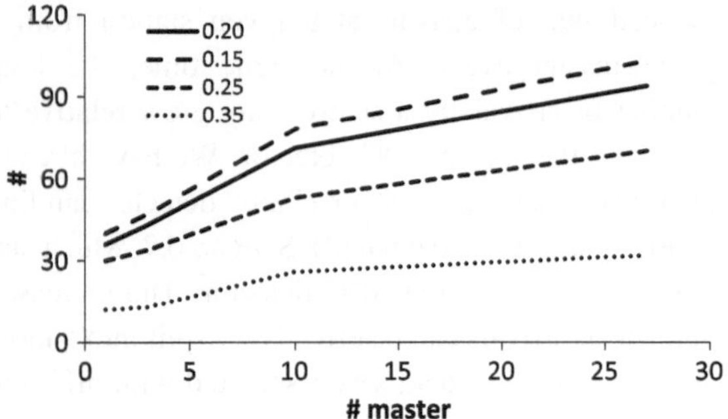

Figure 11
The number of new events found by various sets of master events for different CC thresholds: 0.15, 0.20, 0.25, 0.35

The number of new events grows with the size of master event set. For $CC_{tr} = 0.20$, there are 36 new events for the best master event. For 27 masters, the number of new events is 94. This is a natural trend—since the cross-correlation coefficient falls exponentially with distance, a denser master grid should find more events by virtue of proximity. Ultimately, one might design an iterative procedure to find all possible events (Harris and Dodge, 2011). First, all (reliable) REB events are used as masters to find new events. After an interactive review, these new events are included in the REB and then used as masters to find the next portion of new events. The process is repeated before there is no new event found. To enhance this procedure, one may combine the best templates from different but well-correlated events. For example, one may use as templates stations YKA and MKAR from event 1 and stations ARCES and FINES from event three to create a synthetic event. Moreover, it is possible to combine waveforms from different events to create synthetic waveforms, as proposed by Harris and Paik (2006).

For an area covered by seismic events spaced by less than, say, 30 km (continuous coverage), one needs only one reliable event to start this iterative procedure which will end up in a complete XSEL (or REB) bulletin. The REB events distribution shown in Fig. 1 assumes that the IDC likely needs only one or a few events to start the process which will find all reliable REB events and reorder them according to the number of cross correlations. This reordering also means accurate relocation with the best located events (e.g., the events with ground truth coordinates) defining the absolute locations and confidence ellipses for less successful "cross-correlation neighbors". As we have demonstrated, there exist unreliable (in cross-correlation sense) REB events. A good portion of these events could be re-built with cross-correlation arrivals replacing the misassociated phases. However, a few REB events cannot be healed by cross-correlation. In addition to the recovered and quality checked REB, one will obtain a set of new reliable REB events which might be as large as the REB itself. For the number of existing REB events and the total length of waveforms obtained since the launch of IDC in 2001, this iterative process needs a supercomputer.

Higher CC thresholds reduce the final new event bulletin (i.e., the XSEL less REB) and likely improve its quality. For $CC_{tr} = 0.35$, there are only 12 new events for the best master and 32 new events for 27 masters. Altogether, for all 141 master events there are 9,707 new events built (69 new events per a master event) for $CC_{tr} = 0.15$ and only 5,984 for $CC_{tr} = 0.35$ (42 per a master event). A cross-correlation coefficient of 0.35 is not often among the arrivals in the XSEL and likely belongs to the closer events with higher magnitudes.

A crucial part of the third task is interactive checking of all new events. We have to draw a distinct line between reliable and poor new events. Only a human can make the final decision on event quality. We started with $CC_{tr} = 0.2$ and the set of ten master events. This set finds 126 from 141 REB events by cross-correlation at three and more stations. The residual 15 events are likely internally inconsistent. There are 71 new events to be reviewed and those with a larger number of reliable stations are of a higher priority.

The station performance can be estimated from Table 6. For each station, it lists the portion of all arrivals above a given CC threshold which was used in the final XSEL. For $CC_{tr} = 0.2$, station NOA has the highest rate of $\sim$57 % of all arrivals (above this threshold) used in the XSEL (126 REB and 71 new events). AKASG, FINES, WRA and GERES are characterized by rates above 30 %. At the same time, the input of PETK and KSRS is negligible. These stations are likely able to detect signals from the events with magnitude above 5.0.

Table 6

The percentage of cross-correlation arrivals used in the XSEL

	0.15	0.2	0.25	0.35
AKASG	47.5	51.4	66.1	81.2
ARCES	26.4	26.9	27.3	31.0
CMAR	13.4	14.3	23.0	52.2
FINES	40.9	40.8	40.8	40.8
GERES	27.6	30.7	42.0	60.1
KSRS	1.8	1.8	1.9	5.4
MJAR	10.3	10.3	12.5	43.6
MKAR	7.2	7.6	8.4	10.7
NOA	24.6	56.6	76.6	82.9
PETK	1.9	1.9	1.9	2.8
SONM	19.4	19.4	19.8	30.7
WRA	34.3	37.4	44.1	57.9
YKA	22.4	22.4	25.8	49.2
ZALV	30.1	29.9	29.1	30.9

An important exclusion from the reliability hierarchy is MKAR. It generates many arrivals but 90 % of them are not used in the XSEL. This might be considered as a signature of poor performance. However, the absence in the XSEL does not mean that an arrival is worthless or bogus. The Gutenberg–Richter law implies that there should be more events with smaller magnitudes which can be detected by two stations or even by one station only. As the closest regional array, MKAR has to find valid signals from these smaller events which cannot meet the IDC event definition criteria. This consideration is also applicable to those arrivals at reliable stations which are not in the XSEL. They may also create many two-station events which we do not review.

There is one event with m_b(IDC) = 3.9 which occurred several minutes after the main shock, when the level of microseismic noise was elevated by secondary phases of the main shock and big aftershocks. This event included only four stations. The biggest event has m_b(IDC) = 4.2 and occurred 2 h after the main shock. The smallest new event has m_b(IDC) = 2.8.

Figure 12 displays original waveforms and the IDC solution for a smaller event with m_b(IDC) = 3.4 which includes seven stations with one auxiliary array BVAR. All signals are clear and reliable. Further work is needed to review other events from the set of 71, and then from the set of 94 as obtained by 27 master events, but even the initial effort has shown a large number of events missed in the REB and a high success rate of the cross-correlation pipeline. Definitely, there should be some bogus events among those 71 or 94 built for CC_{tr} = 0.2. One has to decide on the tolerable rate of false alarms and to tune all defining parameters accordingly.

Interactive analysis is a time consuming procedure, and an experienced lead analyst has reviewed with IDC rules and guidelines only 40 from 71 events. The analyst started with the events with the largest number of stations (see Table 7) but also reviewed several less reliable cases. There were built 37 new REB events (success rate 93 %) and their reviewed IDC solutions are listed in Table 8 with locations, origin times and magnitudes. One of the reviewed events was actually an REB event with m_b(IDC) = 4.4 approximately 2,000 km off the main shock. However, it was big enough to generate quasi-sinusoid waveforms of 20 and more seconds at teleseismic distances. It was wrongly built by cross-correlation as an aftershock because all three primary IMS stations which built this event were close (NOA, FINES and ARCES) and at the same great circle with the main shock and the REB event. This problem was fixed by checking the azimuthal gap and relative magnitude.

Signals for the rest two new events were not found by standard methods of interactive analysis. This situation was expected because cross-correlation is definitely able to find valid signals below the noise level. Therefore, the negative result of interactive review does not imply that these events were wrongly built by cross-correlation and the success rate might actually be higher than 93 %.

In many cases, the number of associated phases, nass, and the number of time defining phases, ndef, in Table 8 is larger than the number of stations in the relevant XSEL events. This difference has two sources. The analyst may add arrivals detected by primary 3-C stations or by auxiliary stations; both types are not used for cross correlation. Secondary arrivals may also be added at all station type. For example, the closest auxiliary station AAK (see Fig. 5) reports Pn- and Lg-phases which are used to constrain IDC locations and confidence ellipses. Auxiliary station BVAR is actually a very sensitive array which also detected a good portion of the aftershocks.

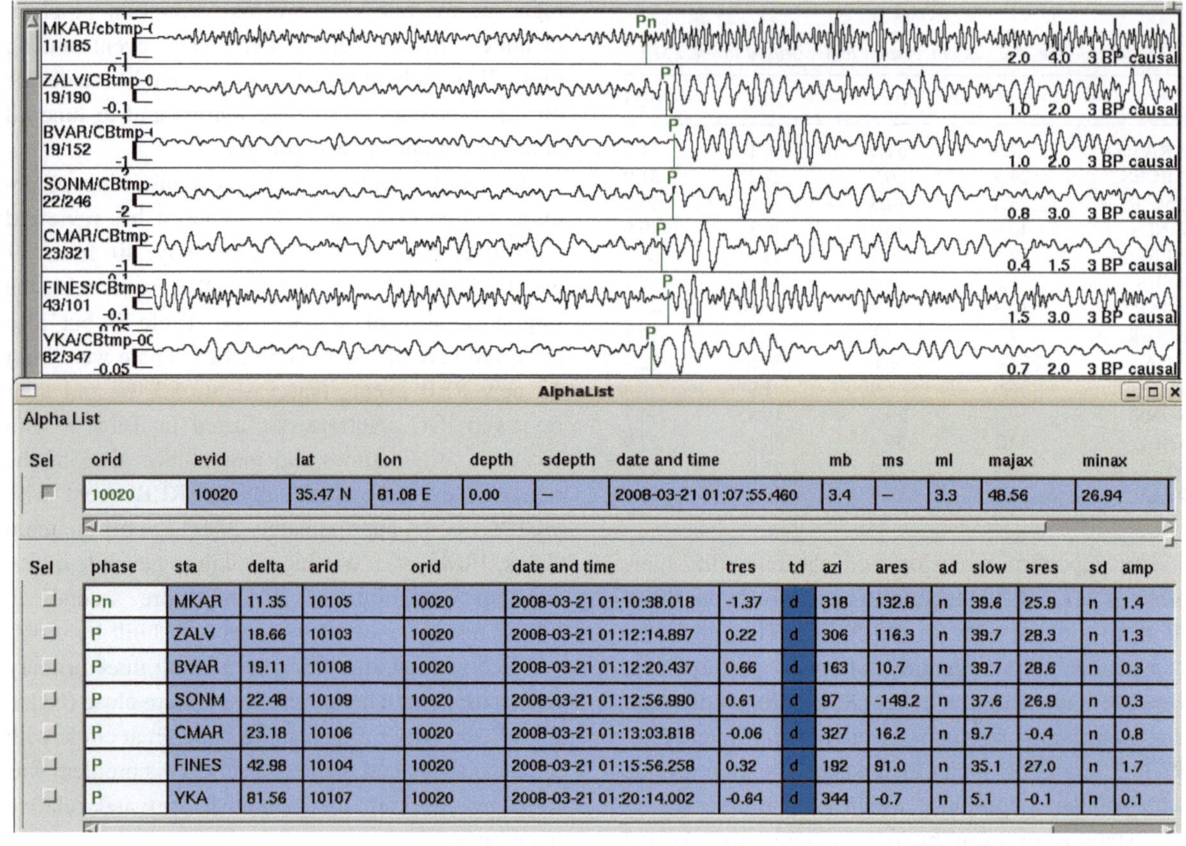

Figure 12

An example of new event built using detections obtained from cross correlation coefficients at six (!) primary stations with m_b (IDC) = 3.4. The event definition criteria are all matched. BVAR is an auxiliary array which was not used as a template but found in interactive analysis. Origin time: 21/03/2008, 01:07:55.46; location: 35.47°N, 81.08°E. Seven waveforms (beams) were filtered with different filters shown in the figure. The list of stations shows the onset times picked by analyst. Azimuths and slownesses for these detections were estimated by automatic procedures from the original waveforms and demonstrate high residuals

All new events in Table 8 were reviewed and thus located using standard IDC software with the parameters of arrivals obtained in standard automatic processing. In many cross-correlation studies, a double-difference (DD) algorithm is used. It is based on the relative travel times from the master and slave event (e.g., SCHAFF *et al.*, 2004). The DD method allows for an accurate location of the slave event relative to the master event which is of importance for many seismological and tectonic applications. For the IDC purposes, absolute locations are of the highest priority. When a master event is mislocated by tens of kilometers (usual values for events with few IMS seismic stations) the relative location introduces systematic errors in all cross-correlated events. An independent location of many events with travel time, azimuth and slowness residuals distributed approximately normally is likely able to provide a more reliable solution on average (KITOV and KOCH, 2007). Therefore, we are going to use the standard IDC location procedure for interactive processing. When all master events are accurately located in absolute terms, one can use them for relative location of slave events.

The fourth task also demands considerable human resources. One has to estimate all defining parameters: CC_{tr}, thresholds for SNR_{CC} and standard SNR together with the lengths of STA and LTA windows, frequency bands and lengths of templates for various seismic phases (e.g., P, Pg, Pn, PKP, Sn, Lg), parameters of F statistic and F_{prob}, the width of the LA association window, the minimum time spacing between arrivals at one station and between events built by the same master event, etc.

Table 7

Number of new events with a given number of stations for various sets of master events

# Masters	1	3	10	27	141
nsta = 3	20	25	46	65	89
nsta = 4	9	10	12	14	19
nsta = 5	6	7	12	10	10
nsta = 6	1	1	1	4	4
nsta = 7	0	0	0	1	1
Total	36	43	71	94	123

Table 8

IDC solutions for 37 new events reviewed by analysts

lat	lon	Day	time	nass	ndef	mb	ML
35.48	80.56	20/03/2008	22:45:32	4	4	3.9	3.5
35.61	81.55	20/03/2008	23:28:20	6	6	3.4	3.0
34.77	81.11	20/03/2008	23:57:10	7	7	3.8	2.8
36.46	81.96	21/03/2008	00:08:12	3	3	2.9	2.7
34.88	80.92	21/03/2008	00:12:07	4	4	4.2	–
34.08	81.28	21/03/2008	00:13:02	4	4	4.1	–
34.91	81.23	21/03/2008	00:15:45	5	5	3.6	3.0
35.31	81.33	21/03/2008	00:41:10	5	5	3.6	3.1
35.47	81.08	21/03/2008	01:07:55	7	7	3.4	3.3
35.35	81.46	21/03/2008	01:15:43	4	4	3.7	–
34.91	82.30	21/03/2008	01:43:01	7	6	3.4	3.0
35.20	80.50	21/03/2008	02:35:33	6	5	3.1	2.9
35.21	81.31	21/03/2008	02:53:49	4	4	3.4	2.2
35.04	80.88	21/03/2008	04:33:13	4	4	3.2	3.1
35.49	81.19	21/03/2008	06:25:59	6	6	3.4	3.5
35.75	81.20	21/03/2008	07:30:36	4	4	3.1	2.8
35.26	80.82	21/03/2008	07:33:52	7	7	3.4	2.5
35.21	80.88	21/03/2008	07:37:01	6	6	3.1	2.9
35.23	81.64	21/03/2008	08:21:26	4	4	3.0	3.0
35.84	80.71	21/03/2008	09:31:23	4	4	3.0	2.9
35.32	81.22	21/03/2008	09:45:47	5	5	3.4	3.0
35.21	80.84	21/03/2008	09:57:07	4	4	3.1	2.6
35.07	82.47	21/03/2008	11:17:15	8	8	3.4	3.0
35.72	81.21	21/03/2008	13:29:53	4	4	2.8	3.2
35.22	81.18	21/03/2008	14:11:55	9	9	3.4	3.2
36.35	81.78	21/03/2008	21:13:34	7	6	2.9	3.2
36.02	81.69	21/03/2008	21:37:47	6	5	2.8	2.9
37.01	80.18	21/03/2008	21:57:17	3	3	3.2	2.9
35.30	81.10	21/03/2008	22:43:02	7	7	3.5	2.8
35.39	80.65	22/03/2008	07:37:56	6	6	3.2	2.8
35.61	81.20	22/03/2008	21:53:18	7	7	3.5	2.8
35.63	81.77	23/03/2008	02:08:54	11	9	3.4	3.1
35.85	80.71	23/03/2008	10:36:31	5	5	3.0	2.9
35.37	81.29	23/03/2008	10:46:12	6	6	3.4	2.8
35.43	81.12	24/03/2008	00:03:08	4	4	3.5	2.8
35.21	80.70	24/03/2008	09:38:48	7	7	3.8	3.4
35.75	80.99	24/03/2008	17:52:46	5	4	3.0	2.9

The problem of optimal thresholds is a typical task for machine learning. Here, we report a few principal relationships based on 45,585 arrivals associated with at least one event built by cross correlation with all 141 REB as master events. First is the probability for an arrival at a given station to be associated with an event as a function of CC. Table 9 lists the cumulative portion of arrivals associated with at least one event below given CC levels. For AKASG, a half of associated arrivals have $|CC| > 0.35$. For ARCES, only arrivals with $|CC| > 0.5$ have the probability to be associated above 50 %. We excluded PETK from the list—it has only 27 associated arrivals, likely for the biggest events. It is worth noting that low CC does not preclude an arrival to be correctly associated with an XSEL event. Using Table 9 and Fig. 6 one can define sound station-dependent thresholds of CC.

Table 9

Cumulative share of arrivals associated with at least one event below given |CC| levels

	0.2	0.25	0.3	0.35	0.4	0.45	0.5
AKASG	0.01	0.13	0.32	0.51	0.66	0.77	0.84
ARCES	0.00	0.03	0.05	0.11	0.25	0.42	0.55
CMAR	0.00	0.11	0.31	0.49	0.65	0.76	0.84
FINES	0.00	0.01	0.04	0.10	0.18	0.28	0.38
GERES	0.00	0.05	0.11	0.20	0.31	0.41	0.52
KSRS	0.02	0.12	0.31	0.50	0.60	0.64	0.74
MJAR	0.00	0.07	0.20	0.37	0.53	0.68	0.79
MKAR	0.00	0.04	0.12	0.23	0.38	0.55	0.71
NOA	0.02	0.26	0.48	0.64	0.74	0.82	0.88
SONM	0.00	0.01	0.05	0.11	0.18	0.25	0.33
WRA	0.01	0.20	0.39	0.54	0.66	0.75	0.81
YKA	0.00	0.05	0.17	0.33	0.48	0.60	0.70
ZALV	0.00	0.00	0.01	0.02	0.07	0.14	0.25

Figure 13 depicts frequency distribution of RM and logα residuals counted in 0.1-wide (dimensionless unit) bins as obtained from all XSEL events built by all 141 master events. For an event, the residuals are the absolute deviations of individual estimates from the average. We have already introduced a tentative RM threshold of 0.7. In the lin-log scale, the number of residuals falls slightly faster than exponentially with only a few residual between 0.6 and 0.7. This observation validates our preliminary assumption on the RM fluctuations for reliable events. One can also consider a smaller threshold. As expected, the set of RM residuals has smaller mean value (0.17) and standard deviation (0.14) compared to those for logα: 0.19 and 0.16, respectively. Individually, the standard deviation varies in a narrow band between 0.16 for AKASG to 0.26 for WRA.

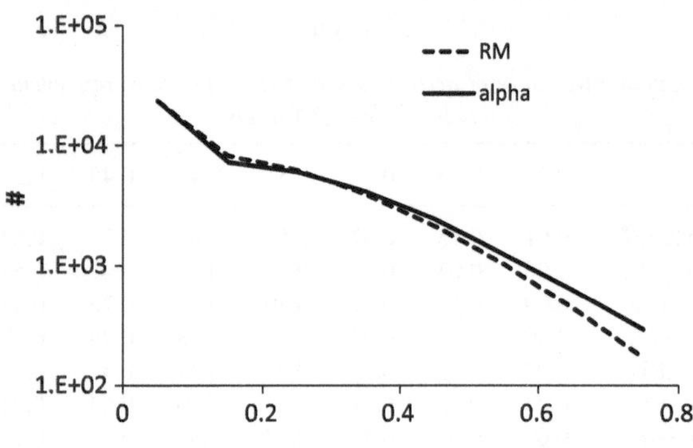

Figure 13

Frequency distribution of RM and logα residuals counted in 0.1-wide dimensionless unit bins as obtained from all XSEL events built by all 141 master events. In the lin-log scale, the number of residuals falls slightly faster than exponentially with only a few residual between 0.6 and 0.7. The residuals of relative magnitude have a smaller standard deviation 0.14 versus 0.16 for logα

5. Conclusion

We have developed and tested a set of procedures allowing automatic detection and event building in accordance with the IDC event definition criteria for REB events. These procedures are arranged in a unique line, in parallel to the current IDC pipeline, and it is possible to issue an automatic XSEL (REB-ready) bulletin when all data from primary stations are available at the IDC. Therefore the XSEL is an analogue of SEL1 but with REB quality.

The core of detection and phase identification consists in cross-correlation of incoming waveforms from primary array stations with a predefined set of waveform templates from accurately prepared master events. A number of standard and newly introduced parameters are estimated for all arrivals. Then the arrivals obtained from one master event at different stations are associated in XSEL events in the local association procedure. The event building procedure includes testing for dynamic consistency between arrivals as based on relative magnitudes.

The obtained XSEL events were compared to the existing REB and new events are reviewed by an experienced analyst. The results of cross-correlation have validated the consistency of approximately 90 % of the studied REB events, i.e., more than three REB aftershocks has been found by each of these events when used as master events. Approximately 10 % of the REB events do not correlate well with these 90 % or with each other. They have to be reconsidered in order to reveal the reason of inconsistency. We have also added 37 new events to the existing REB with dozens more event hypotheses to be reviewed.

All in all, we have demonstrated that several techniques based on cross-correlation are able to significantly reduce the detection threshold of seismic sources in one specific region, and thus worldwide, and to improve the reliability of IDC arrivals and events by a more accurate estimation of defining parameters. The completeness of the REB could be significantly improved (and will be improved in the future) in automatic processing while fitting the event definition criteria. Moreover, there are several options on the way to the completeness: one can always balance the density of master events and computer resources.

The rate of false alarms has also been reduced by several powerful filters compared to the proportion of rejected events in SEL3. The most efficient filter is cross-correlation between waveforms, which separates signals from a small area around a given master event and the whole outer space. Despite its higher sensitivity to seismic sources around a given master event, cross-correlation trims away 99.9 % of the overall arrival flux in routine IDC processing as irrelevant to local association. Only qualified arrivals are considered for association. The travel times from the master event to three or more primary stations allow constraining the origin times for new events in a narrow range of a few seconds. The LA process also includes relative magnitudes to resolve possible dynamic inconsistency between associated arrivals. The RM is a reliable characteristic of relative sizes with low scattering. Two additional filters, *f–k* analysis and *F* statistics, are also based on correlation traces and demonstrate excellent performance, and the former provides accurate estimates of (pseudo-) azimuth and slowness.

The analysts' experience with the cross-correlation event list shows that the workload associated with interactive processing might be reduced by a factor of two or even more. By definition, all event hypotheses in the XSEL have to match the EDC and an analyst has only to review the involved phases according to the IDC rules without formulating additional event hypotheses. It is worth noting that currently the analyst has to observe clear signals above the noise level. This is not a mandatory requirement for cross-correlation, however. It is possible to find and use reliable signals below the noise level (SCHAFF, 2008).

The workload is also reduced because one does not need to search corroborating arrivals at other primary IMS stations, where no signals were found using cross-correlation, and to exercise alternative event hypotheses with varying station configurations. (The probability to find any arrivals extra to those obtained using cross-correlation for two neighbouring events is very small.) In the current version of interactive processing, this station permutation is the most time consuming operation. It is usually associated with smaller, and thus most frequent, events in the SEL3, many of them not matching the EDC. It also involves repeating location of many hypotheses with different station/phase combinations. As an additional benefit, waveform cross correlation is a precise tool for quality check, for both arrivals and events.

Major improvements in automatic processing and interactive review achieved by cross correlation are illustrated by an aftershock sequence of a large continental earthquake. Exploring this event, we have described schematically further steps in the development of a processing pipeline parallel to the existing IDC one in order to improve the quality of the REB together with the reduction of detection threshold. In order to use the cross-correlation pipeline for global monitoring one has to build a global grid of master events. Before the earth surface is covered by master events, the current IDC processing pipeline can be used to monitor the areas where the REB does not provide any master events.

Our example shows that cross-correlation can easily cope with automatic processing of mid-sized aftershock sequence, especially in new areas. Using only one big REB event found by automatic processing we are able to iteratively recover the whole sequence for the following five days. Moreover, we have found several REB events which should be reconsidered because they are believed to include misassociated seismic phases, which have no correlation with all master templates. Such arrivals would have been screened out in cross-correlation processing. For the largest earthquakes, the area of aftershocks may cover tens of thousands square kilometres, and the distance between remote aftershocks may reach several hundred kilometers. In this situation, the set of master events has to cover the entire area with a regular spacing. In some cases, there are no REB events in mid-size areas which are surrounded by seismically active regions. One may try to find new events using cross-correlation and templates at the borders of these blank areas and/or build synthetic templates by extrapolation of the existing templates. Fully synthetic templates are also an option with theoretical time delays at individual sensors of arrays stations and synthetic waveforms or natural waveforms from similar tectonic environments.

Taking into account the results of our preliminary empirical study, one can conclude that cross-correlation can significantly reduce the detection threshold for aftershock sequences of nuclear tests worldwide using the IMS seismic network. An additional improvement can be achieved when continuous waveform data from regional stations are available. At regional distances, the duration of correlated signals may be between 10 s (Pn) and 60 s (Lg) which allows gathering substantial integral energy as expressed by higher correlation coefficients. In the best case, the detection threshold might be reduced by an order of magnitude relative to the standard threshold guaranteed by the IMS network.

There are approximately 80,000 events with non-zero depths. However, we would not propose to introduce master events with depth. There are several reasons. First, cross-correlation is a tool to search for smaller events, whose depths cannot be determined reliably and therefore fixed to zero in line with the CTBTO's focus on potential explosions. Secondly, nuclear explosions can only be shallow, and zero-depth master events are therefore the best choice for cross-correlation—they will not build deep events because of origin times' (i.e., arrival times less travel times associated with the surface masters) mismatch in the LA. This is an effective way to screen out the events not related to nuclear test monitoring. Thirdly, cross-correlation may be a useful tool to detect depth dependent phases and thus to constrain the depth. Our limited experience also shows that for such depth-defining phases such as pP and sP, cross-correlation could also improve detection and characterization because of their similarity to the primary P-wave. However, this technique works better for events in the lower crust and below when the surface reflected phases are sufficiently separated in time from the primary phase. For shallow events, cross-correlation cannot distinguish between P-wave coda and surface reflections.

Cross-correlation requires significant resources for real-time computations. With the tentative algorithms and non-optimized software used in this study we have estimated that one standard processor may run 40 master events. The regionalized approach allows calculating all master events in parallel, and thus the total number of processors is proportional to the number of master events. Under the cross-correlation framework, the number of master events has to be estimated iteratively by refining the REB and then XSEL. Previously, we have estimated the number of master events in seismically active areas as ~2,000. This is a very optimistic figure. As a conservative estimate, we would increase the above figure by an order of magnitude and assume the total number of master events as ~20,000, including all isolated events. Then, the number of processors for real-time calculations is ~500. When all algorithms and software are optimized one will likely be able to reduce the number of processors.

Acknowledgments

The authors are grateful to their colleagues at the IDC John Coyne, Jeffrey Given and Robert Pearce for help and encouragement. We are also thankful to Kirill Sitnikov and Gadi Turyomurugyendo for building several XSEL events. The authors would like to thank the anonymous reviewers for their valuable comments and suggestions to improve the paper.

References

Baisch, S., L. Ceranna, and H.-P. Harjes (2008). *Earthquake cluster: what can we learn from waveform similarity?*, Bull. Seismol. Soc. Am., *98*:6, 2806–2814.

Bobrov, D. and I. Kitov (2011). Analysis of the 2008 Chinese earthquake aftershocks using cross correlation, *Proceedings of monitoring research review 2011: Ground-Based Nuclear Explosion Monitoring Technologies*, pp 811–821, Tucson, Arizona.

Coyne, J., D. Bobrov, P. Bormann, E. Duran, G. Haralabus, I. Kitov, P. Grenard, and Yu. Starovoit (2012). Chapter 17: CTBTO: goals, networks, data analysis and data availability. In: New manual of seismological practice observatory, ed. P. Bormann (forthcoming), doi:10.2312/GFZ.NMSOP-2_ch17.

Douze, E. J. and S. J. Laster (1979). *Statistics of semblance*, Geophys. *4412*, 1999–2003.

Gibbons, S. J. and F. Ringdal (2006). *The detection of low magnitude seismic events using array-based waveform correlation*, Geophys. J. Int. *165*:149–166.

Gibbons, S. J. and F. Ringdal (2012). Seismic monitoring of the north korea nuclear testsite using a multichannel correlation detector, *IEEE transactions on geoscience and remote sensing*, *50*:5, 1897–1909.

Harris, D. B. and T. Paik (2006). Subspace detectors: efficient implementation, Lawrence Livermore National Laboratory technical report UCRL-TR-223177.

Harris, D. B. and D. A. Dodge (2011). *An autonomous system of grouping events in a developing aftershock sequence*, Bull. Seismol. Soc. Am. *101*:2, 763–774.

Kitov, I. and K. Koch (2007). On ground truth events reported in Sweden: assessment of the IDC location calibration data, *Proceeding of the 2007 EGU general assembly, geophysical research abstracts*, *9*, 07689.

Richards, P., F. Waldhauser, D. Schaff, and W.-Y. Kim (2006). *The Applicability of Modern Methods of Earthquake Location*, Pure and Applied Geophysics, *163*, 351–372.

Schaff, D. P. (2008). *Semiempirical statistics of correlation-detector performance*, Bull. Seismol. Soc. Am. *98*, 1495–1507.

Schaff, D. P., G. H. R. Bokelmann, W. L. Ellsworth, E. Zanzerkia, F. Waldhauser, and G. C. Beroza (2004). *Optimizing correlation techniques for improved earthquake location*, Bull. Seismol. Soc. Am., *94*:2, 705–721.

Schaff, D. P. and P. G. Richards (2011). *On finding and using repeating events in and near China*, J. Geophys. Res., 116: B03309, doi: 10:1029/2010/B007895.

Schaff, D. P. and F. Waldhauser (2005). *Waveform cross correlation based differential travel-time measurements at the northern California Seismic Network*, Bull. Seismol. Soc. Am. *95*:2446–2461.

Schaff, D. P. and F. Waldhauser (2010). *One magnitude unit reduction in detection threshold by cross correlation applied to Parkfield (California) and China seismicity*, Bull. Seismol. Soc. Am., *100*:6, 3224–3238.

Selby, N. (2010). *Relative location of the October 2006 and May 2009 DPRK announced nuclear tests using international monitoring system seismometer arrays*, Bull. Seismol. Soc. Am., *100*:4, 1779–1784.

Waldhauser, F. and D. P. Schaff (2008). *Large-scale cross correlation based relocation of two decades of northern California seismicity*, J. Geophys. Res., 113.

(Received November 30, 2011, revised November 10, 2012, accepted November 21, 2012, Published online December 21, 2012)

Pure Appl. Geophys. 171 (2014), 469–484

DOI 10.1007/s00024-013-0646-1

Pure and Applied Geophysics

How to Invert Multi-Band, Regional Phase Amplitudes for 2-D Attenuation and Source Parameters: Tests Using the USArray

W. Scott Phillips,[1] Kevin M. Mayeda,[2] and Luca Malagnini[3]

Abstract—We inverted for laterally varying attenuation, absolute site terms, moments and apparent stress using over 460,000 Lg amplitudes recorded by the USArray for frequencies between 0.5 and 16 Hz. Corner frequencies of Wells, Nevada, aftershocks, obtained by independent analysis of coda spectral ratios, controlled the tradeoff between attenuation and stress, while independently determined moments from St. Louis University and the University of California constrained absolute levels. The quality factor, *Q*, was low for coastal regions and interior volcanic and tectonic areas, and high for stable regions such as the Great Plains, and Colorado and Columbia Plateaus. *Q* increased with frequency, and the rate of increase correlated inversely with 1-Hz *Q*, with highest rates in low-*Q* tectonic regions, and lowest rates in high-*Q* stable areas. Moments matched independently determined moments with a scatter of 0.2 NM. Apparent stress ranged from below 0.01 to above 1 MPa, with means of 0.1 MPa for smaller events, and 0.3 MPa for larger events. Stress was observed to be spatially coherent in some areas; for example, stress was lower along the San Andreas fault through central and northern California, and higher in the Walker Lane, and for isolated sequences such as Wells. Variance reduction relative to 1-D models ranged from 50 to 90 % depending on band and inversion method. Parameterizing frequency dependent *Q* as a power law produced little misfit relative to a collection of independent, multi-band *Q* models, and performed better than the omega-square source parameterization in that sense. Amplitude residuals showed modest, but regionally coherent patterns that varied from event to event, even between those with similar source mechanisms, indicating a combination of focal mechanism, and near source propagation effects played a role. An exception was the Wells mainshock, which produced dramatic amplitude patterns due to its directivity, and was thus excluded from the inversions. The 2-D *Q* plus absolute site models can be used for high accuracy, broad area source spectra, magnitude and yield estimation, and, in combination with models for all regional phases, can be used to improve discrimination, in particular for intermediate bands that allow coverage to be extended beyond that available for high frequency P-to-S discriminants.

[1] Los Alamos National Laboratory, Los Alamos, USA. E-mail: phillipsfive505@comcast.net

[2] Weston Geophysical, Lexington, USA.

[3] Istituto Nazionale di Geofisica e Vulcanologia, Rome, Italy.

1. *Introduction*

An ability to predict amplitudes of regional phases has practical importance in many areas of seismology, including explosion monitoring and the analysis of seismic hazards. This study is motivated by the effort to extend teleseismic techniques of explosion discrimination and yield estimation to the high frequency regional phases that must be examined when events are small. In particular, regional discrimination relies on relative P- versus S-wave, or cross-spectral amplitude ratios (e.g., Pomeroy, 1982). The amplitudes used to form these ratios must be corrected for path and site effects in order to base discrimination, to the best of our abilities, on source differences alone.

Sereno (1988) pioneered techniques to isolate source and path effects on regional phase spectra in Scandinavia, in support of monitoring efforts. More recently Taylor (1998), Walter (2001), and Taylor (2002) refined models for use in what they term magnitude and distance amplitude correction (MDAC). Similar efforts have proven effective in the hazards community (e.g. Malagnini, 2007). These models relied, initially, on frequency dependent but regionally uniform attenuation; however, much effort has been placed in the development of two-dimensional (2-D) station-centric amplitude correction surfaces (Taylor, 1998; Phillips, 1998, 1999; Rodgers, 1999, 2002), and of tomographic models of attenuation (e.g. Singh, 1983; Campillo, 1987, 1993; Xie, 1990; Mitchell, 1997; Baumont, 1999; Phillips, 2000, 2001, 2005; Sandvol, 2001; Ottemoller, 2002; Taylor, 2003; Mayeda, 2005; Pei, 2006; Xie, 2006; Zor, 2007; Phillips and Stead, 2008; Ford, 2009; Pasyanos, 2009) that can used to predict regional phase amplitudes and determine MDAC parameters.

The impact of higher dimensional attenuation models is obviously crucial to the estimation of source spectra, magnitude and yield using regional phase amplitudes; however, their importance to regional phase discrimination is less obvious at first glance. This is because we expect attenuation to be, for example, high for tectonic areas and low for shield areas for all phases, P and S, crust and mantle, and these correlations will cancel the effects of 2-D attenuation models once discriminant ratios are formed. To test this (Phillips, 1998) created 2-D correction surfaces for station LZH in central China for Pn, Pg, Sn, and Lg phases, and a range of short period bands. Cross-validation studies demonstrated scatter reduction in P-to-S, crustal and mantle phase discriminants for a broad area earthquake population that included platform and tectonic areas. In particular, intermediate band discriminant scatter was reduced (2–6 Hz), which promised increased coverage for discrimination procedures, otherwise generally best above 6 Hz. Later studies hinted at why scatter was reduced, in spite of correlated earth structure. Phillips (2001) used multi-phase tomographic inversions for western China and surrounding regions to show that models were indeed correlated, but the S-wave variations were stronger than the P-wave variations, leading to scatter reduction when 2-D corrections were applied. Recent, higher resolution results for Asia (Phillips, 2010) show that attenuation is less correlated between phases than originally shown by (Phillips, 2001), making 2-D corrections that much more important. For this study, we consider the benefits of 2-D path correction on discriminants as established by previous work, and explore a range of inversion techniques that might be used to obtain the amplitude models from which 2-D corrections can be generated.

The next step in this development is to invert for attenuation, site and source parameters simultaneously, using multi-band data. In this study we describe an inversion procedure that solves for 2-D attenuation, and absolute source and site effects for various parameterizations. The USArray provides tremendous regional phase data and this study builds on previous work that determined 1-Hz attenuation, with focus on resolution and correlation with regional geology (Phillips and Stead, 2008). The crux of the inversion is controlling the tradeoff between attenuation and stress, which we perform using constraints on corner frequencies, which can be provided by spectral ratio studies using coda (Mayeda, 2007), and direct waves (Fisk, 2007) applied to a cluster of events of many sizes. In the following we will describe data reduction and inversion methodology, application to Lg amplitude data recorded by the USArray, and present inversion results. Moment constraints from SLU and UCB as well as corner frequency constraints for Wells, Nevada, aftershocks (Mayeda, 2010) will be used to constrain absolute levels and break the attenuation-stress tradeoff. Techniques are simple to extent to multi-phase inversion, which adds no fundamental trade-off issues. The breadth and density of the USArray allows the visualization of spatial variations in amplitude residuals that indicates azimuthal source and near source focusing effects of variable strength, particularly source directivity.

2. Data Reduction

We obtained western US event origins from the Preliminary Determination of Epicenters weekly and monthly catalogs from January 2004 through July 2009 (Fig. 1). Events were selected by limiting to one per quarter-degree bin based on potential station coverage, which was estimated by counting the stations active at the time of the event within a circle of radius determined by a crude magnitude-distance detection criterion. Because data were requested in stages, we occasionally add additional events to each bin; all events were kept. Origin information was then used to request broad-band, vertical component, data segments, and instrument responses from the Data Management Center of the Incorporated Research Institutions for Seismology (IRIS). Data selection procedures are identical to those used by Phillips and Stead (2008), and an additional 2 years of data are included in this study. A record section for a southern Oregon event is shown in Fig. 2, in which we note the predominance of crustal (Pg and Lg) relative to

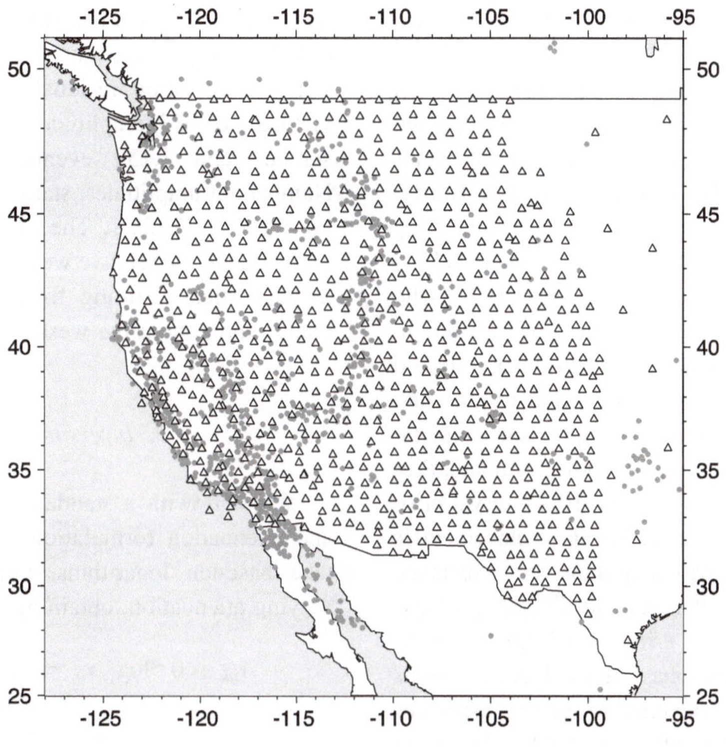

Figure 1

Stations (*triangles*) and events (*dots*) used in the tomography study. Station, response and waveform data were obtained from the IRIS Data Management Center. Event hypocenters were obtained from the USGS Monthly and Weekly bulletins. Events were chosen from a 0.25° grid to maximize potential station coverage

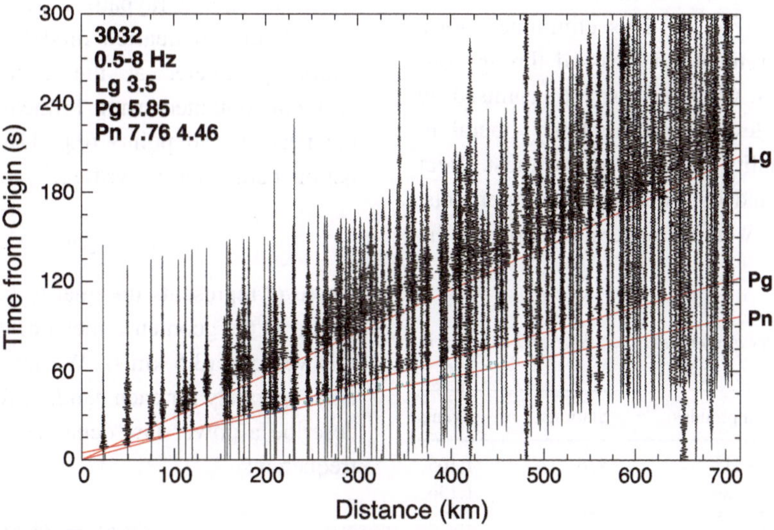

Figure 2

Vertical component, short period seismograms for a southern Oregon event. Some traces have been removed to improve clarity. Pass band, and parameters describing a layer-over-half-space travel time model are indicated in the *upper left*, *red lines* represent model arrivals, and associated phase names are noted on the *right*. *Blue symbols* represent Pn arrival picks made by the authors, *dark blue* are high quality picks, while *cyan* are fair quality picks. Pn arrivals were not determined for most traces due to poor signal to noise. Note the dominance of crustal phases in this section, typical for many areas of the western US

mantle (Pn and Sn) phases typical of western US regional waveforms.

To obtain regional phase amplitudes, we first corrected for instrument response, then bandpassed waveforms into nine overlapping octave width bands between 0.5 and 16 Hz (Table 1) using a Butterworth, four-pole, causal filters, and windowed the Lg using group velocities 3.6–3.0 km/s, prior to taking RMS measurements (similar to HARTSE, 1997, who used 2-pole 2-pass filters). No picks were used to adjust the measurement windows. We also collected background noise measurements prior to the first arrival, and at the end of the waveform segment, taking the lower of the two as our noise estimate. Pre-phase noise was collected over the 10-s interval prior to the measurement window. Window placement is shown in Fig. 3, along with amplitude versus distance plots for three events that sample diverse geology. We set a signal-to-noise criterion of 2.0 for background and 1.0 for pre-phase noise levels. Signals that passed were then converted to spectral estimates using Parseval's theorem (TAYLOR, 2002) to maintain consistency with a monitoring community standard. We further restricted to distances between 100 and 1,000 km. Following coda wave quality control methods described by (PHILLIPS, 2008a), we limited to amplitudes that crudely fit 1-D models compiled for gridded station sets (Fig. 4). This eliminated noise and Sn coda measurements that passed through our signal-to-noise filters, which can be compromised by time-varying noise levels, and are fairly liberal to begin with. This quality control step had most effect on high frequency and long distance measurements. An additional 1–2 % were trimmed after applying the tomographic method described in PHILLIPS and STEAD (2008) to data from each band, based on a three standard deviation residual criterion. This process yielded 465,633 amplitudes for analysis from 827 stations and 1,139 events, magnitudes 2.1–5.9. Numbers of amplitudes, stations and events per band are listed in Table 1. The abundance of Lg data is partly due to the relative weakness of Sn and Sn coda, noted earlier, resulting from an attenuating upper mantle throughout the western US.

Table 1

Band limits, and numbers of amplitudes, stations and events for each

Band limits (Hz)	Amplitudes	Stations	Events
0.5–1.0	49,411	816	1,105
0.75–1.5	70,789	826	1,136
1.0–2.0	74,175	826	1,138
1.5–3.0	68,425	827	1,138
2.0–4.0	60,507	826	1,134
3.0–6.0	49,469	818	1,127
4.0–8.0	41,630	816	1,124
6.0–12.0	29,299	809	1,091
8.0–16.0	21,928	802	1,069

3. *Inversion Methods*

We start with a standard source, site, spreading and attenuation formulation for spectral amplitude, take base-ten logarithms, and allow for laterally varying attenuation, obtaining:

$$A_{ijk} + 0.5\log_{10}x_{ij} = S_{ik} + T_k + R_{jk} - \log_{10}e\Sigma_l \delta x_{ijl}\alpha_{kl}, \quad (1)$$

where indices i, j, k, and l represent source, site, band, and raypath segment, respectively, S represents the source spectral amplitude, T is the source-to-Lg transfer function, R is the relative site effect, x is the great circle distance between source and receiver, δx is a raypath segment used to integrate through the attenuation model, and α is the attenuation parameter at the center of the raypath segment, obtained using bilinear interpolation from the nearest grid points (e.g. UM, 1987). The attenuation parameter is written:

$$\alpha = \omega/2Qv = \pi f/Qv, \quad (2)$$

where Q represents the quality factor, f is frequency (taken as the geometric mean of the band limits), and v is velocity (3.5 km/s). We invert for 2-D attenuation separately for each band, or for 2-D Q_0 and η that describe a power law dependence of attenuation on frequency:

$$Q(f) = Q_0 f^{\eta} \quad (3)$$

Our assumptions include great circle ray paths, a spreading rate of 0.5 (YANG, 2002), isotropic source radiation, a specific source-scaling model, and a regionally uniform and depth independent transfer

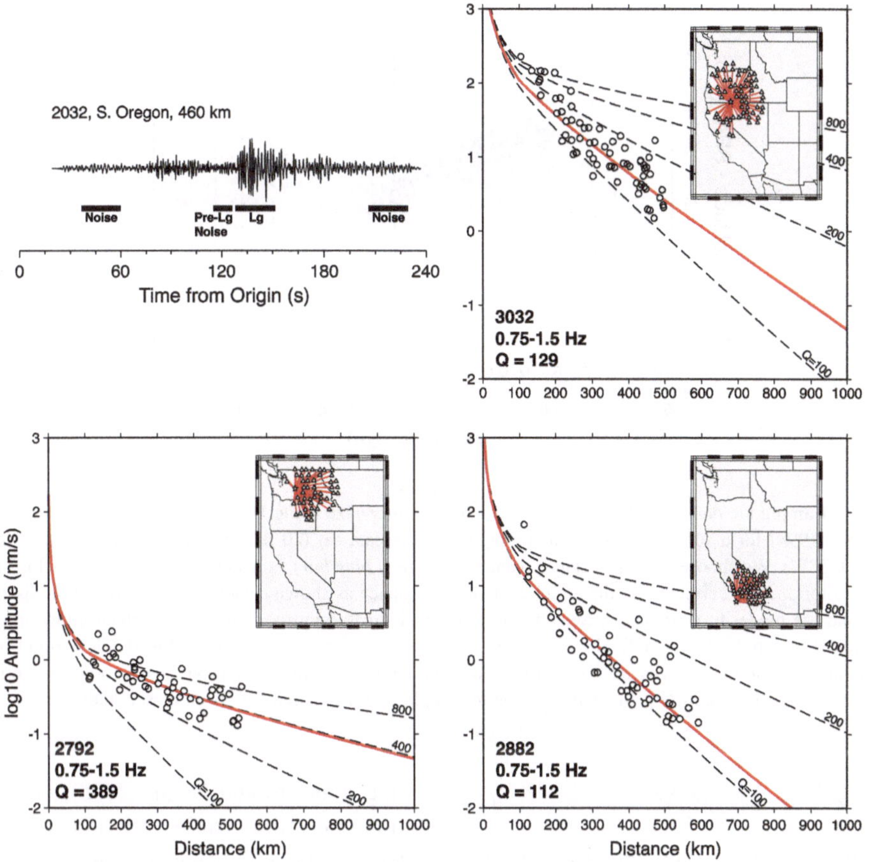

Figure 3
Amplitude measurement and decay with distance. The seismogram is from the southern Oregon event (Fig. 2). *Bars* represent Lg and noise windows. 1-Hz amplitude data for three events are shown in the remaining panels. The southern Oregon event is in the *upper right*, data from a central Washington event display high Q behavior in the *lower left*, and data from a southern California event display low Q behavior in the *lower right*. Best-fit Q are represented by the *red curves* and values are noted in the *lower left* of each panel, while Q for selected values are shown by the *dashed* and *labeled curves*. Site and 2-D attenuation effects contribute to the data scatter. Thumbnail maps show the raypath distributions *(red)* for each event

function. The site and transfer terms cannot be resolved separately, and we solve for their sum, which has been termed the absolute site effect (MALAGNINI, 2004). The transfer function includes the excitation of the regional phase amplitude, which depends on the physics of regional phase generation, such as the range of ray parameters that combine to form the phase, as well as the quality of the spreading model, and details of the measurement procedure, described above. Following the MDAC formulation of WALTER (2001), we assume a BRUNE (1970) source model:

$$S(f) = M_0 \Big/ \left(1 + (f/f_0)^2\right), \tag{4}$$

with scaling described by:

$$2\pi f_0 = (K\sigma/M_0)^{1/3}, \tag{5}$$

$$K = 16\pi/\beta_s^2 \left(R_p^2 \zeta^3 \Big/ \alpha_s^5 + R_s^2/\beta_s^5\right), \tag{6}$$

where M_0 is moment, f_0 is corner frequency, σ is apparent stress (WYSS, 1970), α_s and β_s are estimates of near source compressional and shear velocities (6 and 3.5 km/s), respectively, R_p and R_s are average compressional and shear radiation (0.44 and 0.6), respectively, and ζ is the ratio of compressional to shear corner frequencies, assumed to be 1.0. Further, to allow for non-constant stress scaling (WALTER, 2001) write apparent stress as:

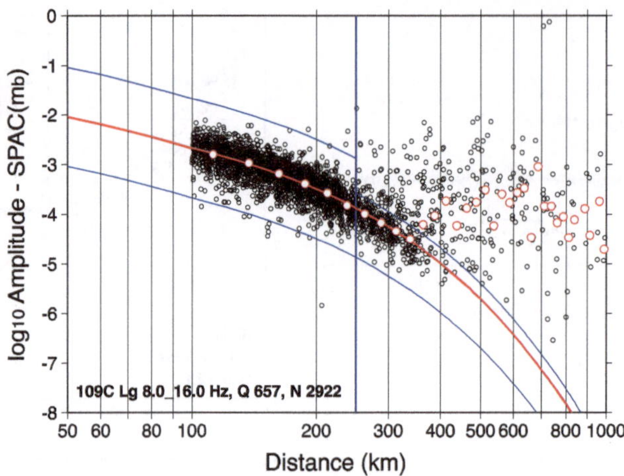

Figure 4
Initial quality control of amplitude data. Amplitudes are corrected for source size using USGS magnitudes and a magnitude-based scaling model (SPAC, TAYLOR, 1998) for a southern California sub-region of the data set (all data for stations with latitude less than 35°N, and longitude less than −115°W). We choose a cutoff distance manually (*vertical blue bar*), and a *Q*-plus-site-term model is fit to data below that distance using an L_1 criterion. Site corrected results are plotted versus distance as shown (*black circles*), along with distance-binned median levels (*red circles*) and the model fit (*red curve*). Selection criteria are indicated by the *blue curves*, which allow an order of magnitude deviation from the model, except for high amplitudes above the cutoff distance, for which we only allow 0.3 deviation from the model. High source-corrected amplitudes at long distances are noise or coda measurements that slip through our signal-to-noise filters, mixed with good data from high *Q* paths that we hope to sample with measurements from closer stations

$$\sigma = \sigma' \left(M_0/M_0'\right)^{\psi}, \qquad (7)$$

where ψ is related to scaling rate, zero for constant stress, cube root scaling, and 0.25 for non-constant stress, quarter root scaling.

We solve a number of inverse problems based on these equations for combinations of 2-D attenuation, 2-D power law exponent, absolute site terms, and individual source terms or event moment and apparent stress. For most effective regularization, and to provide positivity constraints, we re-parameterize and solve for natural logarithms of attenuation, power law exponent, moment and stress. After including regularization constraints, we linearize with respect to a starting model, use analytic partial derivatives, and solve iteratively using LSQR (PAIGE, 1982).

Regularization is critical to the solution of these problems, and we describe in detail here. First, attenuation and its power law exponent, when used, are regularized with first difference spatial constraints. We note that regularizing the logarithm of attenuation is more natural than inverting for and regularizing attenuation itself. The latter technique is commonly employed as the problem is linear (e.g. PHILLIPS and STEAD, 2008), but such regularization constraints cause variations in low *Q* regions to be more heavily penalized, and, conversely, unnatural peaks in *Q* are often obtained in high *Q* regions. This is especially apparent in synthetic tests, and inverting for logarithmic attenuation provides a superior match in such cases. When inverting for 2-D attenuation for each band (instead of using the power law exponent), we apply a second-difference regularization between bands at each spatial grid point. These constraints are lightly weighted to encourage power-law like behavior in areas of poor high frequency coverage, but were turned off for the tests presented here. We also tried damping attenuation in an attempt to match mean levels determined by inversions applied to each band individually, and thus help control the attenuation-stress tradeoff. This controlled the tradeoff poorly, and for results presented here we turn these constraints off.

When inverting for individual, frequency dependent source terms, we damped the sum of source term changes from the starting model to zero for each band. When inverting for source parameters, we damped the sum of moment changes from the starting model to zero for independent moment events, and damped stress changes from the starting model to zero for the independent corner frequency events. Starting source models were based on measured moments and corner frequencies, and were filled in using magnitudes and a standard scaling law for uncalibrated events. This allowed the inversion to stray from the independent moments, as long as the mean level stayed the same, but allowed no variation from the corner frequency measurements. Independent moments were obtained from publicly available lists created by the University of California, Berkeley, and St. Louis University based on regional waveform modeling. Independent corner frequency measurements were obtained from a coda spectral ratio study of the Wells, Nevada, sequence (MAYEDA, 2009). Constraining Wells corner frequencies controlled the tradeoff between attenuation and stress, and greatly facilitated the simultaneous inversion using multiple band amplitudes. Not all stations recorded the Wells events, but the source calibration is expected to extend to regions that are well resolved by crossing raypaths, which includes all but the edge regions where Q-source and Q-site tradeoffs remain. Finally, we regularized stress to follow a scaling law. The weight on this constraint could be varied to obtain a best-fit uniform apparent stress, or allow stress to vary event by event, following FEHLER and PHILLIPS (1991). In this study, the weight was set lightly to encourage events with unconstrained corners to move towards the mean stress for a region surrounding that event (400 km) based on a scaling rate of $\psi = 0$ (Eq. 7).

Site terms were left free in this study. The source constraints dictate their final levels.

We ran four types of inversions for this study, parameterizing attenuation individually by band, or with the frequency dependent model (Eq. 3), and parameterizing the source individually by band, or with the omega-2 model (Eq. 4); while all inversions parameterize site effects as frequency dependent terms for each station:

1. Invert for 2-D Q for each band, and frequency dependent source terms for each event.
2. Invert for 2-D Q_0, 2-D η (Eq. 3), and frequency dependent source terms for each event.
3. Invert for 2-D Q for each band, and moment plus apparent stress (corner frequency, Eq. 4) for each event, include source constraints.
4. Invert for 2-D Q_0, 2-D η, and moment plus apparent stress for each event, include source constraints.

We also invert for best fit 1-D $Q(f)$, Q_0 and η for Runs 1 and 2, and best fit 1-D $Q(f)$ and uniform site term (each phase and band) for Run 1. These will be used to evaluate the improvement we obtain using the 2-D models.

4. Results

We applied tomographic inversions to USArray amplitude data using a latitude-longitude grid with edges 25° to 51°, and −128° to −95°, respectively, and a grid spacing of 0.5°. We show best-fit uniform Q models from Inversions 1 and 2, in which we allow site terms to vary freely, in Fig. 5. The individual band Q results show a slight upward curvature, which can result from noise at the band extremes, but appears minimal in this case. The uniform Q_0–η result matches the individual band results well. Note

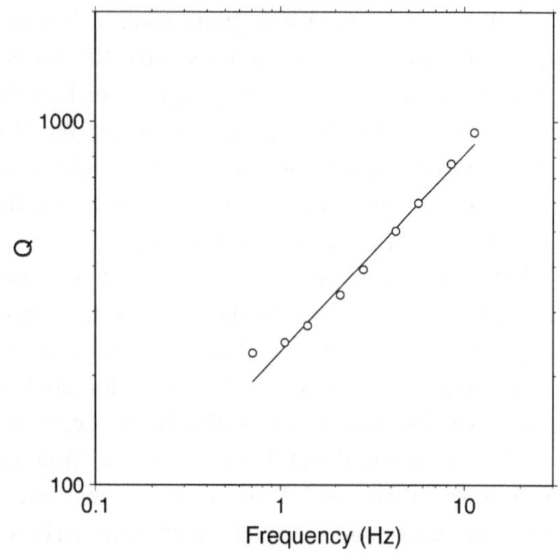

Figure 5
Best-fit uniform Q models. *Circles* represent Q obtained individually for each band, while the *line* represents the best-fit frequency dependent model (Q_0 232, η 0.54). Results are weighted towards regions of high ray density

that the uniform Q results are weighted towards areas with highest raypath density, which tend to be areas of lower Q. We show two 2-D Q images from Inversion 1 in Fig. 6. The Q range is proportionally larger by a factor of two or so for the lower band, which is commonly observed for frequency dependent Q. The more compact range allows the limited color range to display more spatial detail in the high frequency image. We also show Q_0–η images from Inversion 4 in Fig. 7. The η image is plotted using a reversed color bar, which emphasizes the inverse correlation between Q_0 and η.

The 1-Hz Q reported by Phillips and Stead (2008) used a much finer grid interval (0.2°) and covered a smaller area that did not include extreme high Q in the Great Plains, thus the image from that study is somewhat sharper and better defined. We are less concerned with high-resolution detail here; however, basic patterns are the same. Because we invert and regularize logarithmic quantities, high Q levels are muted in this study, for the reasons discussed earlier. Phillips and Stead (2008) noted the correlation between low Q and shear zones (coastal California, Walker Lane) and volcanic areas (Cascades, Yellowstone, San Francisco field, and others surrounding the Colorado Plateau), and between high Q and older, stable crust (Colorado and Columbia plateaus, Central Valley) and intrusive regions (Sierra Nevada, Peninsular Ranges). This study extends 10° further eastward, and we now see high-Q regions such as the Wyoming Basin and Great Plains provinces. We also define a low-Q region encompassing the Colorado Rockies, Rio Grande Rift, and volcanic fields on the southeast edge of the Colorado Plateau.

Q increases at higher frequencies, but regional patterns remain somewhat similar. Low, or flat power law exponents generally correlate with high 1-Hz Q, most obviously for the Great Plains, Colorado Plateaus, Wyoming Basin, and Columbia Plateau, but also for the Sierra-Great Basin, western Mohave, Snake River Plain and Peninsular Ranges. Steep power law exponents correlate with low 1-Hz Q, particularly Pacific coastal regions, the Salton Trough, Rocky Mountains, Rio Grande Rift, Yellowstone, and portions of the Basin and Range.

Moment and corner frequency results from Inversion 4 are shown in Fig. 8. As discussed above, we allow moments to drift from independently determined levels, while retaining their overall level. Outlier moments exist and would be worth looking into further. Otherwise, scatter appears to be 0.2, which is similar to that observed when the same procedure is applied in coda calibration studies. Apparent stress ranges from below 0.1 to above 1 MPa, with means of 0.1 MPa for smaller events and 0.3 MPa for larger events. The slight increase in stress with event size appears to be driven by the data because we regularized the stress to a self-similar model ($\psi = 0$) as described above. We plot apparent stress in map view in Fig. 9. There is much scatter in these results; however, coherent patterns can be seen on small scales in some areas. For example, we observe low stress segments along the San Andreas Fault in central and northern California, with a high stress break north of the central California creeping zone and high stress approaching the Mendocino bend. This contrasts with higher stress in the Walker Lane, along the Garlock fault, and in the Transverse Ranges. Events along the eastern edge of the Basin and Range appear to be lower than average. It is more difficult to see coherent patterns in less seismically active areas; however, a few clusters appear to have coherent stress levels (San Simeon, California; Wells, and Kane Spring, Nevada; Yellowstone, W. Colorado, SW Wyoming, and SW Montana, among others). The "low stress" clusters seen in W. Colorado and SW Wyoming are likely mining areas, and our use of earthquake source models may be inappropriate. Stress is not expected to be well resolved for events outside of the USArray footprint, such as those in Baja California and the Gulf of California. Stress results are sensitive to high frequency noise and may require more intensive quality control on an event-by-event basis before we interpret further.

Site terms are an important part of the inversions, and are found to vary greatly compared to similar studies that rely on permanent, often vault, or borehole deployed, sensors. We believe this is due to the temporary nature of this special array, and a preference for softer sites that allow for more rapid deployment. As a result, site terms reflect very near surface properties and it is difficult to see patterns in site term maps, although subtle differences were

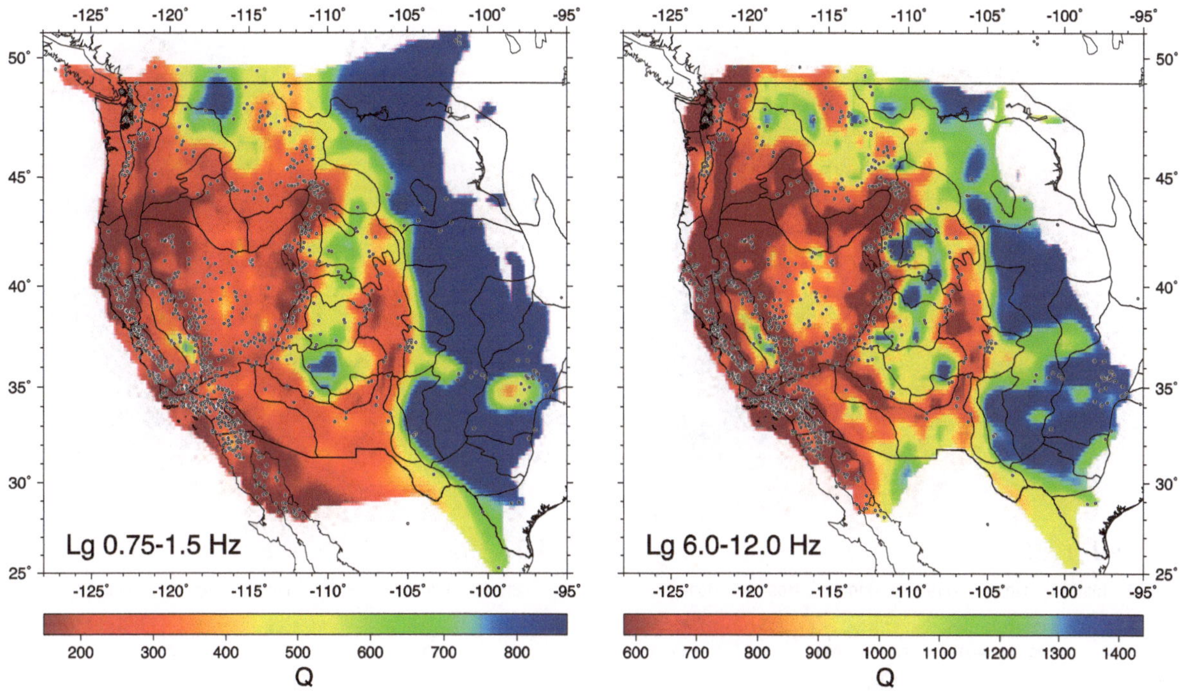

Figure 6
Q for two frequency bands. Results for the 0.75–1.5 Hz band are shown on the *left* and for the 6–12 Hz band on the *right*. *Color bars* differ, we see roughly twice the Q variation at low frequency that we see at high frequency. Geophysical provinces are *outlined* and events used in the inversions are represented by *small dots*. Results are from Inversion 1

noted between Coast Range and Sierra Nevada sites by Phillips and Stead (2008).

RMS residuals from the Inversions 1 through 4 as well as for uniform models are plotted by band in Fig. 10. Residuals for uniform Q and uniform site terms are very large, reaching 0.35 for high bands. Allowing site terms to vary reduces residuals considerably, with RMS peaking just above 0.2, although we expect some Q variation to be absorbed in unrealistic manner by site terms in these runs. Uniform $Q(f)$ (Inversion 1) and uniform Q_0–η (Inversion 2) fits to data are nearly identical. Inversion 1 for 2-D Q displays the lowest RMS, approaching 0.1, but also employs the largest number of free parameters. The number of degrees of freedom is small relative to the number of amplitudes, so we expect the squared RMS to approximate the true residual variance fairly well. In particular, the number of effective free parameters in the attenuation grids, as measured by the trace of the resolution matrix, is normally a small fraction of the total number of grid points used in the inversion (Phillips *et al.*, 2005; Phillips and Stead, 2008). Inversions 2, 3, and 4 increase in misfit, in that order. Interestingly, applying the power-law Q produces very little misfit, relative to applying the omega-squared source model. This is of note because the source model is physically derived while the power law Q is an empirically driven choice. The residuals increase slightly with frequency, which may result from larger proportions of short distance data in the high band populations, with the higher scatter due to the shorter measurement window lengths. Estimation of the band center using the geometric mean may work less well at high frequencies due to source rolloff, and may also be a factor. The highest and lowest bands are fit well by Inversion 1 and less well by the rest, indicating higher noise levels in those bands, or power-law Q and source model breakdown. We favor the former cause, and are cautioned in interpreting Q results for those edge bands as a consequence. We do expect some misfit for mining events, especially if Rg contaminates the Lg measurements, but do not expect much effect on overall scatter due to their small numbers. The outstanding

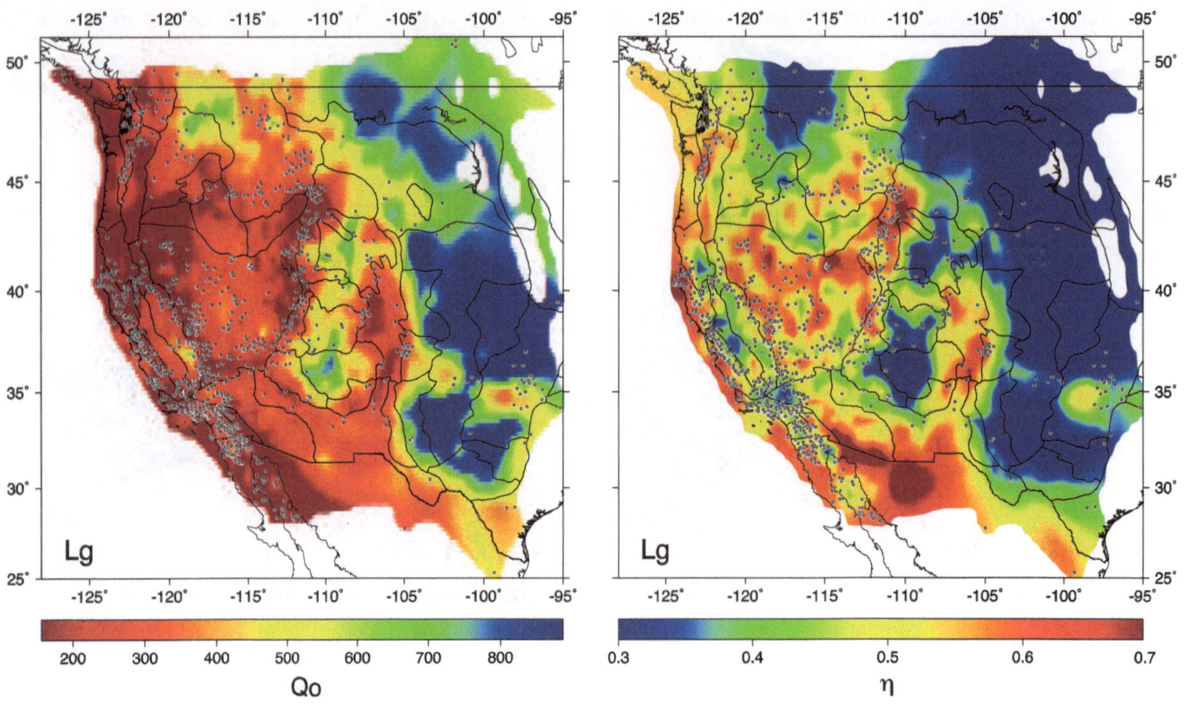

Figure 7
Frequency dependent attenuation. 1-Hz Q (Q_0) is shown on the *left* and the frequency dependent η parameter is shown on the *right*. Geophysical provinces are outlined and events used in the inversions are represented by *small dots*. Results are from Inversion 4

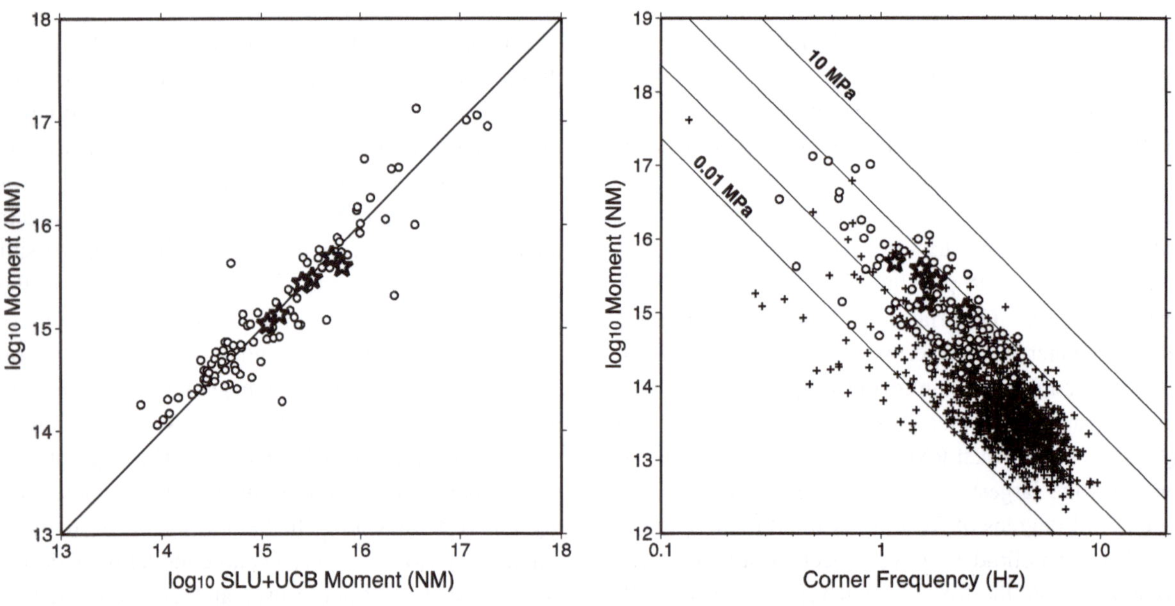

Figure 8
Moment relationship and scaling results. Inversion moments are plotted versus independent values (*circles*) on the *left*. *Stars* represent well-studied Wells aftershocks used to constrain apparent stress (corner frequency). Moment versus corner frequency results are shown on the *right* (+ symbols), with source constraint points shown using the same symbols as on the *left*. Constant apparent stress *lines* are indicated. Results are from Inversion 4

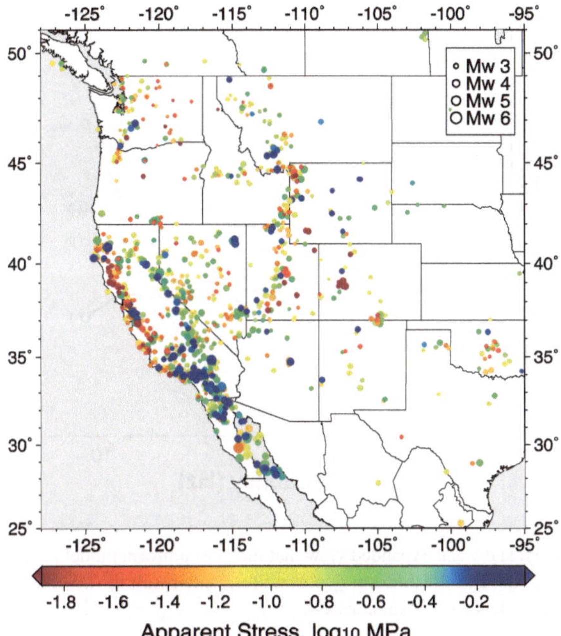

Figure 9
Distribution of apparent stress results. Stress is denoted by symbol color as shown by the *color bar*, and event size is denoted by symbol size as shown in the legend. Results are from Inversion 4

performance of the power-law Q models favors their use with sparse data sets with less resolving power.

The variance reduction between uniform and 2-D models varies by band and between different types of inversions. At worst, reduction is only 50 % for the lowest band, comparing uniform and 2-D results for Inversion 1, allowing site effects free in the uniform Q calculation (1U, Fig. 10). At best, reduction is 90 % comparing high frequency results for Inversion 1, using the uniform Q and uniform site model (1UU, Fig. 10).

5. Discussion

Phillips and Stead (2008) discussed quantitative similarities of their 1-Hz Lg Q image with 1-D regional work of Erickson *et al.* (2004), and results are fairly consistent for higher frequencies as well. Erickson *et al.* (2004) analyzed the same frequency range as this study (0.5–16 Hz) and gave frequency dependent exponents of 0.61–0.72, with errors of 0.15 or so for northern and southern California, the Basin and Range, the Pacific Northwest and the Rocky Mountains. We see from Fig. 7 that the frequency exponent varies between 0.5 to over 0.7 within these regions, but on average, is about 0.6. We avoid the Great Valley and Sierra provinces in taking the northern California average. For southern California, we include low Q coastal and Salton Trough regions, as well as the higher Q Mojave region. For the central US (Erickson *et al.* 2004) observe 1-Hz Q of 640 ($\pm$225) and a frequency exponent of 0.34 ($\pm$ 0.18), similarly for the northeastern US, and our frequency exponents for the Great Plains, and Colorado and Columbia Plateaus are 0.3 or less. These areas are different from the regions analyzed by Erickson *et al.* (2004), so we offer a comparison between generic high Q regions. In all cases, our frequency exponents are 0.1 lower than Erickson *et al.* (2004). These differences are within errors given for their 1-D studies, but the systematic bias could be significant. The 1-D regional results are weighted spatially towards areas of high seismicity, which correlate with lower 1-Hz Q and higher frequency exponent, while our visual averaging from the 2-D map does not take that into account. We have not considered possible biases introduced by differences in measurement techniques, so we are satisfied with a reasonable match.

Although the USArray deployment allows a time-limited sampling of events, we can still benefit from comparisons of our apparent stress results with those of previous studies in the region. Shearer (2006) apply a stacking method to P spectra using 12 years of southern California data, and compute a stress drop based on Madariaga (1976) that we expect to be a factor of 7.5 larger than our apparent stress (based on equating expressions for corner frequency in the two cases). There are no events in common between the two studies. However, we see cases where results are spatially consistent. Both studies observe low stress along the southern San Andreas Fault, near the Salton Sea (0.1 vs. 0.4 MPa, current study listed first, and by "low stress" we refer to the stress range observed in southern California). Stress increases to the west, along the San Jacinto Fault (0.2–1 vs. 0.6–2.5 MPa, both lower in the south), continuing to the Elsinore Fault (0.6–2 vs. 2.5–6.3 MPa). Both studies observe high stress in the wedge between the San Andreas and northern San

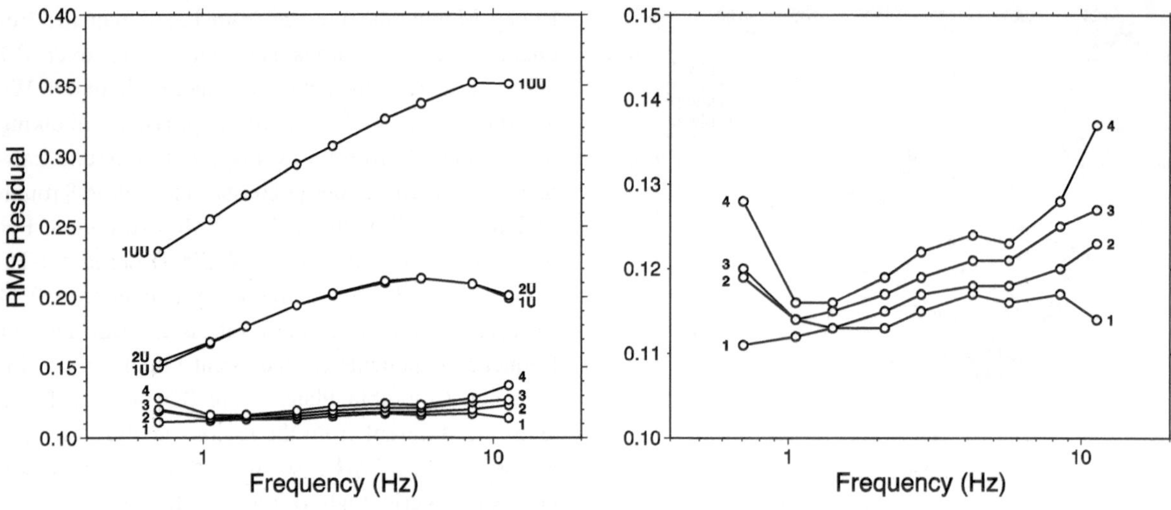

Figure 10
RMS $\log_{10}$ amplitude residuals versus frequency for various inversions. The *left* shows an expanded view that includes uniform model results. Inversions *1–4* are shown, as well as uniform *Q* results for Inversions *1* and *2* (*U*), and the uniform *Q* plus uniform site result for Inversion *1* (*UU*). The scale on the *right* shows greater detail for Inversions *1–4*

Jacinto Faults (1–2 vs. 4–6.3 MPa), and in the Ridgecrest-Coso area, including a cluster to the west (1–2 vs. 4–6.3 MPa). Other low stress areas in common include the San Fernando and Santa Ynez areas (0.2 vs. 0.6–1.6 MPa), and a small cluster northwest of Coso (0.2 vs. 0.4–1 MPa). More detailed comparisons require event-by-event quality control, which we reserve for future efforts. We see a few areas where a match is not achieved, but the small sample used here, the different wavetypes and methods, and lack of common events, lead us to conclude that the results are reasonably consistent. Our small sample precludes broad comparisons of source parameters; however, calibrations based on USArray data for stations with long deployment histories can be used to determine source parameters for past and future events across the region.

The breath and density of the USArray allows us to check our assumptions in unprecedented ways. In particular, our assumption of isotropic Lg radiation can be examined, as the multitude of crossing raypaths should average out any physical effects, and not allow this type of noise to leak into attenuation models. Anisotropic source effects should thus be visible in station-based residual patterns. Plotting residuals by station for selected events from the Wells sequence reveals interesting patterns (Fig. 11). First, the main shock shows a dramatic bimodal distribution of residuals, with high amplitudes to the NNE and low in the opposite direction. The axis of the asymmetry is roughly parallel to the strike of the fault. We believe that rupture directivity drives this pattern (see Dreger *et al.*, 2009; Mendoza and Hartzell, 2009).

Interestingly, an aftershock with focal mechanism similar to the mainshock shows a muted, but spatially anti-correlated pattern of residuals. Spot checks show that envelopes of these events line up nicely, so differences are not due to mislocation, and confirm the higher mainshock amplitudes for northern stations (incidentally even higher for Pg waves) after adjusting for event size using the coda (also see Mayeda, 2010). Patterns for a third aftershock with a strike slip mechanism display a four-lobed pattern, with high amplitude in the expected SH maxima directions, even though we measure the vertical component, indicating a certain amount of SH to SV conversion if this is a focal mechanism effect. In general, patterns are muted, yet spatially coherent, and vary between aftershocks, even though events are from the same source region. The patterns are similar for higher

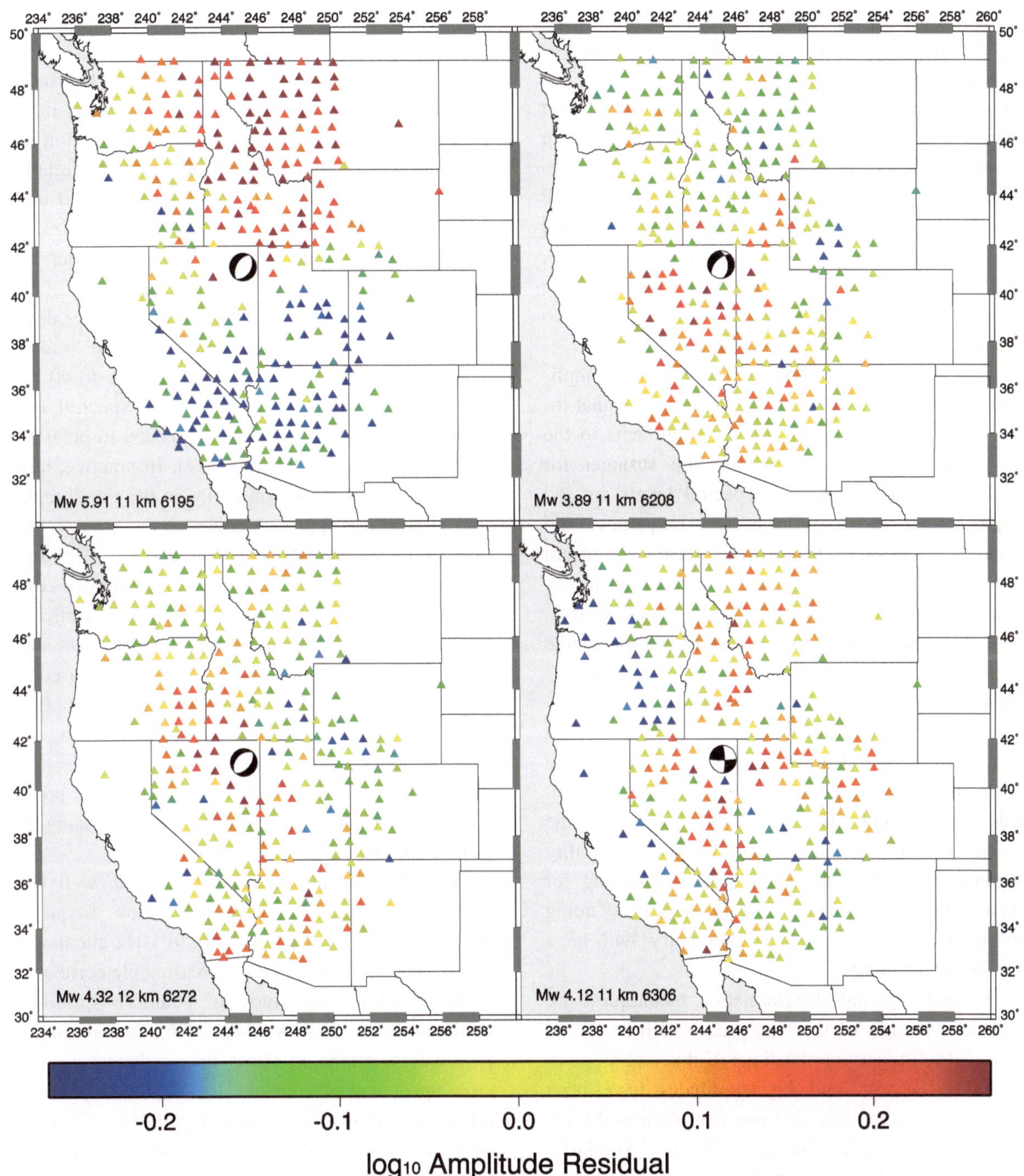

Figure 11

Residual maps for selected Wells events. Log_{10} amplitude residuals for the 0.5–1 Hz band are shown as *colored triangles* for the Wells mainshock (*upper left*) and three aftershocks. *Blue* represents amplitude deficit, while *red* represents amplitude excess, with respect to Inversion 4. SLU source orientation is shown at the Wells location, and corresponding Mw and depth are noted in the *lower left* of each plot. We observe strong directivity effects for the mainshock, weak directivity in the opposite direction for the *upper right* aftershock, spatially coherent but uninterpreted patterns in the *lower left*, typical of most events, and nodal planes expected for SH radiation from a strike-slip event in the *lower right* bands, although even weaker, and observed with fewer stations. We believe azimuthal source effects, including directivity, are partially responsible for these patterns, but that near source, 3-D propagation effects that can vary rapidly across a source region the size of Wells must also play a role. Such mechanisms were proposed for the variation of P waveforms from Semipalatinsk explosions observed across dense networks of Europe (BARKER, 1992), although of much longer range. We note that these effects on Lg would have been very difficult to observe without the USArray.

One effect of the dramatic variation in amplitudes for the Wells mainshock is that our signal-to-noise cut emphasizes the high measurements to the NNW, and because the patterns are stronger for lower bands, low frequency spectral levels are too large, thus decreasing the corner frequency, and apparent stress estimates. For this reason we do not use the Wells mainshock data or constraints in our final inversions. This effect, if more generally true, may also contribute to the poor residuals for source parameter inversions in lower bands (Fig. 10). The coda spectral ratios are less sensitive to the azimuthal source effects (MAYEDA, 2009) and we trust corner frequencies from that analysis, the direct Lg is the issue. To test this further, we binned residuals by azimuth for all events used in the inversions and found that azimuthal variability correlated weakly with event size. Accounting for such effects is a topic for future work, and doing so in areas of sparse station geometry will be a significant challenge.

To test regional discrimination techniques, we will need models of the type presented here for all phases. Preliminary applications of this technique to Asia Pn, Pg, Sn, and Lg amplitudes have been presented in accessible, but non-peer reviewed form (e.g. FISK, 2007; PHILLIPS, 2010). Similarly, Q appears to be transversely isotropic, and preliminary results show 10 % variance reduction for Lg by including those terms (PHILLIPS *et al.*, 2008b). Anisotropic terms may be effective in describing bulk wave propagation behavior along, versus across, major geological structures and phase blockages, as well as in reflecting effects of geological fabric on regional phases.

6. Conclusions

We have shown that USArray data can be used to resolve Lg attenuation and source parameters in the western US. Lg is widely observed in the region due to high mantle attenuation, leading to the relative absence of Sn, and Sn coda. The study relied upon independent moments from waveform studies, as well as independent measurements of corner frequencies from a coda spectral ratio study of the Wells, Nevada, sequence. The independently determined corner frequencies controlled the tradeoff between corner frequency (or apparent stress) and attenuation. Under our assumptions, spectral constraints from only one event are needed to break the tradeoff and set the absolute level. In practice, however, we expect resolution issues in the tomographic inversion as we approach edges of the study region due to Q-site and Q-source tradeoffs. Additional spectral constraints in those areas will improve edge-region model resolution. Further, a large collection of spectral constraints will allow cross-validation studies as are done for travel-time models used in monitoring, and will allow testing of assumptions such as regionally uniform phase excitation. The same issues are important in coda wave studies, and we propose that both coda and direct wave propagation models be constructed in parallel using the same high quality spectral constraints.

We observed attenuation to correlate well with regional geology, and its power-law frequency dependence correlated positively to 1-Hz attenuation levels. Apparent stress showed reasonable correlation with a high-density study of southern California earthquakes by SHEARER *et al.* (2006). Accounting for regional variations in stress may reduce scatter in cross-spectral discriminants. Lg amplitude residuals often showed spatially coherent patterns, of varying strength, reflecting azimuthal source radiation, directivity, or near-source focusing effects. Spatial patterns were not always the same for nearby events, even for those with similar focal mechanisms. Thus, following correction for laterally varying attenuation, scatter in regional distance discriminant ratios will remain for earthquakes due to source and near-source effects. We see hints that these effects can be larger for larger events and lower frequencies, but this will require further effort to substantiate. If so, scatter will be lower for events and frequencies of greatest interest for regional monitoring. We believe that attenuation and source parameter inversions of the type described here can be used to set absolute calibration levels that will allow transportable comparisons of source spectra between regions of study, and should be of great help in isolating test site effects on explosion spectra from regional phases of all types, including coda, to evaluate their potential for broad area yield estimation.

Acknowledgments

This study relied on waveform and ancillary data collected by the Earthscope USArray project. Waveforms we used included contributions from the ANZA Regional, Berkeley Digital Seismograph, Caltech Regional Seismic, Global Seismograph, Western Great Basin, USArray Transportable, US National Seismic, and U. Utah Regional networks. We further acknowledge the Array Operations Facility (NMT), the Array Network Facility (UCSD), and the IRIS Data Mangement Center for efforts to collect and archive USArray data for use by the scientific community. We also thank Robert Herrmann, Douglas Dreger, and students for their timely production of moment tensor results for public consumption. SAC and GMT software were used for processing and display. We greatly appreciate input from two anonymous reviewers. WSP thanks Mark Fisk for discussions about application of source constraints in Asia, and Michael Fehler for introducing the author to source parameter-attenuation inversions many years ago. Publication of this research was supported by the US DOE under contract DE-AC52-06NA25396.

References

Barker, B.W. and J.R. Murphy (1992), *A lithospheric velocity anomaly beneath the Shagan River Test Site: 1. Detection and location with network magnitude residuals*, Bulletin of The Seismological Society of America, *82*, 980–998.

Baumont, D.A, A. Paul, S. Beck, and G. Zandt (1999), *Strong crustal heterogeneity in the Bolivian Altiplano as suggested by attenuation of Lg waves*, J. Geophys. Res., *104*, 20,287–20,305.

Brune, J.N. (1970), *Tectonic stress and spectra of seismic shear waves from earthquakes*, J. Geophys. Res., *75*, 4997–5009.

Campillo, M. (1987), *Lg wave propagation in a laterally varying crust and the distribution of the apparent quality factor in central France*, J. Geophys. Res. *92*: 12604–12614.

Campillo, M., B. Feignier, M. Bouchon, and N. Betoux (1993), *Attenuation of crustal waves across the Alpine range*, J. Geophys. Res., *98*, 1987–1996.

Dreger, D.S., S.R. Ford, and I. Ryder (2009), *Finite-source study of the Wells, Nevada earthquake*, Nev. Bur. Mines and Geol., Reno.

Erickson, D., D.E. McNamara, and H.M. Benz (2004), *Frequency dependent Lg Q within the continental United States*, Bull. Seism. Soc. Am., *94*, 1630–1643.

Fehler, M. C. and W. S. Phillips (1991), *Simultaneous inversion for Q and source parameters of microearthquakes accompanying hydraulic fracturing in granitic rock*, Bull. Seism. Soc. Am., *81*, 553–575.

Fisk, M. and S. R. Taylor (2007), *Robust magnitude and path corrections for regional seismic phases in Eurasia by constrained inversion and enhanced kriging techniques*, in Proceedings of the 29th Monitoring Research Review: Ground-Based Nuclear Explosion Monitoring Technologies, LA-UR-07-5613, Vol. 1, pp. 24–33.

Ford, S.R, W.S. Phillips, W.R. Walter, M.E. Pasyanos, K.M. Mayeda, and D.S. Dreger (2009), *Attenuation tomography of the Yellow Sea/Korean Peninsula from coda-source normalized and direct Lg amplitudes*, Pure and Applied Geophysics, *167*, 1163–1170.

Hartse, H.E., R.R. Taylor, W.S. Phillips, and G.E. Randall (1997), *A preliminary study of regional seismic discrimination in central Asia with emphasis on western China*, Bull. Seism. Soc. Am. *87*, 551–568.

Madariaga, R. (1976), *Dynamics of an expanding circular fault*, Bull. Seism. Soc. Am., *66*, 639–666.

Malagnini, L., K.M. Mayeda, A. Akinci,and P.L. Bragato (2004), *Estimating absolute site effects*, Bull. Seism. Soc. Am., *94*, 1343–1352.

Malagnini, L., K.M. Mayeda, R. Uhrhammer, A. Akinci, and R.B. Herrkmann (2007), *A regional ground motion excitation/attenuation model for the San Francisco region*, Bull. Seism. Soc. Am., *97*, 843–862.

Mayeda, K.M., L. Malagnini, W.S. Phillips, W.R. Walter, and D. Dreger (2005), *2-D or not 2-D, that is the question: A northern California test*, Geophys. Res. Lett., *32*, L12301, doi:10.1029/2005GL022882.

Mayeda, K.M., L. Malagnini, and W.R. Walter (2007), *A new spectral ratio method using narrow band coda envelopes: Evidence for non-self-similarity in the Hector Mine sequence*, Geophys. Res. Lett., *34*, L11303, doi:10.1029/2007GL030041.

Mayeda, K.M., and L. Malagnini (2009), *Apparent stress from coda-derived source ratios: Regional variations, similarities, and differences*, Seism. Res. Lett., *80*:2, 336.

Mayeda, K.M., and L. Malagnini (2010), *Source radiation invariant property of local and near-regional shear-wave coda: Application to source scaling for the Mw 5.9 Wells, Nevada sequence*, Geophys. Res. Lett., *37*, L07306, doi:10.1029/2009GL042148.

Mendoza, C., and S. Hartzell (2009), *Source analysis using regional empirical Green's functions: The 2008 Wells, Nevada, earthquake*, Geophys. Res. Lett., *36*, L11302, doi:10.1029/2009GL038073.

MITCHELL, B.J., Y. PAN, J. XU, and L. CONG (1997) *Lg coda Q variation across Eurasia and its relation to crustal evolution*, J. Geophys. Res., *102*, 22,767–22,779.

OTTEMOLLER, L., N.M. SHAPIRO, S. KRISHNA SING, and J.F. PACHECO (2002), *Lateral variation of Lg wave propagation in southern Mexico*, J. Geophys. Res., *107*, doi:10.1029/2001JB000206.

PASYANOS, M.E., W.R. WALTER, and E.M. MATZEL (2009), *A simultaneous multiphase approach to determine P-wave and S-wave attenuation of the crust and upper mantle*, Bulletin of The Seismological Society of America, *99*, 3314–3325.

PAIGE, S.S. and M.A. SAUNDERS (1982), *Algorithm 583, LSQR: Sparse linear equations and least-squares problems*, Trans Math Software, *8*, 195–209.

PEI, S., J. ZHAO, C.A. ROWE, S. WANG, T.M. HEARN, Z. XU, H. LIU, and Y. SUN (2006) *ML amplitude tomography in North China*, Bull. Seism. Soc. Am., *96*, 1560–1566, doi:10.1785/0120060021.

PHILLIPS, W.S., G.E. RANDALL, and S.R. TAYLOR (1998), *Regional phase path effects in central China*, Geophys. Res. Lett., *25*, 2729–2732.

PHILLIPS, W.S. (1999), *Emperical path corrections for regional phase amplitudes*, Bull. Seism. Soc. Am., *89*, 384–393.

PHILLIPS, W.S., H.E. HARTSE, S.R. TAYLOR, and G.E. RANDALL (2000), *1 Hz Lg Q tomography in central Asia*, Geophys. Res. Lett., *27*, 3425–3428.

PHILLIPS, W.S., H.E. HARTSKE, S.R. TAYLOR, A.A. VELASCO, and G.E. RANDALL (2001), *Application of regional phase amplitude tomography to seismic verification*, Pure Appl. Geophys., *158*, 1189–1206.

PHILLIPS, W.S., H.E. HARTSE, and J.T. RUTLEDGE, *Amplitude ratio tomography for regional phase Q*, Geophys. Res. Lett., *32*, doi:10.1029/2005GL023870, 2005.

PHILLIPS, W.S., R.J. STEAD, G.E. RANDALL, H.E. HARTSE and K. MAYEDA (2008a), Source effects from broad area network calibration of regional distance coda waves, in Advances in Geophysics, 50, Scattering of Short Period Waves in the Heterogeneous Earth, H. Sato and M.C. Fehler, Editors, 319–351.

PHILLIPS, W.S., C.A. ROWE, R.J. STEAD, and D. COBLENTZ (2008b), Mapping Lg Q and anisotropy using the USArray, Proc. 2008 IRIS Workshop, June 4–6, Stevenson, WA, 143.

PHILLIPS, W.S. and R.J. STEAD (2008), *Attenuation of Lg in the western US using the USArray*, Geophys. Res. Lett., *35*, L07307, doi:10.1029/2007GL032926.

PHILLIPS, W.S., X. YANG, R.J. STEAD, M.L. BEGNAUD, and K.M. MAYEDA (2010), Model development for broad area event identification and yield estimation, Proc. 2010 Monitoring Research Review, 486–494.

POMEROY, P.W., W.J. BEST, and T.V. MCEVILLY (1982), *Test ban treaty verification with regional data-a review*, Bulletin of the Seismological Society of America, *72*, S89–129.

RODGERS, A.J., W.R.WALTER, C.A. SCHULTZ, S.C. MYERS, and T. LAY (1999), *A comparison of methodologies for representing path effects on regional P/S discriminants*, Bull. Seism. Soc. Am., *89*, 394–408, 1999.

RODGERS, A.J., and W.R. WALTER (2002), *Seismic discrimination of the May 11, 1998 Indian nuclear test with short-period regional data from station NIL (Nilore, Pakistan)*, Pure And Applied Geophysics, *159*:4, 679–700.

SANDVOL, E., K. AL-DAMEGH, A. CALVERT, D. SEBER, M. BARAZANGI, R. MOHAMAD, R. GOK, N. TURKELLI, and C. GURBUZ (2001), *Tomographic imaging of Lg and Sn propagation in the Middle East*, Pure Appl. Geophys., *158*, 1121–1163.

SERENO, T.J., S.R. BRATT, and T.C. BACHE (1988), *Simultaneous Inversion of Regional Wave Spectra for Attenuation and Seismic Moment in Scandinavia*, J. Geophys. Res., *93*, 2019–2035.

SHEARER, P.M., G.A. PRIETO, and E. HAUKSSON (2006), *Comprehensive analysis of earthquake source spectra in southern California*, J. Geophys. Res., *111*, B06303.

SINGH, S.K., and R.B. HERRMANN (1983), *Regionalization of crustal coda Q in the continental United States*, J. Geophys. Res., *88*, 527–538.

TAYLOR, S.R., and H.E. HARTSE (1998), *A procedure for estimation of source and propagation amplitude corrections for regional seismic discriminants*, J. Geophys. Res., *103*, 2781–2789.

TAYLOR, S.R., A.A. VELASCO, H.E. HARTSE, W.S. PHILLIPS, W.R. WALTER, and A.J. RODGERS (2002), *Amplitude corrections for regional seismic discriminants*, Pure Appl. Geophys., *159*, 623–650.

TAYLOR, S.R., X.N. YANG, and W.S. PHILLIPS (2003), *Bayesian Lg attenuation tomography applied to eastern Asia*, Bull. Seism. Soc. Am., *93*, 795–803.

UM, J., and C. THURBER (1987), *A fast algorithm for two-point seismic ray tracing*, Bulletin of the Seismological Society of America, *77*, 972–986.

WALTER, W.R., and S.R. TAYLOR (2001), A revised magnitude and distance amplitude correction (MDAC2) procedure for regional seismic discriminants: Theory and testing at NTS, report UCRL-ID-146882, Lawrence Livermore Natl. Lab., Livermore, California, http://www.llnl.gov/tid/lof/documents/pdf/240563.pdf.

WYSS, M. (1970), *Stress estimates of South American shallow and deep earthquakes*, J. Geophys. Res., *75*, 1529–1544.

XIE, J.K., and B.J. MITCHELL (1990), *A back-projection method for imaging large-scale lateral variations of Lg coda Q with application to continental Africa*, J. Geophys. Res., *100*, 161–181.

XIE, J.K., Z. WU, R. LIU, D. SCHAFF, Y. LIU, and J. LIANG (2006), *Tomographic regionalization of crustal Lg Q in eastern Eurasia*, Geophys. Res. Lett., *33*, L03315, doi:10.1029/2005GL024410.

YANG, X. (2002), *A numerical investigation of L_g geometrical spreading*, Bulletin of the Seismological Society of America, *92*, 3067–3079.

ZOR, E., E. SANDVOL, J.K. XIE, N. TURKELLI, B.J. MITCHELL, A.H. GASANOV, and G. YETERMISHLI (2007), *Crustal attenuatioin within the Turkish Plateau and surrounding regions*, Bull. Seism. Soc. Am., *97*, 151–161, doi:10.1785/0120050227.

(Received December 15, 2011, revised January 11, 2013, accepted January 15, 2013, Published online August 8, 2013)

Pure Appl. Geophys. 171 (2014), 485–506

DOI 10.1007/s00024-012-0632-z

Pure and Applied Geophysics

Mantle Attenuation Estimated from Regional and Teleseismic P-waves of Deep Earthquakes and Surface Explosions

G. Ichinose,[1] M. Woods,[1] and J. Dwyer[1]

Abstract—We estimated the network-averaged mantle attenuation t^*(total) of 0.5 s beneath the North Korea test site (NKTS) by use of P-wave spectra and normalized spectral stacks from the 25 May 2009 declared nuclear test (mb 4.5; IDC). This value was checked using P-waves from seven deep (580–600 km) earthquakes ($4.8 < M_w < 5.5$) in the Jilin-Heilongjiang, China region that borders with Russia and North Korea. These earthquakes are 200–300 km from the NKTS, within 200 km of the Global Seismic Network seismic station in Mudanjiang, China (MDJ) and the International Monitoring System primary arrays at Ussuriysk, Russia (USRK) and Wonju, Republic of Korea (KSRS). With the deep earthquakes, we split the t^*(total) ray path into two segments: a t^*(u), that represents the attenuation of the up-going ray from the deep hypocenters to the local-regional receivers, and t^*(d), that represents the attenuation along the down-going ray to teleseismic receivers. The sum of t^*(u) and t^*(d) should be equal to t^*(total), because they both share coincident ray paths. We estimated the upper-mantle attenuation t^*(u) of 0.1 s at stations MDJ, USRK, and KSRS from individual and stacks of normalized P-wave spectra. We then estimated the average lower-mantle attenuation t^*(d) of 0.4 s using stacked teleseismic P-wave spectra. We finally estimated a network average t^*(total) of 0.5 s from the stacked teleseismic P-wave spectra from the 2009 nuclear test, which confirms the equality with the sum of t^*(u) and t^*(d). We included constraints on seismic moment, depth, and radiation pattern by using results from a moment tensor analysis and corner frequencies from modeling of P-wave spectra recorded at local distances. We also avoided finite-faulting effects by excluding earthquakes with complex source time functions. We assumed ω^2 source models for earthquakes and explosions. The mantle attenuation beneath the NKTS is clearly different when compared with the network-averaged t^* of 0.75 s for the western US and is similar to values of approximately 0.5 s for the Semipalatinsk test site within the 0.5–2 Hz range.

Key words: Mantle attenuation, deep earthquakes, nuclear explosion monitoring, teleseismic P-waves.

[1] Air Force Technical Applications Center, TTR, 1030 South Highway A1A, Patrick Air Force Base, Brevard, FL 32925, USA. E-mail: gichinose@aftac.gov

1. Introduction

A fundamental question in the estimation of yields of nuclear explosions is the effect of upper mantle attenuation on P-waves at approximately 1 Hz. Estimates of attenuation are needed to calibrate the body-wave magnitude (mb) and also to model teleseismic P-wave spectra for calculating explosion yield. The upper mantle attenuation operator t^* is typically used for this purpose and is measured from the spectral slope of teleseismic P-waves above the corner frequency. Differential t^* studies examine only the ratio of P-wave spectral slopes between two regions to assess if changes to current mb-yield relationships are required; however, absolute t^* measurements are necessary to make more precise yield measurements from either mb (Der and McElfresh, 1980; Der *et al.*, 1985a, b) or teleseismic P-spectra (Burger 1987).

The 2006 and 2009 declared North Korea nuclear tests were conducted in an area that had not been well studied for upper mantle attenuation before 2006, and no history of explosion calibration data were available. Recent studies using differential t^* (M. Woods personal communication) suggest that the upper mantle attenuation beneath the North Korea test site (NKTS) is similar to that of the Semipalatinsk test site (STS) in Kazakhstan. However, a study by Hwang *et al.* (2011) on differential global t^* puts the attenuation beneath NKTS similar to or higher than the western US at the Nevada National Security Site (NNSS), formerly known as the Nevada Test Site. This scenario has major implications because the yields for the 2006 and 2009 declared North Korea nuclear tests would need to be much larger to fit with the higher attenuation NNSS mb-yield formulas rather than the STS formulas. A more focused study on absolute t^* is needed to address whether the mantle attenuation beneath NKTS is similar to that beneath NNSS or STS.

To estimate the mantle attenuation, we use deep earthquakes that occur frequently near the NKTS to partition the absolute t^* into two parts (Fig. 1). $t^*(\mathrm{u})$ is the contribution from the up-going ray from the hypocenter to the surface through the upper mantle, which can be measured from stations that are located above the hypocenter at local-to-regional epicenter distances (<500 km). $t^*(\mathrm{d})$ is the contribution from the down-going ray between the hypocenter to teleseismic distances between approximately 30° and 99°. With a nearby surface explosion, $t^*(\mathrm{total})$ can be measured along a ray path from the source to receivers at teleseismic distances. The sum of $t^*(\mathrm{u})$ and $t^*(\mathrm{d})$ should be equal to $t^*(\mathrm{total})$, because both share nearly coincident ray paths (Fig. 1) and because t^* represents a path-integrated effect of the quality factor Q along a ray. Besides providing an independent validation of t^* measurements for surface explosions, the method will also give us a better understanding of the difference between upper and lower-mantle attenuation.

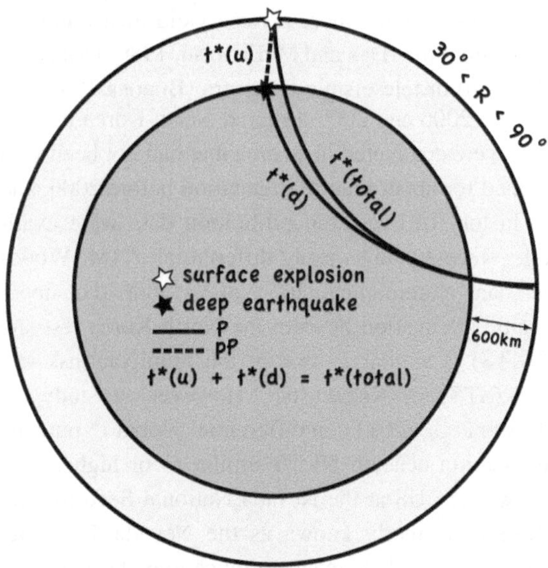

Figure 1
The attenuation $t^*(\mathrm{total})$ measured from a near surface explosion should be approximately equal to the sum of the attenuation $t^*(\mathrm{u})$ measured from a deep earthquake to the surface above the source and the attenuation $t^*(\mathrm{d})$ between the deep earthquake to teleseismic distance

We used deep earthquakes to measure local and global averages of $t^*(\mathrm{u})$ and $t^*(\mathrm{d})$ and check them against the estimated global network average $t^*(\mathrm{total})$ obtained from the declared 2009 North Korea nuclear test. We used P-wave spectra from deep (500–660 km) earthquakes ($M_{\mathrm{w}} > 4.8$–5.5) that frequently occur beneath the Jilin-Heilongjiang, China, region that borders China, Russia, and North Korea. These deep earthquakes are approximately 200–300 km from the NKTS and approximately 200 km from the Incorporated Research Institutions for Seismology (IRIS) seismic station in Mudanjiang, China (MDJ) and IMS station in Ussuriysk, Russia (USRK). The local and teleseismic P-wave displacement spectra were modeled individually and also normalized and stacked over as many as six selected similar deep earthquakes in a cluster. We normalized spectra over as many as 5–22 stations and stacked them for a single event to estimate a network-averaged spectral slope. We also tested $t^*(\mathrm{d})$ using the principle of reciprocity by reversing the source and receiver and estimating $t^*(\mathrm{d})$ from a deep earthquake in the Bismarck Sea region, which was at teleseismic distances to stations around the NKTS. We further checked $t^*(\mathrm{u})$ and $t^*(\mathrm{d})$ values by modeling P and pP for one deep event at an epicentral distance of 43° (Barazangi 1975).

Using deep earthquakes to estimate attenuation has several benefits. At local epicenter distances, waveform data do not have depth phases, refracted rays, or turning rays. Stations in the near field observe only the up-going P and S-wave fields. At teleseismic distances, pP and P travel times are approximately 100 s apart for a source at 600 km, so the pP-phase does not interfere with attenuation estimates on the direct P-phase. Another benefit is that at near-regional epicentral distances attenuation is negligible compared with teleseismic distances; therefore, the trade-off between source corner frequency and attenuation is minimized for near-regional distance P-waves. We use results from a regional moment tensor analysis to determine seismic moment, centroid depth, and radiation pattern independently. These source data are then used as input for modeling local and teleseismic P-wave spectra and should provide more robust and unique estimates for t^*. We also normalize spectra over several stations or events and make spectral stacks to model average local and global t^* values.

A review on absolute t^* measurements from the last four decades returns a wide range of values between very low attenuation t^* to very high values of 1.3 s (e.g. DER, 1977; Table 1). One of the earliest studies by TREMBLY (1968) measured a t^* of 0.96 s ($Q = 450$) from Canada to NNSS. DER (1977) examined spectra from GNOME and SALMON tests in New Mexico and Mississippi at stations across North America for distances less than 30° and found t^* values around 0.4 s ranging between 0.03 and 0.77 s. They also measured teleseismic distance t^* from Novaya Zemlya to North America and found an average value of $\sim$0.2 s ranging between −0.08 and 0.41, much lower than for the NNSS to within North America. In the late 1970s and early 1980s several studies were published on waveform modeling from teleseismic P-waves of nuclear explosions recorded by long-period World-Wide Standard Seismograph Network (WWSSN) instruments. These studies (e.g., BURDICK and HELMBERGER, 1979) typically found that P-wave t^* values of 1–1.3 s yielded reasonable waveform fits. Responses by DER and MCELFRESH (1980) and DOUGLAS and HUDSON (1983) questioned the validity of such large t^* values and stated possible effects of frequency-dependent t^* and source overshoot for the 0.5 s difference in t^* between spectral and waveform estimates. Around this time, researchers recognized that differences in t^* between long-periods and short-periods may be accounted for using frequency-dependent Q based on an absorption band model (e.g., recommendation by CORMIER, 1982). BACHE *et al.* (1985), DER *et al.* (1985a, b, 1986a) and DOUGLAS (1987) published more t^* values for central Asia and NNSS that continued to be below 0.5 s; however, it was clear that differences of approximately 0.2 s existed between NNSS and Kazakhstan, notable in both time-domain waveforms and spectra. These studies typically see signals out to 2–8 Hz, particularly at far-regional distances between approximately 10° and 30°.

The 0.5 s differences in t^* between earlier studies (e.g., DER 1977) and subsequent waveform based studies (e.g., BURDICK and HELMBERGER, 1979) may have been partly because of inadequate bandwidth from older WWSSN instruments (e.g., CHOY and CORMIER, 1986) and also earlier long-period estimates for t^* that were not adjusted for frequency dependence. BURGER (1987) estimated $t^* \approx 0.8$ s from NNSS using short-period WWSSN data in the 0.5–4 Hz band. With frequency-dependent Q models they extended values of t^* out to higher frequencies that were consistent with previous short-period P-wave spectral studies (e.g., DER *et al.*, 1985a, b). CHOY and CORMIER (1986) modeled P-waves from deep earthquakes using modern digital broadband seismograms and placed t^* values from waveform modeling more on a par with short-period spectra. WARREN and SHEARER (2000) estimated an average global t^* value of 0.5 s using 17,836 P and 14,721 PP spectra in the 0.5–0.86 Hz band from an archive of digital data spanning over a decade. HWANG *et al.* (2011) used 190,000 teleseismic P and S-wave spectral ratios up to 0.8 Hz to generate a tomographic map of global differential t^* variations. These differential t^* values at NKTS appear higher than or equal to values near the NNSS, which has implications for yield estimation of the 2006 and 2009 NKTS nuclear explosions given the previously high attenuation estimated at NNSS.

2. *Data*

We used three-component broadband digital waveforms archived at the US National Data Center (US NDC). We also included supplemental waveform data from the Global Seismic Network (USGS/IRIS) and International Monitoring System (IMS) primary and auxiliary stations. We searched for earthquakes with depths greater than 500 km from the US NDC and USGS/NEIC earthquake catalogs in the Jilin-Heilongjiang, China border region and found 18 earthquakes with mb >4.8 between the years 2000 and 2011 (Fig. 2).

Complex source rupture effects must be avoided because they may introduce multiple corner frequencies with intermediate spectral slopes which, if not included in source spectral shape functions, can cause artificially lower t^* values. Ideally, attenuation should be measured from Green functions (impulse response of the Earth); therefore earthquakes with simple impulsive P-wave source time functions are

Table 1

Deep event source data

	Origin time (YYYY/MM/DD HH:mm)	mb	M_w	Depth (km)	% DC	% VR	Strike	Dip	Rake
A	2011/05/10 15:26	5.4	5.50	600	26	91.5	1	10	125
B	2011/01/07 23:34	4.8	4.95	600	74	41.9	52	31	139
F	2009/08/10 12:42	4.8	5.03	580	66	85.8	93	39	170
H	2009/04/18 03:56	4.9	5.09	600	93	79.8	82	22	175
J	2008/10/22 16:18	5.0	4.92	580	95	73.7	67	30	150
N	2004/08/15 15:36	5.0	4.94	580	93	94.3	150	83	88

VR variance reduction, *DC* double couple

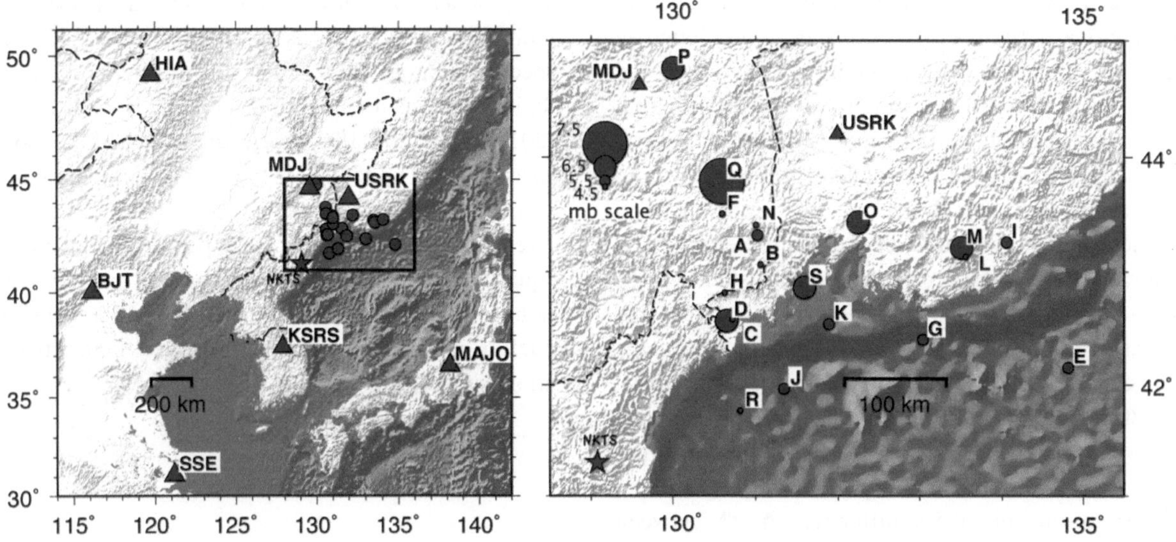

Figure 2
Locations of 18 deep earthquakes selected from the US NDC and USGS/NEIC earthquake catalogs. We selected a cluster of seven events from the catalog on the basis of analysis of P-waves at station MDJ (Table 1). These deep earthquakes are approximately 100–200 km from station MDJ and USRK. These deep events are also 200–300 km from the NKTS

needed. We examined the P-wave complexity of vertical component P-wave displacements at station Mudanjiang, China (MDJ) (Fig. 3) for identifying source finiteness effects. We noted that all of the P-waves were one-sided, but some that were emergent and complex and some that were impulsive and simple. This difference is likely to be because of the large depths of the events, similar focal mechanisms, and close epicenter distances to MDJ. P-waves left the source at nearly the same place on the focal sphere with strong impulsive radiation and identical polarities. The analysis of source time functions, typically requiring the deconvolution of P-waves with empirical Green functions, was not necessary in this unique case. We used the MDJ P-wave displacements as a proxy for source time functions and identified directly viable empirical Green functions for attenuation analysis. Figure 3 shows the first 10 s after arrival of the P-wave. Seven events with mb <5.5 (Table 1) had simple, short durations (approximately 0.8–1 s) and single pulse-like P-waves. Several earthquakes, with magnitudes greater than 6 had P-wave pulse durations two or more seconds longer in duration (events D, O, P, Q, R, and S) and in some cases had two or more sub-events (events O and Q).

The seismic moment and P-wave radiation pattern are important constraints needed in modeling P-wave amplitude spectra. We inverted the regional

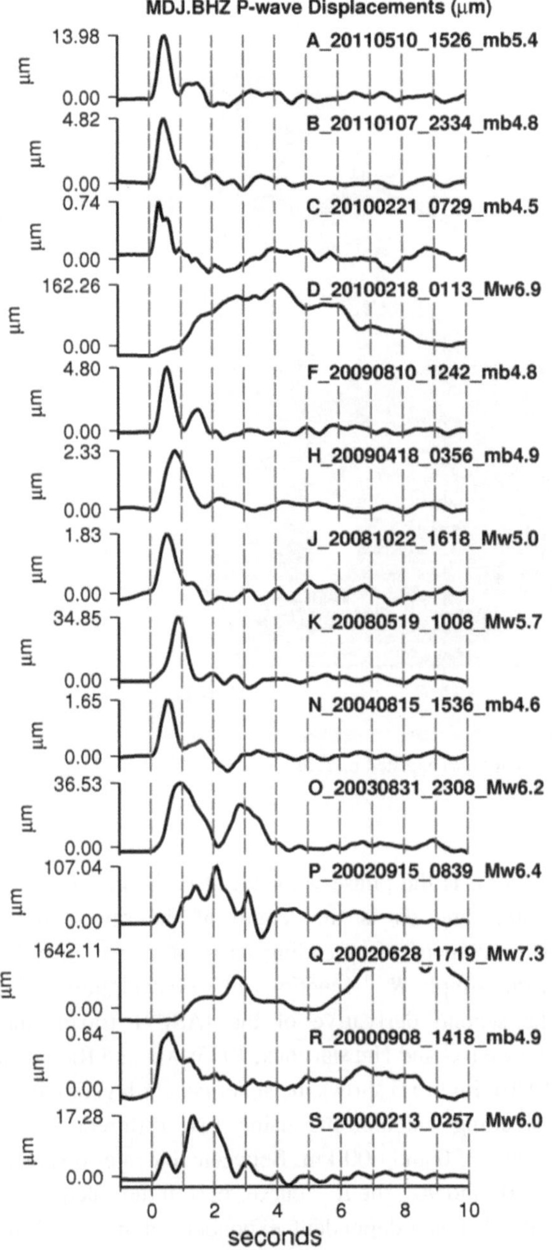

Figure 3
Vertical component, instrument corrected P-wave displacements recorded at station MDJ (100–200 km epicenter distance) from 14 deep earthquakes (>580 km depth). Simple short (<1 s) duration pulse-like P-waves were observed for earthquakes labeled *A*, *B*, *C*, *F*, *H*, *J*, and *N*. These earthquakes all had mb < 5.5 and seem to have ruptured as simple sources. The earthquakes with mb > 6 have longer durations (>2 s) or have multiple sub-event ruptures

distance long-period waves using the moment tensor inversion method (Dreger and Helmberger, 1993; Ritsema and Lay, 1995; Ichinose *et al.*, 2003). The frequency-wave number (*f*-k)-integration method (Zeng and Anderson, 1995) was used to compute Green functions in 10 km depth increments between 10 and 660 km depth. We used the IASP91 one-dimensional layered Earth velocity and density model (Kennett and Engdahl, 1991) down to a depth of 700 km but simplified to only 80 layers. The three-component broadband regional data were instrument corrected for displacement and filtered between 20 and 100 s with a three-pole, two-pass Butterworth filter. We inverted for the point source equivalent deviatoric moment tensors for a range of depths and origin times by shifting the data in 1-s increments about the reported origin time. The solution with the highest variance reduction was selected. Moment tensor solutions were estimated for 6 of the 14 deep earthquakes in the cluster (Fig. 4; Table 1). We also picked the P-wave first motions at the regional stations and computed take-off angles using the same 1D layered Earth model. The first motion polarities are plotted on to the focal sphere for each event and checked to make sure that the moment tensor mechanisms were consistent with the first motion polarities. It is interesting to note that the 10 May 2011 earthquake has a large (80 %) compensated linear vector dipole (CLVD) component (Julian *et al.*, 1998). A full-moment tensor including isotropic Green functions resulted in a low 10 % isotropic component; a large (70 %) CLVD component remained, however. This suggests that the CLVD is not a result of pure implosive source or double-couple shear faulting but possibly from volume contraction because of metastable olivine to beta-spinel mineral transformation within the subducting slab (Kirby *et al.*, 1991).

3. Methods

The earthquake P-wave spectral amplitudes $U(\Delta, f, M_o, R_p)$ can be modeled using linear filter theory, which represents a convolution of a source term $S(f, M_o, R_p)$, an attenuation term $\mathbf{A}(f)$, and a geometrical spreading term $\mathbf{G}(\Delta)$ where Δ is the hypocenter distance (i.e., $\Delta = (R^2 + h^2)^{0.5}$), f is frequency, M_o is the seismic moment, and R_p is the radiation pattern coefficient.

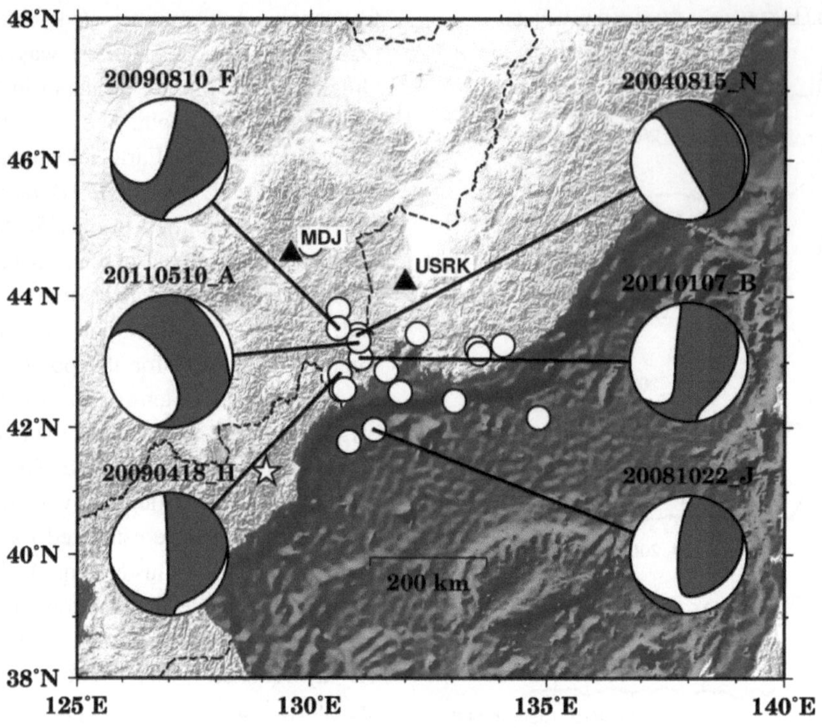

Figure 4
Moment tensor solutions from the six deep earthquakes listed in Table 1

$$U(\Delta, f, M_o, R_p) = S(f, M_o, R_p) \cdot A(f) \cdot G(\Delta)$$

We use an omega-square (ω^2) P-wave source spectral shape function model (Brune, 1970). The source term is:

$$S(f, M_o, R_p) = \frac{R_p F_s M_o}{4\pi\rho\alpha^3 \left(1 + \left(\frac{f}{f_c}\right)^{2n}\right)^{\frac{1}{2}}}$$

where $\boldsymbol{\rho}$ is the density, $\boldsymbol{\alpha}$ is the P-wave velocity near the source, $\boldsymbol{f_c}$ is the corner frequency, and $\boldsymbol{F_s}$ is the free surface amplification correction (e.g., 2 for SH-plane-wave). A value of 2 for $\boldsymbol{\eta}$ results in a high frequency fall-off equivalent to an ω^2 source spectrum. The attenuation term $\boldsymbol{A(f)}$ is $\mathbf{exp}(-\pi\ f\ t^*)$, where the attenuation operator t^* represents the path averaged $\boldsymbol{Q^{-1}}$ along a ray path $\boldsymbol{s}$, i.e.,

$$t^* = \int_s \frac{ds}{cQ}$$

where $\boldsymbol{c}$ is the phase velocity, $\boldsymbol{Q}$ is the mean quality factor, and $\mathbf{ds}$ is the length of the ray segment. The geometrical spreading term $\boldsymbol{G(\Delta)}$ was modeled using simple $\boldsymbol{R^{-n}}$ functions fit to calculations using the second derivative of the IASP91 travel-times (Langston and Helmberger, 1975; Aki and Richards, 1980). Figure 5 shows the spherical earth geometrical spreading calculations using two different source depths of 0 and 600 km. Between epicenter distances of 30° and 90°, the R^{-n} curves best fit the calculations with distance-dependent exponents of $n = 1.13$ for shallow seismic events and $n = 1.05$ for deep earthquakes.

For explosions, we used the Mueller and Murphy (1971) model that also assumes a ω^2 high-frequency decay. We selected granite emplacement data (Stevens and Day, 1985) for NKTS. Figure 6 shows example explosion P-wave displacement spectra using four different emplacement conditions for a 2.4 kT yield (Park, 2009) buried at 300 m depth. The

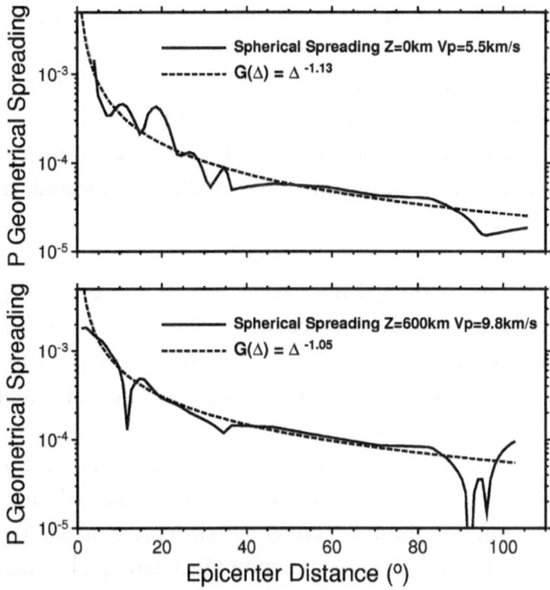

Figure 5
Spherical earth geometrical spreading calculations using the second derivative of the IASP91 travel times with respect to distance. (*Top*) Calculation made for a surface event. (*Bottom*) Calculation made for a hypocenter depth of 600 km. We fit the calculations using a simple distance-dependent geometrical spreading term with an exponent of −1.13 for the shallow events and −1.05 for deep earthquakes

spectra are computed at a distance of 6,671 km without attenuation and then again with a t^* of 0.5 s.

4. *Analysis and Results of t*(u)*

We estimated the upper-mantle attenuation $t^*(u)$ from local-distance P-wave spectra using deep (500–660 km) and moderate-sized ($M_w > 4.8$) earthquakes occurring beneath the Jilin-Heilongjiang, China, region near NKTS (Table 1). We then estimated the lower mantle attenuation $t^*(d)$ from P-wave spectra at teleseismic distances using the same cluster of deep earthquakes. We measured the attenuation t^*(total) from the 25 May 2009 declared nuclear test using a set of 22 teleseismic-distance global network stations. The sum of $t^*(u)$ and $t^*(d)$ from the deep earthquakes should be equal to the t^*(total) estimated for the P-wave of the surface explosion because they share similar paths. Finally, we analyzed two shallow ($M_w \sim 5.6$) earthquakes on and near the NNSS and compared the network-averaged t^* estimates. from NNSS with NKTS.

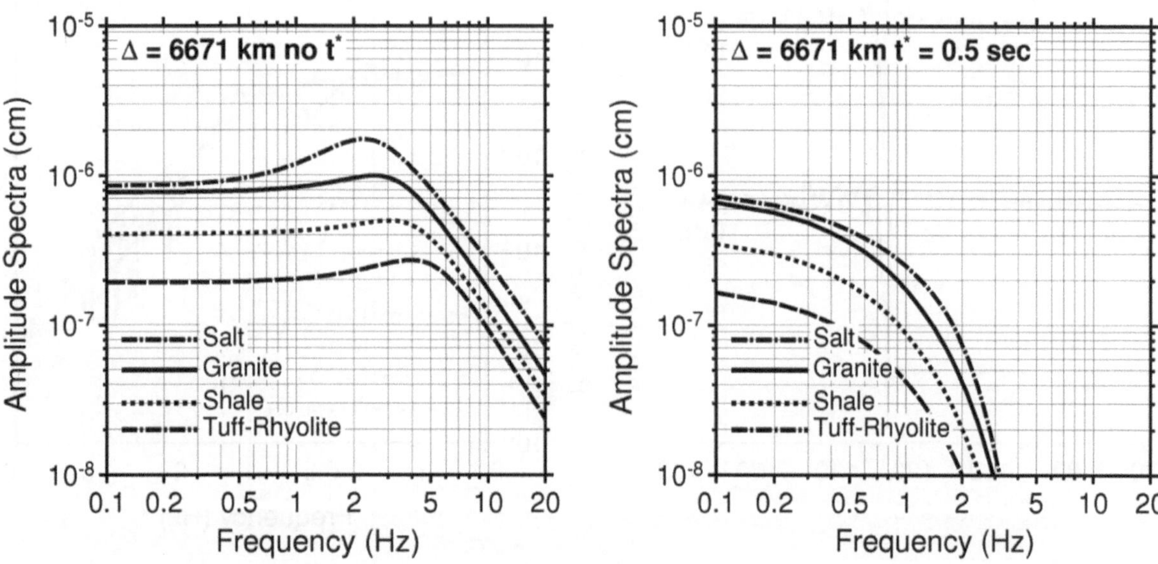

Figure 6
Example 2.4 kT Mueller–Murphy (1971) P-wave explosion source spectra corrected for geometrical spreading at a distance of 60°. We used granite, shale, tuff, and salt emplacement data from Stevens and Day (1985). *Left P*-wave spectra are calculated for the four emplacement data without t^* attenuation. *Right* we applied a t^* of 0.5 s

We first analyzed the most recent deep event at station MDJ. Figure 7 shows the instrument-corrected three-component displacement seismograms recorded by station MDJ at an epicentral distance of 184 km. The 10 May 2011 mb 5.4 (M_w 5.5) earthquake was located at a depth of 600 km based on depth-phases and regional moment tensor modeling. The seismograms have very simple characteristics with an impulsive P-wave on the vertical component and impulsive S-wave arrivals on the horizontal components. We then calculated signal spectra from 5-s windows around the vertical component regional P-wave and noise spectra from a 5-s window ending 1 s before the P-wave arrival. A 10 % Hanning taper was applied to the time window, which was then instrument-corrected to displacement. The data were zero-padded before FFT and then spectra were smoothed using a moving three-point mean averaging window. The seismic moment and radiation pattern-correction coefficient were based on results from an independent moment tensor analysis (Table 1). We then computed a suite of synthetic earthquake spectra for a variety of corner frequencies (f_c) and t^* values, and visually selected a best-fit f_c of 0.85 Hz and a t^* of 0.1 s for the 10 May 2011 M_w 5.5 event (Fig. 7).

A cluster of six deep earthquakes was selected because the earthquakes were closely located in space, had similar faulting mechanisms, and had impulsive single-pulse P-wave displacements recorded at station MDJ (Figs. 3, 8). These deep earthquakes were at epicentral distances between 100 and 300 km from MDJ (Fig. 2). We normalized the displacement spectra, canceling out the different event sizes (seismic moments) so that the spectra could be stacked together for estimating the average high-frequency spectral slope, with normalized logarithmic amplitudes across the six earthquakes. We found from analysis of the individual earthquakes that these spectra all have similar corner frequencies and t^* values, which is consistent with the time domain observations given that the P-waves all had single pulse shapes with very common pulse durations. The

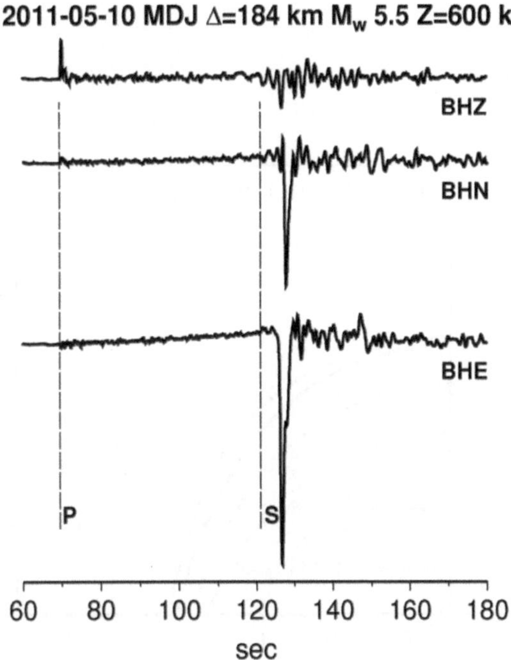

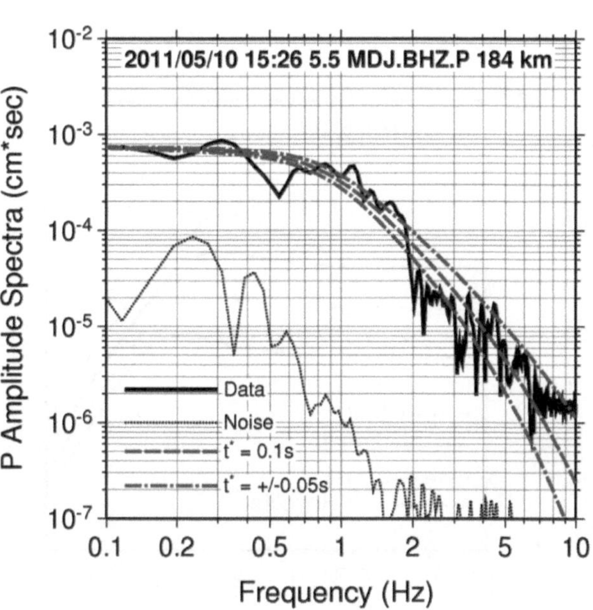

Figure 7

Left instrument-corrected three-component displacement seismograms recorded from station MDJ at an epicenter distance of 184 km from the 10 May 2011 M_w 5.5 earthquake located at a depth of 600 km. *Right* the displacement amplitude spectra were taken from a 5-s window around the vertical component P-wave and noise spectra from a 5-s window ending 1 s before the P-wave arrival. We selected a best-fit corner frequency of 0.85 Hz and a t^* of 0.1 ± 0.05 s

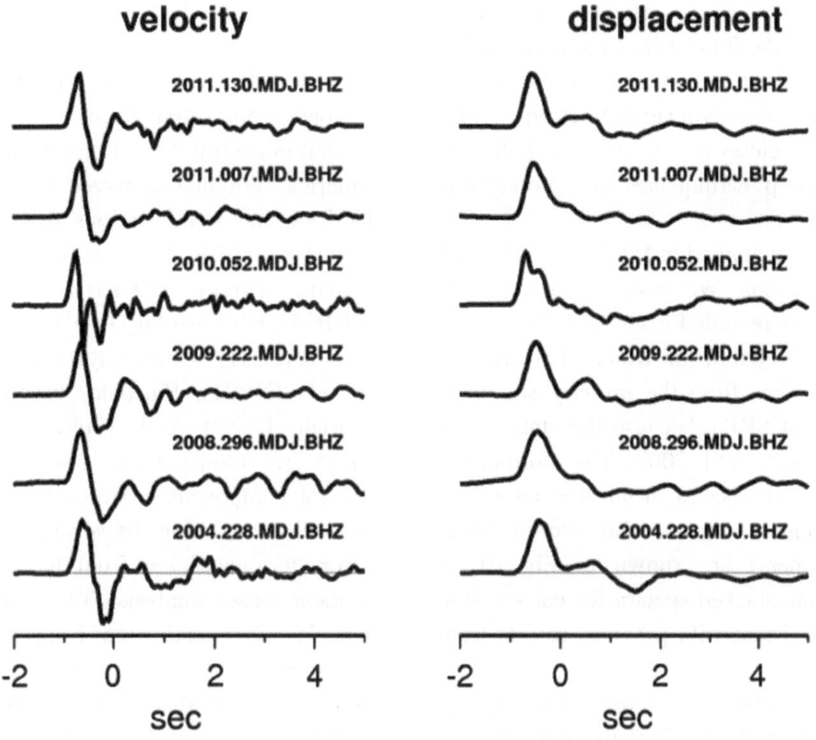

Figure 8
Vertical component P-waves for six deep earthquakes instrument-corrected to velocity (*left*) and displacement (*right*)

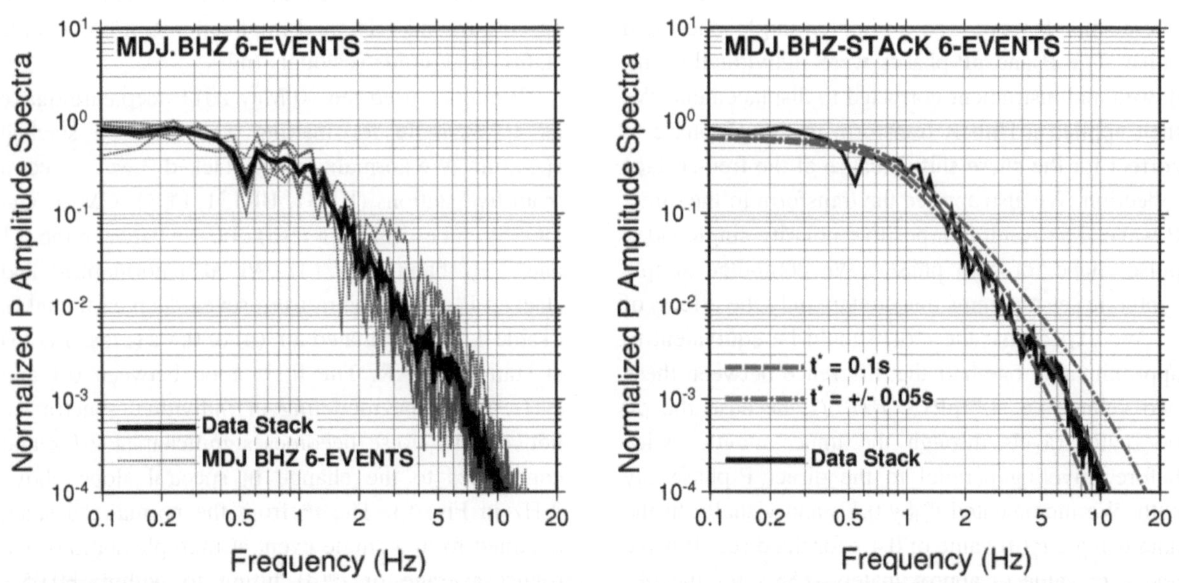

Figure 9
Normalized displacement spectra of vertical component P-waves (Fig. 8) from six deep earthquakes recorded at station MDJ. We stacked the spectra and made a $\log_{10}$ average from the six earthquakes. We found from individual analysis of each event that these spectra all have similar corner frequencies and similar t^* values. The spectra were compared with a model using t^* values of 0.3 and 0.4 s

average t^* across the source region to MDJ was estimated at 0.1 s in the 0.1–10 Hz band (Fig. 9). We expect the attenuation to be low because the epicentral distance to MDJ is only a small fraction (<1/10) of the distance to teleseismic stations and the ray paths from the deep earthquakes are propagating vertically to MDJ.

The same analysis applied to MDJ was performed on two nearby stations. We stack P-wave spectra from six earthquakes recorded from the IMS primary station at Wonju, Republic of Korea (KSRS), and only three earthquakes from the primary station at Ussuriysk, Russia (USRK), because the station was not fully operational until 2007. The normalized P-wave displacement spectra from the Ussuriysk Array (USA0B-center element) and Wonju Array (KS31-center element) are shown in Fig. 10. A t^*(u) of 0.1 s fit the stacked spectra for each station up to 5–7 Hz, consistent with previous results from MDJ.

As a check on t^*(u) and t^*(d) estimates, we modeled time-domain P and pP-phase pulse shapes. We selected the 10 May 2011 deep earthquake recorded at the Karatau, Kazakhstan (KKAR.KZ-network archived at IRIS) at an epicenter distance of 43.2° (Fig. 11). This station was selected because the teleseismic P-wave was a simple, single one-sided pulse. The P and pP-phases were individually windowed and instrument-corrected to displacement. We then applied a Hilbert transform to the pP-phase to correct for the phase shift because of the free-surface reflection. We also applied the transform to the direct P-wave. The result is a positive polarity single-sided pulse shape for both phases. We computed simple synthetic seismograms that contain only the effects of t^*. We expect that the t^*(u) should be equivalent to approximately one-half the difference between these two values, i.e., (t^*(pP) − t^*(P))/2, because the pP ray path travels through the upper mantle twice before traveling parallel to the direct P-phase ray path. We incremented t^* by 0.1 s and visually fit the data using a t^*(d) value of 0.4 s for the direct P-wave and a t^* value of approximately 0.55 s for the pP-phase. The resulting t^*(u) estimated using this method is (0.55–0.4)/2 = 0.075 s, which is similar to the t^*(u) = 0.1 s obtained by use of P-wave spectra from local distance stations.

5. *Analysis and Results of t*(d)*

The t^*(d) is estimated from P-waves recorded by stations at epicentral distances between 30° and 90° located in central Asia, Europe, Australia, and North America. We first analyzed individual P-waves at 10 IMS and GSN stations or arrays in Akbulak, Kazakhstan (ABK31), Alice Springs, Australia (AS31), Keskin, Turkey (BR131), Bucovina, Romania (BUR31), Chiang-Mai, Thailand (CM31), Flin-Flon, Canada (FL31), Karatau, Kazakhstan (KKAR), Mina, Nevada (NV31), Pinedale, Wyoming (PD31), and Sharjah, United Arab Emirates (UOSS). For the arrays, we modeled the P-wave spectra from the vertical component broadband sensor of the center array element, using the seismic moment and radiation pattern coefficient from the regional long-period moment tensor analysis (Table 1). The corner frequencies were estimated from modeling of near-source spectra. Figure 12 shows example spectra from six stations that fit reasonably with t^*(d) values of 0.4 s between 0.1 and 2 Hz. The t^* variations appear visually to be less than ±0.1 s. Noise spectra merge with the spectra of the data window at 2 Hz, changing the slope in the signal spectra. Later we will discuss the fit between the observed and modeled spectra around 1 Hz and implications on body wave station magnitude residuals (mb).

We again used the 10 May 2011 deep earthquake as an example for making a normalized stacked spectrum. We normalized and stacked P-wave spectra from five stations (KK31, BR131, FL31, CM31, and UOSS) and estimated a t^*(d) of 0.4 s between the 0.1 and 2 Hz band (Fig. 13). We also normalized and stacked P-wave spectra from seven deep earthquakes (Table 1) and estimated a t^*(d) of 0.4 s (Figs. 14, 15) at station KKAR. The fit is good between 0.1 and 2 Hz. From the modeling of individual spectra we noticed that noise becomes significant at 2-Hz and contributes to the change in spectral slope above 2 Hz in Fig. 13. The t^* from the normalized stack obtained from a single event at multiple stations is a robust average of t^*(d) fitting to within ±0.05 s (Fig. 13); however, the slopes from the individual stations seem to have variance of the order of ±0.1 s. The normalized spectral stack across the seven deep earthquakes at KKAR in Fig. 15 is interesting

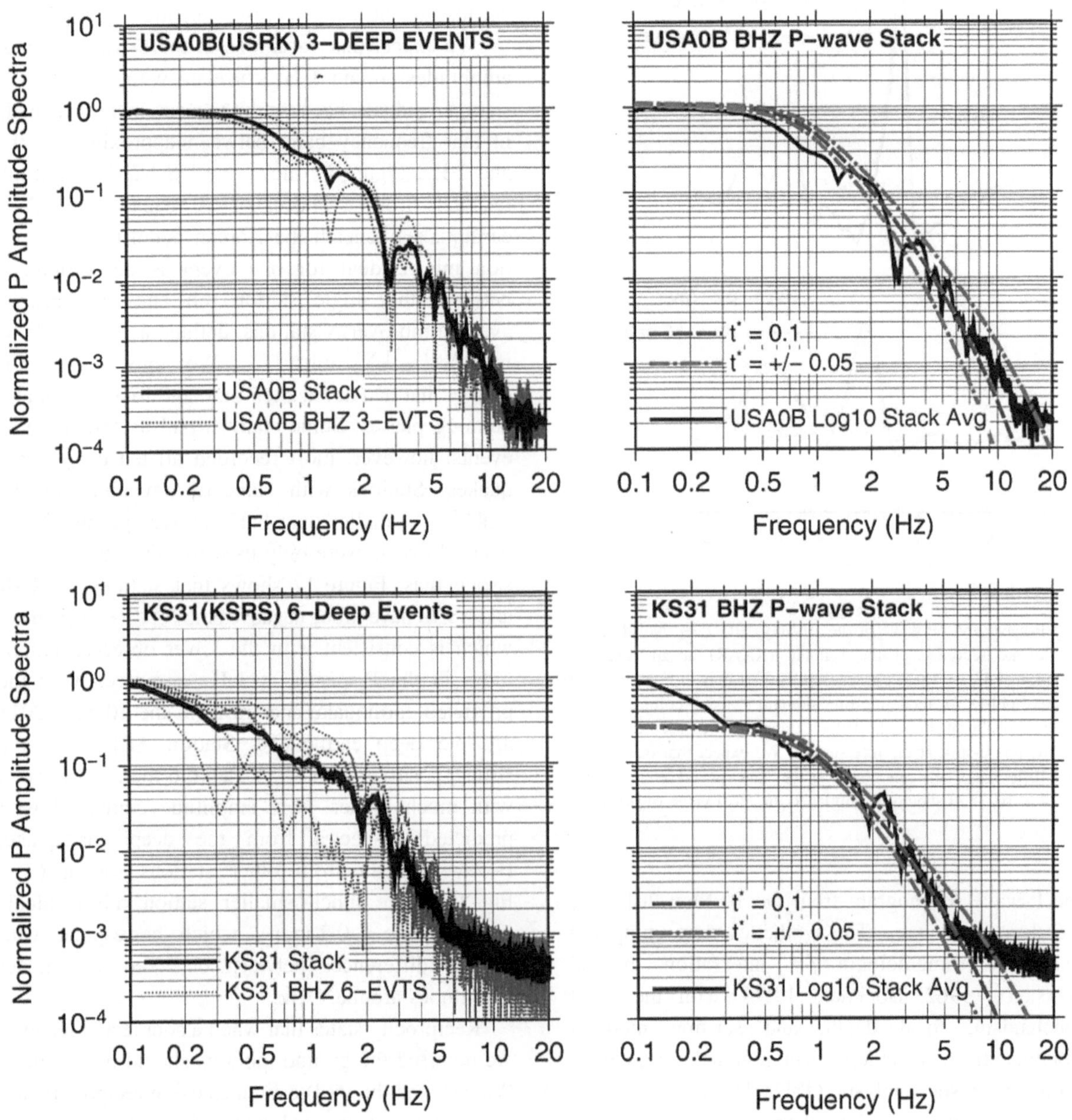

Figure 10
Left Normalized P-wave displacement spectra from the Ussuriysk Array (USA0B-center element) and Wonju Array (KS31-center element) from 3–6 deep earthquakes. *Right* the normalized spectra were stacked to produce $\log_{10}$-averaged spectra (see Fig. 9)

because with proper binning of events with the same depth and magnitude, t^* values can be consistent with a network-averaged t^*.

Time-domain and frequency-domain methods for estimating attenuation should yield similar results of the same frequency bands are used in both modeling methods. We checked the t^*(d) measurements at station KKAR with synthetic seismograms computed using a range of t^* values and comparing these with the observed data. Figure 16 shows a time window of

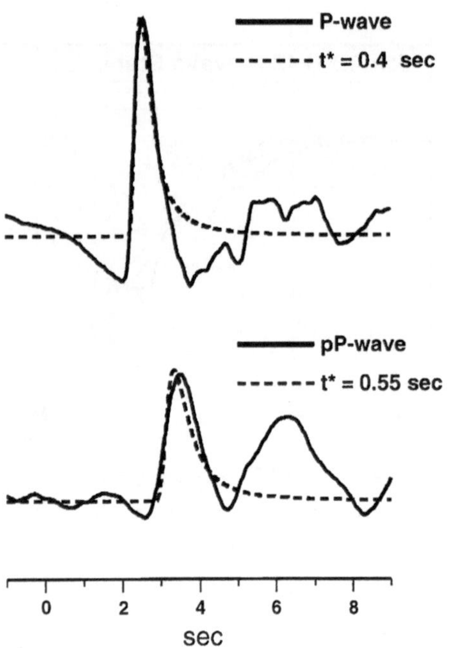

Figure 11
The P-wave and pP-wave of the 10 May 2011 deep North Korea earthquake recorded at the Karatau (KKAR) at an epicenter distance of 43.2°. These phases were Hilbert transformed to create positive pulses for modeling time-domain t^* using complex Q. In the case of the pP-wave the Hilbert transform was used to correct for the phase-shift at the free surface. We expect that one-half of the difference between the P-wave and pP-wave t^* (0.15/2 = 0.075 s) should be similar to the t^*(u) value of 0.1 s

the P-wave seismogram from the 10 May 2011 M_w 5.5 deep earthquake. The synthetics were computed using the time-domain *f–k* integration method (Haskell, 1962; Bouchon, 1976), with the focal mechanism, focal depth, and seismic moment obtained from an independent regional long-period moment tensor analysis (Table 1) being used as inputs to compute the synthetics. The first motion P-wave polarity is consistent with the piercing point on the focal sphere, which is far enough away from the nodal plane to be impulsive (Fig. 16). We convolved an attenuation operator with the synthetics for five different t^* values (0.1, 0.3, 0.4, 0.5, and 0.8 s). The P-wave amplitude and pulse duration best fit the synthetics computed using a t^* of 0.4 s. This time-domain t^* value is equal to the t^*(d) obtained from KKAR P-wave spectra and network-averaged spectra. This is consistent with results shown in Fig. 11 but in this case the data were not phase shifted because we fit using wave-propagation synthetics.

We expect that variations in the P-wave spectral amplitudes around 1 Hz correlate with average MB station residuals, because mb measurements are made in that frequency band between approximately 0.7 and 2 Hz. Figure 17 shows the US NDC station mb residuals from a cluster of as many as ten deep earthquakes beneath the Korean peninsula. The station mb residual for one event is the difference between the single station mb and the network average mb over N stations, defined as $\delta \mathrm{mb}^{(\mathrm{station})} = \mathrm{mb}^{(\mathrm{station})} - \frac{1}{N}\sum_1^N \mathrm{mb}^{(\mathrm{station})}$. We only use earthquakes that had at least ten time and four magnitude-defining phases. Some stations had as few as three events; however, most recorded all ten deep earthquakes. Stations with standard deviations of the $\delta \mathrm{mb}^{(\mathrm{station})}$ smaller than half the average $\delta \mathrm{mb}^{(\mathrm{station})}$ over all events were only used to show stations with strong bias. Figure 17 shows that stations BURAR and ASAR have strong negative $\delta \mathrm{mb}^{(\mathrm{station})}$ (<-0.5 m.u.), which is consistent with the lower observed amplitudes in the P-spectra at 1 Hz compared with the predicted earthquake spectra from the 10 May 2011 M_w 5.5 deep earthquake. Station ABKAR has a strong positive $\delta \mathrm{mb}^{(\mathrm{station})}$ ($>+0.6$ m.u.), consistent with much higher 1 Hz amplitude compared with predicted P-spectra from the deep earthquake. P-wave spectra from other stations shown (e.g., Fig. 12) have much smaller station mb residuals ($-0.2 < \delta \mathrm{mb} + 0.2$ m.u.), which correspond with smaller differences between observed and model amplitudes around 1 Hz.

Reciprocity states that when a source and receiver are reversed the ground motions should be identical. We checked the t^*(d) estimates using reciprocity and reversed the deep earthquakes beneath the Korean peninsula with surface receivers (i.e., MDJ, KS31, and USRK). On the receiver end we reversed them by using a 600 km deep 8 June 2011 earthquake (M_w 5.5) in the Bismarck Sea region recorded from (MDJ, KS31, and USRK) at teleseismic distances between approximately 46° and 52°. We made a normalized stack of P-wave spectra from the three stations and fit a t^*(d) of 0.4 s between the 0.1 and 2 Hz band (Fig. 18) with a range of ± 0.1 s. The network-averaged lower-mantle attenuation measured from

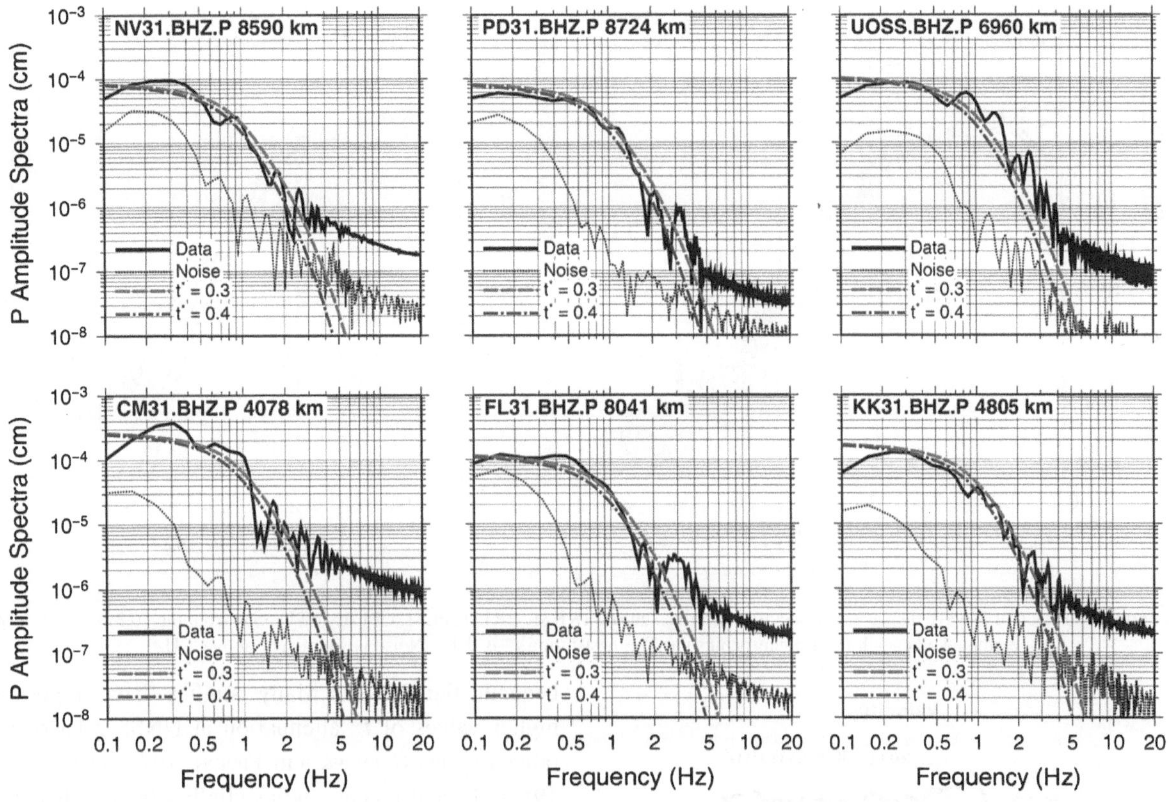

Figure 12
P-wave displacement and noise spectra in cm from 10 s windows at Mina (NV31), Pinedale (PD31), Sharjah (UOSS), Chiang Mai (CM31), Flin-Flon (FL31) and Karatau (KK31). The predicted spectra from t^* values shown are at 0.3 and 0.4 s

teleseismic P-spectra of sources deep beneath the Korean peninsula are consistent with measurements made with a near reversal of the source and receiver.

6. *Analysis and Results of t*(total)*

To measure the attenuation t^*(total) from the 25 May 2009 declared North Korea nuclear test (IDC/CTBTO mb 4.5), we first computed Mueller-Murphy (1971) source spectra using an assumed yield of 2.4 kT based on analysis by Park (2009). We then examined single station P-wave displacement spectra using the explosion source model at stations ABK31, MK31, NV31, and PD31 (Fig. 19). These spectra fit reasonably well using the source model and attenuation t^*(total) values between 0.4 and 0.5 s with some variability between station spectra and noise in the 0.5–3.0 Hz band.

To cancel the source term and average out any site biases, we normalized and stacked the P-wave spectra from 22 stations located at teleseismic distances distributed around the world. The best-fit network-averaged t^*(total) is estimated to be 0.5 s between the 0.5 and 3 Hz band (Fig. 20) where there is best separation between signal and noise amplitudes. The signal in the spectrum typically merges with the noise spectrum around 2 Hz for the deep earthquakes but the explosions merge with the noise spectra around 3 Hz at approximately the same Fourier spectral amplitude of approximately 10^{-8} m s (10^{-6} cm s in the earthquake plots).

7. *Comparison of Attenuation between the Korean Peninsula and Western US*

One of the main purposes of this study was to compare the mantle attenuation beneath NKTS to that

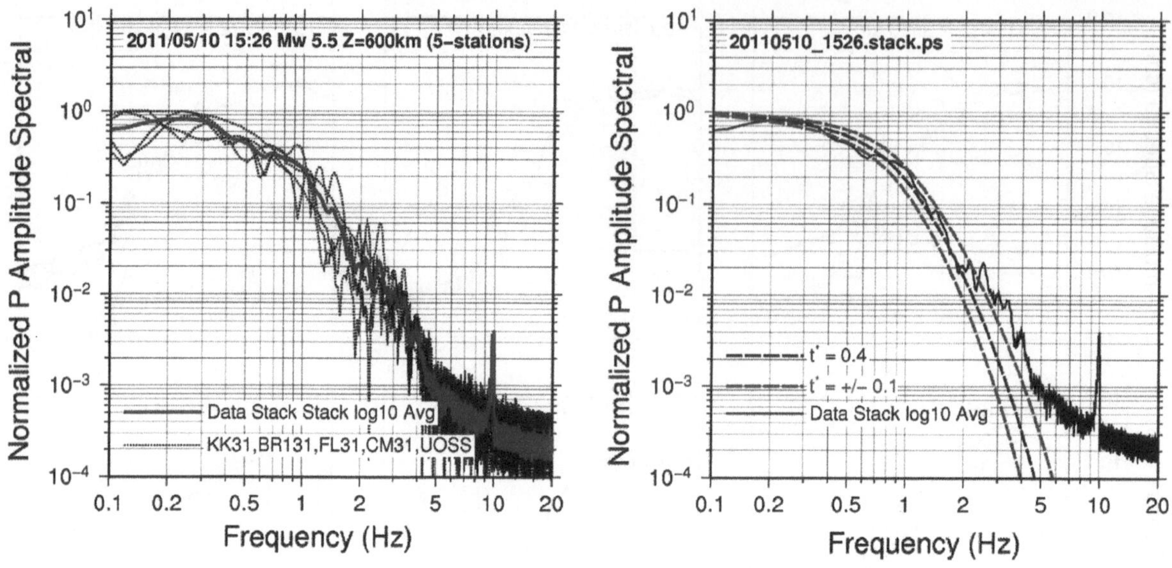

Figure 13
A normalized stack of P-wave displacement spectra from five stations (KK31, BR131, FL31, CM31, and UOSS) from the 10 May 2011 deep earthquake beneath the Korean peninsula. The best fitting t^* is 0.4 s. Noise is significant above 2 Hz

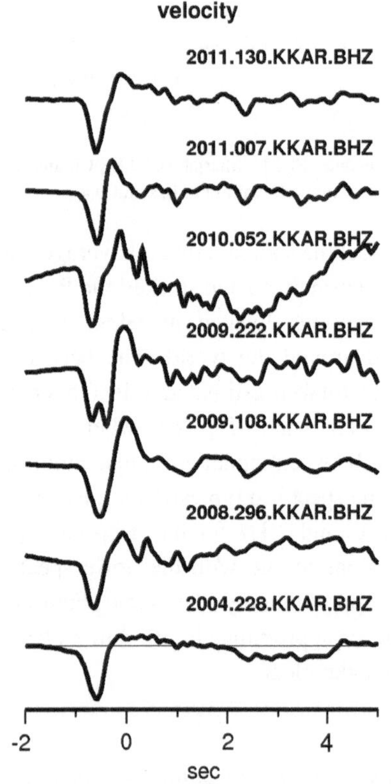

Figure 14
P-waves from seven deep earthquakes at station KKAR.KZ, in velocity amplitudes. These P-waves were used to create a normalized stacked spectrum

beneath the NNSS. Many studies have reported higher values of t^* attenuation at NNSS relative to other regions (Frasier and Filson, 1972; Der *et al.*, 1985a, b) and recent studies conflict on whether the attenuation beneath the Korean peninsula is more like STS or NNSS (Hwang *et al.*, 2011). We applied the same analysis on two shallow earthquakes in the Basin and Range region near NNSS: the 1992 Little Skull Mountain (M_w 5.57) earthquake on the test site and the 1999 Scotty's Junction, Nevada (M_w 5.62) earthquake approximately 100 km to the northwest. We used the moment tensor solutions from Ichinose *et al.* (2003) to model the teleseismic P-wave spectra individually at stations from each event using the same analysis procedures. We measured a range of t^* values between 0.5 and 1.0 s in the 0.1–1 Hz range. The signal typically merged with the noise spectrum around 1 Hz at the same spectral amplitude of approximately 10^{-6} cm s, lower than the 2 Hz intersection seen at most stations from the deep earthquakes. We then normalized and stacked 22 P-wave spectra from GSN stations for the 1999 earthquake and only six stations for the 1992 earthquake. We measured a network-averaged t^* from the spectral slopes of the normalized spectra shown in Fig. 21. These results suggest that the network-averaged t^* for NNSS between 0.1 and 1 Hz is 0.75 s

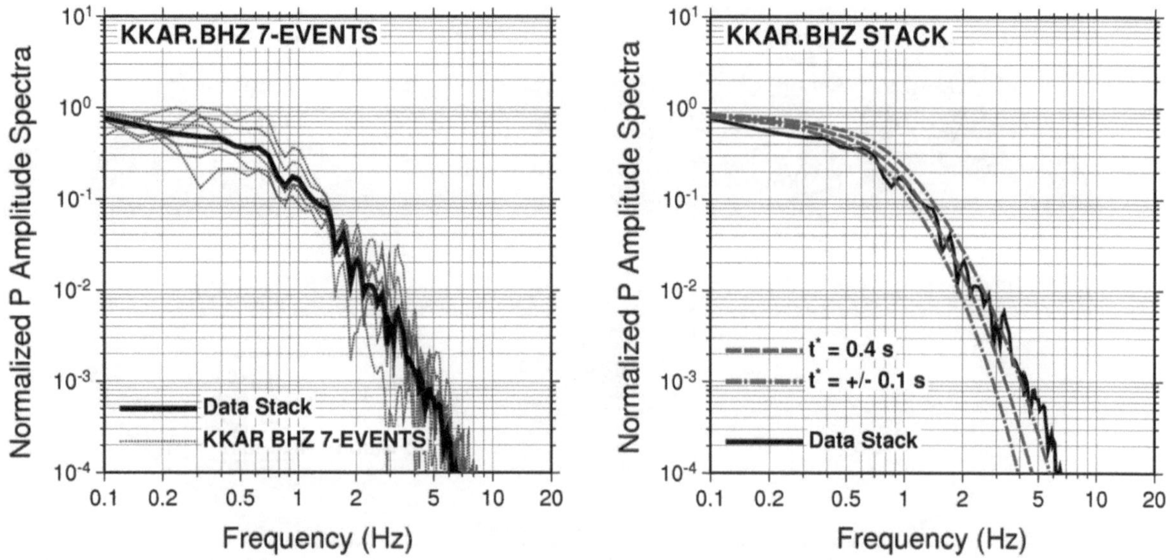

Figure 15
Left single-event-normalized spectra of seven deep earthquakes beneath the Korean peninsula. *Right* a $\log_{10}$ average normalized spectrum. The stack average spectrum fits best with a t^* of 0.4 s up to 2 Hz

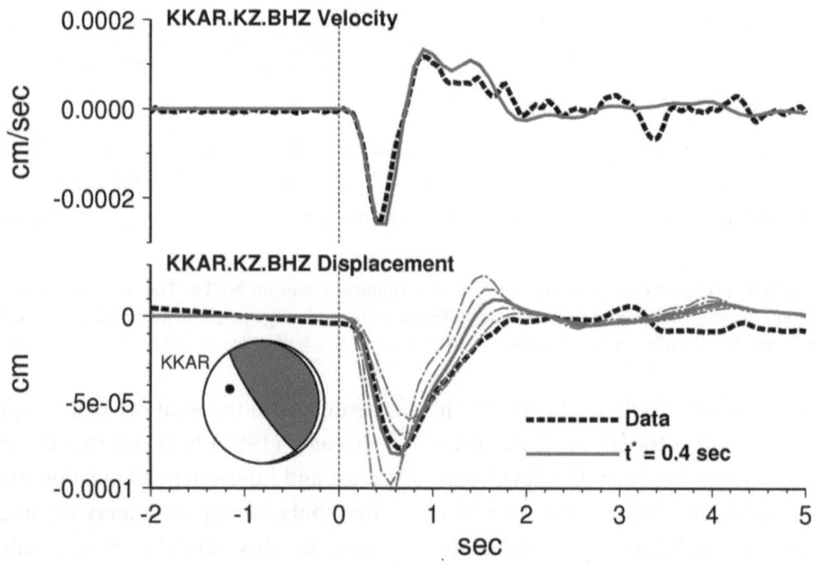

Figure 16
P-wave velocity (*top*) and displacement seismogram (*bottom*) of the 10 May 2011 M_w 5.5 deep earthquake beneath the Korean peninsula recorded at station Karatau (*KKAR.KZ*). We use the depth, mechanism, and seismic moment to calculate the synthetic P-wave seismogram. The seismograms were convolved with a range of t^* operators of 0.1, 0.3, 0.4, 0.5, and 0.8 s. A t^* of 0.4 s best fits the observed P-wave amplitude and pulse duration

for both the 1992 and 1999 earthquakes. There is substantially more variability in the t^* measurements of ±0.25 s from the 1999 earthquake than for the earlier 1992 earthquake because of a larger amount and distribution of stations, however this spread is approximately the same order as that measured from

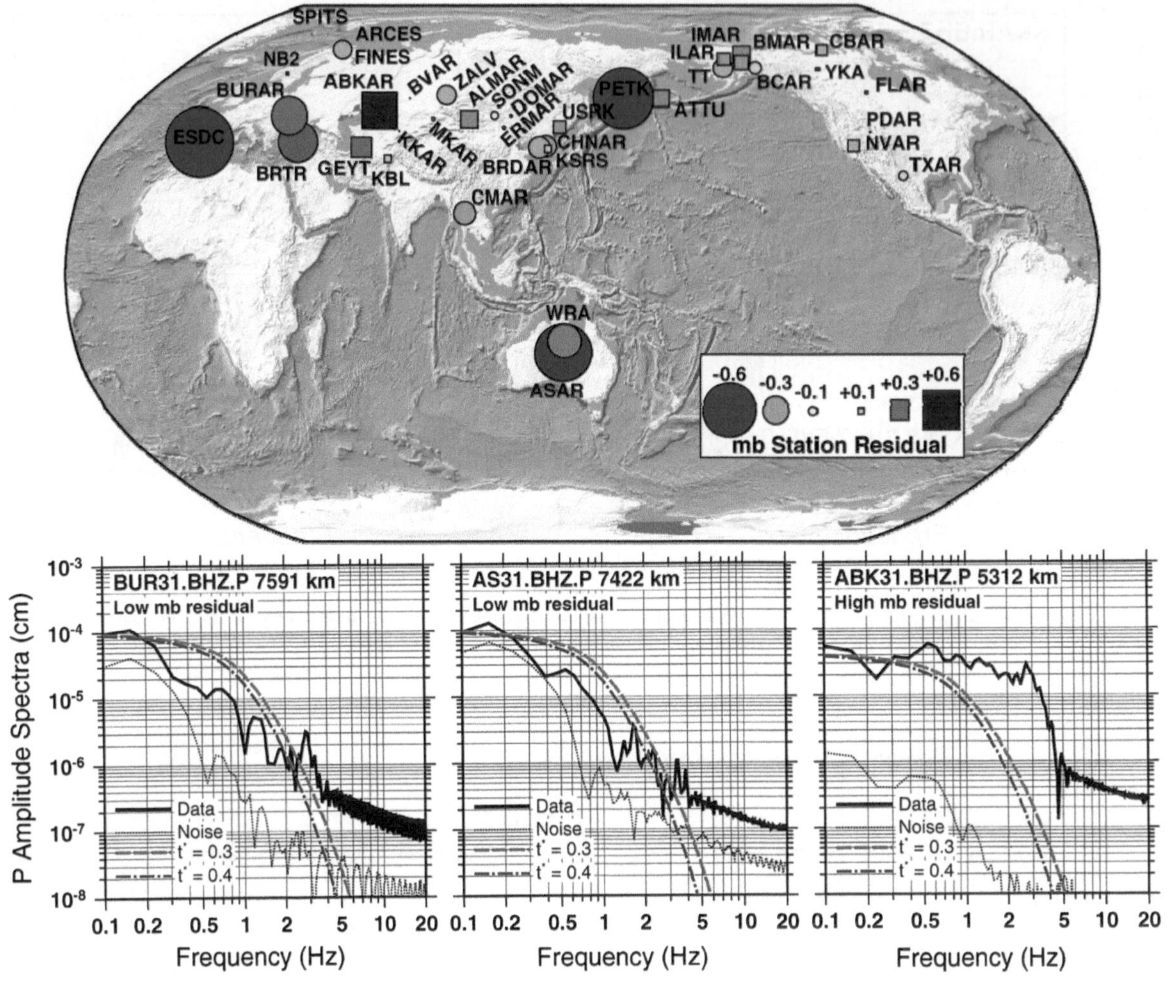

Figure 17
A map of mb station residuals averaged over as many as ten deep earthquakes beneath NKTS. The *insets* show two stations (BUR31 and AS31) with over-predicted P-wave spectral amplitudes around 1 Hz consistent with large negative mb residuals (>−0.5 m.u.) and one station (ABK31) with under-predicted 1-Hz P-wave spectral amplitudes consistent with large positive mb station residual (>+0.6 m.u.)

the 22 stations used in the 2009 declared North Korean nuclear test (Fig. 21). Based on these measurements and comparisons between the level and ratio of noise and signal amplitudes, the network-averaged attenuation beneath NNSS is approximately 0.25 s higher than the network-averaged attenuation of 0.5 s at the NKTS.

8. Discussion

There are several key points to make on the basis of the results of this study. The contribution of attenuation in the upper mantle beneath the Korean peninsula to the total ray path is approximately 0.1 s. Cormier (1982) believed that the contribution of the crust and lithosphere attenuation to the entire ray path may only be approximately 0.1–0.2 s, which is relevant to this study's t^*(u) results. Time domain modeling of Hilbert-transformed teleseismic pP and P-phases for differential attenuation results in values approximately consistent with t^*(u) values measured from local epicenter distance P-wave spectra. Differences in the size and polarity of the average station mb residuals are related to differences between the observed and predicted P-wave spectra. This seems to suggest mb residuals are not the effect of the seismic source and are likely to be partly due to localized site

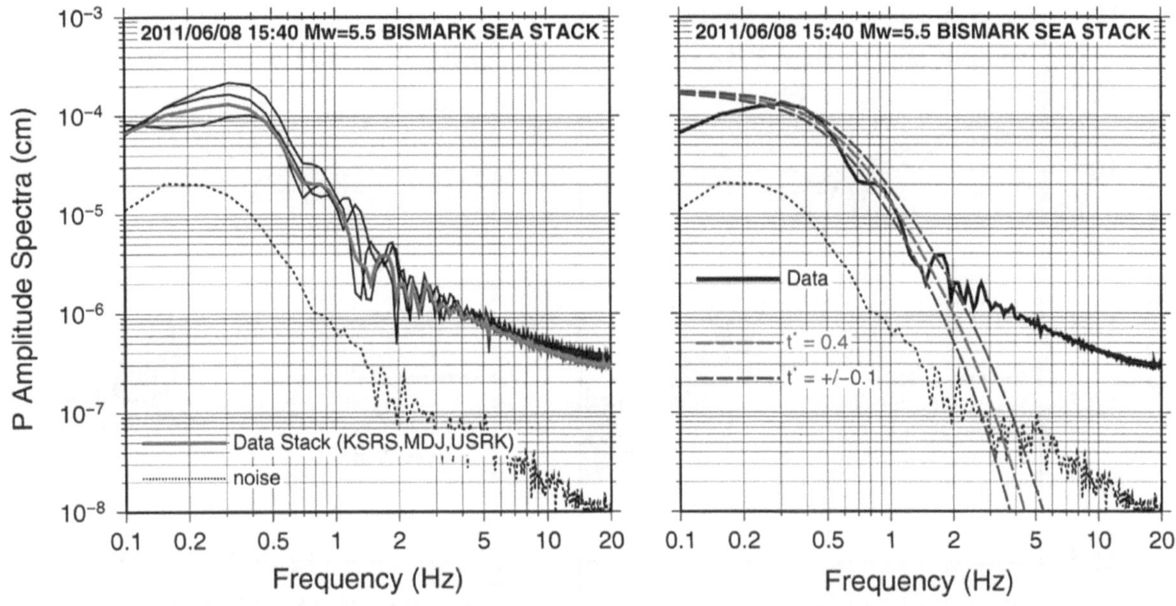

Figure 18
MDJ, KSRS, and URSK P-wave normalized spectral stack from the deep Bismarck Sea event

effects or upper-mantle attenuation. Application of reciprocity, with the reversal between the source and receiver, results in consistent average attenuation values for t^*(d). The implication here is that reciprocal t^* measurements can be made in new aseismic regions with receivers near target source areas.

Independent source variables, f_c and M_o, were estimated from regional spectra and long-period moment tensors. Measuring these values regionally where the attenuation is low avoids any trade-offs typically encountered trying to estimate f_c and t^* uniquely from teleseismic P-waves. We used moderate-sized earthquakes that had corner frequencies <1 Hz. The selection of t^* has little effect on predicted amplitude around f_c for regional distances. In addition to advantages of using deep regional waves, we also sorted out complex events on the basis of P-wave complexity. We avoided earthquakes with complex long duration rupture that typically introduce intermediate corner frequencies and result in artificially lower attenuation estimates. If we underestimated the f_c then that would result in overestimates of attenuation; however, overestimating the f_c will not have a drastic effect on attenuation estimates because we are already below 0.7–2 Hz frequency band of interest in the stacked teleseismic P-spectra.

The sum between local average t^*(u) of 0.1 s and the network-averaged t^*(d) of 0.4 s is equal to the network-averaged t^*(total) of 0.5 s from the 2009 declared North Korea nuclear test. The theory of t^* states that the total attenuation along a ray is the sum of the attenuation from each segment. With some simple algebra, that suggests that the t^* of only the lower mantle below 600 km depth is approximately 0.3 s. This value seems to be approximately constant beyond the distance of 30°. The contribution of attenuation due to passage through the upper-mantle above 600 km is only 0.1 s near NKTS. With only one pass through the upper-mantle the total ray-path t^* is 0.4 s and with two passes for a shallow seismic source t^* is 0.5 s. The use of nearly coincident deep earthquakes and near-by surface explosions enabled separation of amounts of attenuation between the upper and lower-mantle for teleseismic P-waves in the 0.1–2 Hz band.

It is important to place the results from this study in the context of previous research. That requires detailed examination of the differences between absolute t^* measurements from previous studies by Der and McElfresh (1977) and Der *et al.* (1985a, b, 1986a, b). The reasons for these differences can then later be unified on the basis of different regional

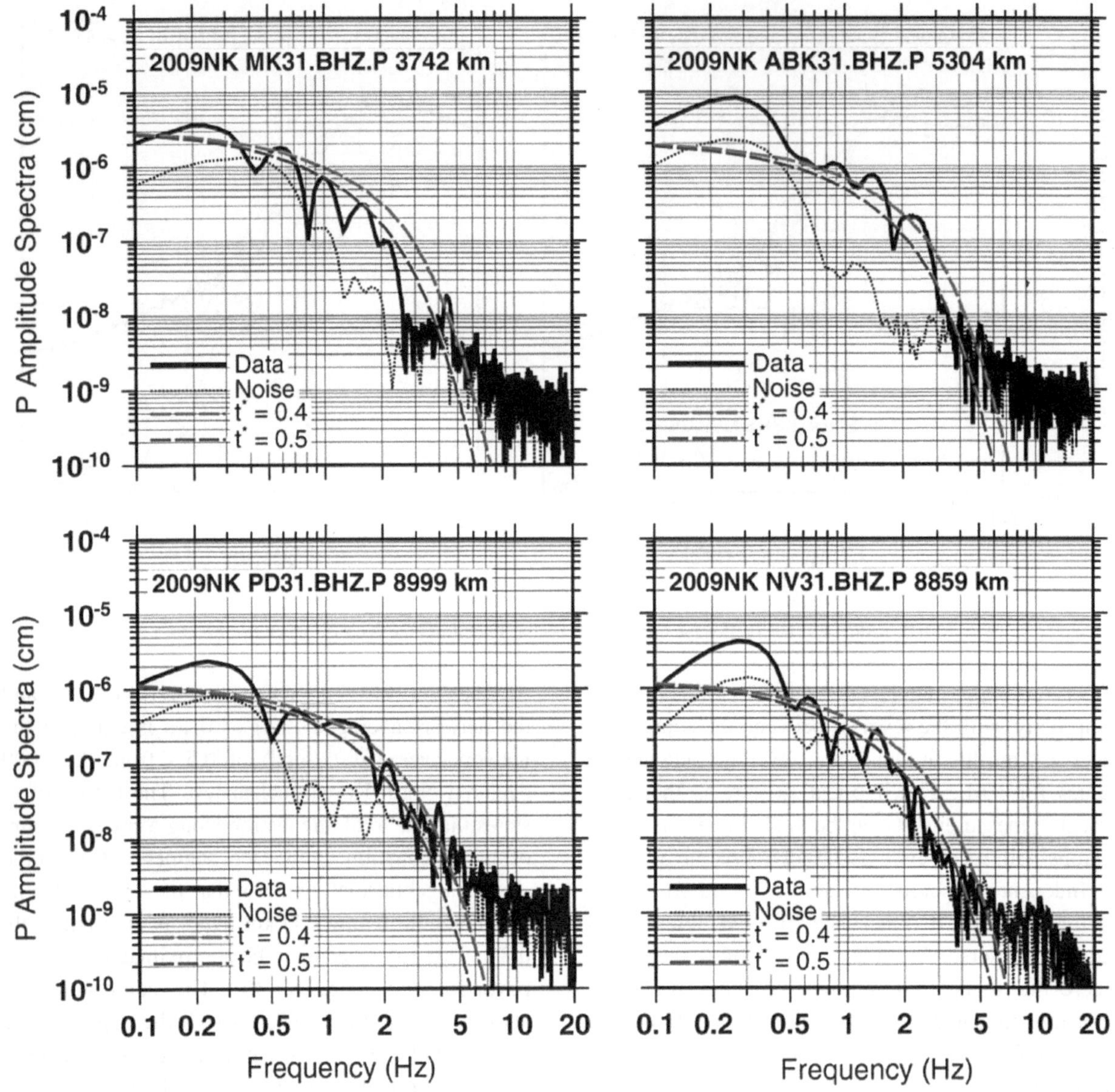

Figure 19
Teleseismic P-wave spectra from stations MK31, NV31, ABK31, and PD31 from the 2009 nuclear test

frequency-dependent Q models. Der *et al.* (1985a, b) reported t^* values that were separated into test-site terms and station terms. Additionally they reported apparent t^* values, $\bar{t}$, that can be adjusted to t^* by use of the formula:

$$\bar{t}^* = t^* + f\left(\frac{\mathrm{d}t^*}{\mathrm{d}f}\right)$$

where f is frequency and ($\mathrm{d}t^*/\mathrm{d}f$) is the t^* versus frequency slope. This slope is negative, therefore apparent t^* values are lower than t^* measurements. Figure 22 illustrates an example of frequency-dependent t^* attenuation based on two approximate frequency-dependent Q models for NNSS and STS.

Der and McElfresh (1977) and Der *et al.* (1985a, b, 1986a, b) reported values for apparent t^* at NNSS using far-regional (10°–30°) P-waves recorded in North America. They also use far-regional P-waves at NORSAR for STS. Waveform data at closer distances enables the measurement of signal amplitudes above

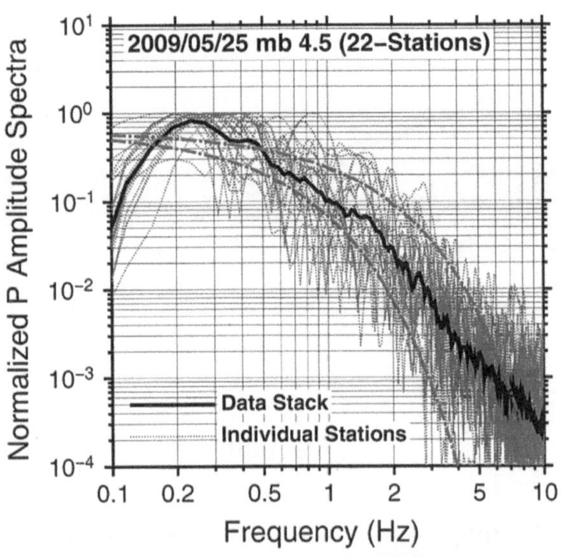

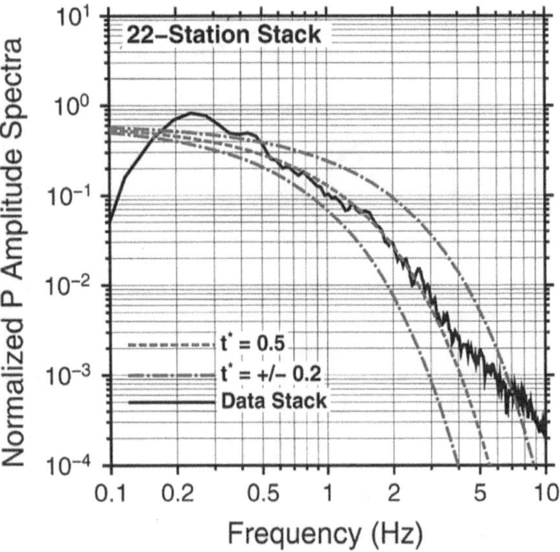

Figure 20
Teleseismic P-wave normalized stacked P-wave spectra from 22 stations

the noise beyond the 1–3 Hz upper limit range compared with this study. Another caveat was that DER *et al.*, (1977) used only explosions in their analysis and those may have higher corner frequencies than earthquakes, allowing for the enhanced excitations at higher frequencies. Based on the differences of this study's bandwidth (0.1–2 Hz) and short-period t^* estimates by DER and MCELFRESH (1977) between 1 and 8 Hz, we will make adjustments to t^* for a common 1 Hz frequency band.

From DER *et al.* (1985a, b; Table 1. Errata), the apparent t^* for test sites terms are 0.07 s at STS (Shagan River, Balapan, and Degelen Mountain) and 0.34 s for NNSS (Pahute Mesa, Climax Stock, and Yucca Flat). We averaged the DER *et al.* (1985b) apparent t^* stations terms to produce an approximate network average of 0.17 (±0.1) s using 22 stations. The sum between the test-site and average station terms for absolute apparent t^* is 0.51 s at NNSS and 0.24 s at STS. For proper comparison of t^* we use a generic frequency-dependent attenuation relationship to convert from $\bar{t}$ to t^* (Fig. 22). The DER *et al.* (1985a, b) apparent t^* attenuation at NNSS of $\bar{t} = 0.5$ s is adjusted for frequency-dependent Q to $t^* = 0.8$ s based on dt^*/df slope of -0.1 s/Hz. We use the same for STS and adjust the $\bar{t}$ value from 0.24 s to $t^* = 0.54$ s. The DER *et al.* (1985a, b) apparent t^* values are found to be only 0.05 s larger than measured in this study for NNSS after adjustments. After adjustments, the STS t^* of 0.54 s is also only 0.04 s larger than the network-averaged t^*(total) of 0.5 s we measured for NKTS. That, also, is consistent with reciprocal t^* measurements at Karatau, Kazakhstan, from NKTS.

9. Conclusions

We used the P-waves from the 25 May 2009 (mb 4.5; IDC) declared nuclear test to measure the absolute t^* of 0.5 s within the frequency band 0.1–3 Hz. This was measured using spectra from the vertical component P-waves from 22 GSN and IMS stations distributed across Asia and North America. We then checked this t^* using six deep earthquakes beneath NKTS. With these deep earthquakes, we split the t^*(total) ray path into two parts: a t^*(u) that represents the attenuation of the up-going ray from the deep hypocenters to the surface receivers and t^*(d) that represents the attenuation along the down-going ray to teleseismic receivers. The sum between t^*(u) of 0.1 s and the t^*(d) of 0.4 s is equal to the t^*(total) of 0.5 s from the 2009 declared North Korea nuclear test. Because the tests on absolute t^* using deep earthquakes and surface explosions are consistent, we can then make an assessment of the amount

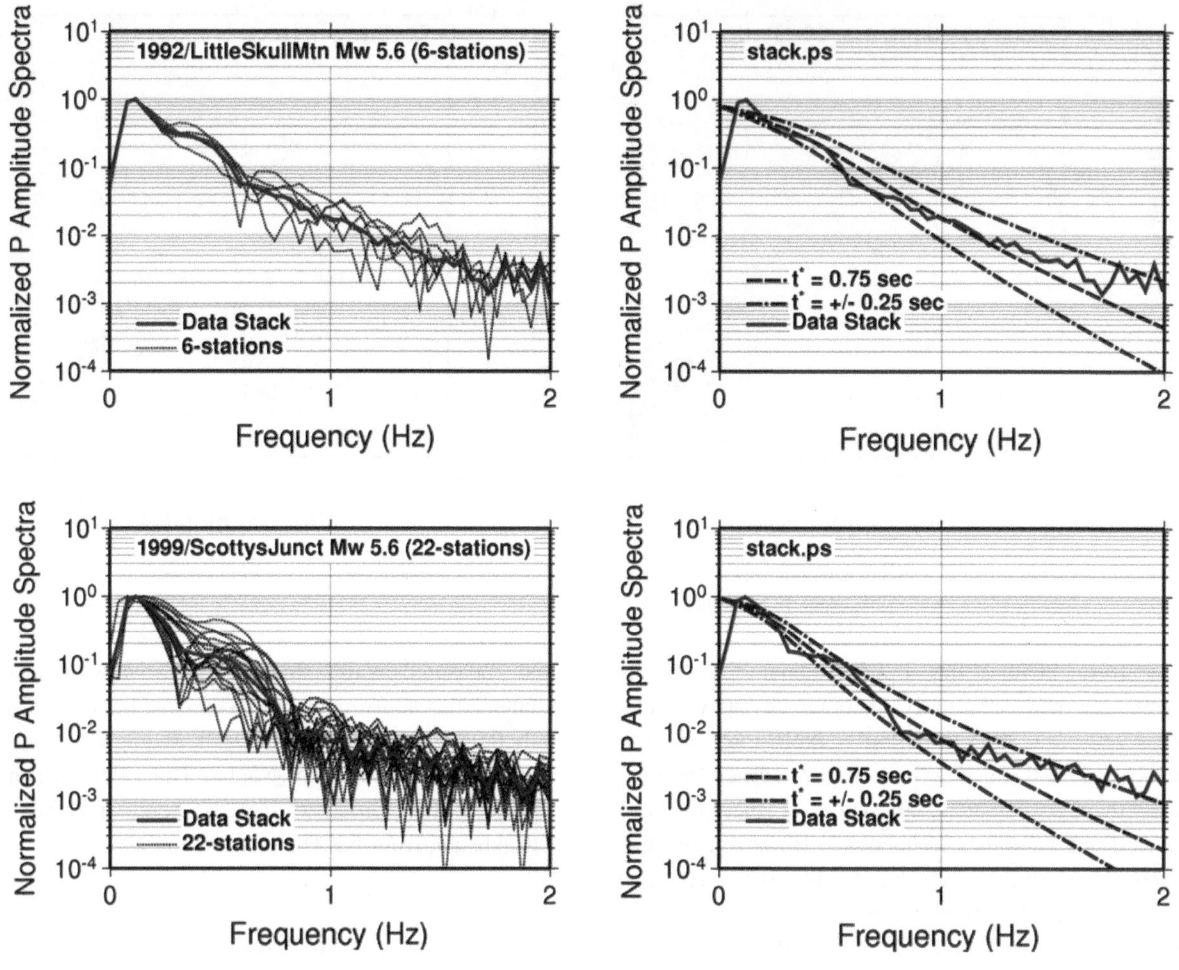

Figure 21
Teleseismic P-wave spectra from two NNSS shallow earthquakes. *Left* normalized P-wave spectra. *Right* stack average spectra

by which different ray path segments contribute to the total amount of attenuation. The theory of t^* states that the total attenuation along a ray is the sum of the attenuation from each segment. We conclude that the t^* of the lower mantle below 600 km depth is approximately 0.3 s (approximately equivalent to a Q of 1,123). This value seems to be constant beyond the distance of 30° from the NKTS. The contribution of attenuation because of passage through the upper-mantle above 600 km is only 0.1 s ($Q \approx 996$). With only one pass through the upper-mantle the total ray-path t^* is 0.4 s and with two passes for the shallow seismic source t^* is 0.5 s. These values represent global network averages for sources beneath the NKTS between the 0.1 and 2–3 Hz band and the variations in P-wave spectra seen at several stations around the 1 Hz level are consistent with average mb station residuals.

The main purpose of this study was to compare the mantle attenuation beneath NKTS to that beneath the NNSS. We applied the same analysis to the NKTS for two shallow earthquakes in the Basin and Range, the 1992 Little Skull Mountain (M_w 5.57) earthquake on the NNSS, and the 1999 Scotty's Junction, Nevada (M_w 5.62) earthquake approximately 100 km to the northwest. We measured a range of t^* values between 0.5 and 1.0 s in the 0.1–1 Hz range. We then normalized and stacked 22 P-wave spectra from GSN stations for the 1999 earthquake and only six stations for the 1992 earthquake. We conclude that the

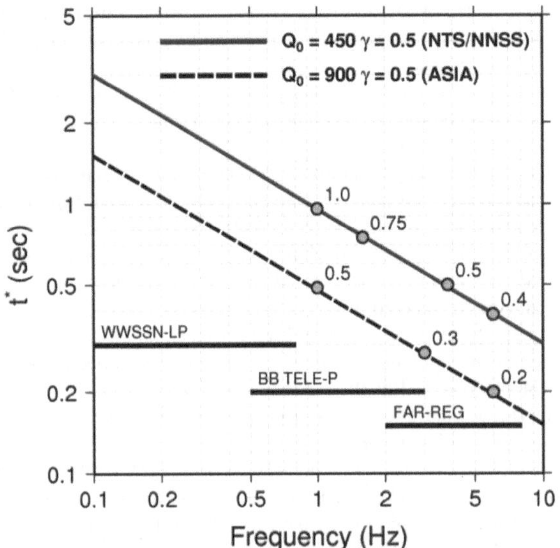

Figure 22
An example demonstration of the effect of frequency-dependent t^* for two frequency-dependent Q(P) relationships using typical values for lower-attenuation Asia (STS and NKTS) and higher-attenuation NNSS. The *bars* represent ranges of expected bandwidth with various datasets used in previous studies

network-averaged t^* for NNSS is 0.75 s from both the 1992 and 1999 earthquakes. On the basis of these measurements and comparisons between the level and ratio of noise and signal amplitudes, the average attenuation beneath NNSS is approximately 0.25 s higher than the network-averaged attenuation of 0.5 s at the NKTS. We also suspect that the attenuation beneath STS is similar to NKTS, based on the analysis in this study and on measurements made by DER *et al.* (1985a, b) adjusted by use of appropriate frequency-dependent t^* models.

Acknowledgments

The views expressed in this article are those of the authors and do not necessarily reflect the position or policy of the United States Air Force or the United States government. The facilities of the IRIS Data Management System and, specifically, the IRIS Data Management Center, were used for access to waveform and metadata required in this study. The IRIS DMS is funded through the National Science Foundation and, specifically, the GEO Directorate through the Instrumentation and Facilities Program of the National Science Foundation under Cooperative Agreement EAR-0552316. Global Seismographic Network (GSN) is a cooperative scientific facility operated jointly by the Incorporated Research Institutions for Seismology (IRIS), the United States Geological Survey (USGS), and the National Science Foundation (NSF).

REFERENCES

AKI, K. and P. RICHARDS (1980). Quantitative Seismology, Freeman and Co., New York.

BACHE, T. C., MARSHALL, P. D. and L. B. BACHE (1985). *Q for teleseismic P waves from Central Asia*, J. Geophysics. Res., *90*(B5), 3575–3587.

BARAZANGI, M., W. PENNINGTON and B. ISACKS (1975). *Global Study of Seismic Wave Attenuation in the Upper Mantle Behind Island Arcs Using pP Waves*, J. Geophys. Res. *80*(8), 1079–1092.

BOUCHON, M. (1976). *Teleseismic body wave radiation from seismic source in a layered medium*, Geophys. J. R. Astr. Soc., *47*, 515–530.

BRUNE, J. (1970). *Tectonic stress and the Spectra of Seismic Shear Waves from Earthquakes*, J. Geophys. Res., *75*(26), 4997–5009, doi:10.1029/JB075i026p04997.

BURDICK, L. J. and D. V. HELMBERGER (1979). *Time functions appropriate for nuclear explosions*, Bull. Seismol. Soc. Am. *69*(4), 957–973.

BURGER, R. W., T. LAY and L. J. BURDICK (1987). *Average Q and yield estimates from the Pahute Mesa test site*, Bull. Seismol. Soc. Am. *77*(4), 1274–1294.

CHOY, G. L. and V. F. CORMIER (1986). *Direct measurement of the mantle attenuation operator from broadband P and S waveforms*, J. Geophys. Res. *91*(B7), 7326–7342.

CORMIER, V. F. (1982). *The effect of attenuation on seismic body waves*, Bull. Seismol. Soc. Am. *72*(6), S169–S200.

DER, Z. A. and T. W. MCELFRESH (1977). *The Relationship Between Anelastic Attenuation and Regional Amplitude Anomalies of Short-Period P Waves in North America*, Bull. Seismol. Soc. Am. *67*(5), 1303–1317.

DER, Z. A. and T. W. MCELFRESH (1980). *Time Domain Methods, The Values of tp* and ts* in the Short-Period Band and Regional Variations of the Same Across the United States*, Bull. Seismol. Soc. Am. *70*(3), 921–924.

DER, Z., T. MCELFRESH, R. WAGNER and J. BURNETTI (1985a). *Spectral characteristics of P waves from nuclear explosions and yield estimation*, Bull. Seismol. Soc. Am. *75*(2), 379–390.

DER, Z., T. MCELFRESH, R. WAGNER and J. BURNETTI (1985b). *Errata for Spectra characteristics of P Waves from nuclear explosions and yield estimation*, Bull. Seismol. Soc. Am. *75*(4), 1222–1223.

DER, Z. A., A. C. LEES and V. F. CORMIER (1986a). *Frequency dependence of Q in the mantle underlying the shield areas of Eurasia, Part III: The Q model*, Geophys. J. R. astr. Soc., *87*, 1103–1112.

DER, Z. A., A. C. LEES, V. F. CORMIER and L. M. ANDERSON (1986B). *Frequency dependence of Q in the mantle underlying the shield areas of Eurasia, Part I: analysis of short and intermediate period data*, Geophys. J. R. astr. Soc., *87*, 1057–1084.

Douglas, A. and J. A. Hudson (1983). *Comments on "Time Functions appropriate for nuclear explosions," by L. J. Burdick and D. V. Helmberger and "Seismic source functions and attenuation from local and teleseismic observations of the NTS events Jorum and Handley," by D. V. Helmberger and D. M. Hadley*, Bull. Seismol. Soc. Am. *73*(4), 1255–1264.

Douglas, A. (1987). *Differences in upper mantle attenuation between the Nevada and Shagan River test sites: Can the effects be seen in P-wave seismograms*, Bull. Seismol. Soc. Am. *77*(1), 270–276.

Dreger, D. S., and D. V. Helmberger (1993). *Determination of source parameters at regional distances with single station or sparse network data*, J. Geophys. Res., *98*, 8107–8125.

Frasier, C. W. and J. Filson (1972). *A Direct Measurement of the Earth's Short-Period Attenuation along a Teleseismic Ray Path*, J. Geophys. Res. *77*(20), 3782–3787.

Haskell, N. A. (1962). *Crustal reflection of plane P and SV waves*, J. Geophys. Res. *67*, 4751–4767.

Hwang, Y. K., J. Ritsema and S. Goes (2011). *Global variation of body-wave attenuation in the upper mantle from teleseismic P wave and S wave spectra*, Geophys. Res. Lett. *38*, L06308, doi: 10.1029/2011GL046812.

Ichinose, G. A., J. G. Anderson, K. D. Smith, and Y. Zeng (2003). *Source parameters of eastern California and western Nevada earthquakes from regional moment tensor inversion*, Bull. Seismol. Soc. Am., *93*, 61–84.

Julian, B. R., A. D. Miller and G. R. Foulger (1998). *Non-double couple earthquakes 1: Theory*, Reviews of Geophysics, *36*(4), 525–549.

Kennett, B. L. N. and E. R. Engdahl (1991). *Travel-times for global earthquake location and phase identification*, Geophys. J. Int., *122*, 429–465.

Kirby, S. H., W. B. Durharm, and L. A. Stern (1991). *Mantle phase changes and deep-earthquake faulting in subducting lithosphere*, Science, *252*, 216–225.

Langson, C. A. and D. V. Helmberger (1975). *A procedure for modeling shallow dislocation sources*, Geophys. J. R. Astron. Soc. *42*, 117–130.

Mueller, R. A. and J. R. Murphy (1971). *Seismic characteristics of underground nuclear detonations. Part I. Seismic spectrum scaling*, Bull. Seism. Soc. Am. *61*(6), 1675–1692.

Park, J. (2009). The North Korean nuclear test: What the seismic data says, Bulletin of the Atomic Scientists, http://www.thebulletin.org/node/7071.

Ritsema, J., and T. Lay (1995). *Long period regional wave moment tensor inversion for earthquakes in the western United States*, J. Geophys. Res., *100*, 9853–9864.

Stevens, J. L. and S. M. Day (1985). *The physical Basis of mb:Ms and Variable Frequency Magnitude Methods for Earthquake/Explosion Discrimination*, J. Geophys. Res., *90*(B4), 3009–3020.

Trembly, L. D. and J. W. Berg (1968). *Seismic Source Characteristics from Explosion-Generated P Waves*, Bull. Seismological Soc. Am. *58*(6), 1833–1848.

Warren, L. M. and P. M. Shearer (2000). *Investigating the frequency dependence of mantle Q by stacking P and PP spectra*, J. Geophys. Res. *105*(B11), 25,391–25,402.

Zeng, Y., and J. G. Anderson (1995). *A method for direct computation of the differential seismograms with respect to the velocity change in a layered elastic solid*, Bull. Seismol. Soc. Am., *85*, 300–307.

(Received November 24, 2011, revised October 19, 2012, accepted November 29, 2012, Published online January 22, 2013)

Pure Appl. Geophys. 171 (2014), 507–521

DOI 10.1007/s00024-012-0623-0

Pure and Applied Geophysics

Seismic Source Characteristics of Nuclear and Chemical Explosions in Granite from Hydrodynamic Simulations

Heming Xu,[1] Arthur J. Rodgers,[1] Ilya N. Lomov,[1] and Oleg Y. Vorobiev[1]

Abstract—Seismic source characteristics of low-yield (0.5–5 kt) underground explosions are inferred from hydrodynamic simulations using a granite material model on high-performance (parallel) computers. We use a non-linear rheological model for granite calibrated to historical near-field nuclear test data. Equivalent elastic P-wave source spectra are derived from the simulated hydrodynamic response using reduced velocity potentials. Source spectra and parameters are compared with the models of Mueller and Murphy (Bull Seism Soc Am 61:1675–1692, 1971, hereafter MM71) and Denny and Johnson (Explosion source phenomenology, pp 1–24, 1991, hereafter DJ91). The source spectra inferred from the simulations of different yields at normal scaled depth-of-burial (SDOB) match the MM71 spectra reasonably well. For normally buried nuclear explosions, seismic moments are larger for the hydrodynamic simulations than MM71 (by 25 %) and for DJ91 (by over a factor of 2), however, the scaling of moment with yield across this low-yield range is consistent for our calculations and the two models. Spectra from our simulations show higher corner frequencies at the lower end of the 0.5–5.0 kt yield range and stronger variation with yield than the MM71 and DJ91 models predict. The spectra from our simulations have additional energy above the corner frequency, probably related to non-linear near-source effects, but at high frequencies the spectral slopes agree with the f^{-2} predictions of MM71. Simulations of nuclear explosions for a range of SDOB from 0.5 to 3.9 show stronger variations in the seismic moment than predicted by the MM71 and DJ91 models. Chemical explosions are found to generate higher moments by a factor of about two compared to nuclear explosions of the same yield in granite and at normal depth-of-burial, broadly consistent with comparisons of nuclear and chemical shots at the US Nevada Test Site (Denny, Proceeding of symposium on the non-proliferation experiment, Rockville, Maryland, 1994). For all buried explosions, the region of permanent deformation and material damage is not spherical but extends along the free surface above and away from the source. The effect of damage induced by a normally buried nuclear explosion on seismic radiation is explored by comparing the motions from hydrodynamic simulations with those for point-source elastic Green's functions. Results show that radiation emerging at downward takeoff angles appears to be dominated by the expected isotropic source contribution, while at shallower angles the motions are complicated by near-surface damage and cannot be represented with the addition of a simple secondary compensated linear vector dipole point source above the shot point. The agreement and differences of simulated source spectra with the MM71 and DJ91 models motivates the use of numerical simulations to understand observed motions and investigate seismic source features for underground explosions in various emplacement media and conditions, including non-linear rheological effects such as material strength and porosity.

1. Introduction

Understanding the excitation of seismic energy emerging from underground explosions is key to nuclear monitoring for accurate interpretation of data recorded at long distances. Seismic (elastic) source characteristics have been estimated from subsurface and surface shock-wave and seismic measurements from nuclear explosions (e.g., Werth and Herbst, 1963; Mueller and Murphy 1971; Aki *et al.*, 1974; von Seggern and Blandford, 1972; Patton and Taylor, 2011). Source spectra are typically characterized by their low-frequency (steady-state) level, corner frequency and high-frequency roll-off slope. The low-frequency level is proportional to the seismic moment (or source strength). The corner frequency is determined by the source dimension, related to yield, depth-of-burial and material properties such as seismic wave speed. The high-frequency roll-off slope is related to the elastic source displacement time history and the form of the radial stress (Denny and Johnson, 1991). Because of the non-linear response of earth materials to high-energy loading from nuclear explosions, these characteristic source parameters are closely related to the rheological properties of the shooting medium, especially material strength, porosity and emplacement depth.

Empirical source scaling laws seek to provide predictions of explosion behavior and have so far

[1] Lawrence Livermore National Laboratory, Livermore, CA 94551, USA. E-mail: rodgers7@llnl.gov

been derived from explosion measurements in a limited number of the geological materials and under limited emplacement conditions (e.g., Stevens and Day, 1985; Denny and Johnson 1991). For other materials and emplacement conditions, the scaling behavior is difficult to infer due to the lack of sufficient free-field measurements and/or poor understanding of the non-linear rheological behavior of the emplacement media. Springer (1966) showed that underground explosions in dry porous media result in significantly reduced seismic amplitudes compared to explosions in hard rock. Perret (1972) analyzed a number of underground nuclear events in several geological environments and reported that explosions in dry porous rocks result in reduced peak motions by factors of 7–40 compared to explosions of equal yield in competent high-strength rocks (e.g., granite). Recently Murphy *et al.* (2011) analyzed historical seismic measurements from Nevada Test Site (NTS) explosions in porous materials and found the seismic source scaling for porous material emplacements can be related to those of hard rock by dividing the MM71 source spectrum for the same yield and depth-of-burial (DOB) in competent rock predicted by an inferred frequency-independent porosity reduction factor. On the other hand, Ford *et al.* (2011) found that corner frequencies and spectral slopes are strongly dependent on emplacement material properties. Specifically for NTS explosions in dry porous materials, Ford *et al.* (2011) reported that source spectral shapes have lower corner frequencies and steeper spectral slopes than those predicted by the MM71 model. These discrepancies suggest more controlled experiments and/or physics-based modeling of the non-linear material response to explosive loading may assist in understanding the spectral characteristics in empirical observations.

The effects of the emplacement media on seismic source characteristics of underground explosions are key to accurately infer explosion yield from distant seismic observations and to understand the excitation of the shear waves so crucial to earthquake-explosion discrimination. Materials such as limestone and sandstone are generally weaker and more porous than hard rock, such as granite, and their response to energetic loading is indeed different (e.g., Werth and Herbst, 1963; Perret and Bass, 1975). Material damage (i.e., permanent changes in material properties) caused by explosions might significantly affect the radiated seismic waves in the form of isotropic and compensated linear vector dipole (CLVD) body force systems, and the composite source characteristics have significant implications for source identification and yield estimation (Ben-Zion and Ampuero, 2009; Patton and Taylor, 2011). Numerical modeling of explosions with hydrodynamic simulations provides an important tool to investigate effective elastic source characteristics in a variety of geological materials, from hard to porous materials, using non-linear material models calibrated from experimental studies. Early hydrodynamic simulations of underground explosions used one-dimensional modeling (e.g., Butkovich, 1965; Andrews and Shlien, 1972; Rodean, 1971). Advances in numerical techniques and computational resources have enabled ever more realistic simulations of explosion shock-waves in earth materials (e.g., Antoun *et al.*, 1999; Stevens and Xu, 2010; Rougier *et al.*, 2011). It is now possible to perform hydrodynamic simulations of explosions on high-performance (parallel) computers in realistic material models in two- and three-dimensions at sufficient resolution to cover the frequencies of interest for seismic monitoring (well above to 10 Hz) and computational domains spanning the non-linear to linear elastic regions. While it remains difficult to model motions from the non-linear, near-source region to distant (regional and teleseismic) seismic stations, one-way coupling strategies are able to pass motions from hydrodynamic to linear elastic calculations (e.g., Stevens and Xu, 2010; Rodgers *et al.*, 2011).

The objective of this study is to perform hydrodynamic simulations of underground nuclear and chemical explosions in a well-characterized hard rock (granite in this study) and to extract seismic source characteristics from the waveforms in the near-field using reduced velocity potentials. We focus on granite because this material has very low porosity (which simplifies the non-linear response) and has been well studied with hydrodynamic modeling. We compute P-wave source spectra and compare results with predictions from the widely used Mueller and Murphy (1971) and Denny and Johnson (1991) models. The variation of source properties with yield and scaled DOB are considered. We also investigate the differences between chemical and nuclear source explosions of the same yield. The interaction between the shock-wave and the free surface results in permanent deformation and material damage. We investigate the seismic radiation from underground explosions by modeling the computed response with elastic Green's functions convolved with the derived source waveform. This study is intended to illustrate a methodology to assess the source characteristics for underground explosions using hydrodynamic simulations. The methodology demonstrated in this study may be applied to other materials, including those with lower material strengths and higher porosity which are expected to undergo more severe non-linear effects (e.g., shear failing, bulking, compaction under dynamic loading) and whose non-linear constitutive behaviors have been calibrated from laboratory experiments (e.g., VOROBIEV *et al.*, 2007; VOROBIEV, 2008).

2. *Hydrodynamic Modeling of Underground Explosions in Granite*

Explosions in granite such as the PILEDRIVER and HARDHAT nuclear tests at NTS have been widely investigated (e.g., ANTOUN *et al.*, 1999, 2001, 2004; ANTOUN and LOMOV, 2003; STEVENS *et al.*, 2003; ROUGIER *et al.*, 2011). In this study we use a constitutive model for granite derived from these data that includes the effects of bulking, yielding and damage on the material response under impact loading conditions (ANTOUN *et al.*, 1999). The equivalent elastic material properties (compressional and shear wave speed and density) for our hydrodynamic calculations are provided in Table 1. The hydrodynamic response is simulated with GEODYN, an Eulerian finite volume code with adaptive mesh refinement (LOMOV *et al.*, 2003). GEODYN can be run in one-, two- or three-dimensions. Figure 1 shows the spherically symmetric one-dimensional (1D) GEODYN simulation responses (radial component peak velocity and displacement) for whole-space models compared to measurements from historical nuclear explosions in granite, specifically the PILEDRIVER, HARDHAT and SHOAL US tests, and tests at foreign sites in the Former Soviet Union (Degelen) and French Test Site in Algeria (Hoggar). The simulation results are seen to be in good agreement with the observed peak velocities and displacements from historical experiments. Note that observed motions for other materials are much more scattered than the result shown in Fig. 1.

Using the same constitutive model we performed similar calculations for a two-dimensional (2D) axisymmetric domain including the free surface. Gravity is included in our calculations except where explicitly stated. In the calculations shown in this study we assume granite properties are uniform in the solid half-space; while this is not entirely realistic, it greatly simplifies comparison with other source models. Extension of the present methodology to cases with vertical gradients in material properties is left to future studies. The explosion time-histories are simulated in different ways for nuclear and chemical sources. For nuclear explosions, the energy is deposited in the source volume as an ideal gas with the mass of the device and an instantaneous temperature and pressure rise (e.g., ANTOUN *et al.*, 2001; ROUGIER *et al.*, 2011). For chemical explosions, we use the Jones–Wilkens–Lee (JWL) equations of state, assuming the properties of a heavy ammonium nitrate fuel oil (ANFO) mix (LEE *et al.*, 1973). In this study we considered explosive yields in units of TNT mass-energy equivalent in the range 0.5–5 kiloton (kt) or 500,000–5,000,000 kg. All explosions are fully tamped, that is, explosives are in contact with hard rock material.

Table 1

Material model parameters for GEODYN simulations and MUELLER and MURPHY (1971, MM71) model following STEVENS and DAY (1985)

Model	P-wavespeed (m/s)	S-wavespeed (m/s)	Density (kg/m^3)
GEODYN	5600	3200	2600
MM71/SD85	5350	2975	2650

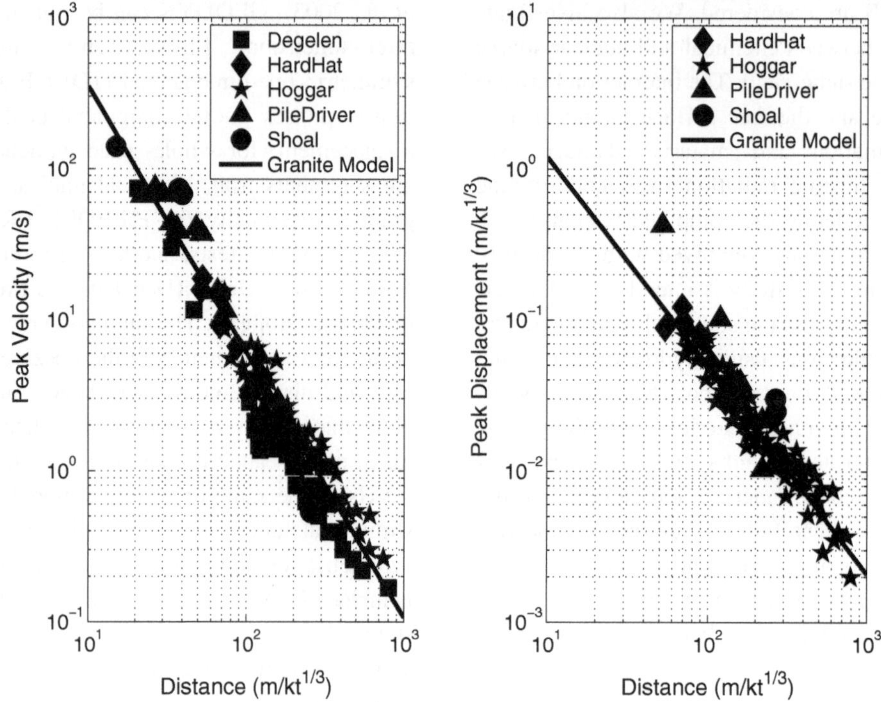

Figure 1
Comparison of the observed peak velocities (*left, symbols*) and displacement (*right, symbols*) from historical experiments in granite versus scaled range with predictions from GEODYN simulations for the granite model (*solid line*)

3. Estimation of the Seismic P-Wave Source Spectra

The source spectra are derived from the simulated motions by calculating reduced velocity potentials, RVP's (Werth and Herbst, 1963; Haskell, 1967; Rodean, 1971; Aki *et al.*, 1974). The time-domain velocity waveforms are taken from the near-field region at marker points where motions are elastic (permanent plastic deformations are less than 1 %).The source spectra are approximated by taking the Fourier transform of the RVP's, accounting for the source-marker point distance and elastic compressional wavespeed and converting to displacement spectra. There are several important factors to consider in this analysis: (1) the reduced displacement or velocity potentials (RDP or RVP, respectively) are defined in the context of elastic motions for an isotropic disturbance in a uniform whole-space with spherical symmetry (Haskell, 1967; Rodean, 1971); (2) gravity with a half-space model makes the wave radiation cylindrically symmetric rather than spherically symmetric; and (3) the presence of the free surface above the explosion results in a reflected tensile wave (Murphy, 1991) which interferes with the primary wave. Because explosions are conducted near the Earth's surface, it is essential to include free-surface effects explicitly. Gravity is omnipresent and causes strong depth dependence in lithostatic pressure, so it must also be considered explicitly.

In order to investigate the effects of these factors on inferred source properties, we compared the time-domain RVP's for a uniform whole-space without gravity and that for a 2D half-space simulation with gravity with the marker point taken below the source, relative to the free-surface. We found good agreement between the time-domain direct P-waves in these two cases, which indicates that we can isolate the direct outgoing P-wave and minimize the interference of a reflected tensile wave. In the following, we estimate P-wave source spectra from the RVP's taken from 2D half-space simulations with gravity.

Simulated source spectra are compared to the MM71 source model (Stevens and Day, 1985) using the elastic material properties in Table 1 at several yields assuming normal scaled depth-of-burial (SDOB) of 122 m/kt$^{1/3}$ following standard containment practice

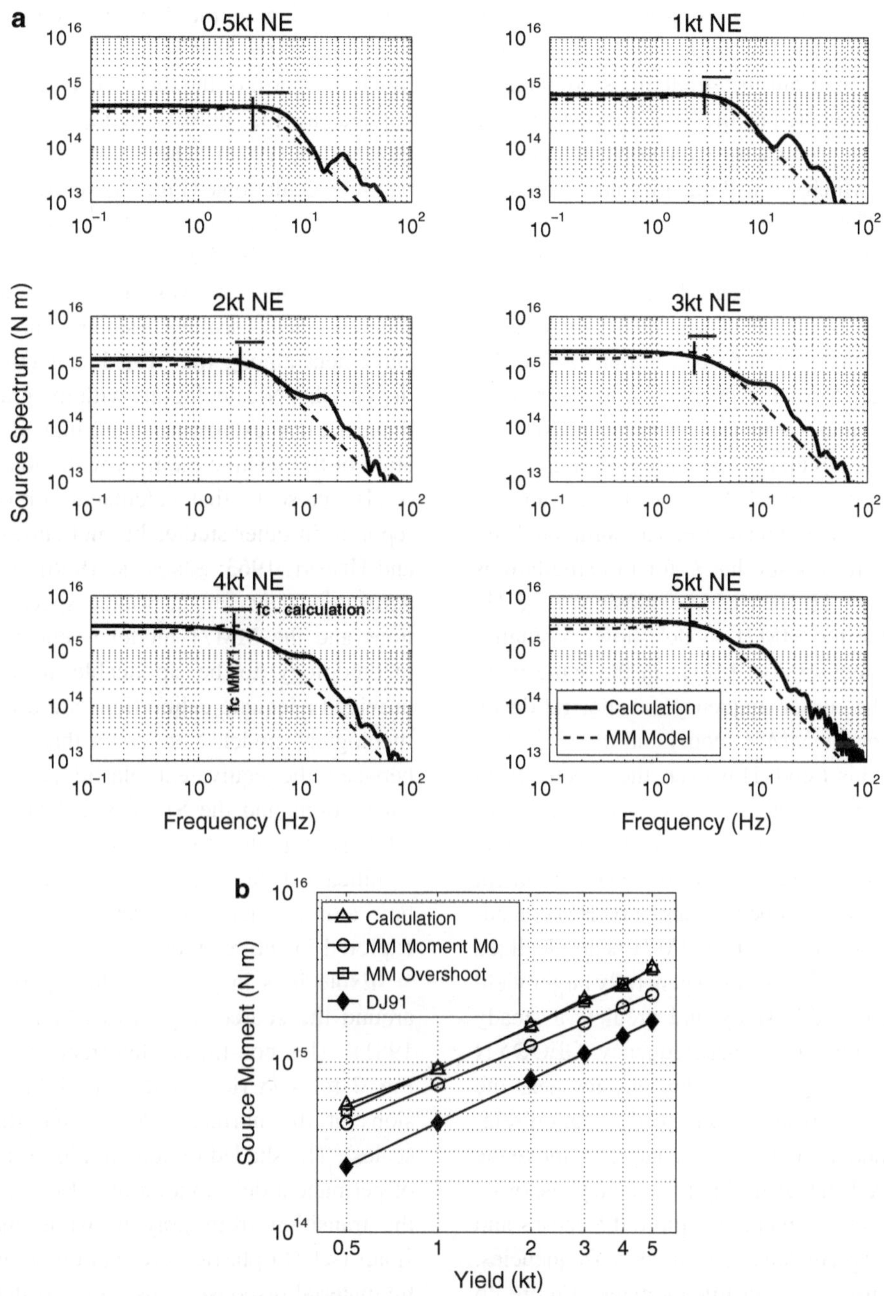

Figure 2

a Comparison of the derived source functions (*solid lines*) with the Mueller–Murphy model (*dashed lines*) for normally buried nuclear explosions at 0.5, 1, 2, 3, 4, 5 kt. In each panel we indicate the corner frequency, f_c, predicted by MM71 (*vertical line*) and the range inferred from the simulated spectra (*horizontal line*). **b** Scaling of seismic moment with yield for GEODYN calculations (*triangles*), MM71 model (*circles*), MM71 model using the overshoot spectral levels (*squares*) and the DJ91 model (*diamonds*)

(U.S. Congress Office of Technology Assessment, 1989). We found results are not very sensitive to small variations in material properties such as those in Table 1. The results for nuclear explosions with yields of 0.5, 1, 2, 3, 4 and 5 kt and at normal SDOB are shown in Fig. 2a. Solid lines represent the source spectra from GEODYN calculations with the granite model and dashed lines represent the MM71 model for each yield and DOB. The spectra agree reasonably well with notable differences. The low-frequency levels (related to seismic moment) from the simulations are slightly higher than that predicted by the MM71 model and the simulations do not produce the overshoot predicted by MM71, both of which we will quantify and discuss in detail below. The corner frequency, f_c, is usually defined as the frequency at which the low-frequency asymptote or seismic moment level intersects the higher frequency asymptote. The simulations produce spectra that do not conform to a simple low-order parametric model. Corner frequencies from the MM71 model (vertical lines) and the range of possible f_c inferred from the simulated spectra (horizontal lines) are indicated in each panel. We find that corner frequencies are quite consistent between the simulated and MM71 spectra. It appears that f_c for the simulations decreases from $\sim$6 Hz for a 0.5 kt explosion to $\sim$3 Hz for 5 kt and that the MM71 model shows a smaller decrease from about 3.0 to 1.8 Hz in the same yield range. Above the corner frequency the spectral decay (slope) is f^{-2} for the MM71 model and the simulated spectra follow this trend. However, the spectra from simulations show complexity above the corner frequency, probably due to non-linear effects at the source. These variations are always below the seismic moment level and at the high-frequency range (10 Hz) considered in nuclear monitoring. The MM71 is an idealized model derived from explosions of generally larger yield than considered in this study and predicts a steady spectral decay above the corner frequency. GEODYN calculations include more complex physics near the source and this may result in enhanced radiation relative to MM71 and oscillations at frequencies above the corner that are reflected in the high-frequency response. Perhaps further research could explore the causes and consequences of predicted energy at these frequencies.

We now consider the seismic moments. Figure 2b shows the inferred seismic moments, M_0, from the simulations (triangles) and MM71 model (circles) and the seismic moment corresponding to the overshoot spectral level from the MM71 model (squares), as well as the moments predicted by the Denny and Johnson (1991, DJ91) model (diamonds). It is seen that the seismic moments inferred from our simulations are slightly higher (about 25 %) than the MM71 model predictions and lie closer to the moments corresponding to the MM71 overshoot levels. The overshoot in the MM71 model is associated with the functional form of the theoretical pressure at the peak applied at the elastic radius (von Seggern and Blandford, 1972). The spectra from the hydrodynamic calculations (Fig. 2a) show little overshoot, presumably due to the functional form of the pressure time-history. The DJ91 model gives about half the moment values of MM71. Note that the seismic moment–yield scaling (slopes) are in excellent agreement among the MM71 and DJ91 models and the GEODYN calculations with the granite model for these nuclear explosions at normal SDOB. The variation of the seismic moment with yield seen in Fig. 2 is also close to that calculated for other materials reported in other studies but not shown (e.g., Werth and Herbst, 1963; Springer, 1966). Such reasonable agreement between the source spectra from simulations and the MM71 model is encouraging for this methodology to be applicable to other materials provided adequate calibrations are used to define the rheological model. Note again that small differences between the equivalent elastic parameters for our calculations and the Stevens and Day (1985) methodology for the MM71 model do not result in significant differences in the P-wave source spectra, but emplacement material properties could be explored in future studies.

Explosions in granite induce permanent damage around the source (e.g., Butkovich, 1965; Rodean, 1981). The non-linear deformations are shown in Fig. 3 for 1 kt (left) and 5 kt (right) nuclear explosions at the normal SDOB (note different length scales). The shaded regions in Fig. 3 indicate degrees of permanent deformation on a logarithmic scale and the transition from gray to white indicates where small (<1 %) plastic deformations transition to elastic material response. This region is determined from the calculations to have experienced shear failure and to have permanent deformation. The non-linear region extends horizontally and vertically from the source to approximately 160 m for 1 kt and 260 m for 5 kt, and extends much further along the free surface. The results are consistent with observed transition boundaries (e.g., Perret, 1972) and calculations (e.g., Butkovich, 1965; Rougier *et al.*, 2011). Also note that the length scale of the transition point

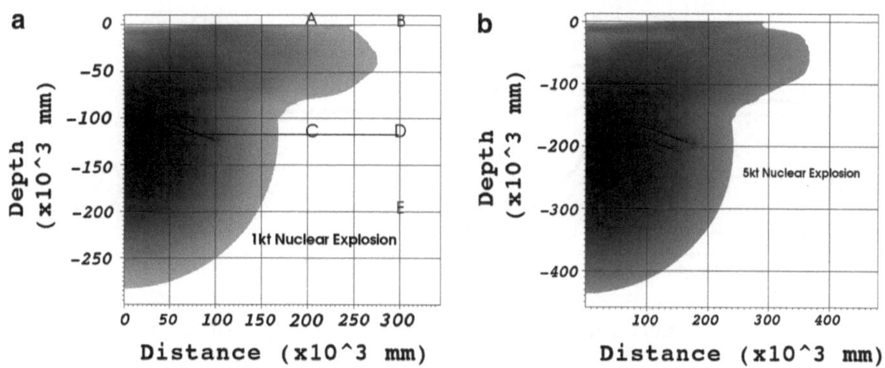

Figure 3

Non-linear deformation regions for 1 kt (**a**, *left*) and 5 kt (**b**, *right*) nuclear explosions at normal burial depths. The *shaded regions* indicate degrees of permanent deformation on a logarithmic scale and the transition from *gray* to *white* indicates where small plastic deformations transition to elastic material response. The markers *A–E* are used later in discussing the motions

from plastic to elastic behavior scales roughly as yield to the 1/3 power.

In the absence of a free surface or for deeply overburied explosions the damage zone should be spherically symmetric, but in the presence of the free surface it is noted from Fig. 3 that the damage zone shapes for the normally buried explosions are nearly spherical at and below the shot depth but become conical near the surface due to the interaction between the reflected waves and the primary shock waves. The aspherical distribution of the non-linear deformation in Fig. 3 indicates that the effect of the free surface is likely to be more complicated than the effect of elastic waves simply interacting with the free surface (i.e., P-to-P reflection and P-to-S conversion). The marker points A–E on the left plot in Fig. 3 are used below to examine the effect of material damage.

4. *Effect of Depth-of-Burial*

Seismic source excitation is dependent on the emplacement depth as well as the material properties (e.g., MM71; Perret and Bass, 1975; Denny and Johnson, 1991). DOB effects seismic excitation via lithostatic pressure because material compression with depth increases material strength and motions at depth are resisted by higher confining pressures. We inferred P-wave source spectra for 1 kt nuclear explosions detonated at SDOB of 0.5, 1.0, 1.5, 2.1, 2.9 and 3.9 (DOB 61, 122, 183, 256, 354 and 476 m, respectively) in our granite model. The source spectra from our calculations are shown in Fig. 4a (solid lines) with each panel comparing the spectra predicted by the MM71 model (dashed lines) at a given SDOB. Seismic moment decreases as SDOB increases for both the MM71 model and our simulations. However, the moments decrease more rapidly with SDOB than the MM71 model predicts. We quantify the effect of SDOB on seismic moment by comparing the moments from our simulations with those for those predicted by the MM71 and DJ91 models in Fig. 4b. This figure indicates that the calculations predict moments in agreement with the MM71 model for SDOB 1.5–3. However, across the range of SDOB considered, our hydrodynamic simulations predict more rapid variation of moment with SDOB than either the MM71 or DJ91 model predicts. Like the comparison of seismic moment with yield shown in Fig. 2b, the DJ91 model consistently predicts lower seismic moments than the MM71 model or our hydrodynamic calculations.

5. *Comparison of 1 kt Nuclear and Chemical Explosions*

The differences between seismic energy emerging from nuclear and chemical explosions are important for interpreting data from nuclear tests and conventional calibration experiments. The fundamental differences between the two explosion types include the initial energy density and the rate of energy

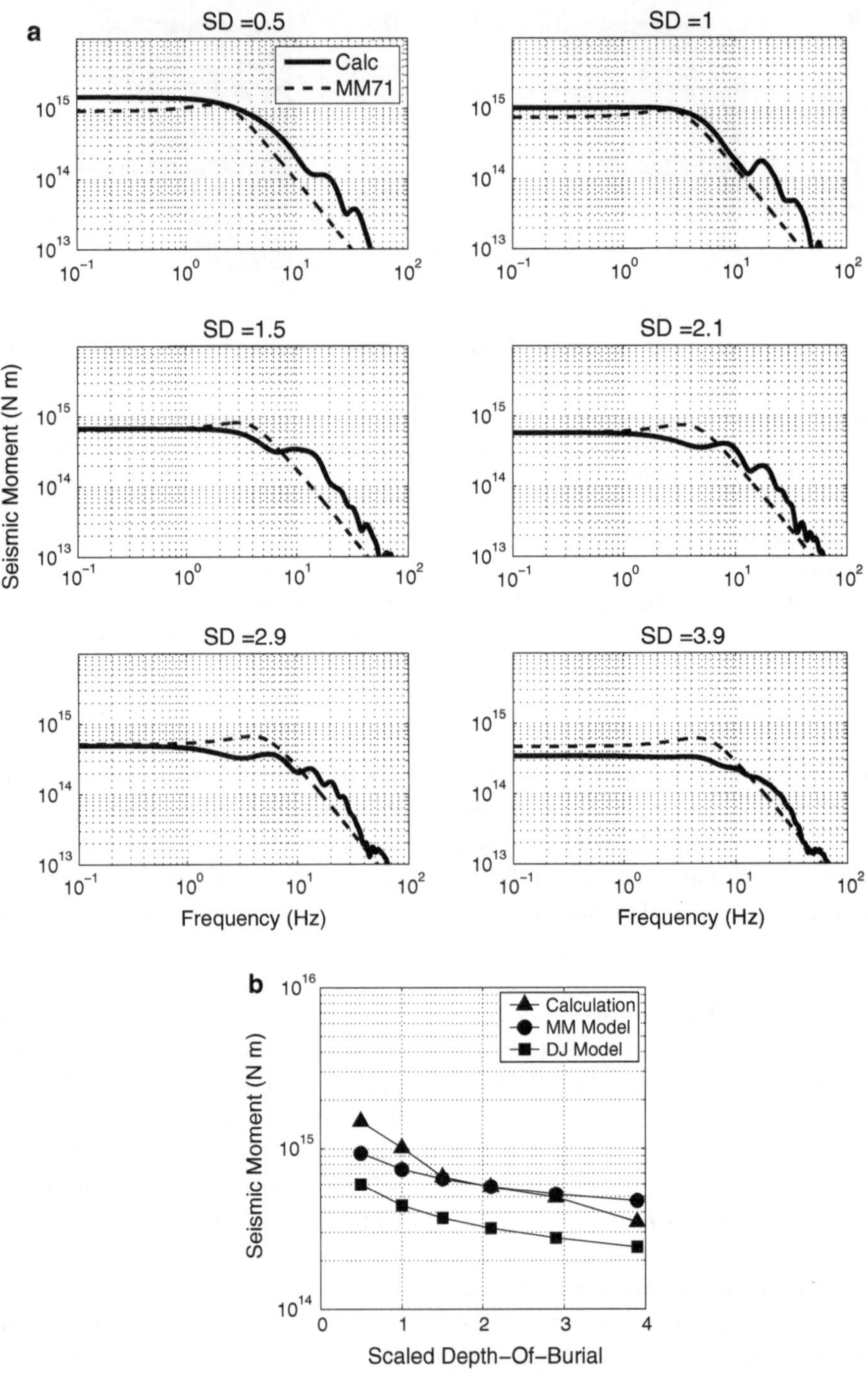

Figure 4

Effect of scaled depth-of-burial (SDOB) on the source spectra for 1 kt nuclear explosions. **a** The source spectra from simulations (*solid lines*) and the corresponding MM71 model results (*dashed lines*) are shown for comparison. **b** Comparison among seismic moments calculated, MM71 model and DJ model for different SDOBs

released during the explosion (Goldstein and Jarpe, 1994). Reports based on the 1993 non-proliferation experiment (NPE) and nearby NTS nuclear explosions indicate that the chemical source spectral amplitude is enhanced relative to that of nearby nuclear explosions by about a factor of two (Denny,

1994; GLENN and GOLDSTEIN, 1994; GOLDSTEIN and JARPE, 1994; WALTER *et al.*, 1994). STUMP *et al.* (1999) showed that there is no apparent spectral shape differences between the nuclear and single-fired chemical source in the bandwidth of the data from 0.36 to 100 Hz at the local sites recording the NPE. In this study we examine the differences in the source spectra from a 1 kt nuclear explosion and 1 kt (TNT energy equivalent) heavy ANFO explosion in granite. The nuclear explosion source representation is an instantaneous energy deposition, as described above, while the chemical explosion is simulated with the JWL equations (LEE *et al.*, 1973) as a spherical charge ignited at the center. The resulting velocity time-histories and source spectra are shown in Fig. 5. The source strengths for 1 kt nuclear and chemical explosions are denoted by solid and dashed lines, respectively. In the time-domain the chemical explosion motions are higher amplitude and longer duration than the nuclear explosions. In the frequency domain, the chemical explosion has a higher low-frequency level by about a factor of two and lower corner frequency than the nuclear explosion. These differences are due to the differences of the energy release rate and not the material properties, which are identical in the two simulations. The source spectral amplitudes and shapes are in good agreement with the derived RVPs from nuclear and chemical explosions at Rainer Mesa at NTS reported by GOLDSTEIN and JARPE (1994), although not duplicated here. Note that the final cavity pressures for these simulations were calculated to be about 50 MPa for the 1 kt chemical explosion and only 12 MPa for the 1 kt nuclear explosion. The final cavity radius can be estimated from the steady-state (long time) density distribution of the hydrodynamic calculations. We inferred these and found the approximate cavity radius, R_c, is larger for the chemical explosion (11.3 m), corresponding to higher RDP and seismic moment, than for the nuclear explosion (9 m). These values support our observation that the predicted P-wave seismic moments are approximately a factor of two higher for chemical compared with nuclear explosions. Seismic moment is related to the final cavity, under the assumption of incompressible rock, by $M_0 = \rho\alpha^2 V_c$ (MURPHY, 1991) where ρ is the density, α is the compressional wave speed and V_c is the

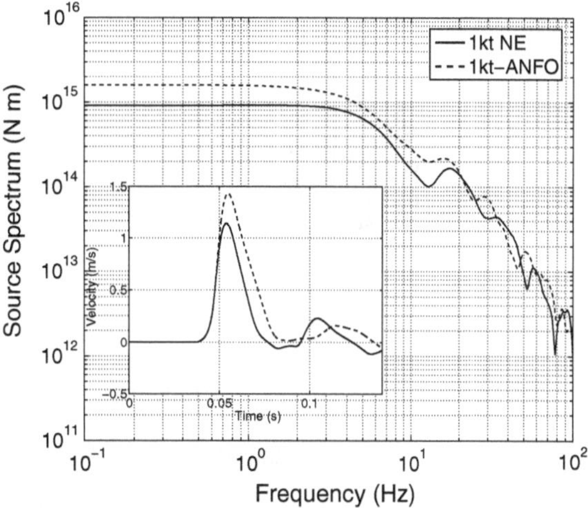

Figure 5
Comparison of source spectra from 1 kt nuclear and 1 kt chemical explosions (both TNT equivalent). The *inset* shows the waveforms in time domain

final cavity volume. Assuming the cavity is spherical, the volume V_c is proportional to R_c^3. Thus, the ratio of moments for chemical compared to nuclear explosions of equal yield in the same medium is equal to the ratio of final cavity radii to the third power. In our case this is $(11.3/9.0)^3 = 1.98$, or quite nearly the factor of 2 seen in Fig. 5. We also report (but do not show) that for spherically symmetric (1D) cases without gravity, the calculations also show that the chemical explosions are more efficient in generating the compressional waves, that is, they have higher seismic moments. We notice the discrepancy between M_0 calculated from the final cavity volume and the seismic moments indicated in Fig. 5 and such discrepancy also was seen in the realistic explosion scenario (AKI *et al.*, 1974). The likely reasons might include that the medium in our calculations is not incompressible.

6. Investigation of the Elastic Radiation from the Non-Linear Damage Zone

As seen in Fig. 3 the non-linear damage region is not spherical but extends further out from the source along the surface. This 2D pattern is rotated around the axis of symmetry resulting in an inverted conical shape of the damage zone above the shot point.

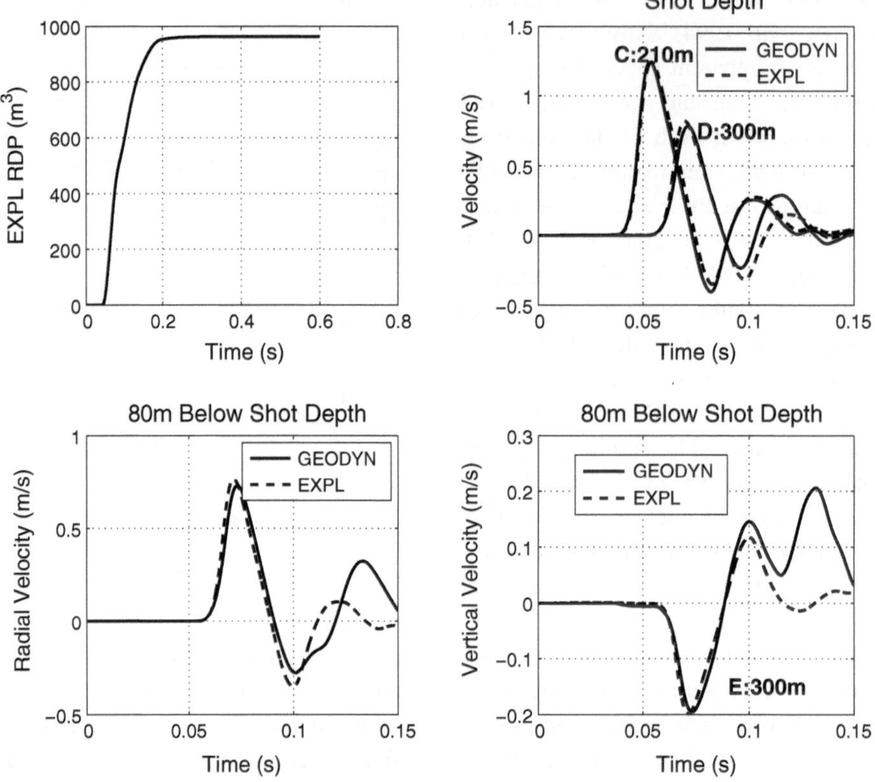

Figure 6

Elastic modeling of the hydrodynamic response for points indicated in Fig. 3a. *Top left* reduced displacement potential for a 1 kt nuclear explosion. *Top right* comparison of the GEODYN calculation and solutions using the Green's function and the derived explosion source function (EXPL) at radial distances of 210 m (**c**) and 300 m (**d**) at the shot depth. *Bottom* comparison of the GEODYN calculation and solutions using the Green's function and the derived source function on radial (*bottom left*) and vertical (*bottom right*) components at radial distances of 300 m (**e**) at a depth of 80 m below the shot point

Damage likely impacts the radiated seismic waves and the relationship of damage and seismic radiation is actively investigated (Ben-Zion and Ampuero, 2009; Patton and Taylor, 2011; Rougier *et al.*, 2011).

Gravity (lithostatic pressure) is not included in the calculations described below to make a straightforward comparison between hydrodynamic and elastic motions to assess the effect from the damage near the source and near the surface. We found that near-source damage such as shown in Fig. 3a is only slightly changed without the gravity from the case with gravity. As described above, the reduced displacement potential is derived from the velocity time-history at a location at the shot depth beyond the non-linear region (210 m, location C in Fig. 3a) and is shown in the top left plot of Fig. 6. The elastic Green's function in the half space (equivalent elastic medium, compressional wavespeed, $v_P = 5{,}600$ m/s, density, $\rho = 2{,}600$ kg/m^3, attenuation quality factor $Q_P = 200$ in this study) is calculated using the wave-number integration method (Luco and Apsel, 1983) at locations A–E for an isotropic source and the source time function is approximated by the reduced displacement potential time-history. The waveforms at locations A–E are obtained by convolving the Green's functions with the RDP with a factor of $4\pi(\lambda + 2\mu) = 4\pi\rho\alpha^2$, where λ and μ are elastic Lame material constants (Aki *et al.*, 1974). Figure 6 shows the comparison of the resulting waveforms (dashed lines) and the GEODYN solutions (solid lines) at the shot depth (locations C and D, top right: radial) and 80 m below (location E, lower panels with radial and vertical components). At the shot level (locations C and D) the P-waves calculated from GEODYN are closely reproduced by the equivalent elastic solutions

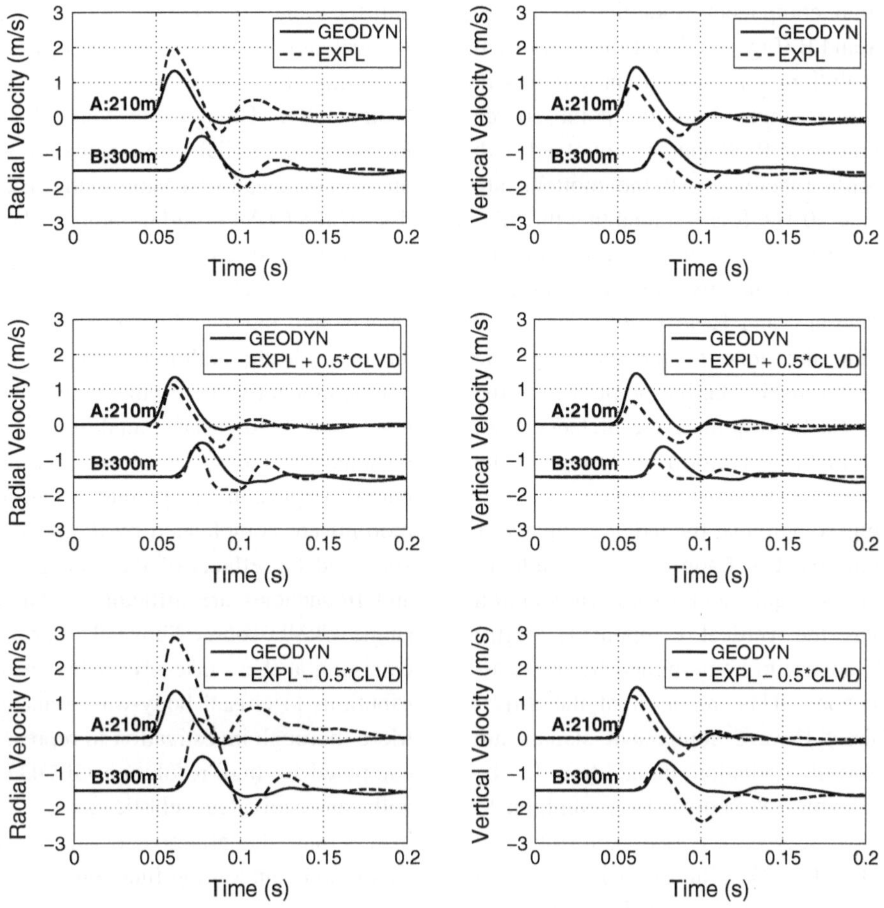

Figure 7

Comparison of the GEODYN solutions and the solutions from a composite source at radial distances of 210 m (**a**) and 300 m (**b**) at the surface. *Top* the source is a simple explosion source. *Middle* composite source is an explosion plus a vertically oriented CLVD with half the explosion moment. *Bottom* composite source is an explosion minus a CLVD with half the explosion moment. Note that CLVD source is located half-way between the explosion depth and the surface and is simultaneous with the isotropic explosion

for an explosion (isotropic) source (EXPL), though after three quarter cycles the difference is clearly visible at location D, likely due to the reflected phases from the free surface. At 80 m below the shot depth the P-waves from the hydrodynamic and equivalent elastic isotropic source are also in good agreement for the first three quarters of the cycle. The differences between the hydrodynamic and elastic responses below the shot point are also due to reflected phases from the free surface, which are excluded in the Green's functions at these locations. When surface reflections are included in the elastic Green's functions the fit to the motions after the first half cycle of the direct P-wave are further degraded. The close agreement of the direct P-wave motion from the hydrodynamic calculations and the elastic response clearly validates that at these locations the motions are indeed elastic and the amplitudes are well represented by an isotropic explosion of equivalent seismic moment and source-time function given by the RDP. The damage zone near the surface contributes only modestly to these direct P-waves, resulting in significant differences only after the first ¾ cycle. Note that we did calculations including gravity and found that it does not impact the direct P-waves and the damage zone is found to show little change compared to calculations without gravity.

At surface locations, however, we see much larger differences between the hydrodynamic calculations and the solutions using the elastic Green's functions convolved with the RDP. Figure 7 (top panels) shows vertical and radial component velocities at surface marker points A and B (Fig. 3a). Solid lines represent the GEODYN solutions and dashed lines represent the response for a single isotropic point source (EXPL) obtained by the RDP source convolved with the elastic Green's functions. All the motions are low-pass filtered at 50 Hz. It is assumed that the first-cycle of the waveform on the surface represents the primary particle motion directly from the source as suggested by TREMBLY and BERG (1966). Note that point A is inside the zone of permanent plastic deformation as seen in Fig. 3a, but point B is in the region of purely elastic response. We included both points to look into the waveform evolution on the surface from the plastic to elastic regimes.

It is interesting to note that the radial components (left panels) from GEODYN appear to more attenuated relative to the equivalent explosion solution (EXPL), whereas the vertical components (right) from GEODYN appear to be amplified by the near-surface damage zone. This implies that the differences in the direct P-wave component signals are generated by seismic radiation in addition to the simple isotropic explosion source. It is common to include a secondary source for buried explosions as a vertically oriented CLVD with extension along a vertical axis and in compression horizontally (STEVENS *et al.*, 2003; YANG and BONNER, 2009; PATTON and TAYLOR, 2011). In order to investigate this possibility, we added the elastic contributions to the motion assuming: (1) CLVD source at the half burial depth (61 m) (note the extra damage is close to the surface, Fig. 3); and (2) the CLVD source has the same time history as the explosion source with no delay relative to the isotropic explosion (EXPL). The far-field elastic solutions from the composite (EXPL + CLVD) source are a simple linear superposition of the two source components. The middle plots in Fig. 7 show that comparison of the GEODYN solutions and the solutions from one composite source at location A and B. This source is modeled as an explosion plus a CLVD with half the explosion moment. It is seen that the radial components for this source are closer to the GEODYN solution, while the vertical component amplitudes are further reduced. The bottom plots correspond to the composite source of an isotropic explosion minus a CLVD source of half the explosion moment in an attempt to boost the vertical amplitudes. This source would have compression along the vertical axis and extension horizontally. Though the vertical amplitudes are increased, the radial components are strongly altered due to horizontal extension than the case with addition of the CLVD (middle plots). The subtraction of the vertically oriented CLVD source degrades the fit to the waveforms at the shot level and below (not shown) and thus would not seem a plausible mechanism while the addition of the CLVD source does not significantly affect the waveforms at the shot level and below. Our analyses of the near-field responses at the surface from this simple normally buried explosion in granite suggest that surface motions are complicated by the near-source damage zone and the effects of the damage at these ranges and frequencies are difficult to characterize with a single CLVD source above the shot point. Motions emerging at these take-off angles are important contributions to crustal body-waves and surface waves. More thorough investigation to represent the motions computed by high-resolution GEODYN simulations with an equivalent elastic source might provide insight into the effective elastic radiation moment tensor and source-time function.

7. Discussion and Conclusion

In this study, seismic source spectra of underground nuclear explosions are inferred from velocity time-histories computed with hydrodynamic simulations with a well-calibrated non-linear material model for granite. We compared P-wave source spectra from our hydrodynamic simulations with the predictions of the MUELLER and MURPHY (1971, MM71) model and with seismic moments from the model of DENNY and JOHNSON (1991, DJ91). Source spectra computed with hydrodynamic simulations for normally buried nuclear explosions are in reasonably good agreement with those predicated by the MM71 model, though we found our hydrodynamic simulations resulted in higher moments than both MM71 and DJ91 predict. The scaling of seismic moment with yield for normally buried explosions is found to be consistent with both the MM71 and DJ91 models.

The effect of SDOB on seismic moment is stronger for our hydrodynamic calculations than that predicted by the MM71 and DJ91 models. Computed source moments are higher by about a factor of two for chemical explosions than for the nuclear explosions of the same yield due to the difference in energy release rate. The relative difference is in good agreement with the observations at the NTS site (Denny, 1994; Goldstein and Jarpe, 1994). The relative energy transfer efficiency might correlate with the emplacement material properties such as material inhomogeneities and porosity, and the enhancement of the chemical explosion source is likely be different for porous materials (e.g., Rodgers *et al.*, 2011).

We attempt to quantify the contributions from the damage around the source and near the free surface for normally buried explosions in the near field. It is noted that damage can be a significant factor affecting radiated seismic waves (e.g., Ben-Zion and Ampuero, 2009; Patton and Taylor, 2011). We use the derived source functions from a 1 kt nuclear explosion convolved with elastic Green's functions to obtain the response at nearby receivers at and below the surface. The comparison between the non-linear calculation and elastic response from an isotropic explosion source shows that the primary waves at steep (downward) takeoff angles are well approximated with the volumetric source (which certainly contains the contribution of damage just around the explosion source), but at shallower angles the primary waves are complicated by near-surface damage. This is consistent with the shape of the damage zone (Fig. 3). In the calculations with composite elastic sources, a simple addition/subtraction of a CLVD with half the explosion moment cannot match the primary near-field surface motions on the both the vertical and radial components simultaneously, though in the far field the addition of the CLVD of half the explosion moment appears to better approximate some observations (Stevens *et al.*, 2003) and a significant impact of Rayleigh wave radiation from a CLVD source on the surface magnitudes is demonstrated (Patton and Taylor, 2011). These studies analyzed data at much farther distances and at lower frequencies. It is possible that an equivalent CLVD source has a different time history and/or cannot be approximated by a point-source halfway between the free surface and shot point. Further investigation of the equivalent elastic source for seismic radiation in the near-field is required.

The volumetric source function derived from the calculations using the RVP certainly consists of the contribution of the damage induced in granite around the explosion source in addition to the explosions itself. The damage increases the coupling between the source and the granite and consequently enhances the radiated waves. To illustrate the effect of sustained material damage generated by explosions on the radiated seismic waves, we performed an additional simulation with the same computational grid but the granite is approximated with an elastic model without shear failure. The resulting motions for the non-linear and linear elastic material models (both have a yield of 1 kt) are compared at two points at the shot depth (C and D in Fig. 3a) in Fig. 8. The solid lines and dashed lines represent the waveforms due to the explosion in non-linear (with damage) and linear elastic (without damage), respectively. Note that the solutions without damage are enlarged by a factor of four to match the amplitudes of those with damage, that is, the damage greatly amplifies the radiated waves due to the weakening of the material properties (Ben-Zion and Ampuero, 2009). Also the waveforms without damage are enriched in higher frequencies, as occurs for anelastic attenuation (absorption). Provided that density undergoes little change, the equivalent Lame's constants in the non-linear granite model might sustain an up to 75 % average reduction in the near field. The significant reduction in the equivalent Lame's constants is important in generating non-spherical components when the upgoing waves reflect from the free surface as a tensile wave and cause more damage. Hydrodynamic modeling must properly account for both the damage caused by compressional and tensional loading. The total effect of damage is strongly dependent on the emplacement conditions and material properties. Further investigations of this type will likely provide insight into how material damage impacts seismic radiation.

The granite material model used in this study has been calibrated with the historical observations and the agreement of the inferred nuclear source spectra with the empirical MM71 model motivates the applications to other materials such as limestone and

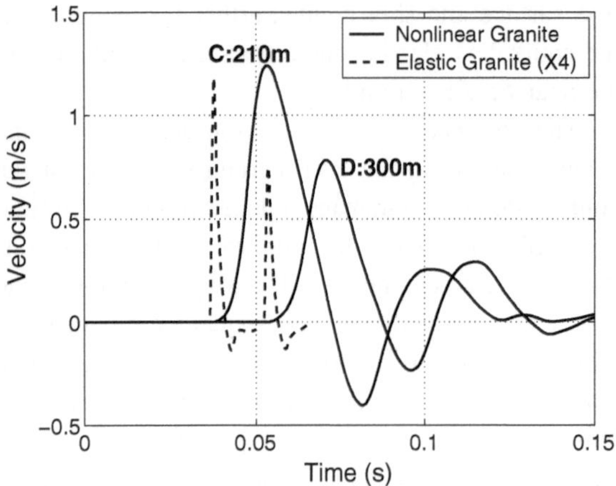

Figure 8
Comparison of the waveforms with (*solid lines*) and without (*dashed lines*) damage in the granite model at the shot depth. Note the waveforms without damage are enlarged by a factor 4

sandstone. These materials are porous and the effect of porosity should be investigated with a series of parametric calculations. The seismic motions for porous media would be reduced relative to those of the competent rock (e.g. PERRET, 1972; MURPHY *et al.*, 2011), and the spectral shapes would further be changed relative to the MM71 model (FORD *et al.*, 2011). The non-linear models simulating the behaviors of such porous materials have been developed (VOROBIEV *et al.*, 2007; VOROBIEV 2008). Though computationally intensive even on today's high performance parallel clusters, such detailed hydrodynamic simulations and resulting source spectra following the strategy in this study will certainly advance monitoring capabilities in a wider variety of realistic geological settings.

Acknowledgments

Discussions with and comments from Bill Walter, Sean Ford and Karl Koch and reviews by the editors and two anonymous referees greatly improved the manuscript. We thank Lew Glenn and Tarabay Antoun for the historical granite explosion data and insightful discussions. We are grateful to the Institute for Scientific Computing Research (ISCR) at LLNL for a Computing Grand Challenge allocation to undertake these calculations. Simulations were performed on the SIERRA Linux cluster operated by Livermore Computing. Funding for this project was provided by the National Nuclear Security Administration, Office of Defense Nuclear Nonproliferation Research and Development. This work performed under the auspices of the US Department of Energy by Lawrence Livermore National Laboratory under Contract DE-AC52-07NA27344. This is LLNL contribution LLNL-JC-519253.

REFERENCES

AKI, K., M. BOUCHON and P. REASONBERG (1974). *Seismic source function for an underground nuclear explosion*, Bull. Seism. Soc. Am., *64*, 131–148.

ANDREWS, D. J. and S. SHLIEN (1972). *Propagation of underground explosion waves in the nearly elastic range*, Bull. Seism. Soc. Am., *62*, 1691–1698.

ANTOUN, T. H., O. Y. VOROBIEV, I. N. LOMOV, L. A. GLENN (1999). Simulations of an underground explosion in granite, *Proceedings of the 11th American Physical Society Topical Conference on Shock Compression of Condensed Matter*, Snowbird, UT June 27–July 2, 1999.

ANTOUN, T. H., I. N. LOMOV and L. A. GLENN (2001). Development and application of a strength and damage model for rock under dynamic loading, *Proceedings of the 38th U.S. Rock Mechanics Symposium, Rock Mechanics in the National Interest*, D. Elsworth, J. Tinucci and K. Heasley (eds.), A. A. Balkema Publishers, Lisse, The Netherlands, 369–374.

ANTOUN, T. H. and I. N. LOMOV (2003). Simulation of a spherical wave experiment in marble using multidirectional damage model, *Proceedings of the 13th American Physical Society Topical Conference on Shock Compression of Condensed Matter*, Portland, OR July 20-25, 2003.

ANTOUN, T. H., I. N. LOMOV and L.A. GLENN (2004). *Simulation of the penetration of a sequence of bombs into granitic rock*, Int. J. Impact Eng., **29**, 81–94.

BEN-ZION, Y. and J.-P AMPUERO (2009), *Seismic radiation from regions sustaining material damage*, Geophys. J. Int., *178*, 1351–1356, doi:10.1111/j.1365-246X.2009.04285.x.

BUTKOVICH, T. R. (1965). *Calculation of the shock wave from an underground nuclear explosion in granite*, J. Geophys. Res., *70*, 885–892.

DENNY, M. (1994). Introduction and Highlights, 1-1, *Proceeding of Symposium on the Non-Proliferation Experiment*, Rockville, Maryland.

DENNY, M.D and L. R. JOHNSON (1991). The explosion seismic source function: models and scaling laws reviewed, *Explosion Source Phenomenology*, p. 1-24, Eds. S. R Taylor, H. J Patton and P.G Richards, Geophysical Monograph, 65.

FORD, S., W. WALTER, S. RUPPERT, E. MATZEL, T. HAUK, and R. GOK (2011). Toward an empirically-based parametric explosion spectral model, *Proceedings of the 2011 Monitoring Research Reviews*, Tucson, Arizona.

GLENN, L and P. GOLDSTEIN (1994), The influence of material models on chemical or nuclear-explosion source functions, 4-68, *Proceeding of Symposium on the Non-Proliferation Experiment*, Rockville, Maryland.

GOLDSTEIN, P. and S. JARPE (1994), Comparison of chemical and nuclear-explosion source spectra from close-in, local and regional seismic data, 6-98, *Proceeding of Symposium on the Non-Proliferation Experiment*, Rockville, Maryland.

HASKELL, N. A. (1967). *Analytic approximation for the elastic radiation from a contained underground explosion*, J. Geophys. Res., *66*, 2937.

LEE, E., M. FINGER and W. COLLINS (1973), JWL equations of state coefficients for high explosives, Lawrence Livermore National Laboratory Technical Report UCID-16189.

LOMOV. I. N., T. H. ANTOUN, J. WAGONER and J. RAMBO (2003). Three-dimensional simulation of the Baneberry nuclear event, *Proceedings of the 13th American Physical Society Topical Conference on Shock Compression of Condensed Matter*, Portland, OR July 20 25, 2003.

LUCO, J.E and R.J. APSEL (1983). *On the Green's functions for a layered half-space. Part I*, Bull. Seismo. Soc. Am., *73*, 909–929.

MUELLER, R. and J. MURPHY (1971). *Seismic characteristics of underground nuclear detonations Part 1: Seismic source scaling*, Bull. Seism. Soc. Am., *61*, 1675–1692.

MURPHY, J.R (1991), Free-field seismic observations from underground nuclear explosions, *Explosion Source Phenomenology*, 25-33, Eds. S. R Taylor, H. J. Patton and P.G Richards, Geophysical Monograph, 65.

MURPHY, J., T. J BENNETT, and B. BARKER (2011), An analysis of the seismic source characteristics of explosions in low-coupling dry porous media, 524-534, *Proceedings of the 2011 Monitoring Research Review*, Tucson, Arizona.

PATTON, H. J. and S. R. TAYLOR (2011), *The apparent explosion moment: Inferences of volumetric moment due to source medium damage by underground nuclear explosions*, J. Geophys. Res., 116, B03310, doi:10.1029/2010JB007937.

PERRET, W. R. (1972), *Seismic-source energies of underground nuclear explosions*, Bull. Seismo. Soc. Am, *62*, 763–774.

PERRET W. R. and BASS R. C. (1975). Free-field ground motion induced by underground explosions, Sandia National Laboratory Report No. SAND74-0252.

RODEAN, H. (1971). *Nuclear-Explosion Seismology*, U.S. Atomic Energy Commission, Oak Ridge, TN.

RODEAN, H (1981). Inelastic processes in seismic wave generation by underground explosions, in *Identification of Seismic source – Earthquake or Explosion*, E. S. Husebye and S.Mykkelveit (Eds.), P. 97–189.

RODGERS, A. J., H. XU, I. N. LOMOV, N. A. PETERSSON, B. SJOGREEN, O. Y. VOROBIEV and V. CHIPMAN (2011). Improving ground motion simulation capabilities for underground explosion monitoring: coupling hydrodynamic-to-seismic solvers and studies of emplacement conditions, *Proceeding of 2011 Monitoring Research Review*, Tucson, AZ.

ROUGIER, E., H. J. PATTON, E. E. KNIGHT and C. R. BRADLEY (2011), *Constrains on burial depth and yield of the 25 may 2009 north korean test from hydrodynamic simulations in a granite medium*, Geophys. Res. Lett., *38*, L16316, doi:10.1029/2011gl048269.

Springer, D.L. (1966), *P-wave coupling of underground nuclear explosions*, Bull. Seismo. Soc. Am, *56*, 861–876.

STEVENS, J.L and DAY, S. M. (1985). *The physical basis of mb:Ms and variable frequency magnitude methods for earthquake/explosion discrimination*, J. Geophys. Res., *90*, 3009–3020.

STEVENS, J.L. and H. XU, (2010), Wave propagation from complex 3D sources using the representation theorem, 519-528, *Proceeding of Monitoring Research Review*, Orlando, FL.

STEVENS, J.L., G.E. BAKER, H. XU, T.J. BENNETT, N. RIMER and S.M. DAY (2003), The physical basis of Lg generation by explosion sources, 456-465, *Proceeding of the 25th Seismic Research Review*, Tucson, Arizona.

STUMP, B. W., D. C. PEARSON and R. REINKE (1999), *Source comparisons between nuclear and chemical explosions detonated at Rainier Mesa, Nevada Test Site*, Bull. Seismo. Soc. Am, *89*, 409–422.

TREMBLY, L. D. and J. W. BERG (1966), *Amplitudes and energies of primary seismic waves near the Hardhat, Haymaker and Shoal nuclear explosions*, Bull Seismol. Soc. Am., 56,643–653.

U.S. Congress, Office of Technology Assessment (1989).*The Containment of Underground Nuclear Explosions,* OTA-ISC-414, U.S. Government Printing Office, Washington, DC.

VON SEGGERN, D. and R. BLANDFORD (1972), *Source time functions and spectra for underground nuclear explosions*, Geophys. J. R. Soc, *31*, 83–97.

VOROBIEV O.YU, LIU B.T, LOMOV I.N, TARABAY T.H (2007) *Simulation of penetration into porous geologic media*, Int. J. of Impact Eng., *34*, 721–731.

VOROBIEV, O. (2008). *Generic strength model for dry jointed rock masses*, Int. J. of Plasticity, *24*, 2221–2247.

WALTER, W. R., K. MAYEDA and H. J. PATTON (1994), Regional seismic observations of the Non-Proliferation Experiment at the Livermore NTS Network, 6-193, *Proceeding of Symposium on the Non-Proliferation Experiment*, Rockville, Maryland.

WERTH, G. and R. HERST (1963). *Comparison of amplitudes of seismic waves from nuclear explosions in four mediums*, J. Geophys. Res., *68*, 1463–1475.

YANG, X. and J. L. BONNER (2009), *Characteristics of chemical explosive sources from time-dependent moment tensors*, Bull Seismol. Soc. Am., *99*, 36–51.

(Received December 23, 2011, revised October 26, 2012, accepted November 3, 2012, Published online November 24, 2012)

Pure Appl. Geophys. 171 (2014), 523–535
© 2012 Springer Basel AG
DOI 10.1007/s00024-012-0591-4

Pure and Applied Geophysics

Testing Event Discrimination over Broad Regions using the Historical Borovoye Observatory Explosion Dataset

MICHAEL E. PASYANOS,[1] SEAN R. FORD,[1] and WILLIAM R. WALTER[1]

Abstract—We test the performance of high-frequency regional P/S discriminants to differentiate between earthquakes and explosions at test sites and over broad regions using a historical dataset of explosions recorded at the Borovoye Observatory in Kazakhstan. We compare these explosions to modern recordings of earthquakes at the same location. We then evaluate the separation of the two types of events using the raw measurements and those where the amplitudes are corrected for 1-D and 2-D attenuation structure. We find that high-frequency P/S amplitudes can reliably identify earthquakes and explosions, and that the discriminant is applicable over broad regions as long as propagation effects are properly accounted for. Lateral attenuation corrections provide the largest improvement in the 2–4 Hz band, the use of which may successfully enable the identification of smaller, distant events that have lower signal-to-noise at higher frequencies. We also find variations in P/S ratios among the three main nuclear testing locations within the Semipalatinsk Test Site which, due to their nearly identical paths to BRVK, must be a function of differing geology and emplacement conditions.

Key words: Nuclear explosion monitoring, regional discrimination, attenuation, event identification.

1. Introduction

High-frequency (>2 Hz) regional P/S ratios are an effective discriminant between closely-spaced earthquakes and explosions (e.g. WALTER *et al.*, 1995; TAYLOR, 1996; HARTSE *et al.*, 1997; KIM *et al.*, 1997; RODGERS and WALTER, 2002; TAYLOR *et al.*, 2002; BOTTONE *et al.*, 2002; WALTER *et al.*, 2007). The application of this discriminant to broad regions, however, has been hampered by large variations in the amplitudes of phases due to propagation effects in the crust and upper mantle. Attenuation models, which can account for much of the observed variability of regional phase amplitudes, have been shown to improve earthquake–explosion discrimination using these phases (e.g. PASYANOS and WALTER, 2009; PHILLIPS *et al.*, 2009), at least in limited areas. While regional high-frequency data has become more widely available in recent years, evaluating the broad area impact of these models is challenging, both because most nuclear explosions were conducted only at designated test sites and because the number of nuclear tests has diminished markedly since most nations signed the 1996 Comprehensive Nuclear-Test-Ban Treaty (CTBT).

The newly available deglitched Borovoye archive data (BAKER *et al.*, 2009) allows us to test regional discriminants, and improvements to them, over a broad region of central Asia. This dataset is an archive of digital seismograms derived from regional waveforms recorded at the Borovoye Observatory, northern Kazakhstan, over a thirty-year period beginning in 1966 (RICHARDS *et al.*, 1992). While seismograms from the observatory have been available to Western scientists since 2001 (KIM *et al.*, 2001), the data contained a large number of glitches and did not include instrument responses for a significant number of events, rendering it unusable for absolute amplitude studies. A joint effort by scientists at Lamont-Doherty Earth Observatory (LDEO) of Columbia University and Los Alamos National Laboratory (LANL) on deglitching the data and determining instrument responses (KIM and EKSTROM, 1996) has been completed, and the data made openly available to outside researchers.

The Borovoye dataset is very useful for testing discriminants using regional P/S ratios. Not only has this station recorded dozens of nuclear explosions at

[1] Lawrence Livermore National Laboratory, Livermore, USA. E-mail: pasyanos1@llnl.gov

the Semipalatinsk Test Site (STS), approximately 700 km away, it has also recorded several Soviet peaceful nuclear explosions (PNEs) which were widely dispersed throughout the territories of the former Soviet Union and are distributed about the recording station. Many explosions in this Borovoye dataset were still unusable due to dropouts, clipped waveforms, or poor signal-to-noise ratio. Still, the number and distribution of the events are sufficient to provide a good test of our ability to use the discriminant over broad regions.

In this paper, we will review the methods we have been using to make regional amplitude measurements, form discriminants, and tomographically invert for an attenuation model. While previous studies have noted geospatial variability in the phase amplitudes (e.g. Murphy *et al.*, 2001) and the characteristics of deeply-buried explosions (e.g. Murphy *et al.*, 1996a), we have modeled amplitude differences as variations in attenuation structure, and used this to make corrections to the observed phase amplitudes. We then apply these methods to recordings made at the Borovoye Observatory to test regional earthquake–explosion discrimination over broad areas, and find that 2-D attenuation corrections can significantly improve event discrimination.

2. *Method*

The basic dataset for both the discriminant and the attenuation tomography are the amplitude measurements of regional phases (Pn, Pg, Sn, Lg) in a series of narrow passbands. Arrival times of all the regional phases were made by hand, as structural complexity prohibits us from reliably using group velocity windows. Amplitude measurements from explosions were made on the three systems (KOD, SS, TSG) in the deglitched Borovoye archive data (Baker *et al.*, 2009; Kim *et al.*, 2010). As only explosions were available in the archive, amplitudes for earthquakes were made on the modern instrument at BRVK. BRVK is part of the IDA (International Deployment of Accelerometers) component of the Global Seismographic Network (GSN) and the data is available from IRIS (Incorporated Research Institutes for Seismology) at http://www.iris.gov. The modern instrument is located at the same physical site as the historic stations. Although it might seem that we are mixing apples and oranges by comparing amplitudes from explosions on the archived data with amplitudes from earthquakes from the modern instrument, all measurements are made in absolute amplitudes (displacement) and the differences should not matter, as long as the instrument responses are correct. We have found no issue with the responses.

The regional P/S discriminant can be manifested in a number of forms, but usually as Pn/Lg, Pg/Lg, or Pn/Sn. For propagation through the platforms (e.g., Kazakh Platform, West Siberian Platform) of central Asia, Pn/Sn appears to be the best combination of phases to use, as both Pn and Sn propagate efficiently in the upper mantle out to long distances, where most of the natural seismicity occurs. Another consideration is the frequency band where the amplitude measurements are made. Higher-frequencies are generally thought to discriminate better (e.g., Walter *et al.*, 1995; Hartse *et al.*, 1997), but both lower signal and higher noise conspire to produce poor signal-to-noise ratios at these frequencies for a number of events.

We initially considered amplitude ratios in passbands of 0.5–1, 1–2, 2–4, 4–6, 6–8, and 8–10 Hz, although we found that we had significantly fewer amplitude measurements passing signal-to-noise in the 0.5–1 Hz passband and above 6 Hz. Also, for the same reasons, the attenuation tomography has markedly poorer coverage in this relatively aseismic region above 6 Hz. We will therefore be mainly considering the 2–4 and 4–6 Hz passbands.

The regional phase amplitudes are also used in an attenuation tomography. LLNL has developed an amplitude inversion method that utilizes an MDAC (magnitude distance amplitude correction) source model (Walter and Taylor, 2001) and has been applied in a simultaneous multi-phase (Pn, Pg, Sn, Lg) amplitude tomography (Pasyanos *et al.*, 2009). Inverting all phases simultaneously allows us to determine consistent attenuation, site, and source terms for all phases, and eliminates non-physical inconsistencies among them. Although we are concentrating on the Pn and Sn phases, which are primarily sensitive to the attenuation structure of the upper mantle, these phases have crustal legs and the amplitudes of the Pg and Lg phases provide additional constraints on the crustal attenuation structure.

From the 2009 study, we have expanded our tomography from the Middle East to cover more of Eurasia, including central Asia. A map (Fig. 1a) shows the dominance of structural platforms (e.g. Kazakh Platform, West Siberian Platform, Russian Platform) across the region. Most seismicity occurs to the south, although only the events to the south and southeast are at regional distances from station BRVK. Figure 1b shows a map of Pn paths. Note the excellent coverage in the south where seismicity is high, and much sparser coverage in the north where stations like BRVK and ARU (Arti, Russia) primarily observe events to the south. Because of the poor prediction of their S-wave amplitudes, known explosions like the PNEs are not currently used in the tomography, although their future incorporation (through the use of an explosion S-wave model) might significantly improve our coverage of this region. Figure 1c shows a map of upper mantle Qp, which is what the Pn paths shown in Fig. 1b are primarily sensitive to. We find high Q (low attenuation) in the platforms to the north, with lower Q (high attenuation) in the tectonic regions to the south. We find very low Q in the Tian Shan, but high Q under the Tarim Basin. The Ural Mountains do not have sufficient coverage to be observed relative to the adjacent platforms. Figure 1d shows the path-averaged attenuation for Pn from BRVK to points in our model. Because the Q near BRVK is high, paths to Borovoye from the whole region tend to have high path-averaged Qs, although there is an important gradient from high Q to moderate Q for events to the southwest.

3. Regional Discrimination

The simplest discriminant (referred to here as "raw") is to take the ratios between the P and S wave amplitudes. Normally, the result is plotted on the log scale, where earthquakes tend to have low or negative values (large S-wave amplitude relative to P), while explosions have high values (small relative S-wave amplitudes). Figure 2 shows maps of presumed earthquakes (circles) and explosions (stars) recorded at station BRVK in the 2–4 Hz (Fig. 2a) and 4–6 Hz (Fig. 2b) passbands. One of the first things to see is that the vast majority of earthquakes recorded at regional distances (approximately between the two concentric circles on the maps) from BRVK occur to the southeast of the station. Events occurring outside of this region may be due to man-made activity rather than natural seismicity, a point we will return to later. Notice as well the large cluster of explosions at the STS, as well as the Meridian PNEs to the south, the Batholith-2 PNE to the southwest, and the Hellum-1 (sometimes referred to as Helium-1) PNE to the northwest (Table 1).

Very clearly, based on the raw amplitudes, the discriminant does not work well, having negative ratios in the 2–4 Hz passband for the STS explosions and high positive values for many earthquakes, especially at far regional distances. In the 4–6 Hz passband, the explosions have higher P/S ratios, but there is great overlap in the populations. At the higher frequency, we also have fewer earthquakes (78) that pass the signal-to-noise criteria, as compared to the 2–4 Hz band (112 earthquakes). When plotted as a function of distance (Fig. 2c, d), it is clear that strong trends with distance do not allow the raw discriminant to be effectively used over broad areas.

Some of the trends shown in Fig. 2c, d are due to 1-D variations, such as different Q and, perhaps, geometrical spreading of the Pn and Sn phases. If so, they can be modeled and the effects removed. We do this by modeling the earthquakes for a magnitude distance amplitude correction (MDAC) that takes into account both 1-D variations in propagation (attenuation and geometrical spreading) and magnitude, which affects the corner frequency of an event. We assume the geometrical spreading of Pn and Sn are the same, and a power-law frequency-dependent Q is modeled as:

$$Q(f) = Q_o f^{\eta}, \tag{1}$$

where Q_o is the attenuation parameter Q at 1 Hz, η its power-law frequency-dependence, and f is at the quarter frequency of the high and low passbands (e.g., 2.5 Hz for 2–4 Hz). For Pn and Sn, we find $Q_o = 370$, $\eta = 0.38$ for Pn and $Q_o = 410$, $\eta = 0.45$ for Sn. We then perform a grid search of all earthquake amplitudes recorded at BRVK for Q_o, η, and

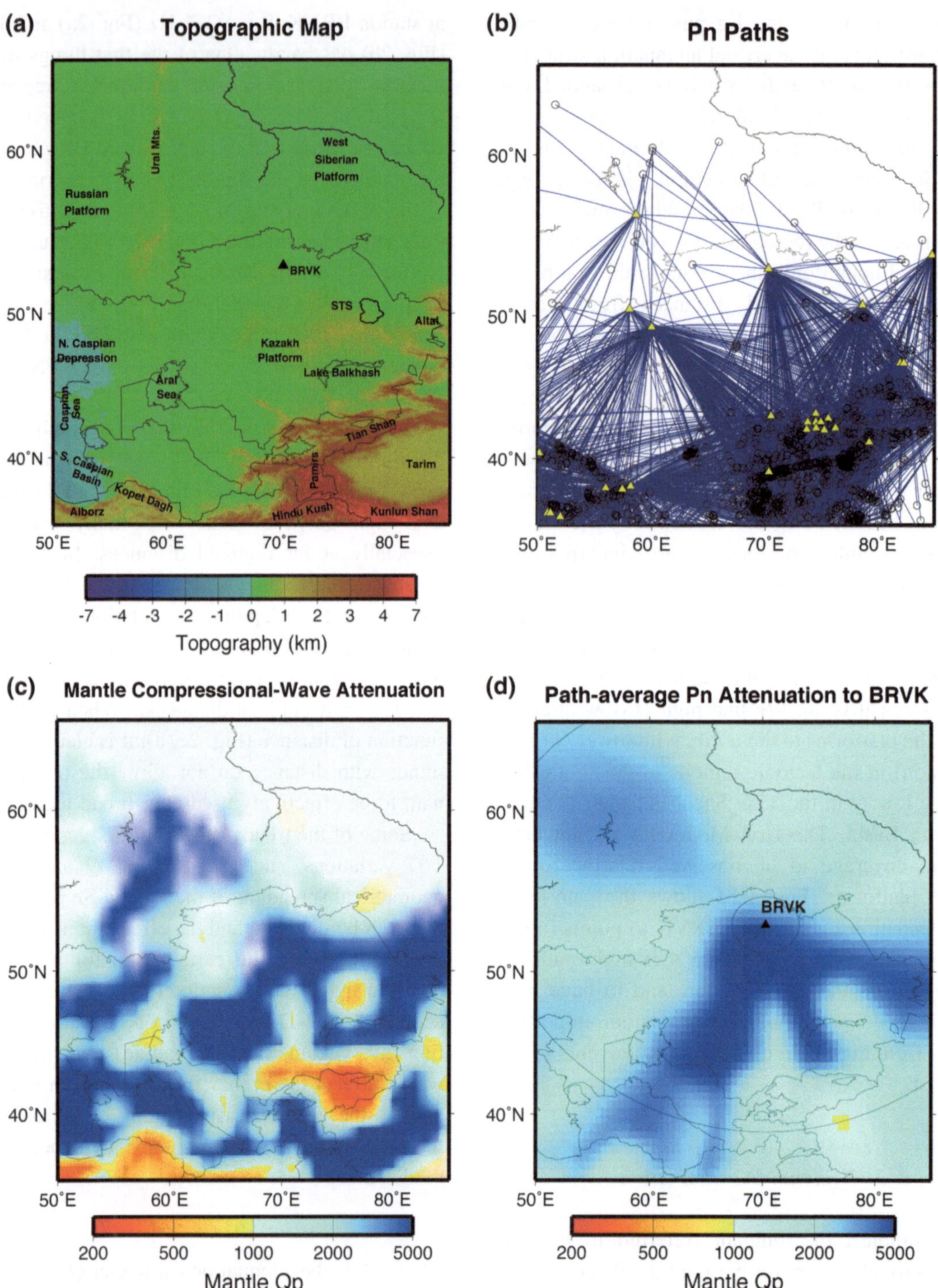

Figure 1

Maps of the study area in Central Asia. **a** Map showing topography and tectonic provinces. Location of the Borovoye station is indicated by *triangle*. STS indicates the Semipalatinsk Test Site. **b** Path map of amplitude measurements for Pn at 2–4 Hz. Events are indicated as *circles*, and stations as *triangles*. **c** Map of the upper mantle attenuation (Qp) at 2–4 Hz from multi-phase amplitude tomography (Pasyanos *et al.*, 2009). **d** Map of path-average Pn attenuation to station BRVK from map points

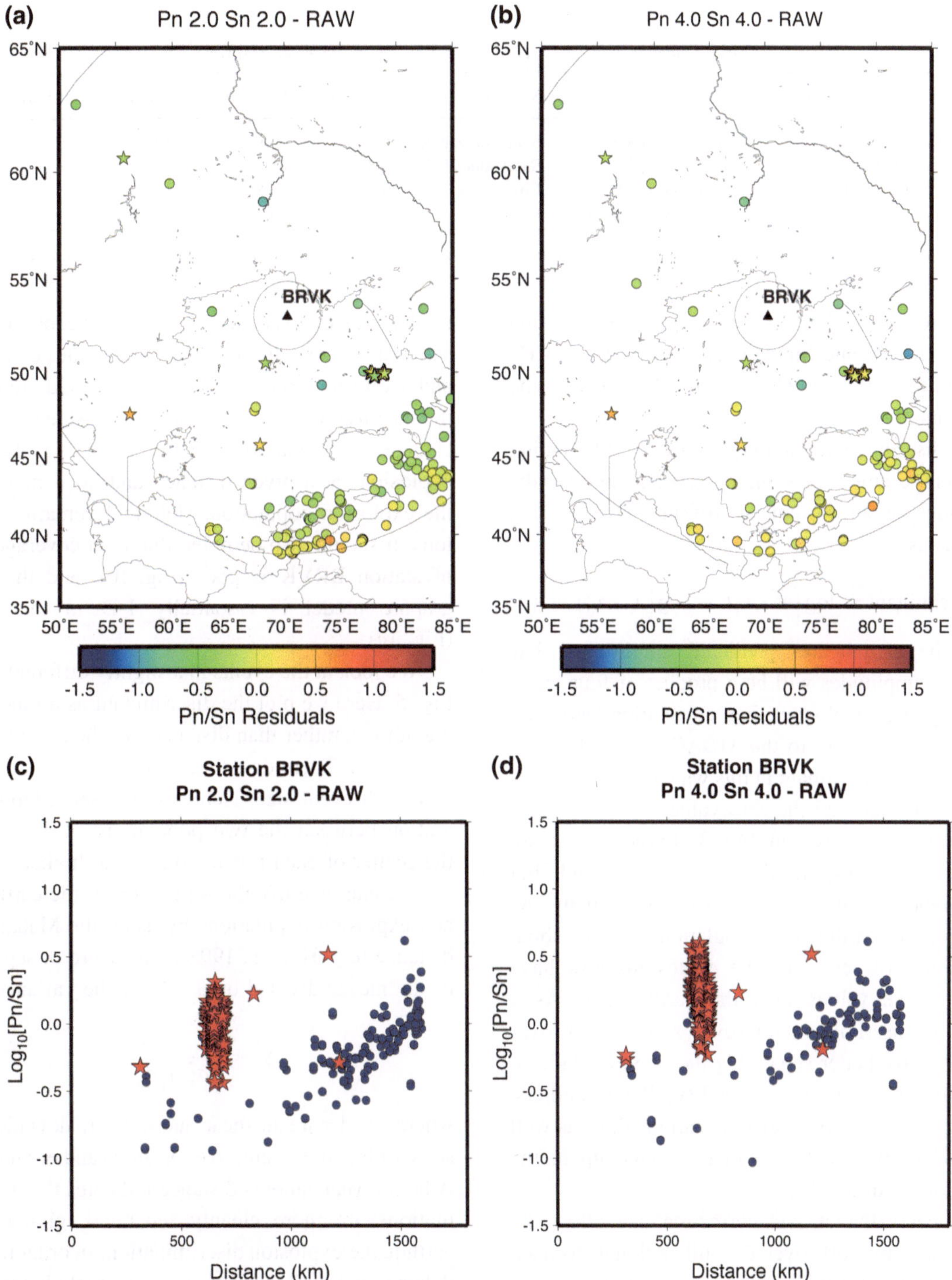

Figure 2
Raw discrimination results for Pn/Sn discriminant. **a** Map view of discriminant at 2–4 Hz. **b** Map view of discriminant at 4–6 Hz. **c** Discriminant as a function of distance at 2–4 Hz. **d** Discriminant as a function of distance at 4–6 Hz. In all plots, explosions are indicated by *stars*, and presumed earthquakes as *circles*

Table 1

Peaceful nuclear explosions recorded at regional distances at Borovoye

PNE name	Date (mm/dd/yyyy)	Location	Medium	Yield (kt)	Depth (m)	Scaled depth-of-burial (m/kt$^{1/3}$)
Meridian-1	8/23/1973	50.527° N 68.323° E	Sandstone/shale	6.3	395	214
Meridian-2	9/19/1973	45.758° N 67.825° E	Sandstone/shale	6.3	615	333
Hellum-1	9/2/1981	60.60° N 55.70° E	Limestone	3.2	2,088	1,417
Batholith-2	10/3/1987	47.60° N 56.20° E	Salt	8.5	1,002	491

All values in the table are from SULTANOV *et al.* (1999), except scaled depth-of-burial, which was calculated from yield and depth

site corrections. This results in $Q = 523$ and $Q = 619$ for Pn and Sn, respectively, in the 2–4 Hz band and $Q = 653$ and $Q = 806$ for Pn and Sn in the 4–6 Hz band.

The discriminant can then be improved by taking the observed amplitudes and correcting them for the amplitudes of an earthquake (of that size, at that location) as

$$\text{Discriminant} = \log\left[A^{\mathrm{P}}/A^{\mathrm{S}}\right] - \log\left[A_{\mathrm{o}}^{\mathrm{P}}/A_{\mathrm{o}}^{\mathrm{S}}\right], \quad (2)$$

where the o indicates predicted amplitudes. The predicted amplitudes will be a product of four terms (source, geometrical spreading, attenuation, and site), all defined according to the MDAC model. By normalizing the discriminant in this way, the earthquakes should scatter around the zero value.

Results are shown in Fig. 3. Large trends are removed, especially in the 2–4 Hz passband, but some trends in the earthquakes still remain. As expected, the earthquake population scatters about the ordinate of zero, and the explosions now have more positive values than the earthquakes. As a result, the discriminant improves significantly over the raw ratios, better separating the two populations. There are also some obvious and significant outliers, both for a few events labeled as earthquakes, as well as the Hellum-1 explosion that we take up in the discussion section below.

It appears that a 1-D correction can allow the discriminant to work over the full regional distance range. Can we use information on the lateral attenuation structure of the earth to further improve our discriminant? We test this by replacing the predicted amplitudes described in Eq. (2) with amplitudes where the attenuation term depends on the path. The results are shown in Fig. 4. Most of the trends in the earthquake population that still existed using the 1-D model are removed and, with the exception of the same outlier events, separate cleanly into earthquake and explosion populations. As expected, removing the increase of P/S amplitudes with distance does not change the separation between earthquakes and explosions at a given distance and, as a member of the explosion population, Hellum-1 remains anomalous. It should also be noted that path coverage north of station BRVK is poor (Fig. 1b), and the tomographic model is essentially 1-D in this region (Fig. 1c).

We look at the events in a slightly different way in Fig. 5. Here we plot the discriminant as a function of magnitude, rather than distance, for the two bands we considered. There do not seem to be any trends in the data with event size. Here, we can see a larger separation between the two populations at 4–6 Hz, but the scatter of each population is also higher.

We can quantify the separation of the earthquake and explosion populations by using the Mahalanobis distance (e.g. HARTSE, 1998), a measure of separation of the means divided by a sum of the variances.

$$\Delta^2 = \frac{(\bar{\mu}_x - \bar{\mu}_q)^2}{\sigma_x^2 + \sigma_q^2}, \quad (3)$$

where $\bar{\mu}$ and σ are the mean and standard deviation, and subscripts x and q refer to explosions and earthquakes. A larger Mahalanobis distance indicates that the populations are more cleanly separated, giving better earthquake/explosion discrimination. In order to avoid "cherry picking" the data, we have included the Hellum-1 explosion and the anomalous earthquake labeled events near the open-pit mines. As a reference point with the raw Pn/Sn amplitude ratios we find a Δ^2 of 0.28 for the 2–4 Hz band and Δ^2 of 1.00 for the 4–6 Hz band. In order to estimate a misclassification rate, we can translate a Mahalanobis distance value into an

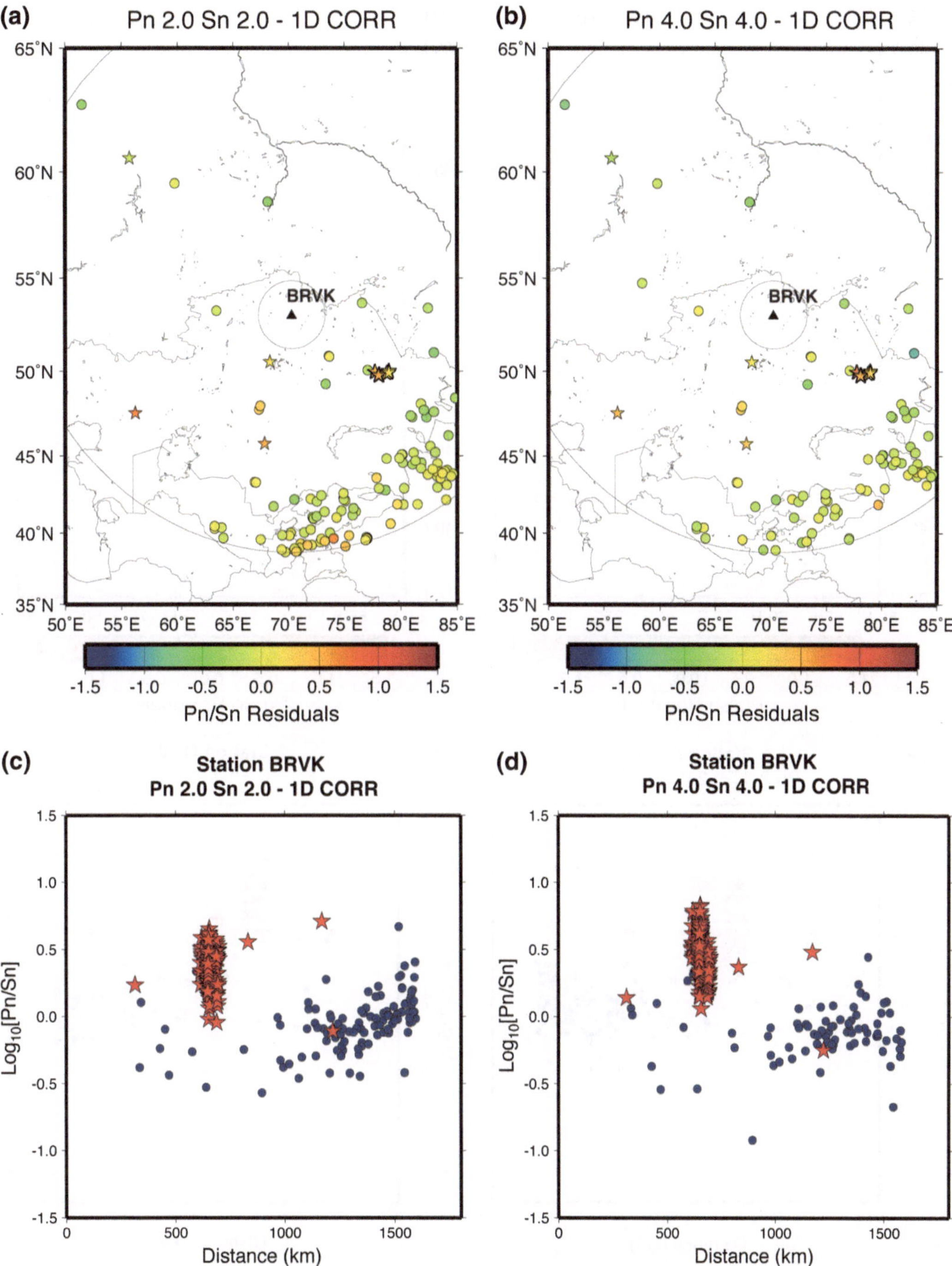

Figure 3
Discrimination results for Pn/Sn discriminant corrected for 1-D propagation effects. **a** Map view of discriminant at 2–4 Hz. **b** Map view of discriminant at 4–6 Hz. **c** Discriminant as a function of distance at 2–4 Hz. **d** Discriminant as a function of distance at 4–6 Hz. In all plots, explosions are indicated by *stars*, and presumed earthquakes as *circles*

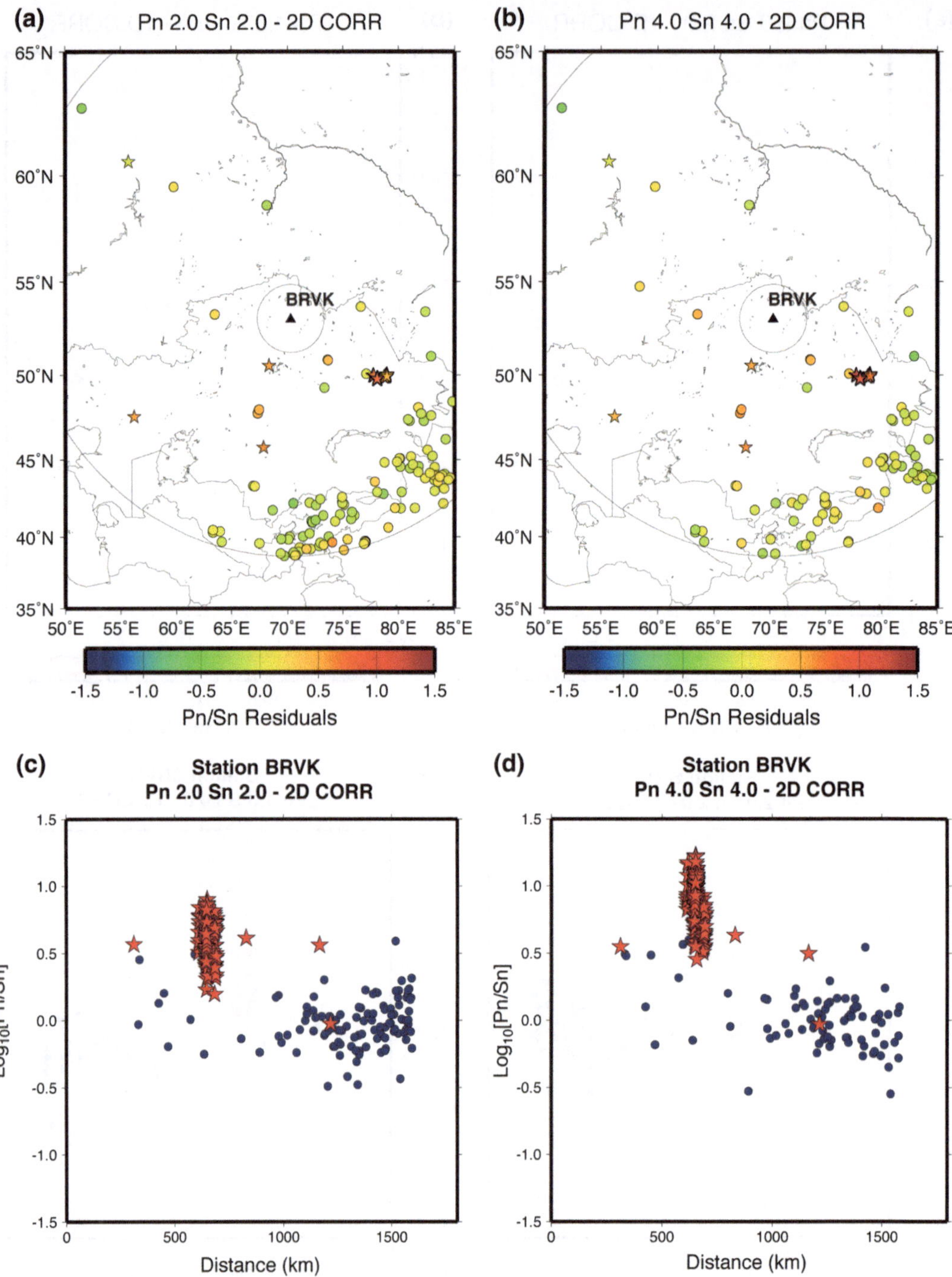

Figure 4

Discrimination results for Pn/Sn discriminant corrected for lateral propagation effects. **a** Map view of discriminant at 2–4 Hz. **b** Map view of discriminant at 4–6 Hz. **c** Discriminant as a function of distance at 2–4 Hz. **d** Discriminant as a function of distance at 4–6 Hz. In all plots, explosions are indicated by *stars*, and presumed earthquakes as *circles*

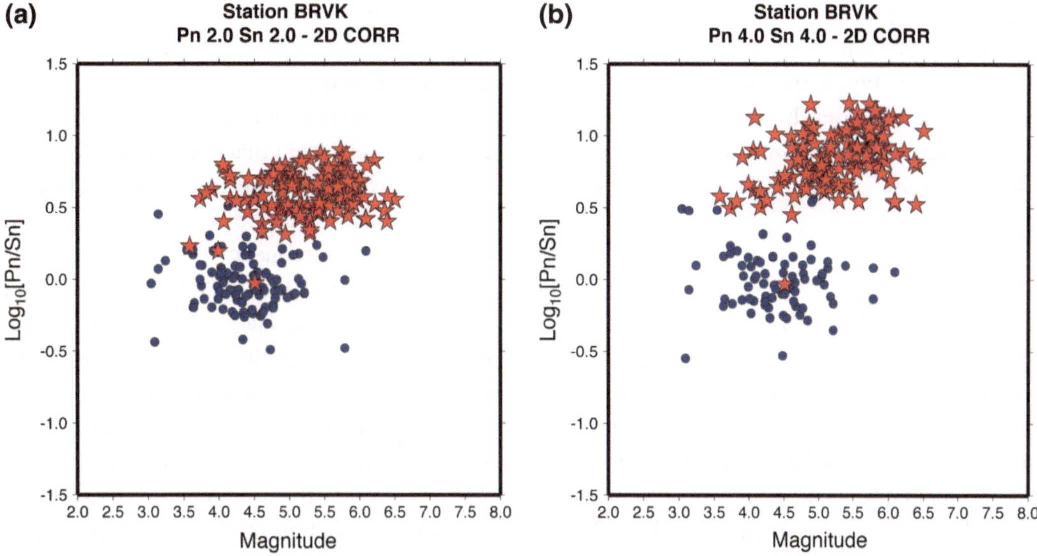

Figure 5
Discrimination results for Pn/Sn discriminant corrected for lateral propagation effects. **a** Discriminant as a function of magnitude at 2–4 Hz. **b** Discriminant as a function of magnitude at 4–6 Hz. In all plots, explosions are indicated by *stars*, and presumed earthquakes as *circles*

equiprobable point (EP), which is the point where the rate of earthquakes misclassified as explosions matches the rate of explosions that are misclassified as earthquakes, according to the formula:

$$\mathrm{EP} = \Phi\left(-\frac{\Delta}{2}\right), \tag{4}$$

where Φ is the normal cumulative probability distribution. This results in equiprobable points of 39.6 % for the 2–4 Hz band and 30.9 % for the 4–6 Hz band. For the 1-D corrections, we find a Δ^2 of 2.46 for the 2–4 Hz band and Δ^2 of 4.50 for the 4–6 Hz band, giving us equiprobable points of 21.6 and 14.4 % for 2–4 Hz and 4–6 Hz bands, respectively. After applying the 2-D corrections from the attenuation model, we find Δ^2 of 10.06 and 9.35 for the two bands, giving us equiprobable points of 5.6 and 6.3 %. A summary of these results is provided in Fig. 6. It appears, then, that the lateral attenuation corrections provide a larger improvement at lower frequencies. This is probably due to the better attenuation model in the lower frequency band, but potentially also due to more consistent and predictable amplitude variations at these frequencies. The use of reliable discriminants at lower frequencies should allow us to examine smaller, distant events that have lower signal-to-noise at higher frequencies.

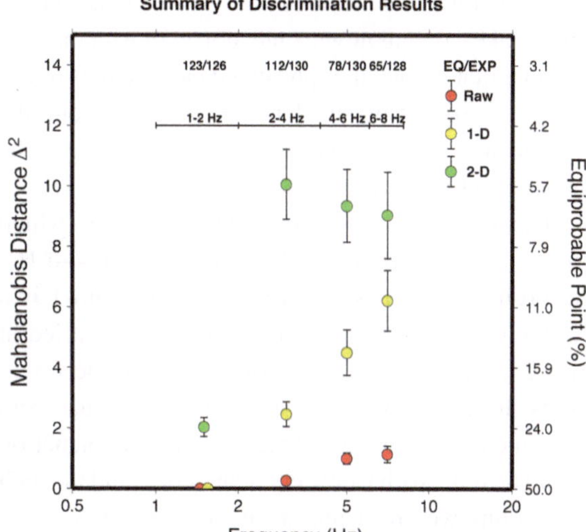

Figure 6
A summary of the discrimination results in this study showing the Mahalanobis distance Δ^2 (*left axis*) and equiprobable point (*right axis*) at 1–2, 2–4, 4–6, and 6–8 Hz for raw amplitude ratios (*red*), and 1-D (*yellow*) and 2-D (*green*) path corrected amplitude ratios. *Errors bars* are the standard deviations of the discriminant values determined using bootstrapping. Numbers of earthquakes and explosions observed in that frequency band are provided at the *top* of the figure

In order to determine the reliability of these results, we employ a bootstrapping method. For each discriminant, the earthquake and explosion populations are resampled with replacement hundreds of times, and statistical analysis performed on the new populations. This is illustrated in Fig. 6 by the bars of standard deviation associated with each discrimination value. Even given the variation from statistical sampling, the improvements from using 2-D corrections is significant.

4. Discussion

Two sets of outliers stand out in our Pn/Sn results, using both 1-D and 2-D path corrections. The first is the set of two presumed earthquake events just south of BRVK, at about the same epicentral distance as STS, which both classify as explosion-like. These events are at the location of a known large open-pit mine (Kevin Mackey, personal communication) and may not be natural earthquakes. While all of the presumed earthquakes in this study come from multiple catalogs where they are labeled as either earthquakes or unknown (and not explosions), it is possible that a few other presumed earthquakes in our dataset, particularly smaller events away from the seismically active areas to the south, might be unidentified mine-related events.

The second outlier is the Hellum-1 PNE, which comes out as earthquake–like at 2–4 and 4–6 Hz. According to Sultanov *et al.* (1999), Hellum-1 is a 3.2 kton explosion in limestone that is overburied at over 2.0 km (see Table 1). Examination of the waveforms shows a very strong Sn shear-wave signal from this nuclear explosion. One possible explanation would be if the path were very anomalous, and the path corrections we are using are invalid. Murphy *et al.* (2001) noted significant differences between PNE waveforms to the north and those to the northwest of BRVK, and Bottone *et al.* (2001) noted some PNE 6–8 Hz P/S ratios were small in the region. However, this seems unlikely to be the cause of the Hellum-1 low P/S values observed here, since we have four earthquakes nearby in the Ural Mountains and another northwest of Hellum-1, each of which have clear high frequency Pn and Sn phases constraining the path correction to BRVK reasonably well.

A second and more interesting possible explanation for the low Hellum-1 PNE P/S value might be related to its relatively small size and great depth. It is known that the explosion P-wave corner frequency depends upon depth, as described in the widely used model of Mueller and Murphy (1971). Fisk (2006, 2007) has postulated that explosion P/S values can be modeled using the Mueller–Murphy P-wave model, where the S-wave are calculated by scaling the S-wave corner frequency by the ratio of the S to P-wave velocity. Taylor (2009) has proposed that this S-wave corner frequency effect might be due to rock damage from the explosion and that it might persist even for very overburied events. In Fig. 7, we show that using this explosion model predicts that the P/S value for Hellum-1 is expected to be much lower at 2–4 and 4–6 Hz than the other PNEs measured here.

However, if this model is correct, then some of the small explosions observed at the Semipalatinsk Test Site might also be expected to have similarly small P/S values, which is not observed. For example, the smallest Semipalatinsk explosions we have measured have m_b values near 3.7, which using the Murphy (1996) magnitude-yield formula of $m_b = 0.75 \log(W) + 4.45$ (where W is yield in ktons), and would imply a yield of about ~ 0.1 kt (Khalturin *et al.*, 2001). There is only very limited published information about the depth of STS nuclear tests, and based on that Israelsson (1994) estimates a very rough relationship of $\text{DOB} = 90(W)^{1/3}$ m. This implies a depth of burial of about 40 m for a 0.1 kt and a depth of about 500 m for 150 kt. There are reasons to think that some of the small tests might be deeper than this relationship implies, for example, due to some of them occurring in tunnels with variable overburden (e.g. Khalturin *et al.*, 2001). In Fig. 7 we show the explosion model P/S ratios for Semipalatinsk from 150 to 0.1 kt at a range of depths (assumed for our purposes to be in granite). If some of the small Semipalatinsk explosions were buried at depths of about 200 m, we would expect them to be outliers at 2–4 and 4–6 Hz as well. Given that the Hellum-1 PNE is the only significant explosion outlier, it would imply that if this model of P/S ratios is correct, the smallest explosions at Semipalatinsk are not significantly overburied. The physical mechanisms that generate explosion S-waves and our ability to model their P/S values for arbitrary depths and geologies remains an area of research.

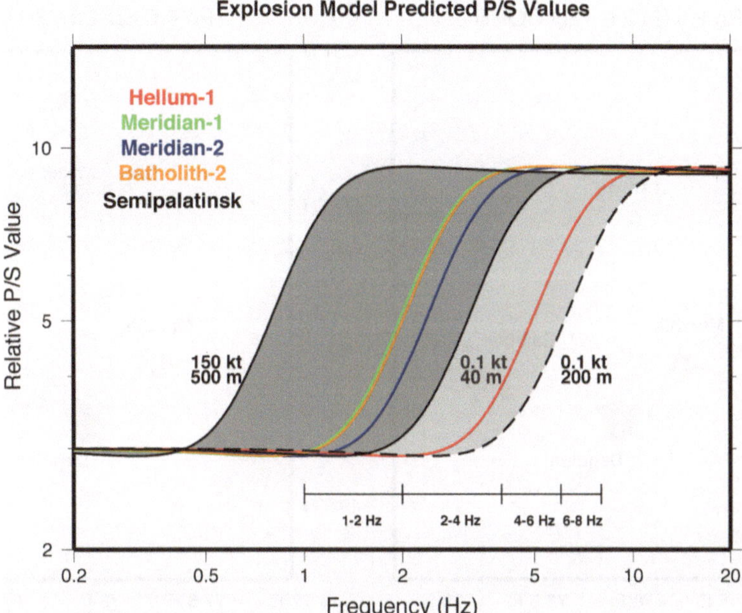

Figure 7
Model of explosion P/S values using the Mueller and Murphy (1971) P-wave model and the Fisk (2006, 2007) conjecture S-wave model for the PNEs and Semipalatinsk explosions. As the depths of the STS explosions are not known, a range of possible values is shown. *Heavier shading* indicates the region covered by the Israelsson (1994) depth of burial scaling, while the *lighter shading* show values when the smaller events are overburied

As an example of the effects of emplacement, Fig. 8 shows a close-up of STS, where we can see the separation of explosions into several locations at Balapan, Degelan, and Murzhik. Events at all of these locations have effectively the same propagation path to BRVK, so variations will solely be due to source effects. We find several interesting things. First, the variation in the discriminant values from STS is only slightly lower than the variations from the distributed earthquakes which, besides unmodeled propagation effects, can also have additional variations from depth and radiation pattern. This means that, in the absence of large numbers of explosions, using the variation in the earthquake population would be a reasonable estimate. Secondly, we find significant variations among the locations, with events at Balapan having lower P/S ratios than events at Degelan and Murzhik. The regions differ from each other in terms of both geology and emplacement conditions. In Degelen, the mountain is a pluton, and tests were carried out in tunnels and conducted in granite or porphyry. It seems possible that the tunnel tests might have a larger range in terms of scaled depth of burial. In contrast, at Balapan and Murzhik, the explosions were carried out in shafts. At Balapan, the rocks are generally tuffs or clastic rocks like conglomerate, sandstone, aleurolite, siltstones, and shales, while tests at Murzhik were carried out in clastic rock (Khalturin *et al.*, 2001).

5. Conclusions

Using a combination of events from the historic deglitched Borovoye archive dataset and more recent events from the modern instrument, we find that high-frequency P/S amplitudes can be used to reliably identify earthquakes and explosions in central Asia. There is good discrimination between earthquakes and explosions both at the STS and for PNEs located throughout the region. More generally, we conclude that seismic discrimination is applicable over broad regions as long as propagation effects are properly accounted for. Applying 1-D and 2-D corrections markedly improves the performance of high-frequency P/S discriminants. Attenuation models that can account for the observed variations in

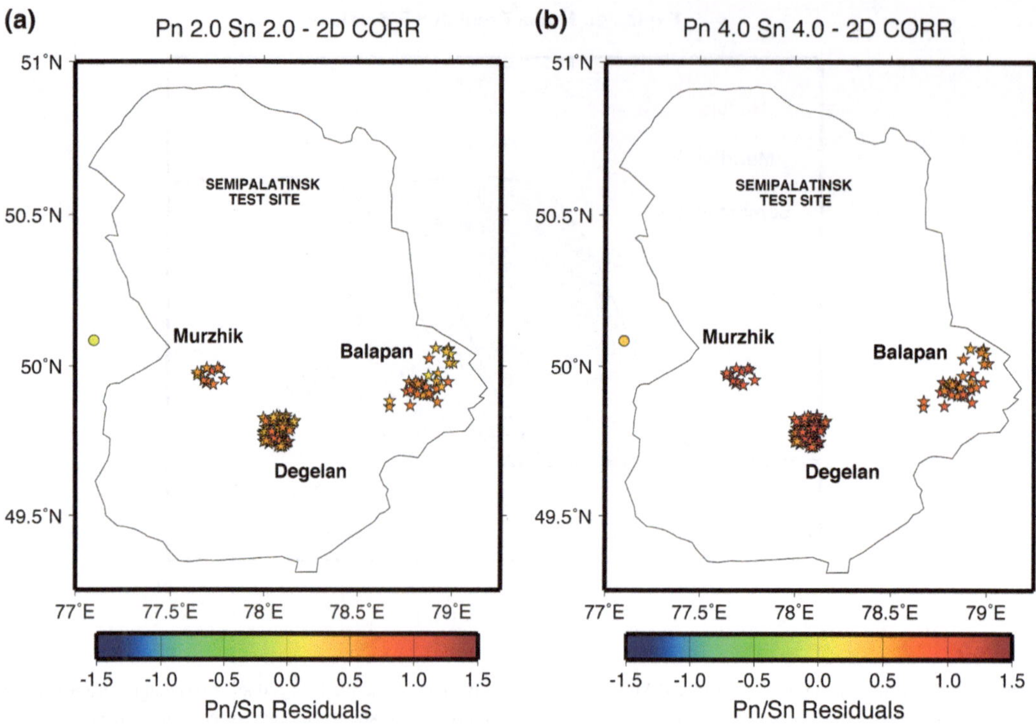

Figure 8
Discrimination results for Pn/Sn discriminant focused on the Semipalatinsk Test Site. **a** Map view of discriminant at 2–4 Hz. **b** Map view of discriminant at 4–6 Hz

amplitudes can further enhance our ability to identify events as earthquake or explosions, particularly in complex regions with stronger lateral variations than the region considered here. Accounting for lateral attenuation variations can also improve the discriminant at lower frequencies, where more events and smaller events can be observed. Finally we note that using multiple frequency bands, including those over 6 Hz, may provide the best means identify the full range of possible explosions including small and/or overburied events such as Hellum-1.

Acknowledgments

This work would not be possible without the tireless efforts of researchers at the Lamont-Doherty Earth Observatory and Los Alamos National Laboratory in compiling, deglitching and determining the instrument responses for thirty years of Borovoye data. We also thank the many people who worked at the Borovoye Observatory over the years to enable the decades of acquisition of digital recordings of seismic ground motion. This data is accessible at http://www.ldeo.columbia.edu/res/pi/Monitoring/Arch/BRV_arch_deglitched.html. We thank Paul Richards for providing boundaries of the Semipalatinsk Test Site (http://www.ldeo.columbia.edu/~richards/Semi.boundaries.html). We thank Terri Hauk and Stan Ruppert for maintaining the LLNL Seismic Research Database, Eric Matzel for making many of the amplitude measurements used in this study, and Alan Sicherman for his assistance with statistical analysis. We also thank Doug Dodge and Mike Ganzberger for the Regional Bodywave Amplitude Processor (RBAP), the tool used to make our amplitude measurements. This work was performed under the auspices of the US Department of Energy by Lawrence Livermore National Laboratory under contract DE-AC52-07NA27344. This is LLNL contribution LLNL-JRNL-516095.

References

Baker, D., W.-Y. Kim, H. Patton, G. Randall, and P. Richards (2009). Improvements to a major digital archive of seismic waveforms from nuclear explosions: The Borovoye seismogram archive, in Proceedings of the 2009 Monitoring Research Review: Ground-Based Nuclear Explosion Monitoring Technologies, LA-UR-09-05276, 12–21.

Bottone, S., M.D. Fisk, and G.D. McCartor (2001). Regional seismic event characterization using Bayesian calibration, Mission Research Corporation report, MRC-R-1621, 87 pp.

Bottone, S., M.D. Fisk, and G.D. McCartor (2002). *Regional seismic event characterization using a Bayesian formulation of simple kriging*, Bull. Seism. Soc. Am., *92*, 2277–2296.

Fisk, M.D. (2006). *Source spectral modeling of regional P/S discriminants at nuclear test sites in China and the former Soviet Union*, Bull. Seism. Soc. Am., *96*, 2348–2367.

Fisk, M.D. (2007). *Corner frequency scaling of regional seismic phases for underground nuclear explosions at the Nevada Test Site*, Bull. Seism. Soc. Am., *97*, 977–988.

Hartse, H., S.R. Taylor, W.S. Phillips, and G.E. Randall (1997). *A preliminary study of regional seismic discrimination in Central Asia with an emphasis on Western China*, Bull. Seism. Soc. Am. *87*, 551–568.

Hartse, H.E., R.A. Flores, and P.A. Johnson (1998). *Correcting regional seismic discriminants for path effects in Western China*, Bull. Seism. Soc. Am., *88*, 596–608.

Israelsson, H. (1994). *Analysis of historical seismograms – root mean square Lg magnitudes, yields, and depths of explosions at the Semipalatinsk Test Range*, Geophys. J. Int., *117*, 591–609.

Khalturin, V.I., T.G. Rautian, and P.G. Richards (2001). *A study of small magnitude seismic events duing 1961–1989 on and near the Semipalatinsk Test Site, Kazakhstan*, Pure Appl. Geophys., *158*, 143–171, doi:10.1007/PL00001153.

Kim, W.-Y. and G. Ekstrom (1996). *Instrument responses of digital seismographs at Borovoye, Kazakhstan, by inversion of transient calibration pulses*, Bull. Seism. Soc. Am., *86*, 191–203.

Kim. W.-Y., V. Aharonian, A.L. Lerner-Lam, and P.G. Richards (1997). *Discrimination of earthquakes and explosions in southern Russia using regional high frequency three-component data from the IRIS/JSP Caucasus network*, Bull. Seism. Soc. Am., *87*, 569–588.

Kim, W.-Y., P.G. Richards, V. Adushkin, and V. Ovtchinnikov (2001). Borovoye digital archive for underground nuclear tests during 1966–1996, Lamont-Doherty Earth Observatory Report, http://www.ldeo.columbia.edu/Monitoring/Data/Brv_arch_ex/brv_text_table.pdf.

Kim, W.-Y., P.G. Richards, D. Baker, H. Patton, and G. Randall (2010). Improvement to a major digital archive of seismic waveforms for nuclear explosions, Final Report on Contract AFRL-RV-HA-TR-2010-1024, http://www.ldeo.columbia.edu/res/pi/Doc/AFRL-RV-HA-TR-2010-1024_07C0004_Final.pdf.

Mueller, R.A. and J.R. Murphy (1971). *Seismic characteristics of underground nuclear detonations: Part I. Seismic scaling law of underground detonations*, Bull. Seism. Soc. Am., *61*, 1675–1692.

Murphy, J.R. (1996). Types of seismic events and their source descriptions, in Monitoring a Comprehensive Test Ban Treaty, E.S. Husebye and A.M Dainty, eds., 225–245.

Murphy, J.R., D.D. Sultanov, B.W. Barker, I.O. Kitov, and M.E. Marshall (1996). Application of Soviet PNE data to the assessment of the transportability of regional discriminants, Phillips Laboratory Report PL-TR-96-2290.

Murphy, J.R., I.O. Kitov, B.W. Barker, and D.D. Sultanov (2001). *Seismic source characeristics of Soviet peaceful nuclear explosions*, Pure Appl. Geophys., *158*, 2077–2101.

Pasyanos, M.E. and W.R. Walter (2009). *Improvements to regional explosion identification using attenuation models of the lithosphere*, Geophys. Res. Lett., *L14304*, doi:10.1029/2009GL038505.

Pasyanos, M.E., W.R. Walter, and E.M. Matzel (2009). *A simultaneous multi-phase approach to determine P-wave and S-wave attenuation of the crust and upper mantle*, Bull. Seism. Soc. Am., *99*, doi:10.1785/0120090061.

Phillips, W.S., X. Yang, G.E. Randall, H.E. Hartse, and R.J. Stead (2009). Regional phase attenuation tomography for central and eastern Asia in Proceedings of the 2009 Monitoring Research Review: Ground-based Nuclear Explosion Monitoring Technologies, LA-UR-09-05276, 536–546.

Richards, P.G., W.-Y. Kim, and G. Ekstrom (1992). *Borovoye Geophysical Observatory, Kazakhstan*, EOS Trans. Am. Geophys. Union, *73*, 201–202, doi:10.1029/91EO00161.

Rodgers, A.J. and W.R. Walter (2002). *Seismic discrimination of the May 11, 1998 Indian nuclear test with short-period regional data from station NIL (Nilore, Pakistan)*, Pure Appl. Geophys., *159*, 679–700, doi:10.1007/s00024-002-8654-6.

Sultanov, D.D., J.R. Murphy and Kh.D. Rubinstein (1999). A seismic source summary for Soviet peaceful nuclear explosions, Bull. Seism. Soc. Am. *89*, 640–647.

Taylor, S. (1996). Analysis of high-frequency Pg/Lg ratios from NTS explosions and western U.S. earthquakes, Bull. Seism. Soc. Am., *86*, 1042–1053.

Taylor, S., A. Velasco, H. Hartse, W.S. Phillips, W.R. Walter, and A. Rodgers (2002). *Amplitude corrections for regional discrimination*, Pure. App. Geophys., *159*, 623–650.

Taylor, S.R. (2009). *Can the Fisk conjecture be explained by rock damage around explosions*? Bull. Seism. Soc. Am., *99*, 2552–2555, doi:10.1785/0120080332.

Walter, W.R. and S.R. Taylor (2001). A revised magnitude and distance amplitude correction (MDAC2) procedure for regional seismic discriminants: theory and testing at NTS, LLNL UCRL-ID-146882, http://www.llnl.gov/tid/lof/documents/pdf/240563.pdf.

Walter, W.R., K. Mayeda, and H.J. Patton (1995). *Phase and spectral ratio discrimination between NTS earthquakes and explosions Part 1: Empirical observations*, Bull. Seism. Soc. Am. *85*, 1050–1067.

Walter, W.R., E. Matzel, M.E. Pasyanos, D.B. Harris, R. Gok, and S.R. Ford (2007). Empirical observations of earthquake and explosion discrimination using P/S ratios and implications for the sources of explosion S-waves, in Proceedings of the 29th Monitoring Research Review, LA-UR-07-5613, 684–693.

(Received November 29, 2011, revised August 28, 2012, accepted September 5, 2012, Published online September 20, 2012)

Pure Appl. Geophys. 171 (2014), 537–547

DOI 10.1007/s00024-012-0627-9

Pure and Applied Geophysics

Sources of Error and the Statistical Formulation of M_S: m_b Seismic Event Screening Analysis

D. N. Anderson,[1] H. J. Patton,[1] S. R. Taylor,[2] J. L. Bonner,[3] and N. D. Selby[4]

Abstract—The Comprehensive Nuclear-Test-Ban Treaty (CTBT), a global ban on nuclear explosions, is currently in a ratification phase. Under the CTBT, an International Monitoring System (IMS) of seismic, hydroacoustic, infrasonic and radionuclide sensors is operational, and the data from the IMS is analysed by the International Data Centre (IDC). The IDC provides CTBT signatories basic seismic event parameters and a screening analysis indicating whether an event exhibits explosion characteristics (for example, shallow depth). An important component of the screening analysis is a statistical test of the null hypothesis H_0: explosion characteristics using empirical measurements of seismic energy (magnitudes). The established magnitude used for event size is the body-wave magnitude (denoted m_b) computed from the initial segment of a seismic waveform. IDC screening analysis is applied to events with m_b greater than 3.5. The Rayleigh wave magnitude (denoted M_S) is a measure of later arriving surface wave energy. Magnitudes are measurements of seismic energy that include adjustments (physical correction model) for path and distance effects between event and station. Relative to m_b, earthquakes generally have a larger M_S magnitude than explosions. This article proposes a hypothesis test (screening analysis) using M_S and m_b that expressly accounts for physical correction model inadequacy in the standard error of the test statistic. With this hypothesis test formulation, the 2009 Democratic Peoples Republic of Korea announced nuclear weapon test fails to reject the null hypothesis H_0: explosion characteristics.

1. The Comprehensive Nuclear-Test-Ban Treaty (CTBT)

Between 1945 and 1996, the United States (US), the Union of Soviet Socialist Republics (USSR), the United Kingdom (UK), France, and China executed a total of 2,398 nuclear weapon explosions (see Mikhailov 1999), including the *Little Boy* and *Fat Man* bombs used by the US in World War II to force the surrender of Japan. The US test *Baker* was detonated in 1946 near the Bikini atoll 27.5 m in the ocean and was observed at seismic stations around the globe. This and other subsequent tests provided scientific evidence that underground nuclear explosions could be detected and potentially characterised with global seismic stations. In the context of nuclear weapon test monitoring, one of the most seismically significant series of tests was executed in 1956 by the UK at the temporary Maralinga test site in Australia. The 1956 Maralinga experiments were by design dual use. In addition to nuclear weapon development data provided by successful detonation, these experiments provided significant scientific characterisation of the Western Australian crust and amplified the call for seismic research relevant to nuclear weapon test monitoring. The 1957 US experiment *Rainier* was the first fully underground nuclear explosion. *Rainier* was detected by about 50 seismic stations around the globe; however, it was confused with earthquakes at some stations (Bolt 1976).

Discussions of a comprehensive ban on nuclear weapon testing, with participation from the US, UK, USSR, France, Canada, Czechoslovakia, Romania, and Poland, occurred over the summer of 1958 in Geneva. These discussions, fundamentally about eliminating the development of nuclear arsenals through a ban on test explosions, were also motivated by concerns of radioactive fallout because most nuclear explosion tests up to that time were above ground. A key finding from these discussions was the need for a global network of state-of-the-art seismic stations (specifically arrays) and associated scientific research leading to the operational capability to monitor for underground nuclear explosions (see Bolt 1976 and Dahlman and Israelson 1977). The 1958 Geneva conference also offered a cautiously

[1] Los Alamos National Laboratory, Los Alamos, NM, USA. E-mail: dand@lanl.gov

[2] Rocky Mountain Geophysics, Los Alamos, NM, USA.

[3] Weston Geophysical, Lexington, MA, USA.

[4] AWE Blacknest, Brimpton, Reading, UK.

optimistic conclusion that techniques to identify seismic events could be developed.

In 1961, US deployment of stations comprising the World Wide Standard Stations Network (WWSSN) began and by 1966, 112 stations were reported operational. This deployment was one component of the US Advanced Research Projects Agency (ARPA) project code named Vela. Project Vela had three main sub-projects: code name Uniform for research and development (RD) to monitor underground explosions, code name Sierra for RD to monitor above ground explosions and code name Hotel for RD to develop satellite detection systems. Vela was funded to enable the verification of the Limited Test Ban Treaty prohibiting the US, USSR, and UK from conducting underwater, atmospheric, and outer space nuclear explosion tests. In addition, Vela provided advanced RD to enable subsequent treaties. A wealth of scientific discoveries on the structure of the earth resulted from WWSSN seismic observations. World seismicity maps from the WWSSN laid the groundwork for the theory of plate tectonics. Deployment of the WWSSN was a significant first step in real-time seismic monitoring for underground nuclear explosions.

The subsequent Threshold Test Ban Treaty (TTBT) prohibited the US and USSR from conducting nuclear weapons tests with yields exceeding 150 kilotons of TNT. The TTBT is still in force today between the US and the Russian Federation. Under the TTBT seismic waves with a propagation path largely in the earth's mantle (teleseismic waves) are analysed to answer three core questions: Where is the seismic event located? What is the source type of the event (explosion or natural)? How large is the event?

The CTBT, in basic obligation, is simple and direct. Article I reads:

1. Each State Party undertakes not to carry out any nuclear weapon test explosion or any other nuclear explosion, and to prohibit and prevent any such nuclear explosion at any place under its jurisdiction or control.
2. Each State Party undertakes, furthermore, to refrain from causing, encouraging, or in any way participating in the carrying out of any nuclear weapon test explosion or any other nuclear explosion.

Protocol for the CTBT calls for the implementation of an International Monitoring System (IMS) and associated International Data Centre (IDC) with the responsibility to receive, collect, process, analyse, report on, and archive data from IMS facilities. Protocol also directs the IDC to provide an event screening service to the treaty signatories, the technical details of which are specified in the CTBT operational manual.

2. *Seismic Event Screening*

Seismic energy is generated by earthquakes, volcanoes, mining and oil exploration explosions, natural fracturing of large rocks, large above-ground explosions, and nuclear weapon tests. A seismic waveform is a measured transient time series of this energy with distinct segments (phases) generated through reflection and refraction with the earth's structure (see Fig. 1). The early arriving segments are called primary phases (P-phases) and the later arriving segments are called secondary phases (S-phases). P-phases are composed of compressional waves. S-phases are composed of slower shear waves. The segment with onset labelled R in Fig. 1 is the Rayleigh surface wave phase. The energy in this phase, relative to the energy in the initial P-phase, is the basis for an IDC event screening hypothesis test.

The path and distance between event and stations are different and if the phase energy measurements from each station could be accurately corrected for path effects, the measurements would represent energy at the source. Magnitudes are empirical measurements of phase energy in $\log_{10}$ units that include corrections for path effects and signal processing filter effects. The M_S magnitude is calculated using an amplitude from the R-phase segment of the filtered waveform. Station magnitudes are averaged to estimate an event (network) magnitude. The m_b magnitude is the average of corrected amplitudes from the P-phase segment of raw station waveforms, filtered near 1 Hz. m_b is computed with a large

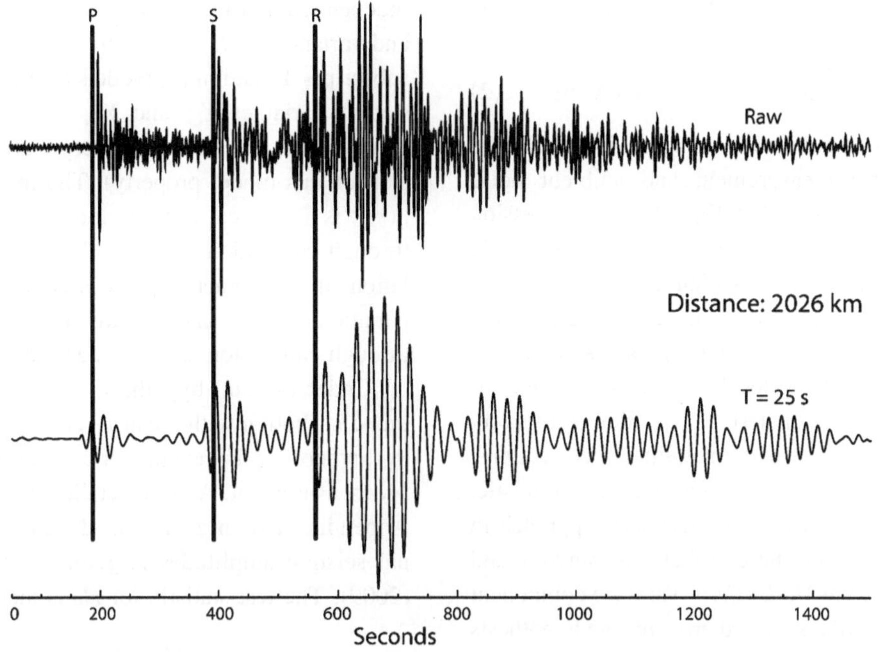

Figure 1
Illustration of the onset of P, S, and Rayleigh (R) phases in a seismic waveform, both as raw data (*top*) and low-pass filtered (*bottom*) at 25 s. A station M_S is calculated using an amplitude from the R-phase segment of the filtered waveform. m_b is an average of corrected amplitudes from the P-phase segment of raw station waveforms, filtered near 1 Hz. m_b is computed with a large network of stations observing an event and is assumed fixed

network of stations observing an event and is assumed fixed. Many surface wave corrections and scales have been developed during the past century. Most notable are scales with distance corrections (GUTENBERG 1945; VANUEK *et al.* 1962; VON SEGGERN 1977; HERAK and HERAK 1993; REZAPOUR and PEARCE 1998; STEVENS and MCLAUGHLIN 2001; BORMANN *et al.* 2009) and scales with corrections for filter effect, distance, and path (MARSHALL and BASHAM 1972; RUSSELL 2006). With calibration analysis, the hypothesis formulation developed in this article is applicable to all commonly used M_S calculations. In general, the model for computed station M_S is

$$M_S = \log_{10}(\text{amplitude}) + \text{path} + \text{distance} + \text{filter effect}. \tag{1}$$

The Rayleigh wave magnitude M_S is made relative to m_b with the regression-like formulation $M_S - \eta(m_b)$. (the analogous network Rayleigh wave magnitude is $\widehat{M_S} - \eta(m_b)$). In contrast, the IDC test statistic is essentially Eq. (15) of the Appendix-a direct extension of the regression formulation. The IDC provides CTBT signatories basic seismic event parameters including event location and depth, measures of event size (magnitudes), and a screening analysis for events with m_b greater than 3.5.

The physical basis of the $M_S - \eta(m_b)$ discriminant is quite mature (see DOUGLAS *et al.* 1971; STEVENS and DAY 1985) and is based on the physics that for a given m_b, a shallow earthquake excites relatively more surface-wave energy than a single-point explosion. However, deep earthquakes like explosions have small M_S for their m_b. This means that single-point underground explosions and deep earthquakes will usually fail to reject the null hypothesis of explosion characteristics in the $\widehat{M_S} - \eta(m_b)$ screening analysis.

IDC seismic event screening includes a statistical test of the null hypothesis that a seismic event has explosion characteristics, with a very small probability of incorrectly rejecting this hypothesis. This conservative IDC event screening analysis retains events of no concern as well as explosions in an event bulletin provided to CTBT signatories.

The probability model of M_S corrected for m_b is

$$\begin{aligned} Y &= M_S - \eta(m_b) \\ &= \mu + \textit{Model Error} + \textit{Station Noise} \end{aligned} \tag{2}$$

where $\eta(m_b)$ is a model of the magnitude and *Station Noise* represents measurement and ambient noise with zero mean. Note that Eq. (2) is a regression model of M_S versus m_b embedded into a simple random effects analysis of variance model.

Model Error is a zero mean random effect that varies from event to event and represents correction model inadequacy from local effects such as inaccurate depth and local material properties, filter/path/distance corrections, and magnitude corrections. These effects are in fact physical and deterministic, yet realistically unknown. The technical approach in Eq. (2) is to model these effects as random and properly include the *Model Error* variance component in calculations of the standard error for the hypothesis test. The variance component for *Model Error* decreases with improved corrections and physical theory, and the term for *Station Noise* in the standard error is reduced through station averaging. Importantly, station averaging cannot reduce *Model Error*.

The variance component for *Model Error* is equal across both the null and alternate hypotheses as is the variance component for *Station Noise*. The differences in the model for the two hypotheses is represented through differences in the term μ. For IDC screening analysis a simple linear regression formulation, $\eta(m_b) = \beta \times m_b$, is used for both the null and alternate hypothesis.

From the conceptual model, define the random variable $Y_{ijk} = M_{Sijk} - \beta\, m_{bj}$ to be the M_S magnitude residual for null and alternate hypotheses $H_{i\,=\,0,\mathrm{A}}$, and event j and station k. The linear model representation of Eq. (2) is then

$$Y_{ijk} = \mu_i + E_j + \epsilon_{ijk} \quad j = 1,2,\ldots, m_i \\ k = 1,2,\ldots, n_{ij}. \tag{3}$$

With the assumption of unbiased source models for H_0 and H_A the expected value is

$$\mathcal{E}\{Y_{ijk}\} = \begin{cases} \mu_0 & \text{for } H_0(i=0) \\ \mu_A & \text{for } H_A(i=\mathrm{A}). \end{cases} \tag{4}$$

The E_j are modelled as independent normal random variables with zero mean and variance τ^2. The ϵ_{ijk} are independent normal random variables with zero mean and variance σ^2. E_j and ϵ_{ijk} are independent across all subscripts. Equation (3) models a correlation between station variables Y_{ijk} and $Y_{ijk'}$ as $\tau^2/(\tau^2 + \sigma^2)$ (see ANDERSON *et al.* 2009 for interpretation and discussion of this model property). The model component $\eta(m_b)$ is a physical correction and is assumed known through calibration, in particular for the IDC formulation the parameter β is assumed known. The parameters μ_i, τ^2 and σ^2 are also assumed known through calibration analysis, demonstrated in Sect. 3.

Under the null hypothesis, $\widehat{M_S} - \beta m_b$ has a normal distribution with mean μ_0 and variance $\tau^2 + \sigma^2/n_{M_S}$, where n_{M_S} is the number of stations observing an M_S measurement. A more detailed derivation is in the Appendix, and an analogous development for regional seismic amplitudes is given in ANDERSON *et al.* (2009). The test statistic for H_0 is then

$$Z_{\widehat{M_S}} = \frac{\widehat{M_S} - (\mu_0 + \beta m_b)}{\sqrt{\tau^2_{M_S} + \sigma^2_{M_S}/n_{M_S}}}. \tag{5}$$

Noted above, the IDC test statistic is essentially Eq. (15) with calibrated uncertainty values derived from IDC operations—a direct extension of the formulation of Eq. (5). The objective of this article is the development and demonstration of an event screening hypothesis test with a more sophisticated standard error. Both Eqs. (5) and (15) are viable options for screening analysis. Comparison of the two would require application to an appropriate subset of IDC operational data and subsequent assessment in terms of accuracy and utility to the CTBT and signatories.

Omitting the term E_j in Eq. (3) implies that the magnitude residual at a station is μ_i plus station noise. This model formulation is fundamentally inconsistent with seismic observation. The standard error of $\widehat{M_S}$ with E_j removed from Eq. (3) is $\sigma^2_{M_S}/n_{M_S}$ $(\tau^2 = 0)$ and decreases as the number of stations observing an event increases. By not including the term E_j, inadequacies in corrections (filter effect, distance, path, and magnitude) are not accounted for in the theoretical model representation of a magnitude, and this bias can never be diminished with a network average calculation. Also, the lower bound of a magnitude standard error, derived from Eq. (3), is non-zero and therefore consistent with realistic seismic observation.

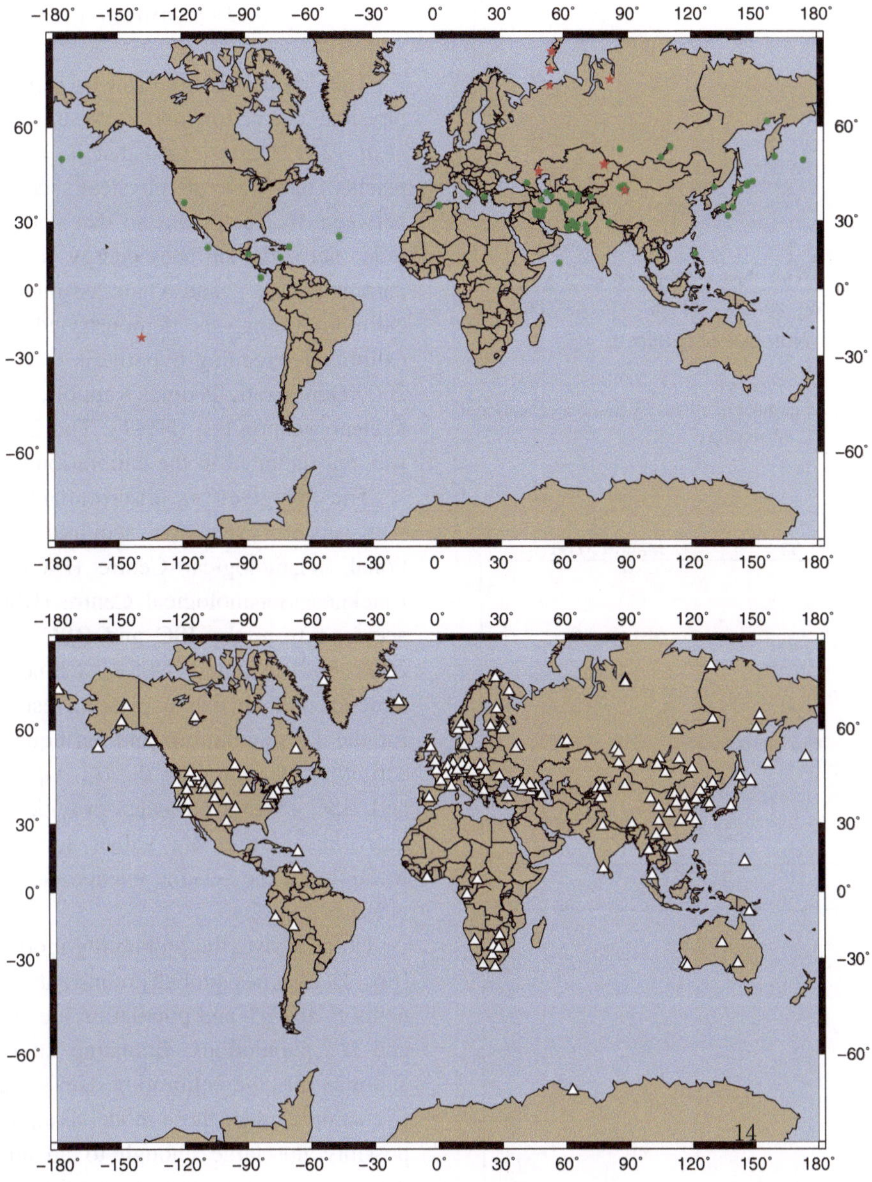

Figure 2
Seismic event and station locations. Explosions (testing sites) are *red stars*, otherwise *green circles*; stations are *triangles*

Another important property of this model is that station magnitude residuals for an event are correlated. This correlation $(\tau^2/(\tau^2 + \sigma^2))$ implies that large adjustment E_j increases correlation between stations because this random adjustment is applied to all stations observing an event, that is, the station residuals probabilistically move together. Small adjustment E_j implies the correction model is good and is conceptually equivalent to stations with incoherent noise. Small adjustment E_j also implies τ^2 is small and the standard error of $\widehat{M_S}$ is fundamentally reduced through network averaging.

Physical correction theory will never be able to adjust amplitudes for all local systematic/physical effects. As discussed previously, these local effects are modelled as random, moving out to all stations (and therefore station M_S values) inducing correlation between station M_S values for an event. As physical

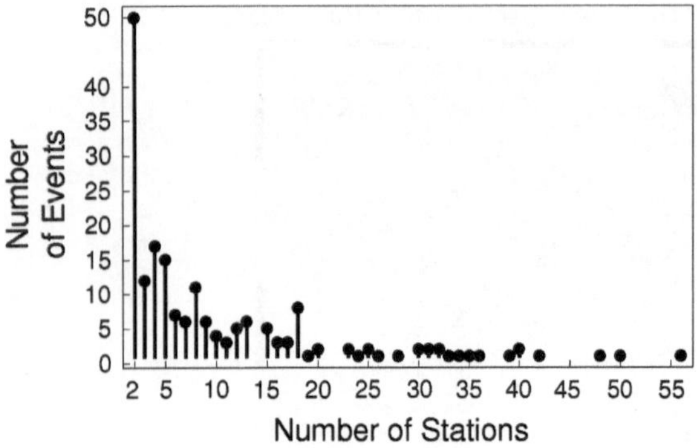

Figure 3
Calibration data summary: number of events by number of stations observing

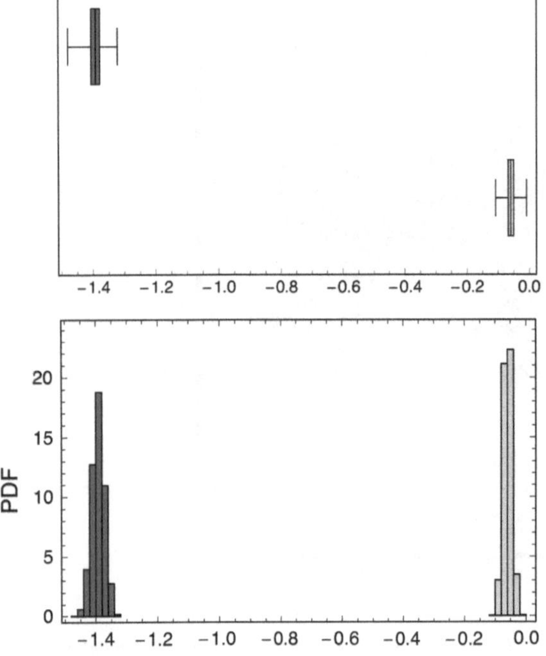

Figure 4
Histograms and box plots of bootstrap means μ_0 and μ_A. (H_0 population—*dark grey*, H_A population—*light grey*)

correction models improve, the model inadequacy terms E_j will be small, giving small values of τ, and the correlation between station M_S values will be small. In the limit, this conceptually gives station M_S values that have been perturbed with incoherent noise.

3. *Demonstration Analysis*

The model component $\eta(m_b) = \beta \times m_b$ is a physical correction and is assumed known. Selby *et al.* (2012) determined that $\beta = 1$ adequately represents the general physical scaling relationship between M_S versus m_b, so that corrected-for-magnitude network surface energy is $\widehat{M_S} - m_b$. The parameters μ_i, τ^2 and σ^2 are assumed known through calibration analysis demonstrated below, and the calibrated screening hypothesis test is applied to the 2009 Democratic Peoples Republic of Korea (DPRK) nuclear weapon test (NWT). The 2009 DPRK NWT was not included in the calibration analysis.

The M_S given m_b discriminant is demonstrated with seismic event data acquired from the International Seismological Centre (ISC) and the AWE Blacknest Seismological Centre (BSC). The events acquired from the ISC and BSC spanned 1964 to 2000 and included USSR, Chinese, and French underground nuclear explosion tests (59 explosions for the H_0 population) and earthquakes and mining activity (129 events for the H_A population). The ISC and BSC event catalogues provide the m_b and the individual station M_S values for each event. The locations of the seismic events and stations are provided in Fig. 2.

Noted above, the probability model of $M_S - \beta\, m_b$ (Eq. 3) assumes globally constant variance components τ^2 and σ^2, and population means μ_i for both H_0 and H_A populations. Bootstrap calibration analysis assumes that the calibration data set is a representative sample under these model assumptions, although possibly not large enough to adequately represent the extremes necessary to confidently estimate the variance components. Calibration analysis is accomplished with bootstrap sampling of corrected M_S values, $M_S - \beta\, m_b$, and the associated event and station indices. The calibration data included $59 + 129 = 188$ calibration events each observed by varying numbers and locations of stations. A summary is given Fig. 3. For example, one event had 56 stations observing and another had 50 stations observing. Also, 12 events had 3 stations observing and 50 events had 2 stations observing (at least two stations were required for an M_S calculation in the analysis). All events with stations observing gave

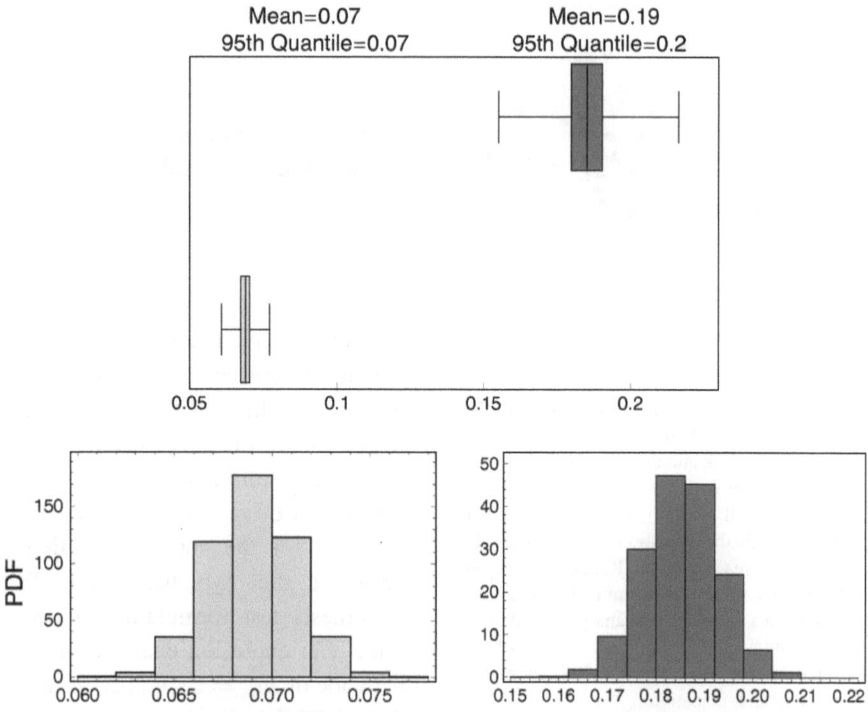

Figure 5
Histograms and box plots of bootstrap variance components (τ^2—*dark grey*, σ^2—*light grey*)

Table 1

Seismic stations and locations observing the 2009 DPRK NWT

ID	Lat.	Lon.	Elev.		Ms	ID	Lat.	Lon.	Elev.		Ms
AAK	42.64	74.49	1.65	Kyrgyzstan	3.89	ASAJ	44.12	142.59	0.20	Japan	2.85
BJI	40.04	116.18	0.04	China	3.83	BVAR	53.02	70.38	0.42	Kazakhstan	3.80
CD2	30.91	103.76	0.63	China	4.23	DAVOX	46.78	9.88	1.83	Switzerland	3.87
FX1	52.88	173.16	0.25	Alaska	3.06	GNI	40.15	44.74	1.61	Armenia	4.03
GTA	39.41	99.81	1.34	China	4.23	HHC	40.85	111.56	1.17	China	4.51
JNU	33.12	130.88	0.54	Japan	2.91	KMI	25.12	102.74	1.94	China	4.16
KSH	39.52	75.92	1.31	China	4.54	KSRS	37.45	127.92	0	South Korea	2.68
LZH	36.09	103.84	1.56	China	4.17	MDJ	44.62	129.59	0.25	China	4.04
MJAR	36.54	138.21	0.42	Japan	3.12	MKAR	46.79	82.29	0.61	Kazakhstan	3.23
NJ2	32.05	118.85	0.05	China	3.74	PETK	53.11	157.70	0.40	Russia	2.99
QIZ	19.03	109.84	0.23	China	4.08	SONM	47.84	106.39	1.42	Mongolia	3.68
SPITS	78.18	16.37	0.32	Norway	3.71	USRK	44.20	132.00	0	Russia	2.33
WMQ	43.82	87.70	0.90	China	4.39	XAN	34.03	108.92	0.63	China	3.61
ZALV	53.95	84.82	0.23	Russia	3.88						

1,906 total event/station records, and so each bootstrap sample had 1,906 randomly selected records. A total of 5,000 bootstrap samples were taken in the calibration analysis.

For each bootstrap sample, the means μ_0 and μ_A were computed. Histograms and basic box plots of these values are given in Fig. 4. The grand mean, -1.39, of the 5,000 H_0 bootstrap means is the calibration value for the test statistic Eq. (5). Also, for each bootstrap sample, the variance components for the one-way random effects model Eq. (3) were computed. The bootstrap variance component estimates are summarised with histograms in Fig. 5. This collection of 5,000 bootstrap variance components provides a technically reasonable

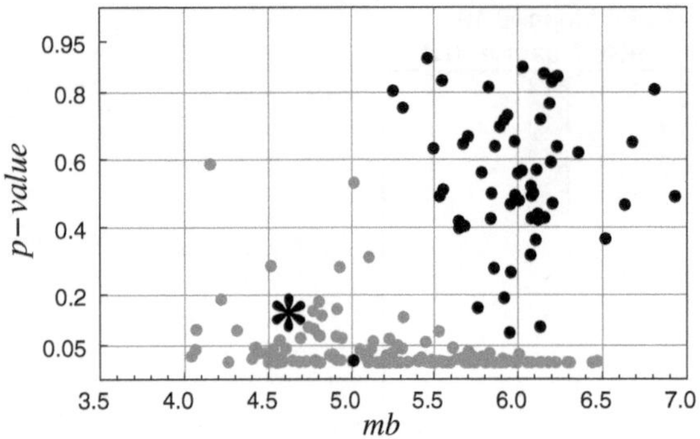

Figure 6
Apparent performance; hypothesis test p-value plot for the calibration data using parameter values from calibration analysis. The explosion (H_0) population is black and the alternate hypothesis (H_A) population is *grey*. The 2009 DPRK p-value is plotted as a *star*. The screened explosion (small p-value) is a peaceful nuclear explosion (PNE) designed to study the formation of cavities in salt. The test was in the Azgir, Guyez region of Kazakhstan on 30 September 1977 (see Mikhailov 1999). The event had two stations observing in the ISC/BCS data used in this analysis, with one anomalously high. Several cold war PNEs exhibited unusually high surface wave energy, and Patton and Taylor (2008) suggest a theoretical explanation

approach to inflate the calibration variance components to more conservative values—the 95th quantile for this analysis. For σ^2, the 95th quantile is 0.07, and for τ^2, the 95th quantile is 0.20.

The 2006 and 2009 DPRK announced NWTs had large network M_S relative to m_b (see Bonner *et al.* 2008; Patton and Taylor 2008 for research on the 2006 DPRK NWT). The ISC/BSC data used in this analysis did not have M_S measurements for the 2006 DPRK NWT. For the 2009 DPRK NWT, 27 M_S measurements were reported ($n_{M_S} = 27$), listed in Table 1, and the event $m_b = 4.62$. From the calibration analysis above, and the recommendation by Selby *et al.* (2011) that $\beta = 1$, the test statistic for H_0 is then

$$Z_{\widehat{M_S}} = \frac{\widehat{M_S} - (-1.39 + 4.62)}{\sqrt{0.20 + 0.07/27}}. \qquad (6)$$

From Table 1, $\widehat{M_S} = 3.687$ giving a "fail to reject H_0" test statistic of $Z_{\widehat{M_S}} = 1.02$ (p-value $= 0.15$). Figure 6 is p-value versus m_b plot for the calibration data, using calibration parameter values (apparent performance), and the p-value for the 2009 DPRK NWT.

Importantly, if the model inadequacy term E_j is not included in the development of the screening hypothesis test, the test statistic for H_0 is of the form

$$Z_{\widehat{M_S}} = \frac{\widehat{M_S} - (\mu_0 + \beta m_b)}{\sqrt{\sigma^2_{M_S}/n_{M_S}}}.$$

Applying the same analysis steps, H_0 is rejected with this test statistic (p-value ≈ 0).

4. Summary

Network averaging ($\widehat{M_S}$) cannot reduce bias in a magnitude correction model. A new potential formulation of the CTBT M_S seismic event screening hypothesis test is developed and demonstrated in this article. The formulation properly partitions total error into *Model Error* and *Station Noise*, and this partition provides for the correct reduction of the standard error of the hypothesis test. With the correct hypothesis test formulation, a decision of "fail to reject H_0: explosion characteristics" is made with a network of $n_{M_S} = 27$ global seismic stations for the 2009 DPRK nuclear weapon test.

Acknowledgments

The authors acknowledge the support of Ms. Leslie A. Casey and the National Nuclear Security Administration Office of Nonproliferation and Treaty Verification Research and Development for funding this work. This work was completed under the auspices of the US Department of Energy by Los Alamos National Laboratory under contract DE-AC52-06NA24596. We thank Dr. Dmitry Storchak, Director of the International Seismological Centre, for his support in the acquisition of the data used in this article. We also thank Dr. Ronan Le Bras, Head of the Software Integration Unit at the International Data Centre, for providing important context in regard to event screening.

Appendix: M_S: m_b Screening Formulation

Established signal processing treats amplitudes as lognormal distributed and therefore magnitudes (path corrected log amplitudes) are normally distributed.

The conceptual representation of the proposed model is

$$\begin{aligned} Y &= \log_{10}(\text{path corrected amplitude}) - \eta \\ &= \mu + \textit{Model Error} + \textit{Station Noise} \end{aligned} \tag{7}$$

where η is a physics model of the magnitude. *Model Error* is a zero mean random effect that varies from event to event and represents model inadequacy from effects such as depth, focal mechanism, local material property, and apparent stress variability. *Station Noise* represents measurement and ambient noise, also with zero mean. Specifically for this development, under H_0: explosion characteristics the source function is η_0 and $\mu = 0$, and the alternate hypothesis (H_A) model is also η_0, however with $\mu = \mu_A$. With the assumption of unbiased source functions, Eq. (7) implies the expected value of Y is

$$\mathcal{E}\{Y\} = \begin{cases} 0 & \text{for } H_0 \\ \mu_A & \text{for } H_A \end{cases} \tag{8}$$

In the development it is assumed that good signal processing practice has been applied giving high confidence in the quality of the observed amplitudes and calculated magnitudes.

For the statistics formulation of Eq. (7), define the random variable Y_{ijk} to be a magnitude residual for hypotheses $H_i = 0$, A and event j and station k. The linear model representation of Eq. (7) is then

$$\begin{aligned} Y_{ijk} = \mu_i + E_j + \epsilon_{ijk} \quad & j = 1, 2, \ldots, m_i \\ & k = 1, 2, \ldots, n_{ij}, \end{aligned} \tag{9}$$

where $\mu_0 = 0$. Analogous to Eq. (7), Eq. (9) reads Y_{ijk} equals a hypothesised source constant μ_i plus a random model inadequacy (event) adjustment E_j plus a station noise adjustment ϵ_{ijk}. Equation (9) is a standard random effects linear model (see SCHEFFE 1959; SEARLE 1971; SEARLE *et al.* 1992).

The E_j are modelled as independent normal random variables with zero mean and variance τ^2. The ϵ_{ijk} are independent normal random variables with zero mean and variance σ^2. E_j and ϵ_{ijk} are independent across all subscripts. This assumption is consistent with model error being uncorrelated with station noise. All model parameters are assumed to be known. As an example, for two stations and three events the statistical properties of E_j and ϵ_{ijk} are succinctly written as

$$\begin{pmatrix} E_1 \\ \epsilon_{(i1)1} \\ \epsilon_{(i1)2} \\ E_2 \\ \epsilon_{(i2)1} \\ \epsilon_{(i2)2} \\ E_3 \\ \epsilon_{(i3)1} \\ \epsilon_{(i3)2} \end{pmatrix} \text{ is normal } (\underline{0}, \Sigma),$$

$$\Sigma = \begin{pmatrix} \tau^2 & 0 & 0 & 0 & 0 & 0 & 0 & 0 & 0 \\ 0 & \sigma^2 & 0 & 0 & 0 & 0 & 0 & 0 & 0 \\ 0 & 0 & \sigma^2 & 0 & 0 & 0 & 0 & 0 & 0 \\ 0 & 0 & 0 & \tau^2 & 0 & 0 & 0 & 0 & 0 \\ 0 & 0 & 0 & 0 & \sigma^2 & 0 & 0 & 0 & 0 \\ 0 & 0 & 0 & 0 & 0 & \sigma^2 & 0 & 0 & 0 \\ 0 & 0 & 0 & 0 & 0 & 0 & \tau^2 & 0 & 0 \\ 0 & 0 & 0 & 0 & 0 & 0 & 0 & \sigma^2 & 0 \\ 0 & 0 & 0 & 0 & 0 & 0 & 0 & 0 & \sigma^2 \end{pmatrix}, \tag{10}$$

where $\underline{0}$ denotes a zero mean vector and Σ is the covariance matrix of the model error components.

$$\begin{pmatrix} Y_{i11} \\ Y_{i12} \\ Y_{i21} \\ Y_{i22} \\ Y_{i31} \\ Y_{i32} \end{pmatrix} = \begin{pmatrix} \mu_i \\ \mu_i \\ \mu_i \\ \mu_i \\ \mu_i \\ \mu_i \end{pmatrix} + W \begin{pmatrix} E_1 \\ \epsilon_{(i1)1} \\ \epsilon_{(i1)2} \\ E_2 \\ \epsilon_{(i2)1} \\ \epsilon_{(i2)2} \\ E_3 \\ \epsilon_{(i3)1} \\ \epsilon_{(i3)2} \end{pmatrix} \text{ is normal } (\underline{\theta}, \Omega), \text{ where}$$

$$\underline{\theta} = \begin{pmatrix} \mu_i \\ \mu_i \\ \mu_i \\ \mu_i \\ \mu_i \\ \mu_i \end{pmatrix}, \quad W = \begin{pmatrix} 1 & 1 & 0 & 0 & 0 & 0 & 0 & 0 & 0 \\ 1 & 0 & 1 & 0 & 0 & 0 & 0 & 0 & 0 \\ 0 & 0 & 0 & 1 & 1 & 0 & 0 & 0 & 0 \\ 0 & 0 & 0 & 1 & 0 & 1 & 0 & 0 & 0 \\ 0 & 0 & 0 & 0 & 0 & 0 & 1 & 1 & 0 \\ 0 & 0 & 0 & 0 & 0 & 0 & 1 & 0 & 1 \end{pmatrix}, \tag{11}$$

and the covariance matrix Ω is

$$\Omega = W \Sigma W^{\mathrm{T}} = \begin{pmatrix} \tau^2+\sigma^2 & \tau^2 & 0 & 0 & 0 & 0 \\ \tau^2 & \tau^2+\sigma^2 & 0 & 0 & 0 & 0 \\ 0 & 0 & \tau^2+\sigma^2 & \tau^2 & 0 & 0 \\ 0 & 0 & \tau^2 & \tau^2+\sigma^2 & 0 & 0 \\ 0 & 0 & 0 & 0 & \tau^2+\sigma^2 & \tau^2 \\ 0 & 0 & 0 & 0 & \tau^2 & \tau^2+\sigma^2 \end{pmatrix}. \tag{12}$$

This two station and three event example is easily generalised to events $j = 1, 2, \ldots, m_i$ and stations $k = 1, 2, \ldots, n_{ij}$. For hypothesis i and a single event j, denote the $1 \times n_{ij}$ vector of magnitude residuals as $\underline{Y}'_{ij} = (Y_{ij1}, Y_{ij2}, \ldots, Y_{ijn_{ij}})$. Then $\underline{Y}_{ij}$ is multivariate normal with $1 \times n_{ij}$ mean vector $\underline{\theta}'_{ij} = (\mu_i, \mu_i, \ldots, \mu_i)$ and $n_{ij} \times n_{ij}$ covariance matrix

$$\Omega_{ij} = \begin{pmatrix} \tau^2 + \sigma^2 & \tau^2 & \tau^2 & \cdots & \tau^2 \\ \tau^2 & \tau^2 + \sigma^2 & \tau^2 & & \tau^2 \\ \tau^2 & \tau^2 & \ddots & & \vdots \\ \vdots & & & \tau^2 + \sigma^2 & \tau^2 \\ \tau^2 & \cdots & \tau^2 & \tau^2 & \tau^2 + \sigma^2 \end{pmatrix} \tag{13}$$

and the network magnitude residual is $\widehat{Y}_{ij} = \underline{1}'\underline{Y}_{ij}/n_{ij}$ is normal with mean μ_i and standard error $\tau^2 + \sigma^2/n_{ij}$.

Conceptually extending the example given in Eq. (10) to two magnitudes gives a block diagonal covariance matrix with the (1,1) block given by Σ in Eq. (10) and the (2,2) block for another magnitude equivalent to Σ in Eq. (10) with notationally different variance components. Introducing a correlation ρ (covariance) in the off-diagonal blocks between the E terms provides the model to calculate the standard error for a hypothesis test.

Let Y_{ijk} be the magnitude residual for m_b and Y^*_{ijk} the magnitude residual for M_S. A test statistic, with both M_S and m_b random, is constructed from network (average) magnitude residuals $\widehat{Y}_{ij}$ and $\widehat{Y}^*_{ij}$. Under H_0, the network magnitude residual $\widehat{m_b}$ is normal with mean $\mu_{m_b} = 0$ and standard error $\tau^2_{m_b} + \sigma^2_{m_b}/n^2_{m_b}$ and $\widehat{M_S}$ is normal with mean $\mu_{M_S} = 0$ and standard error $\tau^2_{M_S} + \sigma^2_{M_S}/n^2_{M_S}$. The standard error of the M_S versus m_b test statistic is then

$$\mathrm{SE}_{\widehat{M_S} - \widehat{m_b}} = \sqrt{\tau^2_{m_b} + \frac{\sigma^2_{m_b}}{n_{m_b}} + \tau^2_{M_S} + \frac{\sigma^2_{M_S}}{n_{M_S}} - 2\,\rho\,\tau_{m_b}\,\tau_{M_S}} \tag{14}$$

for both H_0 and H_A, and the test statistic is

$$Z_{\widehat{M_S} - \widehat{m_b}} = \frac{\widehat{M_S} - \widehat{m_b}}{\sqrt{\tau^2_{m_b} + \sigma^2_{m_b}/n_{m_b} + \tau^2_{M_S} + \sigma^2_{M_S}/n_{M_S} - 2\,\rho\,\tau_{m_b}\,\tau_{M_S}}}. \tag{15}$$

The $Y_{ijk} = M_{Sijk} - \eta(m_{bj})$ model is analogous to Eq. (9) with m_{bj} assumed known. Specifically,

$$Y_{ijk} = \mu_i + E_j + \epsilon_{ijk} \quad j = 1, 2, \ldots, m_i \quad k = 1, 2, \ldots, n_{ij}, \tag{16}$$

with $\eta(m_b)$ as the source model, and equivalent assumptions as with Eq. (9) about the error terms. A common formulation for both hypotheses is $\eta(m_b) = \beta \times m_b$. However, more sophisticated formulations are possible (see PATTON and TAYLOR 2008). With the assumption of unbiased source models for H_0 and H_A the expected value is

$$\mathcal{E}\{Y|m_b\} = \begin{cases} \mu_0 & \text{for } H_0 \\ \mu_A & \text{for } H_A \end{cases} \tag{17}$$

The test statistic for H_0 is then

$$Z_{\widehat{M_S}} = \frac{\widehat{M_S} - (\mu_0 + \beta m_b)}{\sqrt{\tau^2_{M_S} + \sigma^2_{M_S}/n_{M_S}}}. \tag{18}$$

REFERENCES

ANDERSON, D. N., W. R. WALTER, D. K. FAGAN, T. M. MERCIER, and S. R. TAYLOR (2009). *Regional multi-station discriminants: Magnitude, distance and amplitude corrections and sources of error*. Bull. Seism. Soc. Am. *99*, 794–808.

BOLT, B. (1976). Nuclear explosions and earthquakes: The parted veil. San Francisco: W. H. Freeman and Company.

BONNER, J. L., R. B. HERRMANN, and M. HARKRIDER, D. G. PASYANOS (2008). *The surface wave magnitude for the 9 October 2006 North Korean nuclear explosion*. Bull. Seism. Soc. Am. *98*, 2598–2506.

BORMANN, P., R. LIU, Z. XU, K. REN, L. ZHANG, and S. WENDT (2009). *First application of the new IASPEI teleseismic magnitude standards to data of the China national seismograph network*. Bull. Seism. Soc. Am. *99*, 1868–1891.

DAHLMAN, O. and H. ISRAELSON (1977). Monitoring Underground Nuclear Explosions. Amsterdam: Elsevier-North Holland.

DOUGLAS, A., J. A. HUDSON, and V. K. KEMBHAVI (1971). *The relative excitation of body and surface waves by point sources*. Geophys. J. Roy. Astr. Soc. *23*, 451–460.

GUTENBERG, P. (1945). *Amplitudes of surface waves and the magnitudes of shallow earthquakes*. Bull. Seism. Soc. Am. *35*, 3–12.

HERAK, M. and D. HERAK (1993). *Distance dependence of M_s and calibrating function for 20 second Rayleigh waves*. Bull. Seism. Soc. Am. *83*, 1881–1892.

MARSHALL, P. D. and P. BASHAM (1972). *Discrimination between earthquakes and underground explosions employing an improved M_s scale*. Geophys. J. R. Astr. Soc. *29*, 431–458.

MIKHAILOV, V. N. (1999). Catalog of Worldwide Nuclear Testing. New York: Begell-Atom.

PATTON, H. J. and S. R. TAYLOR (2008). *Effects of shock-induced tensile failure on $m_b - M_s$ discrimination: Contrasts between historic nuclear explosions and the North Korean test of 9 October 2006*. Geophys. Res. Lett. *35*, L14301.

REZAPOUR, M. and R. G. PEARCE (1998). *Bias in surface-wave magnitude M_s due to inadequate distance correction*. Bull. Seism. Soc. Am. *88*, 43–61.

RUSSELL, D. R. (2006). *Development of a time-domain, variable-period surface-wave magnitude measurement procedure for application at regional and teleseismic distances, Part I: Theory*. Bull. Seism. Soc. Am. *96*, 665–677.

SCHEFFE, H. (1959). The Analysis of Variance. New York: John Wiley & Sons.

SEARLE, S. R. (1971). Linear Models. New York: John Wiley & Sons.

SEARLE, S. R., G. CASELLA, and C. E. MCCULLOCH (1992). Variance Components. New York: John Wiley & Sons.

SELBY, N. D., P. D. MARSHALL, and D. BOWERS (2012). m_b:M_s event screening revisited. Bull. Seism. Soc. Am. *102*, 88–97.

STEVENS, J. L. and S. M. DAY (1985). *The physical basis of the m_b:M_s and variable frequency magnitude methods for earthquake/explosion discrimination*. J. Geophys. Res. *90*, 3009–3020.

STEVENS, J. L. and K. L. MCLAUGHLIN (2001). *Optimization of surface wave identification and measurement*. Pure. App. Geophys. *158*, 1547–1582.

VANUEK, J., A. *Zatopek*, V. KARNIK, Y. RIZNICHENKO, E. SAVERENSKY, S. SOLOV'EV, and N. SHEBALIN (1962). *Standardization of magnitude scales*. Bull. (Izvest.) Acad. Sci. U.S.S.R., Geophys. Ser. *2*, 153–158.

VON SEGGERN, D. (1977). *Amplitude distance relation for 20-second Rayleigh waves*. Bull. Seism. Soc. Am. *67*, 405–411.

(Received November 30, 2011, revised November 11, 2012, accepted November 21, 2012, Published online February 16, 2013)

Pure Appl. Geophys. 171 (2014), 549–559

DOI 10.1007/s00024-012-0618-x

Pure and Applied Geophysics

Visual Event Screening of Continuous Seismic Data by Supersonograms

BENJAMIN SICK,[1] MARCO WALTER,[1] and MANFRED JOSWIG[1]

Abstract—We present a new visualization method for human inspection of seismic data called supersonograms, which maximizes the amount of time and stations visible on screen while retaining the possibility to detect short and low-signal to noise ratio (SNR) signals. This visualization approach is integrated into a seismological software suite used in the seismic aftershock monitoring system (SAMS) of Comprehensive Nuclear-Test-Ban Treaty Organization (CTBTO) on-site inspections (OSI) to detect suspicious events eventually representing aftershocks from an underground nuclear explosion (UNE). During an OSI, huge amounts of continuous waveform data accumulate from up to 50 six-channel mini-arrays covering an inspection area of 1,000 square kilometers. Sought-after events can have magnitude as low as M_L −2.0 and duration of just a few seconds, which makes it particularly hard to discover them in large, noisy datasets. Therefore, the data visualization is based on nonlinearly scaled, noise-adapted spectrograms, i.e., sonograms, which help to distinguish weak signal energy from stationary background noise. Four single-trace sonograms per mini-array can be combined into supersonograms, since the array aperture is small and sonograms suppress differences of local site noise, allowing an analyst to check array-wide signal coherence quickly. In this paper, we present the supersonograms and the software on the basis of a dataset from a creeping, inhabited landslide in Austria where the same station layout is used as in an OSI. Detected signals are fracture processes in the sedimentary landslide, i.e., slidequakes, with M_L −0.5 to −2.5 between July 2009 and July 2011. These signals are comparable in magnitude and duration to expected weak UNE aftershocks.

Key words: OSI, SAMS, passive method, signal processing, seismic.

1. Introduction

The seismic aftershock monitoring system (SAMS) of an on-site inspection (OSI) has to cover events within an area of up to 1,000 square kilometers (CTBTO 1996). To meet these criteria, the technique of nanoseismic monitoring (JOSWIG 2008) is used and up to 50 seismic mini-arrays are installed for monitoring. A total of 40 inspectors are allowed in an OSI, and an inspection can last up to 60 days with a maximum extension of 70 days (CTBTO 1996). All data processing must be done on-site by the inspectors, and since event detections may inform the ongoing inspection, it is important to analyze the seismic data as fast as possible. Only a few studies on aftershocks due to explosions are available; an overview can be found in FORD and WALTER (2010). JARPE *et al.* (1994) found that explosions from chemical and nuclear tests have similar aftershock rates but that magnitudes of aftershocks from explosions were smaller relative to aftershocks from earthquakes, assuming similar magnitude of the source event. Furthermore, manual processing of the continuous waveforms is essential. State-of-the-art automatic detection algorithms are not useful in the OSI scenario as signatures of expected events are unknown a priori and signals from many different noise sources would lead to too many false-positive automatic detection picks. Training of sophisticated automatic processing tools to support manual analysis in such conditions is complicated and is the topic of current research. Seismological datasets from an OSI are large because of the many stations but short in the sense of recorded time. An automatic algorithm would have to take into account the many unknowns, e.g., station-specific characteristics such as geology, weather influence, and typical signals occurring in an inspection area (IA) at specific stations. Depending on regulations by the inspected state party (ISP), the geology, and the time available, stations cannot be buried deeply and station locations cannot always be optimal (e.g., sediments instead of solid rock). This makes them especially exposed to local noise sources. Noise sources and regulations, e.g., traffic by military

[1] University of Stuttgart, Institute for Geophysics, Azenbergstrasse 16, 70174 Stuttgart, Germany. E-mail: benjamin.sick@geophys.uni-stuttgart.de

vehicles, might even be introduced deliberately by the ISP to compromise the measurements. An algorithm would have to be tuned separately for each mini-array with very little data (data accumulate after station deployment) and no real events (aftershocks of a UNE). Manual analysis enables expert analysts with many years of experience with seismic signals to take into account all of these influences and possible error sources and make the decision to investigate a signal further or not. They are also the ones who deploy the stations, and thus they know in great detail which noise sources might influence the stations (e.g., a river, road, or train track nearby, that the station is on a hill and more exposed to wind gusts and rain, etc.). Events can be very sparse, and one missed event could be the only hint of an UNE. The limited resources and tough time schedules led to the development of the new visualization technique of supersonograms, especially suited for manual processing of large datasets with very low-SNR events. The supersonograms are incorporated into a new software suite called NanoseismicSuite.

1.1. Nanoseismic Monitoring

The method of nanoseismic monitoring fills the gap between passive seismics and microseismic networks (Joswig 2008). Data acquisition is based on the application of seismic mini-arrays, which are suited for azimuth determination of an incoming signal and have been used in numerous studies. Seismic arrays have a long history in CTBT monitoring (e.g., Ringdal and Husebye 1982; Ringdal 1990), and much work has been done on fundamental array design (e.g., Hauberich 1968; Mykkeltveit *et al.* 1983; Harjes 1990). However, sparse arrays with three or four stations are rarely considered (e.g., Suyehiro 1967; Ward and Gregersen 1973; Chiu *et al.* 1991; Kvaerna and Ringdal 1992; Kennett *et al.* 2003), although they offer great improvements for automated processing at minor investment costs (e.g., Sokolowski and Miller 1967; Joswig 1990, 1993b). As the mini-arrays lead or navigate to the source of a signal, we established the term seismic navigating system (SNS), which consists of three vertical seismometers and one three-component seismometer for monitoring. Each mini-array is arranged with the three-component station in the center and the three one-component vertical stations in a surrounding tripartite array. Depending on the epicentral distance of expected signals, the array aperture is usually between around 50 and 200 m. Use of these mini-arrays combines array processing with three-component processing and thus provides calculation of back azimuths, apparent velocities, and particle motion. Array processing as time shifting and stacking is not yet integrated into the standard monitoring process. Furthermore, the mini-arrays are designed for fast and easy installation by two persons, which is crucial in an OSI. Nanoseismic monitoring focuses on forensic seismology (Zucca 1998; Douglas 2007) with manual screening of data and sought-after events just above the ambient noise level. Apart from OSI, further applications are sinkhole monitoring (Wust-Bloch and Joswig 2006), active fault mapping (Häge and Joswig 2009), and monitoring of slope dynamics (Walter *et al.* 2009, 2012; Walter and Joswig 2009).

1.2. Heumoes Slope Example Dataset

Unfortunately, results of the Integrated Field Exercise 2008 (IFE08) in Kazakhstan where seismic data were recorded continuously by 30 seismic small arrays for 15 days cannot be presented in this study due to formal restrictions. The advantages of event screening of seismic data by analysis of supersonograms are explained exemplarily using seismic data recorded permanently by three mini-arrays at Heumoes slope, Austria. The creeping Heumoes slope is situated in the Vorarlberg Alps, Austria, around 25 km south of Bregenz (Fig. 1). The inhabited slope extends ~1.5 km in east–west and ~600 m in north–south direction and shows movement rates of a few centimeters per year at the surface (Lindenmaier *et al.* 2005). A permanent network consisting of three mini-arrays was installed from July 2009 for 2 years to observe slope-related signals caused by movement of the slope (Fig. 1, Walter and Joswig 2008; Walter *et al.* 2011). The observed seismic signals with magnitudes between $M_L = -0.7$ and $M_L = -2.5$ are generated by brittle deformation of the unstable slope material, i.e., fracture processes or "slidequakes" (Gomberg *et al.* 2011). As the Heumoes slope is inhabited, the seismic recordings are dominated by anthropogenic noise sources. Differentiation

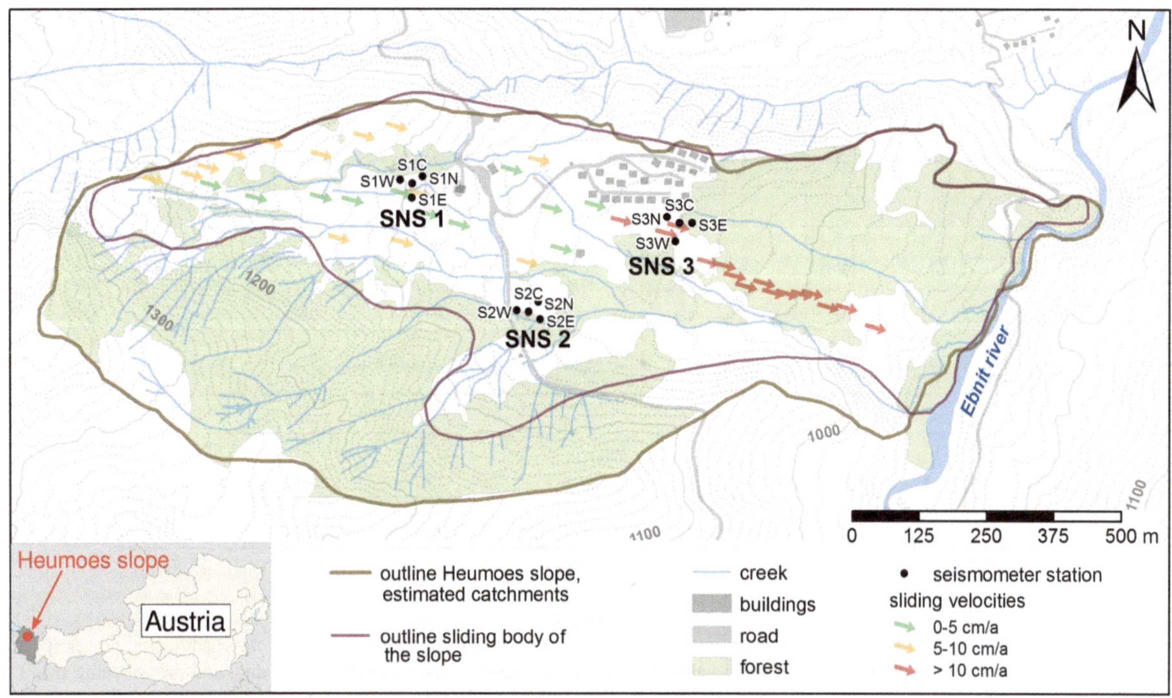

Figure 1
Heumoes slope, sliding velocities, and location of the three seismic mini-arrays (SNS1, SNS2, and SNS3) and their stations (modified after Lindenmaier *et al.* 2005)

between slope-related signals and anthropogenic or natural noise transients by noise forensics is essential (Douglas 2007).

2. *Supersonogram Event Screening and Classification*

Automatic processing of seismic data is not practicable at the moment because sought-after signals in nanoseismic monitoring have very low SNR and are often only detected at single mini-arrays. Furthermore, comprehensive training of an automatic detector would be difficult because fracture processes in the given slope are infrequent and signal patterns can vary significantly depending on the origin of the event. Automatic picking algorithms would either miss crucial events (false negatives) or create an abundance of events coming from the various noise sources on the slope (false positives). An example can be found in Spillmann *et al.* (2007), where only ~0.0034 % of the 66,409 triggered events were slope related. Research into automatic processing of such events is ongoing.

Manual processing on the other hand allows an experienced seismologist with deep knowledge of the setting and noise sources of the given area to set potential events in a broader context and thus eliminate false positives while even finding events disturbed by noise bursts. The need for manual screening of continuous data encouraged the development of a new software suite capable of displaying large datasets on screen while preserving the capability to detect very weak and short-lasting events. The usually applied seismograms (time domain) were no longer sufficient, and instead energies of seismic data are visualized in the form of spectrograms (time–frequency domain). The spectrograms are enhanced by multiple signal processing steps and are called sonograms (Joswig 1993a).

2.1. *Sonogram Calculation Steps*

The sonogram calculation steps are essential for understanding of the resultant supersonograms and are explained here after Joswig (1993a, 1995).

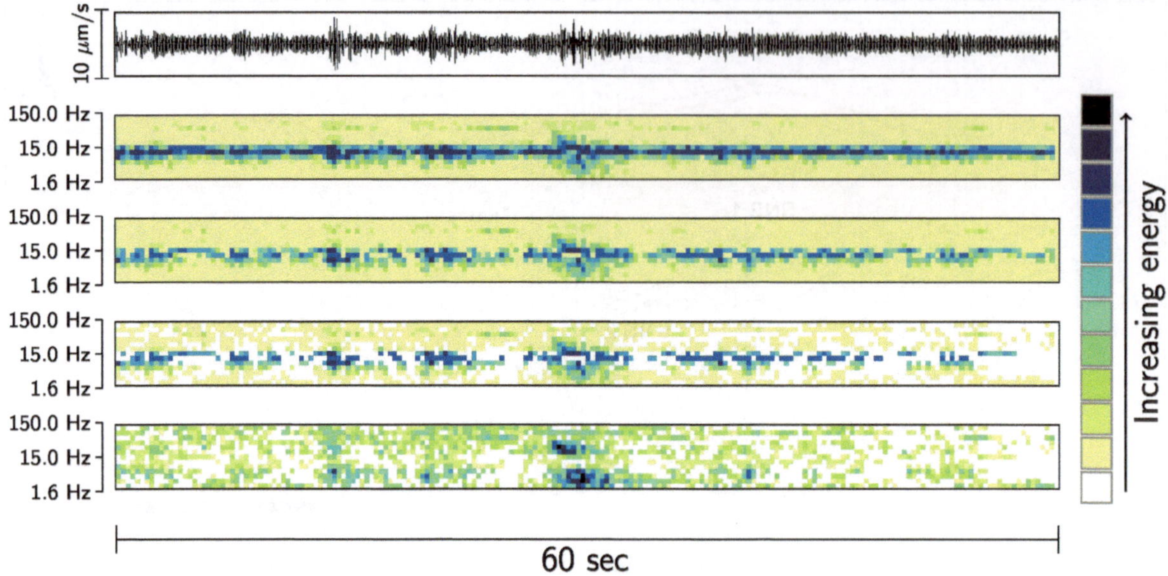

Figure 2
Processing steps of sonogram calculation for a local earthquake in a noisy environment, from *top* to *bottom*: seismogram, power spectral density spectrogram with logarithmic amplitudes and half-octave frequency bands, noise adaptation, blanking, and prewhitening (M_L 1.0, distance 7.7 km, 2010/11/05 15:02:20 UTC)

The time signal is processed by short-time Fourier transformation (STFT). It is split into segments of 256 samples $x(\tau)$, which are tapered with a $\sin^2(\tau)$ windowing function and transformed by fast Fourier transformation (FFT) to $X(\omega)$ with segment overlap of approximately 50 %.

The resulting spectrograms are based on the power spectral density (PSD) from the STFT and are filtered in 13 half-octave wide passbands (Eq. 1).

$$A(\omega, t) = \sum_{\text{half-octave}} X(\omega)X(\omega)^*. \quad (1)$$

The amplitudes of the resultant time–frequency matrix are scaled logarithmically. If a log-normal noise distribution is assumed, the logarithmic-scaled spectrogram has a Gaussian distribution of noise with mean $\mu(\omega)$ and variance $\sigma(\omega)$. Noise adaptation has to be outlier resistant, and a more robust solution is to use the median $M(\omega) = M_{50}$ instead of the mean and $S(\omega) = M_{75} - M_{50}$ instead of the variance. $M(\omega)$ and $S(\omega)$ are both calculated for each frequency band, allowing individual adaptation. For noise adaptation, we subtract $2^{M(\omega)}$ from our signal and allow only values grater than $2^{M(\omega)+S(\omega)}$ (blanking). Therefore, we eliminate disturbing artifacts from our signal visualization (Eq. 2).

$$D(\omega, t) = \begin{cases} \log_2\left(A(\omega, t) - 2^{M(\omega)}\right), & A(\omega, t) > 2^{M(\omega)+S(\omega)} \\ 0, & \text{else} \end{cases} \quad (2)$$

For further scaling by prewhitening we define the log noise variance (Eq. 3).

$$N_D(\omega) = \log_2 2^{M(\omega)+S(\omega)} - 2^{M(\omega)}, \quad (3)$$

where $D(\omega, t)$ and $N_D(\omega)$ are each rounded to the nearest integer value to suppress fine-grain amplitude differences less than $\sqrt{2}$.

Subtracting the frequency-resolved noise variance N_D from the logarithmic-scaled energy distribution performs a prewhitening where the significance of any local energy spot is rated and therefore color-coded as a multiple of the background noise variance (Eq. 4). This interpretation can be seen in the formulation of Eq. 5.

$$\mathrm{SONO}(\omega, t) = \begin{cases} D(\omega, t) - N_D(\omega), & A(\omega, t) > 2^{M(\omega)+S(\omega)} \\ 0, & \text{else} \end{cases} \quad (4)$$

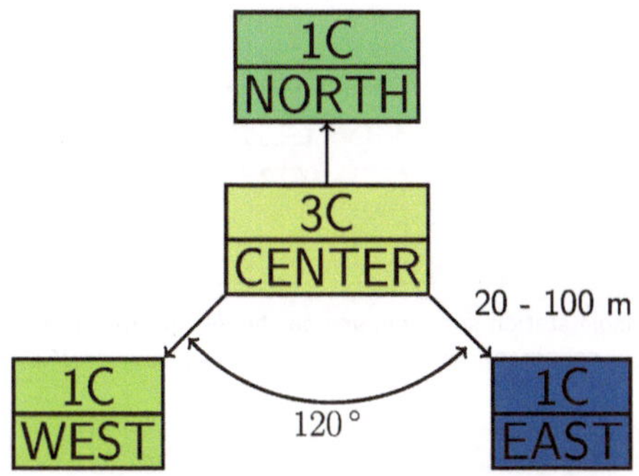

(a) Layout of mini-array with one three-component (3C) central station and three vertical component (1C) stations.

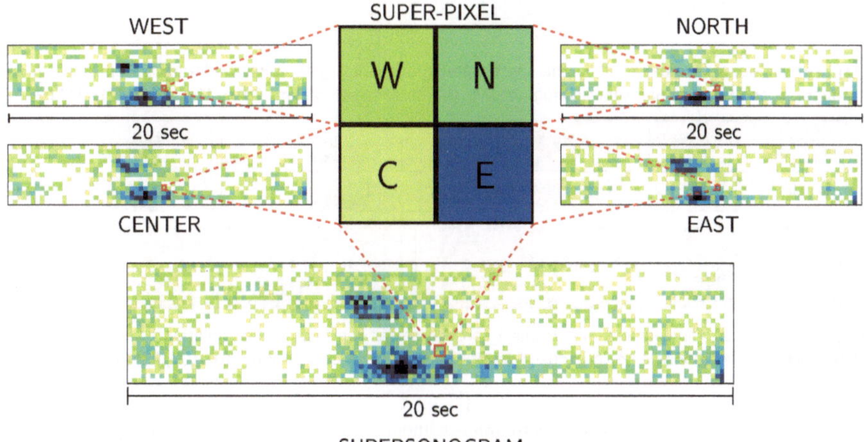

(b) Compilation of super-sonogram from four sonograms of a mini-array (same event as in Figure 2). The pixels of each single sonogram create the "super-pixels" of the super-sonogram.

Figure 3
Layout of mini-array and compilation of supersonogram

$$\mathrm{SONO}(\omega,t) = \begin{cases} \log_2\left(\frac{A(\omega,t)-2^{M(\omega)}}{2^{M(\omega)+S(\omega)}-2^{M(\omega)}}\right), & A(\omega,t) > 2^{M(\omega)+S(\omega)} \\ 0, & \text{else} \end{cases} \tag{5}$$

Sonograms were originally developed for automatic pattern recognition (Joswig 1990) but are perfectly capable of assisting analysts in manual screening and identification of very small-scale events, especially in noisy environments. Sonograms filter disturbing stationary background noise, and the half-octave band division of the frequency bands is based on human perception of frequencies, which enhances manual detection. Event patterns are prominent even when analyzing large datasets with varying noise conditions. Amplitudes are visualized with a specially developed color scale which emphasizes contrasts (Fig. 2). Sonogram scaling makes use of values above the saturation of the color scale, which can result in black areas for strong events. For detection purposes, it is enough to recognize strong events as such; further analysis of, e.g., amplitude ratios, for detailed classification of events can be done with seismograms with scaling on the maximum values. The individual processing steps for sonogram calculation are illustrated in Fig. 2 on the basis of a weak local earthquake near the Heumoes slope. The elimination of disruptive dominating narrowband noise which is permanently present can be seen prominently in Fig. 2.

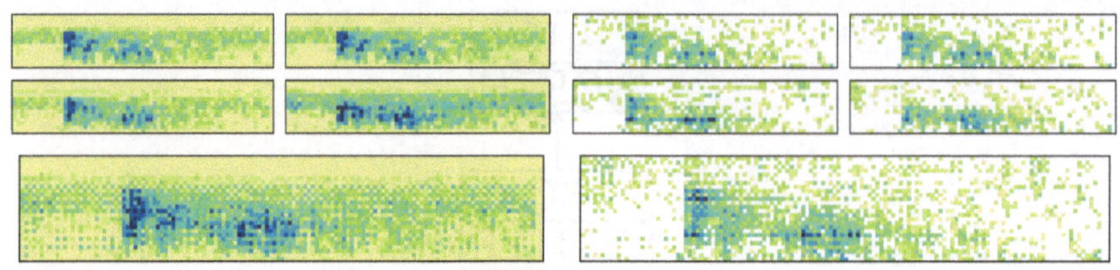

(a) Significantly different single station spectrograms on the left get much more similar through sonogram processing which makes super-sonogram creation possible (M_L 1.2, distance 24.02 km, 2010/04/02 03:14:43 UTC).

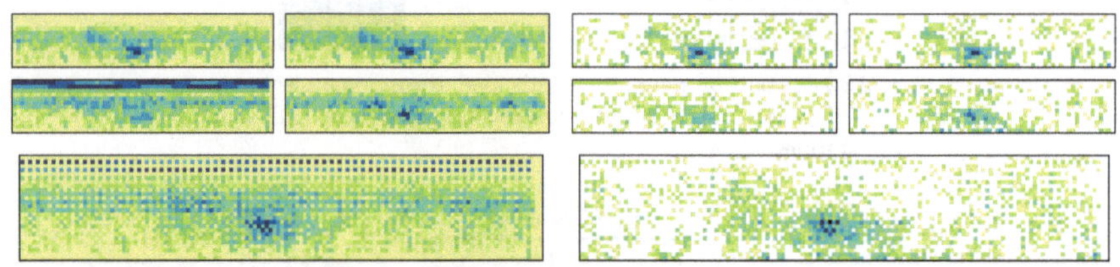

(b) The almost not visible event in the spectrograms on the left gets visible by lifting it from noise in the sonograms and the resulting super-sonogram (M_L 0.7, distance 18.55 km, 2010/10/27 04:35:15 UTC).

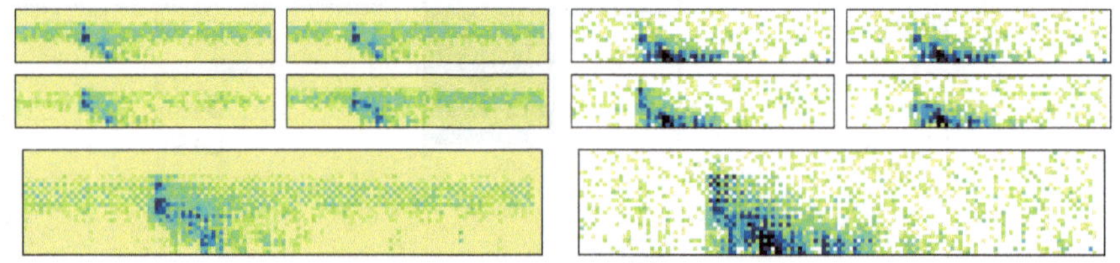

(c) Example of an improvement of a P-onset through better contrasts and completion of energies of lower frequencies of the impulsive i.e. broadband onset (M_L 1.0, distance 4.94 km, 2010/03/23 15:58:50 UTC).

Figure 4

Comparison of supersonogram compilation of ordinary spectrograms without signal-enhancing sonogram steps (*left*) versus sonograms (*right*). Figures consist of the four spectrograms/sonograms in the order given in Fig. 3b plus the resulting supersonogram at the *bottom*. Signal length of all examples is 20 s

2.2. *Supersonograms*

Single stations of one mini-array are within 200 m distance, which makes it possible to combine the four vertical traces of one mini-array into a so-called supersonogram. The combination is done by using "superpixels" at each time and frequency position. The horizontal traces of the three-component stations are used later for more detailed interpretation of signals. Each "superpixel" of the supersonogram consists of four pixels, each from one vertical trace of the mini-array. Figure 3 shows how pixels of four ordinary sonograms create one "superpixel," which is then used in the supersonogram.

Supersonogram visualization provides additional advantages over array processing methods, which were primarily introduced to estimate back azimuth and apparent velocity. The visualization aids in the pure detection and differentiation of events from noise with fast coherence checks and acts as preprocessing before events are inspected more thoroughly by beam-forming and localization; e.g., array-wide signal coherence can be checked very fast by looking at just one trace. Incoherent signals show up as spotted patterns, while coherent ones create consistent areas of similar color. Furthermore, the amount of data which can be displayed on one screen increases significantly. Supersonograms can be displayed on screen with a very small dimension and yet events are still prominent to an analyst. Additionally, faulty or very noisy data from single stations can be recognized immediately in comparison with other stations of the same array.

Tests with common spectrograms (Fig. 4) show that the combination of traces from different stations into one trace is only possible with the signal processing steps of the sonogram calculation. On

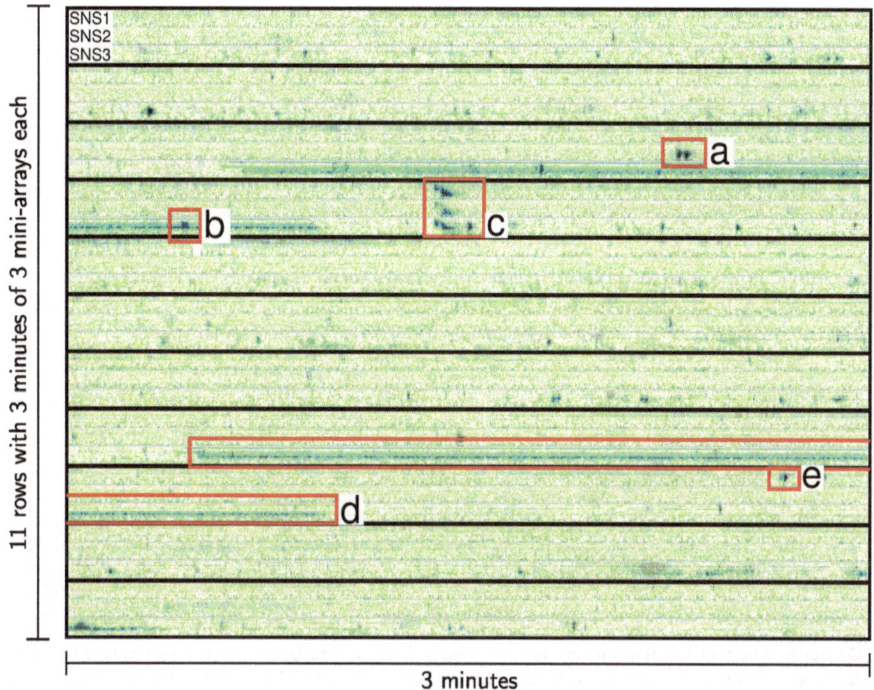

Figure 5

Screenshot of the software module SonoView with supersonograms of three mini-arrays of a larger timespan with recurring anthropogenic noise (start 2009/10/21 00:00:00 UTC). The screenshot shows 11 rows, each containing 3 min of supersonograms of all three mini-arrays. **a** Multiple frost heave events at SNS2. **b** Frost heave event in noise of SNS3. **c** Local earthquake (M_L0.3, distance 10.0 km) with frost heave event on SNS3 *right* after earthquake. **d** Anthropogenic noise transient caused by a pump installed near SNS3. **e** Frost heave event at SNS1

the other hand, if regular spectrograms are combined to form a supersonogram, the varying noise conditions at each single station show up dominantly and obfuscate events (Fig. 4a). Low-SNR signals are not visible without the enhancements (Fig. 4b). With sonograms, the onset times of events get significantly clearer by better contrasts of pre- and post-onset time signals as well as, e.g., the extension of low-frequency parts of an impulsive, i.e., broadband, onset (Fig. 4c).

2.3. *Signal Classification*

Sonograms allow an analyst to classify events by multiple factors partly known from conventional seismogram analysis. Energies of different frequencies, amplitudes, signal duration, and different seismic onsets of phases are the main classification attributes. Additionally, the direct visualization of frequency contents allows, e.g., immediate recognition of moving signal sources. Signals of sources approaching a station become more broadband while those of receding sources become more narrowband, as can be seen in Fig. 6c for a snowcat moving on the Heumoes slope. By the use of supersonograms in a large-timespan display, events can be furthermore analyzed by periodicity and number of stations, and most importantly the event can be put in a broader context by looking at the data of other stations and other points of time with the same scaling. Anthropogenic noise can be identified more easily and excluded based on, e.g., repetition of signals of the same energy or/and a constant repeating frequency as shown in Fig. 5 for a water pump installed at the Heumoes slope. Regional and teleseismic seismicity need to be classified in an OSI to exclude it from UNEs, while local seismicity can be the result of aftershocks of an UNE (Fig. 6b). In the case of landslide monitoring, slidequakes and other slope-related events are of particular interest. Figure 6a shows one of the many registered events in the slope which is classified to be a slidequake. All examples show how different types of seismic sources create

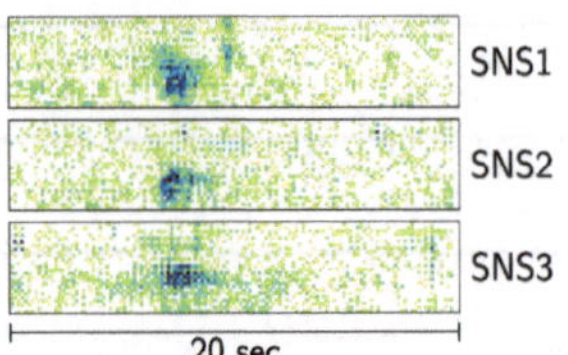

(a) Super-sonograms of slidequake at all mini-arrays (M_L − 1.8, distance 0.2 km from SNS1, 0.33 km from SNS2 and 0.35 km from SNS3, 2010/05/22 04:16:05 UTC).

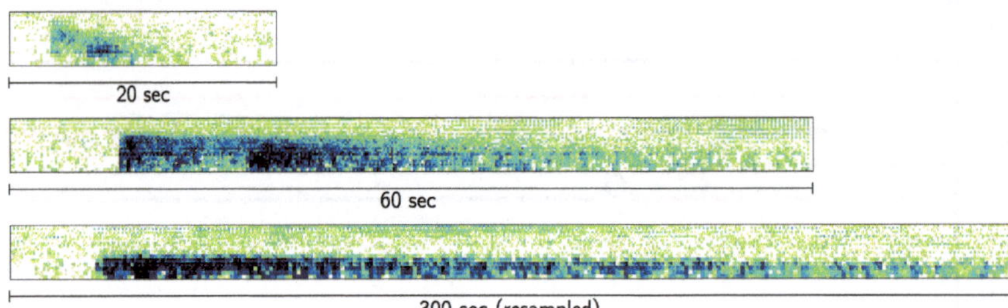

(b) Super-sonograms of natural seismicity at SNS1. From top to bottom: local earthquake (M_L 0.9, distance 17.5 km, 2010/10/28 14:00:35), regional earthquake (M_L 2.4, distance 68 km, 2009/11/02 12:14:40 UTC), teleseismic earthquake (Haiti region, 5 minutes super-sonogram resampled from 400 Hz to 100 Hz, 2010/01/12 22:04:05).

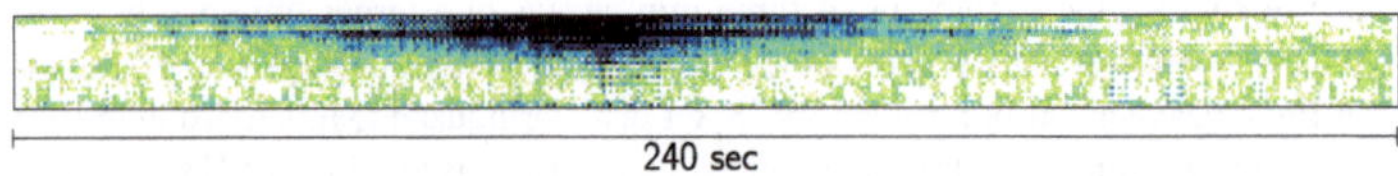

(c) Super-sonogram with characteristic "cigar" shape of moving noise source (here snowcat on Heumoes slope, 4 minutes super-sonogram resampled from 400 Hz to 100 Hz, 2011/02/01 19:55:15 UTC).

Figure 6
Examples of natural seismicity and noise

specific patterns in the supersonograms and therefore help an analyst in manual screening.

3. *Software*

The supersonogram technique has been integrated into a software suite called NanoseismicSuite, which consists of four modules: SeisServ, SonoView, TraceView, and HypoLine (Fig. 7). A basic description of each software module is given here; more information on the software can be found on the nanoseismic monitoring webpage (http://www.nanoseismic.net).

3.1. *SeisServ*

SeisServ reads seismic data and metadata in Center for Seismic Studies (CSS) or MiniSEED format from files or an Oracle database. It provides these data to the other modules and allows editing of the metadata, e.g., the geometry of seismic stations.

3.2. *SonoView*

After loading the data, SonoView is the first application to be used in a typical event screening scenario. It visualizes supersonograms in a manner that maximizes the data visible on one screen. An

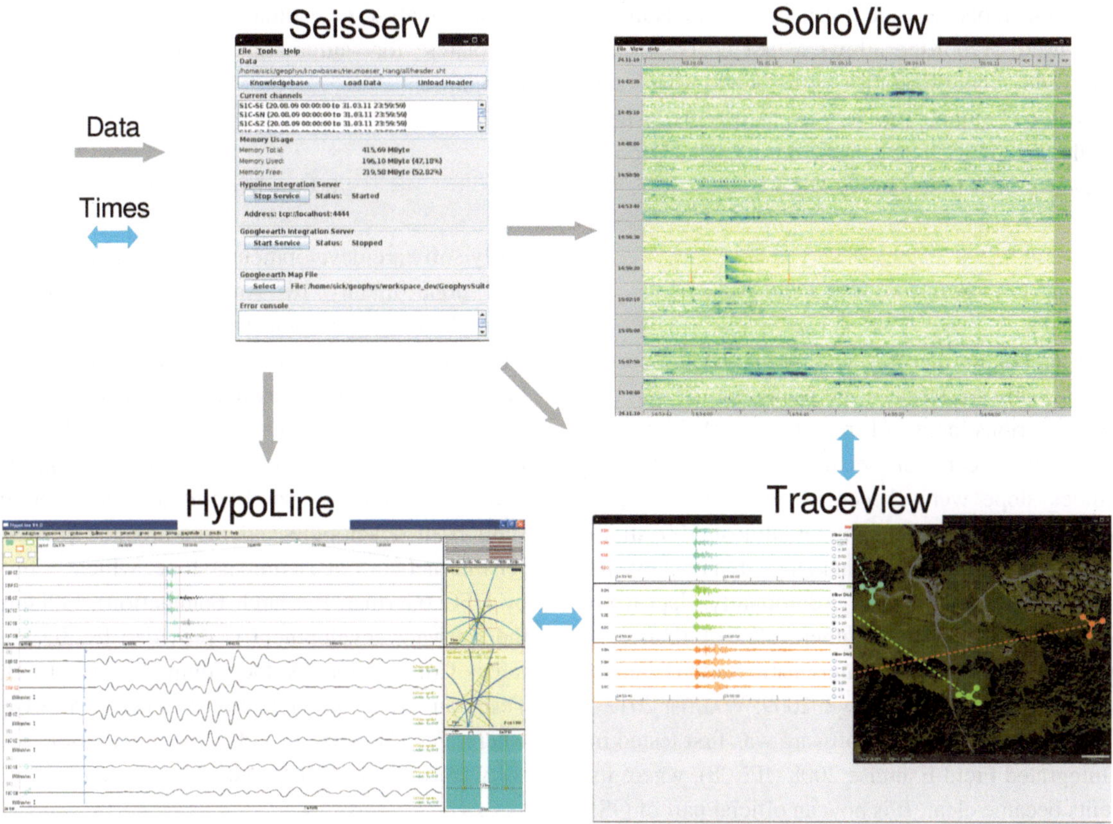

Figure 7
NanoseismicSuite: overview of components and interfaces. *Arrows* indicate data and timing interfaces. SeisServ provides data to SonoView, TraceView, and HypoLine, which synchronize current screening times with each other

arbitrary amount of mini-arrays and time spans can be loaded. An analyst can scroll quickly through the continuous data in SonoView and mark suspicious events for further processing steps.

3.3. *TraceView*

Detected events from SonoView can be further analyzed in TraceView, which visualizes the seismograms of these events together with a map of the measurement area with locations of the mini-arrays. It provides a two-dimensional neighborhood overview of mini-arrays, which cannot be provided by the one-dimensional listing of SonoView. TraceView shows the seismograms of the currently selected mini-array and the five adjacent mini-arrays based on geographic context. Basic filters and scalings can be applied to the seismograms, and georeferenced images of the measurement area can be shown on the map (e.g., satellite images).

3.4. *HypoLine*

The last application in the processing pipeline of the NanoseismicSuite is HypoLine, which is used for localization and magnitude estimation of events. Accurate three-dimensional underground models of the measurement area are often not known a priori, and localization is done by time difference of arrival (TDOA) hyperbolae and S-P distance circles based on one-dimensional velocity models. At the moment, HypoLine supports processing of data from up to six mini-arrays, which it gets from TraceView. This subset of mini-arrays is no restriction for weak events because they are only visible at the surrounding stations. For further processing of single events, tools such as Geotools, Seisan, Pitsa or SeismicHandler can be used. HypoLine allows a first coarse localization and identification with interactive and graphical techniques for very weak events, where the influence of each parameter on event location is displayed in real time (Joswig 2008).

4. Conclusions and Discussion

The technique of supersonograms and the corresponding software tools of the NanoseismicSuite are used in various areas of monitoring of low-SNR events. Two years of seismic data recorded at Heumoes slope were processed with the software, which allowed the screening of almost 100 slidequakes and a multitude of other slope dynamic events (Walter *et al.* 2009, 2011). Additionally, data from another landslide in the southern alps of France is being processed with the NanoseismicSuite (Walter *et al.* 2009; Walter and Joswig 2009; Walter 2012). For CTBTO purposes, the software was first tested in the Integrated Field Exercise 2008 (IFE08), where its benefits became clear. It is now an official part of OSI SAMS and regularly used in CTBTO training cycles to train OSI SAMS team members, e.g., in the OSI advanced training course 2nd training cycle (AC2TC), where the software was used in training with IFE08 data as well as in the field in Hungary, where data recorded from the field campaign were analyzed directly. Other areas of usage are monitoring of induced seismicity (Häge *et al.* 2012) and sinkholes (Wust-Bloch and Joswig 2006).

For future development it is planned to integrate automatic detection algorithms into the software to assist the manual processing by visual indications. Especially the field of pattern recognition provides promising algorithms (Joswig 1996), and supersonograms are predestined to be used here. Datasets can be screened partly manually, and detected events can be used as templates to train automatic detection algorithms. Research with self-organizing maps (also known as Kohonen maps, Kohonen 2001) to cluster events by unsupervised learning is ongoing. Self-organizing maps provide the capability to create an overview of the existing seismicity by prototypes of each event type. These prototypes can be used by a human analyst to estimate local seismicity and as templates for automatic detection with pattern recognition.

Acknowledgements

Early software development of the NanoseismicSuite was done mainly by Andreas Poszlovszki, who created the modular object-oriented application basis which allowed the further extensive development by the main author into an application which is used now in many areas. Matthias Guggenmos and Andreas Eisermann also contributed code. The authors thank Patrick Blascheck and Eberhard Claar for their support in developing and installing the permanent seismic network at Heumoes slope, which is funded by DFG (German Research Foundation). Funding of the software development is provided by the Preparatory Commission for the Comprehensive Nuclear-Test-Ban Treaty Organization (CTBTO Prep Com), where Peter Labak provides the integration of the software into the OSI SAMS environment.

References

Chiu, J. M., Steiner, G., Smalley Jr., R., and Johnston, A. C. (1991), *PANDA: A simple, portable seismic array for local- to regional-scale seismic experiments*, Bull. Seism. Soc. Am. *81*, 1000–1014.

CTBTO, Comprehensive Nuclear-Test-Ban Treaty (Preparatory Commission for the Comprehensive Nuclear-Test-Ban Treaty Organization, 1996).

Douglas, A. (2007), *Forensic seismology revisited*, Surv Geophys. *28*, 1–31.

Ford, S. R., and Walter, W. R. (2010), *Aftershock Characteristics as a Means of Discriminating Explosions from Earthquakes*, Bull. Seism. Soc. Am. *100*, 364–376.

Gomberg, J., Schulz, W., Bodin, P., and Kean, J. (2011), *Seismic and geodetic signatures of fault slip at the Slumgullion Landslide Natural Laboratory*, J. Geophys. Res. *116*, 20 pp.

Häge, M., and Joswig, M. (2009), *Spatiotemporal characterization of interswarm period seismicity in the focal area Nový Kostel (West Bohemia/Vogtland) by a short-term microseismic study*, Geophys. J. Int. *179*, 1071–1079.

Häge, M., Blascheck, P., and Joswig, M. (2012), *EGS hydraulic stimulation monitoring by surface arrays - location accuracy and completeness magnitude: the Basel Deep Heat Mining Project case study*, J. Seismol.

Harjes, H.-P. (1990), *Design and siting of a new regional array in Central Europe*, Bull. Seism. Soc. Am. *80*, 1801–1817.

Haubrich, R. A. (1968), *Array design*, Bull. Seism. Soc. Am. *58*, 977–991.

JARPE, S., GOLDSTEIN, P., and ZUCCA, J. J., *Comparison of the non-proliferation event aftershocks with other Nevada Test Site events, UCRL-JC-117754. In Non-proliferation Experiment Symposium*, Rockville, Maryland, 19-21 April 1994.

JOSWIG, M. (1990), *Pattern recognition for earthquake detection*, Bull. Seism. Soc. Am. *80*, 170–186.

JOSWIG, M. (1993), *Single-trace detection and array-wide coincidence association of local earthquakes and explosions*, Comput. Geosci. *19*, 207–221.

JOSWIG, M. (1993), *Automated seismogram analysis for the tripartite BUG array: an introduction*, Comput. Geosci. *19*, 203–206.

JOSWIG, M. (1995), *Automated classification of local earthquake data in the BUG small array*, Geophys. J. Int. *120*, 262–286.

JOSWIG, M. (1996), *Pattern recognition techniques in seismic signal processing, In Proceedings of the 2nd Workshop on Application of Artificial Intelligence Techniques in Seismology and Engineering Seismology 12*, 1996, 37–56.

JOSWIG, M. (2008), *Nanoseismic monitoring fills the gap between microseismic networks and passive seismic*, First Break *26*, 117–124.

KENNETT, B. L. N., BROWN, D. J., SAMBRIDGE, M., and TARLOWSKI C. (2003), *Signal Parameter Estimation for Sparse Arrays*, Bull. Seism. Soc. Am. *93*, 1765–1772.

KOHONEN, T. (2001), *Self-organizing maps, Springer* Ser. Inf. Sci. *30*, 501 pp.

KVAERNA, T., and RINGDAL, F. (1992), *Integrated array and three-component processing using a seismic microarray*, Bull. Seism. Soc. Am. *82*, 870–882.

LINDENMAIER, F., ZEHE, E., DITTFURTH, A., and IHRINGER, J. (2005), *Process identification at a slow-moving landslide in the Vorarlberg Alps*, Hydrological Processes *19*, 1635–1651.

MYKKELTVEIT, S., ÅSTEBØL, K., DOORNBOS, D. J., and HUSEBYE, E. S. (1983), *Seismic array configuration optimization*, Bull. Seism. Soc. Am. *73*, 173–186.

RINGDAL, F., and HUSEBYE, E. S. (1982), *Application of arrays in the detection, location, and identification of seismic events*, Bull. Seism. Soc. Am. *72*, 201–224.

RINGDAL, F. (1990), *Introduction to the special issue on regional seismic arrays and nuclear test ban verification*, Bull. Seism. Soc. Am. *80*, 1775–1776.

SOKOLOWSKI, T. J., and MILLER, G. R. (1967), *Automated epicenter locations from a quadripartite array*, Bull. Seism. Soc. Am. *57*, 269–275.

SPILLMANN, T., MAURER, H., GREEN, A. G., HEINCKE, B., WILLENBERG, H., and HUSEN, S. (2007), *Microseismic investigation of an unstable mountain slope in the Swiss Alps*, J. Geophys. Res. *112*, 25 pp.

SUYEHIRO, S. (1967), *A search for small, deep earthquakes using quadripartite stations in the Andes*, Bull. Seism. Soc. Am. *57*, 447–461.

WALTER, M., and JOSWIG, M. (2008), *Seismic monitoring of fracture processes generated by a creeping landslide in the Vorarlberg alps*, First Break *26*, 131–136.

WALTER, M., and JOSWIG, M., *Seismic characterization of slope dynamics caused by softrock-landslides: The Super-Sauze case study*. In MALET, J.-P., REMAÎTRE, A., BOOGARD, T. (Eds), *Proceedings of the International Conference on Landslide Processes: from geomorphologic mapping to dynamic modelling*, Strasbourg (CERG Editions, 2009), 215–220.

WALTER, M., NIETHAMMER, U., ROTHMUND, S., and JOSWIG, M. (2009), *Joint analysis of the Super-Sauze (French Alps) mudslide by nanoseismic monitoring and UAV-based remote sensing*, First Break *27*, 75–82.

WALTER, M., WALSER, M., and JOSWIG, M. (2011), *Mapping rainfall-triggered fracture processes, and seismic determination of landslide volume at the creeping Heumoes slope*, Vadose Zone J. *10*, 487–495.

WALTER, M., ARNHARDT, C., and JOSWIG, M. (2012), *Seismic monitoring of rockfalls, slide quakes, and fissure development at the Super-Sauze mudslide, French Alps*, Eng. Geol. *128*, 12–22.

WARD, P. L., and GREGERSEN, S. (1973), *Comparison of earthquake locations determined with data from a network of stations and small tripartite arrays on Kilauea Volcano, Hawaii*, Bull. Seism. Soc. Am. *63*, 679–711.

WUST-BLOCH, G. H., and JOSWIG, M. (2006), *Pre-collapse identification of sinkholes in unconsolidated media at Dead Sea area by 'nanoseismic monitoring' (graphical jackknife location of weak sources by few, low-SNR records)*, Geophys. J. Int. *167*, 1220–1232.

ZUCCA, J. J. (1998), *Forensic seismology supports the Comprehensive Test Ban Treaty, Science & Technology Rev.*, LLNL, CA., Sept. 1998, 4–11.

(Received October 17, 2011, revised October 2, 2012, accepted October 10, 2012, Published online November 30, 2012)

Pure Appl. Geophys. 171 (2014), 561–573

DOI 10.1007/s00024-012-0617-y

Pure and Applied Geophysics

Suppression of Periodic Disturbances in Seismic Aftershock Recordings for Better Localisation of Underground Explosions

FELIX GORSCHLÜTER[1] and JÜRGEN ALTMANN[1]

Abstract—For precise localisation of a potential underground nuclear explosion, the Comprehensive Nuclear-Test-Ban Treaty Organization, during an on-site inspection, can set up seismic sensors to find the very small signals from aftershocks. These signals can be masked by periodic disturbances from, for example, helicopters. We present a new method to characterise every such disturbance by the amplitude, frequency and phase of the underlying sine in the time domain using a mathematical expression for its Hann-windowed discrete Fourier transform. The contributions of these sines are computed and subtracted from the complex spectrum sequentially. Two examples show the performance of the procedure: (1) synthetic sines superposed to a coal-mine induced event, orders of magnitude stronger than the latter, can be removed successfully, (2) removal of periodic content from the signals of a helicopter overflight reduces the amplitude by a factor 3.3 when the frequencies are approximately constant. The procedure cannot yet cope with peaks that change frequency too fast, for example by the Doppler effect when passing, and with peaks that lie too close to each other. Improvement to solve these problems seems possible.

Key words: Seismic, localisation of underground explosions, spectral analysis, periodic noise reduction.

1. *Introduction*

The Comprehensive Nuclear-Test-Ban Treaty Organization (CTBTO) builds up and maintains a global network of sensors to detect every underground explosion with a yield of 1 kT TNT-equivalent or better. Teleseismic detections result in localisation uncertainties on the order of 10 km. For a more precise determination of the hypocentre of the explosion and to find further indicators as to whether the explosion was a nuclear one, the CTBTO can carry out an on-site inspection (OSI) in the area of interest if the country is a Treaty party. A seismic aftershock monitoring system (SAMS) can be placed at the surface to detect the very small vibrations produced by relaxations in the rock around the cavity. However, helicopters and vehicles used by the inspectors, noise from existing infrastructure in the country, or even intended disturbance attempts can generate seismic signals which can mask the weak aftershock signals.

Many man-made noise sources (engines, etc.) are of a periodic nature, and airborne sound can couple into the ground. Periodic signals show up as peaks in the frequency spectrum. The weak signals of aftershocks, on the other hand, are of a pulsed shape, and their spectrum is broadband. With the Fourier transform (which converts a certain interval of time-domain data into its spectrum) these properties give the opportunity to distinguish between disturbing periodic noise and impulse-type aftershock events (Fig. 1). We investigate whether the disturbing peaks can be characterised and subtracted from the superposed spectrum, so that the broadband content of the impulsive event remains.

Removing periodic noise can be done by traditional methods. If the noise peaks lie above or below the frequency range of interest, a simple low- or high-pass filter can remove them effectively. If the disturbing spectral peaks overlap with the signal spectrum—which is often the case if the latter is broadband—one can use notch filters individually tuned to the frequencies contained in the disturbance. This requires finding these frequencies and is not adapted to the different strengths of the peaks. Different methods for noise reduction in seismic data are described in ROBINSON and TREITEL (1980), BUTTKUS (2000), PETER BORMANN (2009). To better remove the peaks, we are investigating a new method that exactly

[1] TU Dortmund, Dortmund, Germany. E-mail: gorschlueter @e3.physik.tu-dortmund.de; altmann@e3.physik.tu-dortmund.de

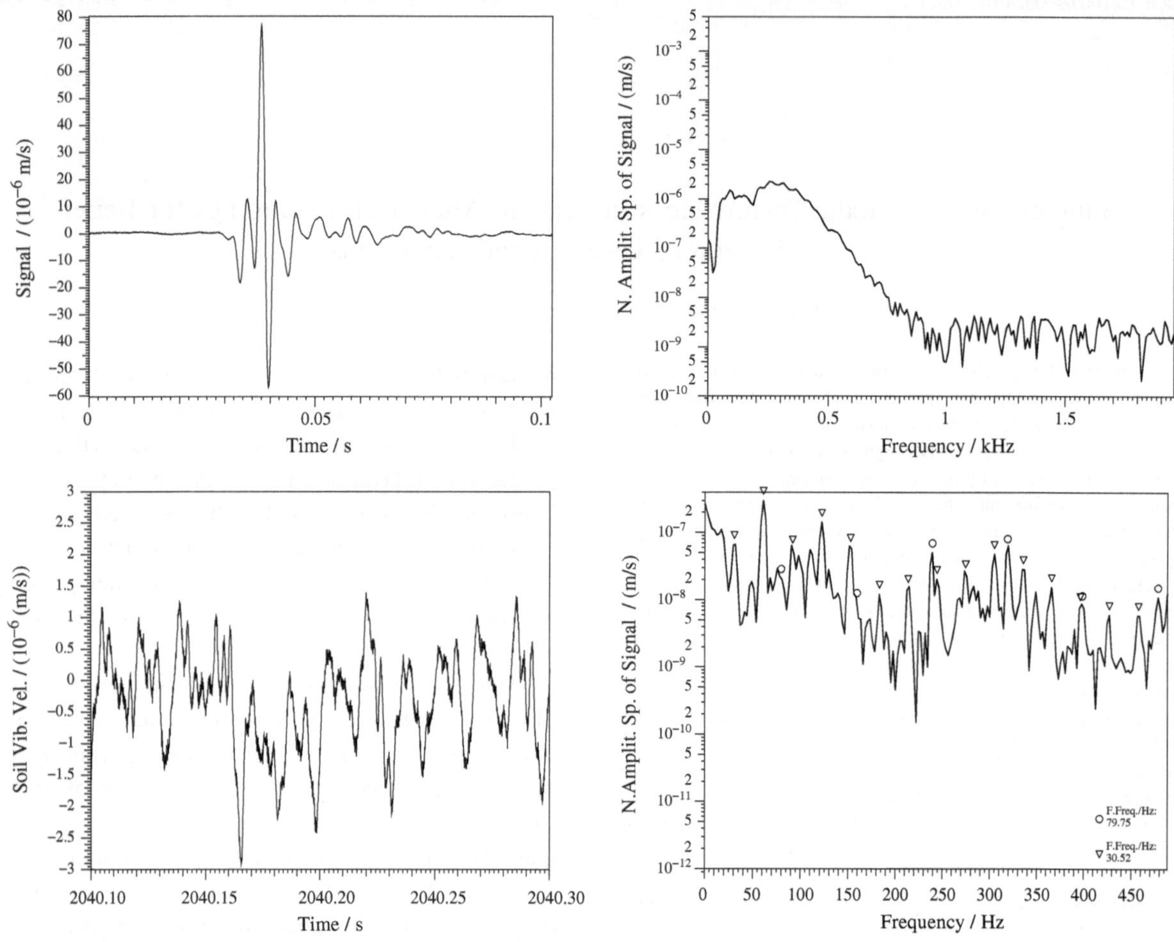

Figure 1
Demonstration of periodic and impulse events: seismic signal (*left*) and spectrum (*right*) of a firecracker at about 100 m distance from a geophone (*top*) and the seismic signal of a flying helicopter at about 1 km distance (*bottom*). In the spectra of the helicopter two harmonic series can be found stemming from the main (*triangles*) and the tail (*circles*) rotor, respectively

characterises the peaks by amplitude, frequency and phase and subtracts their single spectra one after the other from the complex spectrum of the given time series.[1] The method will be explained in mathematical detail elsewhere. It promises to complement existing methods of spectral estimation, in particular for finding frequencies contained in a signal, in the time domain, such as the autoregressive moving average (ARMA) model, or in the spectral domain, such as the averaged periodogram or minimum variance spectral estimation (e.g. Kay 1988; Oppenheim and Schafer 1999). Its first step can be seen as an extension of the periodogram technique by finding the peaks of spectral power, but for accurate determination of the phase then it works in the complex domain where averaging becomes meaningless. We assume that the sines in the time series have constant frequency during the time interval for one spectrum, but that the amplitude, frequency and phase can change from interval to interval, so that again averaging techniques cannot be applied. Here we report on the application of our peak-identification and -subtraction method to the problem of periodic disturbances in seismic aftershock measurements as they are stipulated by the CTBT for OSIs.

The next section describes the analytical expression for the discrete spectrum of a single sine and the

[1] First results have been published in a conference poster: Jürgen Altmann and Felix Gorschlüter (2009).

algorithm used to fit it to the data. Section 3 shows examples of an application of the peak fitting and subtraction algorithm, and Sect. 4 discusses the results.

2. Background and Theory

2.1. Analytical Expression of the Complex Spectrum of a Monofrequent Sine

We assume that in each spectrum the periodic content does not change in time. This is often at least approximately fulfilled over the time interval used for one discrete spectrum. A signal consisting of periodic contributions can be expressed by a superposition of sine functions. Each such sine has an amplitude A_0, a frequency ν_0 and a phase ϕ_0, its continuous time course is

$$s(t) = A_0 \sin(2\pi\nu_0 t + \phi_0) \tag{1}$$

Three facts need to be taken into account (Brigham 1988):

- Real data are gained by analogue-digital-converters (ADC) by sampling the continuous signal $s(t)$ with a certain rate (e.g., the CTBTO uses 500 Hz when performing OSI exercises); thus the data are a sequence of discrete values. This is equivalent to multiplying $s(t)$ with an equidistant Dirac comb
- Real signals can only be handled for a finite duration, and they can change over time, so only a short interval of data is transformed into its spectrum. Mathematical this means multiplication with a rectangle function
- Because of the first two items mentioned the discrete spectrum of a sine does not consist of two δ functions at the frequency of the sine and its negative [as the Fourier transform of the continuous $s(t)$ would be] but depends on the position of the frequency ν_0 with respect to the equidistant comb of discrete frequencies of the spectrum. To reduce the ensuing spectral leakage $s(t)$ is multiplied with a window function (Hann window in our case)

Multiplications in the time domain are equivalent to convolutions in the frequency domain causing a multiplicity of terms. After the convolution of the terms described in Table 1 the analytical expression of the spectrum of a single, monofrequent sine becomes:[2]

$$\begin{aligned} G(\nu) = i\frac{A_0}{4\sqrt{3}N}\Bigg[&\sin\left(\pi T(\nu-\nu_0)\right)\bigg(2\cot\left(\pi\Delta t(\nu-\nu_0)\right) \\ &\qquad - \cot\Big(\pi\big(\Delta t(\nu-\nu_0) - \frac{1}{N}\big)\Big) \\ &\qquad - \cot\Big(\pi\big(\Delta t(\nu-\nu_0) + \frac{1}{N}\big)\Big)\bigg)\cdot e^{i\left(\pi T(\nu-\nu_0)-\phi_0\right)} \\ &- \sin\left(\pi T(\nu+\nu_0)\right)\bigg(2\cot\left(\pi\Delta t(\nu+\nu_0)\right) \\ &\qquad - \cot\Big(\pi\big(\Delta t(\nu+\nu_0) - \frac{1}{N}\big)\Big) \\ &\qquad - \cot\Big(\pi\big(\Delta t(\nu+\nu_0) + \frac{1}{N}\big)\Big)\bigg)\cdot e^{i\left(\pi T(\nu+\nu_0)+\phi_0\right)}\Bigg] \end{aligned} \tag{2}$$

The numerical factor arises from the normalisation (periodogram and additionally Hann window). In the discrete Fourier transform (DFT), in order to create a discrete spectrum the result of the three steps in Table 1 is sampled in the spectral domain by

[2] The derivation will be published elsewhere.

Table 1

List of the components used to get the analytical expression of a sine with the properties described above

	Time domain	Frequency domain
Sine	$s(t) = A_0 \sin(2\pi\nu_0 t + \Phi_0)$	$S(\nu) = i\frac{A_0}{2}[\delta(\nu+\nu_0) - \delta(\nu-\nu_0)]e^{i\Phi_0\frac{\nu}{\nu_0}}$
Rectangle	$\Pi(\frac{t}{T} - \frac{1}{2})$	$Te^{-i\pi T\nu}\mathrm{sinc}(T\nu)$
Hann window	$\frac{1}{2} - \frac{1}{2}\cos(\frac{2\pi t}{T})$	$\frac{1}{2}\delta(\nu) - \frac{1}{4}\delta(\nu+\frac{1}{T}) - \frac{1}{4}\delta(\nu-\frac{1}{T})$
Dirac comb	$\frac{1}{\Delta t}\mathrm{III}(\frac{t}{\Delta t}) = \frac{1}{\Delta t}\sum_{k=-\infty}^{\infty}\delta(\frac{t}{\Delta t} - k)$	$\mathrm{III}(\Delta t \cdot \nu) = \sum_{k=-\infty}^{\infty}\delta(\Delta t \cdot \nu - k)$

By Fourier theory, the terms on the right have to be convolved. Note that only the expression for the rectangle does not consist of δ functions, so the overall result is a sum of infinite repetitions of the sinc functions. The δ functions of the Fourier transform of the Hann window are so close to each other ($\pm\Delta\nu = \pm\frac{1}{T}$) that summing up these three sinc functions (by convolution) results in a broadened peak at the frequencies $\pm\nu_0$

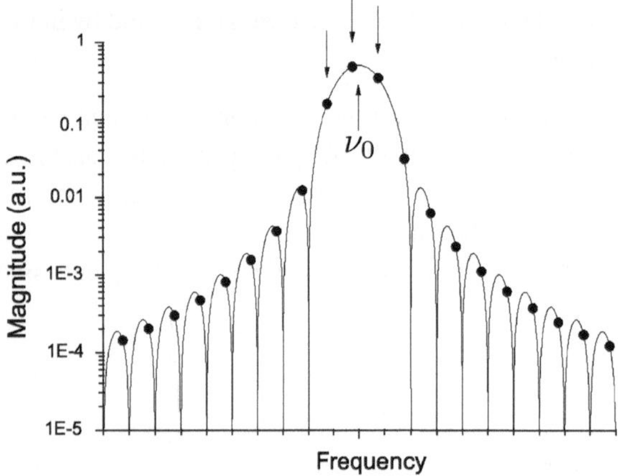

Figure 2
Plot of the absolute magnitude of $G(\nu)$ (Eq. 2) which is the theoretical spectrum of one sine, region around the peak. For a given sampling rate and number of samples the position of the discrete frequency values (*dots*) is fixed. The centre frequency ν_0 in function $G(\nu)$ determines the values at the discrete frequencies. One example: If ν_0 is exactly equal to one of the discrete frequencies, all magnitude values except for the highest three become zero

multiplication with a Dirac comb of spacing $\Delta\nu$ where $\Delta\nu = \frac{1}{T}$ is the inverse of the duration $T = N \cdot \Delta t$ with N the number of samples used as input, and Δt is the sampling period. This corresponds to convolving the time-domain result with the inverse transform, a Dirac comb of spacing T, producing infinite repetition that can also be seen as cyclic. If one takes Eq. (2) at the discrete frequencies $n \cdot \Delta\nu$, one gets the values of the discrete spectrum (Fig. 2). In a real recording several such contributions with different frequency, amplitude and phase are superposed, plus possibly additional contributions, for example from broadband noise. The additional sines need not be harmonics of a fundamental frequency.

2.2. *Fitting Peak Parameters to a Given Spectrum and Peak Subtraction*

First the discrete Fourier transform is applied to a section of the data because the algorithm works in the spectral domain. In order to find the relevant peaks, the next step is to find all the local maxima in the power spectrum and reject from them all the candidates that do not fulfil certain criteria. This is done to avoid false-positive fittings (especially in this experimental state of the algorithm where the thresholds are not yet optimised) and to save computing time by using the knowledge that the fitting will not be satisfactory.

- The line width must not exceed a certain value. It is defined as the sum of the magnitudes (absolute values)[3] of the highest value of a peak and its two neighbours divided by the magnitude of the highest value. The result depends on the position of the frequency relative to the grid of discrete frequencies and ranges from 2.0 to 2.2 for a monofrequent sine. For real signals we use an empirical threshold of 2.6 in order to tolerate noise and small frequency shifts which of course lead to line broadening
- Fitting candidates must be stronger than the general frequency-dependent background of the respective sensor and the specific background of the actual spectrum in the vicinity of the peak by certain factors (here 4.0 and 20 in power, respectively). The specific background is modelled by a

[3] Here and in the following we use "magnitude" in the mathematical, not the seismological sense.

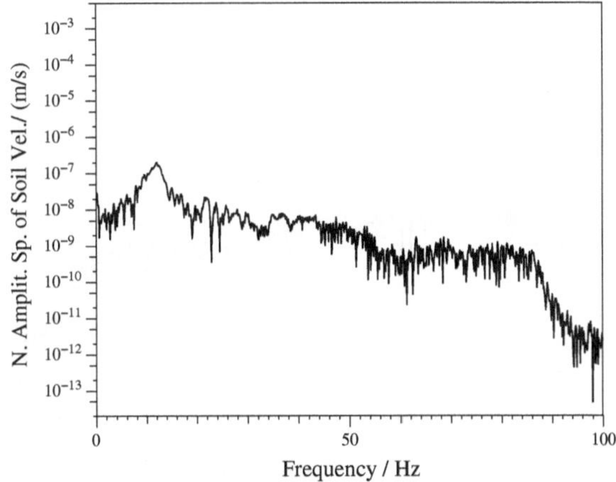

Figure 3
Magnitude spectrum of a seismic event in a coal mine

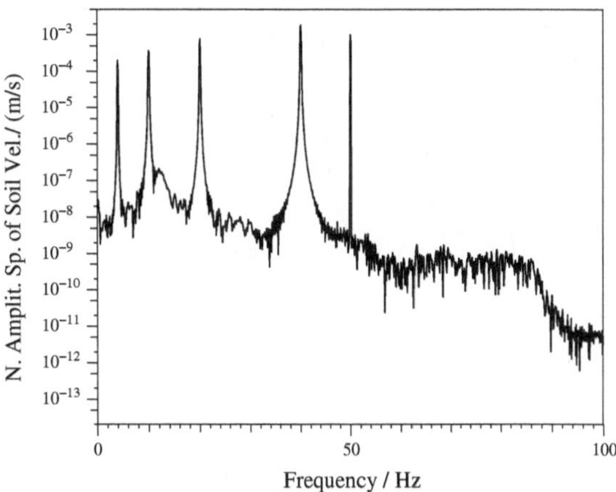

Figure 4
Magnitude of sum spectrum after superposing the complex spectra of Fig. 3 with artificial sines with amplitudes 10^3–10^4 times the amplitude of the seismic event. The peak at 50 Hz is narrow because the sine frequency nearly coincides with a discrete frequency

piecewise linear curve representing the valleys of the spectrum (here consisting of 20 parts)

- At each side, two discrete frequency values closest to the zero as well as to the Nyquist frequency are skipped to avoid complications with the local-maximum search and by mirrored peak components

After sorting the remaining candidates by amplitude, the algorithm is fed one by one with peaks of decreasing amplitude. Each single peak is mainly formed by three to four samples (Fig. 2). These samples have most of the power and thus are relatively least influenced by flanks of peaks at other frequencies or by the underlying signal. For each peak the three samples with highest spectral amplitudes are used. The expression $G(\nu)$ is fitted to their complex values by simultaneously varying the peak parameters amplitude A_0, frequency ν_0 and phase ϕ_0 by the non-linear Levenberg–Marquardt method (Press *et al.* 1989). Start values for the three parameters are gained from the three samples:

- The start amplitude is the sum of the highest three magnitudes of a peak times $\sqrt{2}$
- The start frequency is the weighted mean frequency of the peak. The three frequencies with the highest magnitude are multiplied with their corresponding magnitude values, summed up and divided by the sum of the three magnitude values
- The start phase is calculated as: $\frac{\pi}{2}$ plus the phase of the value with highest magnitude (the 2 π-arctangent of the complex number) minus $1.35 \cdot \pi \cdot a$

where a is the decimal fraction (non-integer part) of the start frequency in units of $\Delta\nu$.

These expressions have been derived from consideration of Eq. 2. The one for the phase was adapted according to the experiences made with many different spectra. The factor 1.35 was set after numerous tests; this is not the ideal solution, but it suffices for finding an adequate start value. After the fitting process the main criterion to decide whether or not a line has been fitted successfully is the normalised sum χ^2 of the squared deviations between the three complex samples of the discrete spectrum and the analytical expression. If a peak is to be subtracted, the complex spectrum of a peak with the gained parameters, computed with $G(\nu)$, is subtracted from the given spectrum, and the procedure is repeated with the next-highest peak, etc., until all candidates have been processed. Because the $G(\nu)$ spreads over the whole spectrum, peaks influence each other so criteria of neighbouring peaks might change after subtraction of a peak. This could result in peak candidates becoming valid which had been rejected before. To avoid this, one can redo all the steps to this point after a successful fitting and subtraction of a peak. Of course this greatly increases the processing time. For both examples in this work (the coal-mine-induced event with artificial lines and the helicopter overflight) the algorithm was restarted.

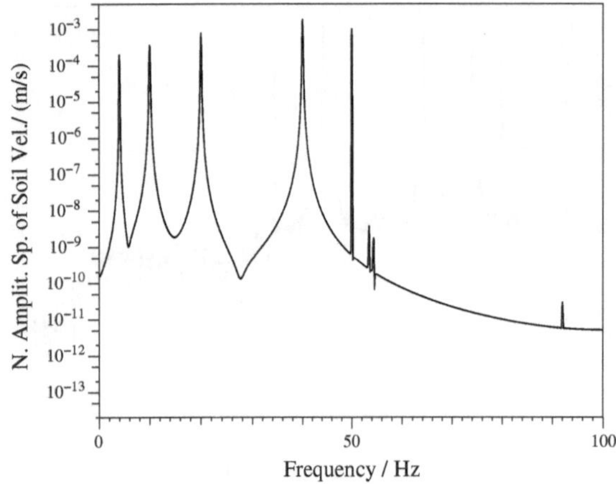

Figure 5
Magnitude spectrum of fitted and subtracted sines. Note that three sines were subtracted that had not been added before (53.4, 54.3, 92.0 Hz)

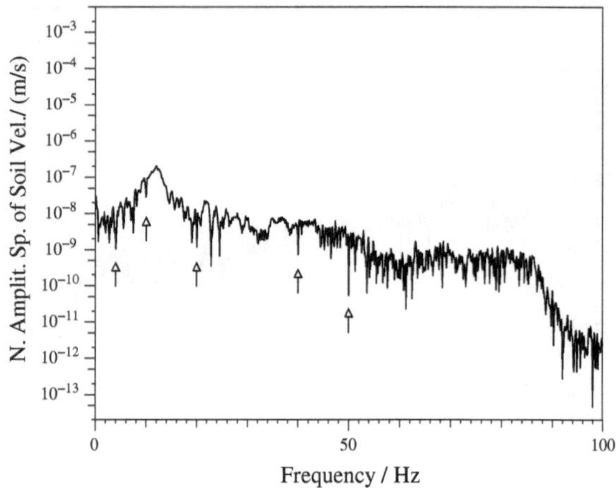

Figure 6
Magnitude spectrum remaining after peak fitting and subtraction from the superposed spectrum of Fig. 4. The frequencies of the artificial lines (respectively of the notches) are marked with *arrows*

The last step is to inversely transform the remaining spectrum back to the time domain and divide the result by the window function. If a longer set of time-domain data is to be cleaned of peaks, the process is carried out for several time intervals with some overlap; $\frac{1}{4}$ overlap proved to be appropriate to reduce margin errors from small window values. After spectrum processing and inverse Fourier transform the overlapping parts of neighbouring intervals are averaged using a sliding linear weighting scheme.

3. Results of the Peak Fitting and Subtraction Algorithm

3.1. Real Seismic Event Superposed with Artificial Sine Waves

As a first demonstration we take a coal-mine-induced event in the area of Hamm-Herringen, Germany (Bischoff *et al.* 2010)[4] as a model for a substantially non-periodic aftershock signal (spectrum Fig. 3, signal Fig. 7) and superpose it with artificial sine functions with different frequencies, amplitudes and phases as an example of periodic disturbances. The sampling rate for these data is 200 Hz and the segment length is only 15 s. To get a good spectral resolution we take a single spectrum with 2,048 samples, corresponding to 10.24 s of the time domain. By adding these pure sine functions (spectrum in Fig. 5) to the seismic event (spectrum in Fig. 3) one gets the signal and spectrum shown in Figs. 8 and 4, respectively. In this example, the sine amplitudes are 10^3–10^4 times the event amplitude (Table 2).

After addition of extremely strong artificial peaks, the seismic event is completely masked (see magnitude scales in Figs. 7, 8). The peak fitting results in sine parameters that are impressively close to the original ones (Table 2). As a consequence, their spectral contributions are subtracted nearly fully. The event again becomes the strongest component in the data, and its shape is reproduced well (Fig. 9). The remaining (periodic) differences from the original data (Fig. 7) are mainly caused by the spectral power which is lost in the notches at the peak positions (see Figs. 6, 10). The reasons for this are non-ideal peak parameters due to small contributions from other peaks, and of course by the underlying broadband signal. Lower magnitudes at certain frequencies in the broadband spectrum of an impulse event cause periodic disturbances similarly to higher ones. As the algorithm is fed with the superposed data, it has no

[4] Data kindly provided by Monika Bischoff and Sebastian Wehling-Benatelli (Institut für Geologie, Mineralogie und Geophysik; Ruhr-Universität Bochum). Measured: 1 February 2007, 19:18:28 UTC during the "HAMNET" acquisition period. Location depth is 1,056 m, distance between source and sensor 1,441 m, Mercalli intensity: −0.48, vertical component of a 3D-velocity sensor.

Table 2

Parameters of all lines (Fig. 5, left: input, right: found lines)

Input (exact)			Found		
Frequency (Hz)	Amplitude (mm/s)	Phase (rad)	Frequency (Hz)	Amplitude (mm/s)	Phase (rad)
4	0.5	0	3.9999997529	0.5000212055	−0.00001572
10	1	1	10.000016047	1.0001643250	0.99938260
20	2	2	20.000000164	2.0000225443	1.99999110
40	5	3	40.000000293	5.0000028117	2.99998852
50	2.5	4	49.999999850	2.4999975712	4.00000328
			53.404826315	9.0516943955e-006	4.38969166
			54.323507210	4.9913225312e-006	0.59328760
			91.996010077	7.7939926521e-008	0.35863423

The three lines at the bottom are categorised as valid lines and subtracted by mistake—these have not been added before and thus their subtraction from the spectrum leads to a manipulation of the data which should be avoided

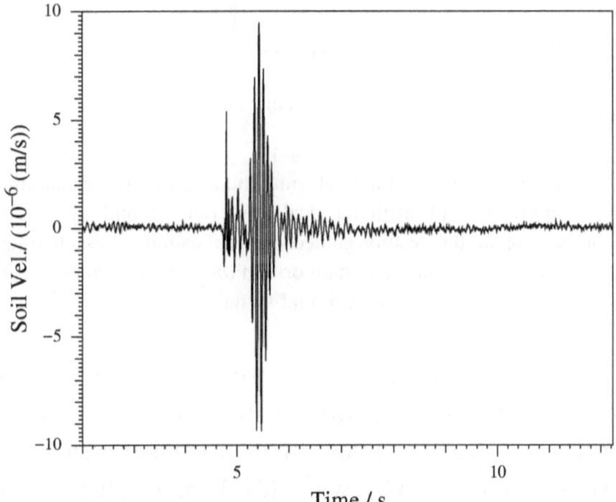

Figure 7
Seismic signal in the time domain (associated with the spectrum in Fig. 3). Note that the peak-to-peak value is $1.9 \cdot 10^{-5}\ \frac{m}{s}$

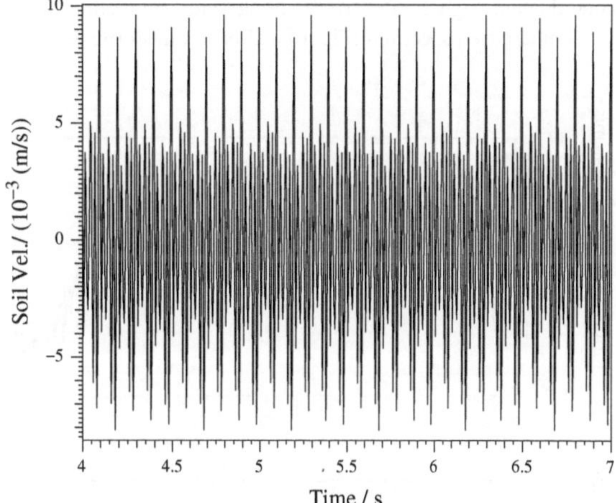

Figure 8
Time-domain signal after addition of artificial peaks (associated to spectrum of Fig. 4). Note that the peak-to-peak value is $1.8 \cdot 10^{-2}\ \frac{m}{s}$

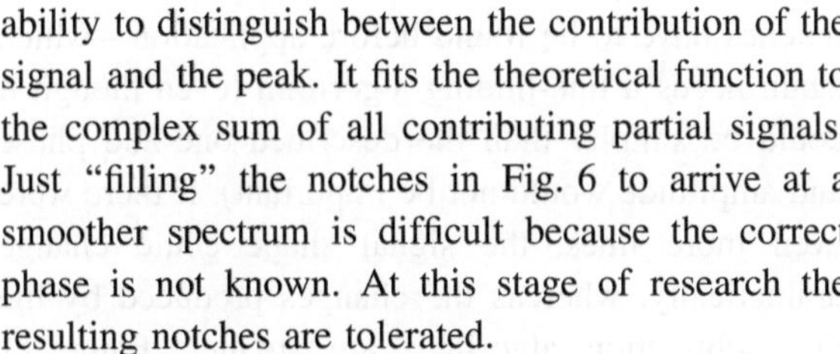

ability to distinguish between the contribution of the signal and the peak. It fits the theoretical function to the complex sum of all contributing partial signals. Just "filling" the notches in Fig. 6 to arrive at a smoother spectrum is difficult because the correct phase is not known. At this stage of research the resulting notches are tolerated.

3.2. *Comparison with Notch-Filtered Data*

If periodic noise with constant frequencies disturbs the data it is common to use notch filters to suppress it and thus to enhance the signal-to-noise ratio. As the amplitude of our artificial sines (Sect. 3.1) is strong, the values from spectral leakage even in the far flanks need to be suppressed (only at the 50 Hz sine is the width narrow, Fig. 4) the notches need to have a certain width. A compromise is to be made: On the one hand an increased width leads to a larger suppressed bandwidth which influences the original data, on the other hand if the width is too small, parts of the peak remain in the signal. The Blackman filter used provides only −74dB attenuation in the stopband—after a single application the peaks are markedly reduced but still remain significantly stronger than the surrounding frequencies (compare with peaks in Fig. 4, divide peak heights by ≈5,000). We carried out tests with different filter

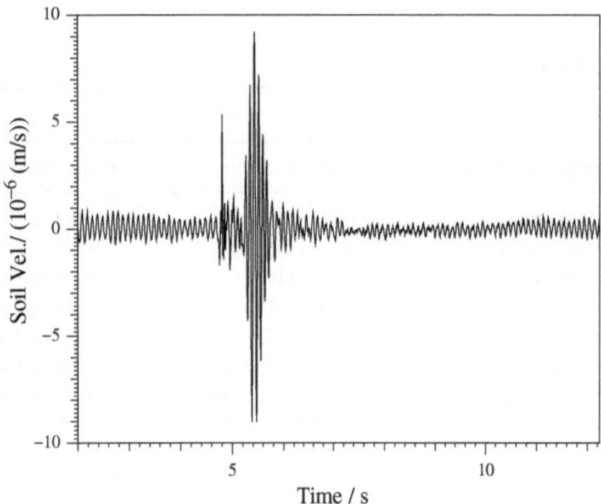

Figure 9
Time-domain signal after peak fitting and subtraction in the superposed spectrum (Fig. 6) and back transformation

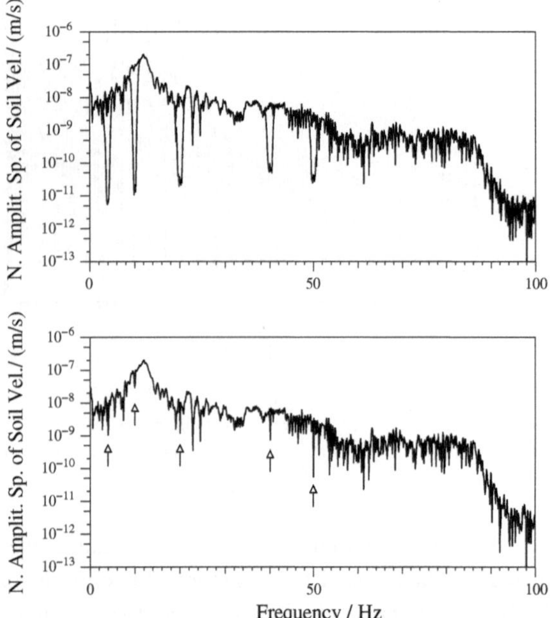

Figure 10
Spectra of the coal-mine-induced event (1,024 values from 2,048 signal samples) after addition and suppression of artificial peaks. Comparison of the results gained by application of notch filters (*top*) and the peak-subtraction algorithm (*bottom*), in both figures the *upper line* is the original spectrum (without artificial sines) which is substantially congruent except for frequencies close to the suppressed peaks. The filter notches are located exactly at the frequencies of the sines

kernels [between 1.2 and 2.4 Hz notch-filter width (between the two points where the magnitude is reduced to one half)] and single and double

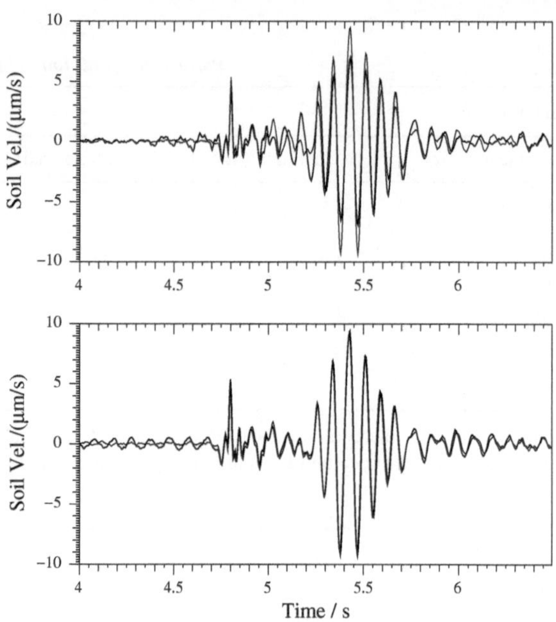

Figure 11
Time-domain signal of the coal-mine-induced event after addition and suppression of artificial peaks, section around the event. Comparison of the results gained by application of notch filters (*top*) and the peak-subtraction algorithm (*bottom*), the *thin lines* are the original signal

application. The best results were gained by double application of notch filters with a width of 1.6 Hz, that is 16.384 times the frequency step of $\Delta\nu \frac{1}{N\Delta t} = 0.0977$ Hz, with 801 kernel values i.e. 4 s—these are shown in Figs. 10 and 11.

It is possible that a better adjustment of the parameters of the notch filters could lead to further improvement of its performance. Nevertheless, this approach is less flexible as the frequencies of the notches have to be found before application—which again needs a line-finding algorithm (even though it could be simpler than the described one and phase and amplitude would not be important). If there were even more lines, the signal shape could change dramatically, whereas the changes produced by the line subtraction algorithm are smaller. Figure 11 shows that the P onsets are clearly reproduced by both approaches. The peak-subtraction algorithm lets the event shape largely unchanged; a small oscillation is added before the event. The notch filters change the signal shape more strongly: the surface wave starts to increase earlier, reaches lower amplitudes and decreases more slowly. In addition, a spurious signal of increasing amplitude appears before the P onset. With the line subtraction, on the other hand, the oscillation before the event is stationary. The influence of these changes on the characterisation of the event needs to be reviewed by an aftershock analyst and hence are not commented by us.

3.3. Real Seismic Helicopter Data

Helicopter data are good examples for showing how the algorithm behaves if real data are processed. The engine of a helicopter typically is a turbine which produces turbulent airflow, resulting in non-periodic signals. This turbine typically powers one main and one tail rotor, having a certain ratio of revolution rates and consisting of blades cutting the air which produces strong periodic signals.

Signals used in this section were recorded by us during a test measurement of the OSI Division of the CTBTO near Varpalota, Hungary, in September 2011.[5] Ground velocity was measured by 4.5 Hz geophones; the sample rate was 10 kHz. For this analysis we use $N = 8{,}192$ samples (that is, 0.8192 s for each spectrum). This is a tradeoff: on the one hand, large values mean good spectral resolution and thus good separation of neighbouring peaks; on the other hand, peak broadening increases if there is a frequency shift in this time interval which yet cannot be handled by the fitting algorithm. The spectrogram of 90 s of a helicopter overflight is shown in Fig. 12. Note that there is significant power (far) below 250 Hz, the Nyquist frequency applicable at the SAMS during OSI exercises of the CTBTO. In the beginning of the time interval shown, the helicopter approaches the sensor so the frequencies received (white horizontal lines) appear to be higher than corresponding to the revolution rates of the rotors and their harmonics (Doppler shift). The velocity component towards the sensor decreases and becomes negative after the helicopter passed the point of closest approach to the sensor, resulting in lower frequencies from the departing helicopter. At long distances the velocity component towards the sensor is almost constant so that the lines produced by the helicopter become horizontal again. Because of the long duration, an overlap between adjacent time intervals of $\frac{1}{4}$, that is 2,048 samples, is used. Figure 13 shows the frequencies and amplitudes of the successfully subtracted peaks for each spectrum. The resulting spectrogram is shown in Fig. 14. Figure 15 contains quantitative information about the spectral power that was reduced by peak subtraction. Shown is the relative difference in the spectral power sums before and after subtraction, normalised to the power sum of the original spectrum. As the power sum of a spectrum approximately equals the mean-square value of the signal in the time domain, this is a relative measure of the remaining signal strength in that time interval. Small values mean small differences in the total spectral power. As there are no circumstances known that the algorithm ever increases the power sum (by peak subtraction with wrong phase or whatever) this occurs if no peaks are subtracted. Except for the centre region, the mean-square value in the time domain is reduced by 70–90 %, corresponding to multiplication by a factor 0.3 to 0.1—that is, the root-mean-square value is reduced with a factor 0.5 to 0.3. Because this example has strong periodic content, there is much to subtract; remaining are mainly broadband noise and sines with smaller amplitude, except at the centre where the higher-frequency sines change frequency too rapidly to be represented by Eq. 2 and thus are not subtracted. Two figures are given to show extreme cases. Figure 16 corresponds to a time with very constant lines and shows good results. The rms value is reduced by a factor about 2.5, the peak-to-peak value by a factor 3.3. As the algorithm fits a monofrequent sine to the data, results get worse if frequencies change in time during the analysed interval, as shown in Fig. 17. Here the rms value is reduced with a factor 0.88, the peak-to-peak one with 0.93.

In comparison to the example with few artificial sines in Sect. 3.1, in the good example the peak-to-peak value has been reduced by only a factor 3.3, and the remaining amplitude is still relatively high. If only a simple amplitude criterion were used to detect

[5] The measurements were done for the Preparatory Commission for the Comprehensive-Nuclear-Test Ban Treaty Organization (CTBTO), Vienna, under Contract No. 2011-1260. We thank the On-Site Inspection Division of the CTBTO for the good co-operation.

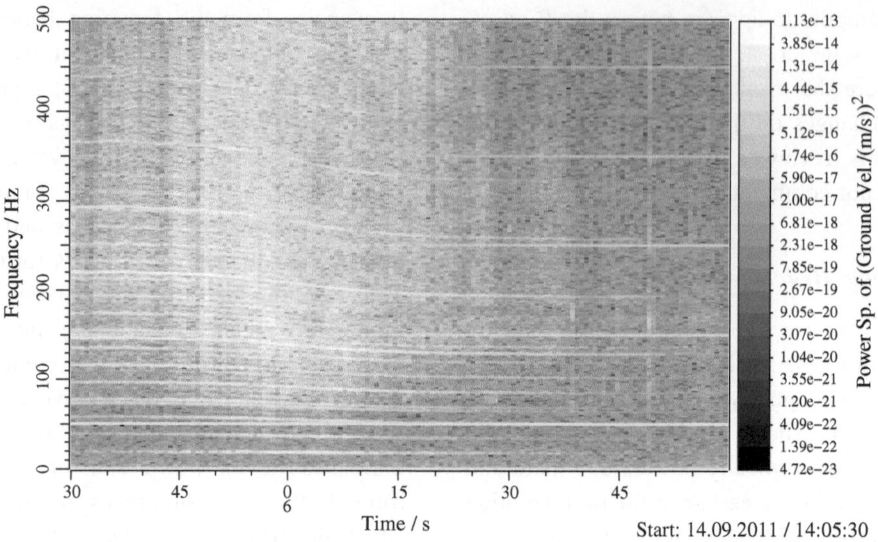

Figure 12
Sequence of 145 power spectra with 8,192 samples of a helicopter overflight. The sampling rate is 10 kHz, here the lowest 500 Hz are shown. There are three harmonic series of frequencies, one is mains-line hum probably produced by a nearby power generator (constant 50, 150, 250 Hz, etc.), one stems from the main rotor (multiples of around 6.0 Hz, every third is stronger) and one is from the tail rotor (fundamental frequency around 68 Hz). The variation of the Doppler shift, as the source moved with a velocity component towards the sensor, passed the point of closest approach (at about 14:06:00), and moved away from it again, is easy to recognise

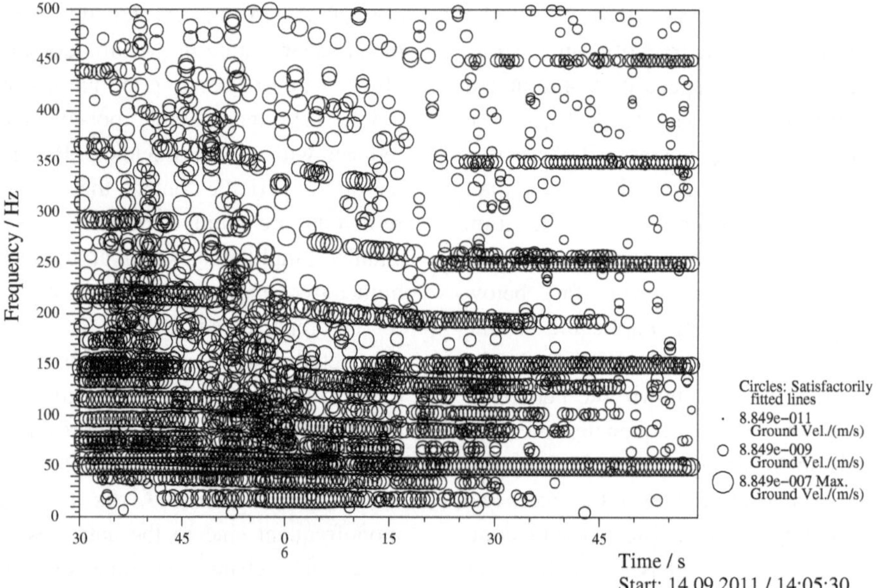

Figure 13
Sequence of results of the peak fitting and subtraction algorithm. Every *circle* stands for a successfully subtracted peak (with an amplitude of at least 10^{-4} times the amplitude of the strongest peak). Because the algorithm fits the theoretical function of a pure sine (which has a frequency that does not change in time) results get worse in times of changing frequencies. When the helicopter reaches its closest point to the sensor the Doppler shift passes through zero quickly and changes to lower frequencies afterwards. The line starting at 220 Hz shows a gap from 5:50 to 6:00. During these 10 s the frequency shifts by 9 Hz from 217 to 208 Hz. The line with the highest frequency that is reliably subtracted starts at 97 Hz and is shifted down to 93 Hz

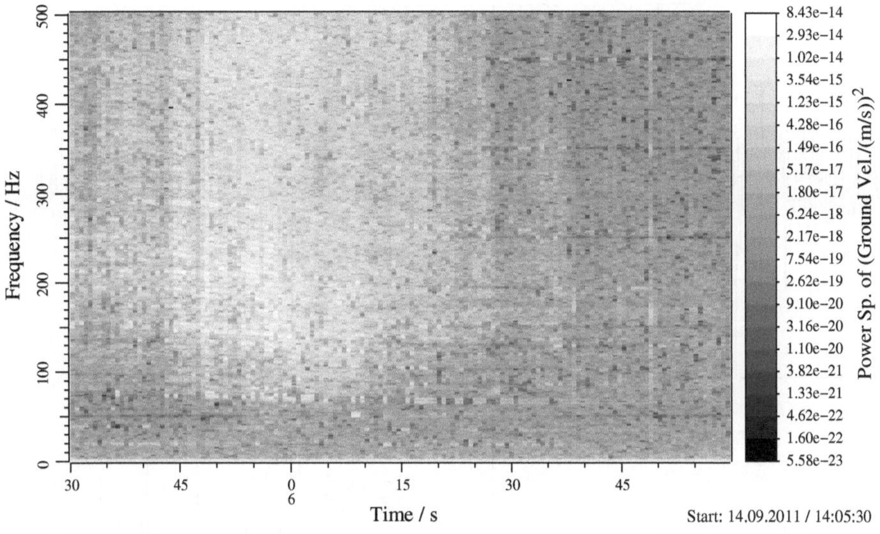

Figure 14
Same sequence as Fig. 12 but after subtraction of the peaks shown in Fig. 13. Most of the peaks are reduced successfully, at some regions one can even recognise that the *bright lines* changed to ones *darker* than the surrounding; these are the notches mentioned in Sect. 3.1 (see Fig. 6)

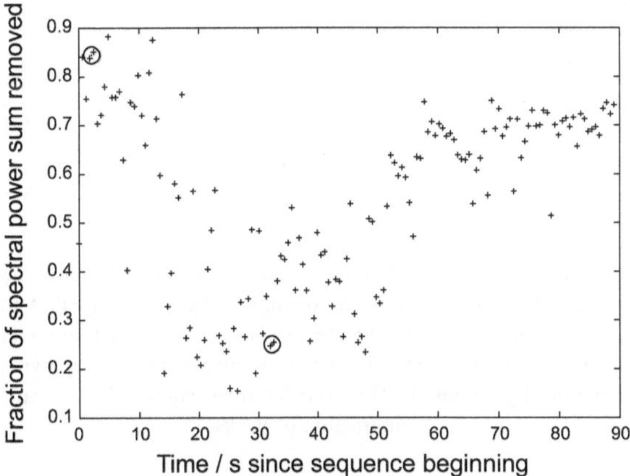

Figure 15
Degree of peak removal, same time interval as in Figs. 12– 14. The differences in the spectral power sums before and after subtraction are normalised to the power sum of the original spectrum. Values of 0.7–0.9 mean that ≈70–90 % of the mean-square value in the time domain is reduced, corresponding to a reduction by 0.5–0.3 in root-mean-square value. The time-domain data of the marked intervals are shown in the Figs. 16 and 17

a pulse masked in the original signal, the pulse would have to be relatively strong. Further research will look into improved procedures for peak removal. This example allows estimation of the tolerated frequency rate of change. As the frequency shift is proportional to the frequency, in times of changing frequencies the fitting gets worse for higher-order harmonics. Considering the line starting at 220 Hz, peak fitting is interrupted during 10 s. During that time its frequency shifts by 9 Hz which means (linear shift assumed) that a frequency shift in the order of $0.9\,\frac{\mathrm{Hz}}{\mathrm{s}}$ in this case (or generalised: $0.6\,\frac{\Delta\nu}{T}$) is too strong to be handled. On the other hand, the line with the highest frequency that is reliably subtracted starts at 97 Hz and is reduced in 10 s by 4 Hz, which means that the algorithm is able to handle frequency shifts of at least $0.4\,\frac{\mathrm{Hz}}{\mathrm{s}}$, respectively $0.3\,\frac{\Delta\nu}{T}$.

4. Discussion and Outlook

The peak-finding and peak-subtraction procedure described is a new method of cleaning signals and spectra from unwanted periodic noise. The algorithm fitting the analytical expression of a monofrequent peak to the maxima of a spectrum works well if the frequencies are not too close and do not change in time. If the peaks do not interfere with each other strongly, artificial sine functions added to real data and to pure noise can be subtracted successfully even if the sine amplitudes are higher by orders of magnitude. This is the case when the frequencies differ by more than about $10\Delta\nu$. In real helicopter signals, 70–90 % reduction of the mean-square value can be

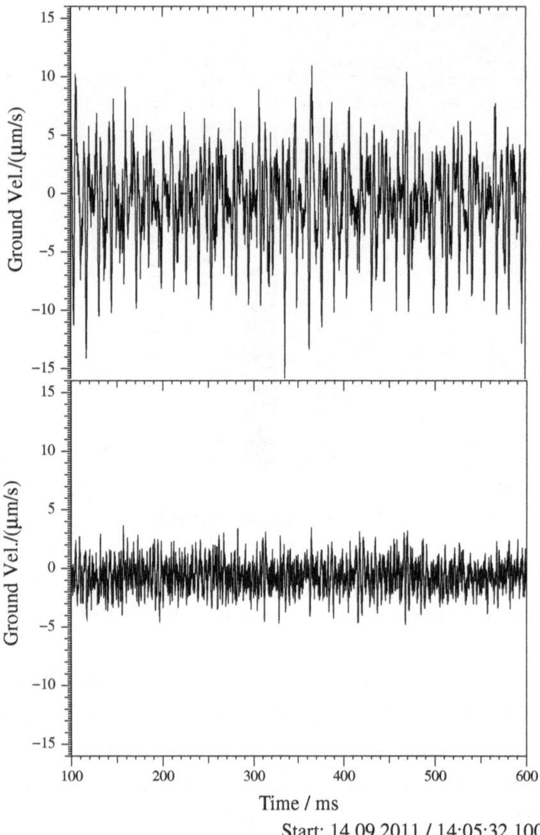

Figure 16
0.5 s of the time-domain signal of the helicopter overflight when the algorithm gives good results (computed by inverse FFT of the 4th and 5th spectrum with spectral-power-sum reduction factors 84 and 85 %, see Fig. 15, 2.1 s, corresponding to rms-value reduction by about 60 %). *Top* original signal, *bottom* after subtraction of peaks

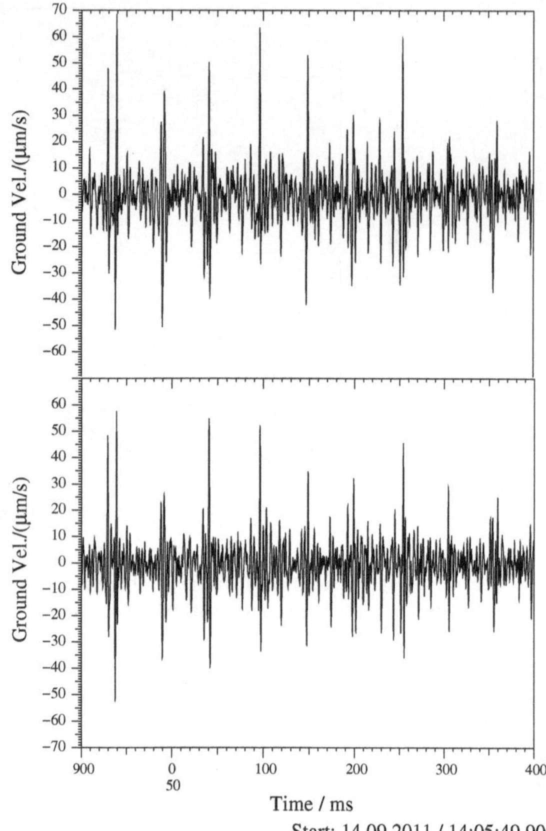

Figure 17
0.5 s of the time-domain signal of the helicopter overflight when the algorithm gives bad results (computed by inverse FFT of the 32th and 33th spectrum with spectral-power-sum reduction factors 22 and 21 %, see Fig. 15, 19.9 s, corresponding to rms-value reduction by about 12 %). *Top* original signal, *bottom* after subtraction of peaks

achieved. The remaining signal mainly consists of non-periodic components (probably produced by turbulent airflow) and of peaks strongly varying in time. If the frequency change stays below about 0.3 times the spectral frequency step during the time used for one spectrum, the theoretical expression represents the peak shape well enough and subtraction can succeed. On-going research is devoted to three problems:

- Including the small contributions of distant peaks by trying to optimise, in a second stage, the parameters of all found peaks simultaneously
- In case of close peaks, when the expression for a single one is no longer usable, optimise the parameters of two in parallel
- Investigate how cases can be handled when the frequency changes faster than the limit above, e.g., by time-varying Doppler shift

The procedure promises to become a flexible tool for removing unwanted periodic disturbances from seismic records, including the possibility to select which peaks to remove if some are of interest but they are disturbed by others.

At the moment our procedure works on a single time series. It is expected that expanding it to array signals will improve the results. The least that one expects is better differentiation between spectral peaks that occur consistently at several sites from those that are due to random fluctuations. Beyond that, signals cleaned from unwanted periodic disturbances should give better results in all forms of array processing.

Acknowledgments

We thank the OSI Division of the CTBTO for their good cooperation. Further thanks go to Sebastian Wehling-Benatelli (Institut für Geologie, Mineralogie und Geophysik; Ruhr-Universität Bochum, Germany) and Monika Bischoff (formerly Institut für Geologie, Mineralogie und Geophysik, Ruhr-Universität Bochum, Germany, now Bundesanstalt für Geowissenschaften und Rohstoffe, Hannover, Germany) for providing the data and the interpretation of the HAMNET project which was part of the Collaborative Research Centre 526 "Rheology of the Earth" at Ruhr-University Bochum and funded by the German Research Foundation (DFG). Finally we are grateful for the comments from two anonymous reviewers that contributed much to improving the article.

References

Jürgen Altmann and Felix Gorschlüter. *Removing Periodic Noise to Detect Weak Impulse Events*. Poster presented at International Scientific Studies (ISS 09) Conference, Vienna, 10–12 June, 2009. Available from http://www.ctbto.org/fileadmin/user_upload/ISS_2009/Poster/SEISMO-08J%20%28Germany%29%20-%20Jurgen_Altmann%20and%20Felix_Gorschluter.pdf (16. Dec. 2011).

Monika Bischoff, Alpan Cete, Ralf Fritschen and Thomas Meier. *Coal mining induced seismicity in the Ruhr Area, Germany*. Pure Appl. Geophys., *167*:63–75, 2010.

Peter Bormann. *Seismic signals and noise*. pages 1 – 34. Deutsches GeoForschungsZentrum GFZ, Potsdam, 2009.

E. Oran Brigham. *Fast Fourier Transform and Applications*. Prentice-Hall, Englewood Cliffs NJ, 2nd edition, 1988.

Burkhard Buttkus, editor. *Spectral Analysis and Filter Theory in Applied Geophysics*. Springer, 2000.

Steven M. Kay. *Modern Spectral Estimation: Theory and Applications*. Prentice-Hall, Englewood Cliffs NJ, 1988.

Alan V. Oppenheim and Ronald W. Schafer. *Discrete Time Signal Processing*. Prentice-Hall, Englewood Cliffs NJ, 2nd edition, 1999.

William H. Press, Saul A. Teukolsky, William T. Vetterling and Brian P. Flannery. *Numerical recipes in Pascal: the art of scientific computing*. Cambridge University Press, 1st edition, 1989.

Enders A. Robinson and Sven Treitel, editors. *Geophysical Signal Analysis*. Prentice-Hall, Englewood Cliffs NJ, 1980.

(Received January 5, 2012, revised October 8, 2012, accepted October 10, 2012, Published online January 1, 2013)

Pure Appl. Geophys. 171 (2014), 575–585

DOI 10.1007/s00024-012-0586-1

Pure and Applied Geophysics

The Influence of Spatial Filters on Infrasound Array Responses

David J. Brown,[1] Curt A. L. Szuberla,[2] David McCormack,[3] and Pierrick Mialle[1]

Abstract— A spatial filter is often attached to a microphone or microbarometer in order to reduce the noise caused by atmospheric turbulence. This filtering technique is based on the assumption that the coherence length of turbulence is smaller than the spatial extent of the filter, and so contributions from turbulence recorded at widely separated ports will tend to cancel while those of the signal of interest, which will have coherence length larger than the spatial dimensions of the filter, will be reinforced. In this paper, the plane wave response for a spatial filter with an arbitrary arrangement of open ports is determined. It is found that propagation over different port-to-sensor distances causes out-of-phase sinusoids to be summed at the central manifold and can lead to significant amplitude decay and phase delays as a function of frequency. The determined spatial filter plane wave response is superimposed on an array response typical of infrasound arrays that constitute the International Monitoring System infrasound network used for nuclear monitoring purposes. It is found that signal detection capability in terms of the Fisher Statistic can be significantly degraded at certain frequencies. The least-squares estimate of signal slowness can change by up to 1.5° and up to 10 m/s if an asymmetric arrangement of low and high frequency spatial filters is used. However, if a symmetric arrangement of filters is used the least-squares estimate of signal slowness is found to be largely unaffected, except near the predicted null frequency.

Key words: Infrasound, spatial filter, pipe filter, array response.

1. Introduction

Wind noise impinging on an acoustic sensor is dependent on the size of the turbulent eddy whose scale depends on the wind speed (see, e.g., Walker and Hedlin 2010). An arrangement of pipes distributed over a spatial area with a number of ports sampling the atmosphere and connected to a single common sensor can be used to mitigate against the effects of wind noise if the spatial scale of the filter system is larger than those of the turbulent eddy and smaller than those of the acoustic signal of interest. Walker and Hedlin (2010) provide a thorough review of the nature of acoustic noise due to wind turbulence and common mitigation procedures.

Although effective at reducing the contribution of turbulent wind noise on recordings taken from acoustic sensors, it has been established that spatial filters can also have a deleterious influence on the recorded signal of interest. Hedlin *et al.*, (2003) discuss the amplitude attenuation as a function of frequency that will occur as a result of out of phase summing at the sensor due to different propagation path lengths through the various ports. In addition, impedance mis-matches within the acoustic filter system generate internally reflected waves that cause resonances to occur (Alcoverro 2002). Indeed, Alcoverro (2002) using an analogue in electrical circuit theory shows that it is possible to determine the frequency response of a spatial filter system provided the design specifications are known accurately, however, the analysis presented in Alcoverro (2002) assumes vertically incident waves.

In this paper we determine how the plane wave response of a spatial filter impacts the infrasound array response in terms of signal detectability and slowness determination. It should be noted that the analysis presented here does not consider the internal resonances caused by impedance mis-matches internal to the acoustic filter system. It has been shown (Alcoverro 2002) that this resonance will also introduce significant phase shifts and needs to be

[1] International Data Centre, Comprehensive Nuclear-Test-Ban Treaty Organization, Preparatory Commission, Vienna International Centre, PO Box 1200, 1400 Vienna, Austria. E-mail: David.Brown@ctbto.org
[2] Geophysical Institute, University of Alaska Fairbanks, 903 Koyukuk Drive, Fairbanks, AK 99775-7320, USA.
[3] Geological Survey of Canada, Natural Resources Canada, 1 Observatory Crescent, Ottawa, ON K1A 0Y3, Canada.

considered in a complete analysis of the spatial filter response.

In Sect. 2 we derive the plane wave response for infrasound spatial pipe filter systems, and in Sect. 3 determine how it modifies the infrasound array response. Section 4 investigates the effect of the modified array response on the Fischer Statistic and hence the signal detectability, and Sect. 5 determines the effect on the measured back azimuth in terms of the least squares estimate. Section 6 determines the effect on the least squares estimate of signal slowness when low-frequency and high-frequency spatial filters are distributed asymmetrically among the sensors. In Sect. 7 the influence of the spatial filter system on measured signal slowness is simulated using artificial signals that have been contaminated with a realistic pink noise model.

2. *The Spatial Filter Plane Wave Response*

Elementary considerations show that the addition of two sinusoids $A_0 \sin(\omega(t-t_0))$ and $A_1 \sin(\omega(t-t_1))$ produces a third sinusoid $A \sin(\omega(t-\tau))$ such that

$$A_0 \sin(\omega(t-t_0)) + A_1 \sin(\omega(t-t_1)) = A \sin(\omega(t-\tau)) \tag{1}$$

with $A = \sqrt{(A_0 \cos\omega t_0 + A_1 \cos\omega t_1)^2 + (A_0 \sin\omega t_0 + A_1 \sin\omega t_1)^2}$ and $\tau = \frac{1}{\omega}\arctan\left(\frac{A_0 \sin\omega t_0 + A_1 \sin\omega t_1}{A_0 \cos\omega t_0 + A_1 \cos\omega t_1}\right)$.

When a monochromatic signal passes through the N-ports of a spatial filter each port presents to the sensor a copy of the signal with a phase shift when compared to the signal recorded on an unadorned (i.e., filterless) reference sensor at the same location. The phase shift is due to the time delay caused by the different propagation distances and speeds travelled by the signal from each port to the sensor when compared to that of the reference. When the time delay for each port has been determined the resultant signal at the sensor due to the summation of the signal from each port can be found by repeated use of the sinusoidal addition rule given in Eq. 1.

Consider a spatial filter consisting of 4 lengths of pipe in the form of a cross with high-impedance ports placed every metre along each pipe, as shown in Fig. 1. Assume also a monochromatic signal propagating in the $+x$ direction.

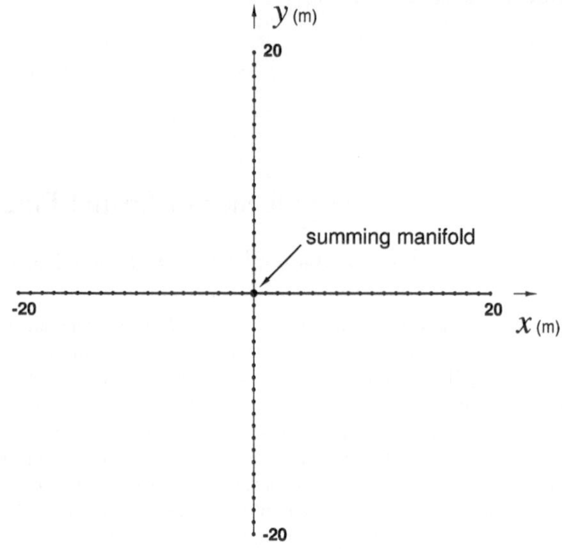

Figure 1
Spatial filter in the form of a cross with the summing manifold at the centre with a port placed every metre

Acoustic propagation through the spatial filter system consists of a through-air (TA) component and a through pipe (TP) component. We will assume here that all TA propagation is at the trace velocity V of the signal across the array, and all TP propagation is at a speed c determined by Kirchhoff's transmission line model (Kirchhoff 1868) for sound propagation in a cylindrical conduit, which is presented in Benade (1968) as Eq. 7. If we assume time zero to correspond to the arrival of a surface of constant phase at port 1, the port with the least value of the x-coordinate, then the time difference between the arrival of the surface at the sensor via the TP path to that of the reference is $\Delta t_1 = \frac{|x_1|}{V} - \frac{D_1}{c}$ where D_1 is the TP travel distance from port 1 to the sensor. More generally, the delay for a signal travelling via port k is

$$\Delta t_k = \frac{|x_1|}{V} - \left(\frac{D_k}{c} + \frac{|x_k - x_1|}{V}\right). \tag{2}$$

Kirchoff's analysis treats the propagation of acoustic waves inside a cylinder from the point of view of transmission line theory where a propagation constant $\Gamma(\omega) = \sqrt{(R + i\omega L)(G + i\omega C)}$ is defined as a function of angular frequency ω in the usual manner. Here, R and L are the real and imaginary parts of the impedance, and G and C the real and imaginary parts of the admittance.

For the acoustic problem under consideration these parameters are expressed in terms of the molecular properties of air and the radius a of the cylinder. The reader is directed to BENADE (1968) for the exact form of these expressions.

The attenuation constant α and phase constant β are determined according to the usual transmission line formalism as $\Gamma(\omega) = \alpha(\omega) + i\beta(\omega)$ with the phase velocity $c = \frac{\omega}{\beta}$.

Here, the attenuation constant α is defined to be $A_x = A_0 e^{-\alpha x}$, where A_0 is the amplitude at a reference point, and A_x is the amplitude a distance x from the reference.

Tabulated values of the attenuation constant α and phase velocity c as a function of frequency are shown in Table 1 for various pipe radii a, assuming a local sound speed of 343 m/s. These results show that the attenuation constant α is small enough in all cases to be considered to be zero for the remainder of this paper.

The amplitude and phase as a function of frequency for the filter shown in Fig. 1 has been computed and is shown in Fig. 2 for an internal pipe radius $a = 1$ cm. Here, the amplitude and phase are provided by Eq. 1, and the relative time delay given by Eq. 2.

The corresponding results for an internal pipe radius of 5 cm is shown in Fig. 3.

Unexpectedly the radius 1.0 cm and radius 5.0 cm curves show good agreement. In both cases for a 0° incident wavefront the phase decreases steadily to around minus 60° at 5 Hz where it stays constant out to 10 Hz, the 45° incident wavefront exhibiting only a fairly small deviation in the phase information. A predicted amplitude attenuation occurs for increasing frequencies, being more pronounced for the 0° incident wavefront above 5 Hz. In both cases, a predicted amplitude attenuation of around 60 % occurs at 5 Hz. The effects are less marked in all cases as the trace velocity increases.

A spatial filter configuration commonly used at infrasound arrays is the rosette-style filter shown in Fig. 4.

The plane-wave response for both the high-frequency (18 m diameter) and low-frequency (70 m radius) configurations is shown in Fig. 5 and are similar to those discussed in HEDLIN *et al.*, (2003).

The response for the 18 m filter shows a linear phase shift with frequency, being the same for all incident trace velocities, achieving around −70° at 5 Hz. The amplitude attenuation is fairly innocuous, being around 90 % at 5 Hz, and deceasing as the

Table 1

Acoustic velocity c and attenuation constant α as a function of frequency for sound propagation in a cylinder for three values of the internal radius a

Frequency (Hz)	$a = 1.0$ (cm)		$a = 2.0$ (cm)		$a = 5.0$ (cm)	
	c (m/s)	α (/m)	c (m/s)	α (/m)	c (m/s)	α (/m)
0.01	109.5	0.00005	192.0	0.00002	274.5	0.00001
0.03	173.6	0.00008	251.1	0.00003	298.2	0.00001
0.07	225.5	0.00011	277.3	0.00004	312.0	0.00001
0.1	243.3	0.00012	285.4	0.00005	316.6	0.00002
0.3	278.9	0.00017	306.4	0.00007	327.2	0.00003
0.7	296.9	0.00023	318.0	0.00011	332.5	0.00004
1.0	303.4	0.00027	321.8	0.00013	334.2	0.00005
3.0	318.8	0.00045	330.4	0.00022	337.8	0.00009
7.0	326.7	0.00067	334.6	0.00033	339.6	0.00013
10.0	329.3	0.00080	336.0	0.00039	340.2	0.00015

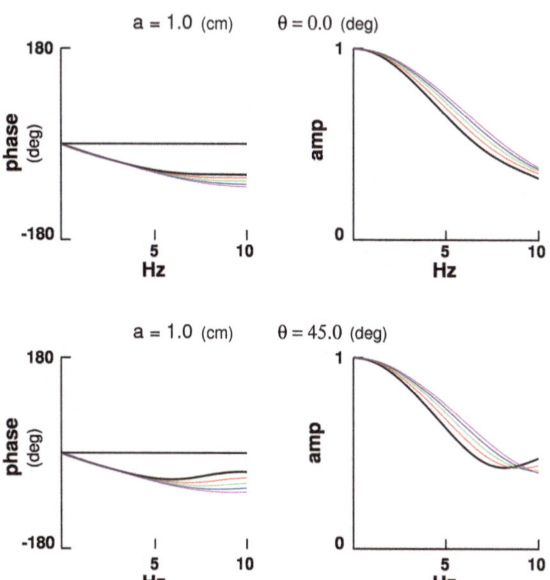

Figure 2
Phase and amplitude variation as a function of frequency for the spatial filter shown in Fig. 1 with internal pipe radius = 1 cm. Results for five values of the trace velocity V are indicated: *black line* $V = 340$ m/s; *red line* $V = 390$ m/s; *green line* $V = 440$ m/s; *blue line* $V = 490$ m/s; *pink line* $V = 540$ m/s, and for two values of the incident angle $\theta = 0.0°$ and $\theta = 45.0°$ from propagation in the $+x$ direction

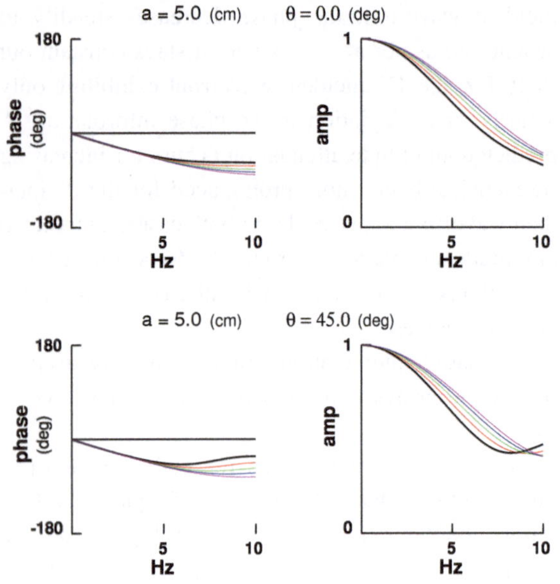

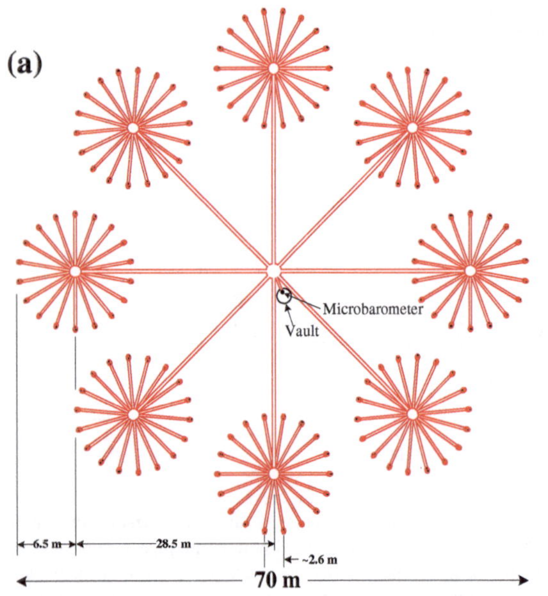

Figure 3
Phase and amplitude variation as a function of frequency for the spatial filter shown in Fig. 1 with internal pipe radius = 5 cm. Results for five values of the trace velocity V are indicated: *black line* $V = 340$ m/s; *red line* $V = 390$ m/s; *green line* $V = 440$ m/s; *blue line* $V = 490$ m/s; *pink line* $V = 540$ m/s, and for two values of the incident angle $\theta = 0.0°$ and $\theta = 45.0°$ from propagation in the $+x$ direction

trace velocity increases. The presence of the 70 m filter is significantly distorting the signal. A 180° phase shift and complete amplitude annulment is observed at the characteristic frequency, which is defined by the TA propagation speed divided by the diameter of the spatial filter, which for 340 m/s propagation speed is 4.85 Hz.

3. Modified Array Response

Inclusion of phase and amplitude information of the kind presented in Sect. 2 will likely alter the array response. In this section we determine the modified array response function for an array with spatial filters attached to the sensors.

It can be shown (see, e.g. Kennett 2002) that a plane wave with slowness **s** and angular frequency ω impinging on an infrasound or seismic array with N sensors is modulated by the array response function

$$S(\mathbf{p}-\mathbf{s},\omega)=\sum_{j=1}^{N}e^{i\omega(\mathbf{p}-\mathbf{s})\cdot\mathbf{r}_j} \tag{3}$$

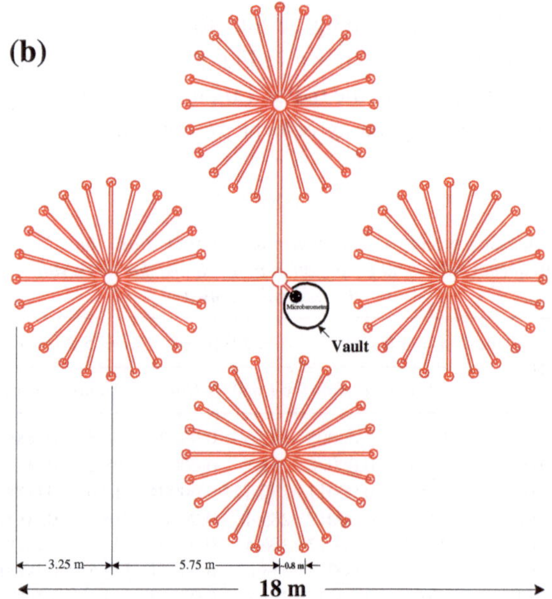

Figure 4
The 'rosette' spatial filter design in common use at International Monitoring System (IMS) style infrasound arrays (Christie 1999). **a** Low frequency 144 port 70 m diameter filter. **b** High frequency 96 port 18 m diameter filter

where $\mathbf{r}_j$ is the position vector of the jth sensor, and $\mathbf{p}$ is a point in the slowness plane, so that for the broadband process $\psi(\mathbf{x},t) \sim \int_{\omega_1}^{\omega_2} A(\omega)e^{i\omega(t-s\cdot\mathbf{x})}d\omega$, the normalized signal power P as a function of slowness is given by Eq. 4.

$$P(p,s) = \left| \int_{\omega_2}^{\omega_1} A(\omega)S(p-s,\omega)d\omega \right|^2 \tag{4}$$

For the present analysis we assume that $A(\omega)$ is described by the gaussian function $A(\omega) = e^{-(\omega-\omega_0)^2}$ about a central frequency ω_0, thus avoiding purely monochromatic signals, which may tend to prematurely drive an array into spatial aliasing as frequency is increased. Furthermore, we assume that amplitude spectrum is discretely sampled by a finite number of frequency pickets $\omega_m = \omega_0 2^{\frac{m}{M}}$ for $m = \ldots, -2, -1, 0, 1, 2, \ldots$ where M is the number of frequency pickets per octave. In this study we assume that $\omega_m = \omega_0 2^{\frac{m}{9}}$ for $m = -2, -1, 0, 1, 2$.

With these constraints, the normalized power as a function of slowness is given by

$$P = \left| \sum_{m=-2}^{2} A(\omega_m) \sum_{j=1}^{N} e^{i\omega(\mathbf{p}-s)\cdot r_j} \right|. \tag{5}$$

Application of spatial filters of the kind considered in Sect. 2, requires the amplitude correction $B(\omega)$ and phase shift $\varphi(\omega)$ be included in the expression for the array response. This now becomes the modified array response

$$\hat{S}(p-\mathbf{s},\omega) = \sum_{j=1}^{N} B_j(\omega)e^{i[\omega(p-s)\cdot\mathbf{r}_j-\phi(\omega)]} \tag{6}$$

The modified array response given by Eq. 6 has been computed for the test array shown in Fig. 6. This array consists of eight sensors in the form of a low-frequency outer triangle with an inverted high frequency triangle in the centre such that the two inner most sensors are co-located. The low-frequency sensors are equipped with a 70 m rosette filter and the high-frequency sensors are equipped with a 18 m rosette filter of the sort displayed in Fig. 4.

Figure 7 shows the array response (Eq. 3) for an un-adorned array compared to the modified array

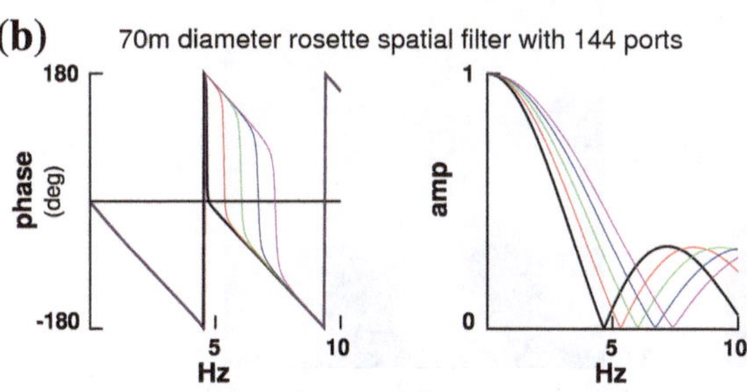

Figure 5
Phase and amplitude variation as a function of frequency for the rosette filters shown in Fig. 4. **a** 18 m diameter. **b** 70 m diameter. Results for five values of the trace velocity V are shown: *black line* $V = 340$ m/s; *red line* $V = 390$ m/s; *green line* $V = 440$ m/s; *blue line* $V = 490$ m/s; *pink line* $V = 540$ m/s), propagation is in the $+x$ direction

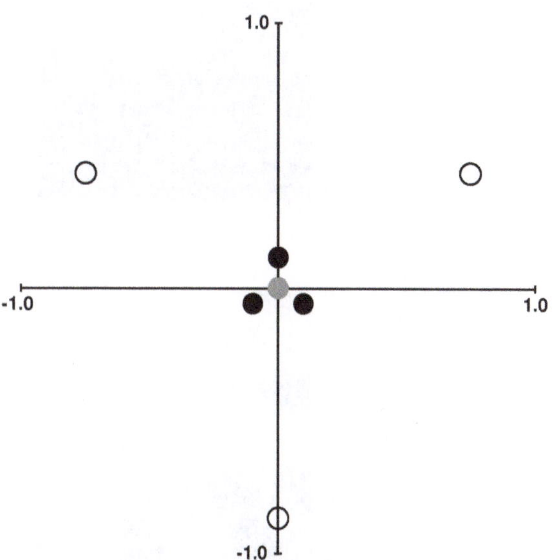

Figure 6
Theoretical eight-element infrasound array with two types of spatial filter, *circles* indicate sensor locations. *Open circles* indicate a 70 m diameter 144-port rosette low-frequency spatial filter. *Black shaded circles* indicate 18 m diameter 96 port rosette high frequency spatial filter. *Grey shaded circles* indicate co-located sensors, one with a small rosette and one with a large rosette. Indicated distances are in km

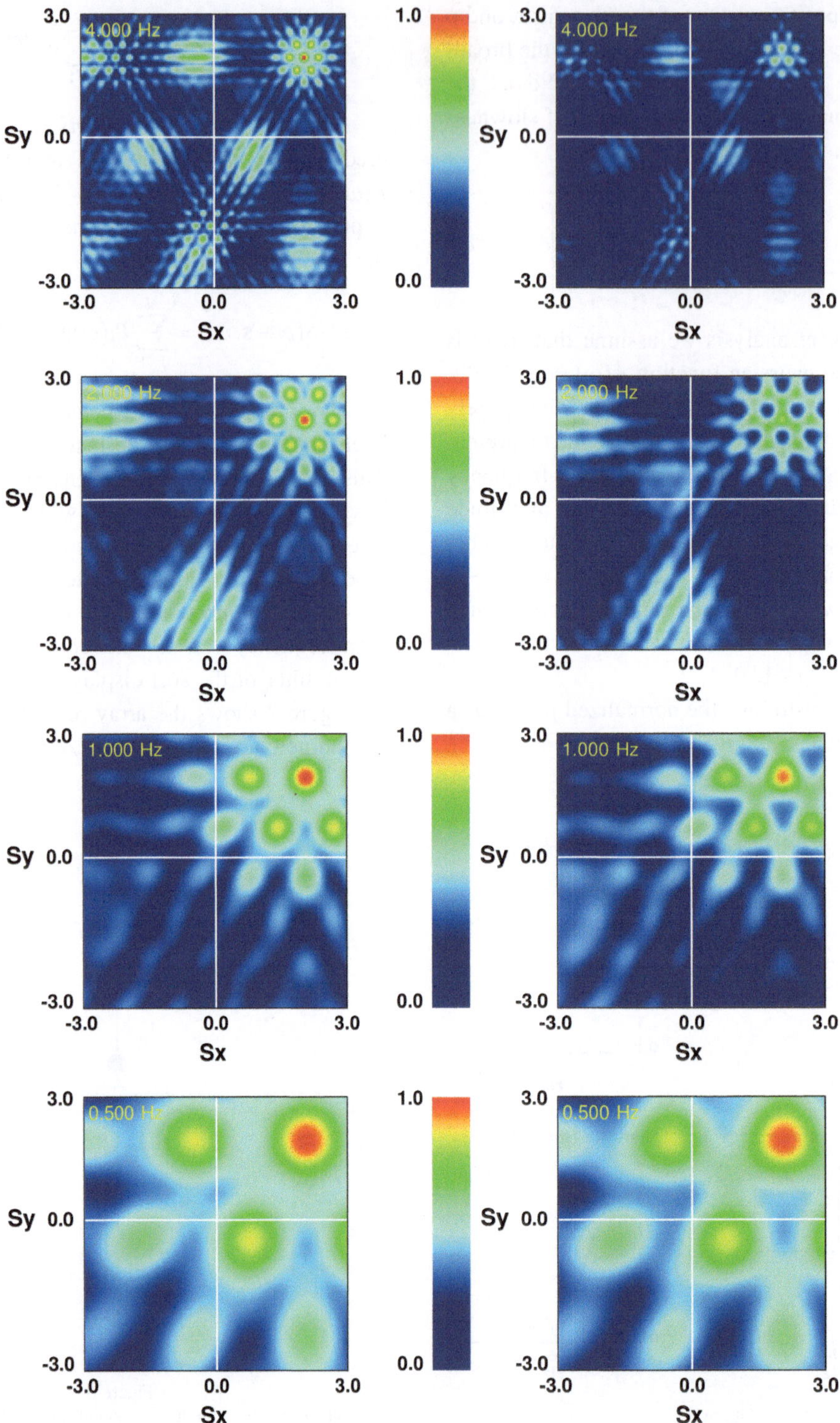

Figure 7

Unmodified (*left panels*) and modified (*right panels*) normalized array power as a function of slowness for various frequencies for the test array shown in Fig. 6. The signal is assumed to be arriving from the northeast (i.e., 45° from North) at the local sound speed, i.e., there is no vertical component

response (Eq. 6) for the arrangement sensors and filters for the test array in Fig. 6.

As expected, the modified array response has had a fairly significant effect. This is largely due to the presence of the 70 m spatial filters. The reduction in amplitude is obvious at higher frequencies. The reduction in main lobe amplitude compared with the sidelobes can be significant at higher frequencies, and this is likely to drive the array into spatial aliasing at lower frequencies. To more properly investigate the affect on the array response we need to determine the affect on the measured Fischer Statistic (F-stat) and the least-squares estimate of signal slowness and its variance.

4. *Effect on Measured F-stat*

An un-adorned or filterless array will cause our test signal ψ to register an infinite F-stat, since by definition ψ represents a purely plane wave. To prevent such infinities from occurring we introduce a slight distortion of the wavefront such that the array response function becomes the modified form

$$S(p-s,\omega)=\sum_{j=1}^{N}\mathrm{e}^{i\left[\omega(p-s)\cdot\mathbf{r}_j+\theta_j\right]} \quad \text{where } \theta_j=2\pi M\times \mathrm{rand}()_j \tag{7}$$

where rand() is a random number between 0 and 1, and M is a multiplier used to set the measured F-stat to a predetermined value, such as 100, for the frequency dependent incident waveform.

The F-stat in the frequency domain can be determined using Shumway's formula (Shumway 1971)

$$F(p,s)=(N-1)\frac{P(p,s)}{P_T-P(p,s)} \tag{8}$$

where $P_T=\left|\int_{\omega_2}^{\omega_1}A(\omega)\mathrm{d}\omega\right|^2$ is the total signal + noise power.

For the array and filter configuration shown in Fig. 6, and the plane wave ψ impinging upon it, the measured F-stat is modified by the spatial filters as shown in Table 2.

For the theoretical infrasound array and spatial filter arrangement shown in Fig. 6 the predicted affect on the measured F-stat can be significant for signals above 1.0 Hz, although the effect seems to be more pronounced for more coherent signals with higher F-stat; the less coherent signals are not affected to the same extent. Fortunately the broadband character and generally lower frequency nature of the signals of interest for which the International Monitoring System (IMS) arrays are designed will reduce somewhat the deleterious effect of the spatial filters. These results are of course dominated by the inclusion of the 70 m spatial filters, which was never intended for use at the higher frequencies.

Table 2

The effect of spatial filters on theoretical F-stat values for various frequencies

Frequency (Hz)	F-stat (no spatial filter)	F-stat (spatial filter)
4.0	100.0	2.5 ± 0.1
	50.0	2.5 ± 0.2
	20.0	2.3 ± 0.4
	10.0	2.0 ± 0.3
	5.0	2.0 ± 0.2
2.0	100.0	8.9 ± 1.6
	50.0	9.0 ± 1.6
	20.0	8.2 ± 2.2
	10.0	5.4 ± 1.5
	5.0	4.3 ± 1.4
1.0	100.0	32.5 ± 9.7
	50.0	24.2 ± 8.0
	20.0	15.0 ± 3.4
	10.0	8.9 ± 2.3
	5.0	4.9 ± 1.3
0.5	100.0	68.9 ± 20.0
	50.0	40.4 ± 10.2
	20.0	17.9 ± 3.5
	10.0	10.0 ± 1.5
	5.0	5.3 ± 0.7
0.1	100.0	98.1 ± 8.2
	50.0	49.4 ± 3.1
	20.0	19.9 ± 0.9
	10.0	10.0 ± 0.3
	5.0	5.0 ± 0.2

The uncertainty is determined by taking 20 random numbers in Eq. 7 each of which cause the desired F-stat in the 'no spatial filter' case (column 2) to be achieved and then taking the standard deviation of the determined F-stat in the case in which a spatial filter was applied (column 3)

5. *Modification to the Least Squares Estimate of Signal Slowness*

Assuming our array of n-sensors is distributed in 'd-dimensions', where d is either 2 or 3, we can construct a $N \times d$ matrix X of unique inter-sensor separations. A vector τ_o of observed inter-sensor time delays corresponds to the arrival of a plane wave with d-dimensional vector slowness s through the relationship

$$\tau_o = Xs + \varepsilon \tag{9}$$

where $\boldsymbol{\varepsilon}$ is the error in measuring the time-delays.

The solution in $\mathbf{s}$ is found by minimizing the error $R^2 = \varepsilon^T\varepsilon$, where T indicates matrix transposition.

The least-squares solution to Eq. (9) is found to be (see, e.g., Rao 1973)

$$\hat{s}_o = C^{-1}X^T\tau_o \tag{10}$$

where $C = X^TX$.

In order to accommodate the inclusion of spatial filters a correction factor $\boldsymbol{\mu}$ needs to be subtracted from the vector of observed inter-sensor time delays, so that the 'correct' vector $\boldsymbol{\tau}_C = \boldsymbol{\tau}_O - \boldsymbol{\mu}$ is determined. Note that if the same spatial filter is applied to each sensor then the vector $\boldsymbol{\mu} = \mathbf{0}$ because the same phase delay is applied to each sensor implying that the inter-sensor delays remain unchanged.

With spatial filters attached to each sensor Eq. 10 implies that the least-squares estimate on slowness is given by $\hat{s}_o = C^{-1}X^T(\tau_c + \mu)$, from which the correct slowness vector can be determined to be

$$\hat{\mathbf{s}}_C = \hat{\mathbf{s}}_O + \delta\mathbf{s} = \hat{\mathbf{s}}_O - C^{-1}X^T\boldsymbol{\mu} \tag{11}$$

A Cramer–Rao lower bound estimate of the variance in $\boldsymbol{\tau}_O$ is given by (Szuberla and Olson 2004)

$$\hat{\sigma}^2_{\tau_o} = \frac{\tau_o^T(I - R)\tau_o}{N - r} \tag{12}$$

where $R = XC^{-1}X^T$, r is the rank of R, and I is the identity matrix. Eq. 11 can be written in terms of the corrected time delays as

$$\begin{aligned}\hat{\sigma}_{\tau_O} = \hat{\sigma}_{\tau_C} + \boldsymbol{\mu}^T(I - R)\boldsymbol{\tau}_C + \boldsymbol{\tau}_C^T(I - R)\boldsymbol{\mu} \\ + \boldsymbol{\mu}^T(I - R)\boldsymbol{\mu}\end{aligned} \tag{13}$$

It can be seen that without accommodating the time delays due to the spatial filters the variance on $\boldsymbol{\tau}$ is being over estimated by the factor

$$A = \boldsymbol{\mu}^T(I - R)\boldsymbol{\tau}_C + \boldsymbol{\tau}_C^T(I - R)\boldsymbol{\mu} + \boldsymbol{\mu}^T(I - R)\boldsymbol{\mu} \tag{14}$$

The affect on least-squares estimate of slowness for signals recorded on an infrasound array with the sensor and spatial filter configuration shown in Fig. 6 can be estimated via Eqs. 11 and 14. An internal pipe diameter of 1-cm is assumed for the pipe filters in the following analysis.

The array shown in Fig. 6 (assumed to be two-dimensional), being an eight-element array, has 28 possible sensor pairs implying both the vector $\boldsymbol{\tau}_o$ of observed inter-sensor time delays and the correction vector $\boldsymbol{\mu}$ is 28-dimensional. Since the array consists of two groups of four sensors with the same spatial filter applied, the vector $\boldsymbol{\mu}$ contains 12 elements that are identically zero and 16 identical non-zero elements. The non-zero element of $\boldsymbol{\mu}$ is shown in Table 3 as a function of frequency together with components of the slowness correction vector $\delta\mathbf{s}$.

The correction vector $\delta\mathbf{s}$ is found to be negligible at all frequencies and so the change in measured back azimuth and trace velocity will be of no consequence, and since the vector $\boldsymbol{\mu}$ occurs in each term on the right-hand side of Eq. 14 it also follows that the magnitude of the correction A to the estimate of the variance $\sigma_{\boldsymbol{\tau}}$ to the vector $\boldsymbol{\tau}$ will also be negligible.

6. Asymmetric Array

It may perhaps be anticipated that the null result is due to the symmetric nature of the arrangement of the

Table 3

Estimates of the maximum element of the time correction vector $\boldsymbol{\mu}$ to the observed inter-sensor time delay vector, and the correction δs to the least squares estimate of slowness as a function of frequency for the test array shown in Fig. 6, and the asymmetric counterpart shown Fig. 8

Freq (Hz)	max{μ} (s)	$\delta\mathbf{S}$ (s/km) symmetric array (Fig. 6)	$\delta\mathbf{S}$ (s/km) asymmetric array (Fig. 8)
0.1	0.106	$\begin{bmatrix} 9.3 \times 10^{-7} \\ \approx 0 \end{bmatrix}$	$\begin{bmatrix} 3.5 \times 10^{-2} \\ 6.1 \times 10^{-2} \end{bmatrix}$
0.5	0.089	$\begin{bmatrix} 7.8 \times 10^{-7} \\ \approx 0 \end{bmatrix}$	$\begin{bmatrix} 2.9 \times 10^{-2} \\ 5.1 \times 10^{-2} \end{bmatrix}$
1.0	0.086	$\begin{bmatrix} 7.5 \times 10^{-7} \\ \approx 0 \end{bmatrix}$	$\begin{bmatrix} 2.8 \times 10^{-2} \\ 4.9 \times 10^{-2} \end{bmatrix}$
2.0	0.083	$\begin{bmatrix} 7.2 \times 10^{-7} \\ \approx 0 \end{bmatrix}$	$\begin{bmatrix} 2.7 \times 10^{-2} \\ 4.7 \times 10^{-2} \end{bmatrix}$
4.0	0.081	$\begin{bmatrix} 7.1 \times 10^{-7} \\ \approx 0 \end{bmatrix}$	$\begin{bmatrix} 2.7 \times 10^{-2} \\ 4.6 \times 10^{-2} \end{bmatrix}$

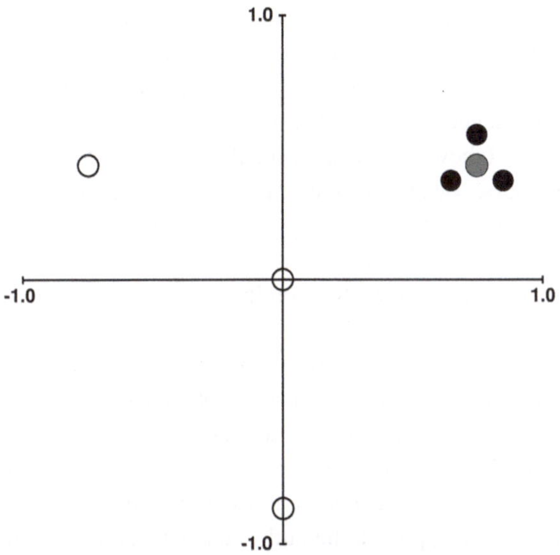

Figure 8
Theoretical eight-element infrasound array with two types of spatial filter, *circles* indicate sensor locations. *Open circles* indicate a 70 m diameter 144-port rosette low-frequency spatial filter. *Black shaded circles* indicate 18 m diameter 96 port rosette high frequency spatial filter. *Grey shaded circles* indicate co-located sensors, one with a small rosette and one with a large rosette. Indicated distances are in km

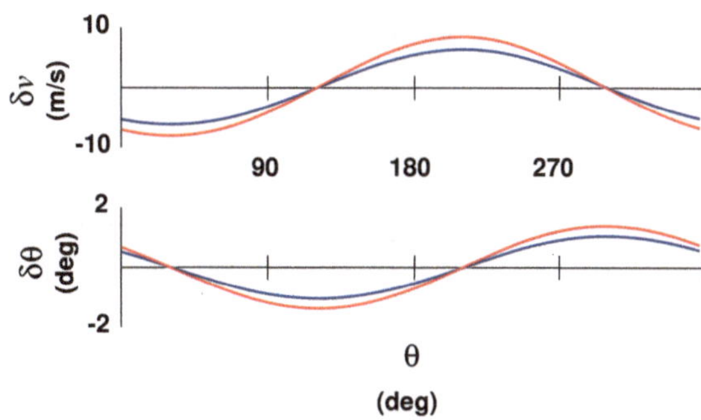

Figure 9
Estimated change in the least squares estimate of wave speed δv and back azimuth $\delta\theta$ as a result of applying the asymmetric arrangement of low and high frequency spatial filters shown in Fig. 8 (*red line* 0.1 Hz, *blue line* 4.0 Hz). A wave speed of 343 m/s was assumed

spatial filters in the array, i.e. the small-radius filters placed symmetrically around the centre sensor may reduce the bias in the measured slowness. A second simulation has been performed in which the high-frequency, small-radius spatial filters are placed around one of the corner sensors, as shown in Fig. 8.

The results are shown in Table 3. The change in the slowness correction vector is significantly larger this time. The change in the back azimuth, $\delta\theta$, and change in measured speed, δv as a function of signal back azimuth, θ, are shown in Fig. 9. A speed of propagation of 343 m/s was assumed.

With this asymmetric arrangement of spatial filters a maximum deviation of 1.5° in the least squares estimate of back azimuth is predicted, together with a corresponding maximum deviation of 10 m/s in the least squares estimate of wave speed.

7. Simulation

In order to test the theoretical predictions, of Sect. 5, a simulation of slowness estimation was

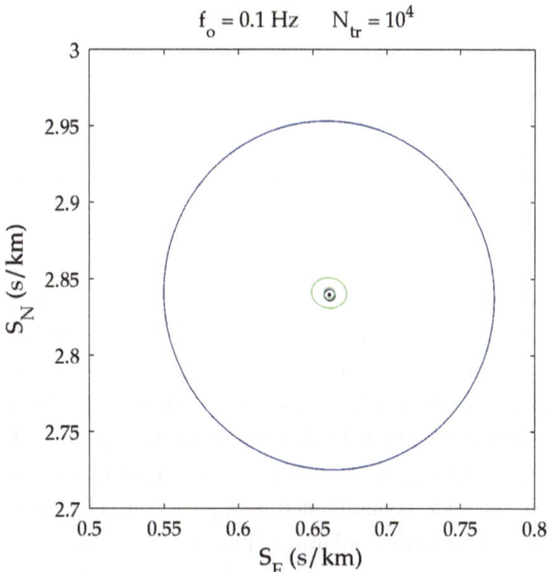

Figure 10
Simulation results for wave packets at 0.1 Hz. The nominal slowness, corresponding to 13.1° back azimuth at 343 m/s trace velocity is shown in *black*. The bare, seven-sensor array is shown in *blue*, while the eight-element array equipped with spatial filters is shown in *green*. Confidence ellipses are calculated at the 95 % level for 10^4 trials at 10 dB SNR, as described in the text

arranged for the array depicted in Fig. 6. An ensemble of 10^4 quasi-monochromatic wave packets corresponding to the frequencies given in Table 1 was prepared. These packets comprised roughly seven full oscillations in a Hanning-tapered window. The packets were then phase shifted to correspond to a planar arrival, from 13.1° back azimuth with a trace velocity

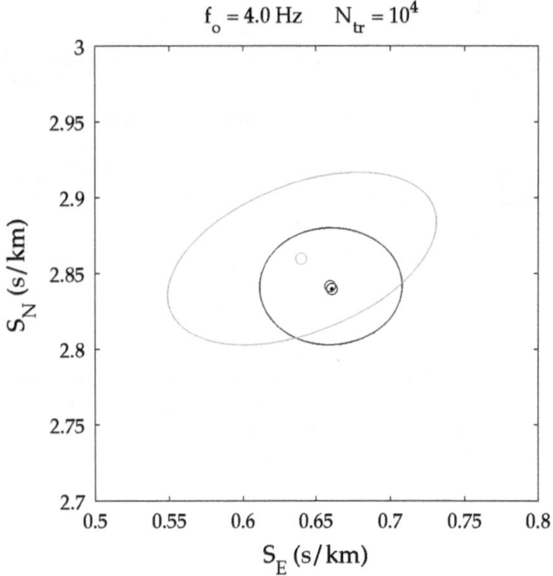

Figure 11
Simulation results for wave packets at 4.0 Hz. The nominal slowness, corresponding to 13.1° back azimuth at 343 m/s trace velocity is shown in *black*. The bare, seven-sensor array is shown in *blue*, while the eight-element array equipped with spatial filters is shown in *green*. Confidence ellipses are calculated at the 95 % level for 10^4 trials at 10 dB SNR, as described in the text

of 343 m/s, at each inlet port or sensor. This set of arrival parameters was selected because it is not a function of any symmetry in neither the array nor pipe inlet geometries. Each packet was then sampled at 20 Hz (consistent with CTBT-type infrasound arrays). To simulate noise akin to a realistic infrasound station, each example was contaminated with pink noise (1/f amplitudes, uniformly randomized phases) at 10 dB SNR.

The array shown in Fig. 6 was treated in two different respects: (a) with no pipe array, such that there were seven bare sensors (the co-located center elements would be identical) exposed to the noisy packets, and (b) simulating the effects of the pipe array by further phase shifting the packet at each inlet to correspond with its propagation down the pipes ($a = 1.0$ cm was used here) at speeds consistent with Table 1. Each sensor response in the latter case was built by summing respective inlet-phase-shifted packets. For each of the two treatments and frequencies, a 95 % confidence ellipse was constructed for the slowness estimates of the noisy packets. These results are shown in Figs. 10 and 11, for two of the frequencies given in Table 3.

In Fig. 10 we show the results for packets at 0.1 Hz. The results at relatively low frequencies such as this are consistent with our intuition, in that the confidence ellipse for the array equipped with spatial filters (shown in green) is smaller than that of the bare sensors, owing to the incoherent noise reduction afforded by the summing of each inlet's signal at the sensor. The results are also largely unbiased at this relatively high SNR (nominal slowness shown in black).

The results for 4 Hz, shown in Fig. 11 are markedly different, and perhaps not predicted so well by the theory for this configuration of sensors and inlets. The slowness deviation is of order 10^{-2} s/km. The advantage of incoherent noise reduction is largely lost as we approach the null frequency, as seen by the nearly equal areas of each ellipse. In this case, the inlet-equipped result also exhibits a pronounced bias in slowness that would correspond with 12.6° back azimuth and 341 m/s trace velocity estimates. The uncertainties in these estimates are still relatively small, of order 1° and 1 m/s, so other effects (e.g., winds between the source and array) may easily dominate.

8. Conclusions

The plane wave response for several spatial filter designs, typical of those in use at IMS infrasound stations, has been determined. The analysis incorporated internal acoustic velocities that were specified by the Benade (1968) analysis for sound propagation in a conduit. The determined amplitude and phase response as a function of frequency were incorporated in the usual array response for an arrangement of infrasound sensors. It was found that the incorporation of spatial filters at infrasound stations can have a significant deleterious affect on the signal detectability in terms of the measured Fischer statistic, particularly at frequencies above 2 Hz. Signals with dominant frequency below 1 Hz have only a minor reduction in their detectability. The incorporation of mixed spatial filter types, i.e., spatial filters of different size, and so different phase delays, at infrasound sensors in the same array, can have significant impact on the measured back azimuth and signal speed in term of the least-squares estimate if an asymmetric arrangement of spatial filters is used. However, if a symmetric arrangement of spatial filters is used then there is little impact on the least-squares estimate of signal slowness. Simulation of semi-realistic signals shows that some effects are more pronounced than predicted. The uncertainties arising from these effects are still relatively small compared to other, naturally-caused, perturbations in signal parameter estimation.

Acknowledgments

Part of this work was carried out when the first author was a visiting scientist at the Geological Survey of Canada, Natural Resources Canada in 2002.

Disclaimer The views expressed in this paper are those of the authors and do not necessarily reflect those of the Preparatory Commission.

References

ALCOVERRO, B. 2002. *Frequency Response of Noise Reducers*. Proc. of the Infrasound Technology Workshop, De Bilt, The Netherlands, 28–31 October.

BENADE, A.H. 1968. *On the Propagation of Sound Waves in a Cylindrical Conduit*. J. Acous. Soc. Am. *44* No. 2, pp. 616-623.

CHRISTIE, D.R. 1999. *Wind-noise reducing pipe arrays*. Report IMS-IM-1999-1. International Monitoring System Division, Comprehensive Nuclear-Test-Ban Treaty Organization, Vienna, Austria, 22 pp.

HEDLIN, MAH, ALCOVERRO, B, and D'SPAIN, G. 2003. *Evaluation of rosette infrasonic noise-reducing spatial filters*. J. Acoust. Soc. Am. *117*, 1880–1888.

KIRCHHOFF, G. (1868) *On the influence of heat conduction in a gas on sound propagation*. Ann, Phys. Chem. *134*, 177–193.

KENNETT, B.L.N. 2002. *The Seismic Wavefield Volume II: Interpretation of seismograms on regional and global scales*. Cambridge University Press.

RAO, C. R. 1973. *Linear Statistical Inference and Its Applications*. Wiley, New York. 625 pp.

SHUMWAY, R.H. 1971. *On detecting a signal in N stationarily correlated noise series*, Technometrics, *13*, 499.

SZUBERLA, C.A.L and OLSON, J.V. 2004. *Uncertainties associated with parameter estimation in atmospheric infrasound arrays*. J. Acoust. Soc. Am. *115* No. 1. pp. 253–258.

WALKER, K.T, and HEDLIN, MAH. 2010. *A Review of Wind-Noise Reduction Methodologies*, in Infrasound Monitoring For Atmospheric Studies. Springer, New York. pp. 141–182.

(Received September 30, 2011, revised August 20, 2012, accepted August 24, 2012, Published online September 21, 2012)

Pure Appl. Geophys. 171 (2014), 587–597

DOI 10.1007/s00024-012-0576-3

Pure and Applied Geophysics

Sources of Error Model and Progress Metrics for Acoustic/Infrasonic Analysis: Location Estimation

STEPHEN ARROWSMITH,[1] DAVID NORRIS,[2] ROD WHITAKER,[1] and DALE ANDERSON[1]

Abstract—How well can we locate events using infrasound? This question has obvious implications for the use of infrasound within the context of nuclear explosion monitoring, and can be used to inform decision makers on the capability and limitations of infrasound as a sensing modality. This paper attempts to answer this question in the context of regional networks by quantifying current capability and estimating future capability using an example regional network in Utah. This example is contrasted with a sparse network over a large geographical region (representative of the IMS network). As a metric, we utilize the location precision, a measure of the total geographic area in which an event may occur at a 95 % confidence level. Our results highlight the relative importance of backazimuth and arrival time constraints under different scenarios (dense vs. sparse networks), and quantify the precision capability of the Utah network under different scenarios. The final section of this paper outlines the research and development required to achieve the estimated future location precision capability.

Key words: Infrasound, event location, nuclear explosion monitoring.

1. Introduction

Infrasound has been used for decades as a sensor modality to monitor atmospheric nuclear tests. Within the last ten years, the deployment of the international monitoring system (IMS) infrasound network under the comprehensive nuclear-test-ban treaty (CTBT), in addition to successes from prototype regional networks, is highlighting the additional value of infrasound as a supplemental tool for monitoring underground tests. With the drive towards monitoring low yield tests at regional distances, infrasound has the potential to play a key role in identifying mining blasts and other manmade events that become prevalent at these low seismic magnitudes. In addition, infrasound can provide constraints on source depth, which can be critical for reliably estimating the yield from seismic data and for source identification purposes.

Within the context of treaty monitoring, it is important to communicate to decision makers the capability of different technologies. Recent papers (LE PICHON *et al.* 2009; GREEN and BOWERS, 2010) have quantified the detection capability of the IMS infrasound network in terms of yield thresholds. However, there has not yet been any attempt to quantify what is achievable with regional networks of infrasound arrays, or to formally quantify localization uncertainty. With the recent deployment of a prototype infrasound network in Utah, it is possible to provide a preliminary assessment of the capability. While the capability will depend on the network configuration (e.g., number and density of stations), which will differ from network to network, there is value in providing an assessment of the Utah network for decision makers as an example scenario. The results will both quantify the current state-of-art capabilities in localization, as well as project what might be achievable in the future by leveraging research advances.

The purpose of this paper is to outline a formal approach for assessing network location precision for decision makers. Previous studies at IMS scales have assessed location accuracy using some measure of the azimuthal gap. However, in treaty monitoring applications the question of more relevance is how large a geographic area a given test is constrained within. For onsite inspection activities, areas of up to 1,000 km^2 are permitted. We discuss a methodology for estimating location precision for a regional infrasound

[1] Los Alamos National Laboratory, Los Alamos, USA. E-mail: sarrowsmith@gmail.com

[2] Applied Physical Sciences, Arlington, VA 22203, USA.

network, then illustrate both existing capability and an estimate of future capability using the Utah infrasound network as an example. Finally, we outline the research needed to move from the first to the second scenario.

This paper uses the Bayesian infrasound source location (BISL) (MODRAK *et al.*, 2010) to formally assess location precision. The advantage of BISL as compared with other techniques for this purpose is that both measurement error and model error are formally accounted for, and prior constraints can be readily folded into the solution. As a forward method, BISL contrasts with the standard inverse approach that is used for infrasound location (BROWN *et al.*, 2002a; CERANNA *et al.*, 2009). Such inverse techniques are based on Geiger's method (GEIGER, 1912). For using azimuthal information in addition to arrival time constraints, BRATT and BACHE (1988) outline a formulism that has been used by researchers in the infrasound community (CERANNA *et al.*, 2009). BROWN *et al.* (2002b) and EVERS *et al.* (2007) implemented methods where bearing intersections were weighted by the sine of the angle of intersection. An innovative approach for event location that uses a space–time approach has also been developed (SZUBERLA and ARNOULT, 2011; SZUBERLA *et al.*, 2009) but is not considered in this study because it is applicable for near-source recordings and not to events recorded at regional and global distances.

2. Methodology

2.1. The Bayesian Infrasonic Source Locator: A Recap

MODRAK *et al.* (2010) introduced the Bayesian infrasonic source location (BISL) method for localizing infrasound events using a regional network of infrasound arrays. BISL uses both arrival time and back azimuth constraints to provide credibility bounds (analogous to confidence bounds in frequency statistics) on event location. This probabilistic approach accounts for uncertainties in both measurement and model error relevant to the infrasound location problem. We can represent the uncertainties associated with each parameter by:

$$\sigma(\theta)_{\text{total}} = \sigma(\theta)_{\text{measured}} + \sigma(\theta)_{\text{model}}$$
$$\sigma(\phi)_{\text{total}} = \sigma(\phi)_{\text{measured}} + \sigma(\phi)_{\text{model}}$$

where, θ represents back azimuth and ϕ represents arrival time.

As a crude estimate of the total uncertainty in back azimuth, $\sigma(\theta)_{\text{total}}$, we can make no correction for back azimuth deviation due to wind bias, and simply resort to empirical observations of back azimuth deviations to define the uncertainty. As a first order enhancement we can expect that propagation modeling would allow us to correct the azimuths for wind bias and reduce the corresponding model uncertainty. For arrival times, we may assume that phase is unknown and different at each array; therefore $\sigma(\phi)_{\text{model}}$ should be sufficiently large to capture the range of possible arrival times at the arrays. Similarly, as a first-order enhancement, we can utilize a distance-dependent probability density function on group velocity, with additional levels of sophistication that may include azimuthal and temporal dependence. Such an approach is currently being explored by MARCILLO *et al.* (2012). These choices and level of sophistication affects the subsequent model error.

These uncertainties are implemented in the calculation of the likelihood, assuming Gaussian distributed errors, over all possible locations (latitude, longitude), origin times (t_0), and group velocities (v). Following MODRAK *et al.* (2010), the likelihood is calculated from:

$$P(d|m) = \prod_{i=1}^{n} \Theta_i(\theta_i|m)\Phi_i(t_i|m)$$

where,

$$\Theta_i(\theta_i|m) = \frac{1}{\sqrt{2\pi\sigma_\theta^2}} \exp\left[-\frac{1}{2}\left(\frac{\gamma_i}{\sigma_\theta}\right)^2\right]$$
$$\Phi_i(t_i|m) = \frac{1}{\sqrt{2\pi\sigma_\phi^2}} \exp\left[-\frac{1}{2}\left(\frac{\varepsilon_i}{\sigma_\phi}\right)^2\right],$$

represent the individual likelihood components for the backazimuths and arrival times, respectively. With $d_i = d_i(x_0, y_0, x_i, y_i)$ as the distance from each hypothetical source to the i'th array, and for Cartesian coordinates, the residual terms are:

$$\gamma_i \equiv \theta_i - \arctan\left(\frac{y_i - y_0}{x_i - x_0}\right)$$
$$\varepsilon_i \equiv t_i - \left(t_0 + \frac{d_i}{v}\right).$$

Location solutions are computed from the multivariate posterior probability distribution of location parameters. A uniform prior on the group velocity is used in our initial development of BISL (that is, the group velocity can vary over some range, typically chosen to represent the possible range of infrasonic group velocities at a particular distance scale). The limitation of the initial development, as outlined in MODRAK *et al.* (2010), is the assumption that the group velocity is the same for each array in the model. In practice, different phases will typically be observed in different directions due to wind, which introduces anisotropy not accounted for in our preliminary model. At present, we account for the model anisotropy through the standard deviation in arrival time, which is implemented in the likelihood equations. As long as this standard deviation is set sufficiently large to capture the possible range of arrival times for different phases, the model error is adequately accounted for. Future development of BISL will include construction of source-path specific priors for group velocity including priors based on atmospheric predictions of group velocity with associated errors. These improvements should better account for model error through explicitly accounting for anisotropy and by incorporating different priors to different stations (this is discussed in the last section of this paper).

2.2. *A Location Metric*

Event location capability should ideally be estimated by the precision of a location estimate (this is the area of uncertainty of the location ellipse or polygon). If both measurement and model errors are adequately estimated, the location ellipse or polygon should enclose the actual event location on most occasions (the credibility or confidence level at which the ellipse or polygon is estimated should define the percentage of times the solution will not enclose the location). Previous studies of infrasound location capability, which have focused on the IMS network, have represented location capability by an estimate of the accuracy based on the azimuthal separation between arrays for a given source location (e.g., LE PICHON *et al.*, 2009; GREEN and BOWERS, 2010). Location accuracy (expressed in km) can be differentiated from precision (expressed in km^2). Within the context of the CTBT, where onsite inspection activities can occur inside an area of 1,000 km^2, precision is a more useful measure (assuming, of course, that the model uncertainties are appropriately defined).

Using BISL as our location technique, precision can be calculated in a straightforward manner by calculating the area enclosed by a location polygon at a specified credibility. These polygons are derived from the posterior distribution of location parameters, where the posterior integrating constant is derived with numerical integration, with the initial model formulation. In this study, we use a credibility of 0.95 for calculating the polygons, because this corresponds to a typical level at which confidence regions are calculated. Three different scenarios are considered: (1) a best case scenario where each event is detected at every array, (2) a 'typical' northern hemisphere winter scenario where the stratospheric wind jet blows towards the east, and (3) a 'typical' northern hemisphere summer scenario where the stratospheric wind jet blows towards the west. For scenarios 2 and 3 we assume that no arrivals are observed at distances greater than 200 km in the counter-wind direction (i.e., no stratospheric arrivals are observed in the counter-wind direction). However, within 220 km, tropospheric arrivals are observed 50 % of the time, regardless of direction. Although these assumptions are simplistic, the resultant maps provide a more realistic assessment of the location capability than the simulation shown in Fig. 1, and also provide a sense of the effect of seasonal variations.

3. *The Utah Infrasound Network*

The Utah infrasound network was deployed in two phases. In 2006, three infrasound arrays were installed around Great Salt Lake to record infrasound signals from rocket motor detonations at the

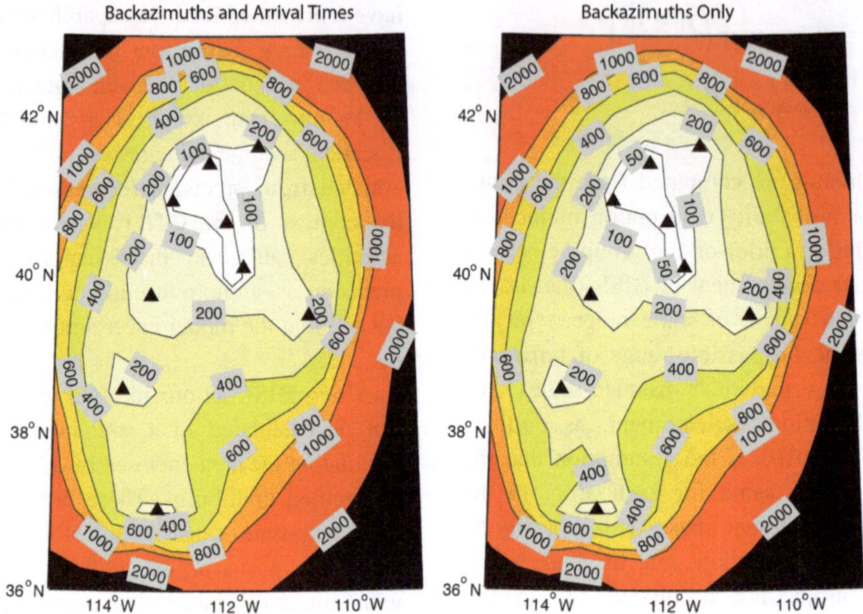

Figure 1
Simulation results showing location uncertainty (in km^2) as a function of location for the Utah infrasound network. Based on the arguments outlined in the paper, these values are representative of current localization capability

Utah Test and Training Range (UTTR) (STUMP *et al.*, 2007). In 2008–2009, the network was extended to include six new arrays in order to study infrasound from small earthquakes in the intermountain seismic belt. The full 9-array network is an excellent test bed for research into the capability of high-density regional infrasound networks. The locations of the arrays in the network are provided in Table 1 and shown as black triangles in the figures.

Table 1
Locations of arrays in the Utah infrasound network

Array name	Latitude	Longitude
BGU	40.9204	−113.0309
BRP	39.4727	−110.7409
EPU	41.3901	−112.4099
FSU	39.7196	−113.3900
HWU	41.6071	−111.5642
LCM	37.0109	−113.2444
NOQ	40.6526	−112.1186
PSU	38.5332	−113.8555
WMU	40.0795	−111.8310

4. *Results*

4.1. *The Utah Infrasound Network*

The location precision estimates have been computed for two scenarios. The first scenario represents an estimate of current capability. The second scenario attempts to assess what might be possible in the future given further research (we discuss the research that is required to fulfill these goals as outlined in the following section). Each scenario is tested and applied on the Utah infrasound network (representative of a spatially-dense network).

Under the first scenario (current operational capability) the following assumptions are made. First, for back azimuth we assume that the existing sum of measurement and model error, $\sigma(\theta)_{\text{total}}$, at any given array can be represented by a standard deviation of 3°. This value is consistent with recent empirical observations of tropospheric and stratospheric signals from ground-truth explosions at the Hawthorne Army Ammunition Depot in Nevada (Negraru, Pers. Comm.) in addition to azimuth deviations observed from measurements of earthquakes in the western

USA (Mutschlecner and Whitaker, 2005). Second, we assume that the identification of phase is non-unique, and therefore that group velocity could vary from 0.28–0.34 km/s (neglecting thermospheric returns). For the size of the Utah network this could result in a spread as large as $\sim$180 s (this value is calculated from the maximum distance between all pairs of arrays and the subsequent difference in travel time for the mean group velocity of 0.31 km/s and the slowest group velocity of 0.28 km/s). Setting $\sigma(\theta)_{\text{total}}$ equal to 100 s, it is clear that, for this network configuration, back azimuth dominates the location capability (Fig. 1). On this basis, the most significant improvements to current capability can be made by reducing the back azimuth model error. The measurement error for back azimuth, typically quoted as $\sim$0.5° but dependent on the array configuration and processing parameters, is small in comparison to model error (Szuberla and Olson, 2004). Of course, this conclusion is based on the Utah network, which is a relatively dense infrasound network. For sparse networks, the importance of arrival time on the solution should be greater.

As shown in Fig. 1, location precisions of 50 km^2 are possible in the densest portion of the network. The whole network region is enclosed by the 400 km^2 contour, indicating that any event within this region will have a location precision better than this value. As one moves away from the region spanned by the network, the location precision degrades as expected. Most notably, the location precision does not improve by adding arrival time constraints with the large model error used. For comparison, under the second scenario (future capability), we assume that a realistic attainable improvement in the total back azimuth uncertainty is $\sigma(\theta)_{\text{total}} = 1.5°$ and that a realistic attainable improvement in arrival time estimation is 20 s (Hedlin *et al.*, 2011). We expect that these improvements will be primarily driven by the reduction of model error. The resultant simulations, shown in Fig. 2, can be compared with the simulations shown in Fig. 1. The improvement in localization precision is dramatic, with the network now enclosed by a precision of 50 km^2. Additionally, it is clear that although the dominant effect is back azimuth, the inclusion of arrival times does improve the location capability at these smaller model errors.

The simulations of typical summer and winter scenarios, using estimates of present capability, are shown in Fig. 3. These plots represent averages of ten realizations for each season (as discussed above, each realization has a random distribution of observations, with 50 % of arrays detecting signals inside 220 km). These results provide a more realistic estimate of location precision, and how it differs at different times of the year. Location precisions of <400 km^2 are possible close to individual arrays, with a more typical location precision of $\sim$800 km^2 surrounding most of the network. During the summertime, since the stratospheric winds blow towards the west in the northern hemisphere mid-latitudes, the network does not detect events at distances greater than 220 km to the west. This scenario is reversed during winter when the stratospheric winds at these latitudes blow towards the east.

4.2. The IMS Network: A Test for the European Region

For comparison to small regional scales, consider a large region with spatially limited stations. For this case, arrival times are much more important, as is illustrated for the two example events in Fig. 4. The top panel in Fig. 4 represents current *operational capability* ($\sigma(\phi)_{\text{total}} = 1{,}000$ s to account for any phase at any array, and $\sigma(\theta)_{\text{total}} = 3^{\circ}$) while the bottom panel represents future capability ($\sigma(\phi)_{\text{total}} =$ 100 s to account for any phase at any array, and $\sigma(\theta)_{\text{total}} = 1.5°$). We note that by operational capability we refer to automatic or real-time methods. The improvements suggested by the future scenario maps require enhancements to the BISL algorithm, in addition to enhancements in propagation models, as outlined below.

5. Research and Development Needed for Location Improvements

This paper argues that, for dense deployments like the Utah network, key enhancements to location capability will come through propagation modeling that can reliably predict the azimuthal deviation caused by the propagation of infrasound from source

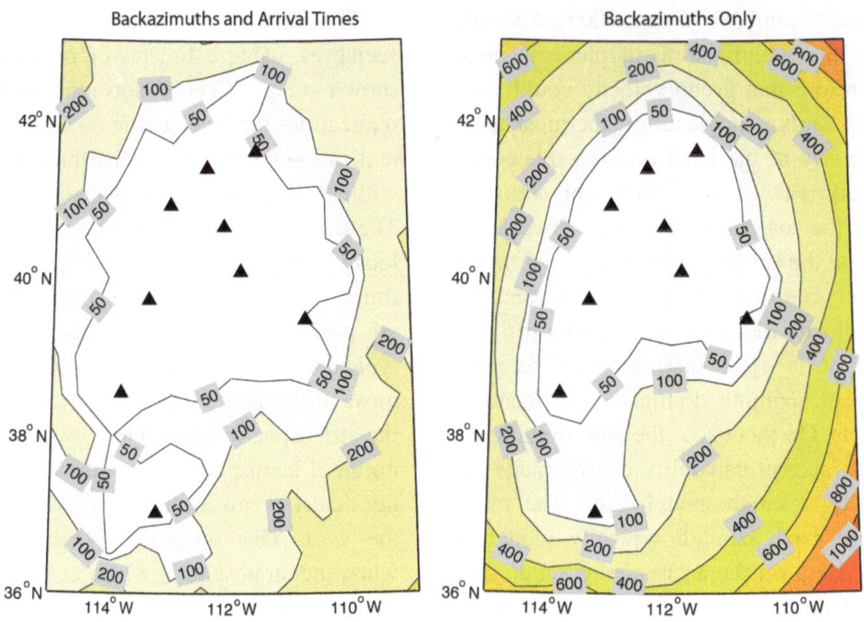

Figure 2
Simulation results showing potential localization uncertainty (in km^2). Based on the arguments outlined in the paper, these values are representative of potential future localization capability

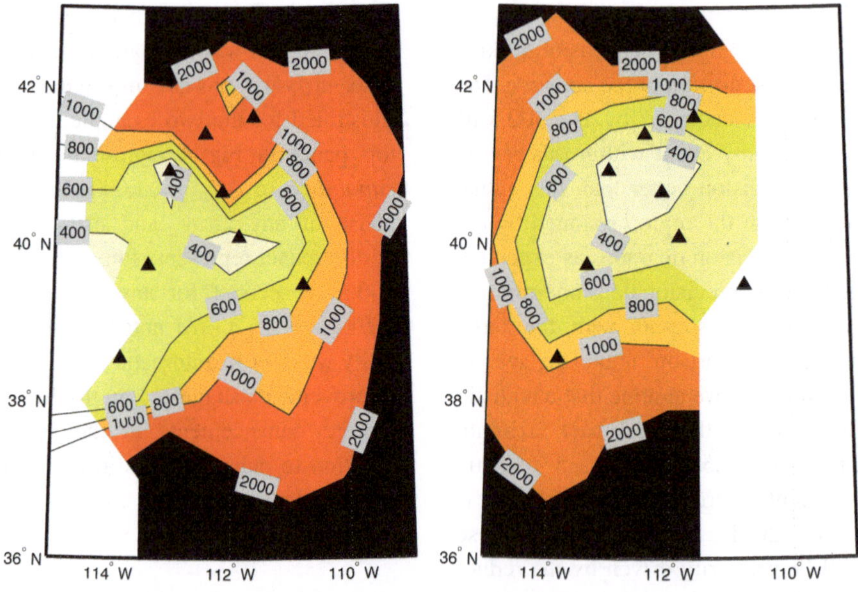

Figure 3
Summer (*left*) and winter (*right*) maps using $\sigma(\theta)_{\text{total}} = 3°$ and $\sigma(\phi)_{\text{total}} = 100$ s. *Colors* represent the location precision in km^2. Areas *shaded white* denote regions where ≤ 1 station would detect an event, and therefore localization using BISL is not possible

to receiver. The prior on azimuth is constructed by embedding physical azimuthal predictions into the appropriate error model. For sparse network configurations, however, enhancements to the localization algorithm are needed that can incorporate physics-based priors and account for propagation anisotropy. The steps required to meet these enhancements can be broken down into three categories, which are

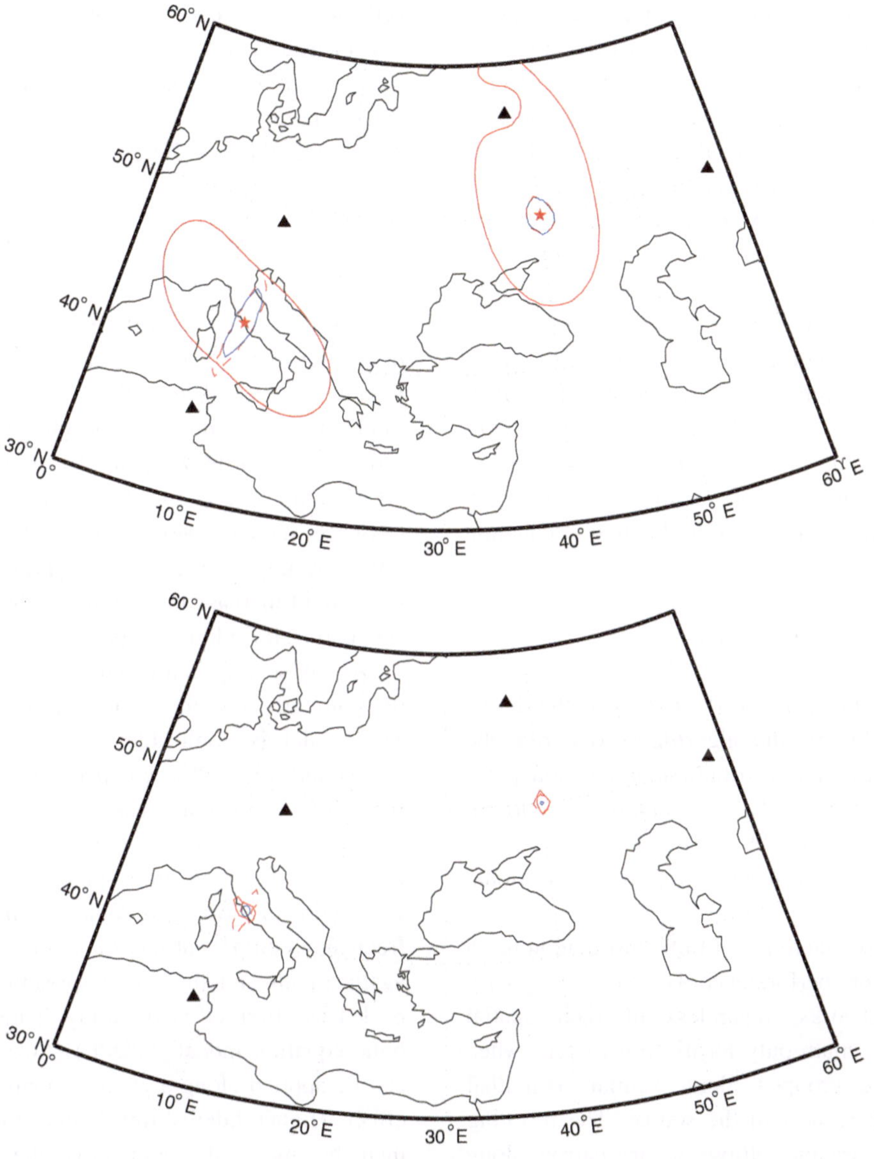

Figure 4
Location simulations for two hypothetical events in Europe with four IMS arrays

discussed separately below. A final major enhancement can be made through combining seismic and infrasonic data for the combined geolocation of seismo-acoustic events; this is discussed as a fourth category.

5.1. *Improvements to the Mathematics of the Location Framework*

The current implementation of BISL assumes the same group velocity to each array in the network. Although the Bayesian prior accounts for the unknown group velocity, it does not adequately account for different phase arrivals at different arrays; this scenario is currently accounted for by increasing the standard deviation in arrival times to include a model error component. Model anisotropy may be a second-order effect at global distances where stratospheric returns are the predominant arrival, and such arrivals are typically observed along similar azimuths in the direction of stratospheric wind. However, at regional distances, where tropospheric returns are common, model anisotropy is more pronounced. Further research and development of BISL is required to enable the inclusion of more sophisticated model predictions, incorporating state-of-the-art propagation models and atmospheric specifications, in the form of Bayesian priors (Marcillo *et al.*, 2012). The Utah network simulations suggest that back azimuths are the dominant effect on the location precision; however, for global scale monitoring using the IMS network, where perhaps only two arrays will detect an event, the inclusion of arrival times will be essential. In this scenario, refinements to the existing BISL framework are expected to lead to dramatic improvements in location precision. The key refinement needed is the inclusion of an array-specific group velocity; in other words ε_i in the likelihood equation would equal:

$$\varepsilon_i = t_i - \left(t_0 + \frac{d_i}{v_i}\right).$$

Included in this framework advance is the development of BISL-specific algorithms to derive the multivariate posterior distribution of location parameters, specifically a Gibb's sampling algorithm (Casella and George, 1992). This development will enable the use of BISL in a near-real-time operational setting.

Another case where travel time information has a large impact on performance is sparse or poorly distributed networks, regardless of their spatial extent. Poor azimuth-only localizations occur when the arrays are grouped along similar azimuthal directions with respect to the source. The resulting localization uncertainty ellipse is very narrow along its major axis, elongated along the direction to the arrays. The quantification of this effect is achieved using geometric dilution of precision (GDOP) (Yarlagadda *et al.*, 2000; Levanon, 2000). GDOP metrics are derived by quantifying the sensitivity of source localization to array measurements, and good GDOP is supported when the arrays span a variety of azimuthal directions.

In the case of poor GDOP, travel time information can supplement the azimuthal data and result in significantly improved localization precision. This improvement occurs because the uncertainly ellipses for travel-time only localizations are elongated orthogonal to that for azimuth. Therefore, when the two measurements are combined, the localization is more tightly constrained, with an uncertainty ellipse that is more circular in nature. [As an example, see Norris and Gibson (2002)].

5.2. Improvements to Propagation Models, Atmospheric Specifications

For predicting arrival times and back azimuth deviations at regional distances (<2,000 km), 3D ray tracing is state-of-the-art. To first order, existing implementations do a reasonable job at predicting arrivals beyond the zone of silence with state-of-the-art 4D atmospheric specifications (e.g., the Ground-to-Space model, Drob *et al.*, (2003)). There are however, improvements to the propagation models that would increase the fidelity of the predictions by accounting for additional physics.

As ray tracing is a geometrical approximation, it does not account for frequency dependent effects. The model is applicable as long as the acoustic wavelength is small in comparison with the physical scales of the medium (Kinsler *et al.*, 1982; Jensen *et al.*, 1994). As this assumption breaks down, the resulting effect for infrasound propagation is that we begin to observe geometric dispersion. Each frequency component takes a slightly different propagation path and there is a spreading of the wave packet as observed at the array. Time-domain parabolic equation models (TDPE) (Norris *et al.*, 2007) can be applied along a given azimuth to predict this effect. Higher fidelity travel time predictions could then be made by comparing the observed and predicted waveforms, where the same phase arrival (peak picking) algorithm would be applied.

Although the TDPE approach can be used for travel times, it does not account for deviations along a given azimuth, which the 3-D ray model quantifies with the azimuthal deviation metric. Research could be pursued to account for these effects using full-wave models. The approach could either be based on exploring 3-D implementations, or in developing hybrid N by 2D solutions.

Another area of research is in the prediction of the propagation paths (phases). The well-established paths of interest are capped at the top of the troposphere and stratosphere. Quantifying their stability, strength, and evolution can be captured with the high-resolution atmospheric specification proposed below, coupled with Monte Carlo simulations. An additional propagation path may also be relevant: infrasonic energy gets trapped in the stratopause and "rides" this height for an extended distance before ultimately escaping and refracting back down to the ground. The trapping occurs as the energy bounces between layers of inhomogeneities driven by gravity waves. This path is significant because the group velocities are much larger than those for a stratospheric path. More research is needed in further quantifying the condition needed and frequency to which this additional propagation path occurs.

For the atmospheric specification, at distances inside 200 km, the addition of fine-scale structure is typically required in order to predict observed arrivals. The key research required, associated with improvements to propagation models and atmospheric specifications *for location purposes*, is related to the incorporation of this fine-scale structure and its effect on the prediction of travel times and back azimuth deviations. Research is needed on the implementation of such methods for the prediction of Bayesian prior PDF's on azimuthal deviations and arrival times.

Recent work has illustrated the value in incorporating spectral models for gravity waves in propagation simulations (Kulichkov, 2004; Gibson *et al.*, 2009). With the use of such stochastic models, Monte Carlo methods are needed to adequately sample the solution space. Due to the dynamic nature of the atmosphere, the solution space is spanned both spatial and temporally. Predictions that capture the relevant scales are on the order of 0.1–1 degree in space and hourly in time. Obviously this modeling fidelity must be balanced with the resulting payoff and computation loads they introduce. The modeling turn-around time for analysis and operational groups must also be considered. With these issues in mind, research in optimization of the front-to-end modeling chain would also be beneficial.

In all cases, these more sophisticated group velocity and back azimuth models are predictions of stochastic atmospheric processes and therefore have error. Embedding these predictions into the appropriate error model provides the priors for the advanced framework.

5.3. Validation of Propagation Models, Atmospheric Specifications

Perhaps the most neglected area of research to date has been the adequate validation of propagation models using large ground-truth datasets (comprising 1,000's of arrivals). In part, this has been a result of the limited infrasound data openly available to researchers. Recent work with USArray data has illustrated how spatially-aliased most infrasound networks are, and how USArray is enabling such validation for the first time (Hedlin *et al.*, 2010). Unfortunately, USArray comprises single acoustic elements and therefore cannot be used to validate model predictions of back azimuth deviations. To address this limitation, research deployments like the Utah network are essential for gathering sufficient ground-truth data with which to properly test models. Such deployments are in their infancy, and little research has been done to date on this aspect. However, based on the findings presented in this paper, such work is critical to validate the model implementations, which will ultimately improve the localization capability that can be obtained using regional infrasound networks.

5.4. Seismo-acoustic Location

The complementary nature of infrasound and seismic data argues for combining both datasets for simultaneous localization of seismo-acoustic events. Seismic data provide better traveltime constraints whereas infrasound back azimuths are superior owing to the relatively low lateral heterogeneity in the atmosphere. Che *et al.* (2009) demonstrate the advantage of combining both technologies by demonstrating how a combined seismo-acoustic location is more accurate, compared to both seismic and infrasound locations, for a ground-truth mining explosion in Korea. Pinsky *et al.* (2012) report on highly accurate seismo-acoustic locations for a series of explosions in Israel. These papers show promise but there has been little research on how to best combine seismic and infrasound data for localization.

Towards this goal, we plan to explore an extension of the BISL framework discussed in this paper through the inclusion of seismic traveltime constraints.

6. Conclusions and Future Research

We have discussed metrics for regional infrasound monitoring, and how these might be improved with improvements to existing atmospheric specifications and propagation models. Similar metrics are needed to guide the development of atmospheric specifications and infrasound propagation models (e.g., how well one can predict the arrival times and amplitudes of signals), but these are not the focus of this paper. Ultimately, for the CTBT monitoring problem, the key parameters are detection, location, and yield estimation. The authors are currently developing acoustic/infrasonic yield estimation methods that require good source location for path corrections. By providing metrics to quantify the capability of infrasound technology in estimating these parameters, the community can better provide decision makers with the tools they need to understand both existing and potential capability.

Acknowledgments

We thank David Green for his comments and suggestions on a draft of this manuscript and two anonymous reviewers for their constructive feedback. We also thank Leslie Casey for proposing this manuscript and for funding this work. This work was completed under the auspices of the U.S. Department of Energy by Los Alamos National Laboratory.

References

Bratt, S.R. and Bache, T.C., 1988. *Locating Events with a Sparse Network of Regional Arrays*, Bull. Seism. Soc. Am., *78*, 780–798.

Brown, D.J., Katz, C., Le Bras, R., Flanagan, M.P., Wang, J. and Gault, A.K., 2002a. *Infrasonic Signal Detection and Source Location at the Prototype International Data Center*, Pure appl. geophys., *159*, 1081–1125.

Brown, P., Whitaker, R.W., Revelle, D. and Tagliaferri, E., 2002b. *Multi-station infrasonic observations of two large bolides: signal interpretation and implications for monitoring of atmospheric explosions,* Geophys. Res. Lett., *29*. doi:10.1029/2001GL013778.

Casella, G. and George, E., 1992. *Explaining the Gibbs sampler*, The American Statistician, *46*, 167–174.

Ceranna, L., Le Pichon, A., Green, D.N. and Mialle, P., 2009. *The Buncefield explosion: a benchmark for infrasound analysis across Central Europe*, Geophys. J. Int., *177*, 491–508.

Che, I.-Y., Shin, J.S. and Kang, I.B., 2009. *Seismo-acoustic location method for small-magnitude surface explosions,* Earth Planets Space, *61*, e1-e4.

Drob, D.P., Picone, J.M. and Garces, M., 2003. *Global morphology of infrasound propagation*, J. Geophys. Res., *108*. doi: 10.1029/2002JD003307.

Evers, L.G., Ceranna, L., Haak, H.W., Le Pichon, A. and Whitaker, R.W., 2007. *A Seismoacoustic Analysis of the Gas-Pipeline Explosion near Ghislenghien in Belgium*, Bull. Seism. Soc. Am., *97*, 417–425. doi:10.1785/0120060061.

Geiger, L., 1912. *Probability method for the determination of earthquake epicenters from the arrival time only*, Bull. St. Louis. Univ., *8*, 60–71.

Gibson, R., Drob, D.P. and Broutman, D., 2009. *Advancement of Techniques for Modeling the Effects of Fine-scale Atmospheric Inhomogeneities on Infrasound Propagation.* in 2009 Monitoring Research Review, Tucson.

Green, D.N. and Bowers, D., 2010. *Estimating the detection capability of the International Monitoring System infrasound network,* J. Geophys. Res., *115*. doi:10.1029/2010JD014017.

Hedlin, M.A.H., De Groot-Hedlin, C.D. and Drob, D.P., 2011. *A Study of Infrasound Propagation Using Dense Seismic Networks American Geophysical Union Fall Meeting, Abstract A31A-0037*, Bull. Seism. Soc. Am., Submitted.

Hedlin, M.A.H., Drob, D.P., Walker, K.T. and De Groot-Hedlin, C.D., 2010. *A study of acoustic propagation from a large bolide using a dense seismic network*, J. Geophys. Res., *115*. doi: 10.1029/2010JB007669.

Jensen, F., Kuperman, W., Porter, M., Schmidt, H., 1994. *Computational Ocean Acoustics, AIP Press, Woodbury, NY, Sec. 3.3.1.*

Kinsler, L, Frey, A., Coppens, A. and Sanders, J. 1982. *Fundamentals of Acoustics, 3rd. Ed., Wiley, NY, Sec. 5.13.*

Kulichkov, S.N., 2004. *Long-range propagation and scattering of low-frequency sound pulses in the middle atmosphere*, Meteor. and Atmos. Phys., *85*, 47–60. doi:10.1007/s00703-003-0033-z.

Le Pichon, A., Vergoz, J., Blanc, E., Guilbert, J., Ceranna, L., Evers, L.G. and Brachet, N., 2009. *Assessing the performance of the International Monitoring System's infrasound network: Geographical coverage and temporal variabilities*, J. Geophys. Res., *114*. doi:10.1029/2008JD010907.

Levanon, N., 2000. *Lowest GPS in 2-D scenarios*, IEE Proc.-Radar, Sonar Navig., *147* (3), 149–155.

Marcillo, O., Arrowsmith, S.J., Whitaker, R.W. and Anderson, D.N., 2012. *Enhancements to the Bayesian Infrasound Source Location Method.* in 2012 Monitoring Research Review, Albuquerque, NM.

Modrak, R.T., Arrowsmith, S.J. and Anderson, D.N., 2010. *A Bayesian framework for infrasound location*, Geophys. J. Int., *181*, 399–405. doi:10.1111/j.1365-246X.2010.04499.x.

Mutschlecner, J.P. and Whitaker, R.W., 2005. *Infrasound from earthquakes*, J. Geophys. Res., *110*. doi:10.1029/2004JD005067.

Norris, D., Bhattacharyya, J. and Whitaker, R.W., 2007. *Development of Advanced Propagation Models and Application to the study of Impulsive Infrasonic Events.* in 29th Monitoring Research Review, edn 2007. Denver, CO.

NORRIS, D. and GIBSON, R., 2002. *InfraMAP Enhancements: Environmental/Propagation Variability and Localization Accuracy of Infrasonic Networks.* in 24th Seismic Research Review, pp. 809–813, Ponte Vedra Beach, Florida.

PINSKY, V., GITTERMAN, Y., BEN HORIN, Y. and ARROWSMITH, S.J., 2012. *Seismo-acoustic analysis for series of ammunition demolition explosions at Sayarim, Israel.* in European Geophysical Union General Assembly, Vienna, Austria.

STUMP, B., BURLACU, R., HAYWARD, C., PANKOW, K.L., NAVA, S., BONNER, J., HOCH, S., WHITEMAN, D., FISHER, A., KIM, T.S., KUBACKI, R., LEIDIG, M., BRITTON, J., DROBECK, D., O'NEILL, P., JENSEN, K., WHIPP, K., JOHANSON, G., ROBERSON, P., READ, R., BROGAN, R. and MASTERS, S., 2007. *Seismic and Infrasound Energy Generation and Propagation at Local and Regional Distances: Phase I—Divine Strake Experiment.* Air Force Research Laboratory.

SZUBERLA, C.A.L. & ARNOULT, K., 2011. Locating explosions, volcanoes, and more with infrasound, *Physics Today, 64*, 74–75. doi:10.1063/1.3580503.

SZUBERLA, C.A.L. and OLSON, J.V., 2004. *Uncertainties associated with parameter estimation in atmospheric infrasound arrays*, J. Acoust. Soc. Am., *115*, 253–258.

SZUBERLA, C.A.L., OLSON, J.V. and ARNOULT, K., 2009. *Explosion localization via infrasound*, JASA Express Letters, *126.* doi: 10.1121/1.3216742.

YARLAGADDA, R., ALI, I., AL-DHAHIR, N. and HERSHEY, J., 2000. *GPS GDOP metric*, IEE Proc.–Radar, Sonar Navig., *147* (5), 259–264.

(Received November 29, 2011, accepted August 12, 2012, Published online September 20, 2012)

Pure Appl. Geophys. 171 (2014), 599–619

DOI 10.1007/s00024-012-0575-4

Pure and Applied Geophysics

GT0 Explosion Sources for IMS Infrasound Calibration: Charge Design and Yield Estimation from Near-source Observations

Y. GITTERMAN[1] and R. HOFSTETTER[1]

Abstract—Three large-scale on-surface explosions were conducted by the Geophysical Institute of Israel (GII) at the Sayarim Military Range, Negev desert, Israel: about 82 tons of strong high explosives in August 2009, and two explosions of about 10 and 100 tons of ANFO explosives in January 2011. It was a collaborative effort between Israel, CTBTO, USA and several European countries, with the main goal to provide fully controlled ground truth (GT0) infrasound sources, monitored by extensive observations, for calibration of International Monitoring System (IMS) infrasound stations in Europe, Middle East and Asia. In all shots, the explosives were assembled like a pyramid/hemisphere on dry desert alluvium, with a complicated explosion design, different from the ideal homogenous hemisphere used in similar experiments in the past. Strong boosters and an upward charge detonation scheme were applied to provide more energy radiated to the atmosphere. Under these conditions the evaluation of the actual explosion yield, an important source parameter, is crucial for the GT0 calibration experiment. Audio-visual, air-shock and acoustic records were utilized for interpretation of observed unique blast effects, and for determination of blast wave parameters suited for yield estimation and the associated relationships. High-pressure gauges were deployed at 100–600 m to record air-blast properties, evaluate the efficiency of the charge design and energy generation, and provide a reliable estimation of the charge yield. The yield estimators, based on empirical scaled relations for well-known basic air-blast parameters—the peak pressure, impulse and positive phase duration, as well as on the crater dimensions and seismic magnitudes, were analyzed. A novel empirical scaled relationship for the little-known secondary shock delay was developed, consistent for broad ranges of ANFO charges and distances, which facilitates using this stable and reliable air-blast parameter as a new potential yield estimator. The delay data of the 2009 shot with IMI explosives, characterized by much higher detonation velocity, are clearly separated from ANFO data, thus indicating a dependence on explosive type. This unique dual Sayarim explosion experiment (August 2009/January 2011), with the strongest GT0 sources since the establishment of the IMS network, clearly demonstrated the most favorable westward/eastward infrasound propagation up to 3,400/6,250 km according to appropriate summer/winter weather pattern and stratospheric wind directions, respectively, and thus verified empirically common models of infrasound propagation in the atmosphere.

[1] Geophysical Institute of Israel, POB 182, 71100 Lod, Israel. E-mail: yefim@seis.mni.gov.il

Key words: Controlled surface chemical explosion, IMS infrasound station calibration, yield estimation, airblast secondary shock delay, infrasound propagation.

1. Introduction

The International Monitoring System (IMS) comprises an infrasound network (currently of 60 stations) for nuclear tests monitoring via recording of low-frequency acoustic waves emitted from remote sources placed on the Earth surface or in the atmosphere (e.g., CHRISTIE *et al.*, 2001). To improve monitoring of explosion sources, i.e., detection, identification, location and yield estimation, the infrasound stations should be calibrated. The best procedure includes fully controlled large-scale on-surface explosions producing strong acoustic signals that can be observed at large distances. A well-established calibration experiment should include measurements of ground truth zero (GT0) information: detonation time, accurate GPS coordinates, precise TNT equivalent yield, and extensive seismic and acoustic observations at near-source zone and also at local/regional distances.

A number of large nuclear and chemical (up to 8 kT) on-surface and near-surface tests were conducted in the 1970–1990's, many of them at White Sands Military Range (WSMR) (Website http://GlobalSecurity.org). Unfortunately, all of them were conducted prior to the establishment of IMS infrasound stations, starting in 1996. In the last decade, a very few controlled explosions were conducted providing well-documented observations of infrasound waves at large distances beyond 1,500 km (CHRISTIE, 2005), which were recorded at IMS stations. The remarkable Watusi experiment was conducted recently at the Nevada Test Site (Bhattacharyya *et al.*, 2003), where the high explosives (HE) charge (19 tons of TNT equivalent) was detonated in a cylindrical container that was partially above the ground level. The acoustic signals of this relatively small explosion were observed at several IMS infrasound stations up to 2,165 km. Two controlled ground-level explosions (5 and 27 tons of ammunition) were conducted at the Woomera test facility in Australia, where clear infrasound signals were recorded at 470–1,260 km at different azimuths (Brown *et al.*, 2003).

Strong surface explosions in NW Russia (presumably due to demolition of old ammunition) were reported in a recent NORSAR study (Ringdal, 2005). Clear acoustic signals were observed at both infrasound and seismic sensors of Apatity and ARCES arrays placed at $\sim$250 km from the source. Combined seismic and infrasound signal processing showed that location estimations based on infrasonic detections match closely the standard seismic data locations, but a joint seismic/infrasound location procedure implies assigning (to the distinct infrasound arrivals dataset) of reliable weights which cannot be chosen without GT0 events. Bolide explosions were used to study the capabilities and limitations of source location procedures based on infrasound travel times and azimuth deviations derived from ray tracing formulations (Garces *et al.*, 2002). Seismic and infrasound data from construction explosions near Seoul were jointly used to produce a combined event location (Stump *et al.*, 2002). However, the location accuracy of used procedures in all these cases cannot be reliably estimated without GT0 explosions.

The propagation features of infrasonic waves from atmospheric explosions with yields <1 kT in the distance range of more than 1,500 km are still poorly understood. Such experiments may provide considerable insight into the detection capability. Until recently no GT0 explosions, relevant for IMS infrasound calibration, and no IMS stations recorded signals from GT0 sources were presented anywhere in the Eastern Mediterranean/Middle East (EM–ME) region.

To improve infrasound monitoring in the EM–ME region, a number of large-scale surface controlled explosions were conducted in August 2009 and January 2011 by the Geophysical Institute of Israel (GII) in different weather and wind conditions, which were supported by the US Army Space and Missile Defence Command (SMDC) and PTS CTBTO (Gitterman, 2010, 2011). The collaborative efforts of several organizations provided this creative, cost-efficient venture, improving monitoring and verification of the Comprehensive Test Ban Treaty (CTBT) in the region. The explosions were realized at Sayarim Military Range (SMR), Negev desert, in the site of the regular demolition of outdated ammunition by Israel Defense Forces (IDF).

The main goals of the explosion experiments—recording infrasound signals at several IMS stations, and experimental demonstration of seasonal (summer/winter) variation in infrasound propagation at far-regional distances—were reached. Dense seismo-acoustic network of portable and permanent stations provided good datasets at near-field and local ranges. We analyze here air-shock and acoustic records aiming to understand observed unique blast effects, determining wave parameters that are better suited for explosive yield estimation, and deriving empirical relations between the yield and the relevant wave parameters.

This paper covers analysis and interpretation of the explosion parameters and near field recordings and provides a reference to other papers devoted to Sayarim experimental explosions, in-particular: long distance propagation and modeling of the infrasound signals (Fee, 2012), and seismo-acoustic energy partitioning at near-source and local distances (Bonner, 2012).

2. *Charge Design*

Major design conceptions, elaborated by GII, were maximal concentration of explosives in the charge assembled from numerous similar units and increased energy release to the atmosphere—in order to provide more distant infrasound observations (Fig. 1). Special attention was given to details of charge design and configuration to ensure the understanding of observed explosion effects and energy generation. A similar design was used in both experiments, however main agent explosives were rather different (Table 1).

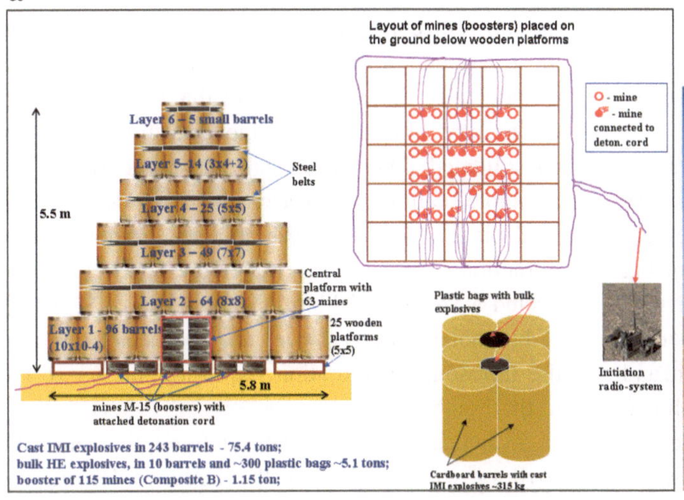

b

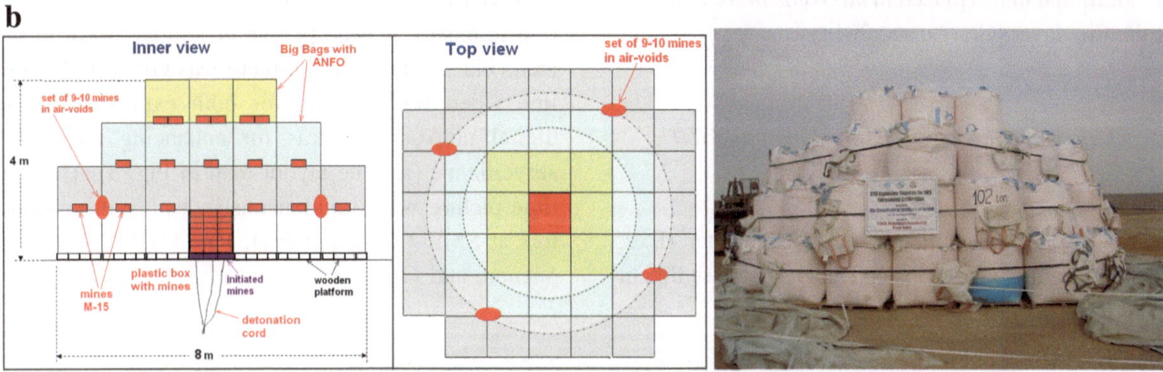

Figure 1
Charge design of main Sayarim shots in August 2009 (**a**), and in January 2011 (**b**)

Table 1

Parameters of explosives used in large-scale Sayarim surface explosions

Event	Date	Main agent	Density (g/cm^3)	Velocity of detonation, VOD (m/s)	Charge unit (kg)	Charge booster
Ex1	26/08/09	Cast IMI	1.54–1.67	7,130–7,980	Barrel 315	Mines M-15
Ex2	24/01/11	Bulk ANFO	0.80–0.81	2,400	Big bag 870	(each one 10 kg composition B)
Ex3	26/01/11					

The large-scale surface explosion Ex1 in August 2009 was characterized by (Fig. 1a): (a) big barrels ($\sim$315 kg each) filled up to the top with strong cast HE, provided by Israel Military Industries Ltd. (IMI); (b) minimal air voids between the barrels by filling the voids with plastic bags containing HE bulk explosives, and between six charge layers; (c) nearly compact pyramidal shape (measured dimensions: base 5.8 × 5.9 m and height 5.5 m); (d) a strong booster composed of several mines M-15, placed on the ground upside down, to provide the upward detonation and additional upward cumulative effect, and 63 mines on the central platform of the first layer; (e) multiple-point initiation scheme to ensure reliable detonation (Gitterman, 2010). The IMI explosives consist of a mixture of different recuperated HE: TNT and gun powder as major agents, smaller amounts of Composition B and RDX, and some other components.

Two explosions, Ex2 and Ex3, conducted in January 2011 have slightly different features (Fig. 1b): (a) primary agent ANFO mixture (94 % ammonium nitrate and 6.0 % fuel oil), placed in waterproof big bags, supplied by Explosives Manufacturing Industries (1997) Ltd., Israel (EMI); (b) nearly hemispherical shape (measured actual dimensions: base 8 × 8 m and height 4 m); (c) single mines M-15 beneath each bag (double mines for the upper layer) for reinforcement of the detonation wave front propagating upward; (d) additional mines placed in air-voids in four corners of the first layer (Gitterman, 2011).

3. *Conducting of the Explosions and GT0 Parameters*

All charges in summer (Ex1) and winter (Ex2, Ex3) experiments were placed on the soft sediment surface at the same site (Fig. 2). In the explosion area the subsurface layer of ~0.5–1 m consists of soft and loose sediments, with consolidated sediments below. Geologically the subsurface media is presented by Quaternary alluvial conglomerates, underlined by consolidated limestone, chalk and chert rocks. The available shallow subsurface velocity model is shown in Table 2. The model was obtained from a seismic refraction survey at Sayarim Valley, where GII conducted a Seismic Calibration Explosion of 32.5 tons in boreholes (Gitterman *et al.*, 2005), not far from the Infrasound Calibration Explosion site (~16 km to the South).

In January 2011, per IDF safety requirements, the territory within 6 km of the large explosion was surveyed by helicopter about 2 h before the detonation to ensure that no people or animals were in the area (Fig. 2). In addition, there was GII installed online video broadcasting for both explosions using digital network cameras for enhancing the safety surveillance (see the layout map in Fig. 3). The real time picture was simultaneously displayed on monitors at SMR Command Post, PTS CTBTO office in Vienna, and GII.

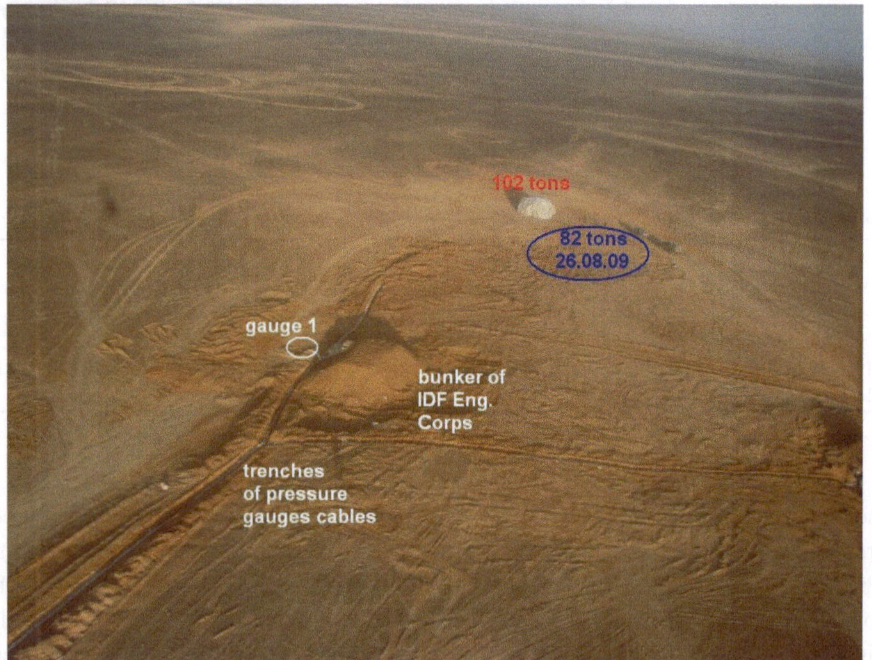

Figure 2
Helicopter view of the large explosion site (morning of 26 January 2011). The previous 82-ton shot place is shown, with distance between the charge centers ~30 m

Table 2

Velocity model of the shallow subsurface at Sayarim Valley

Layer no.	Depth interval (m)	Interval velocities, V_p (m/s)	Averaged layer thickness (m)	Average V_p (m/s)
1	0 down to 5–7	1,130–1,300	0–6	1,215
2	5–7 down to 17–20	1,650–1,720	6–18.5	1,685
3	17–20 down to 85–100	1,930–2,020	18.5–92.5	1,975
4	Deeper than 85–100	3,750–4,120	>92.5	3,935

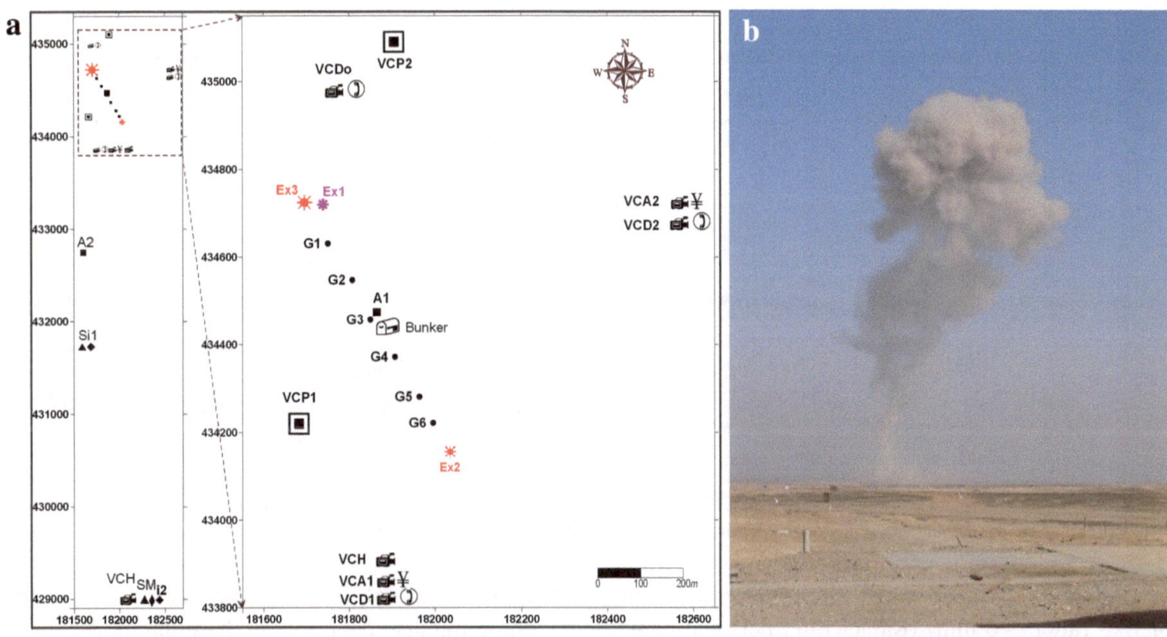

Figure 3

The Ex2 and Ex3 experiment layout (winter 2011): location of the explosions (and the 2009 shot Ex1) and near-source measurement systems: pressure gauges (G1–G6), accelerometers (A1, A2), home and speed video-cameras, and monitoring safety cameras that were broadcasting the real time video through the internet (**a**); view of the dust and gases column from the Command Post at 6 km for Ex3 after 6 min (**b**)

Table 3 presents origin (detonation) time (OT), GPS coordinates (accuracy 4–5 m), total weight of all explosives, altitude of the charge and near-surface air temperature during the detonation. The OT values were based on an electric circuit attached to the detonator with an appropriate PC-based system that includes GPS time. In addition, we checked the timing via analysis of raw data files from the GII data acquisition system and verification by near-source accelerometric records. The day and time (~0930 hours local time) for the 2009 shot Ex1 were chosen due to favorable wind conditions for infrasound propagation (Garces *et al.*, 2009; Bowman *et al.*, 2009). In 2011 the smaller 10-ton shot Ex2 was conducted as late as possible in the afternoon (~1518 hours local time) given the short daytime in January and the large 100-ton shot Ex3 was conducted as early as possible in the morning, to check the day variability of weather conditions.

4. Near-source Observations

GII deployed various measuring and observation systems at close distances including pressure gauges, accelerometers, regular and speed video-cameras. The experiment layout of the explosion and GII recording systems, similar for both experiments, is shown on Fig. 3.

The maximal height of the dust and gases column was estimated using recordings of a video camera:

Table 3

Parameters of large-scale surface explosions, at Sayarim Military Range

#	Date	Nominal charge weight (kg)	Detonation time (GMT)	Altitude (m)	Air temperature (°C)	Latitude (°N)	Longitude (°E)
Ex1	26/08/09	81,664	06:31:54.00	558	35	30.00057	34.81351
Ex2	24/01/11	10,240	13:17:53.80	546	24	29.99555	34.81668
Ex3	26/01/11	102,080	07:17:42.44	558	13	30.00064	34.81324

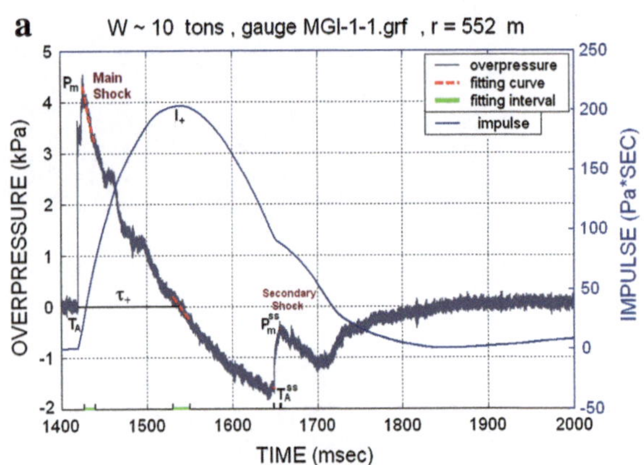

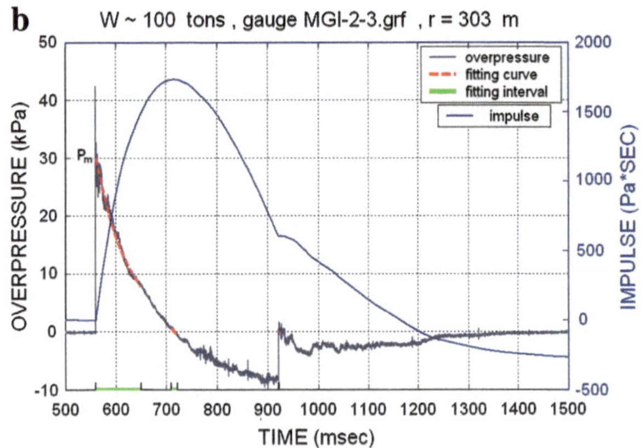

Figure 4
Sample records for Ex2, gauge G1 (filtered) (**a**), and Ex3, gauge G3 (**b**). Exponential fitting curves (*red*) and intervals (*green*) are also shown. Note a significant difference of the maximal recorded amplitude and the peak fitting overpressure value, for Ex3

~1,250 m for Ex2 (after ~4 min), and ~2,750 m for Ex3 (after ~6 min) (Rafael Ltd., personal communication) (see Fig. 3b).

5. *High-pressure Measurements*

High-pressure sensors are required to evaluate the efficiency of the charge design and energy generation of ANFO explosives and estimate the energy released by the explosions, as the TNT equivalent yield of the charge. Furthermore, the records of air-shock waves are used to analyze and explain some interesting blast phenomena. The IDF team installed six pressure gauges XTL-190-5G/50A/100A along a line in the distance range 100–600 m (Fig. 3a). The gauges with disc type baffles were mounted on steel rods ~1.5 m above the surface providing side-on free-field overpressure measurements with sampling rate of 2 MHz.

We measured accurately basic parameters of recorded air-blast waves:

- Time of arrival T_A^{ms}, peak pressure P_m^{ms}, positive phase impulse I_+, positive phase duration τ_+—for the main shock;
- parameters (T_A^{ss}, P_m^{ss})—for the secondary shock (see below in details).

Samples of the air-shock overpressure records with explanations of the measured parameters are presented in Fig. 4, together with calculated shock-wave impulse (blue curves).

The records of the small shot Ex2 are contaminated by high-frequency noise (at frequency about 107.5 kHz), that hampered accurate measurement of the parameters. Therefore, we applied the 33.33 kHz low pass filter (Fig. 5), thus improving the measurement accuracy.

Due to some irregularities in the high-pressure time history, high-frequency noise and spikes, the waveform was approximated by 2–3 exponent fit curves, for more reliable estimation of (P_m, τ_+) values for the main shock and (T_A, P_m) values for the

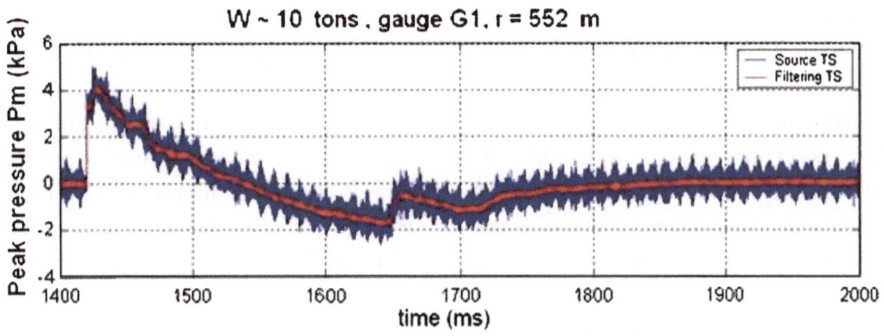

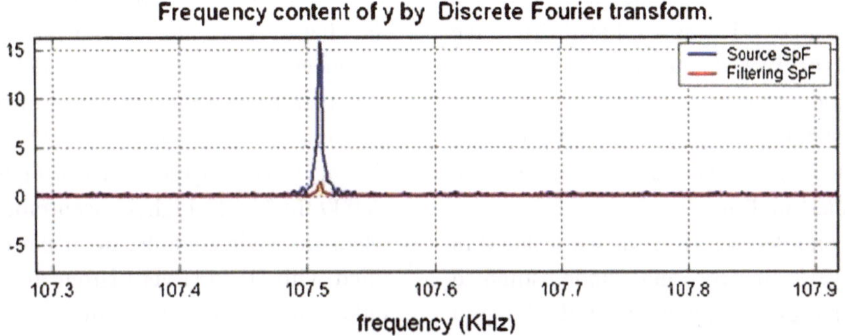

Figure 5
Application of low pass 33.33 kHz filtering to a sample record for Ex2

Table 4

Accurately measured basic air-blast wave parameters for Ex2 and expected (standard) values (in brackets) as obtained by BECv4 procedure for the charge 10.3 ton ANFO, actual altitude and air temperature

Distance, r (m)	Arrival time, T_A^{ms} (ms)	Peak over-pressure, P_m (kPa)	Positive phase duration, τ_+ (ms)	Impulse, I_+ (kPa ms)	Secondary shock (SS)		
					T_A^{ss} (ms)	Delay, Δt (ms)	P_m^{ss} (kPa)
351	863.08 (857.8)	7.14 (6.94)	103.60 (117.2)	331.82 (356.4)	1,072.06	209.0	1.81
452	1,127.57 (1,148.2)	5.40 (5.05)	107.51 (126.0)	252.39 (278.4)	1,347.46	219.9	1.45
552	1,419.14 (1,440.1)	4.31 (3.88)	119.32 (132.8)	203.64 (228.5)	1,649.14	230.0	1.17

secondary shock (Fig. 4). Then we calculated the secondary shock delay $\Delta t = T_A^{ss} - T_A^{ms}$. Sample measured values for the explosion Ex2 are presented in Table 4.

The air-blast measured parameters were found enlarged (P_m) or reduced (τ_+, I_+), for all gauges and all three explosions, compared to expected (standard) values obtained by the DDESB Blast Effects Computer, Version 4.0 (BECv4), an Excel template (Swisdak, 2000) (see Table 4). This procedure provides easy computation of a wide variety of free-field air-blast parameters (in English or Metric units), for different explosives and charges (including on-surface sources), in broad pressure, charge weight and distance ranges. The parameters are calculated by reliable empirical scaling relationships based on numerous fully-controlled experimental explosions of hemispherical charges (e.g., Sadwin and Swisdak, 1970). Air-blast estimations can be obtained also by commonly used ConWep computation procedure (Conwep, 1997), but only for sea level conditions, whereas BECv4 takes into account also an altitude (or atmospheric pressure) and the temperature (in some cases the difference is significant).

6. Estimation of TNT Equivalent Yield from Air-blast Wave Data

The TNT equivalent estimation is based on the mentioned BECv4 procedure, using the actual altitude and air temperature at the explosion site (see Table 3). The accurately measured values of peak over-pressure, positive phase duration and impulse for the main shock were utilized for calculation of an appropriate yield for each gauge, and then averaged over several gauges, for each specific air-shock wave parameter. The gauge G1 was excluded from the yield estimation for Ex3 due to anomalously low pressure amplitudes (distance 103 m), supposedly because the air-shock propagation may have been affected by the nearby IDF bunker (see Fig. 2). The results are presented in Table 5.

Increased P_m and decreased τ_+ values resulted in appropriately overestimated and underestimated yields. The positive phase impulse I_+ is the integral and stable characteristic of the air-blast wave (unlike the one-point peak amplitude), and we consider the impulse-based estimation as the most reliable, accepted as GT0 parameter of Sayarim calibration explosions (shown bold in Table 5), and used in the following developing of charge-scaled relationships. Note that the impulse-based yield estimation is close to the average of estimations by both parameters (P_m and τ_+), because the impulse is about proportional to their multiplication.

The same blast wave parameters and a similar multi-station procedure were applied to charge estimation for track bomb explosions, concluding that the most reliable and accurate acoustic wave property (for yield estimation) is the impulse of the airblast (Koper *et al.*, 2002).

The obtained TNT yield estimation for explosion Ex1 is about 20 % more than the nominal weight of the charge. There are several possible factors for the enhanced air-blast energy and the appropriate enlarged TNT yield: (1) strong IMI explosives with high detonation velocity ~7,500 m/s on the average for the whole charge, ~10 % higher than for TNT (6,900 m/s); (2) high concentration of explosives, when most air-voids between the charge units were filled with HE; (3) very strong booster and multiple initiation scheme; (4) upward detonation of the charge.

Note that the ratio of yield estimations based on peak over-pressure and maximal impulse is very similar for the two large shots: 152/96 = 1.58 for Ex1 with IMI explosives stronger than TNT, and 119.6/76.8 = 1.56 for Ex3 with ANFO explosives, which is weaker than TNT.

Table 5

Peak pressure, positive phase duration and impulse for the main shock and appropriate estimations of the TNT equivalent yield for the Sayarim explosions

Event	Gauge	Distance, r (m)	Peak overpressure, P_m		Positive phase duration, τ_+		Positive phase impulse, I_+	
			Measured (kPa)	Yield (tons)	Measured (ms)	Yield (tons)	Measured (Pa s)	Yield (tons)
Ex1	G2	197	74.4	164	129.9	33	2,963	101
	G3	294	34.9	157	163.9	52	1,953	93
	G4	394	22.3	163	185.6	59	1,508	95
	G5	509	15.0	155	199.6	57	1,202	98
	G6	611	10.6	123	219.5	67	966	92
	Average			152.4		53.6		**96.0**
Ex2	G3	351	7.14	9.20	103.60	5.1	331.8	7.7
	G2	452	5.40	10.1	107.51	4.5	252.4	7.4
	G1	552	4.31	10.8	119.32	5.6	203.6	7.2
	Average			10.0		5.1		**7.4**
Ex3	G2	203	53.5	106.8	127.4	24.7	2,424.4	72.1
	G3	303	30.16	132.0	156.36	34.4	1,731.5	76.0
	G4	405	18.7	124.0	179.9	42.2	1,314.7	75.8
	G5	513	13.3	122.0	200.54	48.8	1,109.85	82.7
	G6	580	10.9	113.0	206.03	46.5	943.3	77.3
	Average			119.6		39.3		**76.8**

Based on the known standard ANFO to TNT equivalence of about 0.82, the expected equivalent TNT yield values are 8.5 ton for Ex2, and 84.5 ton for Ex3. The obtained yield estimates are a little smaller than the expected values (~13 % for Ex2, and ~9 % for Ex3). We note several possible factors for lower air-blast energy and appropriate reduced equivalent TNT yield: (1) not quite hemispherical shapes (especially for Ex2); (2) inhomogeneous charges (bags and boxes for Ex2, placement of numerous mines with stronger explosives in the ANFO charge body for Ex3); (3) numerous air-voids between the charge units (plastic boxes and big bags) that were not filled by explosives (as was done for Ex1). All these factors caused a non-uniform charge that resulted in observed blast anomalies: jetting, asymmetrical blast fronts, multiple shock-wave phases, turbulence, collisions of wave fronts, and, consequently, in some energy losses.

7. Secondary Shock Effect

In the 2011 ANFO explosions Ex2 and Ex3, a distinct secondary shock (SS) wave was observed at all gauges during the negative phase of the pressure–time curves, showing negative or close to zero peak pressures (Figs. 4, 7a, b). It is similar to SS waves observed for surface 20 and 100 ton ANFO shots in Alberta, Canada (Sadwin and Swisdak, 1970), and opposite to the case of Sayarim 2009 Ex1, where positive SS peak pressures were observed (Fig. 7c) (Gitterman, 2010). The explosion Ex1 comprised cast IMI explosives with higher density and velocity of detonation than ANFO that was used in 2011 (see Table 1). Evidently these strong explosives caused smaller time delays between main and secondary shocks, resulting in positive SS peak pressures.

Clear SS waves were observed also at two Kinemetrics K2 accelerometers (A1 and A2 in Fig. 3a), placed on the surface and subjected to the strong impact of the air shock wave. For the 102 ton shot Ex3, a vertical acceleration ~4 g was measured at the closest station, corresponding to the main air-shock MS arrival (Fig. 6).

In this known, but rarely reported phenomenon, the air-blast wave for any finite chemical explosion source can exhibit numerous repeated shocks of small amplitudes at various times, caused by successive implosion of rarefaction waves from the contact surface between explosion products and the air (Baker, 1973). A higher pressure shock front propagates faster, therefore the time delay between the main shock (MS) and SS phases increases with distance (see Fig. 7a, b), as well as with the charge yield.

Using the charge cubic root scaling law we developed relationships for the scaled delay Dt and the scaled distance R (for estimated TNT equivalent charge W, presented in Table 5):

$$\begin{aligned} \mathrm{Dt} &= \mathrm{Dt}/W^{1/3}\ (\mathrm{s/kg}^{1/3}), \\ R &= r/W^{1/3}\ (\mathrm{m/kg}^{1/3}). \end{aligned} \quad (1)$$

Some differences in air temperature and pressure (altitude) for different explosions are considered to cause minor changes in air-blast parameters, comparable with measurement errors of distance and wave parameters, and are not applied here in the distance and time delay scaling.

We extended the SS dataset, utilizing a broader observation range for more complete analysis of this unique air-blast feature. In the Sayarim 2011 experiment (Ex2 and Ex3) a number of acoustic sensors were deployed by Weston Geophysical Corporation (WGC), the University of Mississippi (UM), and the University of Firenze (UF), at a distance range 1–37 km (Gitterman, 2011), providing good records of SS waves that we included in the analysis.

A linear RMS fit regression curve of the scaled delay Dt versus logarithmic scaled distance R was obtained for two Sayarim ANFO explosions, with a high correlation parameter C^2 (Fig. 8):

$$\begin{aligned} \mathrm{Dt}(\mathrm{s/kg}^{1/3}) &= 0.0057565 \times \log(R) + 0.0032, \\ C^2 &= 0.985, \quad 2 < R < 1{,}000\,\mathrm{m/kg}^{1/3} \end{aligned} \quad (2)$$

In addition, we tried to extend also the source range by checking some acoustic records of the WSMR large-scale ANFO explosions Distant Image (2,214 metric tons ANFO, equivalent to 1,815 tons TNT), and Minor Uncle (2,472 metric tons ANFO, equivalent to 2,027 tons TNT), at distances 28–60 km, presented in Norris (2007). We identified SS phases, measured the delays and scaled for TNT equivalent yield, and found a good agreement with Sayarim ANFO explosions and the fit curve (Fig. 8).

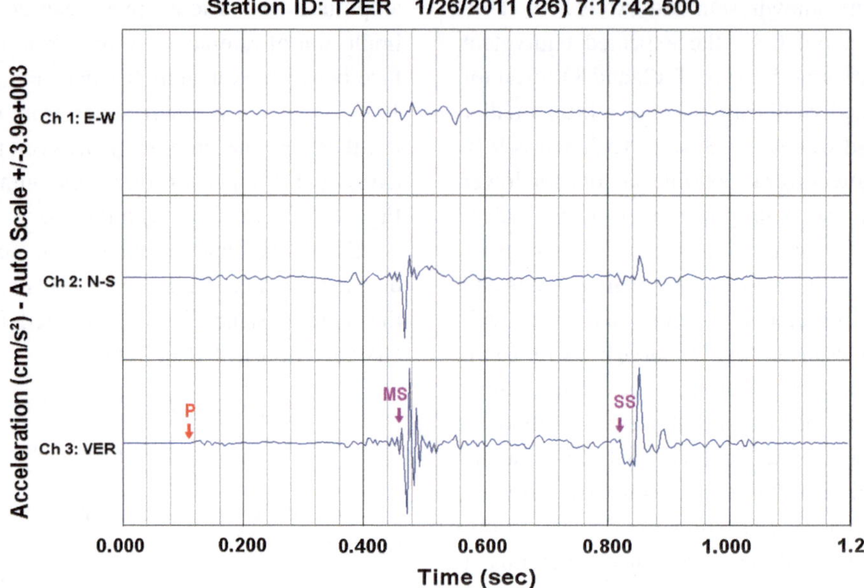

Figure 6
Accelerogram of Ex3 at A1 ($\sim$300 m), showing arrivals for P-waves, and air-blast main (MS) and secondary (SS) shocks

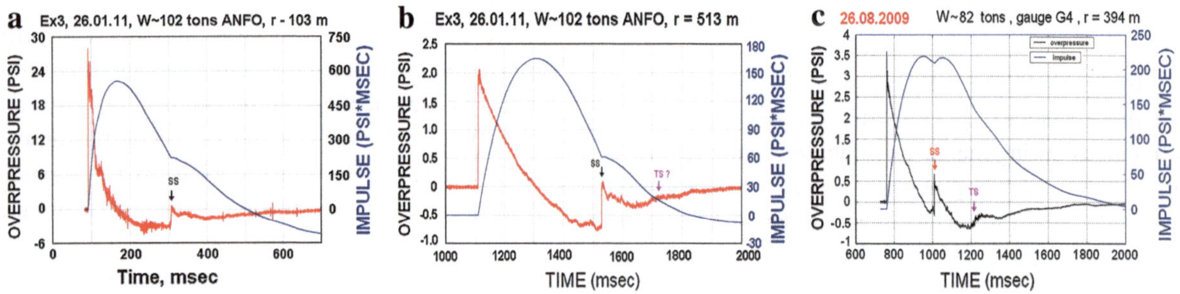

Figure 7
Samples of shock-waves records for Ex3 at different distances (**a**, **b**). Gauge G5 showed SS delay Δt $\sim$0.42 s (**b**), compared to a smaller Δt $\sim$0.24 s, for Ex1, at a similar distance (**c**)

Thus Eq. (2) describes the data over a broad range of charges (10–2,725 tons ANFO) and distances (0.1–60 km).

However the data for the Sayarim 2009 explosion Ex1 with IMI explosives, using records of five high-pressure gauges and also three accelerometers at a distance range 200–610 m, are clearly separated from the ANFO data, and, though showing a linear relationship between the scaled delay and the log-scaled distance, but significantly lower than the fit curve (Eq. 2) (Fig. 8). The IMI explosives are much stronger than ANFO and, as we estimated, more energetic than TNT, having a higher detonation velocity (see Table 1). Supposedly, this factor explains the smaller SS delays.

The obtained results show that there is an option to use the stable and reliable air-blast parameter, SS delay, as a new potential yield estimator. For ANFO shots on dry soft sediment surface, if the SS phase is clearly identified on acoustic or seismic records, then the charge weight can be accurately estimated.

8. *Empirical Scaling Relationship for Basic Air-blast Parameters from Sayarim Explosions*

When analyzing the dependence of the main shock peak pressure in air-shock waves on the scaled distance, in a broad distance range (0.1–10.5 km) we found that the data of all three Sayarim explosions,

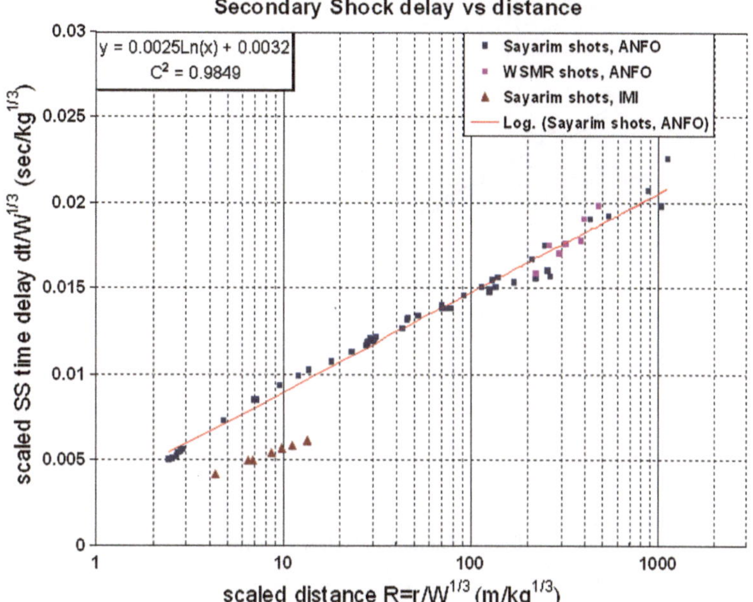

Figure 8
Scaled time delay between MS and SS phases versus scaled distance (calculated for TNT equivalent charge weights)

although of very different ANFO and IMI explosives, show a unified linear relationship in the double log-scale, indicating a power law curve, with a low spreading (Fig. 9). The RMS regression procedure provided a simple power law with a high correlation parameter C^2:

$$P_m(\mathrm{Pa}) = 410,801 \times R^{-1.3461},\ C^2 = 0.98,\quad 2 < R < 500\ \mathrm{m/kg}^{1/3} \tag{3}$$

In the peak pressure regression analysis, we used the same near-source high-pressure and close acoustic records as for the SS delay regression for Sayarim shots, including WGC, UM and UF sensors. The distance range was restricted to 10.5 km, in order to stay in the field of strong air-shock waves ($P_m >$ 100 Pa, approximately corresponds to the threshold of window breakage), and to avoid atmospheric/wind effects on propagation of weak (elastic) infrasound waves. Equation (3) is similar to one of the standard reference empirical relations for the open air detonation (Perkins, 1964): $P_m(\mathrm{Pa}) = 360{,}070 \times R^{-1.38}$, which fits well the Sayarim data (Fig. 9).

For comparison, we present in Fig. 9 also an empiric-analytical relationship derived from the Landau analytical form, that was developed for weak air-shock waves at large distances (Landau, 1945), using measured near-source peak pressures at 200–600 m from Sayarim 2009 shot Ex1 (V. Pergament, personal communication):

$$\Delta P/P_0 = 1/\{R[\log(R/\pi)^{1/2}]^{1/2}\}, \tag{4}$$

where P_0 is normal atmosphere pressure, 10^5 Pa.

Equation (4), which is valid for $R > 3.4$ m/kg$^{1/3}$, demonstrates a good correspondence to Sayarim data, especially at close distances.

We also developed empirical scaling relationships for the main shock basic parameters: peak over-pressure P_m, positive phase duration τ_+ and impulse I_+, using only high-pressure gauges data (i.e., in a narrow distance range 100–600 m). In addition to the charge cubic root scaling (only τ_+ and I_+, using the estimated TNT equivalent charge W, Table 5), we scaled these measured air-blast parameters to the reference atmospheric conditions at the sea level (altitude $H = 0$ m, temperature $T = 15$ °C), following Sachs' scaling procedures (Sachs, 1944; Sadwin and Swisdak, 1970). The results are shown in Fig. 10, the following RMS fit expressions were obtained:

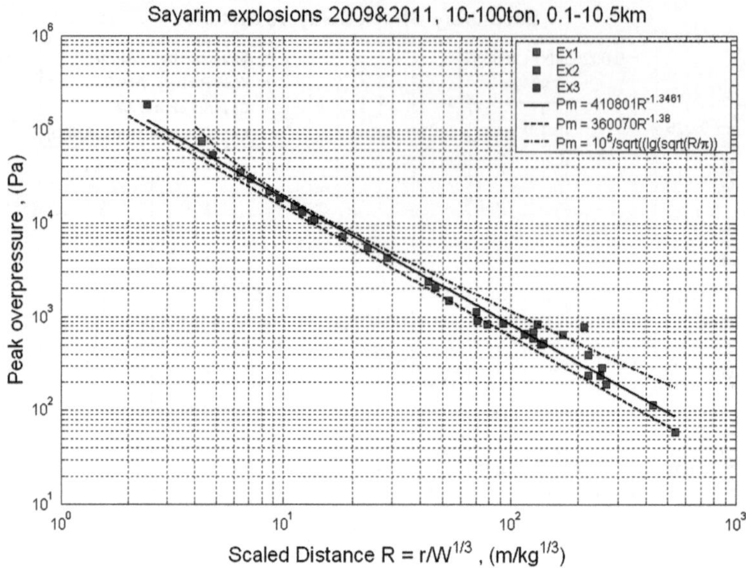

Figure 9
Peak over-pressure versus scaled distance (calculated for TNT equivalent charge weights) for Sayarim explosions

$$P_m(\mathrm{Pa}) = 662{,}764 \times R^{-1.5451} \quad (5)$$

$$\tau_+(\mathrm{ms/kg}^{1/3}) = 1.4063 \times R^{0.46639} \quad (6)$$

$$I_+(\mathrm{Pa\,ms/kg}^{1/3}) = 268{,}343 \times R^{-0.94907} \quad (7)$$

The reference (TNT standard) values, obtained by BECv4 procedure, are also presented (Fig. 10). The scaled observed data demonstrate a high consistency for all three shots with different types of explosives and TNT yield, and correspond well to power fit curves (5–7), except for τ_+ values (Fig. 10b) that seems to fit much better to a polynomial curve (Fig. 10d). There is a discrepancy of the scaled Sayarim data with the reference (by BECv4) values (as appropriate to the analysis of the TNT yield estimation presented before): measured peak pressures are a little higher (Fig. 10a), positive phase durations are lower (Fig. 10b, d), whereas positive impulses show almost the same values as the TNT standard curve (Fig. 10c). This result corresponds, evidently, to impulse-based estimation of yield values, which are crucial for scaling relations.

Note an expected stronger attenuation with distance for high near-source peak pressures, compared with weaker P_m values in a broad local distance range (Eq. 3).

Obtained small spreading and high uniformity for all three shots of the scaled data, especially of the positive impulse, verify the yield evaluation procedure and estimated TNT yields.

9. Audio-visual Observations of Blast Effects

Two home video-cameras, three special safety monitoring cameras (radio and cellular transmitted), and a speed Phantom-3 camera (3,100 frames/s) were placed at different distances and azimuths in the two explosions (Fig. 3a), for recording the unique blast phenomena. Snapshots from home video of the 102-ton shot show an expanding, evidently following the shock wave, short-term ($\sim$0.5 s) white spherical condensation cap, which is clearly visible due to specific air (humidity 61 %, temperature 13 °C) and lighting conditions (Fig. 11). Snapshots of speed video-record for the ten ton shot show non-spherical segments and multiple phases in the air-shock wave front at short times and distances (Fig. 11). Apparently, the same multiple shock-front phases were also observed in the large Ex3 and correspond to multiple peaks in the positive (compression) phase of the air-shock wave recorded by close pressure gauges (Fig. 7a). Supposedly, these front phases/peaks are due to non uniformity

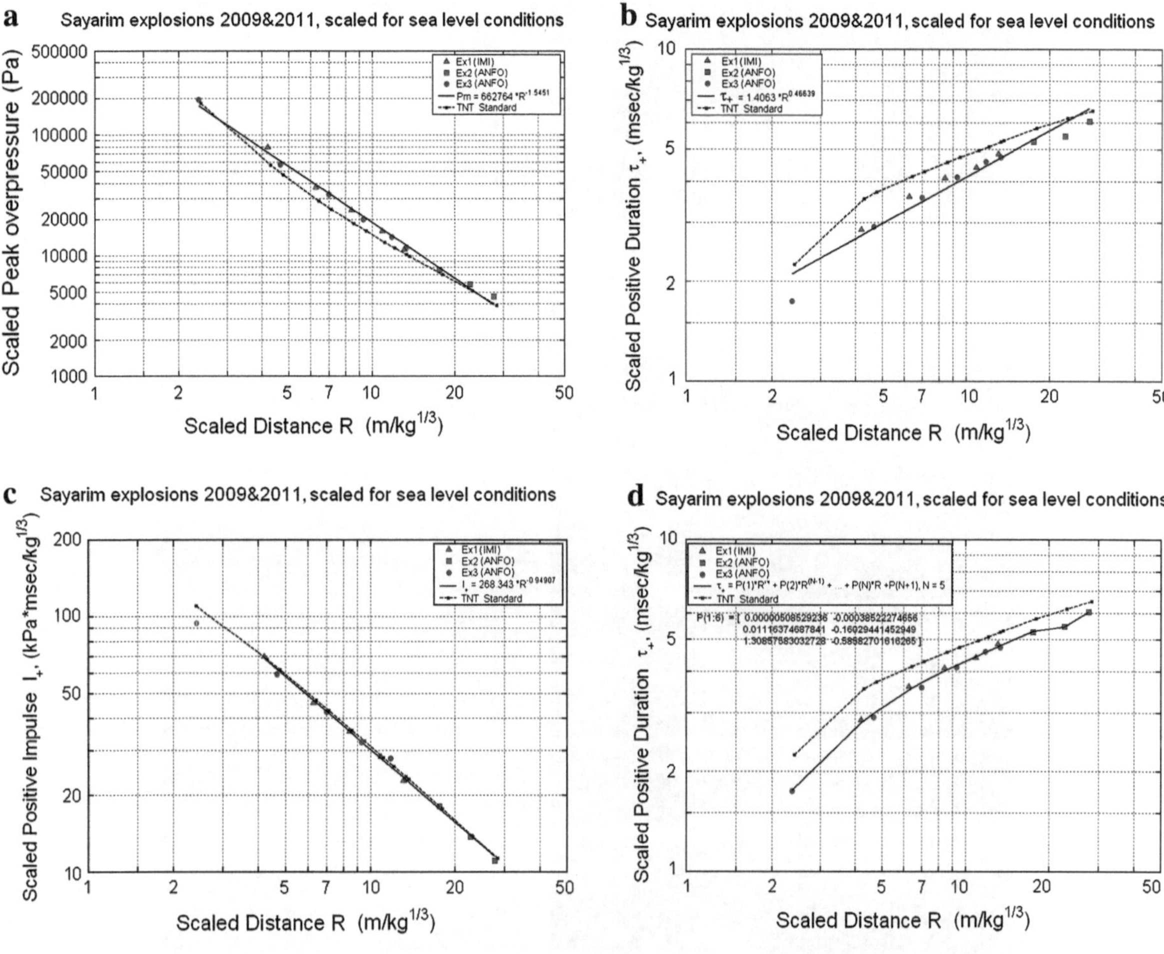

Figure 10

Air-blast data measured for three Sayarim large explosions, scaled to the TNT yield charge and the sea level conditions, show a high uniformity: peak pressure (**a**), positive phase duration (**b**, **d**) and positive phase impulse (**c**). Reference (TNT standard) values, obtained by BECv4 procedure, are also shown

of the charges, and cannot be attributed to any specific charge elements. Away from the explosion, the peaks are merged, and after ~70–100 m, a stable uniform main shock phase is formed.

Many observers of the large Ex3 placed at ~9 km reported hearing two clear "bang" sounds with a delay of less than 1 s, which were interpreted as two separate explosions. Multiple pronounced "bang" sounds can be clearly identified at recordings of the monitoring video-cameras (at several hundreds meters) for both shots Ex2 and Ex3. Initially it was suggested that these calibration shots were not simultaneous as planned, however, detailed analysis of the data showed that these audio-phenomena were caused by secondary shocks in the air-blast wave observed at all records of high-pressure gauges and accelerometers (Figs. 4, 6, 7). These records, along with the yield estimates using the main-shock peak pressures (Table 5) confirm that all explosive materials were fully detonated in the initial 2–3 ms.

10. Crater Observations and Seismic Magnitudes

A regular hemispherical crater was created by the 82-ton IMI explosion in 2009 (Gitterman, 2010). The Survey of Israel conducted accurate GPS crater measurements just after the shot, and provided 2D–3D crater images, and accurate diameter D and depth H estimations (Fig. 12; Table 6).

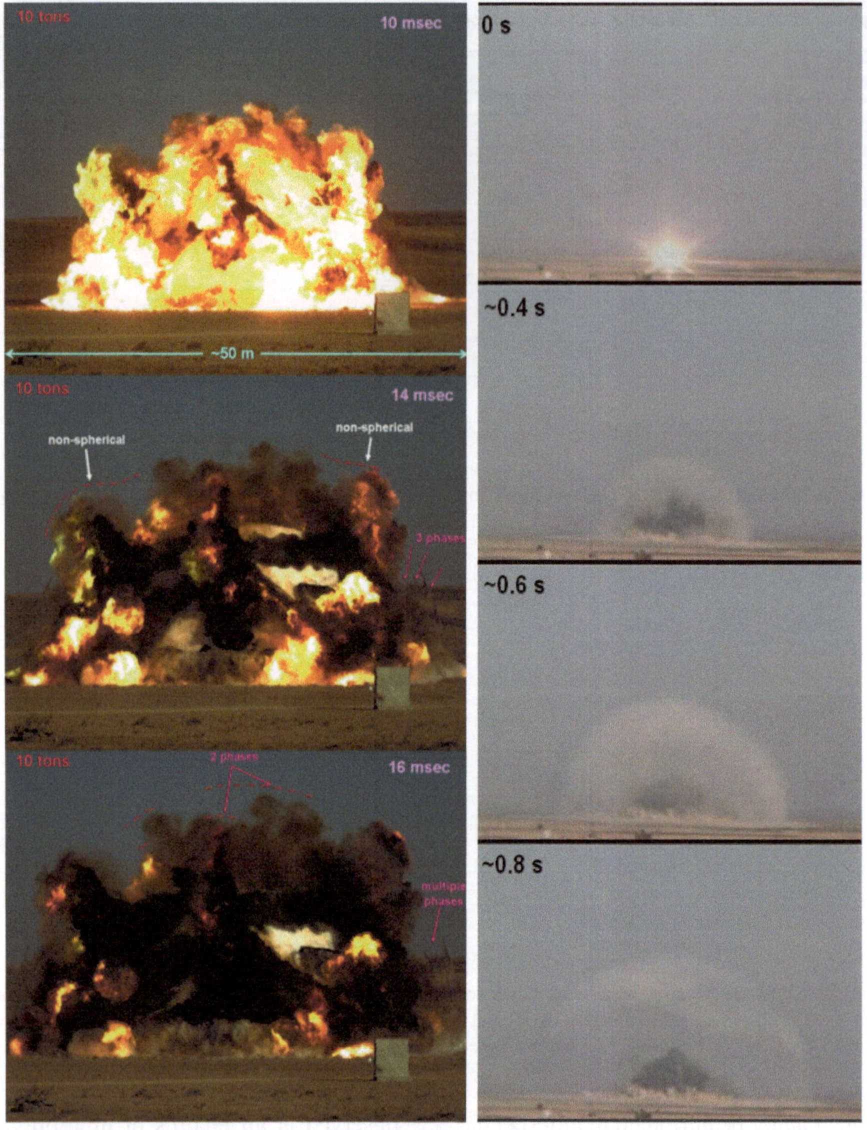

Figure 11
Snapshots from speed video-recording for Ex2 at 360 m (*left*) and from usual video recording at 6 km for Ex3 (*right*), show unique blast effects

Unlike the 2009 explosion, complex irregular-shaped craters were found for the two January 2011 ANFO shots, with step-like walls and a cone of crashed rocks in the center (Fig. 13).

For comparison, we used empirical equations for craters of large-scale explosions (in Russia) on soft soils surface (Adushkin and Khristoforov, 2004):

$$D(\mathrm{m}) = 2 \times 3.36\, W^{0.336};\quad H(\mathrm{m}) = 1.78\, W^{0.316} \qquad (8)$$

and another diameter equation based on 200 large surface shots (Kinney and Graham, 1985):

$$D(\mathrm{m}) = 8 \times W^{1/3}, \qquad (9)$$

where W is TNT equivalent yield in tons.

The crater parameters calculated using Eqs. (8) and (9) are presented in Table 6, jointly with measured values, demonstrating significantly smaller sizes.

A hemispherical surface explosion of 90.7 ton TNT, on alluvium, in the US Army Waterways Experiment (1962), was similar to Sayarim shot Ex1, but created a bigger crater with radius $D = 42.6$ m,

Figure 12
Crater created by the 2009 explosion Ex1 (*top*), and 2D and 3D images provided by the Survey of Israel (*bottom*)

Table 6

Measured crater dimensions, compared with calculated values and local seismic magnitudes M_d

Event	Nominal weight (tons)	TNT equivalent yield (tons)	VOD (m/s)	M_d	Diameter, D (m)			Depth, H (m)	
					Measured	Calculated		Measured	Calculated Eq. (8)
						Eq. (8)	Eq. (9)		
Ex1	81.66 IMI	96.0	~7,500	2.5	30.0	31.1	36.6	5	7.5
Ex2	10.24 ANFO	7.4	2,400	2.2	12.0	13.2	15.6	1.5	3.35
Ex3	102.1 ANFO	76.8		2.8	29.0	28.9	34.0	3	7.0

$H = 6.3$ m (CraterDatabase_V1.3 2004). The ConWep procedure (ConWep, 1997) provided also much larger crater size estimations for Ex3 (102 tons ANFO on dry sandy clay): $D = 41$ m, $H = 13$ m.

The local duration magnitude M_d estimated from the Israel Seismic Network (ISN) records showed two interesting important features (Table 6). Firstly, the magnitude for Ex1 with higher TNT yield is smaller that for Ex3 (though the crater is larger). Evidently, the main reason is the much higher VOD of stronger IMI explosives compared to ANFO. A similar effect was found for small borehole shots NEDE in the USA in 2008, where larger amplitude Rg and Love waves were generated from shots using black powder (VOD 530 m/s) and ANFO emulsion (VOD 5,260 m/s), than shots with Composition B (VOD 8,100 m/s) (Stroujkova, 2012). Secondly, all magnitudes seem smaller than could be expected, based on extensive GII experience in monitoring of numerous detonations of outdated ammunition at SMR. For example, an ammunition shot at SMR in June 2008 (consisted mainly of M-15 mines) with TNT yield about 10 tons, similar to Ex2, but with the downward detonation (the booster mines were on the charge top, as on Fig. 14), produced a higher $M_d = 2.5$ (Gitterman and Hofstetter, 2008; Gitterman, 2009).

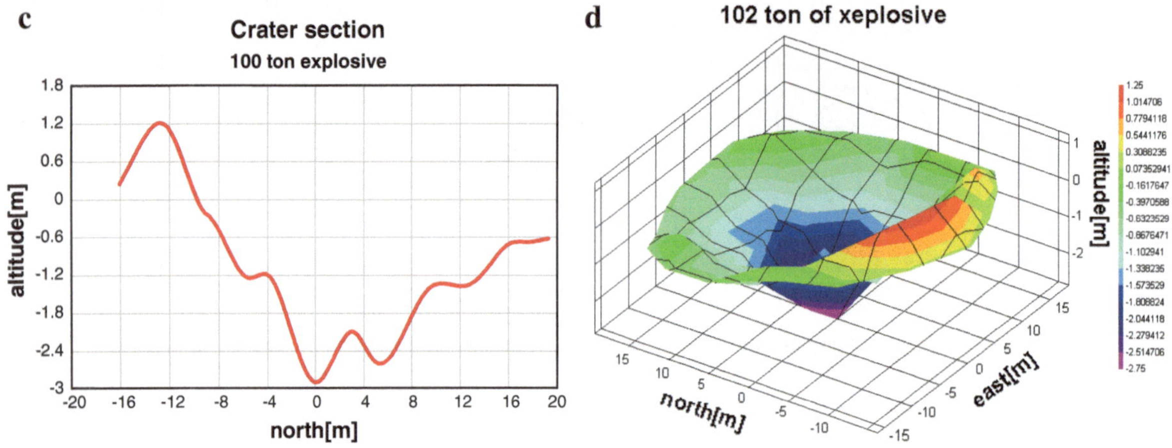

Figure 13
Crater general view for 2011 Sayarim explosions Ex2 (**a**) and Ex3 (**b**), and 2D (**c**) and 3D (**d**) crater images for Ex3, provided by the IDF Engineer Corps team

The results show that crater parameters, the diameter and especially the depth, and seismic magnitudes, for all Sayarim explosions were smaller than the expected values, based on data for previous similar experiments, thus indicating a decreased coupling of explosive energy with the ground, supposedly due to the specific charge design and especially the upward detonation.

A striking evidence for importance of the detonation direction for cratering was found for test explosions at SMR with equal charges of 1 ton of TNT and Composition B, with different detonation direction (GITTERMAN, 2009). For two explosions (separately TNT and Composition B), conducted in June 2008, the boosters (mines) were placed on the top of the charge (Fig. 14a), providing downward detonation. Both explosions produced large deep (~0.5–0.7 m) craters (a little deeper for the explosion with more powerful Composition B) (Fig. 14b). In distinction, the upward detonation was applied to 1 ton (mix of TNT, Composition B and RDX) shot in December 2008, resulting in a very small shallow crater (Fig. 14c, d; Table 7). Evidently a significant difference in the crater size prevail over non-identical charge shape and placement conditions.

The smaller crater for the upward detonation evidences that relatively large and small portions of explosive energy were released to the atmosphere, and penetrated the ground as seismic energy, respectively, compared with downward detonation. It is also confirmed by lower amplitudes of seismic waves from the December 2008 explosion recorded at ISN stations (Fig. 15), and by an appropriately smaller magnitude, compared to June 2008 shots (Table 7). Consequently, cratering effects observed for surface explosions can be considered as an indicator for

Figure 14
Cratering effects for two SMR test shots: downward detonation (*red arrows*) of 1 ton TNT in June 2008 (**a**), that created a large deep crater (**b**), and upward detonation of the same size charge in December 2008 (**c**), that produced a very small shallow crater (**d**) (from GITTERMAN, 2009)

Table 7

Observed crater sizes for 2008 test shots at SMR

Date	Charge (kg) explosives	Detonation direction	M_d	Crater size	
				Diameter (m)	Depth (m)
2008/06/24	1,020 TNT	Down	1.8	2.5–2.6	0.5–0.6
2008/06/25	1,020 CompB	Down	2.0	2.4–2.5	0.6–0.7
2008/12/02	1,040 TNT, CompB&RDX	Up	1.6	1.8–2.0	0.3–0.4

explosion energy partitioning between seismic waves in the ground and acoustic waves in the atmosphere.

11. Discussion

Based on an empirical scaling relationship for the dominant period of infrasound waves, another yield estimation for the 2009 explosion was obtained. The physical basis for this relationship is the increased acoustic transit time of the blast radius with increased yield (REVELLE, 1998). Besides, the dominant period is not influenced much by propagation like peak amplitudes. Utilizing a well-recorded dataset of Israel and Cyprus infrasound stations at a distance range of 150–570 km, where single clear main arrivals were observed, the yield of 128 tons of TNT was roughly estimated based on the average dominant period

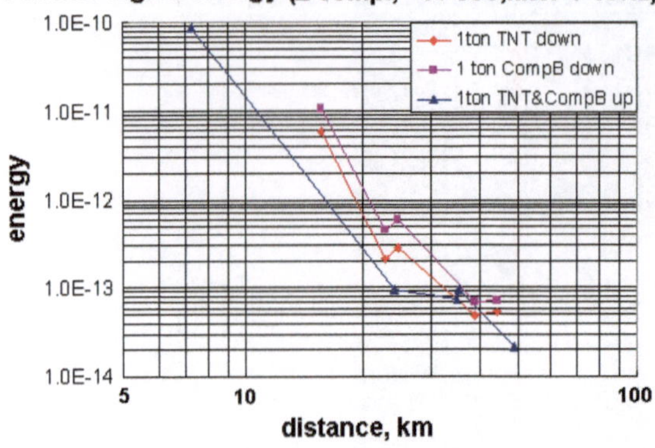

Figure 15

Calculated energy (in counts) of the whole seismic signal (vertical component) from 2008 test shots recorded at five close ISN stations

(GITTERMAN, 2010). This estimation is relatively close to the reliable yield of 96 tons from air-blast measurements (Table 5).

Similar to the near-source pressure gauges, anomalous enhanced peak pressure amplitudes were found at local infrasound stations for the 2009 explosion, indicating a possible upward directivity effect and asymmetric energy radiation to the atmosphere (GITTERMAN, 2010). Possibly, this effect caused an overestimated yield of 0.3–0.5 kT, obtained at two far-regional IMS infrasound stations I26DE and I48TN of the International Data Center (IDC), using wind-corrected amplitudes, as applied to several yield-range-amplitude attenuation laws, including LANL2003 (BROWN, 2009). This large misfit in yield estimation denotes an obvious need for refinement and tuning of the monitoring procedures, using GT0 infrasound sources, to obtain reliable yield values. However, using a new-developed Parabolic Equation-based semi-empirical relation, a reasonable value ~0.1 kT at the dominant frequency ~0.5 Hz was estimated from I48TN data for the 2009 explosion (LE PICHON, personal communication, 2010).

As mentioned before, the air-blast parameter SS delay could be used as a yield estimator if the explosive type is known, based on the developed scaled relationships. On the other hand, the explosive type can be identified and the detonation velocity can be roughly estimated, if the yield is known and SS delays are measured. It seems that atmospheric or surface nuclear explosions do not produce air-blast SS phases (e.g., BRODE, 1956), because a nuclear test provides an instantaneous and point-like source, where parameters of chemical shots (crucial for the SS effect)—charge size, detonation velocity, expanding detonation products (gases)—are not applicable. Then, by a logical extension, it is guessed that the SS phase could be used as a discriminant between a small nuclear and large chemical explosion. Theoretically it is possible under specific conditions, but it can be practical only at close distances where the main shock and especially secondary shock are still observable, and not transformed into numerous phases of acoustic/infrasound wave.

12. Conclusions

Two large-scale shots of about 100 tons, with a pyramid/hemisphere charge design, were successfully conducted by GII in summer 2009 (Ex1) and winter 2011 (Ex2, Ex3). The explosion design was rather complicated and different from the ideal homogenous hemisphere used in similar past experiments. Strong boosters and an upward charge detonation scheme were applied to ensure that more energy is radiated to the atmosphere. Enhanced peak pressures at all distances and smaller craters and seismic magnitude indicated that the developed charge design provided a strong explosion energy generation and necessary energy partition: large portions of energy to the atmosphere and less to the ground as seismic energy. Under these conditions evaluation of the actual explosion yield, as one of the important source parameters, is crucial for the GT0 calibration experiment. The accurate yield estimations were obtained, based on measured positive phase impulse in air-blast waves at near-source distances, and considered as ground truth parameter. Compared to the nominal explosives weight, the estimated TNT equivalent yield was enlarged for the 2009 shot and reduced for the 2011 shots.

Dense near-field observation systems provided valuable abundant data sets of audio-visual, air-shock and acoustic records that were utilized for interpretation of observed unique blast effects and determination of blast wave parameters, suited for

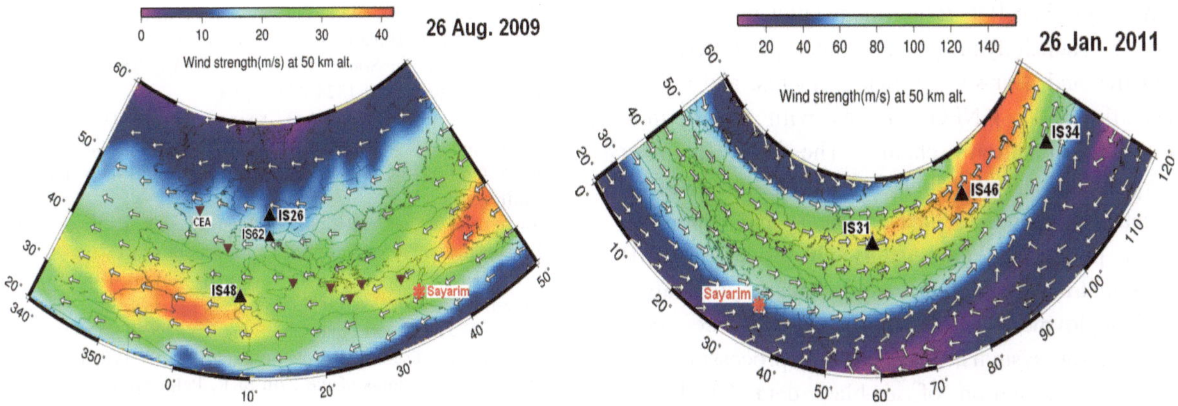

Figure 16
Modeling of stratospheric winds for the pair of large Sayarim shots (Ex1 and Ex3), based on atmospheric specifications from the ECMWF, provided 6-hourly models of 91 layers on a global 0.5° × 0.5° grid, the figure is derived from these models at 50 km altitude (courtesy of L. Evers, KNMI). The explosion site and recorded IMS (*triangle*) and some regional portable (*inverted triangle*) infrasound stations are shown

yield estimation. Empirical functional relationships between basic blast-wave parameters, the yield and the distance were developed, including a novel secondary shock delay relation, that provided a base for a new yield estimator.

The main goal of this dual Sayarim calibration experiment was reached: fully controlled infrasound sources, the strongest since the establishment of the IMS network, were observed at several IMS infrasound stations in Europe, Middle East and Asia, thus establishing the first GT0 infrasound dataset for this region.

In the 2009 experiment, ten portable infrasound arrays were deployed by collaborating institutions from eight countries at regional distances in Europe and EM–ME countries, coordinated by the University of Hawaii and CTBTO teams. In the 2011 experiment, institutions from 20 countries collaborated to set up a dense infrasound network for near-regional observations. All together 20 portable arrays were deployed in 13 countries throughout the EM–ME region by local institutions, coordinated by the CTBTO and UM teams, that provided recording equipment for the far field observations in 2011 (Gitterman, 2011).

The infrasound signals were observed at numerous regional and IMS stations up to 3,400 km to the west/north-west for the summer 2009 shot Ex1, and up to 6,250 km to the east for the winter 2011 shot Ex3—further than expected. The very clear westward/eastward infrasound propagation according to appropriate summer/winter stratospheric wind directions, respectively, was demonstrated (Fig. 16), thus verifying empirically common models of infrasound propagation in the atmosphere.

The complementary smaller Ex2 shot (winter 2011) of 10 tons was conducted during the afternoon 2 days prior to the Ex3 (main 102-ton explosion detonated in the morning), providing additional valuable data for the analysis of charge scaling and infrasound propagation features affected by various atmospheric (wind) conditions.

The unique collected database provides an important contribution in the modeling of long-range infrasound propagation in the atmosphere, detection analysis at different stations, and refining and tuning of IDC evaluation procedures of infrasound monitoring and yield estimation.

Acknowledgments

Many organizations and persons participated in the preparation of Sayarim calibration explosions, measurements and data processing. High-quality explosives in convenient packages were supplied by IMI Ltd. (I. Veksler) for the 2009 experiment, and by EMI Ltd. (Dr. D. Hershkovich) for the 2011 experiment. Elita Security Ltd. (S. Kobi) assembled the 2009 IMI charge with maximal concentration of explosives. The IDF Experiment Division (E. Stempler, Y. Hamshidyan) provided appropriate territory, logistics and near-source measurements, and assembled the 2011 ANFO charges with the optimal initiation/detonation scheme. The IDF Engineer Corps (Z. Savir) contributed to preparation of infrastructure for close-in measurements, and conducted crater parameters measuring and processing. GII personnel helped in logistics procedures, preparation and deployment of numerous near-source local observation systems, and in data processing and graphic presentation of air-blast data (U. Peled, N. Perelman). Thanks to our collaborators Dr. J. Bonner, Weston Geophysical Corporation, Dr. R. Waxler, the University of Mississippi and Dr. E. Marchetti, the University of Firenze, for supplementing data of close seismo-acoustic observations in 2011 experiment. Thanks to V. Pergament of Magnitogorsk State Technical University for assisting in the analysis of the pressure-distance relationship, to Dr. L. Evers, the Royal Netherlands Meteorological Institute (KNMI), for supplementing stratospheric winds modeling. Sayarim experiments were supported by the US Army SMDC in 2009 (M. Pickens), and PTS CTBTO in 2011 (Dr. L. Zerbo and Dr. J. Given). Research work of one of the authors (Y.G.) was supported by the Israel Ministry of Immigrant Absorption. We are thankful to anonymous reviewers for valuable comments, especially relating to additional aspects of the air-blast secondary shock delay application, and the analysis of seismic magnitudes.

References

Adushkin, V. and B. Khristoforov (2004). *Craters of Large-Scale Surface Explosions, Combustion, Explosion, and Shock Waves*, Vol. *40*, No. 6, pp. 674–678.

Baker, W. E. (1973). Explosions in Air, University of Texas Press, Austin and London, 268p.

Bhattacharyya, J., H.E. Bass, D.P. Drob, R.W. Whitaker, D.O. Revelle and T.D. Sandoval (2003). Description and analysis of infrasound and seismic signals recorded from the Watusi explosive experiment, September 2002, Proceedings of the 25th SRR, Tucson, Arizona, September 2003.

Bonner, J., R. Waxler, Y. Gitterman and R. Hofstetter (2012). Seismo-acoustic energy partitioning at near-source and local distances from the 2011 Sayarim explosions in the Negev desert, Israel. Bull. Seis. Soc. Am. (in press).

Bowman J.R., H. Israelsson, G. Shields, M. O'Brien and Y. Gitterman (2009). Detection and Characterization of Infrasound Signals at IMS Stations from Explosions at the Sayarim Military Range, Israel: 2005-2009. Presented at the Infrasound Technology Workshop, Brasilia, Brazil, November 2-6, 2009.

Brode, H. L. Point Source Explosions in Air; The Rand Corp., Research Memo RM-1824-AEC, 1956.

Brown, D., C. Collins and B. Kennet (2003). The Woomera infrasound and seismic experiment, September-October 2002, in Proceedings of Infrasound Technology Workshop, La Jolla, California.

Brown, D., I. Kitov, N. Brachet, P. Mialle and R. Le Bras (2009). Enhancements to the CTBTO infrasound processing system, Presented at the Infrasound Technology Workshop, Brasilia, Brazil, November 2-6, 2009.

Christie, D., B. Kennet and Ch. Tarlowski (2005). Detection of regional and distant atmospheric explosions at IMS infrasound stations, Proceedings of the 27th SRR, Palm Springs, CA, September 2005.

Christie, D., J. Vivas Veloso, P. Campus, M. Bell, T. Hoffmann, A. Langlois, P. Martysevich, E. Demirovic, and J. Carvalho (2001), *Detection of atmospheric nuclear explosions: The infrasound component of the International Monitoring System*, Kerntechnik, *66*, 96–101.

Conwep (1997), a collection of conventional weapons effects calculations from the equations and curves of TM 5-855-1 "Design and Analysis of Hardened Structures to Conventional Weapons Effects". USAE Engineer Research and Development Centre, Vicksburg, MS.

CraterDatabase_V1.3 (2004). http://keith.aa.washington.edu/craterdata/.

Fee, D., R. Waxler, Y. Gitterman, J. Given, J. Coyne, P. Maille, P. Grenard, M. Garces, J. Assink, D. Drob, D. Kleinart, H. Buchanan, and L. Zerbo (2012). Overview of the 2009 and 2011 Sayarim Infrasound Calibration Experiments (in preparation).

Garces M., D. Fee, R. Waxler, C. Hetzer, J. Assink, D. Drob, A. Le Pichon, Y. Gitterman, and R. Hofstetter (2009). The Sayarim Calibration Experiment: Theory and Observations. Presented at the Infrasound Technology Workshop, Brasilia, Brazil, November 2-6, 2009.

Garces, M., C. Hetzer, K. Lindquist and D. Drob (2002). Source location algorithm for infrasonic monitoring, Proceedings of the 24th SRR, Ponte Vedra Beach, Fl, September 2002.

Gitterman, Y., V. Pinsky, A.-Q. Amrat, D. Jaser, O. Mayyas, K. Nakanishi, and R. Hofstetter (2005). Source features, scaling and location of calibration explosions in Israel and Jordan for CTBT monitoring, Isr. J. Earth Sci., 54: 199-217.

Gitterman Y. and R. Hofstetter (2008). Infrasound Calibration Experiment in Israel: Preparations and Test Shots. Proceedings of the 30th Monitoring Research Review, LA-UR-08-05261.

Gitterman, Y., M. Garces, R. Bowman, D. Fee, H. Israelsson, R. Hofstetter, V. Pinsky (2009). Near-source and far-regional observations for Sayarim test explosions. Proceedings of the 2009 Monitoring Research Review: Ground-Based Nuclear Explosion Monitoring Technologies, LA-UR-09-05276, pp. 724-734.

Gitterman, Y. (2010). Sayarim Infrasound Calibration Explosion: near-source and local observations and yield estimation, in Proceedings of the 2010 Monitoring Research Review: Ground-Based Nuclear Explosion Monitoring Technologies, LA-UR-10-05578, Vol. II, pp. 708–719.

Gitterman, Y., J.W. Given, J. Coyne, R. Waxler, J. L. Bonner, L. Zerbo, R. Hofstetter (2011). Large-scale controlled surface explosions at Sayarim, Israel, at different weather patterns for

Infrasound Calibration of the International Monitoring System, in Proceedings of the 2011 Monitoring Research Review: Ground-Based Nuclear Explosion Monitoring Technologies, LA-UR-11-04823.

GlobalSecurity.org, http://www.globalsecurity.org/wmd/ops/testing-effects.htm.

Kinney, G. and K. Graham (1985). Explosive shocks in Air, Springer-Verlag, New York, 9-10.

Koper, K., T. Wallace, R. Reinke and J. Leverette (2002). *Empirical Scaling Laws for Track Bomb Explosions based on Seismic and Acoustic Data*, Bull. Seis. Soc. Am., Vol. *92*, No. 2, 527-542.

Landau, L. (1945). *On shock waves at large distances from the place of their origin*, Applied Mathematics and Mechanics, V. *9*, No. 4, pp.96-103 (in Russian).

Perkins, B. and W. Jackson (1964). Handbook for prediction for air-blast focusing, BRL report no. 1240.

Norris, D., J. Bhattacharyya and R. Whitaker, 2007. Development of advanced propagation models and application to the study of impulsive infrasonic events. Proceedings of the 29th Monitoring Research Review, September 2007.

ReVelle, D., R. Whitaker, W. Armstrong (1998). Infrasound from the El Paso super-bolide of October 9, 1997. LANL Report LA-UR-98-2983.

Ringdal, F. and J. Schweitzer (2005). Combined seismic/infrasound processing: A case study of explosions in NW Russia, NORSAR Sci. Rep. 2-2005, Semiannual Technical Summary, Kjeller, August 2005.

Sachs, R.G. (1944). The dependence of Blast and Ambient Pressure and Temperature. BRL report 466, May 1944.

Sadwin L. and Swisdak, M. (1970). Blast Characteristics of 20 and 100 Ton Hemispherical ANFO Charges. NOL Data Report, NOL TR 70-32, 17 Mar 1970.

Stroujkova, A., J. Bonner, M. Leidig, R. Martin, and P. Boyd (2012). Seismic studies of explosions with different velocities of detonation in Barre granite, *In press, BSSA.*

Stump, B. W., S. M. McKenna, C. Hayward and T-S. Kim (2002). Seismic and Infrasound data and models at near-regional distances, Proceedings of the 24th SRR, Ponte Vedra Beach, Fl, September 2002.

Swisdak M. (2000). DDESB Blast Effects Computer, Version 4.0.

(Received November 29, 2011, revised August 9, 2012, accepted August 12, 2012, Published online September 6, 2012)

Pure Appl. Geophys. 171 (2014), 621–627
© 2012 Springer Basel AG
DOI 10.1007/s00024-012-0494-4

Pure and Applied Geophysics

Modelling of a Single-Channel Beta–Gamma Coincidence Phoswich Detector Using Geant4 for the Conversion Electron Energy Peak Resolution and Beta–Gamma Coincidence Efficiency Improvement

Weihua Zhang,[1] Pawel Mekarski,[1] Marc Bean,[1] Jing Yi,[1] and Kurt Ungar[1]

Abstract—In this study, an optimized single-channel phoswich well detector design has been proposed and assessed in order to improve beta–gamma coincidence measurement sensitivity of xenon radioisotopes. This newly designed phoswich well detector consists of a plastic beta counting cell (BC404) embedded in a CsI (Tl) crystal coupled to a photomultiplier tube. The BC404 is configured in a cylindrical pipe shape to minimise light collection deterioration. The CsI (Tl) crystal consists of a rectangular part and a semi-cylinder scintillation part as a light reflector to increase light gathering. Compared with a PhosWatch detector, the final optimized detector geometry showed 15 % improvement in the energy resolution of a ^{131m}Xe 129.4 keV conversion electron peak. The predicted beta–gamma coincidence efficiencies of xenon radioisotopes have also been improved accordingly.

Key words: Phoswich detector design, Monte Carlo simulation, FWHM, xenon radioisotopes.

1. *Introduction*

In the Comprehensive Nuclear Test Ban Treaty (CTBT) verification regime, the two-dimensional beta–gamma coincidence detector is widely used for radioxenon activity measurement due to its ability to distinguish ^{131m}Xe and ^{133m}Xe and to suppress non-coincident background events as well as for its high sensitivity to the coincident events characteristic of the xenon radioisotopes of interest (Saey and De Geer 2005). The beta–gamma coincidence detector basically consists of two scintillators, i.e. an inorganic scintillator and a plastic scintillator counting cell. In many beta–gamma coincidence systems, beta and gamma radiations are detected in two separate channels, such as the automated radioxenon sampler/analyzer (Reeder *et al.* 1998), the Swedish Automatic Unit for Noble Gas Acquisition (SAUNA) (Ringbom *et al.* 2003) and the automatic portable radiometer of isotopes xenon (Popov et al. 2000). To simplify the coincidence system configuration and calibration procedure, a phoswich, namely PhosWatch (PW5), beta–gamma coincidence detector has been developed, in which both beta and gamma radiations can be detected by one single channel (Hennig and Tan 2006). The original design of the PW5 detector is presented in Fig. 2. As shown in the figure, the detector has a cavity drilled inside the inorganic scintillator, which accommodates the plastic scintillator counting cell (BC404). The embedded plastic cell is filled with xenon gas to be counted. The xenon radioisotope decays by emitting gamma-rays or X-rays in coincidence with beta-particles or conversion electrons (CE), respectively. The thin plastic scintillator is used to absorb low energy beta-particles and CE (less than 400-keV approximately), while the longer range gamma-rays and X-rays are mainly detected by the inorganic scintillator. When a xenon radioisotope emits both CE and X-rays, a beta–gamma coincident signal can be distinguished. The detector enables the same energy (average 30.4 keV) K-X-rays (including K-alpha and K-beta X-rays) emitted from ^{131m}Xe (abundance 53.75 %) and ^{133m}Xe (abundance 54.85 %) to be separated by the well-specified discrete conversion electrons peaks at 129.4 keV (abundance 61.6 %) and 198.7 keV (abundance 62.9 %), respectively (NuDat2.6 2012). The peak width of CE and X-rays is mostly limited by the resolution of the scintillators, but the spatial variations in the light collection efficiency throughout the detector can significantly degrade the overall energy resolution of the detector.

[1] Radiation Protection Bureau of Health Canada, 775 Brookfield Road, AL 6302D1, Ottawa, ON K1A 1C1, Canada. E-mail: weihua.zhang@hc-sc.gc.ca

It was shown previously that the degraded X-ray and CE peak energy resolutions due to light collection efficiency deterioration (light-path obstruction) by the embedded beta counting cell can be minimised by varying the embedded beta counting cell positions inside the CsI (Tl) crystal (ZHANG *et al.* 2010). The previous simulation demonstrated that the PW5 detector configuration obtained by placing beta counting cell furthest away from the photomultiplier tube (PMT) can provide the best overall CE energy peak resolution (ZHANG *et al.* 2010). However, in this configuration some gamma-rays may escape out of the thinner part of CsI (Tl) crystal, which may result in a lower beta–gamma coincidence measurement sensitivity of xenon radioisotopes, especially for ^{135}Xe due to its relatively high energy gamma-ray (249.8 keV). For this reason, it is worthwhile to investigate an alternative phoswich detector design that will have both optimal CE and X-ray energy resolution and beta–gamma coincidence efficiency for all xenon radioisotopes. The design guidance, incorporates the following key principles: single-channel phoswich detector for keeping system simplicity; minimization of lost or obstructed photons by embedding the beta counting cell between the inorganic scintillator and the PMT; availability of a commercial scintillator and ease of manufacturing; and the shape of CsI (Tl) scintillator for maximum light gathering by optical reflection.

2. *Monte Carlo (MC) Simulation for Detector Geometry Design*

An experimentally determined pure ^{131m}Xe standard beta–gamma coincidence spectrum by PW5 detector is illustrated in Fig. 1. The two dimensional histograms (c) and (d) at the bottom are distributions resulting from energy deposition events in the BC404 (beta energy axis) and CsI (Tl) (gamma energy axis) detectors, respectively. The top view of the three dimensional histogram at the top right (b) contains information regarding the beta–gamma coincidence decay events detected by the PhosWatch detector. The histogram is divided into several rectangularly shaped regions of interest (ROI) based on the gamma-ray energy and beta-particle energy calibrations. The boundaries of each ROI rectangular depend on xenon radioisotope decay modes. The gross counts in each ROI are calculated by summing the counts per channel and used to calculate the radioxenon concentration, as listed in Fig. 1a.

As shown in Fig. 1d, the energy resolution of the prototype PW5 [beta counting cell located at geometry centre of CsI (Tl) detector] at ^{131m}Xe 129.4 keV CE peak is experimentally determined by a pure ^{131m}Xe source as 38.0 %, which is expressed as peak full width at half maximum (FWHM) divided by peak center position. Subsequently the optical resolution scale of BC404 scintilltor was defined such that the simulated FWHM at 129.4 keV CE peak matches this experiment's value. The resolution scale was also applied to the following steps in the new phoswich detector geometry design.

There are several steps involved in performing the proposed phoswich detector design and optimization. The first step of simulation design involved the addition of an identical PMT to the face opposite the original PMT window. This additional PMT collected the light directed to the other half of the scintillator crystal as opposed to reflecting it back around the inner scintillator shell. The second step involved altering the geometry of the PW5 detector itself in order to reflect the maximum amount of light possible into the PMT. This involves finding the optimum of CE and X-ray energy resolution and beta–gamma coincidence efficiency by varying the embedded plastic cell positions inside the CsI (Tl) detector, and the analysis of several different types of reflectors on the surface opposite the PMT window.

All the above detector geometry and designs evaluations were performed with MC simulations according to the ^{131m}Xe CE energy peak resolution. The MC simulations were created using the GEANT4 Toolkit (AGOSTINELLI *et al.* 2003). All of the data processing and output was performed through the implementation of AIDA (VINCENZO INNOCENTE 2003) in the simulation. GEANT4 has also been shown to accurately simulate the PW5 detector system, including the optical signals received by the PMT (MEKARSKI *et al.* 2009). The identical simulation methodology has been applied for this study with any changes to geometry or tracking explained. GEANT4 allows the user to define the geometry and physics processes to be

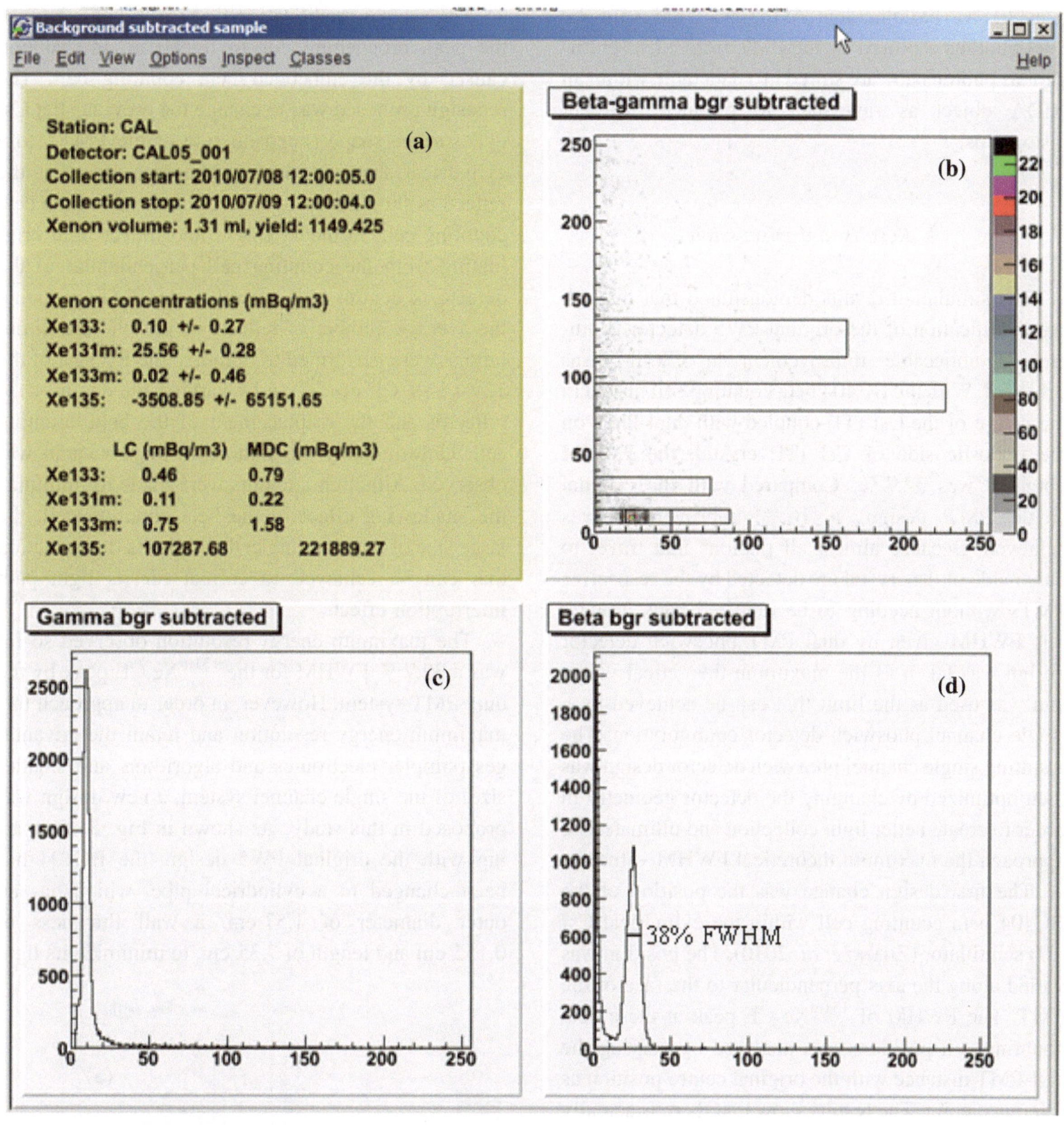

Figure 1

An experimentally determined pure ^{131m}Xe standard beta–gamma coincidence spectrum by PW5 detector. The analyzed radioxenon concentration (**a**); *Top view* of the 3D histogram regarding the beta–gamma coincidence decay events detected by PhosWatch detector (**b**); the 2D histograms of energy deposition events at CsI (Tl) (**c**), and BC404 (**d**)

used in the simulation. In this simulation, both the standard electromagnetic and optical process packages were defined. GEANT4 then handles the transportation and interaction of the simulated particle through the defined geometry. The radioactive decay module (G4RadioactiveDecay.3.2) was used for radioxenon isotopes. For each radioxenon decay event, the Geant4 detector module records the energy deposition in each scintillator and counts the number of photons captured by the PMT from each scintillator, with a time stamp. The time difference between the BC-404 plastic cell and the CsI (Tl) scintillator is recorded, and events occurring within a specified "coincidence-time window" of each other in the respective scintillator are considered to be coincident. At the end of the event, the total photons captured by the PMT from each scintillator as photon pairs are stored in a data table within an AIDA object as raw data to produce different histograms.

3. Results and Discussion

The simulation results demonstrated that through the modification of the original PW5 detector geometry, a noticeable improvement in FWHM was achieved. With the BC404 beta counting cell placed in the centre of the CsI (Tl) coupled with dual PMT on the opposite side of CsI (Tl) crystal, the FWHM obtained was 32.7 %. Compared with the original single PMT design, a 16 % improvement was achieved. Because almost all photons that travel to either side of the crystal are detected by the respective PMTs without needing to be reflected back around, this FWHM given by dual PMT phoswich detector design was taken as the maximum theoretical value, and was used as the limit that can be achieved by a single channel phoswich detector configuration. The resulting single channel phoswich detector design was then optimized by changing the detector geometry in order to create better light collection and ultimately to approach the maximum theoretical FWHM value.

The first design change was the position of the BC404 beta counting cell within the cylindrical CsI (Tl) scintillator (Zhang *et al.* 2010). The position was varied along the axis perpendicular to the face of the PMT. The FWHM of ^{131m}Xe CE peak at these beta counting cell positions was analysed by varying the cell-PMT distance with the original centre position as reference point. The results show that there is actually a decrease in FWHM for a position close to the PMT face. However, for positions farther away from the PMT there is an improvement in the FWHM of the system. The simulation demonstrated that the detector configuration obtained by placing the beta counting cell furthest away from the PMT provides the best detector energy resolution. A 9 % improvement in the FWHM of the 129.4 keV CE of ^{131m}Xe was achieved compared to the FWHM value obtained with the beta counting cell at the CsI (Tl) geometric centre (Zhang *et al.* 2010).

The second modification was to further minimise the peak broadening due to light path interruption caused by the embedded beta counting cell. The redesign proposed was to change the previous flat CsI (Tl) surface shape (opposite side of the PMT), to a cylindrical shaped reflector. The idea was that the reflectors should help to reflect photons from the beta counting cell. In theory this would reflect light originating from the counting cell perpendicular to the detector's central axis directly into PMT, reducing the average number of reflections per photon. Simulations were performed to optimize the FWHM of the 129.4 keV CE of ^{131m}Xe by varying the depth of the reflector and the displacement of the beta counting cell. Unfortunately, no significant improvement was observed. Although attempts were made to minimize the shadowing effect of the beta counting cell, the large size of the counting cell still limits the reduction that can be achieved to correct for its light path interruption effect.

The maximum energy resolution observed so far was a 32.7 % FWHM for the ^{131m}Xe CE peak by the dual PMT system. However, in order to approach this maximum energy resolution and retain the advantages (simpler electronics and algorithms and smaller size) of the single channel system, a new design was proposed in this study. As shown in Fig. 2, comparing with the original PW5 design, the BC404 has been changed to a cylindrical pipe, which has an outer diameter of 1.57 cm, a wall thickness of 0.155 cm and length of 7.35 cm, to minimize its light

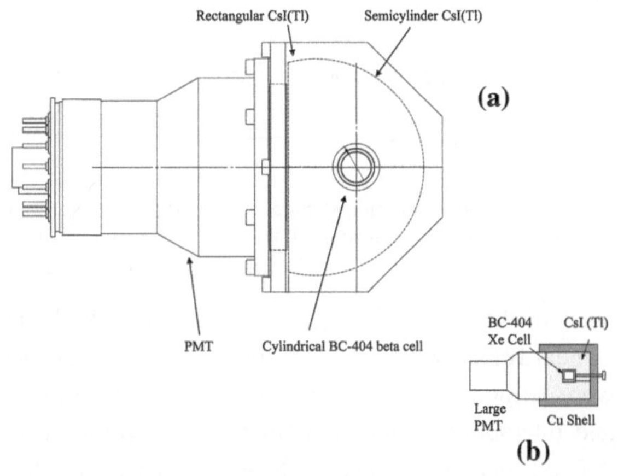

Figure 2
The schematic diagram of proposed detector geometry modeling by Geant4 simulations (**a**); the original design of PW5 (**b**)

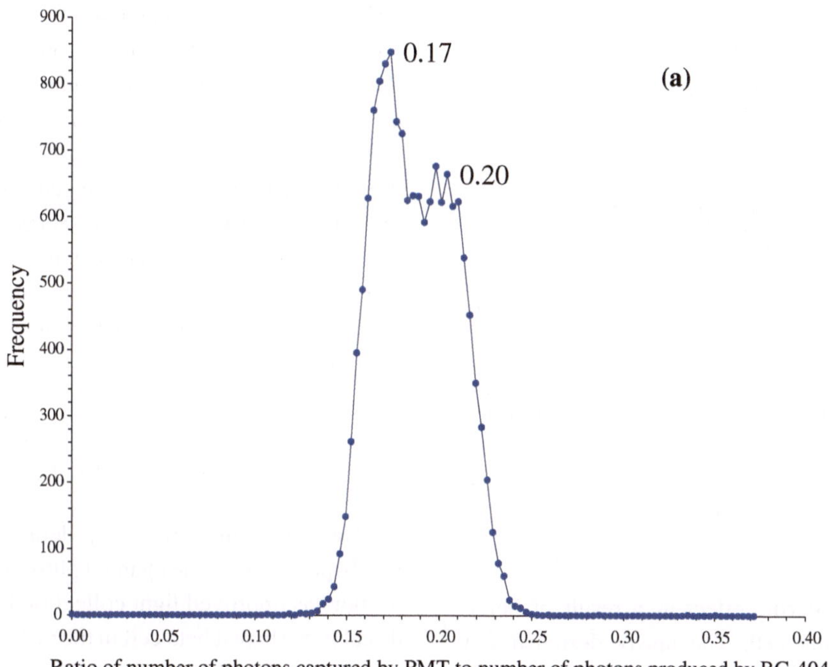

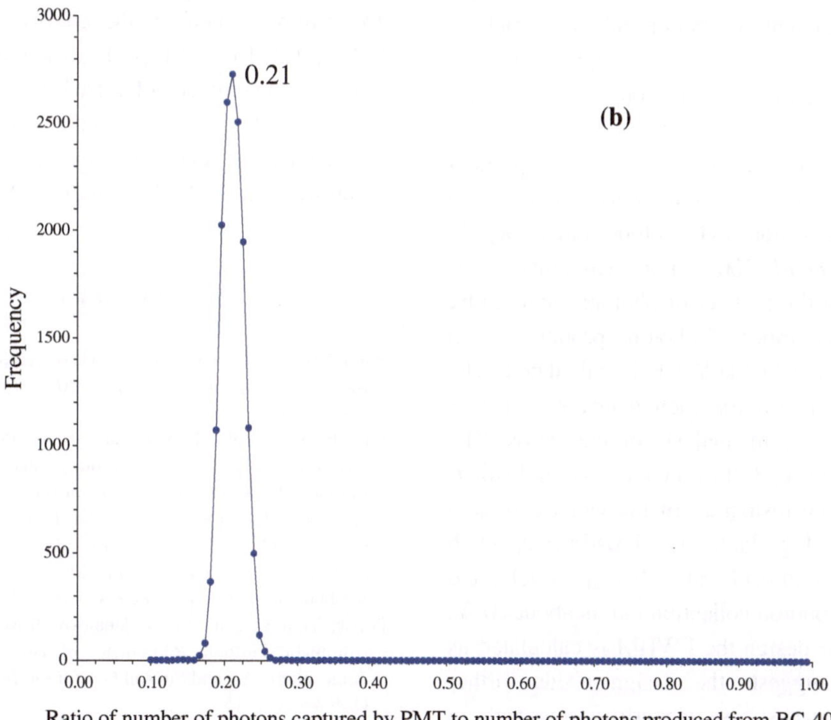

Figure 3
The BC404 photon counting histogram analysis for optimizing detector design parameters, **a** the worst-case scenario **b** the best-case scenario

Table 1

The coincidence detection efficiencies ($\varepsilon_{\beta\gamma}$) of the different x-rays and gamma rays of xenon radioisotopes by Geant4 modeling

Radioxenon	ROI limits, keV			$\varepsilon_{\beta\gamma}$ (%)
	y axis (CsI)	*x* axis (BC404)	ROI index	
^{133}Xe	17.47–42.59	0–347.50	1	89.6 (87.2)
	67.80–98.73	0–396.50	2	72.5 (68.5)
	17.47–42.59	16.25–89.55	3	33.0 (32.6)
	17.47–42.59	239.03–396.50	4	8.8 (8.0)
	17.47–42.59	170.07–396.50	5	24.6 (22.9)
	17.47–42.59	16.25–161.90	6	60.2 (58.4)
^{131m}Xe	17.47–42.59	96.99–161.90	7	78.0 (74.9)
^{133m}Xe	17.47–42.59	170.07–239.03	8	69.6 (65.3)
^{135}Xe	226.58–275.47	0–908.0	9	72.9 (63.8)

Data in brackets is simulated with original PW5

shadowing and scattering effect as a result of being an installed counting cell. The newly designed CsI (Tl) crystal consists of a rectangular component (10.16 × 10.16 × 1.27 cm) and a semi-cylindrical scintillation component (5.08 cm radius), which is used as a light reflector to increase light gathering. These scintillator geometries are commercially available. The readout electronics of the simulated system described in this study are based on the PW5 product package that is currently being beta-tested at the CTBT Canadian xenon radioisotope laboratory in Ottawa (Hennig *et al.* 2008). For each coincidence event, the ratio of the number of photons captured by the PMT to the number of photons produced from BC404 by ^{131m}Xe 129.4 keV CE is calculated. The performance evaluation for each detector design is based on the histogram analysis of the ratios. The histogram plots in Fig. 3 illustrate the best- and worst-case scenarios. The histogram of the worst-case scenario, as shown in Fig. 3a, has two broad peaks which are distributed around 0.17 and 0.20, respectively, and the maximum of photon collection rate is about 40 %. With this detector design the FWHM is calculated as 39.1 %, which suggests the design needs further improvement. The best-case scenario corresponds to the proposed detector design in this study. The histogram is a single peak and narrowly distributed around a mean value of 0.21, as shown in Fig. 3b, and the maximum of the photon collection rate from BC404 can reach as high as 100 %. The FWHM of ^{131m}Xe 129.4 keV CE is calculated as 33.1 % in this best-case scenario. The beta–gamma coincidence detection efficiencies of the four xenon radioisotopes of interest, as listed in Table 1, are also calculated with the simulation results both at the proposed detector design in this study and original PW5 detector geometries using the same BC404 resolution scales. The simulated results in Table 1 indicate that the radioxenon coincidence counting efficiencies by the new detector design are about 5–14 % better than those achieved with original PW5 system.

4. Conclusions

The MC simulation study demonstrated that the newly proposed single channel phoswich well detector design has improved light collection by using a cylindrical pipe shaped beta cell in the CsI (Tl) crystal and a semi-cylindrical light reflector. Compared with PW5 detector, the final optimized detector geometry showed 15 % improvement in the energy resolution of the ^{131m}Xe 129.4 keV CE peak. Based on the simulations scaled to the empirical FWHM the ^{131m}Xe 129.4 keV CE peak, the calculated beta–gamma coincidence efficiencies of xenon radioisotopes are significantly better than those obtained with a PW5 detector.

References

P.R.J. Saey and L.-E. De Geer, Notes on radioxenon measurements for CTBT verification purposes, 2005, Appl. Radiat. Isot., *63* (5–6), 765–773.

P.L. Reeder, T.W. Bowyer and R.W. Perkins, Beta–gamma counting system for Xe fission products, 1998, Journal of Radioanalytical and Nuclear Chemistry, *235*, 89–94.

A. Ringbom, T. Larson, A. Axelson, K. Elmgren and C. Johansson, SAUNA—a system for automatic sampling, processing, and analysis of radioactive xenon, 2003 Nuclear Instruments and Methods in Physics Research A, *508*, 542–553.

Popov, Yuri S., and Yuri V, Dubasov, Russian Federation, 2000 Automatic Portable Radiometer of Isotopes Xe ARIX-02 in Atmospheric Air and Subsoil Gas for On-Site Inspection, CTBT/OSI/WS-6/PR/15.

Hennig, W., Tan, H., Warburton, W.K., McIntyre, J.I., Single-channel beta gamma coincidence detection of radioactive Xenon using digital pulse shape analysis of phoswich detector signals, 2006, IEEE Transactions on Nuclear Science *53* (2), 620–624.

NuDat2.6, 2012. National Nuclear Data Center, information extracted from the NuDat2 database, http://www.nndc.bnl.gov/nudat2/.

ZHANG W., MEKARSKI P. and UNGAR K, Beta–gamma coincidence counting efficiency and energy resolution optimization by Geant4 Monte Carlo simulations for a phoswich well detector, 2010, Appl. Radiat. Isot., *68*(12), 2377–81.

Geant4 Collaboration (S. AGOSTINELLI et al.) Accelerators, Spectrometers, Detectors and Associated Equipment, 2003, Nuclear Instruments and Methods in Physics Research A *506*(3), 250-303.

V. INNOCENTE, L. MONETA, A. PFEIFFER, "Review of the Abstract Interfaces for Data Analysis (AIDA) from a developer's perspective", CERN Version1.0, 2003.

P. MEKARSKI, W. ZHANG, K. UNGAR, M. BEAN, E. KORPACH, Monte Carlo simulation of a PhosWatch detector using Geant4 for xenon isotope beta–gamma coincidence spectrum profile and detection efficiency calculations, 2009, Applied Radiation and Isotopes, *67* (10), 1957–1963.

W. HENNIG, H. TAN, W.K. WARBURTON, A. FALLU-LABRUYERE, K. SABOUROV, M.W. COOPER, J.I. MCINTYRE, and A. GLEYZER, Development of a COTS radioxenon detector system using Phoswich detectors and pulse shape analysis, 2008 Monitoring Research Review: Ground-Based Nuclear Explosion.

(Received October 20, 2011, revised April 3, 2012, accepted April 16, 2012, Published online May 15, 2012)

Pure Appl. Geophys. 171 (2014), 629–644

DOI 10.1007/s00024-012-0581-6

Pure and Applied Geophysics

Analysis of Radionuclide Releases from the Fukushima Dai-Ichi Nuclear Power Plant Accident Part I

G. Le Petit,[1] G. Douysset,[1] G. Ducros,[2] P. Gross,[1] P. Achim,[1] M. Monfort,[1] P. Raymond,[3] Y. Pontillon,[2] C. Jutier,[1] X. Blanchard,[1] T. Taffary,[1] and C. Moulin[1]

Abstract—Part I of this publication deals with the analysis of fission product releases consecutive to the Fukushima Dai-ichi accident. Reactor core damages are assessed relying on radionuclide detections performed by the CTBTO radionuclide network, especially at the particulate station located at Takasaki, 210 km away from the nuclear power plant. On the basis of a comparison between the reactor core inventory at the time of reactor shutdowns and the fission product activities measured in air at Takasaki, especially ^{95}Nb and ^{103}Ru, it was possible to show that the reactor cores were exposed to high temperature for a prolonged time. This diagnosis was confirmed by the presence of ^{113}Sn in air at Takasaki. The ^{133}Xe assessed release at the time of reactor shutdown (8×10^{18} Bq) turned out to be in the order of 80 % of the amount deduced from the reactor core inventories. This strongly suggests a broad meltdown of reactor cores.

Key words: Fukushima Dai-ichi accident, Takasaki, fission products, reactor damages, isotopic ratios, CTBTO.

1. Introduction

The Fukushima I nuclear power plant (NPP) in Japan (also known as Fukushima Dai-ichi) suffered major damage from the 9.0 MW moment magnitude scale categorized earthquake and subsequent tsunami that hit Japan on March 11, 2011. First commissioned in 1971, the plant consists of six boiling water reactors (BWR) managed by the Tokyo Electric Power Company (TEPCO). Since September 2010, unit 3 had been fueled by a small fraction of mixed-oxide (MOX) fuel, in addition to low enriched uranium (LEU). Only three of the six units were reported as nominally operating (Units 1, 2 and 3) at the time of the event. Although the operating reactors were automatically shutdown after the earthquake, the subsequent tsunami disabled the emergency generators required to cool down the reactors and the spent-fuel pools (SFP). It is today accepted that Units 1, 2 and 3 reached a fuel meltdown state, within a time period of <100 h after the occurrence of the earthquake.

Through its involvement in the verification of compliance with the comprehensive nuclear-test-ban treaty (CTBT), CEA has developed expertise and tools in the following fields: data processing chains including atmospheric transport modeling (global and mesoscale), radioactive spectral emission analysis and radiological impact calculations, nuclear reactor physics and nuclear fuel cycle. The originality of the present study is to take advantage of these complementary approaches to analyze the radionuclide releases of the Fukushima NPP accident.

In order to estimate the worldwide impact of the atmospheric radioactive releases and to understand the accident severity, CEA analyzed the data collected by the radionuclide Particulate and Noble gas stations of the comprehensive nuclear-test-ban treaty organization (CTBTO) network and the measurements performed by the laboratories supporting this network.[1] As our institution hosts the French CTBTO certified laboratory we participated in the re-analysis of samples collected by the particulate detection network.

At the time of this publication, numerous papers dealing with the Fukushima accident (Leon *et al.*,

[1] CEA, DAM, DIF, 91297 Arpajon, France. E-mail: gilbert.le-petit@cea.fr

[2] CEA, DEN, Cadarache, France.

[3] CEA, DEN, Saclay, France.

[1] Disclaimer: the views expressed in this publication are those of the authors and do not necessarily reflect the views of the CTBTO Preparatory Commission.

2011; Bowyer *et al.*, 2011; Bolsunovsky and Dementyev, 2011; Sinclair *et al.*, 2011; Zhang *et al.*, 2011; Manolopoulou *et al.*, 2011; Pittauerová *et al.*, 2011; Qiao *et al.*, 2011; Tanimoto *et al.*, 2011; Okada, 2011; Masson *et al.*, 2011; Chino *et al.*, 2011), were released, but none specifically based on the reactor damage assessment from fission products detections in Japan.

In the first part of this publication (Part I) we report observations and interpretations of the Fukushima NPP accident features relying on the chronology and the nature of major nuclide detections at Takasaki (210 km southwest from the Fukushima NPP) where the CTBTO Particulate (JPP38) and Noble Gas (JPX38) Japanese stations are located. Based on direct gamma measurement carried out at JPP38 and on post-analysis of samples sent to CTBTO laboratories, we performed a detailed analysis in order to assess the severity of reactor damages at Fukushima NPP.

This analysis is based on the level of the quantified activity concentrations and on physicochemical characteristics of detected fission products (FPs). Measured activation products (APs) in air linked to major components included in the fuel cladding were also of interest for interpreting reactor core damages.

Accordingly, these elements were used to provide a better confidence in assessed source terms deduced from atmospheric transport modeling calculations. This is dealt with in the second part of the publication (Part II, Achim *et al.*, 2012).

2. *First Data Collections at Takasaki, Japan*

Dates/times mentioned hereafter refer to coordinated universal time (UTC) and sample collection date refers to collection start. Nuclear data are taken from reference (Bé *et al.*, 2008).

As part of the CTBT network, the particulate station located at Takasaki (equipped with a fully automatic RASA type high volume air sampler (Miley *et al.*, 1998) follows a standard sampling-measurement cycle: (1) 24 h particle collection onto a polypropylene filter (standard sampled volume: about 23,000 m^3, down to 4,000 m^3 during the Fukushima event for radiation protection purposes); (2) 24 h sample decay in order to eliminate the largest part of radon progenies; (3) 24 h gamma analysis of the collected sample using a high efficiency HPGe detector. A preliminary spectrum is automatically transferred to the International Data Center in Vienna and to National Data Centers every 2 h. On March 14, 2011 at 06:55 the measurement of a sample collected 2 days earlier started. Figure 1 shows the last preliminary gamma spectrum and the full one. The first one (on the left, acquisition time 71,926 s) exhibits only natural radionuclide signatures although the HPGe energy resolution is drastically altered (>5 keV) owing to cooling problems following a power loss at the station. The second one (on the right, acquisition time 79,100 s) shows gamma rays corresponding to fission products (^{132}Te at 228 keV and ^{131}I at 364 keV) growing in the spectrum during acquisition. The energy resolution of these gamma lines is almost nominal (1.4 keV) suggesting a station's operator intervention. The sudden detection of FPs in the spectrum suggests that contaminated air mass directly entered the station premises on March 15, 2011 at 04:00 ($\pm$1 h). FPs identified in this spectrum of poor quality were the most volatile ones: ^{131}I, ^{132}I, ^{133}I, ^{134}Cs, ^{137}Cs and ^{132}Te. This was the first detection in the CTBTO network of FPs related to the Fukushima accident.

Direct FP activity concentrations derived from the detector calibration curve were not reliable since the measurement geometry differed significantly from the calibrated one. Consequently, in order to provide a rough assessment of air activity concentrations, it was necessary to build a calibration curve in accordance with the actual measurement geometry. This was done considering the maximum free volume surrounding the detector and limited by the shielding (Fig. 2; Nadalut *et al.*, 2010). It has been estimated to about 4.5 l (Forrester, 2011) matching at a rough approximation a Marinelli Baeker geometry. This geometry was simulated using a Monte Carlo code taking into account the HPGe detector characteristics. Considering the spectrum acquisition time corresponding to the first FPs detection (Fig. 1) the activity concentrations for ^{131}I and ^{137}Cs have then, respectively been, estimated to 3,700 $\pm$ 1,000 Bq/m^3 and 400 $\pm$ 100 Bq/m^3 (March 15, 2011 04:00).

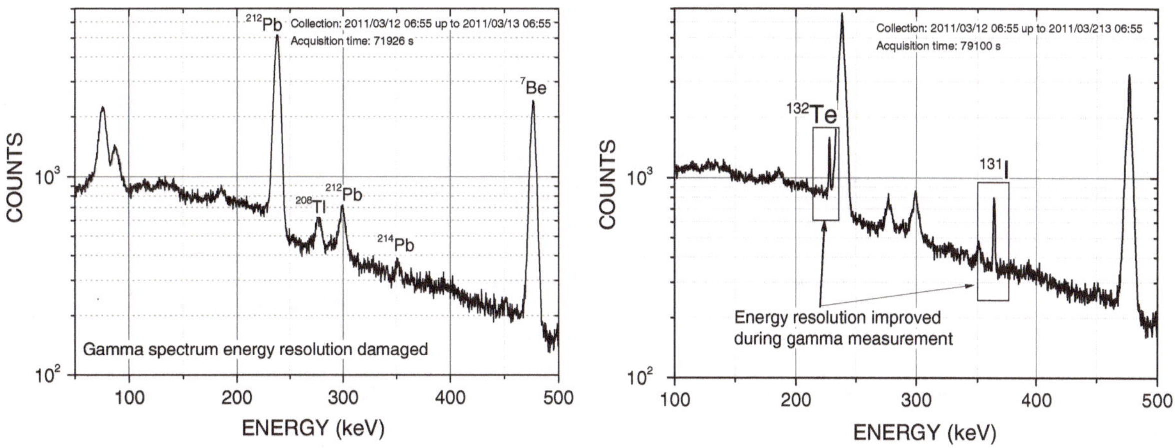

Figure 1
Sample collection start March 12, 2011 at 06:55. *On the left* acquisition time 71,926 s (no fission product in the spectrum). *On the right* acquisition time 79,100 s (FPs visible)

Figure 2
Takasaki particulate station (RASA type). Filter sample assembly processing on the *left* and detail of the detector and shielding configuration on the *right*. Free volume between shielding and detector is assessed to 4.5 l (figure courtesy of CTBTO)

Filter related to the collection of March 13, 2011 06:55, as the previous one, can not contain any FP. ^{131}I, ^{132}I, ^{133}I, ^{134}Cs, ^{137}Cs, ^{129m}Te, ^{131m}Te, ^{132}Te, ^{99}Mo and ^{140}La detected in this spectrum (Fig. 3) were related again to air masses present on the station premises and a probable deposit of radioactive materials onto the sampling and detection units.

The presence of a 1,001 keV gamma line (Fig. 4) was noticed. Count rate of 2.2×10^{-3} counts per second (cps) was similar to the one usually observed for the environmental samples collected at JPP38 prior to the Fukushima accident although the spectrum baseline was about 20 times higher than the standard one. This gamma line was detected at higher count rates (up to 23 cps) for the hottest sample (collection start March 15, 2011) and suggested the presence of ^{234m}Pa (line of ^{234m}Pa at 766 keV interfered with the ^{95}Nb main emission line and the 63 keV line of its direct parent ^{234}Th can not be detected owing to the level of the spectrum baseline at low energy). Usually ^{234m}Pa (and ^{234}Th) measured

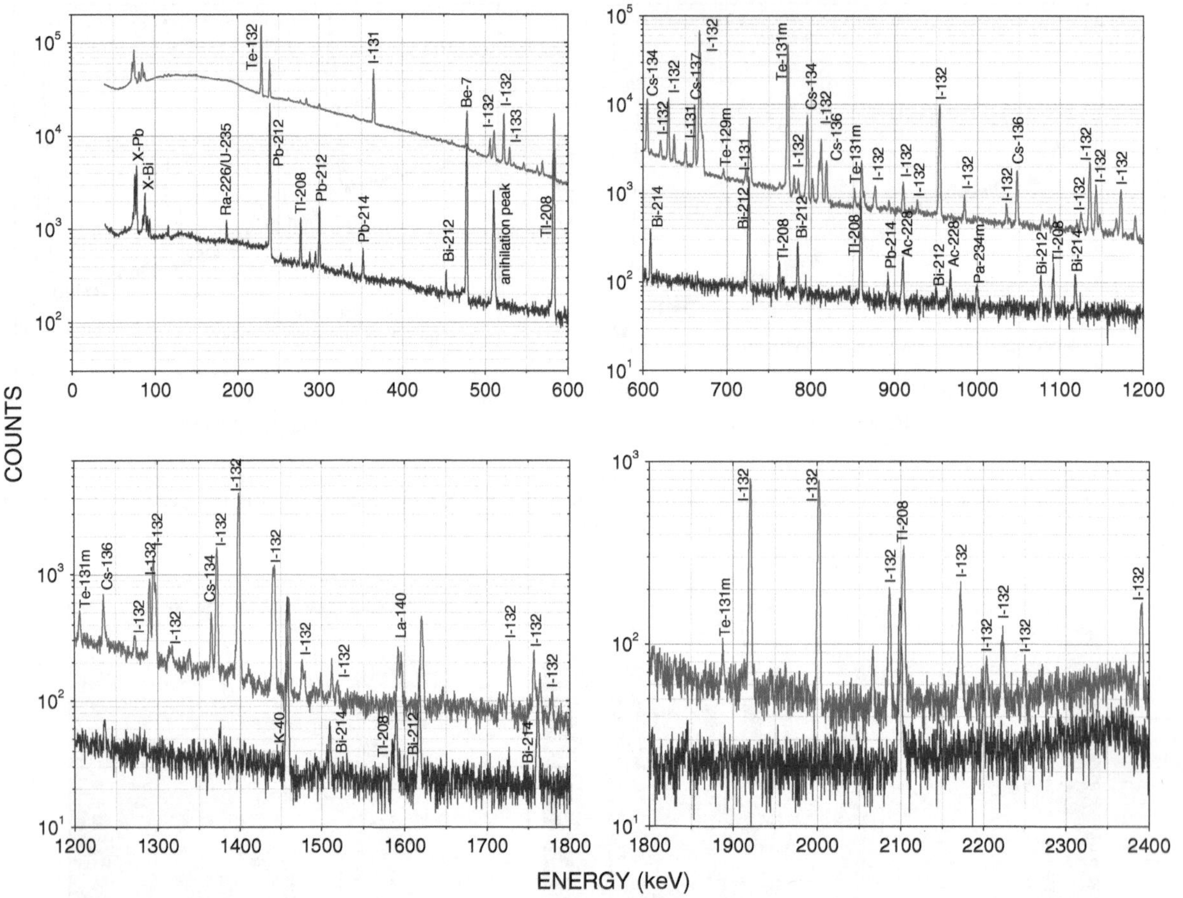

Figure 3
green Sample collection start March 13, 2011 at 06:55 (acquisition time 73,102 s) compared with *blue* sample collected prior to the Fukushima accident (collection start March 10, 2011)

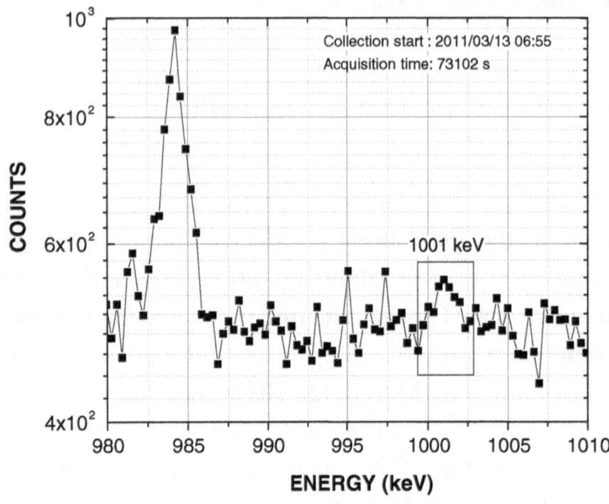

Figure 4
Evidence of a gamma line at 1,001 keV (related to the collection start March 13, 2011 06:55)

by gamma spectrometry implies the presence of ^{238}U (not detectable at an environmental level from its gamma lines) as one of its direct progeny. This detection would have led to a worrying situation in terms of radiological impact.

A detailed analysis of the hottest sample spectra allows for rejecting the presence of ^{238}U. Relying on the gamma spectrum related to the collection of March 22, 2011 (acquisition time: 12,694 s, dead time: 12 %), we determined that the 1,001 keV peak was due to pile-up of 637.0 keV and 364.5 keV gamma lines, both pertaining to ^{131}I disintegration scheme (Fig. 5). Likewise, Fig. 5 clearly shows some other pile-up peaks related to ^{131}I, respectively, at ~649 keV (Σ 364.5, 284.3) and ~729 keV (Σ 364.5, 364.5 and Σ 80.2, 284.3, 364.5).

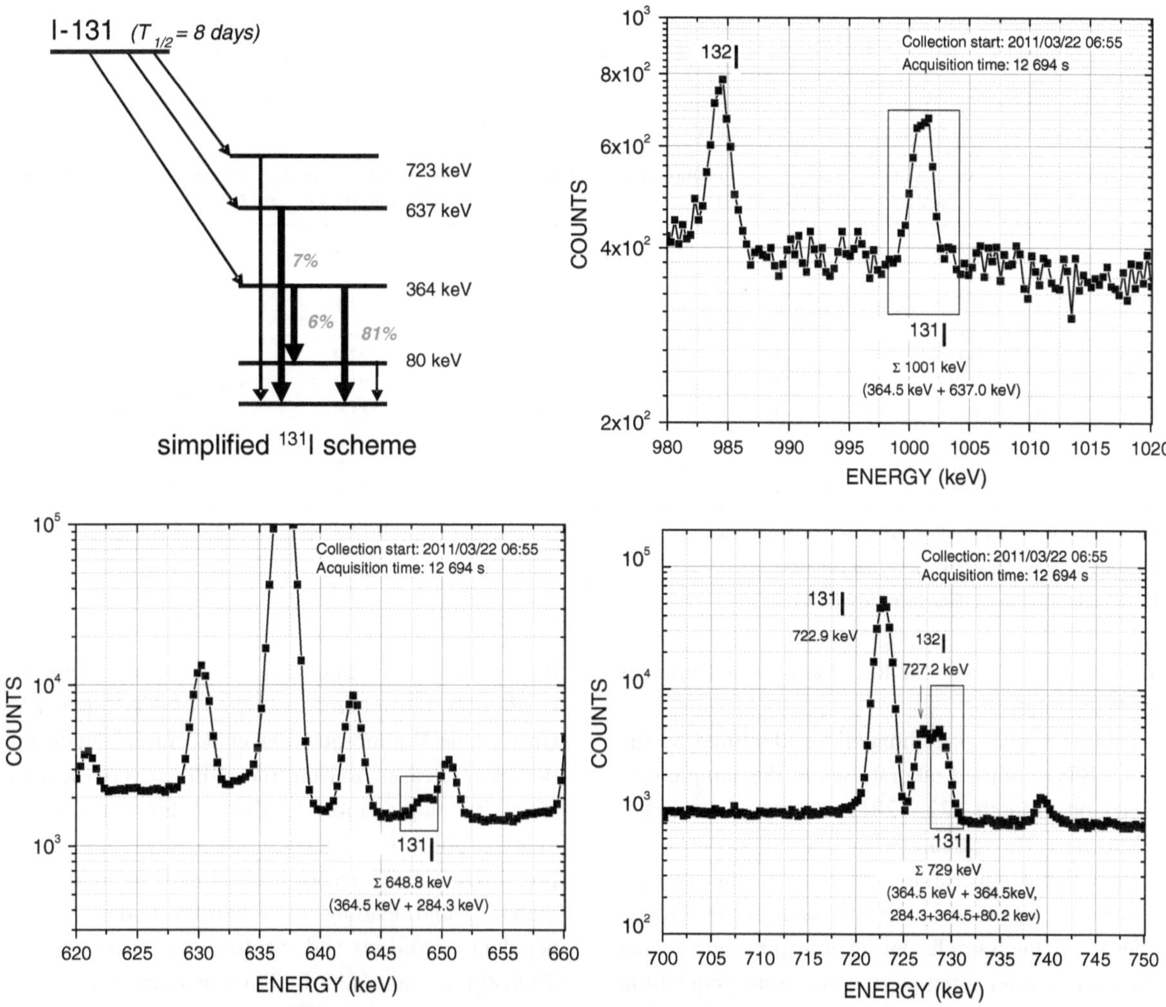

Figure 5

On the upper-left simplified ^{131}I disintegration scheme; *on the upper-right* and *lower left* and *right*, respectively, pile-up peaks related to ^{131}I at 1,001 keV, 649 keV and 729 keV (sample collection start March 22, 2011 06:55 at Takasaki, acquisition time 12,694 s, dead time: 12 %)

3. Tentative Spectral Post-Interpretation in Terms of Reactor Damages

In this publication, focus was put on the gamma emission spectrometry analysis of one of the hottest samples collected at Takasaki on March 22, 2011. Following the CTBTO procedure, this sample has been split into two parts and sent to two CTBTO certified laboratories for in-depth analysis. As expected, this sample exhibits a complex spectrum, with more than 200 gamma lines to be explained, including multi-gamma radioisotopes like ^{134}Cs, ^{136}Cs or ^{132}I, involving true coincidence summing (TCS) and false coincidence summing (pile-up) effects. TCS correction determinations have been performed relying on a specific procedure described in Jutier *et al.* (2007).

Furthermore some radionuclides have to be associated as parent/daughter such as ^{129m}Te/^{129}Te, ^{132}Te/^{132}I, ^{140}Ba/^{140}La (which have been produced during sampling, decay and acquisition of the gamma emission spectrum) with appropriate branching ratios and half-lives to be correctly processed. In order to perform correct identification of the radionuclides, as well as to extract the weakest gamma lines, spectra were interactively reviewed. A customization is needed to build up an efficient gamma library in order to avoid non-identified peaks or, on the contrary, to

Table 1

Average activity concentrations of radionuclides detected and quantified by two CTBTO laboratories related to a sample collected at Takasaki

Takasaki radionuclide station, air collection started March 22 06:55 up to 23 06:55, 2011

Radionuclide	Type	Main gamma line (keV)	Activity concentration Bq/m^3 (Unc: 1σ in %)	Radionuclide	Type	Main gamma line (keV)	Activity concentration Bq/m^3 (Unc: 1σ in %)
^{7}Be	Nat.	477.6	1.35×10^{-3} (5.5 %)	^{95}Nb	FP	765.8	2.88×10^{-4} (5.0 %)
^{234m}Pa[b]	Nat.	1,001.0	1.25×10^{-3} (25 %)	^{99}Mo	FP	140.5	4.11×10^{-3} (16 %)
^{110m}Ag	FP	657.8	6.55×10^{-5} (8.0 %)	^{103}Ru[a]	FP	497.1	4.95×10^{-5} (17 %)
^{140}Ba	FP	537.3	2.86×10^{-4} (12 %)	^{86}Rb[a]	FP	1,076.8	2.82×10^{-4} (15 %)
^{134}Cs	AP	604.7	3.63×10^{-2} (4.0 %)	^{125}Sb	FP	427.9	3.21×10^{-4} (5.0 %)
^{136}Cs	FP	818.5	5.81×10^{-3} (4.3 %)	^{113}Sn[a]	AP	391.7	4.67×10^{-5} (10 %)
^{137}Cs	FP	661.7	3.54×10^{-2} (3.8 %)	^{127}Te	FP	417.9	5.16×10^{-3} (12 %)
^{131}I	FP	364.5	2.18 (4.0 %)	^{129}Te	FP	459.6	3.47×10^{-2} (4.7 %)
^{132}I	FP	667.7	9.98×10^{-2} (4.7 %)	^{129m}Te	FP	695.9	5.49×10^{-2} (4.0 %)
^{140}La	FP	1,596.2	4.60×10^{-4} (5.0 %)	^{132}Te	FP	228.1	1.00×10^{-1} (5.5 %)

Nat natural, *FP* fission product, *AP* activation product

[a] Radionuclide quantified by only one laboratory

[b] False identification as demonstrated in Sect. 2

avoid false allocations due to numerous interferences and/or overlaps.

Table 1 shows all radionuclides identified by the two CTBTO certified laboratories in the sample collected on March 22, 2011 (sampled volume: 19,226 m^3). The calculated activity concentrations and associated uncertainties result from an average between results found by the two laboratories. Owing to decay of the sample during transportation to the laboratories, dead times during spectrum acquisition time are for both laboratories of the order of 3 %, less than the one observed at Takasaki station (11.5 %). This value is low enough to allow appropriate standard corrections. Some isotopes have been detected by only one laboratory (^{103}Ru and ^{86}Rb). This might be explained by a possible sample inhomogeneity (Han *et al.*, 2012).

Among the radionuclides listed in Table 1, apart from ^{7}Be which is a cosmogenic radionuclide, the others are FPs (^{110m}Ag, ^{140}Ba, ^{136}Cs, ^{137}Cs, ^{140}La, ^{131}I, ^{132}I, ^{99}Mo, ^{95}Nb, ^{103}Ru, ^{86}Rb, ^{125}Sb, ^{127}Te, ^{129}Te, ^{129m}Te and ^{132}Te) and APs (^{113}Sn, ^{134}Cs[2]). It is useful to refer to VERCORS (Pontillon *et al.*, 2010; Pontillon and Ducros, 2010a, b) and PHEBUS (Clément, 2005; Clément *et al.*, 2003; Dubourg *et al.*, 2005) programs conducted by CEA and IRSN (Institut de Radioprotection et de Sureté Nucléaire) on FPs releases during the NPP severe accident involving degradation of the reactor cores (Lewis *et al.*, 2008). These experiments rely on irradiated pressurized water reactor fuel samples including their cladding and leading to a corium under different accident conditions and for different burn-ups, up to 70 GWd/t. One of the findings of these experiments has led to categorize FPs in four volatility classes based on the release rate and specific behaviors, including oxygen potential and material chemical interactions observed in situation of a severe accident:

Class 1: Volatile FPs including fission gas (xenon and krypton), iodine, cesium, rubidium, tellurium, cadmium, silver and antimony. Released fraction of these elements is *quasi*-total even before reaching the meltdown state. The release of these elements is accelerated in oxidizing media and slightly delayed for tellurium and antimony.

Class 2: Semi-volatile FPs: molybdenum, barium, rhodium, palladium, and technetium characterized by a large release rate, in certain cases similar to the one of volatile FPs, but very sensitive to the redox conditions and exhibiting an important retention in the upper structure of the reactor vessel.

[2] ^{134}Cs is mainly an activation product formed in the fuel from the reaction: $^{133}\text{Xe} \rightarrow {}^{133}\text{Cs} + \beta^- + \bar{\nu} \rightarrow {}^{133}\text{Cs}(n,\gamma){}^{134}\text{Cs}$. It is often denoted as a secondary fission product.

Class 3: Low volatile FPs: ruthenium, cerium, strontium, yttrium, niobium, lanthanum and europium. They are characterized by a weak release but nevertheless significant (up to 10 % during the fuel degradation). However the release rate can achieve higher values for fuels under high burn-up conditions release of 15 and up to 30 % have been measured for niobium, ruthenium and cerium with 70 GWd/t UO_2 fuel burn-up, or under specific conditions like ruthenium under air. Strong retention on upper structure of the reactor is expected for these FPs.

Class 4: Non volatile FPs: zirconium, neodymium and praseodymium. These FPs do not exhibit any quantifiable release as under the conditions of the VERCORS experiments.

Actinides including uranium, plutonium, neptunium, americium and curium have each their own behavior. However, they can be subdivided into two groups: the former, uranium and neptunium, with releases reaching potentially 10 % and behaviors closer to the group of the low volatile FPs, the latter including plutonium with very weak releases much less at <1 %.

These volatility classes are useful to provide relevant information as they allow identification of FPs measured from releases following a severe nuclear reactor accident to be confirmed or inferred by checking the consistency between the identified FPs and their respective class of behavior. It is especially the case when identification of a radionuclide from a gamma spectrometric measurement is not obvious or consistent. In order to verify the global consistency of the whole set of FPs identified in the sample collected on March 22, 2011, one relies onto normalized activity ratios through an easily identifiable and quantifiable FP, ^{137}Cs (Table 2). The second column of Table 2 shows an estimate of the reactor core inventory (in TBq/tU) provided by the CESAR Code[3]. A standard BWR fuel assembly has been considered taking into account realistic burn-up assumptions (Ducros *et al.*, 2012). Ten days after reactor shutdown, the core inventories depends at first order on the irradiation power for short-lived radionuclides (22 MW/t has been considered) and on the burn-up rate for long-lived radionuclides (30 GWd/t has been considered). A more sophisticated model like the one considered in another study (Kirchner *et al.*, 2012) leads to similar core inventory (<20 % average discrepancy for the whole set of radionuclides listed in Table 2). The fourth column of Table 2 is related to measured activity concentrations in Bq/m^3 (collection start March 22, 2011). Fifth column (column 4/column 2) expresses the signature of the released fraction, related to each FP. The last column gives the ratio between the previous one divided by the ^{137}Cs activity concentration in order to get a FP behavior comparison against cesium.

Table 2 globally shows a good consistency in behavior of the previously defined classes of FPs according to our knowledge of their releases and transportation features. These derived data are of great interest to provide a reliable estimate of the reactor core conditions:

With regard to the volatile FPs (class 1), it is worth noticing that the three cesium isotopes (^{134}Cs, ^{136}Cs, ^{137}Cs) lead to similar ratios with regard to their initial core inventory. Like cesium, isotope ^{86}Rb is an alkaline element and exhibits the same behavior, with a ratio of 91 % of the ^{137}Cs one. This is consistent with the fact that this FP has always showed in the VERCORS experiments the same behavior as ^{137}Cs.

As expected, tellurium isotopes exhibit behavior similar to that of cesium. Ratios of about 2 are probably not significant as tellurium and cesium globally behave in the same way considering both their release from the fuel and their transportation as aerosols. This slight difference in comparison with cesium is probably mainly due to uncertainty on the estimate on their core inventory along with differences in physical–chemical properties. It should be noticed that ^{129}I, even if produced at low level in the reactor core, should be present (even if not detected) in the environment as a daughter of ^{129}Te and ^{129m}Te.

Iodine exhibits a ratio larger than 20 as compared to cesium resulting from a larger volatility as revealed in the VERCORS and PHEBUS experiments.

On the contrary, ^{125}Sb and ^{110m}Ag (Fig. 6), listed as volatile FPs in term of release, seem to behave as rather semi-volatile due to their well-known strong

[3] (Vidal *et al.*, 2006) developed by CEA and was qualified for BWR fuel elements to be reprocessed at the La Hague French centre. It has also been validated from calculations based on the representative BWR EOLE model.

Table 2

FP activity ratios normalized to ^{137}Cs

Radionuclide	Core inventory (TBq/tU)	Type	Activity concentration in air at Takasaki (Bq/m^3)	Released fraction (FR)	Ratio FR/FR_137Cs
^{110m}Ag	9.62×10^1	Volatile	6.55×10^{-5}	6.81×10^{-7}	6.74×10^{-2}
^{140}Ba	2.20×10^4	Alkaline earth	2.86×10^{-4}	1.30×10^{-8}	1.29×10^{-3}
^{134}Cs	4.43×10^3	Alkali	3.63×10^{-2}	8.19×10^{-6}	8.11×10^{-1}
^{136}Cs	6.70×10^2	Alkali	5.81×10^{-3}	8.26×10^{-6}	8.58×10^{-1}
^{137}Cs	3.50×10^3	Alkali	3.54×10^{-2}	1.01×10^{-5}	1.00
^{131}I	9.30×10^3	Halogen	2.18	2.34×10^{-4}	$2.32 \times 10^{+1}$
^{140}La	2.53×10^4	Trivalent	4.60×10^{-4}	1.82×10^{-8}	1.80×10^{-3}
^{99}Mo	3.24×10^3	Transition element	4.11×10^{-3}	1.27×10^{-6}	1.26×10^{-1}
^{95}Nb	3.54×10^4	Transition element	2.88×10^{-4}	8.13×10^{-9}	8.05×10^{-4}
^{95}Zr	3.20×10^4	Tetravalent	$<2.30 \times 10^{-5}$	–	–
^{86}Rb	3.06×10^1	Alkali	2.82×10^{-4}	9.22×10^{-6}	9.13×10^{-1}
^{103}Ru	3.04×10^4	Platinoid	4.95×10^{-5}	1.63×10^{-9}	1.62×10^{-4}
^{106}Ru	1.47×10^4	Platinoid	$<1.82 \times 10^{-4}$	–	–
^{125}Sb	2.83×10^2	More volatile group	3.21×10^{-4}	1.13×10^{-6}	1.12×10^{-1}
^{127m}Te	3.39×10^2	Chalcogen	6.03×10^{-3}	1.78×10^{-5}	1.78
^{129}Te	1.96×10^3	Chalcogen	3.47×10^{-2}	1.77×10^{-5}	1.75
^{129m}Te	1.96×10^3	Chalcogen	5.49×10^{-2}	2.80×10^{-5}	2.77
^{132}Te	3.58×10^3	Chalcogen	1.00×10^{-1}	2.79×10^{-5}	2.76
^{133}Xe	1.45×10^4	Noble gas	$>1.4 \times 10^{+3}$	$>1 \times 10^{-1}$	$>1 \times 10^{+4}$

Column 2: first order of magnitude of reactor core inventory (in TBq/tU) on March 22, 2011 (10 days after the reactor shutdown) for a constant irradiation power of 22 MW/t and a burn up rate of 30 GWd/t, column 4: FP measured activity concentrations (in Bq/m^3) in air at Takasaki, column 5: column 4/column 2, last column : activity ratios normaliezd to ^{137}Cs value

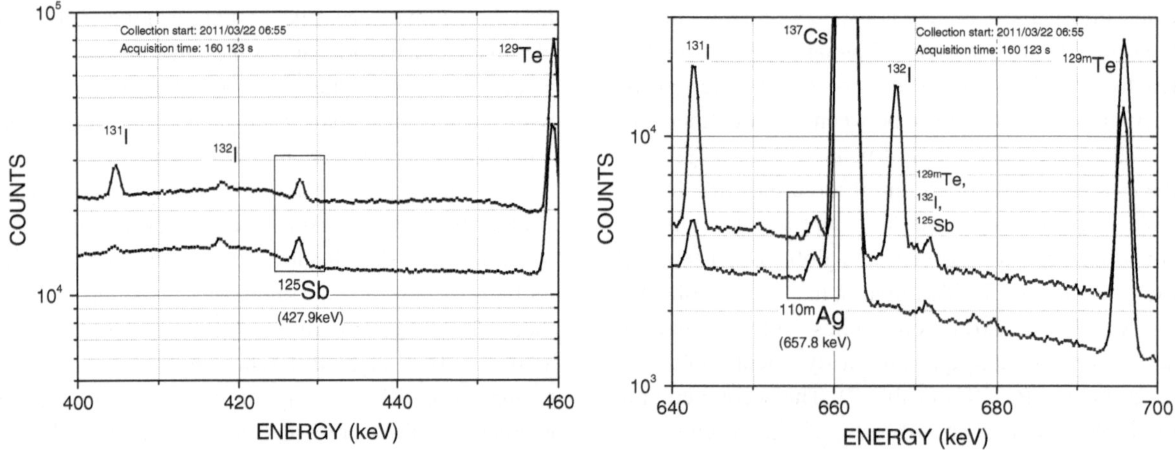

Figure 6

^{125}Sb and ^{110m}Ag detected and quantified by two CTBTO laboratories in the sample collected on March 22, 2011 06:55 at Takasaki. Spectra are shifted for improved visibility

retention within the reactor core structure. Their release rate (normalized to ^{137}Cs), in the order of 7 and 11 % compared to cesium, are consistent with the calculated value for ^{99}Mo (13 %).

Ten days after reactor shutdowns, the core activity inventory ratio ^{111}Ag/^{110m}Ag is about 6. This leads to a gamma line intensity ratio of 0.7 based on respective gamma emission probabilities and detection efficiencies. That leads to a calculated activity concentration for ^{111}Ag of about 4.5×10^{-5} Bq/m^3, well below its minimum detectable activity (1.4×10^{-3} Bq/m^3).

The value related to ^{140}Ba (0.13 %, Table 2) does not completely reflect the release rates measured in VERCORS and PHEBUS experiments, but it has been shown that this FP could be easily released from the fuel element, rather under reducing than oxidizing conditions, with a trend to condense in the upper part of fuel bundle and, consequently, a small amount generally reaches the containment.

As regard to the low volatile FPs (class 3), two key FPs were identified in the collected samples of March 22, 2011 at Takasaki: ^{95}Nb and ^{103}Ru (Table 1; Fig. 7). These elements imply an important damage of the reactor cores with a very probable core melting. ^{140}La is detected as a direct progeny of ^{140}Ba but does not reflect the reactor releases.

Usually ^{95}Nb is associated with its parent nuclide ^{95}Zr (half-life: 64 days, $E_{\gamma} = 756.7$ keV, $I_{\gamma} = 54.3$ %). ^{95}Zr was never detected in the whole set of spectra acquired at Takasaki and the only evidence of ^{95}Nb occurrence in atmosphere was its single gamma line signature at 765 keV ($I_{\gamma} = 99.8$ %). Consequently, owing to the importance of the presence of ^{95}Nb for the understanding of the Fukushima accident, we decided to perform measurements at different times in order to verify from the count rate decay that the apparent radioactive period matched the expected one. Figure 8 shows a very good agreement between the observed half-life (35.3 ± 1.1 days) determined from repeated measurements of the sample dated March 22, 2011 and the tabulated value (34.991 ± 0.006 days).

Compared to ^{137}Cs, ^{95}Nb exhibits a ratio of 0.08 %, which points out a high and prolonged fuel temperature following a probable dewatering of the fuel element bundle. Relying on the VERCORS experiments feedback only, meltdown fuel samples led to significant releases of this isotope. On the other hand, the fact that ^{95}Zr was not detected with ^{95}Nb is also consistent with CEA/IRSN experiments that showed very weak releases of this FP even following a core meltdown. In addition, this observation is in agreement with the disadvantageous vapor pressure constant (KIRCHNER *et al.*, 2012) ratio between oxidizing state ZrO_2 and NbO_2 (0.07). This is consistent with the Minimum Detectable Concentration (MDC) measured for ^{95}Zr (2.30×10^{-5} Bq/m^3) compared to the activity concentration measured for ^{95}Nb (Table 2).

Isotope ^{103}Ru quantified by one CTBTO laboratory in the sample collected on March 22, 2011 at Takasaki (Table 1) emits gamma lines at 497 keV (I_{γ}: 89.5 %) and 610 keV (I_{γ}: 5.64 %). This isotope is quantified at a low level, close to the MDC, from only its most intense gamma line. A second measurement at a later time led to the expected counting rate and allowed us to confirm this isotope which is of a great importance with regard to diagnosis of the reactor core damages. Its very low ratio (0.016 %) normalized to ^{137}Cs (Table 2) is consistent with a volatility

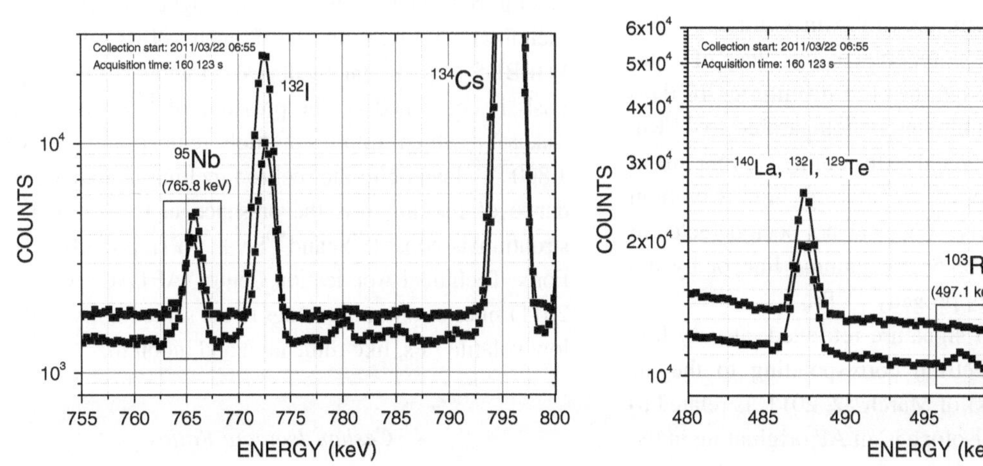

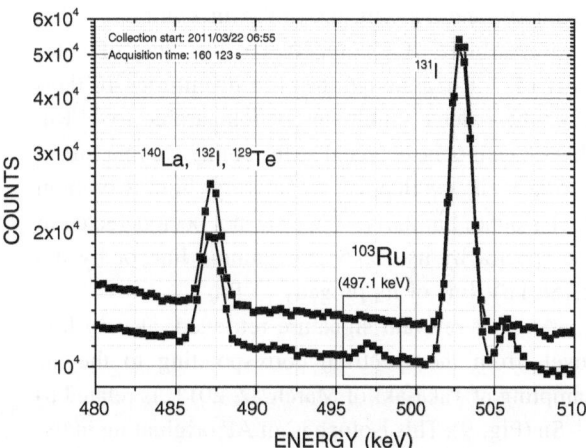

Figure 7
Detection of ^{95}Nb (*on the left*) by two CTBTO laboratories and ^{103}Ru (*on the right*) detected by only one laboratory in the sample collected on March 22, 2011 06:55 at Takasaki. Spectra are shifted for improved visibility

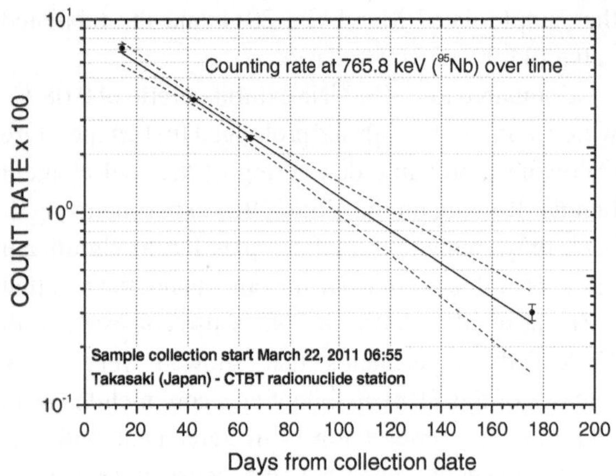

Figure 8
Evolution of the counting rate at 765.8 keV over time; four measurements have been performed in a CTBTO laboratory on a 180 day period following the sampling on March 22, 2011 (fitted function in *bold* and associated confidence band)

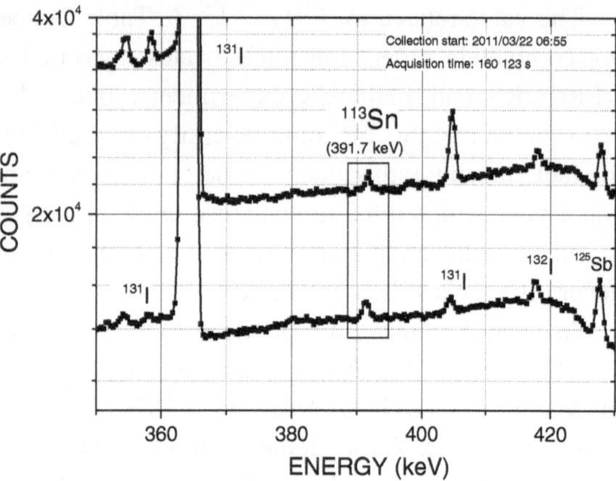

Figure 9
Activation product ^{113}Sn (gamma line at 392 keV) detected and confirmed from two CTBTO laboratories (sample collection start March 22, 2011 06:55 at Takasaki). Spectra are shifted for improved visibility

degree corresponding to overheating and extended fuel melting of the reactor cores and invalidates a potential signature related to dewatering of the SFP 4. Indeed, such assumption would have led to a ratio much higher for ^{103}Ru, close to the one of volatile FPs since it is well-known that in case of SFP drying, air ingress induces high releases of Ru under its tetraoxyde form (RuO_4) with kinetic similar to that the one of volatile FPs such Cs and I (Lewis *et al.*, 1998). Moreover, ^{106}Ru would have also been detected. In addition, all measurements of contaminated water samples within SFP-4 delivered by TEPCO have revealed a ^{131}I/^{137}Cs ratio consistent with the core inventory and not with the SFP-4 inventory.

Regarding ^{141}Ce belonging to the class of volatility of ^{103}Ru, although the core inventories 10 days after the reactor shutdowns is comparable to ^{103}Ru, this FP cannot be detected as the single gamma line at 145 keV is overlapped by a very high Compton background. Similarly, ^{144}Ce has not been detected in the spectra by its 133.5 keV gamma line or by the 2,186 keV line of its progeny ^{144}Pr.

Another relevant signature retrieved, at very low level, from the spectrum corresponding to the air sampling at Takasaki of March 22, 2011, is related to ^{113}Sn (Fig. 9). This isotope is an AP originating in the fuel rod cladding made of Zircaloy-2 alloy and composed of Zr associated to Sn (1.5 %), Fe (0.15 %), Cr (0.1 %) and Ni (0.05 %). Isotope ^{112}Sn has only a natural abundance (in atom) of 1 % but a 0.7 barn thermal cross section (σ_n, γ) which produces ^{113}Sn through the reaction ^{112}Sn(n, γ)^{113}Sn. Other stable tin isotopes with higher natural abundances lead to radioactive isotopes produced at much lower amount owing to a lower cross section.

Detection of ^{113}Sn implies high temperature of fuel rod claddings initiated by the exothermic reaction between Zircaloy and steam. This exothermic reaction is able to heat fuel rod up to 1,800 °C or more.

^{95}Nb and ^{103}Ru in the gamma spectrum related to the collection on March 22, 2011 are confirmed and precisely quantified by the analysis performed at CTBTO laboratories. These key elements suggest reactor core meltdowns relying on the French VERCORS and PHEBUS experiment feedback. Moreover, this diagnosis is confirmed by the presence of ^{113}Sn in air at Takasaki which implies a high temperature, above 1,800 °C, leading to the fuel rod melting. Another evidence of melting was the presence in the air of the strontium isotopes (^{89}Sr and ^{90}Sr + ^{90}Y) reported by Air Force Technical Application Centre (AFTAC) (Lucas, 2011) since strontium belongs to the same family of the low volatile FPs, like ruthenium and niobium.

4. *Cesium Isotopic Ratios*

Relevant information can also be obtained from isotopic ratios involving isotopes with identical chemical behavior. Figure 10 shows isotopic ratios $^{137}Cs/^{134}Cs$ and $^{137}Cs/^{136}Cs$ related to two radionuclide stations of the CTBTO network (located at Sand Point/USA and at Okinawa/Japan). These ratios given for the March–April 2011 period were decay-corrected to March 11, 2011 12:00. The evolution of the isotopic ratios over time is linked to the core inventories of Unit 1, 2 and 3 and does not depend on the dates of FP releases following the different explosions and venting. As shown by Fig. 10, the isotopic ratios $^{137}Cs/^{134}Cs$ are quite constant over time (significantly higher values are due to a local effect) and lead to an average value close to 1 in proper accordance with the ratio ($\sim$0.8 on March 11, 2011) as assessed from the core inventory drawn up on the basis discussed in §III. The constant ratio $^{137}Cs/^{134}Cs$ points out that the radioactive aerosols were released from the same material source. This ratio value fully complies with data published in numerous papers (e.g. KIRCHNER *et al.*, 2012). Although the level of the $^{137}Cs/^{136}Cs$ ratio $\sim$3–4 (when decay corrected to March 11, 2011) and $\sim$2 (at the date of the core inventory), is roughly at the order of magnitude expected from the core inventories ($\sim$5, see Table 2), it can be observed a large variation of the ratios as noticed in several papers (e.g., BIEGALSKI *et al.*, 2011; ENDO *et al.*, 2011). This can be explained by the features of the mixed releases. The BWR Unit 3 is fueled by only some percents of mixed-oxide (MOX) which can not

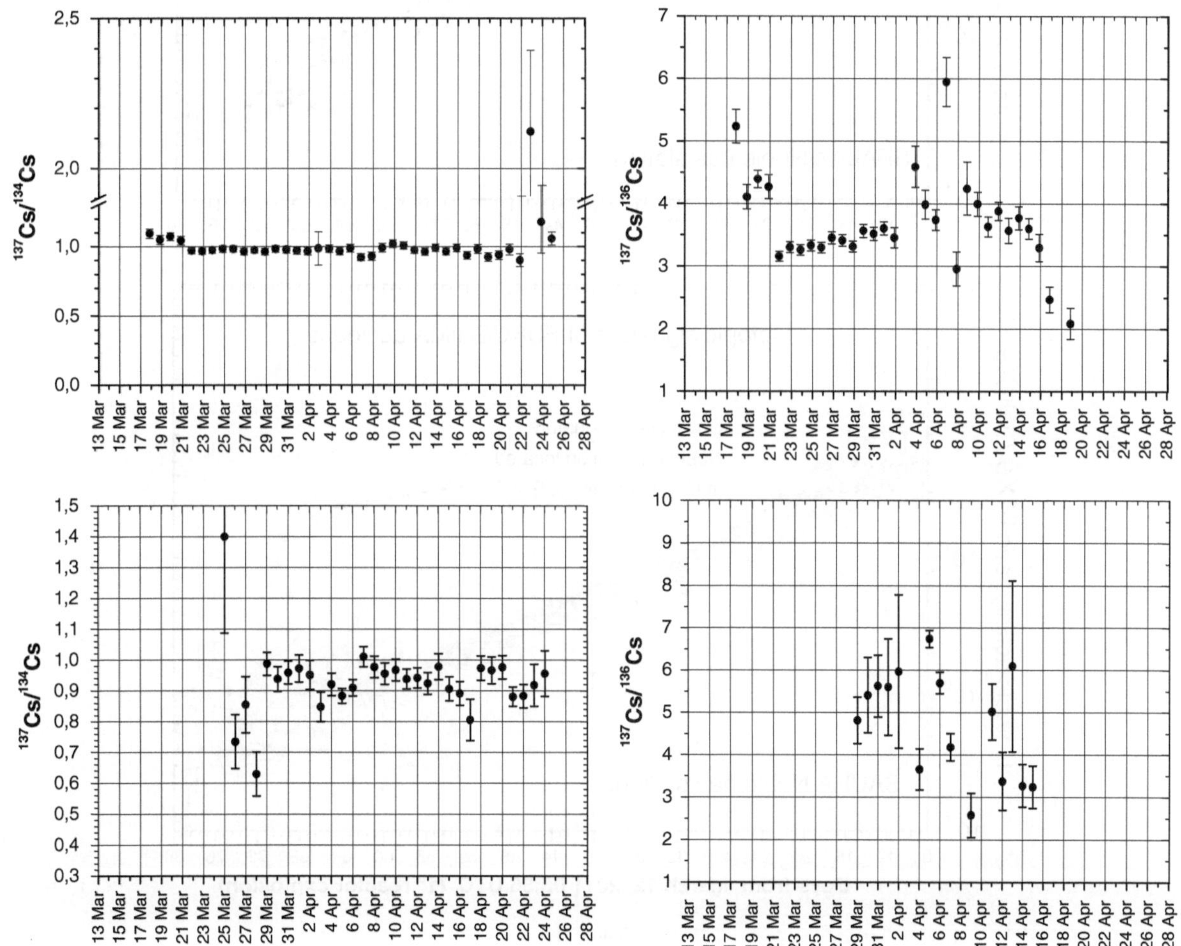

Figure 10

Cesium isotopic ratios ($^{137}Cs/^{134}Cs$ and $^{137}Cs/^{136}Cs$) decay corrected to the reference date March 11, 2011 and related to radionuclide stations of the CTBTO network located at Upper/Sand Point (USA, 55.0 N 160.0 W), Lower/Okinawa (Japan, 26.5 N 127.9 E)

explain the lowest $^{137}Cs/^{136}Cs$ ratios due to a higher fission yield for the shielded nuclide ^{136}Cs. A possible explanation could be linked to releases following the fires related to SFP-4 characterized by the absence of the ^{136}Cs isotope.

5. $^{131m}Xe/^{133}Xe$ Isotopic Ratios

Gaseous FPs, especially ^{131m}Xe, ^{133}Xe, ^{133m}Xe and ^{135}Xe isotopes are potentially detected by CTBTO noble gas stations and are also likely to

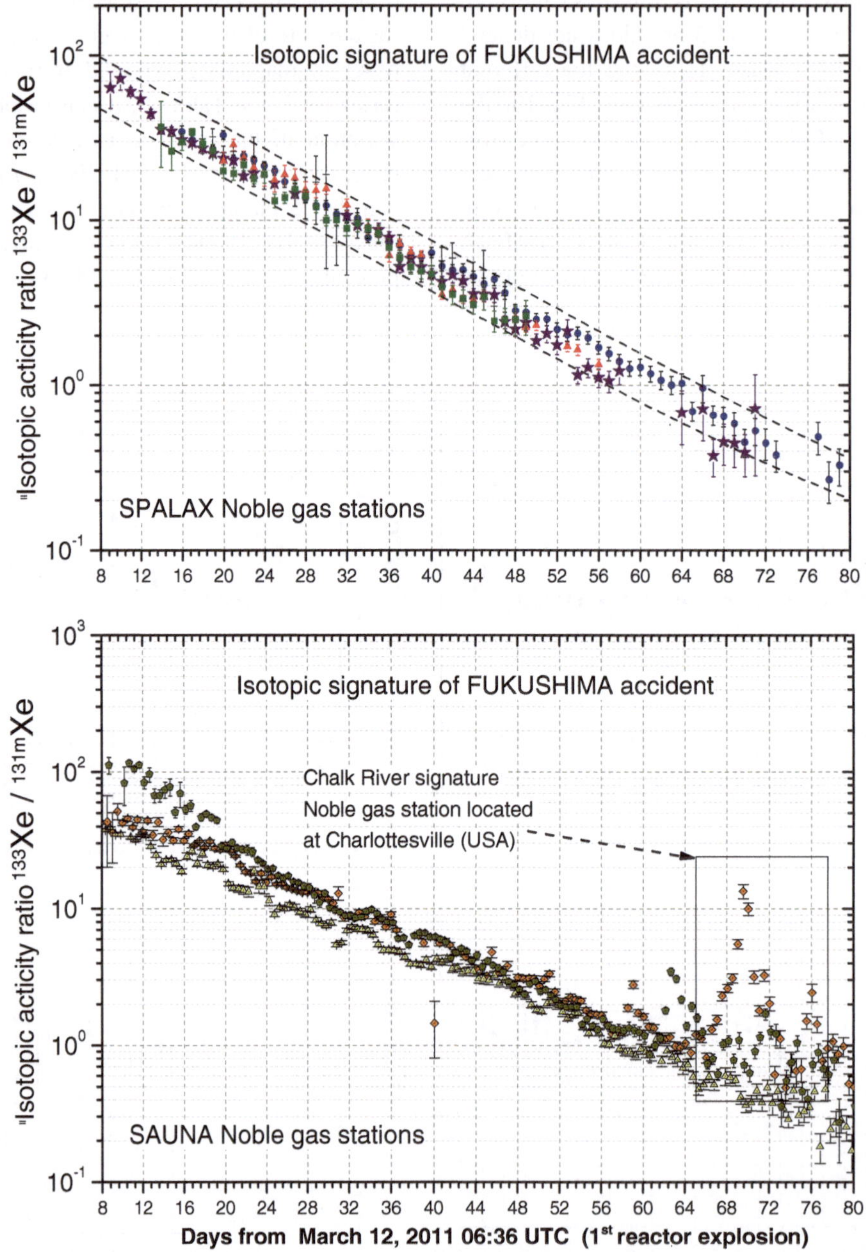

Figure 11

Upper part $^{133}Xe/^{131m}Xe$ isotopic ratio evolution measured over time by SPALAX noble gas stations of the CTBTO network and from French national means, all located in the Northern hemisphere. The bands reflect the assessed uncertainties on radioxenon core inventories. *Lower part* $^{133}Xe/^{131m}Xe$ isotopic ratio evolution measured over time by some SAUNA noble gas stations (Noble Gas station located at Charlottesville exhibits the signature of radioxenon coming from Chalk River laboratory) provide key information. $^{133}Xe/^{137}Cs$ ratio estimate given in Table 2 is based on a comparison of the Fukushima NPP core inventories and measurements performed at Takasaki. It clearly expresses an absence of fission gases retention and therefore reflects the reactor states at the initial time of the accident.

The ^{133}Xe assessed release at the time of reactor shutdown (8×10^{18} Bq, see Part II) turned out to be the order of 80 % of the amount deduced from the reactor core inventories (Table 2). This strongly suggests a broad meltdown of the reactor cores. Upper Fig. 11 shows isotopic activity ratios $^{133}Xe/^{131m}Xe$ with associated uncertainties (confidence level 1σ) measured over time by all SPALAX noble gas stations of the Northern hemisphere within the CTBTO network and the French National one. Lower Fig. 11 presents the same data for some SAUNA noble gas stations pertaining to the CTBTO network. The worldwide atmospheric radioxenon inventory is supposed to reflect the time evolution of the radioxenon related to the reactor core inventories at the time of the explosions. It is assumed that atmospheric radioxenon concentrations are not significantly modified by the decay of iodine isotopes.

Figure 11 shows a clear Fukushima accident signature over a long time period (80 days). The ratio decays over time with an 8.6 days pseudo half-life which is close to the expected one considering the half life of both ^{133}Xe (5.2 days) and ^{131m}Xe (11.9 days) isotopes. This piece of information is valuable in order to assess monitoring capabilities of the CTBTO network in the weeks following the Fukushima accident in case of any relevant other event should have occurred. As an illustration, lower Fig. 11 presents the isotopic activity ratios $^{133}Xe/^{131m}Xe$ over time achieved from the analysis of the SAUNA data station located at Charlottesville (USA) which clearly exhibits a singular signature coming from the Chalk River Laboratory medical isotope production plant (Canada) superimposed to the Fukushima one.

Figure 12 illustrates the method used to determine the activity concentrations for both high and very weak detections from two samplings taken at Yellowknife (Canada) on March 24, 2011 (respectively,

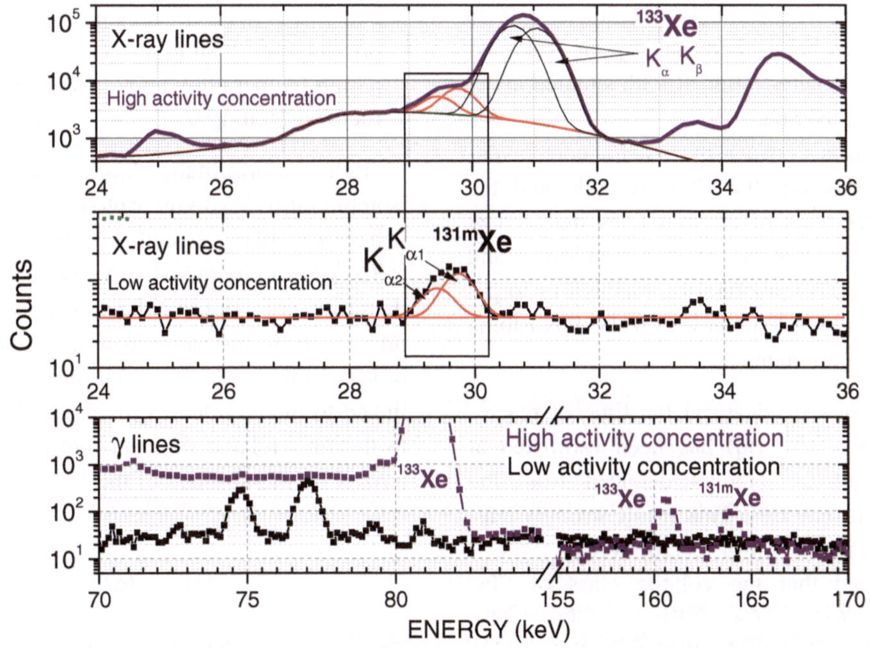

Figure 12
Detection of radioactive xenon isotopes by the SPALAX noble gas station located at *Yellowknife* (Canada) related to two samples collected, respectively, on March 24, 2011 (purple, ^{133}Xe and ^{131m}Xe, respectively, 2,300 ± 30 and 38 ± 2 Bq/m^3) and on May 30, 2011 (*black*, ^{133}Xe and ^{131m}Xe, respectively, 0.8 ± 0.1 and 2.3 ± 0.3 mBq/m^3): ^{131m}Xe is detected both from the 164 keV gamma line (*lower/purple*) and from X-ray at ~30 keV (*upper*) at high activity concentration and from only X-ray at ~30 keV at low activity concentration (*middle/black*)

2,300 ± 30 and 38 ± 2 Bq/m^3 for ^{133}Xe and ^{131m}Xe) and on May 30, 2011 (respectively, 0.8 ± 0.1 and 2.3 ± 0.3 mBq/m^3 for ^{133}Xe and ^{131m}Xe; uncertainties are given at a confidence level 1σ). ^{133}Xe is quantified from its 81 keV gamma line and ^{131m}Xe from its 164 keV gamma line (I_γ: 1.98 %) in case of high concentration. At a very low level, ^{131m}Xe is quantified with a maximum likelihood method (Vivier *et al.*, 2009) from the X-ray multiplet region at 29.4 keV (Xe $K_{\alpha1}$) and 29.7 keV (Xe $K_{\alpha2}$) which exhibits higher probability emissions, respectively, 15.4 and 28.5 %. As ^{133m}Xe was never measured in the whole set of samples, no X-rays interference had to be taken into account.

6. Conclusions

The work presented here is a post-analysis of radionuclide releases originating from the March 2011 Fukushima NPP accident. It illustrates that data collected by CTBTO radionuclide stations and laboratories can be efficiently used for other purposes than detecting nuclear explosions. In particular, we used data collected by radionuclide noble gas stations from the Northern hemisphere and by the Takasaki radionuclide particulate station JPP38, situated about 210 km from Fukushima.

The comparison between the reactor core inventories performed using the CESAR code and the activity concentrations of radionuclides detected at Takasaki 10 days after the reactor shutdowns and quantified by CTBTO laboratories, enabled us to put forward the following hypotheses on the Fukushima Dai-ichi reactor core damages:

- The ^{133}Xe assessed release at the time of reactor shutdowns (8 × 10^{18} Bq) has been estimated at about 80 % of the amount deduced from the reactor core inventories. This result strongly suggests a broad meltdown of reactor cores.
- It was showed that the volatile class of FPs, especially ^{134}Cs, ^{136}Cs, ^{86}Rb, ^{127m}Te ^{129}Te, ^{129m}Te, ^{132}Te cesium and tellurium radioisotopes, exhibited the same behavior as ^{137}Cs (ratios close to 100 % when normalized to ^{137}Cs) which is consistent with the severe nuclear power plant experiments conducted by CEA and IRSN programs (VERCORS and PHEBUS). On the contrary ^{125}Sb and ^{110m}Ag, considered as volatile FP, seem to behave as rather semi-volatile FP due to their well known strong retention within the reactor core structures. Their release rates normalized to that of ^{137}Cs are, respectively, about 7 and 11 %; this is consistent with the value calculated the same way for ^{99}Mo (13 %) as determined in VERCORS and PHEBUS experiments. The value related to ^{140}Ba (0.13 %) does not completely reflect the release rates obtained from VERCORS experiment but complies with PHEBUS program feedback which has shown that this FP could condense in the upper element fuel bundle and consequently a small amount generally reaches the containment.
- Compared to ^{137}Cs, ^{95}Nb exhibits a ratio of 0.08 % which points out a high temperature fuel state under a prolonged time, following a probable dewatering of the fuel element bundles, relying on the CEA experiments (VERCORS). ^{95}Nb, ^{103}Ru measured in air at Takasaki suggest the Fukushima reactor core suffered meltdowns relying on the French VERCORS and PHEBUS experiment feedback. Moreover, this diagnosis is confirmed by the presence of ^{113}Sn in air at Takasaki which implies a high temperature, above 1,800 °C, leading to the fuel rod melting.

Relevant information has also been obtained from isotopic ratios involving isotopes with same chemical behavior, like cesium isotopes. Unlike the ^{137}Cs/^{134}Cs ratios which are constant over time, and, therefore, point out that the radioactive particles were released from the same material source, ^{137}Cs/^{136}Cs ratios exhibit a large variation over time possibly linked to mixed releases involving those consecutive to the fires related to spent-fuel pool No. 4 characterized by the absence of the ^{136}Cs isotope.

Acknowledgments

The authors wish to thank the Comprehensive Nuclear-Test-Ban Treaty Organization for fruitful discussions about the data collected by radionuclide stations of the International Monitoring Network during the Fukushima event. A special thanks should be expressed to the local operators of the CTBTO Takasaki radionuclide stations for maintaining data availability in such exceptional circumstances.

References

Leon J.D., Jaffe D.A., Kasper J., Knecht A., Miller M.L., Robertson R.G.H. and Schubert A.G.,. *Arrival time and magnitude of airborne fission products from the Fukushima, Japan, reactor incident as measured in Seattle, WA, USA.*, J. Environ. Radioact. *102*, 1032-1038, 2011

Bowyer T.W., Biegalski S.R., Cooper M., Eslinger P.W., Haas D., Hayes J.C., Miley H.S., Strom D.J. and Woods V., *Elevated radioxenon detected remotely following the Fukushima nuclear accident*, J. Environ. Radioact. *102*, 681, 2011

Bolsunovsky, A., Dementyev, D. *Evidence of the radioactive fallout in the center of Asia (Russia) following the Fukushima Nuclear Accident.* J. Environ. Radioact. *102*, 1062-1064, 2011

Sinclair L.E., Seywerd H.C.J., Fortin R., Carson J.M., Saull P.R.B., Coyle M.J., Van Brabant R.A., Buckle J.L., Desjardins S.M. and Hall R.M., *Aerial measurement of radioxenon concentration of the west coast of Vancouver Island following the Fukushima reactor accident*, J. Environ. Radioact., (Article in Press), doi:10.1016/j.jenvrad.2011.10.011

Zhang W., Bean M., Benotto M., Cheung J., Ungar K. and Brian Ahier, *Development of a new aerosol monitoring system and its application in Fukushima nuclear accident related aerosol radioactivity measurement at the CTBT radionuclide station in Sidney of Canada*, J. Environ. Radioact. *102*, 1065, 2011

Manolopoulou, M., Vagena, E., Stoulos, S., Ioannidou, A., Papastefanou, C,. *Radioiodine and radiocesium in Thessaloniki, Northern Greece due to Fukushima nuclear accident*. J. Environ. Radioact. *102*, 796-797 (2011)

Pittauerová D., Hettwig B. and Fischer H.W., *Fukushima fallout in Northwest German environmental media*, J. Environ. Radioact. *102*, 877, 2011

Qiao F., Wang G., Zhao W., Zhao J., Dai D., Song Y. and Song Z., *Predicting the spread of nuclear radiation from the damaged Fukushima Nuclear Power Plant, 2011*. Chinese Science Bulletin *56*, 1890, 2011

Tanimoto T., Uchida N., Kodama Y., Teshima T. and Taniguchi S., *Safety of workers at the Fukushima Daiichi nuclear power plant, LANCET* Volume: *377* Issue: 9784, Pages: 2180-2180 Published: JUN-JUL: 2011

Okada, S., *Off-Site Activities regarding the accident at the Fukushima Daiichi Nuclear Power Station and Lesson from Them.* World Engineers' Convention, Geneva, 4-9 September, 2011

Masson O., Baeza A., Bieringer J. *et al.*, *Tracking of Airborne Radionuclides from the Damaged Fukushima Dai-Ichi Nuclear Reactors by European Networks*, Environ. Sci. Technol. *45*, 7670-7677, doi:10.1021/es2017158, 2011

Chino, M., Nakayama, H., Nagai,H., Terada, H., Katata, G., and Yamazawa, H., *Preliminary Estimation of Release Amounts of ^{131}I and ^{137}Cs Accidentally Disharged from the Fukushima Daiichi Nuclear Power Plant into the Atmosphere*. Journal of Nuclear Science and Technology, Vol.*48*, No.7, p. 1129-1134, 2011

Achim, P., Monfort., M., Le Petit., G., Gross., P., Douysset., G., Taffary., T., Blanchard., X., and Moulin., C., *Analysis of Radionuclide Releases from Fukushima Dai-Ichi Nuclear Power Plant Accident Part II.* Pure Appl Geophys. doi:10.1007/s00024-012-0578-1, 2012

Bé, M.C., Chisté, V., Dulieu, C., *CEA-R-6201 ISSN 0429 – 3460 Nucléide-LARA Bibliothèque des émissions X et gamma*, CEA/LNE-LNHB, 129, 2008

Miley, H.S., Bowyer, S.M., Hubbard, T.W., McKinnon, A.D., Perkins, R.W., Thompson, R.C.,Warner, R.A., *A description of the DOE Radionuclide Aerosol Sampler/Analyzer for the Comprehensive Test Ban Treaty*, Journal of Radioanalytical and Nuclear Chemistry, VoL *235*, N° 1-2, 83-87, 1998

Nadalut, B., Barnes, E., Schröttner, T., Le Petit, G., Haquin, G., Davies, A., Greenwood, L., *RASA samples at the Laboratory: Status of the Investigation*, CTBTO Laboratory Workshop, Buenos Aires (Argentina), *5* December, 2010

Forrester, F., Pacific Northwest National Laboratory, P.O. Box 999, MSIN J4-65 Richland, WA 99352 USA, private communication, 2011

Jutier, C., Gross, P., Le Petit, G., *A new synthetic formalism for true coincidence summing calculations*, Nucl. Instrum and Methods A *580*, 1344-1354, 2007

Han, D., Duran, E., Wang, J., *Laboratory Analysis of IMS Samples Related to Fukushima Event*, Fukushima Lessons Learned Workshop, Vienna (Austria), 14-15 June 2012

Pontillon, Y., Ducros, G., Malgouyres, P.P., *Behaviour of fission products under severe PWR accident conditions: the VERCORS experimental program - Part 1: General description of the program*, Nuclear Engineering and Design *240* 1843-185, 2010

Pontillon, Y., Ducros, G., *Behaviour of fission products under severe PWR accident conditions: the VERCORS experimental program - Part 2: Release and transport of fission gases and volatile fission products*, Nuclear Engineering and Design *240*, 1853-1866, 2010

Pontillon, Y., Ducros, G., *Behaviour of fission products under severe PWR accident conditions: the VERCORS experimental program - Part 3: Release of low-volatile fission products and actinides*, Nuclear Engineering and Design *240* 1867-1881, 2010

Clément, B., *The Phebus Fission Product and Source Term International Programs*, Int. Conf. Nuclear Energy for New Europe 2005 Bled, Slovenia, September 5-8, 2005

Clément, B., Hanniet-Girault, N., Repetto, G. *et al.*, *LWR severe accident simulation: synthesis of the results and interpretation of the first Phebus FP experiment FPT0*, Nucl Eng Design *226*, 5-82, 2003

Dubourg, R., Faure-Geors, H., Nicaise, G., Barrachin, M., *Fission product release in the first two PHEBUS tests FPT0 and FPT1*, Nucl Eng Design *235*, 2183-2208, 2005

Lewis, B.J., Dickson, R., Iglesias, F.C., Ducros, G., Kudo, T., *Overview of experimental programs on core melt progression and fission product release behaviour* Journal of Nuclear Materials, *380* p126-143, 2008

Vidal, J.M. *et al.*, "*CESAR: a Code for Nuclear Fuel and Waste Characterization*," Proc. of WM'06 Conference, Tucson, USA, Feb. 26 – March 2, 2006

Ducros, G. *et al.*, *Main lessons learnt from Fission Product release, for the understanding of Fukushima Dai-ichi NPP status*, accepted *in* Trans. Am. Nucl. Soc. (2012 ANS Winter Meeting, San Diego - USA)

KIRCHNER, G., BOSSEW, P., DE CORT, M., *Radioactivity from Fukushima Dai-ichi in air over Europe; part 2: what can it tell us about the accident?* Journal of Environmental Radioactivity, doi:10.1016/j.jenvrad.2011.12.016, 2012

LEWIS, B.J., CORSE, B.J., THOMPSON, W.T., KAYE, M.H., IGLESIAS, F.C., ELDER, P., DICKSON, R., LIU, Z., *Vaporisation of low volatile fission products under severe Candu reactor accident conditions*, Journal of Nuclear Materials, *252*, 235-256, 1998

LUCAS, J., *Airborne sampling following Fukushima reactor accident*, Air Force Technical Application Center, Workshop on Signatures for Medical Isotope Production, Strassoldo (Italy), 13-17, June 2011

BIEGALSKI, S.R. *et al.*, *Analysis of data from sensitive US monitoring stations for the Fukushima Dai-ichi nuclear reactor accident*, Journal of Environmental Radioactivity, doi:10.1016/j.jenvrad.2011.11.007, 2011

ENDO, S., KIMURA, S., TAKATSUJI, T., NANASAWA, K., IMANAKA, T., SHIZUMA, K., *Measurement of soil contamination by radionuclides due to the Fukushima Dai-ichi nuclear power plant accident and associated estimated cumulative external dose estimation*, Journal of Environmental Radioactivity, doi:10.1016/j.jenvrad.2011.11.006, 2011

VIVIER, A., LE PETIT, G., PIGEON, B., BLANCHARD, X., *Probabilistic assessment for a sample to be radioactive or not; application to radioxenon analysis*, Journal of Radioanalytical Nuclear Chemistry doi:10.1007/s10967-009-0315-0, 743-748, 2009

(Received December 14, 2011, revised July 6, 2012, accepted July 18, 2012, Published online September 16, 2012)

Pure Appl. Geophys. 171 (2014), 645–667

DOI 10.1007/s00024-012-0578-1

Pure and Applied Geophysics

Analysis of Radionuclide Releases from the Fukushima Dai-ichi Nuclear Power Plant Accident Part II

PASCAL ACHIM,[1] MARGUERITE MONFORT,[1] GILBERT LE PETIT,[1] PHILIPPE GROSS,[1] GUILHEM DOUYSSET,[1] THOMAS TAFFARY,[1] XAVIER BLANCHARD,[1] and CHRISTOPHE MOULIN[1]

Abstract—The present part of the publication (Part II) deals with long range dispersion of radionuclides emitted into the atmosphere during the Fukushima Dai-ichi accident that occurred after the March 11, 2011 tsunami. The first part (Part I) is dedicated to the accident features relying on radionuclide detections performed by monitoring stations of the Comprehensive Nuclear Test Ban Treaty Organization network. In this study, the emissions of the three fission products Cs-137, I-131 and Xe-133 are investigated. Regarding Xe-133, the total release is estimated to be of the order of 6×10^{18} Bq emitted during the explosions of units 1, 2 and 3. The total source term estimated gives a fraction of core inventory of about 8×10^{18} Bq at the time of reactors shutdown. This result suggests that at least 80 % of the core inventory has been released into the atmosphere and indicates a broad meltdown of reactor cores. Total atmospheric releases of Cs-137 and I-131 aerosols are estimated to be 10^{16} and 10^{17} Bq, respectively. By neglecting gas/particulate conversion phenomena, the total release of I-131 (gas + aerosol) could be estimated to be 4×10^{17} Bq. Atmospheric transport simulations suggest that the main air emissions have occurred during the events of March 14, 2011 (UTC) and that no major release occurred after March 23. The radioactivity emitted into the atmosphere could represent 10 % of the Chernobyl accident releases for I-131 and Cs-137.

Key words: Fukushima Dai-ichi accident, atmospheric transport modeling, source terms evaluation, Cs-137, I-131, Xe-133, CTBTO.

1. *Introduction*

On March 11, 2011, the tsunami induced by the 9.0 magnitude earthquake that occurred east of Japan caused serious damage to the cooling systems of the Fukushima Dai-ichi Nuclear Power Plant (NPP). Due to the lack of cooling, hydrogen and vapor blasts, dewatering of spent fuel rod pools and fires led to the release of radioactive materials into the atmosphere. Although at the time this article was written, the comprehensive understanding of the accident was not fully established, it was accepted that units 1, 2 and 3 reached a fuel fusion state. The detection in the air of fission and activation products by monitoring stations belonging to the International Monitoring System (IMS) of the Comprehensive Nuclear Test Ban Treaty Organization (CTBTO) is likely to provide relevant information on the reactor core damage. CEA relied on these IMS data and on analysis performed by radionuclide laboratories supporting the CTBT network in order to estimate the worldwide distribution of the atmospheric releases of radioactive material and to better understand the accident features. The first part (Part I) of this study is dedicated to observations and interpretations of the accident characteristics relying on the chronology of the major detections and the nature of detected radionuclides (LE PETIT *et al.*, 2012). This linked part (Part II) deals with the Atmospheric Transport Modeling (ATM) at global scale (long range simulations). The objective is to assess the arrival time of radionuclides over IMS stations and to evaluate the quantities released into the atmosphere. We will mainly focus on Cs-137, I-131 and Xe-133 fission products. Hence, the two first radionuclides will be the main contributors to dose and worldwide industrial Xe-133 background could be modified by Fukushima radioxenon release affecting the performances of the CTBT radionuclide monitoring network.

2. *Sequence of the Accident and Observations at Fukushima Dai-ichi Site*

After the automatic shutdown of the reactors, cooling problems were caused by the March 11

[1] CEA, DAM, DIF, 91297 Arpajon, France. E-mail: pascal.achim@cea.fr

tsunami. During venting operations on several units, hydrogen explosions and fires occurred. Table 1 gives the sequence of main events that could have led to the release of radionuclides into the atmosphere (Japanese Nuclear and Industrial Safety Agency, NISA; Okada, 2011; Ministry of Education, Culture, Sports, Science and Technology, MEXT). Four of the six units were damaged as a result of the tsunami. Only units 1, 2 and 3 were in operation at the time of the event. As unit 4 had been already stopped, it was not involved in venting operations and did not suffer an hydrogen explosion. However, it was damaged by the explosion of unit 2 leading to a fire in the spent fuel storage pool. It should be noted that other events than those mentioned in Table 1 could have led to atmospheric emissions, like cooling operations by spraying water from helicopters or terrestrial devices that could have resuspended small quantities of radionuclides.

2.1. *Dose Rate Measurements*

Figure 1 shows the evolution of the dose rates measured during March 2011 on the monitoring points located on the Fukushima Dai-ichi site (TEPCO Press releases). The monitoring points are approximately located 1 km away from the reactors. Measurement of amplitude over time depends on both the amount of material released in the atmosphere and the location of the monitoring stations with regard to the prevailing wind. Since the time series are incomplete, they may not represent all the events that could have occurred. However, measurements show a succession of peaks of dose rate, which can be related to the events presented in Table 1. Releases that led to the highest measured dose rates took place on March 14 and 15 UTC (events 7–11 in the table). After March 25, dose rates decrease continuously with time. Even though they give an indication of release timing, dose rates are difficult to interpret since they do not allow the identification of radionuclides. Hence, dose rates partly result from the contribution of short half-life radionuclides such as I-132 (2.295 h) (Quélo *et al.*, 2011). In addition, dose rate monitoring devices are subject to saturation phenomena, thus making the interpretation of the relative peak amplitudes difficult.

2.2. *Assessment of Atmospheric Cs-137 and I-131 Leakage after March 22*

From March 22, measurements of activity concentrations in Cs-137 and I-131 were conducted at the Fukushima Dai-ichi site by the TEPCO Company (TEPCO Press releases) (Fig. 2, left). These measurements were made daily from mobile sampling

Table 1

Sequence of main identified events that could have lead to atmospheric releases

Date (UTC)	Concerned unit	Events
03/11 05:46		Automatic shutdown of units 1, 2 and 3 due to earthquake
03/12 01:17	Unit 1	1) Venting
03/12 06:36	Unit 1	2) Hydrogen explosion in reactor building
03/12 23:41	Unit 3	3) Venting
03/13 02:00	Unit 2	4) Venting
03/13 20:20	Unit 3	5) Venting
03/14 02:01	Unit 3	6) Hydrogen explosion in reactor building
03/14 15:02	Unit 2	7) Venting
03/14 21:15	Unit 2	8) Explosion. Possible damage of pressure system. Damage of building wall of unit 4 reactor
03/15 00:38	Unit 4	9) Fire occurred in spent fuel cooling pool
03/15 20:45	Unit 4	10) Fire occurred in spent fuel cooling pool
03/15 23:20	Unit 3	11) White smoke generated
03/20 02:00	Unit 3	12) Rise of primary containment vessel pressure
03/21 06:55	Unit 3	13) Greyish smoke
03/23 07:20	Unit 3	14) Black smoke

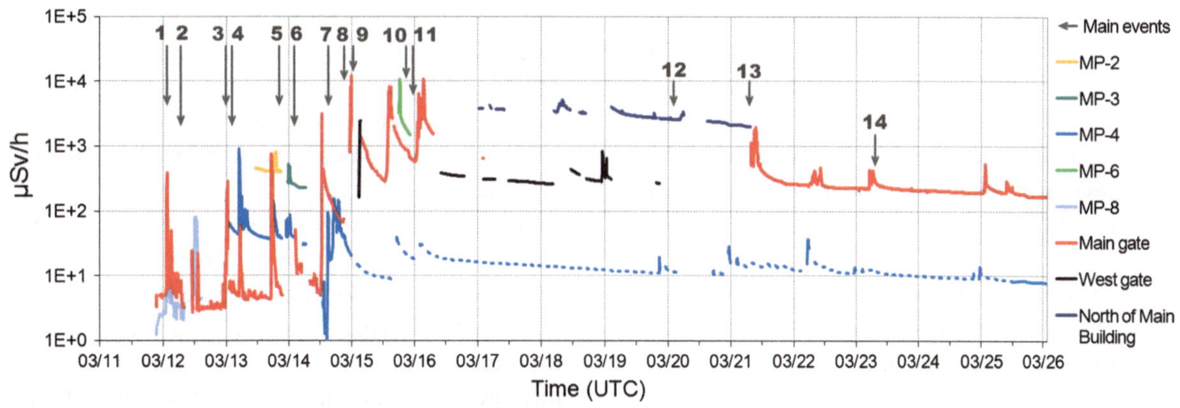

Figure 1
Dose rates measured on the monitoring points located around the Fukushima plant. Events identified in Table 1 are shown by the *arrows*

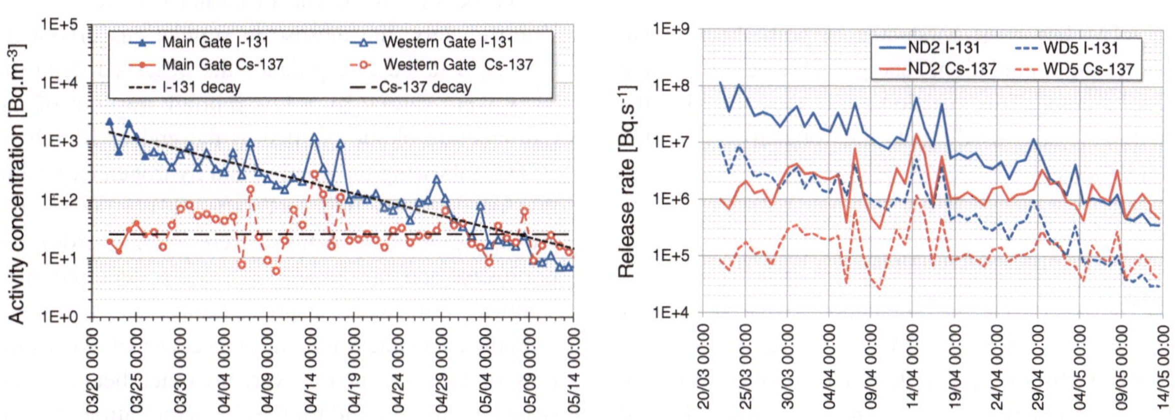

Figure 2
Left activity concentrations measured on the Fukushima Dai-ichi site for I-131 and Cs-137. *Right* estimated release rates (Gaussian formulation) required to match measurements. *ND* normal diffusion; *WD* weak diffusion; 2 and 5 refer to wind speed (m/s)

devices located near the so-called monitoring points "Western Gate" and "Main Gate". Gamma spectroscopy measurements were performed at the Fukushima Dai-ini site. Measurements show that the I-131 activity concentrations decrease with time following approximately the radioactive decay of the radionuclide, with some exceptions, such as the period from April 13 to 19. Concerning Cs-137, measured activity concentrations appear fairly constant, in agreement with the half-life of the radionuclide. This result points out that no major release has been measured by these monitoring devices after March 22 and rather indicates a continuous release of radionuclides into the atmosphere at a constant emission factor.

Figure 2 (right) provides rough estimation of release rates required to obtain measurements presented in Fig. 2 (left). Release rates are calculated from a Gaussian dispersion formulation, assumed to be relevant due to the small distance between the reactors and the monitoring devices. It is assumed that sensors were under the direct influence of releases, which was not necessarily the case in reality. Building effects on dispersion are not taken into account and ground deposition and radioactive decay phenomena are neglected (both are negligible at short distances for considered radionuclides). Calculations are performed for a 20 m high source and for two meteorological conditions assumed to provide reasonable major and minor estimations of release rates: i.e., unstable atmosphere with a 2 m s^{-1} wind speed (Normal Diffusion 2) and a stable atmosphere with a 5 m s^{-1} wind speed (Weak Diffusion 5). By time integration of the release rates over the observation period ($\sim$2 months from March 22), these two conditions lead to releases ranging from 1×10^{13} to 1×10^{14} Bq for I-131 and from 1×10^{12} to 1×10^{13} Bq for Cs-137. The work should be continued to estimate the amounts released during leaks (i.e., between main releases) that may have occurred before March 22.

3. *Elements on Atmospheric Transport Modeling (ATM)*

ATM has been conducted at both the regional (short range simulations) and global scale (long range simulations). This section describes briefly the methodologies used and the objectives of the simulations.

3.1. *Short Range Simulations*

The regional scale simulations have been conducted using MM5 V3.7 (MM5, 2005) and WRF V3.3 (WRF website) mesoscale meteorological models. These well-known systems are parallelized, limited area, non hydrostatic, terrain following and sigma-coordinate models designed to simulate or predict mesoscale atmospheric circulation. NCEP's Global Forecast System (GFS) meteorological data with 6 h and 0.5° resolutions have been used as initialization and boundary conditions (National Centers for Environmental Prediction/GFS website). Among the existing atmospheric dispersion models, the 3D lagrangian particle dispersion model (LPDM) FLEXPART was used (Stohl *et al.*, 1998; FLEXPART homepage). The versions of the model used here are those specifically developed for MM5 (V.6.2) and WRF (Fast and Easter, 2006). It should be noted that in the standard WRF FLEXPART version, wet deposition phenomena are not fully implemented. To correct this, the calculation of precipitation subgrid variability was made available in particular the calculation of the fraction of surface that undergoes precipitation. This fraction is a function of convective and large-scale precipitation and also depends on cloud cover. The method developed is based on the MM5 FLEXPART model, where cloud cover is estimated from the total water contained in the air column between ground and roof level of the calculation domain.

Regional calculations were carried out to simulate, at the scale of a few hundreds kilometers, the dispersion of radionuclides emitted by Fukushima Dai-ichi NPP. In particular, one goal was to simulate dry and wet deposits and to compare results with measured dose rates in the Ukedo river basin located on the northwest region of the Fukushima Dai-ichi site (DOE, 2011). Because of their spatial and temporal resolutions, long range models are not able to reproduce this event (Takemura *et al.*, 2011). Unfortunately, the mesoscale simulations driven by 0.5° GFS failed to reproduce the observed deposits. They were caused by a low rainfall washout of the radionuclide plume emitted at the end of March 14 (UTC) following the venting and hydrogen explosion that occurred on unit 2 (Okada, 2011). While the prevailing wind directions were mainly from the northwest and northeast, the release occurred in a southeast flow during a short duration, which was not reproduced in the simulations because of shortcomings in GFS data. However, no other better large-scale data were found for the characterization of these meteorological conditions (Mathieu *et al.*, 2012). Figure 3 shows an example of a comparison between observations and simulations at the Fukushima Airport (RJSF station, 37.2275N; 140.4281E), located 60 km southwest of Fukushima Dai-ichi NPP. In this example, the WRF mesoscale wind field has a resolution of 1.6 km and 1 h. This figure shows that the southeast wind observed during the early hours of March 15 is not reproduced by the model (similar result has been obtained with MM5).

Several mesoscale simulation attempts were performed by varying the vertical calculation grid resolution and the relaxation coefficients towards the large-scale input data inside and outside the boundary layer for the coarser domain. A possible way to improve the quality of the simulations for this time period would be to proceed to observations assimilation. Satisfactory results have been obtained with the Japanese SPEED and WSPEEDI-II emergency

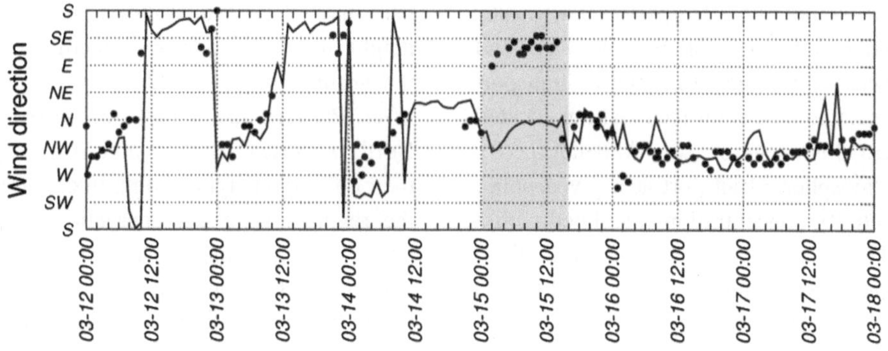

Figure 3
Wind directions observed at the METAR Fukushima airport station (*black dots*) and wind directions simulated using WRF (*black solid line*). The *grey area* corresponds to the period of time during which releases have been transported towards the northwest of Fukushima Dai-ichi plant

response systems (Chino *et al.*, 2011; Katata *et al.*, 2012a, b; Terada *et al.*, 2012). Large-scale data used in these models have higher temporal and spatial resolution than the GFS ones and a large amount of local meteorological observations are assimilated. As our regional simulations are still ongoing, only long range results are presented in this article.

3.2. *Long Range Simulations*

Atmospheric Transport Modeling at global scale has been carried out using the particle dispersion lagrangian model FLEXPART (V.8.2) and the NCEP/GFS meteorological data (http://weather.noaa.gov/pub/SL.us008001/ST.opnl) with 6 h, 0.5° × 0.5° and 1° × 1° resolutions. This ATM system is suitable regarding the spatio-temporal characteristic scales of the problem to be solved. The objective of long range simulations is to assess the arrival time of radionuclides at different monitoring stations located over the globe and evaluate the quantities released into the atmosphere. As previously quoted, we will mainly focus on Xe-133 (half life 5.244 days), I-131 (8.023 days), and Cs-137 (30.05 years) which are volatile fission products (see Part I of the publication).

3.3. *Gas and Particulate Deposition, Particle Size and Emission Height*

During events such as the Fukushima or Chernobyl accidents, gases and particles emitted in the atmosphere can be transported over very long distances and over long periods of time. Typically, the spatio-temporal scales of the problem are of the order of the circumference of the globe and several months. Because of these atmospheric dispersion and transportation scales, radioactive decay, gas-particulate conversion, and dry and wet deposition can drastically affect the behavior of emitted material. It should be noted that the noble gases (such as radioxenon) are not affected by deposition phenomena.

Wet deposition of gases or aerosols on the ground is due to the washout of the radionuclide plume by precipitation (rainfall, snow, etc.). It does not depend linearly on the precipitation rate. It is based on a scavenging coefficient, which depends on the precipitation rate and the considered radionuclide. To calculate the dry deposition of gases and aerosols, a classical approach in the dispersion models is to separate the gravity fallout (settling) and the interaction with soil and vegetation. The total deposition rate is the sum of these two contributions. The deposition due to interaction with the soil is calculated for altitudes between the ground and a reference height (e.g., 15 m). It is usually a function of aerodynamic drag terms produced by vegetation canopies and soil nature. Settling velocity is assumed to be zero for gases but depends on density and diameter for particles. This last point is important because as a FLEXPART particle can not represent several settling velocities, gases and aerosols trajectories must be calculated separately.

Preliminary calculations were performed to estimate the relative influence of particle size and deposition processes on the atmospheric transport efficiencies calculated at several IMS stations located in the northern hemisphere (see Appendix). By efficiency of atmospheric transport at a given point, we mean the activity concentration calculated with a unit release. Figure 4 (left) shows the normalized efficiencies calculated at the locations of 14 IMS stations. Efficiencies are normalized to the maximum reached at the station JPP38-Takasaki. Calculations are carried out with 1 and 0.1 μm particle diameters (Kaneyasu *et al.*, 2012), respectively. The result shows that the variation of particle diameter in this range has a small influence on the calculated behaviour of particles in the flow. Figure 4 (centre) shows the efficiencies obtained by considering on the one hand an inert tracer (i.e., without wet and dry deposition) and an aerosol with 1 μm diameter and dry deposition on the other hand. As expected, dry deposition has a weak effect on submicronic particles during transport. Figure 4 (right) shows the efficiencies calculated by using both an inert tracer and a 1 μm diameter aerosol with wet and dry deposition. With precipitation rates included in employed GFS data, wet deposition appears to have a significant effect on the results.

The height at which the radionuclides are emitted can have a significant influence on their dispersion into the atmosphere. Qiao *et al.* (2011) used a climate model to estimate the long range dispersion of Fukushima releases over a period of 3 months. In particular, they studied the influence of the emission height of radionuclides by selecting a release close to the ground, in the 5,000 m layer and the 10,000 m layer. Particles are transported more rapidly towards North America, Europe and Asia especially when their emission height is high. In our simulations, since the releases were not energetic (excluding cooling pool fires) and the explosions that occurred were controlled, it is considered that releases take place in the 0–200 m layer. As mentioned in the following section, radionuclides can quickly reach the middle and upper troposphere in situations when an updraft occurs.

4. Meteorological Conditions

Figure 5 shows the large-scale meteorological patterns for four dates between March 15 and March 24 (the maps were produced from NCEP/GFS 0.5° data using the GrADS system [http://www.iges.org/grads]). These different patterns present wind vector fields and geopotential height at 500 hPa. Geopotential height is useful to highlight large-scale cyclonic and anticyclonic structures. During this time period, a strong westerly jet stream blew over the Pacific Ocean from Japan to California, and over the Atlantic Ocean from the North American continent to Iceland. The flow was then directed towards the north pole because of a high pressure system over Europe (e.g., March 15 map). The high pressure structure established over Western Europe then led the flow to Central Europe. Because the

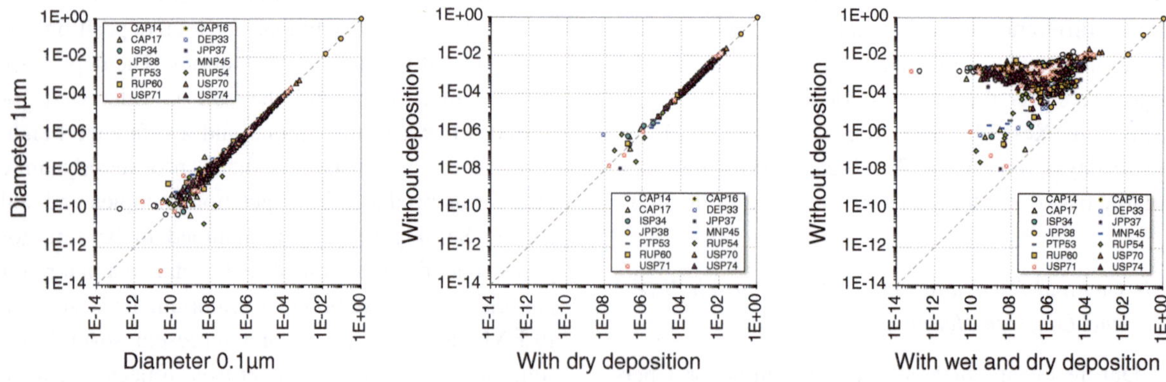

Figure 4
Left influence of particle diameter on transport efficiencies. *Center* influence of dry deposition. *Right* influence of wet deposition

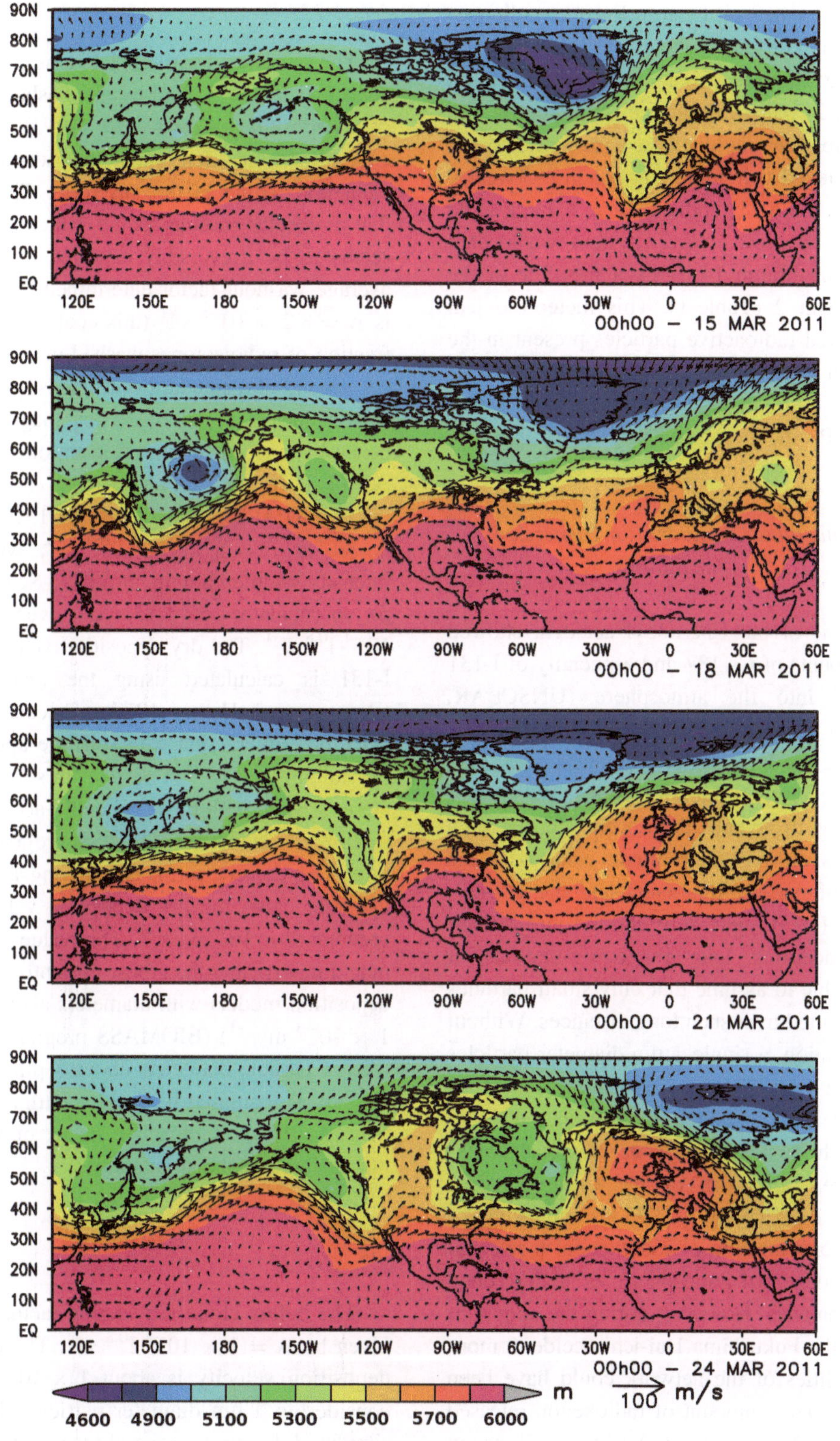

Figure 5
Large-scale wind field at 500 hPa (NCEP/GFS). *Scale color* geopotential height (m)

downdraft was located on Eastern Europe, radionuclides suspended in the atmosphere could have reached Eastern Europe before reaching Western Europe (e.g., March 21 and 24 maps). Radionuclides released into the atmosphere were rapidly transported around the globe, and achieved a circumnavigation in 2–3 weeks. TAKEMURA *et al.* (2011) indicate that a large-scale updraft caused by a low pressure system was located over Japan from March 14 to 15 at the time of the explosion of unit 2 (Table 1). This meteorological situation allowed radioactive particles present in the boundary layer to reach the middle and upper troposphere and the jet stream layer, and finally transported them over the Pacific Ocean in 3 or 4 days.

5. Radionuclides Considered in Dispersion Simulations

The analysis of the Chernobyl accident showed that large amounts of Cs-137 and especially of I-131 were emitted into the atmosphere (UNSCEAR, 2000). In order to compare these quantities with those released during the Fukushima accident, Cs-137 and I-131 are considered in the simulations. Because of the complexity of events that occur during such accidents, it is difficult to estimate what were the distributions of particle diameters and the gas/particulate ratios (I-131) that were emitted into the air. However, concerning long range simulations, it seems reasonable to assume that only small particles may be transported over such long distances. Without further information, a single 1 μm diameter particles distribution for Cs-137 and I-131 was considered. In addition to the Cs-137 and I-131, simulations were performed with Xe-133. Hence, as previously mentioned, in the framework of the Comprehensive Nuclear Test Ban Treaty (CTBT), monitoring of atmospheric concentration of radioxenon is relevant to provide evidence of atmospheric or underground nuclear weapon tests. However, during the couple of months after the Fukushima Dai-ichi accident, monitoring capabilities of the network could have been affected by the large amount of radioxenon released by the accident. The impact of Fukushima radioxenon releases on the worldwide Xe-133 background must also be investigated.

5.1. I-131

The gaseous form of I-131 is supposed to be deposited on the ground by the washout of the plume (wet deposition). CAPUT *et al.* (1993) have conducted experiments to determine the washout factor of elemental iodine in the gas form. The results show that molecular gaseous iodine, which is very reactive, seems to be irreversibly captured in rainwater. The average washout factor determined from experiments is $\Lambda = 8.2 \times 10^{-5}\ \mathrm{s}^{-1}$ (this coefficient indicates the fraction of radioisotopes washed out from 1 m^3 of air per second at a standard rain intensity of 1 mm h^{-1}). This value is of the order of magnitude of those generally accepted in deposition models [e.g., $\Lambda = 7 \times 10^{-5}\ \mathrm{s}^{-1}$ (PITTAUEROVÁ *et al.*, 2011)]. In the FLEXPART model, gaseous I-131 washout factor is set to $\Lambda = 8 \times 10^{-5}\ \mathrm{I}^{0.62}\ \mathrm{s}^{-1}$, where I is the intensity of the rain in mm h^{-1}. Scavenging factor for particulates of I-131 is given by $\Lambda = 1 \times 10^{-4}\ \mathrm{I}^{0.8}\ \mathrm{s}^{-1}$. The dry deposition velocity of gaseous I-131 is calculated using the resistance method (WESELY and HICKS, 1977). Calculated deposition velocities (interaction with soil) are approximately in the range 1×10^{-3}–$1 \times 10^{-2}\ \mathrm{m\ s}^{-1}$. These values are consistent with those generally accepted in impact assessment models (BIOMASS program, 2003). For the I-131 aerosol form, considering a 1 μm particle diameter class, the dry deposition velocity is of the order of $1 \times 10^{-4}\ \mathrm{m\ s}^{-1}$. This value is one order of magnitude smaller than those currently used in particle deposition models with diameters above 1 μm (about $1 \times 10^{-3}\ \mathrm{m\ s}^{-1}$) (BIOMASS program, 2003). In the case of particle diameters above 1 μm, the deposition velocity is increased by gravity settling (SLINN, 1982; SEHMEL, 1980). Radioactive decay of I-131 is taken into account in the simulations.

5.2. Cs-137

The below cloud scavenging factor for Cs-137 is given by $\Lambda = 1 \times 10^{-4}\ \mathrm{I}^{0.8}\ \mathrm{s}^{-1}$. The calculated dry deposition velocity is about $1 \times 10^{-4}\ \mathrm{m\ s}^{-1}$ when considering 1 μm diameter particles. Regarding the simulated durations (several weeks), the influence of radioactive decay of Cs-137 will be negligible on calculated activity concentrations.

5.3. Xe-133

Radioxenons are highly volatile fission products. Noble gases are not affected by wet and dry processes. Radioactive decay of Xe-133 is taken into account in the simulations.

5.4. Iodine Gas/Particle Conversion

The behavior of I-131 in the atmosphere is known to be complex as the gaseous form could gradually evolve into the particulate form. This has been the subject of many studies, particularly following the accident at the Chernobyl plant in April 1986. Uematsu *et al.* (1988) used atmospheric measurements made in Japan and on a boat in the Pacific Ocean to estimate the characteristic conversion gas/aerosol time for I-131. With the assumption that about 60 % of the total I-131 was present in the gas form in the Chernobyl releases, they found an average conversion time of about 2–3 weeks, with a minimum conversion time of about 12 days. The relative uncertainty is estimated to be about a factor of 2. Masson *et al.* (2011) indicate that the I-131 gas/I-131 total (total = gas + particles) ratio measured at the site of the Fukushima Dai-ichi plant from March 22 to April 4, 2011 was 71 ± 11 %. The average ratio measured in Europe until April 12 on a station network is very close, 77 ± 14 %. According to the authors, the similarity of the measured ratios suggests that the conversion gas/particle is small.

To verify this hypothesis, a calculation of atmospheric dispersion was performed by considering a single release which occurred on March 14 from 18 to 24 h (UTC), linked to the explosion on unit 2. In this numerical experiment, the release is assumed to be made up of 70 % of I-131 gas and 30 % of I-131 aerosol and a diameter of 1 μm is chosen for particulate form. The wet and dry deposition processes for gas and particulate forms are those described above. The radioactive decay of I-131 is taken into account, but gas/particle form transfers are not simulated. Assuming that the deposition process is realistically reproduced, the transfer between the gas phase and particulate phase can be considered low if the gas/particle ratio calculated in Europe is close to the initial release one.

Figure 6 (left) presents the activity concentrations obtained at some IMS stations located in Western Europe. The radionuclides released into the atmosphere during the simulated event reached monitoring stations between March 24 and April 13, which is in good agreement with measurements carried out in Europe (Masson *et al.*, 2011). The figure shows that travel time is about 2–4 weeks. Figure 6 (right) shows the simulated gaseous I-131/total I-131 ratios. Considering that the maximum simulated activity concentration is about 4,000 μBq m^{-3} and assuming a detection limit of 0.5 μBq m^{-3}, the average simulated gaseous I-131/total I-131 ratio is about 94 % (95 % considering all the values calculated without taking the detection limit into account). This result may indicate that about 40 % of gaseous I-131 should be converted to particulate phase to match the measured average ratio in Europe (~77 ± 14 %). According to the travel time between Japan and Europe, this suggests that the conversion time is in good agreement with the one proposed by Uematsu *et al.* (1988). However, due to the uncertainties of the simulations and measurements, this work should be continued. Iodine gas/particle conversion is not taken into account in the simulations.

6. Simulations at the JPP38: Takasaki Station

Simulations are carried out using NCEP/GFS global meteorological data with 6 h and 0.5° resolution. Figure 7 shows the activity concentrations calculated at the JPP38 station (see Appendix, Fig. 17) selecting the releases from Table 1. The activity concentrations are normalized to the total. For each event, it is assumed that the release lasts 2 h and that a unit amount of an inert tracer is emitted (without deposition or radioactive decrease). A short duration release is chosen to distinguish the relative contribution of each emission at the monitoring station.

6.1. Cs-137

Due to the relatively short distance between Fukushima Dai-ichi and JPP38 (~210 km), it was considered that the behavior of a tracer can be representative of the behavior of small sized aerosols

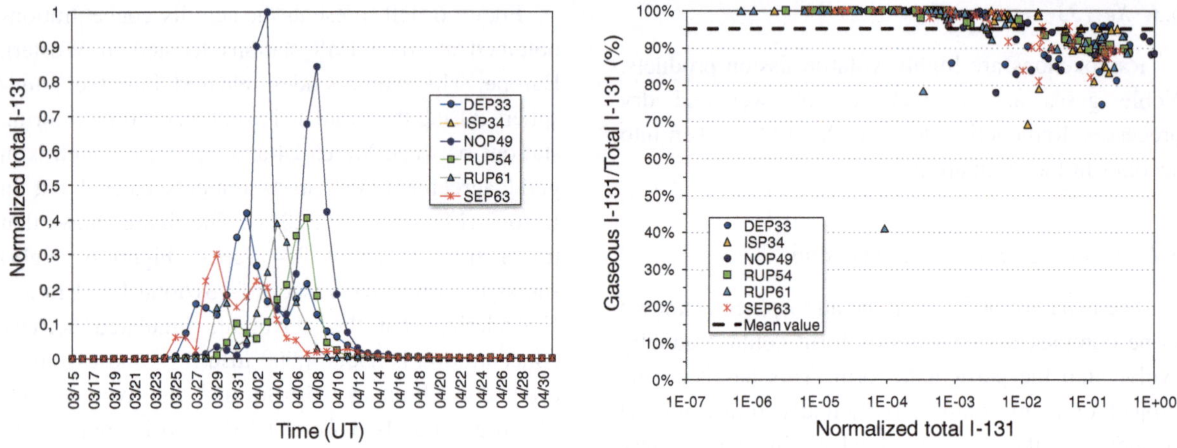

Figure 6
Estimation of iodine gas/particle conversion. *Left* activity concentrations calculated at the location of IMS stations situated in western Europe. *Right* corresponding gaseous/total I-131 ratios

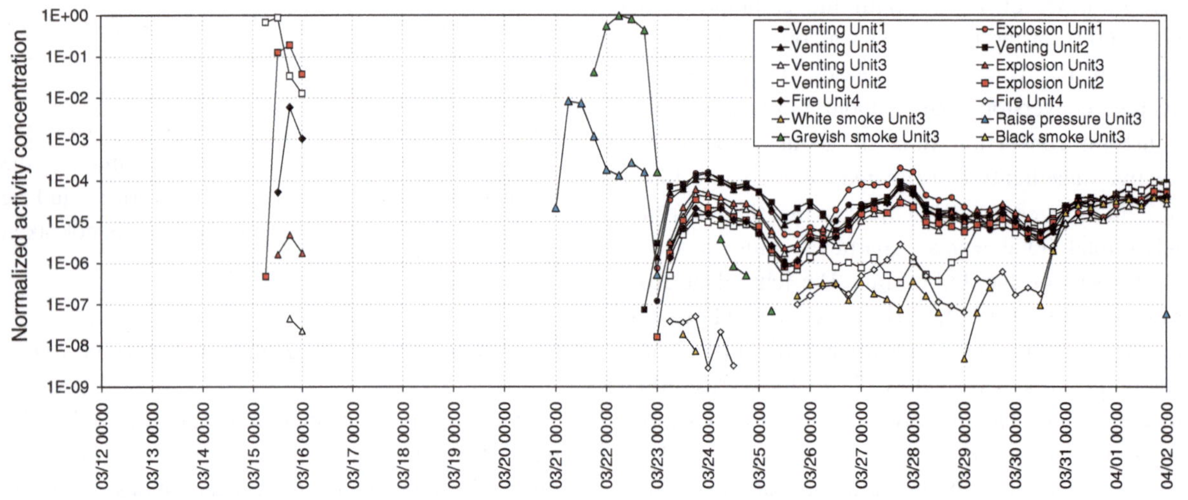

Figure 7
Calculated normalized activity concentrations at the JPP38 Takasaki station considering main events identified in Table 1

of Cs-137 during the first few days of the accident. Results presented above have shown that the behavior of aerosols was insensitive to the particle diameter when it was smaller than 1 μm (Fig. 4). For this particle size, it was shown that dry deposition phenomena were weak. Previous results also highlight that wet deposition was not very effective for the period leading to higher activity concentrations of JPP38. Figure 7 shows that the calculated activity concentrations were almost the same on March 15 and for the period from March 21 to 23. During those days, the northeasterly wind was favorable to the transport of radionuclides towards the station. The figure shows that on March 15, the station was mainly sensitive to releases 7 and 8 (venting and explosion of unit 2) that occurred in the late afternoon of March 14. For the March 21 to 23 period, the station was sensitive to releases 12 and 13 (rise of primary containment vessel pressure and greyish smoke on unit 3) that occurred during March 20 and 21. From March 23, the contribution of other releases is more difficult to assess. From March 30, radionuclides emitted at the beginning of the event were able to travel around the globe and to contribute to the simulated signal.

Figure 8 (left) presents the activity concentrations calculated and observed at JPP38 station for Cs-137.

The activity concentrations calculated are those obtained by considering a continuous release from March 12 to April 1, 2011 and also the total contribution of all short releases (Fig. 7). The release rate is the same in all simulations. Results are normalized by the maximum activity concentration calculated for the period, obtained with the continuous release. The measured activity concentrations are normalized by the maximum observed, reached on March 15, 2011. The dynamics of the measurements is relatively well reproduced by the simulations. On March 15 and for the period from March 21 to 23, the activity concentrations from the simulated puffs and continuous release are substantially equivalent. This indicates that during these days, releases that occurred on late March 14 and during March 20 and 21 have mainly contributed to the measured signal. However, the figure shows significant activity concentrations measured on March 31, which are only reproduced with the continuous release scenario. Differences observed between the continuous release and puffs may reflect the fact that releases which have reached JPP38 may have lasted more than 2 h. This also points out that other events than those identified in Table 1 may have occurred. Figure 8 (right) shows the comparison between measured and simulated activity concentrations. On March 15, a reasonable agreement is obtained by considering a total release of the order of $\sim 5 \times 10^{16}$ Bq on March 14 (releases 7 and 8). For March 21 to 23, a fairly good agreement is found by considering a release of the order of $\sim 2 \times 10^{16}$ Bq on March 20 (release 12) or $\sim 4 \times 10^{14}$ Bq on March 21 (release 13). The total estimated release of Cs - 137 is of the order of 5×10^{16}–7×10^{16} Bq.

In Fig. 8 (right), an additional release of $\sim 3 \times 10^{14}$ Bq was added during March 29 in order to reproduce the observed peak during March 30 and 31. In order to determine the effective release period, a backward calculation was carried out from the monitoring station for those 2 days (Fig. 9, left). Measurements of dose rate on the monitoring points located close to the Fukushima Dai-ichi NPP show a rise of the signal over March 28 and 29 (Fig. 9, right). These observations are in agreement with the hypothesis that an atmospheric release may have occurred during these dates. However, no particular event has been identified in news reports concerning the accident (NISA). Chino *et al.* (2011) have made an estimation of the I-131 and Cs-137 source terms released into the atmosphere between March 12 and early June 2011. To satisfy the agreement between simulation and measurements, the authors have found that an increase of 1.4×10^{14} Bq h^{-1} in Cs-137 was needed between March 29 and 30.

6.2. I-131

Figure 10 (left) presents the activity concentrations calculated and observed at JPP38 station for I-131. The calculation of the dispersion takes into account the radioactive decay of I-131. It was assumed that fires that occurred on unit 4 cooling pools could not lead to I-131 releases as the unit had already been stopped at the time of the tsunami and the spent fuel was too old to release appreciable amount of I-131. As for Cs-137, aerosols are assumed to be small enough to behave as a gas and to neglect the dry deposition phenomena. It was also assumed that wet deposition is not effective for periods during which the main peaks have been measured. Because of the short transport time between the release point and monitoring stations (distance $\sim$210 km), the gas/particle conversion have been neglected. By considering that the measurements at the IMS stations are only representative of the contribution of aerosols, a correction coefficient has to be applied to account for the fact that I-131 releases are also composed of gas. For aerosols, quite good agreement is obtained by considering a total release of the order of 6×10^{17} Bq at the end of the day on March 14 (releases 7 and 8: venting and explosion on unit 2). From March 21 to 23, an agreement is found by considering a release of the order of $\sim 4 \times 10^{16}$ Bq on March 20 (release 12) or $\sim 4 \times 10^{14}$ Bq during March 21 (release 13). An additional release of $\sim 2 \times 10^{14}$ Bq was added during March 29 to match the observed peak during the March 30 and 31 period. An increase in I-131 release rate was also needed between March 29 and 30 in Chino *et al.* (1.8×10^{14} Bq h^{-1}). The estimated total release is about $\sim 6 \times 10^{17}$ Bq. Assuming that the initial releases composition was 70 % gas and 30 % particles, the total I-131 estimated release is found to be $\sim 2 \times 10^{18}$ Bq.

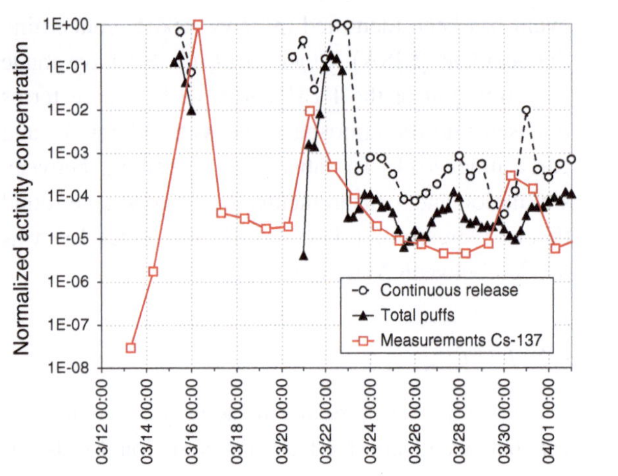

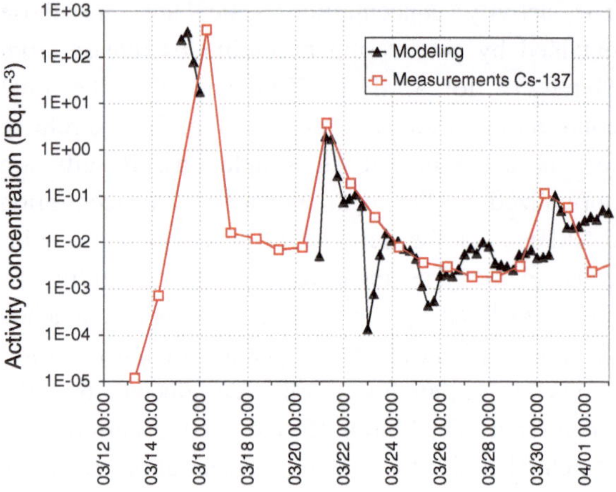

Figure 8

Left normalized Cs-137 activity concentrations measured and calculated at the JPP38 Takasaki station considering a continuous release and main identified events. *Right* Cs-137 activity concentrations measured and simulated using estimated source terms

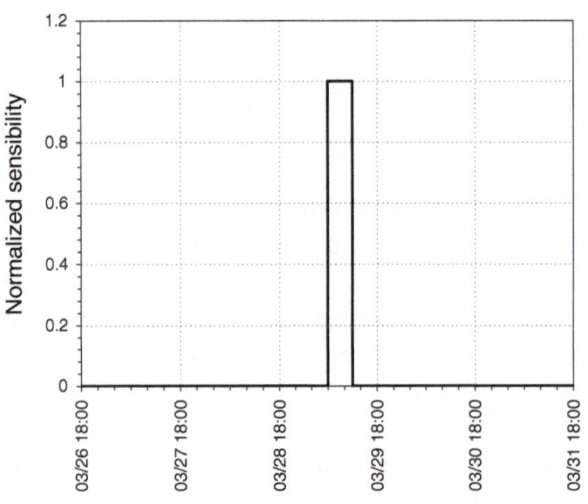

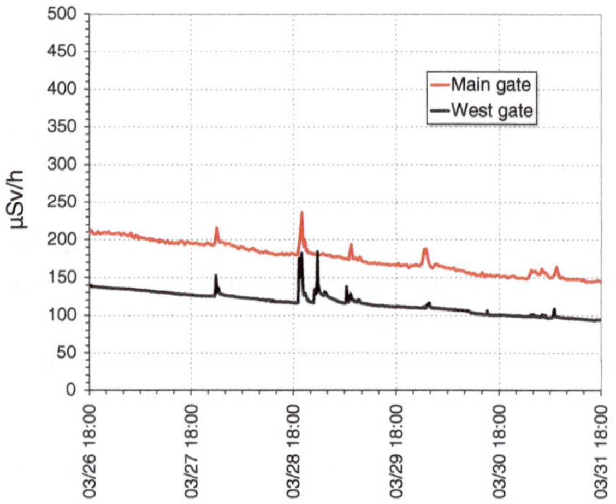

Figure 9

Left sensitivity between the site of Fukushima Dai-ichi site and the JPP38 station for a backward calculation started on March 30 and 31 from the station. *Right* dose rates measured on the monitoring points Main Gate and West Gate between March 26 and 31

Other releases, which have not reached the monitoring station due to meteorological conditions, might have to be added to the estimated I-131 and Cs-137 releases. Uncertainties on calculated source terms are significant and are related in part to the measurements. The I-131 and Cs-137 uncertainties have been assessed at ±25 % (3,700 ± 1,000 Bq m^{-3} and 400 ± 100 Bq m^{-3}, respectively) on March 15 measurements. In addition, after March 15, the JPP38 radionuclide station had been contaminated by the radioactive cloud as shown in Part I of the publication. Uncertainties on source terms are also related to inaccuracies in the wind fields used and may also be due to the dispersion model itself. Hence, the 0.5° wind fields may not be sufficiently resolved to match the JPP38 observations, as this station is close to Fukushima and located in an area where the topography is complex (Appendix, Fig. 17).

7. *Evaluation of Source Terms with Long Range Simulations*

Atmospheric transport of aerosols was carried out using 1.0°. NCEP/GFS data and simulations related to the Xe-133 were performed with 0.5° data. To

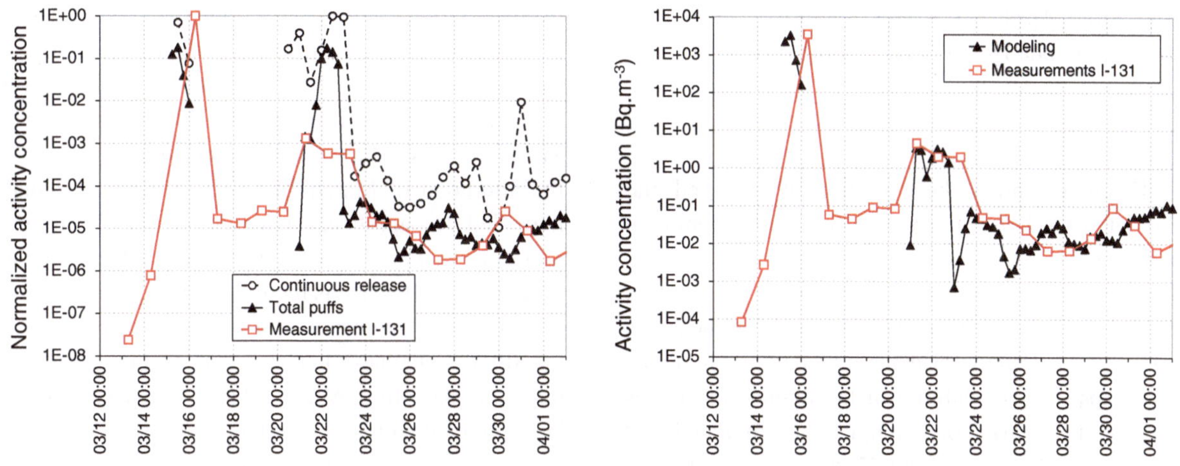

Figure 10
Left normalized I-131 activity concentrations measured and calculated at the JPP38 Takasaki station considering a continuous release and main identified events. *Right* I-131 activity concentrations measured and simulated using estimated source terms

assess the influence of the wind fields resolution on long range simulations, a calculation was performed by simulating the dispersion of an inert tracer continuous release emitted from the Fukushima Dai-ichi site during the period of time from March 12 to May 1, 2011. Calculations were performed with both the 1.0° and the 0.5° data. Figure 11 shows the results at several IMS stations located over the Northern hemisphere (see Appendix, Fig. 18). FA2 and FA5 bands are plotted in the figure to identify the 1.0° results which are in a factor of 2 or 5, respectively, compared to the results obtained at 0.5°. The two simulations appear to be in proper agreement, with FA2 = 87 % and FA5 = 97 %. The worst scores were obtained for the JPP38 station which is the closest to the Fukushima NPP. In this particular case, given the relative short distance between JPP38 and the facility, a better agreement between observations and simulations is obtained as expected with the 0.5° resolution data. For other IMS stations located at a longer distance from Fukushima, 1.0° and 0.5° wind fields are found to lead to substantially similar results.

Based on source terms evaluated from JPP38 station measurements, long range simulations were performed using 1.0° GFS wind fields. Direct calculations were carried out considering the events listed in Table 1. It is assumed that all events can lead to Cs-137 releases. As unit 4 was already shutdown at the time of the tsunami, it is assumed that the reactor

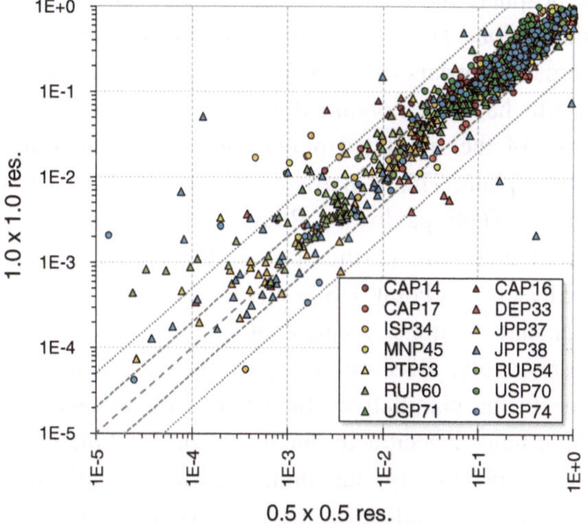

Figure 11
Comparison of activity concentrations simulated on IMS stations using 1.0° and 0.5° resolution GFS wind fields

and spent fuel cooling pools can not lead to I-131 releases. Because Xe-133 is highly volatile, it is considered to be emitted into the atmosphere at the time of the hydrogen explosions of units 1, 2 and 3.

All simulations indicate that the material emitted into the atmosphere mostly remained over the northern hemisphere, showing that the inter-hemisphere exchanges are limited. Only a few minor detections were observed over the southern hemisphere (e.g., for I-131: 1 μBq m^{-3} at FJP26, Nadi,

Fidji, from April 5 to 6 and 7 μBq m^{-3} at PGP51, Kavieng, Papua New Guinea, from April 11 to 12).

7.1. Cs-137

Figure 12 shows the evolution of observed and simulated Cs-137 activity concentrations. Temporal synchronization between observations and simulations is satisfactory. From one station to another, relative activity concentration level differences are not only based on the distance travelled by the plume from Japan (for readability of the figures, the *y* axis may be different). For example, the simulated and observed levels at the MNP45 station (Ulaanbataar, Mongolia) are higher than those obtained at the DEP33 station (Freiburg, Germany). This indicates that the complex transport of radionuclides in the atmosphere has been, in part, well reproduced in the simulations. However, long range transport of Cs-137 aerosols appears difficult to simulate. Hence, poorer results have been obtained for stations located on the edge of the simulated plume and/or close to Fukushima plant. This is particularly the case for RUP60 station (Petropavlovsk, Russia), which is located about 2,200 km northeast from Fukushima and for USP71 (Sand Point, USA) and CAP16 (Yellowknife, Canada) stations located at 62.5N and 55.33N respectively. Bias on transport of Cs-137 may be due to the cumulative effect over long distances of wind fields inaccuracies and/or an underestimation of the dispersion of lagrangian particles with high latitudes. Depending on the intensity of precipitation rates contained in the meteorological input data, the wet deposition can also be locally overestimated or underestimated on the path of the particle cloud. Uncertainties on the initial size distribution of aerosols can also affect the representativeness of the calculated dry deposition. Some of the differences between simulations and observations may also be due to specific phenomena that have not been taken into account, such as re-suspension of part of cesium initially deposited on the ground. Some smaller peaks observed after the main peaks may be due to the resuspension.

At stations where the aerosol behaviour is the most consistent with the observations, the source terms evaluated using the JPP38 detections lead to an overestimation of the long distance simulated detections. Total release estimated from the long range simulations is about 1×10^{16} Bq ($\sim$5–7 times lower than those obtained using the JPP38 measurements). The simulations show that the events that occurred in the afternoon of March 14 UTC on the unit 2 may have led to significant releases of Cs-137 in the atmosphere. For these events, a total release of the order of 6×10^{15} Bq is required to find an agreement with the measurements ($\sim$10 times lower than for the JPP38 calculation). The events on the unit 3 and that took place from March 20 to 23 are also found to be significant with a required total release of the order of 2×10^{15} Bq. Simulations suggest that the hydrogen explosion and/or venting operations of units 1 and 3 (March 12 and 13/14) may have resulted in a total release of 2×10^{15} Bq.

7.2. I-131

Figure 13 shows the evolution of observed and simulated I-131 activity concentrations. To perform the comparison at the IMS stations, simulations were carried out considering only the aerosol fraction of emissions (iodine gas phase is not measured by IMS stations). The agreement between simulations and observations is generally better for I-131 than for Cs-137. This can be attributed to the half-life of I-131 (8.023 days), which leads to a faster depletion of the cloud and thus reduces the relative errors of simulations (due to wet deposition, etc.). The dynamics of the measurements is rather well reproduced by the simulations (note that the amplitude of the *y* axis of USP70 and USP71 stations is different from the other stations). The temporal synchronization between observations and simulations is quite satisfactory. As for the Cs-137, poorer results are obtained for stations located on the edge of the simulated plume and/or near Fukushima: e.g., RUP60 (Petropavlovsk, Russia), CAP16 (Yellowknife, Canada) and USP71 (Sand Point, USA) stations. A general agreement is nevertheless obtained by considering a total I-131 aerosol source term of approximately 1×10^{17} Bq emitted during the different puffs. Assuming that the I-131 aerosol release represents only 30 % of the total emission, the total release (aerosol + gas) would be about 4×10^{17} Bq without considering

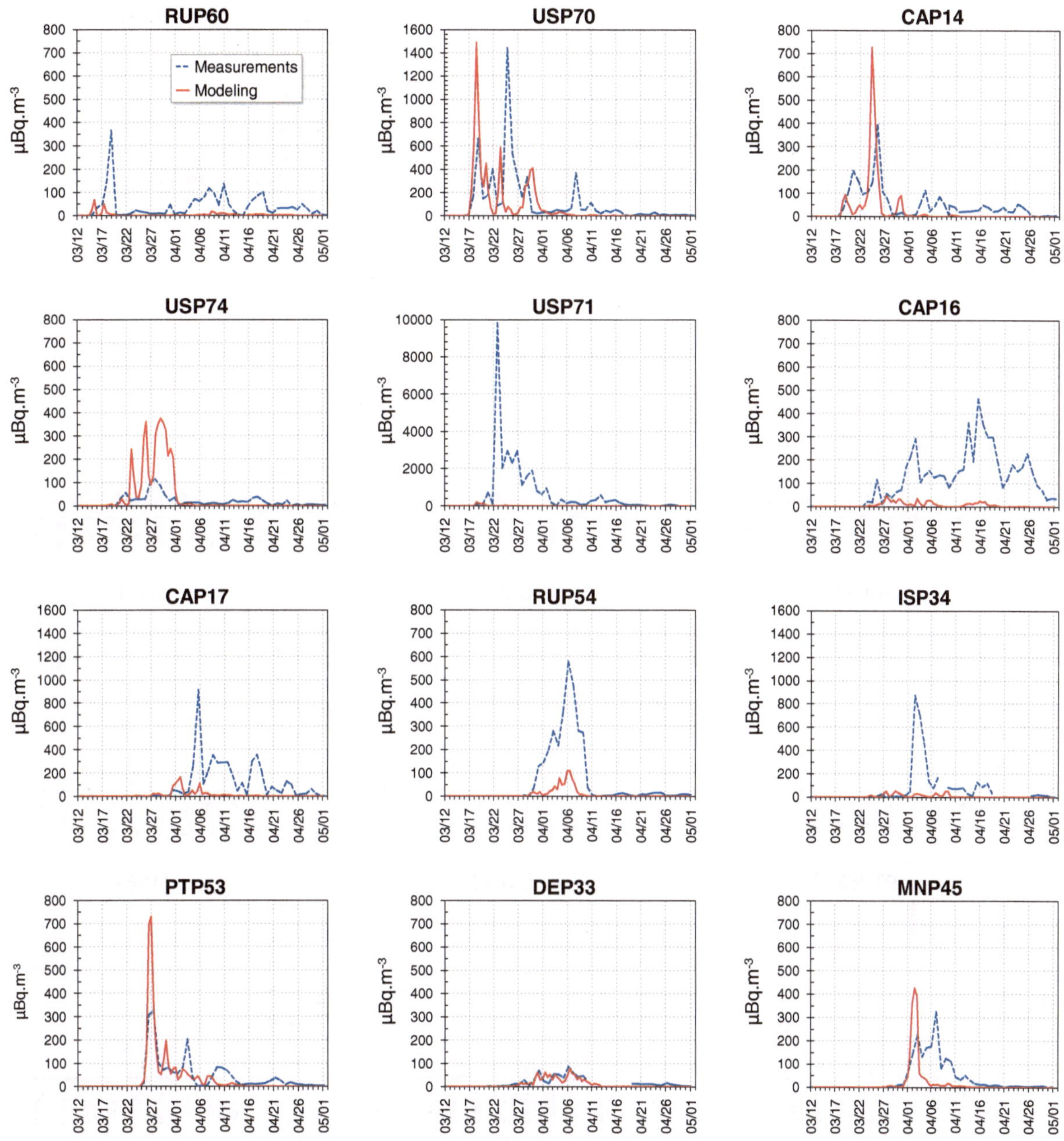

Figure 12
Activity concentrations in Cs-137 measured and simulated from March 12 to May 1, 2011 at some IMS stations. The stations are represented according to the dates of arrival of the plume of radionuclides. *Top left* the station that detected the earliest. *Bottom right* the station that detected the latest

the gas/particle conversion phenomena. As for the Cs-137, a factor ~5 is observed between the estimated total releases from simulations at JPP38 and distant stations. Simulations show that the events which took place in the afternoon of March 14 UTC on the unit 2 led to major releases of I-131 in the atmosphere. For these events, a release of the order of 6×10^{16} Bq is required to find an agreement with measurements (~10 times lower than for the JPP38 calculation). Events on unit 3, which took place from

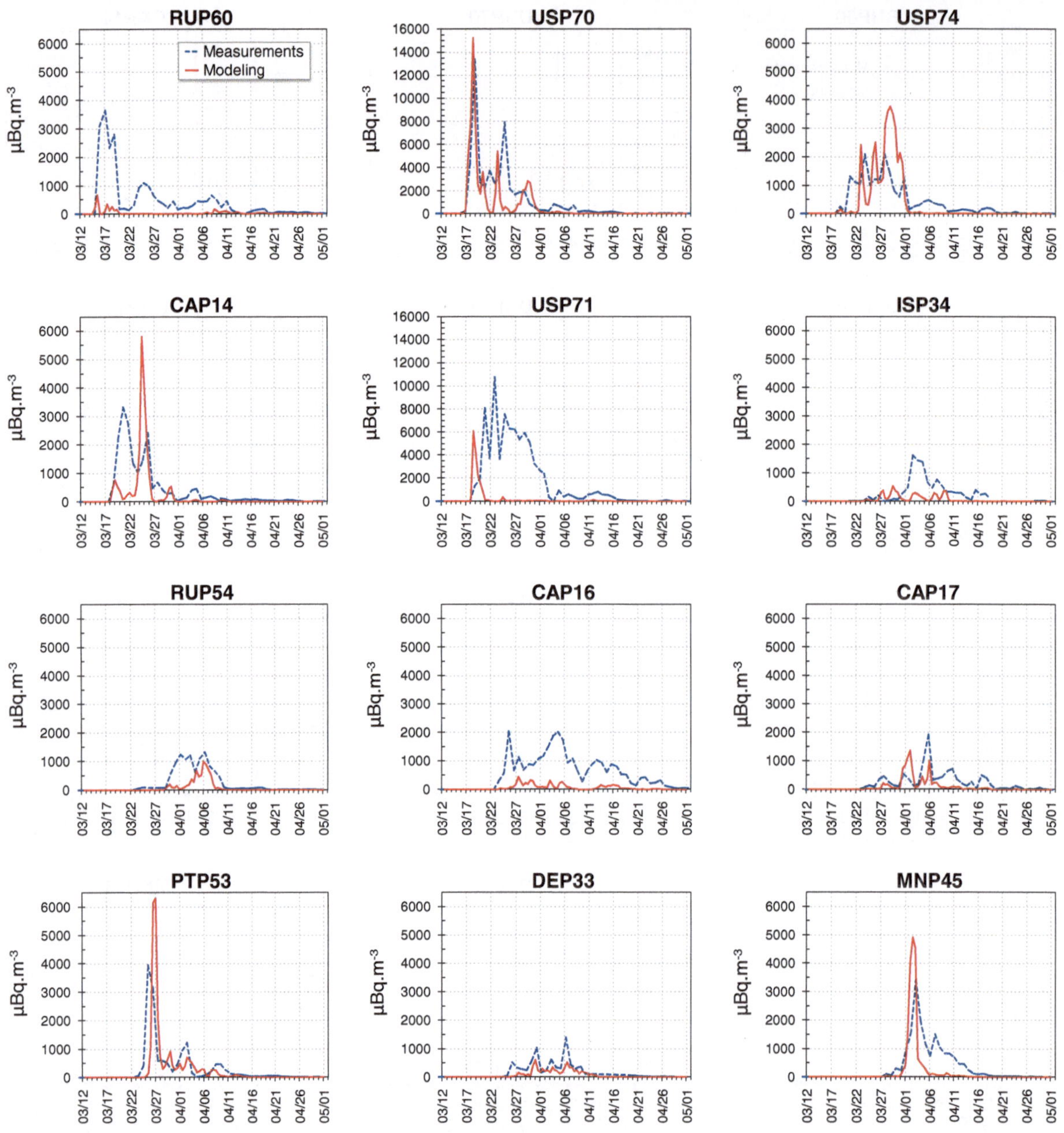

Figure 13

I-131 activity concentrations measured and simulated from March 12 to May 1, 2011 at some IMS stations. The stations are represented according to the arrival dates of the plume

March 20 to 23, are also found to be significant, with a required release of the order of 3×10^{16} Bq. For example, the 1.4 mBq m^{-3} peak measured from the April 5 to 6 UTC at DEP33 station (Freiburg, Germany) seems to be correlated with these emissions. Simulations show that the venting and hydrogen explosion related to unit 1 (March 12) may have led to significant releases of about 2×10^{16} Bq. Simulations suggest that the peak of 13.4 mBq m^{-3} measured from March 17 to 18 UTC by the USP70 (Sacramento, USA) station could be mainly due to these releases.

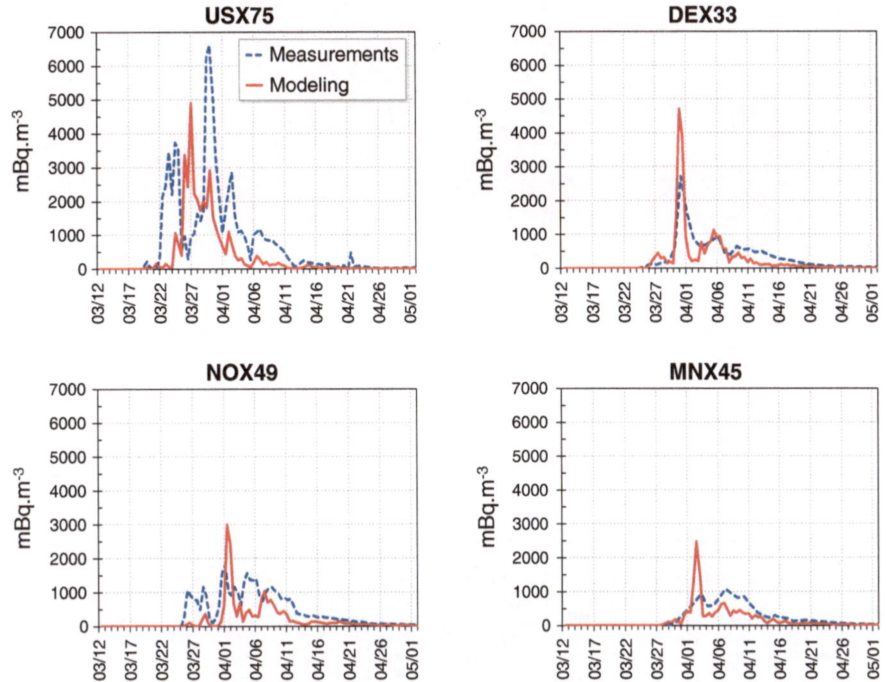

Figure 14
Activity concentrations in Xe-133 measured and simulated from March 12 to May 1, 2011 at some IMS stations. The stations are represented according to the dates of arrival of the plume

7.3. Discussion on Cs-137 and I-131 Results

Releases calculated for Cs-137 and I-131 appear to be much larger than those calculated for the continuous leakage between March 22 and mid-May 2011 (see Sect. 2.2). Continuous leakage calculated from March 22 may represent only 1/100 (Cs-137) and 1/1,000 (I-131) of total releases and they will be neglected.

Total source terms calculated from long range simulations for Cs-137 and I-131 (resp. 1×10^{16} Bq and 1×10^{17} to 4×10^{17} Bq) are in good agreement with those presented in the literature (Chino *et al.*, 2011; Mathieu *et al.*, 2012; Winiarek, 2012) These results suggest that the source terms estimated from the station JPP38 are probably too high. Chino *et al.* (2011) found also that the largest releases took place on early March 15 and that no major release occurred after March 24 which is also consistent with our results. The main releases could be linked to the explosion and the pressure suppression chamber damage of unit 2. However, our simulations suggest that the hydrogen explosions and/or venting operations of units 1 and 3 (March 12 and 13/14) may have also resulted in significant releases (Katata, 2012a). Because of the meteorological situation, these releases seem to have been rapidly blown towards the Pacific Ocean (especially the release of March 12) and may not have been significantly measured by the stations network in the vicinity of the NPP.

It was estimated that the total Cs-137 and I-131 releases emitted into the atmosphere during the Chernobyl accident were 8.5×10^{16} and 1.8×10^{18} Bq respectively. Concerning this event, the release of Cs-137 was estimated to be about 30 % of the core inventory and that of I-131 is estimated to be about 50 % (UNSCEAR). The source terms estimated in our study from long range simulations show that Fukushima releases could represent about 10 % of the Chernobyl emissions for these two isotopes.

7.4. Xe-133

Because of its high volatility, it is estimated that the Xe-133 was mainly emitted into the atmosphere

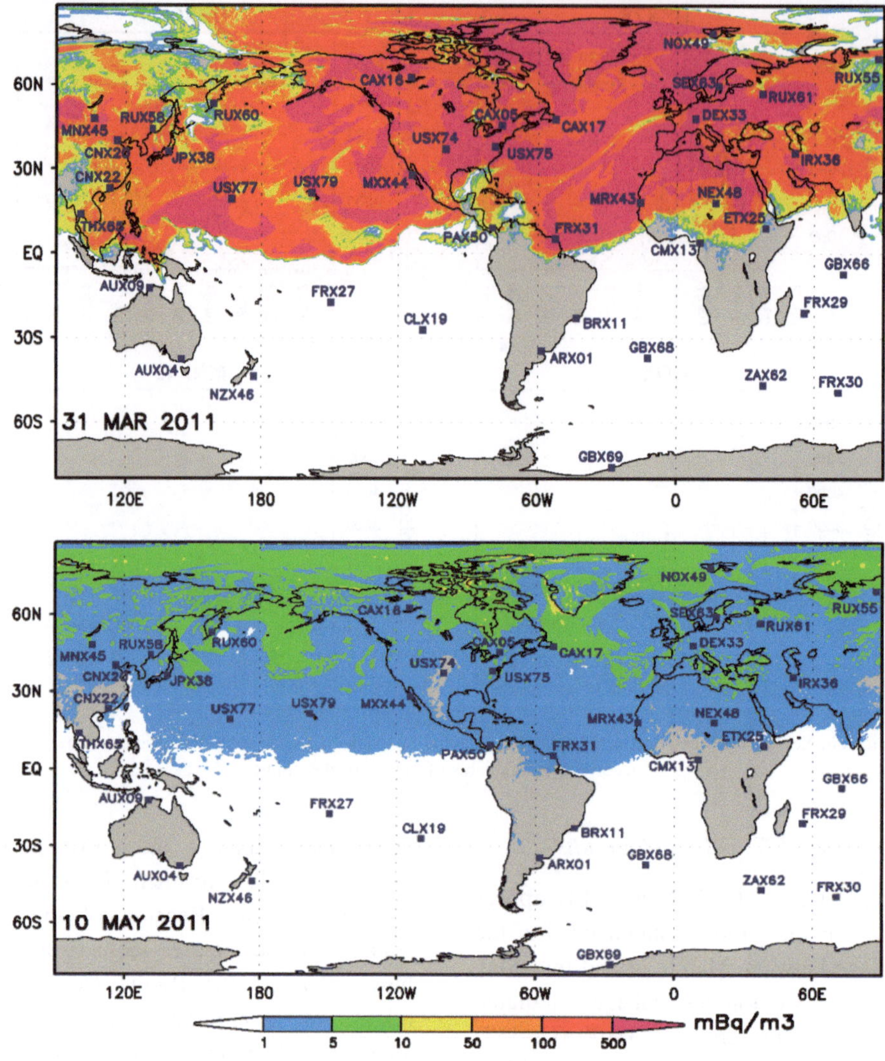

Figure 15

Xe-133 activity concentrations at ground-level for two dates. All IMS noble gas stations are represented on the maps

during the hydrogen explosions of units 1, 2 and 3. Releases are supposed to have occurred over 6 h time periods centered at the time of explosions, which allows us to consider some venting and leakage before and after those moments. Figure 14 shows the activity concentrations (expressed in mBq m^{-3}) measured and calculated at some IMS stations located in the northern hemisphere (see Appendix, Fig. 19). Quite good agreement is obtained considering that $\sim 2 \times 10^{18}$ Bq are released during each puff (total release of $\sim 6 \times 10^{18}$ Bq).

As previously observed, the results are less satisfactory at stations located on the edge of the simulated plume. For other stations, a fairly good temporal synchronization is obtained and the dynamics of the measurements is globally well reproduced. Figure 15 shows the maps of calculated activity concentrations for two selected dates within a 2 month period after the beginning of the releases. Results show that the cloud was dispersed in the whole northern hemisphere and as mentioned previously, that air exchanges between north and south were low (as for the Chernobyl event). The maps suggest a relatively rapid decrease of concentrations, resulting from the dispersion of gas and half-life of Xe-133. Because of the large amount of radioxenon

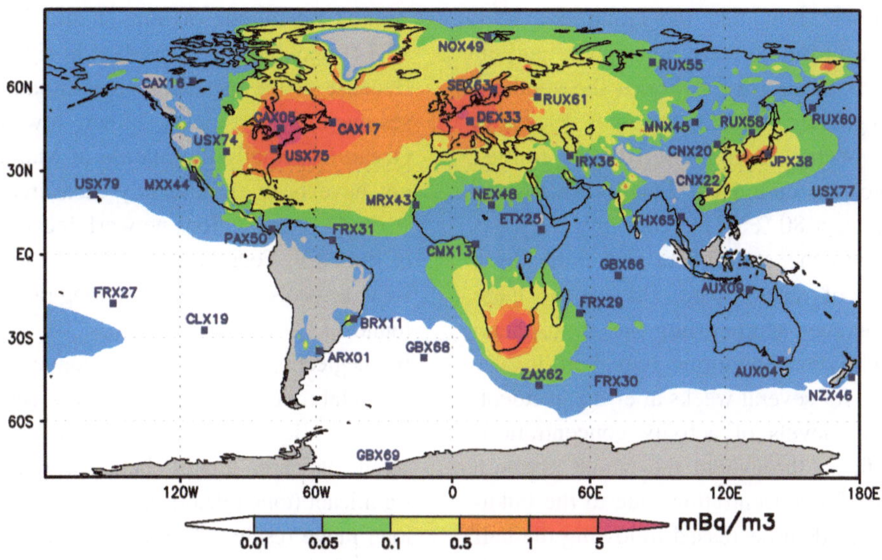

Figure 16
Global average simulated background in Xe-133 due to atmospheric releases of major nuclear facilities during normal operation (ACHIM *et al.*, 2011): Nuclear Power Plants and factories producing radionuclides for medical use (MATTHEWS, 2010; KALINOWSKI and TUMA, 2009). These facilities are mainly distributed in the northern hemisphere (note that Figs. 16 and 17 color scales are differents)

emitted into the atmosphere, high levels of activity concentrations were found that remained locally above 1 Bq m^{-3} for several weeks after the accident. Until early May, levels of activity concentrations remained significant throughout the northern hemisphere. They were above average levels usually observed due to releases of major nuclear facilities during normal operations (Fig. 16). From early May 2011, Xe-133 concentrations due to Fukushima Dai-ichi releases were of the order of magnitude of industrial background. Measurements show that the situation has returned to near normal since June. Xe-131 m/Xe-133 ratios measured over time by the CTBT noble gas network and French national devices show a clear signature of the Fukushima Dai-ichi accident over a long period of time (about 80 days). This information is useful to assess the monitoring capabilities of the network if another significant event occurred within a few weeks after the accident. For example, the USX75 station (Charlottesville, USA) has shown, 2 months after the accident, a typical signature from the Chalk River medical isotope production plant (Canada) that clearly exceeded the Fukushima ratios (see Part I of the publication).

The total estimated source term gives a fraction of core inventory of about 8×10^{18} Bq at the time of reactor shutdown. This result suggests that at least 80 % of the core inventory has been released into the atmosphere and indicates a broad meltdown of reactor core (see Part I of the publication). Total source term is in good agreement with literature (MATHIEU *et al.*, 2012). However our result is lower than in STOHL *et al.* (2012) where the authors found a total release from 12×10^{18} to 18×10^{18} Bq which is higher than the entire estimated Xe-133 inventory. According to the authors, a significant part of Xe-133 released could be due to I-133 decay.

8. *Conclusion*

This part of the publication (Part II) is dedicated to atmospheric transport modeling. Simulations were mainly carried out at global scale by considering Cs-137, I-131 and Xe-133 volatile fission products to assess the arrival time of radionuclides at different IMS stations located on the globe and to evaluate the quantities released into the atmosphere. These analyses are valuable to estimate the Fukushima reactor core damages (Part I of the publication) and to assess the monitoring capabilities of the CTBT network following the accident.

All simulations show that the cloud has been mainly dispersed in the northern hemisphere and air exchanges appear to be very low with the southern hemisphere. Regarding Xe-133, the total release is estimated to be of the order of 6×10^{18} Bq emitted during the explosions on units 1, 2 and 3. This result suggests that at least 80 % of the core inventory has been released into the atmosphere and indicates a broad meltdown of reactor core. Due to the large amount of radioxenon emitted into the atmosphere, levels of activity concentrations remained locally above 1 Bq m^{-3} for several weeks after the accident. Until early May, levels of activity concentrations remained significant throughout the whole northern hemisphere. Xe-133 concentrations due to the Fukushima Dai-ichi accident decreased following the half life of the radionuclide to be of the order of magnitude of industrial background from mid May 2011. Measurements by noble gas stations located in the northern hemisphere show that the situation had returned to near normal during June 2011. The evolution of measured Xe-131 m/Xe-133 ratios shows a clear signature of the Fukushima Dai-ichi accident over a long period of time (about 80 days). The knowledge of activity concentrations and of isotopic ratios over time were essential to maintain the monitoring capabilities of the CTBT network if another major event had arisen in the weeks following the accident.

Regarding Cs-137 and I-131, simulations were performed considering both JPP38 and distant IMS stations measurements. Atmospheric transport modeling results are in a reasonable agreement with measurements on most stations but appear poorer for stations located on the edge of the simulated plume and/or close to Fukushima NPP. Bias on transport may be due to the considered source terms and to the cumulative effect over long distances of inaccuracies in wind fields. Calculations suggest that the main air emissions have occurred on March 14 (explosion and pressure suppression chamber damage of unit 2) and that no major release occurred after March 23. The JPP38 station appeared to be mainly concerned by March 14 and March 20 and 21 releases (rise of pressure of unit 3). The hydrogen explosions and/or venting operations of units 1 and 3 (March 12 and 13/14) may have also resulted in significant releases. Because of the meteorological situation, these releases have been quickly blown towards the Pacific Ocean (especially the release of March 12) and may not have been significantly measured by the JPP38 station and the stations network located in the vicinity of the NPP. Total atmospheric releases of Cs-137 and I-131 aerosols estimated from long range simulation are found to be 1×10^{16} and 1×10^{17} Bq, respectively. By neglecting gas/particulate conversion phenomena, the total release of I-131 (gas + aerosol) could be 4×10^{17} Bq. Emissions estimated using JPP38 measurements are higher but uncertainties could be significant (total releases are $\sim$5–7 larger regarding the long range results). The amounts of Cs-137 and I-131 emitted into the atmosphere during the Fukushima accident could represent 10 % of the Chernobyl accident releases (estimated to be 8.5×10^{16} Bq of Cs-137 and 1.8×10^{18} Bq of I-131).

Acknowledgments

The authors wish to thank the Comprehensive Nuclear-Test-Ban Treaty Organisation for fruitful discussions about the data collected by radionuclide stations of the International Monitoring Network during this event. A special thank you should be expressed to Harry Dupont of the Alten Company for his expertise and for the implementation of dispersion and mesoscale meteorological models.

Appendix

Figure 17 shows the IMS stations (International Monitoring System) closest to the Fukushima Dai-ichi NPP. These stations are JPP38 - Takasaki; Japan ($\sim$210 km) and RUP58 - Ussuriysk; Russia Fed. ($\sim$1,100 km). The Fukushima Dai-ichi site and the city of Tokyo are also indicated on the map. Figures 18 and 19 show the IMS networks of radionuclide (aerosols) and noble gas stations in their September 2011 operational states.

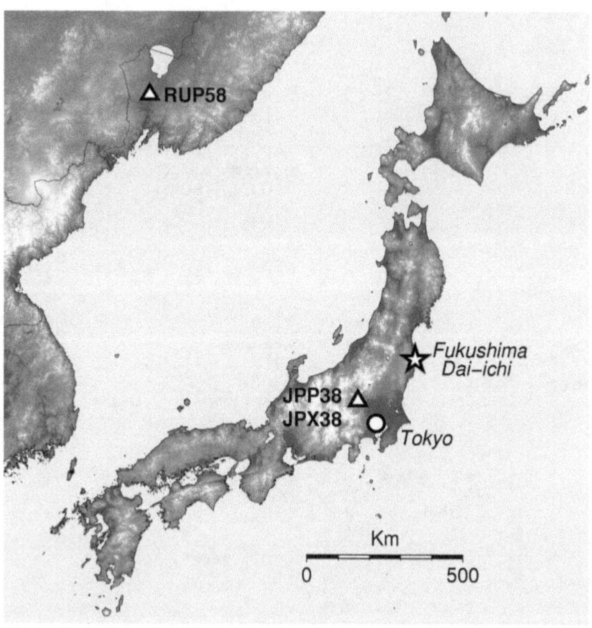

Figure 17
Location of IMS stations (*triangles*) situated close to the Fukushima Dai-ichi plant (*star*)

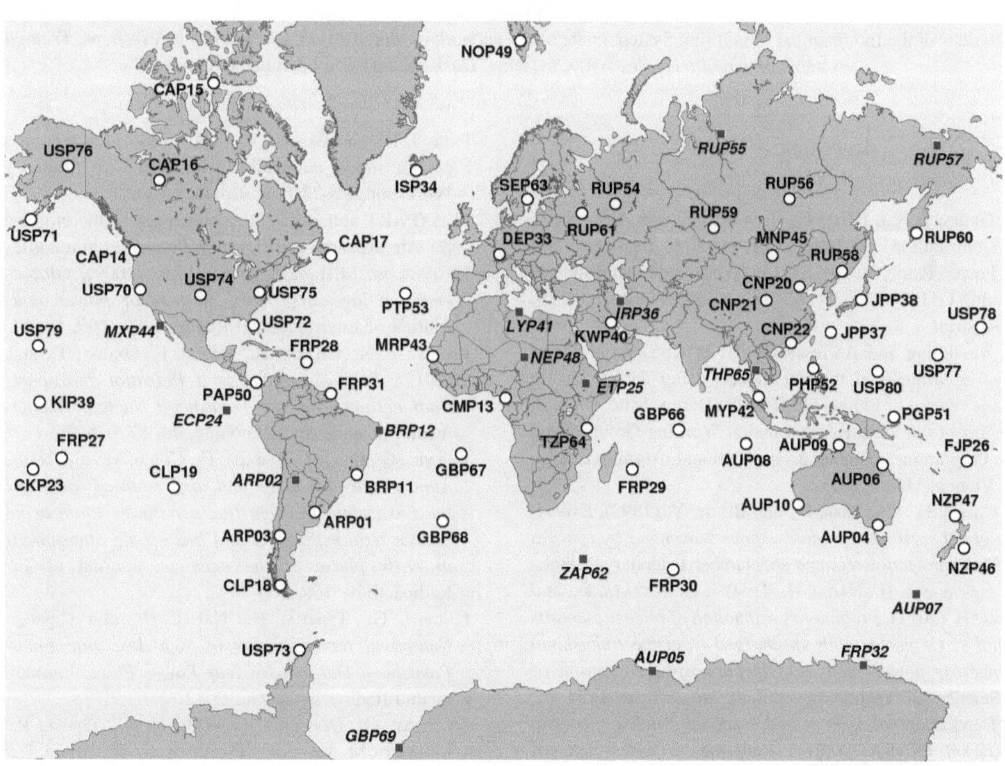

Figure 18
Particulate stations of the International Monitoring System (state of the network in March 2011). *White circles* operational stations. *Dark squares* non operational stations

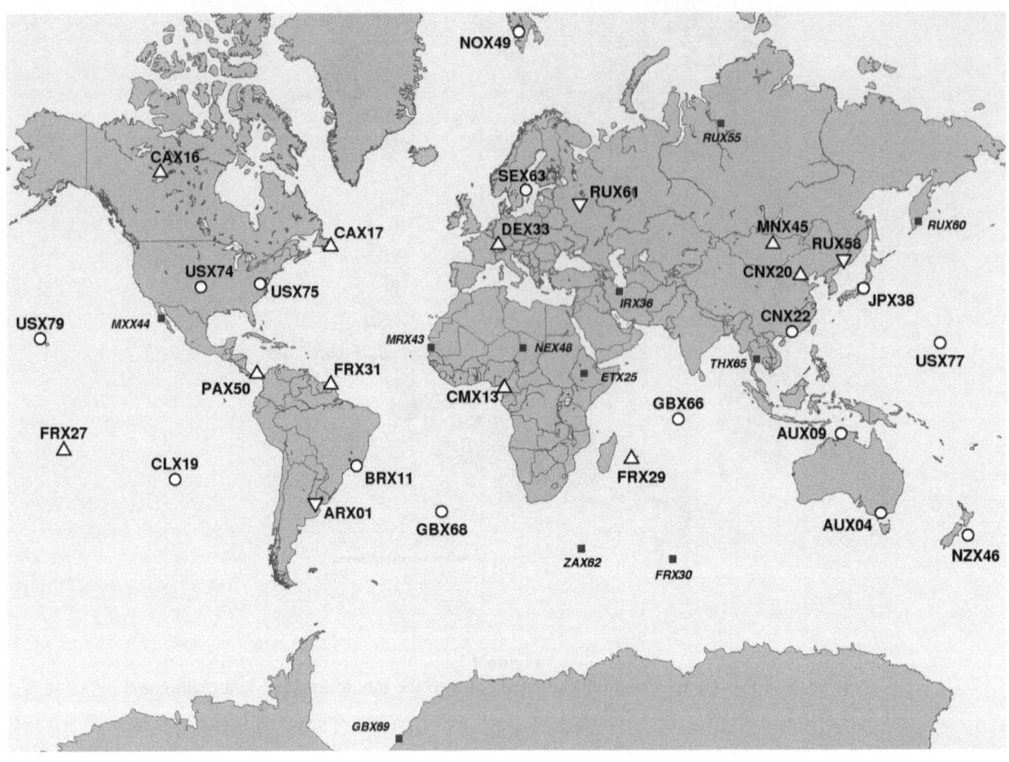

Figure 19

Noble gas stations of the International Monitoring System (state of the network in March 2011). *Circles* SAUNA systems. *Triangles* SPALAX systems. *Inverted triangles* ARIX systems. *Dark squares* non operational stations

References

ACHIM, P., GROSS, P., LE PETIT, G., TAFFARY, T. and ARMAND, P. (2011), Contribution of Isotopes Production Facilities and Nuclear Power Plants to the Xe-133 Worldwide Atmospheric Background. CTBT Science and Technology, June 8 to 10, Vienna, Austria.

BIOsphere Modelling and ASSessment (BIOMASS) programme: Testing of environmental transfer models using data from the atmospheric release of Iodine-131 from the Hanford site, USA, in 1963. Report of the Dose Reconstruction Working Group of the Biomass Programme, Theme 2. International Atomic Energy Agency, Vienna, March 2003.

CAPUT, C., CAMUS, H., GAUTHIER, D., and BELOT, Y. (1993), *Experimental study of washout of iodine vapour scavenged by rain* (in French). Radioprotection Volume *28*, Number 1, January–March.

CHINO, M., NAKAYAMA, H., NAGAI, H., TERADA, H., KATATA, G., and YAMAZAWA, H. (2011), *Preliminary estimation of release amounts of ^{131}I and ^{137}Cs accidentally discharged from the Fukushima Daiichi nuclear power plant into the atmosphere*. Journal of Nuclear Science and Technology, Vol. *48*, No. 7, 1129–1134.

DOE, US Department of Energy and National Nuclear Security Administration (NNSA), (2011) Radiological assessment of effects from Fukushima Daiichi power plant. Radiation monitoring data updated from March 22 to May 13.

FAST, J. D., and EASTER, R. C. A. (2006), Lagrangian particle dispersion model compatible with WRF. 7th Annual WRF User's Workshop, 19–22 June, Boulder, CO.

FLEXPART and FLEXTRA homepage at the Norvegian Institute for Air Research (NILU). http://transport.nilu.no/flexpart.

KALINOWSKI, M.B., and TUMA, M.P. (2009), *Global radioxenon emission inventory based on nuclear power reactor reports*. Journal of Environmental Radioactivity *100*, 58–70.

KANEYASU, N., OHASHI, H., SUZUKI, F., OKUDA, T., and IKEMORI, F. (2012), *Sulfate Aerosol as a Potential Transport Medium of Radiocesium from the Fukushima Nuclear Accident*. Environmental Science & Technology, *46*, 5720–5726.

KATATA, G., OTA, M., TERADA, H., CHINO, M., and NAGAI, H. (2012), *Atmospheric discharge and dispersion of radionuclides during the Fukushima Dai-ichi Nuclear Power Plant accident. Part I: Source term estimation and local-scale atmospheric dispersion in early phase of the accident*. Journal of Environmental Radioactivity *109*, 103–113.

KATATA, G., TERADA, H., NAGAI, H., and CHINO, M. (2012), *Numerical reconstruction of high dose rate zones due to the Fukushima Dai-ichi Nuclear Power Plant*. Journal of Environmental Radioactivity *111*, 2–12.

LE PETIT, G., DOUYSSET, G., DUCROS, G., GROSS, P., ACHIM, P., MONFORT, M., RAYMOND, P., PONTILLON, Y., JUTIER, C., BLANCHARD, X., TAFFARY, T., and MOULIN, C. (2012), *Analysis of radionuclide releases from the Fukushima Dai-ichi Nuclear Power Plant accident, Part I*, Pure Appl. Geophys. doi:10.1007/s00024-012-0581-6

MEXT, Ministry of Education, Culture, Sports, Science and Technology—Japan: Reading of environmental radioactivity level by prefecture, Time series data.

MM5, Modeling System Version 3. PSU/NCAR mesoscale modeling system. Tutorial Class Notes and User's Guide, January 2005.

MASSON, O., *et al.* (2011), *Tracking of Airborne Radionuclides from the Damaged Fukushima Dai-ichi Nuclear Reactors by European Networks.* Environmental Science & Technology, Vol *45*, 7670–7677.

MATHIEU, A., KORSAKISSOK, I., QUÉLO, D., GROËLL, J., TOMBETTE, M., DIDIER, D., QUENTRIC, E., SAUNIER, O., BENOIT, J.P., and ISNARD, O. (2012), *Atmospheric Dispersion and Deposition of Radionuclides from the Fukushima Daiichi Nuclear Power Plant Accident.* Elements, Vol. *8*, 195-200.

MATTHEWS, M. (2010), Workshop on Signatures of Medical and Industrial Isotopes Production (WOSMIP)—A Review. US DOE Report PNNL-19294, February 2010.

NCEP/GFS meteorological data at global scale. http://weather.noaa.gov/pub/SL.us008001/ST.opnl.

NISA, Japanese Nuclear and Industrial Safety Agency: Bulletins on conditions of Fukushima Daiichi Nuclear Power Station. http://www.nisa.meti.go.jp/english/press/index.html.

OKADA, S. (2011), Off-Site Activities regarding the Accident at the Fukushima Daiichi Nuclear Power Station and Lesson from Them. World Engineers' Convention, Geneva, 4–9 September.

PITTAUEROVÁ, D., HETTWIG, B., and FISCHER, W. (2011), *Fukushima fallout in Northwest German environmental media.* Journal of Environmental Radioactivity, *102*, 877–880.

QIAO, F.L., WANG, G.S., ZHAO, W., ZHAO, J.C., DAI, D.J., SONG, Y.J., and SONG, Z.Y. (2011), *Predicting the spread of nuclear radiation from the damaged Fukushima Nuclear Power Plant.* Chinese Science Bulletin Vol. *56*, No. 18, 1890–1896.

QUÉLO, D., GROËLL, J., DIDIER, D., MATHIEU, A., KORSAKISSOK, I., TOMBETTE, M., QUENTRIC, E., BENOIT, J.P., and ISNARD, O. (2011), Atmospheric transport modeling and situation assessment of the Fukushima accident. 15th annual GMU conference on atmospheric transport & dispersion modeling, July 12–14, Fairfax, Virginia, U.S.A.

SEHMEL, G.A. (1980), *Particle and gas dry deposition: a review.* Atmospheric Environment Vol. *14*, pp. 983–1011.

SLINN, W.G.N. (1982), *Predictions for particle deposition to vegetative canopies.* Atmos. Environ., 16, 1785–1794.

STOHL, A., HITTENBERGER, M., and WOTAWA, G. (1998), *Validation of the Lagrangian particle dispersion model FLEXPART against large-scale tracer experiment data.* Atmos. Environ. *32*, 4245–4264.

STOHL, A., SEIBERT, P., WOTAWA, G., ARNOLD, D., BURKHART, J.F., ECKHARDT, S., TAPIA, C., VARGAS, A., and YASUNARI, T.J. (2012), *Xenon-133 and caesium-137 releases into the atmosphere from the Fukushima Dai-ichi nuclear power plant: determination of the source term, atmospheric dispersion, and deposition.* Atmospheric Chemistry and Physics, *12*, 2313–2343.

TAKEMURA, T., NAKAMURA, H., TAKIGAWA, M., KONDO, H., SATOMURA, T., MIYASAKA, T., and NAKAJIMA, T. (2011), *A Numerical Simulation of Global Transport of Atmospheric Particles Emitted from the Fukushima Daiichi Nuclear Power Plant.* SOLA, Vol. *7*, 101–104.

TEPCO Press releases. http://www.tepco.co.jp/en/nu/press/f1-np/index-e.html.

TERADA, H., KATATA, G., CHINO, M., and NAGAI, H. (2012), *Atmospheric discharge and dispersion of radionuclides during the Fukushima Dai-ichi Nuclear Power Plant accident. Part II: verification of the source term and analysis of regional-scale atmospheric dispersion.* Journal of Environmental Radioactivity *112*, 141–154.

UNSCEAR: *Exposures and effects of the Chernobyl accident.* Annex J. Report Vol 2, 456–457, (2000).

UEMATSU, M., MERRIL J.T., PATTERSON T.L., DUCE, R.A., and PROSPERO J.M. (1988), *Aerosol residence time and iodine gas/particle conversion over the North Pacific as determined from Chernobyl radioactivity.* Geochemical Journal, Vol. *22*, 157–163.

WESELY, M.L. and HICKS, B.B. (1977), *Some factors that affect the deposition rates of sulfur dioxide and similar gases on vegetation.* J. Air Poll. Contr. Assoc., *27*, 1110–1116.

WINIAREK, V., BOCQUET, M., SAUNIER, O., and MATHIEU A. (2012), *Estimation of Errors in the Inverse Modeling of Accidental Release of Atmospheric Pollutant: Application to the Reconstruction of the Cesium-137 and Iodine-131 Source Terms from the Fukushima Daiichi Power Plant.* Journal of Geophysical Research, doi:10.1029/2011JD016932.

WRF, Weather Research and Forecasting Model. http://www.wrf-model.org/index.php and http://www.mmm.ucar.edu/wrf/users.

(Received December 8, 2011, revised August 6, 2012, accepted August 12, 2012, Published online September 20, 2012)

Pure Appl. Geophys. 171 (2014), 669–676

DOI 10.1007/s00024-012-0564-7

Pure and Applied Geophysics

Discrimination of Nuclear Explosions against Civilian Sources Based on Atmospheric Radioiodine Isotopic Activity Ratios

Martin B. Kalinowski,[1] Yen-Yo Liao,[1] and Christoph Pistner[2]

Abstract—A global monitoring system for atmospheric radioactivity is being established as part of the International Monitoring System that will verify compliance with the comprehensive nuclear-test-ban treaty (CTBT) once the treaty has entered into force. This paper studies isotopic activity ratios to support the interpretation of observed atmospheric concentrations of ^{135}I, ^{133}I and ^{131}I. The goal is to distinguish nuclear explosion sources from civilian releases. Simulated nuclear explosion releases along with observational data of radioiodine releases from historic nuclear explosions at the Nevada Test Site are compared to simulated light water reactor releases in order to provide a proof of concept for source discrimination based on radioiodine isotopic activity ratios.

Key words: CTBT, environmental monitoring, international monitoring system, isotope activity ratios, radioiodine, radioactivity monitoring, source discrimination, test ban.

1. Introduction

Radioactive iodine can be collected and measured on a daily basis at the 80 radionuclide stations of the International Monitoring System (IMS) that is currently being established (see e.g., Hoffmann *et al.*, 1999; Kalinowski and Schulze, 2002). Though it may be released in gaseous form, iodine tends to attach itself to aerosol particles during its transport through the atmosphere. Due to their half-lives and fission yields the iodine radionuclides ^{135}I, ^{133}I and ^{131}I are relevant for detecting a nuclear explosion (De Geer, 2001).

Besides many atmospheric tests, also underground nuclear explosions were occasionally detected over long distances. For example, ^{131}I was detected in Sweden and related to the venting from the nuclear explosion called Baneberry that was conducted at the Nevada Test Site (NTS) on 18 December 1970 (De Geer, 1991). The underground explosion of the Soviet Union conducted on 2 August 1987 resulted in ^{131}I observations in Denmark, Finland, Germany, Sweden and Norway (Bjurman *et al.*, 1990). Both ^{133}I and ^{131}I were measured in Oslo and several Swedish stations in the aftermath of the underground test at Novaja Zemlja on 26 February 1987 (De Geer, 1991; Bjurman *et al.*, 1987). These reported observations give confidence that there is a significant chance for radioiodine releases from nuclear tests being discovered by IMS stations since they have a much better detection probability than the historic detection capabilities.

Due to its volatility, radioiodine is a likely observable radioactive signature of underground nuclear explosions. Though it is less likely to escape from deep underground than radioxenon, it is much more likely to be released than particle bound radioactivity. However, radioactive iodine may also be released during normal operations of nuclear facilities. Since radioiodine may be detected at stations located downwind to nuclear facilities, proper source characterization is important for determining whether an event is possibly a nuclear explosion. This paper investigates the multiple isotopic ratios for source discrimination similar to the approach successfully taken for radioxenon (Kalinowski and Pistner, 2006; Kalinowski *et al.*, 2010).

In this paper, the combinations of isotopic ratios of the following calculated data sets are used:

1. simulations of various nuclear explosion scenarios and different degrees of fractionation,
2. simulations for light water reactor (LWR) operational cycles.

[1] Carl Friedrich von Weizsäcker Center for Science and Peace Research (ZNF), Beim Schlump 83, 20144 Hamburg, Germany. E-mail: martin.kalinowski@uni-hamburg.de

[2] Öko-Institut e.V., Rheinstraße 95, 64295 Darmstadt, Germany.

In the first section of this paper, the simulations are introduced. The second section uses these simulations to demonstrate the useability of iodine isotopic activity ratios for source discrimination by demonstrating a clear separation of the multiple isotopic activity plot in a reactor and a test domain. The proof of concept is provided by testing the proposed source characterization method against observed radioiodine releases from underground nuclear explosions at Nevada. The quality of these data is discussed in more detail in a companion paper (Kalinowski and Liao, 2012).

2. Simulated Radioiodine Releases

2.1. Simulations of Various Nuclear Explosion Scenarios

A nuclear explosion takes place in a very short time and the nuclear chain reaction ends in much less than a second. At that moment, the initial fission product yield is formed and radioactive decay dictates the subsequent activity changes with time. Three different fission product sources are studied with regard to the fissile material used and the neutron energy. Fission of ^{235}U and ^{239}Pu is simulated at fission neutron energies and ^{238}U with neutrons stemming from fusion using the Bateman equations (Bateman, 1910) implemented in Matlab (Kalinowski, 2011; Liao, 2011). A range of scenarios regarding fractionation are considered. These cover various times of fractionation between two extremes. One extreme assumes that the radioactive iodines are fully removed from their precursors and allowed to decay immediately after the nuclear explosion, in the other extreme all fission products are held together, perhaps in an underground cavity, such that ingrowth from precursors is allowed and the cumulative yield of radioiodine is produced.

The temporal development of radioiodine activities with various times of fractionation after a nuclear explosion with ^{235}U as fissile material can be seen in Fig. 1a and with ^{239}Pu in Fig. 1b both with fission neutron energies. Figure 1c shows the same for ^{238}U as fissile material with fast neutrons.

The comparison of simulations with measurements of radioactive iodine releases at the NTS after underground explosions is presented and discussed in a companion paper (Kalinowski and Liao, 2012).

2.2. Simulations for Light Water Reactors

Simulations of LWR fuel burn-up with 3.2 % enrichment in ^{235}U through three 1-year operational reactor power cycles are conducted to explore the possible radioiodine isotopic signature of nuclear reactor releases under different operational conditions. The methodology is described in Kalinowski and Pistner (2006).

The simulations are performed with the cell burn-up code-system MCMATH (Pistner 2006), a combination of MCNP (Monte Carlo N-Particle Transport Code System) Version 4 (Briesmeister, 2000) and the analysis tool Mathematica (Wolfram, 1999), which are employed in an iterative mode. Mathematica is applied to calculate the change of isotopic compositions due to the neutron interactions within the fuel and including radioactive decay chains. About 60 actinide and 150 fission product isotopes are followed to account for changes in the fuel composition. The Monte Carlo neutron transport code MCNP is used to calculate the neutron energy spectrum and the effective cross-sections at certain burn-up steps for a given isotopic composition of the fuel. MCNP uses continuous energy cross-section libraries based on ENDF/B-VI for all nuclides to calculate the flux spectrum and effective cross-sections. New effective cross-sections are calculated regularly to account for changes in fuel composition. An extensive benchmark of the calculational system in comparison to other available tools (such as ORIGEN2) has been performed (Pistner, 2006), demonstrating the high accuracy of the results.

Fuel and reactor parameters are selected with the intention to represent a typical constellation of the most frequently used reactor type (LWR). Accordingly, the fresh fuel in the burn-up calculations is assumed to be uranium oxide. An enrichment of 3.2 % in ^{235}U is assumed. Three reactor cycles are simulated, each lasting for 317 days at a constant average power density of 38.3 W/g(HM) and thus generating a burn-up per reactor cycle of 11 MWd/kg(HM). The corresponding total flux at full power starts with 2.6×10^{14} s^{-1} cm^{-2} and reaches

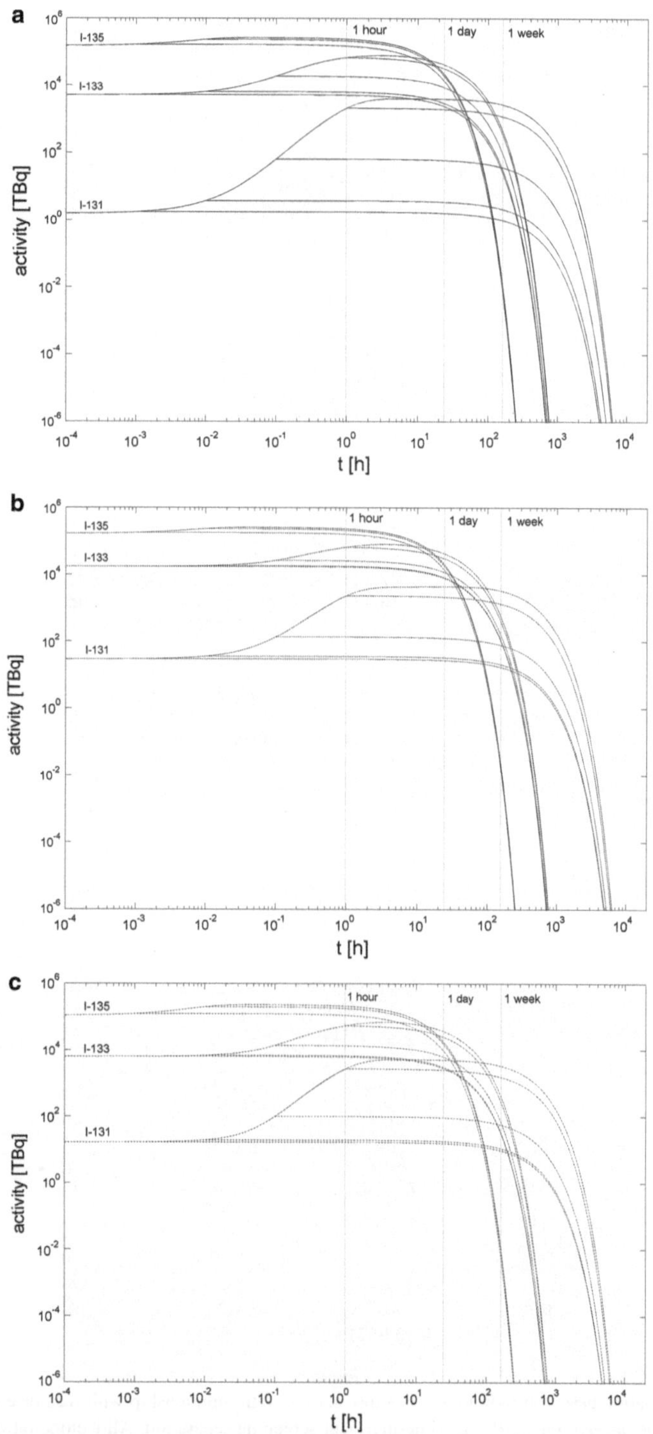

Figure 1
a Temporal development of radioiodine activities with various times of fractionation after a nuclear explosion with ^{235}U as fissile material. The *x*-axis notes the time elapsed since the nuclear explosion occurred. Each group of *curves* applies for one of the three radioiodine isotopes as annotated. In each group, the *uppermost curve* is valid for the case of no fractionation. The *lower curves* separate from the *uppermost curve* by fractionation of radioiodine from all precursors at the following time steps: 0.001, 0.01, 0.1 and 1. **b** Temporal development of radioiodine activities with various times of fractionation after a nuclear explosion with ^{239}Pu as fissile material. **c** Temporal development of radioiodine activities with various times of fractionation after a nuclear explosion with ^{238}U as fissile material with fast neutrons

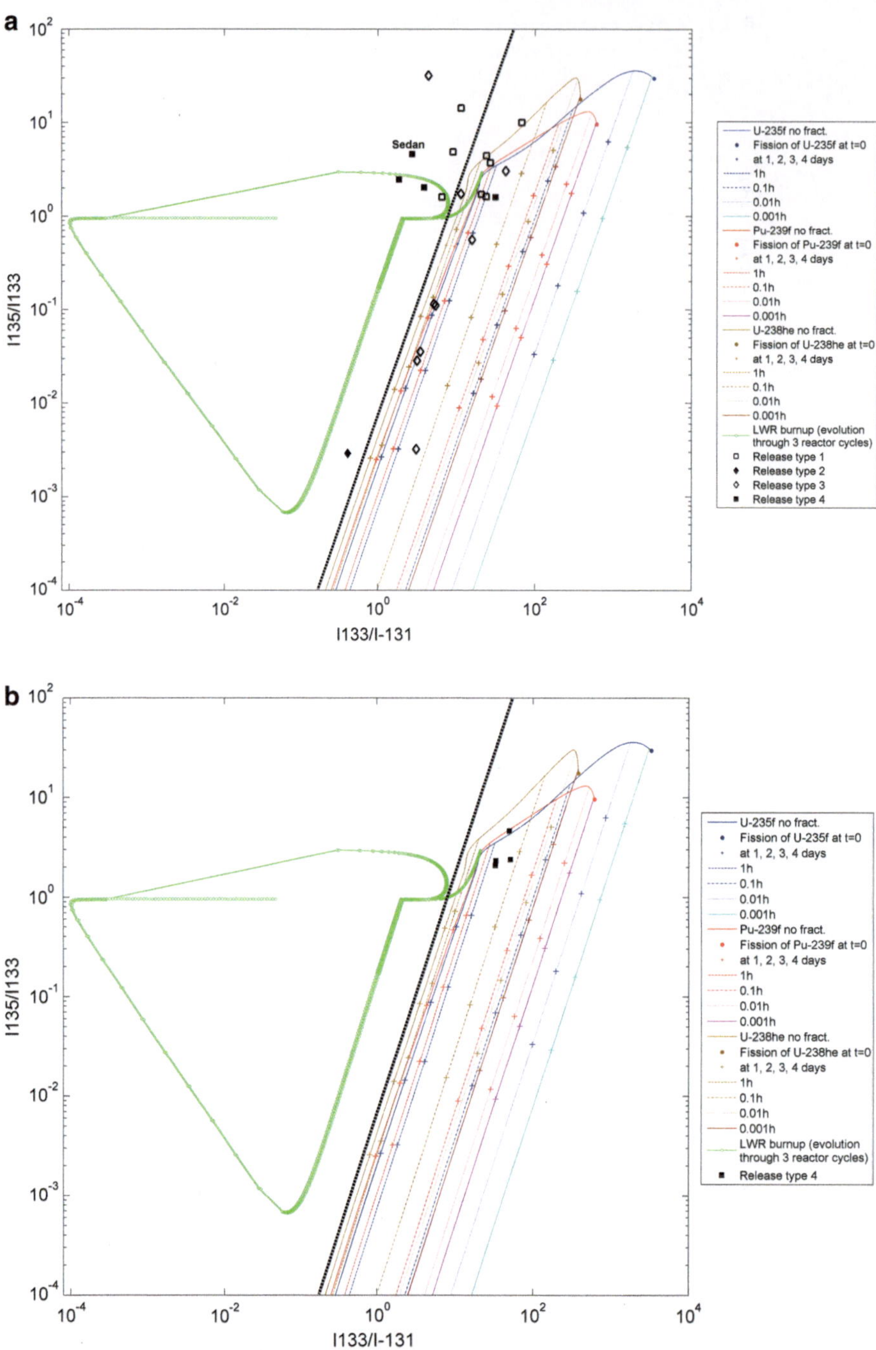

Figure 2

Time-invariant source discrimination based on radioiodine isotopic activity ratio relationship with reactor emission data for the case that all three isotopes are measured. The *dashed line marks* the time-invariant screening separation. All isotope ratio relations found above (i.e., *left to*) this line are definitely irrelevant for CTBT verification purposes, because they cannot be explained by a nuclear explosion. All samples that have ratio dependencies found below (i.e., *right to*) the line might be relevant for CTBT monitoring purposes. The exact location of the separation line is subject to further studies. Both versions of this figure contain observational data from radioiodine releases from nuclear explosions at the Nevada Test Site. While **a** shows all data as reported by SCHOENGOLD *et al.* (1996) for various tests with inconsistent measurement procedures for the different isotopes, **b** has only data consistently measured for all three isotopes after the Sedan explosion and summarized in PLACAK (1963)

4.0×10^{14} s^{-1} cm^{-2} at the end of the third cycle. Between the full power periods, 30-day sequences are assumed for revision and fuel reloading. During these breaks, in the calculations a neutron flux five orders of magnitude below the full power level is assumed, such that radioactive decay dominates changes in fuel composition. The power reduction and revamping are assumed to be instantaneous. In reality, a controlled shut-down extends over 1 or 2 days. The simulations are thus conservative by exaggerating the effect of changes in neutron flux on the isotopic activity ratios. At the end of the third cycle, the low power level is kept for 90 instead of 30 days.

3. *Source Discrimination Based on Radioactive Iodine Isotopic Activity Ratios*

In Fig. 2a and b, two different isotopic activity ratios of radioiodine are shown on the abscissa and ordinate. Two isotopes are used on the abscissa (^{135}I/^{133}I); another combination of two is taken for the activity ratio on the ordinate (^{133}I/^{131}I). For a number of fractionation times (at $t = 0$, 0.001, 0.01, 0.1 and 1 h) the temporal evolution of isotopic activity ratios are entered as trajectories starting from the initial yield ratios for the time zero after the explosion. The simulation curves entered here separate the plot area into two distinctive domains similar as for radioxenon (Kalinowski and Pistner, 2006; Kalinowski *et al.*, 2010). The trajectories for three operational cycles of a nuclear reactor follow a circular pattern in the left half of the plot. The simulation curves for nuclear explosion scenarios remain in the right half of the plane. Only a small tail of the trajectory belonging to the power reactor cycle reaches into the nuclear test domain. However, this applies only for the first 2 weeks of fresh fuel being exposed to the neutron flux. An emission resulting solely from such fresh fuel in a power reactor is highly unlikely. However, this short exposure time is typical for isotopic production with uranium targets. Fortunately, there are only a very few facilities of this type in the world and their location is well known.

In addition to the simulations, Fig. 2a contains all data of radioiodine releases from historic nuclear explosions at the NTS as published by Schoengold *et al.* (1996). Figure 2b has data points reported by Placak (1963) for observations at various locations and times after the Sedan explosion that was conducted at the NTS on 7 June 1962. The quality of these data is analyzed in detail by Kalinowski and Liao (2012). In Fig. 2a the majority but not all of the Nevada data (15 out of 23) are found as expected on the right side of the line separating the reactor and the explosion domain. The few entries on the left side of the line are associated with inconsistently measured isotopes as was revealed by Kalinowski and Liao (2012). The isotopic activity ratio of ^{133}I and ^{131}I for these releases is lower than expected due to the fact that for the shorter lived isotope ^{133}I only the gaseous part of its activity is reported, while for ^{131}I both gaseous and particle bound activities are taken account of. Therefore, the atmospheric concentrations of ^{133}I are underestimated and the isotopic activity ratios with respect to ^{131}I are shifted towards too low values and across the separation line into the reactor domain.

This data inconsistency as found in the report of Schoengold *et al.* (1996) can be avoided when using the original reports. This is demonstrated here with the example of the Sedan explosion that is marked in Fig. 2a and found far in the reactor domain. The original report on resulting off-site environmental radioactivity (Placak, 1963) lists the particle and gas bound activities separately. Therefore, it is possible to use consistently the gaseous phase of iodine only. The resulting data are entered in Fig. 2b. The four available data points are all found clearly in the test domain. The observation with the largest concentrations that was selected by Schoengold *et al.* (1996) and that is marked in Fig. 2a moves to an activity ratio for ^{133}I and ^{131}I of about one order of magnitude higher in Fig. 2b.

This demonstrates a robust method for source discrimination that works even independent from radioactive decay, but it depends on the condition that all three radioiodine isotopes are present in the plume at concentrations above the minimum detectable concentration (MDC). It further depends on the assumption that a single release source is detected. Since the axes are on logarithmic scale, the decay causes the point representing a certain isotopic starting mixture to move on a straight diagonal line towards minus infinity in both dimensions. If iodine precursors are released together with the iodine nuclides and are therefore present in the plume, the resulting curve would be a bent curve that converges with the straight line defined by decay. The direction of this straight line is determined by the decay constant of the three isotopes under consideration.

A separation line is drawn between the reactor and the explosion domain. Any isotopic activity ratio combination on the separation line moves down along that line with the radioactive decay. Any isotopic activity ratio combination off that line remains on the same side of the line forever. Radioactive decay changes the isotopic activity ratios along a line that is parallel to the separation line. For different times of fractionation of radioiodine from their precursors, the isotopic activity ratios move closer to the separation line but they do not cross it.

This decay invariance offers an important quality for source discrimination. It allows for screening out all irrelevant detections because the separation between explosion and non-explosion scenarios is independent from the age of the detected release, i.e., neither the delay before release nor the time for transport through the atmosphere needs to be known.

Figure 2 allows also for a discussion of the capability of only two isotopes to distinguish nuclear test signatures from reactor releases while the third is not detected. This is particularly facilitated by any assumption that can be made on the time of the explosion, i.e., on the delay between generation and detection of the iodine fission products. For this purpose it is useful to consider the + signs in the plot that mark the first, second, third and fourth day after the time of the explosion. The available data related to the Sedan test are all valid for a few hours after the explosion ranging from 2¼ to 12 h. Therefore, the entries in Fig. 2b are all found at the upper end of the simulated curves. In contrast, Fig. 2a contains data points for releases that occurred up to 4 days after the explosion. These points are found deep down along the simulation curves close to the fourth + sign counted from the solid circle at the top of the simulated curves marking the time zero.

The activity ratio of ^{135}I to ^{133}I is larger for nuclear explosions than for any part of the power reactor operations cycle at the time of a fresh release. Due to radioactive decay, this distinctive feature vanishes within 1 day after the explosion because the activity ratio decays below the maximum ratio found in the reactor cycle. The isotopic activity ratio of ^{133}I and ^{131}I initially is higher for nuclear explosions as compared to any power reactor status. In the extreme case of no fractionation, for 1–2 days the explosion signature remains higher than any civilian release with the exception of isotopic production signatures. In the other extreme case of instantaneous complete fractionation, the source discrimination for clearly identifying nuclear explosions against power reactors lasts for at least 5 days. As a result, ^{135}I is not required for source categorization, if the release is detected early enough. Figure 2b demonstrates that the off-site observations reported by Placak (1963) could have been used for determining the source to be a nuclear explosion solely based on the activity ratio of ^{133}I to ^{131}I.

4. *Conclusions and Implications for Monitoring of Nuclear Explosions*

The radioiodine isotopic activity ratio of ^{133}I and ^{131}I is well suited for source discrimination, in particular if early fractionation has occurred and if isotopic production facilities can be ruled out as the source. If the plume is detected early enough, the activity ratio of these two isotopes remains larger than a certain threshold that may conservatively be set to 10 and it can be used with high confidence for source discrimination for all nuclear testing scenarios.

A more robust method for source discrimination that works independently of the delay between release and detection is demonstrated here based on the relationship of two different isotopic activity ratios. This requires three radioiodine isotopes to be quantified. This is very likely for atmospheric nuclear tests and possible even for underground explosions. In the case that certain isotopes are not detected, the method may as well work with substituting the missing concentration by an appropriate value for the MDC. For example, in the likely case that the short-lived ^{135}I is not observed, its MDC can be used as a substitute for the ^{135}I concentration in calculating its isotopic activity ratio with respect to ^{133}I. This establishes an upper bound for this ratio and if this is found to be below the separation line, the measurement must be caused by a nuclear explosion. If it is above the line, it could result from either a reactor or an explosion source. This approach of making use of the MDCs has been presented for radioxenon by KALINOWSKI *et al.* (2010) and is demonstrated to work very effectively by POSTELT (2012).

A special advantage of this multiple isotopic activity method is its independence on the time periods elapsed between generation and release as well as between release and detection, i.e., it is dilution and decay invariant.

A proof of concept for this categorization method is given here by demonstrating it with simulations of nuclear reactors and explosions and by a comparison with observational data of radioiodines released from historic nuclear explosions at the NTS. Further investigations to explore the opportunities for an operational approach are required. In particular, the conditions and limitations are that a suitable combination of isotopes needs to be detected and that the uncertainties of the isotopic activity ratio should not be too large. The detection of all three radioiodine isotopes in an atmospheric sample is very rare. Since the spread of observable radioiodine concentrations has yet to be determined, no generalized quantitative condition can be given for the requirement on their uncertainties. However, it is obvious that the lower the concentration uncertainties are the sharper will a categorization algorithm be that is based on multiple isotopic activity ratios. Background and interfering sources are not a significant issue since the typical sources for the infrequent detection of radioiodine are normally well known and in most cases are radiopharmaceutical production facilities and hospitals with nuclear medicine applications at known locations. It might as well be a severe reactor accident.

Additional categorization methods will be employed. These may be based on a single ratio as well as on absolute concentrations and may involve an outlier analysis to classify a measurement as anomalous in comparison to typical atmospheric background at that detector site. This can be supported by atmospheric transport simulations using information about the known civilian sources.

Acknowledgments

The work of Martin Kalinowski and Yen-Yo Liao was funded by the German Foundation for Peace Research (DSF) and the University Hamburg. Christoph Pistner conducted his research while being a staff member with the Interdisciplinary Research Group in Science, Technology and Security (IANUS) at Darmstadt University of Technology.

REFERENCES

BATEMAN, H. (1910), *The solution of a system of differential equations in the theory of radio-active transformation*, Proc. Cambridge Phil. Soc., (*16*), page 423.

BJURMAN, B., *et al.* (1987), The detection in Sweden of short-lived fission products probably vented from the underground nuclear test at Novaya Zemlya on 2 August 1987. FOA Report C 20673-9.2, Stockholm September 1987.

BJURMAN, B., *et al.* (1990), *The Detection of Radioactive Material from a Venting Undergound Nuclear Explosion*. J. Environ. Radioactivity *11* (1990) 1–14.

BRIESMEISTER, J.F., ed. (2000), MCNP—A General Monte Carlo N-Particle Transport Code. LA-13709-M, Version 4C. Los Alamos, New Mexico: Los Alamos National Laboratory.

DE GEER, L.-E. (1991), *Observations in Sweden of venting underground nuclear explosions*. Paper prepared for Symposium on Underground Nuclear Weapons Testing: Potential Environment Impacts and their Containment. Ottawa, Canada, 23–24 April 1991.

DE GEER, L.-E. (2001), *Comprehensive Nuclear-Test-Ban Treaty: Relevant radionuclides*. Kerntechnik *66* (3), 113–120.

HOFFMANN, W., KEBEASY, R. and FIRBAS, P. (1999), *Introduction to the verification regime of the Comprehensive Nuclear-Test-Ban Treaty*, Physics of the Earth and Planetary Interiors, *113*, 5–9.

KALINOWSKI, M.B. (2011), *Characterisation of prompt and delayed atmospheric radioactivity releases from underground nuclear tests at Nevada as a function of release time*. In: Journal of Environmental Radioactivity, *102*, Issue 9, September 2011, pp. 824–836. doi:10.1016/j.jenvrad.2011.05.006

KALINOWSKI, M.B., AXELSSON, A., BEAN, M., BLANCHARD, X., BOEYER, T.W., BRACHET, G., HEBEL, S., MCINTYRE, J.I., PETERS, J., PISTNER, C., RAITH, M., RINGBOM, A., SAEY, P., SCHLOSSER, C., STOCKI, T.J., TAFFARY, T., UNGAR, R.K. (2010), *Discrimination of nuclear explosions against civilian sources based on atmospheric xenon isotopic activity ratios*. In: Becker, A., Schurr, B., Kalinowski, M.B., Koch, K., Brown, D. (ed.): Recent Advances in Nuclear Explosion Monitoring. Pure and Applied Geophysics Topical Volume 167/4-5. pp. 517–539. doi:10.1007/s00024-009-0032-1.

KALINOWSKI, M.B. and LIAO, Y.-Y. (2012), Isotopic characterization of radioiodine and radioxenon in releases from underground nuclear explosions with various degrees of fractionation. Submitted to Pageoph Topical Volume II on Recent Advances in Nuclear Explosion Monitoring.

KALINOWSKI, M.B. and PISTNER, C. (2006), *Isotopic signature of atmospheric xenon released from light water reactors*, Journal of Environmental Radioactivity, *88* (3), 215–235.

Kalinowski, M.B. and Schulze, J.: *Radionuclide Monitoring for the Comprehensive Nuclear-Test-Ban Treaty*. Journal of Nuclear Materials Management *30*, No. 4, Summer 2002, 57–67.

Liao, Y.-Y. (2011), *Fraktionierung bei der Freisetzung von Leitnukliden für die Entdeckung von unterirdischen Kernwaffentests*. Diplomarbeit, Universität Hamburg, September 2011.

Pistner, C. (2006): *Neutronenphysikalische Untersuchungen zu uranfreien Brennstoffen (Neutronics calculations for inert matrix fuels)*, Ph.D. thesis, Darmstadt University of Technology.

Placak, O.R. (1963), Project Sedan, Final Off-Site Report, U.S. Public Health Service, Report PNE-200F, 25 April 1963.

Postelt, F. (2012), The Potential of Radionuclide Ratios for Spectrum Categorisation Algorithms. Submitted to Pageoph Topical Volume II on Recent Advances in Nuclear Explosion Monitoring.

Schoengold, C.R., DeMarre, M.E. and Kirkwood, E.M. (1996), Radiological effluents released from U.S. continental tests 1961 through 1992, United States Department of Energy—Nevada Operations Office, DOE/NV-317 (Rev.1) UC-702, Las Vegas, August 1996.

Wolfram, S. (1999): The Mathematica Book, 4th ed., Wolfram Media/Cambridge University Press, 1999.

(Received December 2, 2011, revised July 8, 2012, accepted July 20, 2012, Published online August 12, 2012)

Pure Appl. Geophys. 171 (2014), 677–692

DOI 10.1007/s00024-012-0580-7

Pure and Applied Geophysics

Isotopic Characterization of Radioiodine and Radioxenon in Releases from Underground Nuclear Explosions with Various Degrees of Fractionation

Martin B. Kalinowski[1] and Yen-Yo Liao[1]

Abstract—Both radioxenon and radioiodine are possible indicators for a nuclear explosion. Therefore, they will be, together with other relevant radionuclides, globally monitored by the International Monitoring System in order to verify compliance with the Comprehensive Nuclear-Test-Ban Treaty once the treaty has entered into force. This paper studies the temporal development of radioxenon and radioiodine activities with two different assumptions on fractionation during the release from an underground test. In the first case, only the noble gases are released, in the second case, radioiodine is released as well while the precursors remain underground. For the second case, the simulated curves of activity ratios are compared to prompt and delayed atmospheric radioactivity releases from underground nuclear tests at Nevada as a function of the time of atmospheric air sampling for concentration measurements of ^{135}I, ^{133}I and ^{131}I. In addition, the effect of both fractionation cases on the isotopic activity ratios is shown in the four-isotope-plot (with ^{135}Xe, ^{133m}Xe, ^{133}Xe and ^{131m}Xe) that can be utilized for distinguishing nuclear explosion sources from civilian releases.

Key words: Nuclear explosion, test ban, CTBT, isotope activity ratios, radioiodine, radioxenon, fractionation, radioactivity monitoring, source discrimination.

1. Introduction

Due to their half-lives, fission yields and decay radiation the iodine radionuclides ^{135}I, ^{133}I and ^{131}I as well as the xenon radionuclides ^{135}Xe, ^{133m}Xe, ^{133}Xe and ^{131m}Xe are relevant for detecting a nuclear explosion (De Geer, 2001). Though two of the latter are metastable isomers, for convenience, this paper refers to the entities of this quartet as the four relevant xenon isotopes.

Xenon isotopes are the most likely observable radioactive signatures of underground nuclear explosions because xenon is a gas and chemically inert. Radioiodine is volatile and can also be released from underground nuclear explosions. Since it is less likely to escape from deep underground than xenon, however. it is still much more likely to be released than its particle bound precursors in the decay chains, and some degree of fractionation can be expected. This fractionation influences the temporal development of the radioxenon isotopes released from underground into an atmospheric plume, because it depends on the amount of its radioiodine precursor isotopes which are present in the same plume. This paper studies empirically the degree of fractionation and theoretically the impact of various fractionation scenarios on the isotopic activity ratios of radioxenon.

Atmospheric radioiodine and radioxenon are monitored on a daily basis at the radionuclide stations (see, e.g., Kalinowski and Schulze, 2002) of the International Monitoring System (IMS) that is currently being established (see, e.g., Hoffmann *et al.*, 1999; Kalinowski 2006). Though it may be released in gaseous form, during its transport through the atmosphere iodine tends to undergo to some degree chemical reactions and subsequently to attach itself to aerosol particles. It will be collected on particle filters at the 80 radionuclide stations of the IMS network. For radioxenon, suitable sensors will be installed at 40 of these stations with the option of a later expansion to all 80 sites of the radionuclide particle monitoring network.

A previous paper described the temporal development of prompt and delayed atmospheric radioactivity releases from underground nuclear tests at the Nevada Test Site in the United States (Kalinowski, 2011). A subset of this data is used in this paper to investigate the degree of fractionation that occurred with the reported releases of

[1] Carl Friedrich von Weizsäcker Center for Science and Peace Research (ZNF), Beim Schlump 83, 20144 Hamburg, Germany. E-mail: martin.kalinowski@uni-hamburg.de

radioiodine at that site. Other previous papers presented the proof of principle (Kalinowski and Pistner, 2006) and a demonstration (Kalinowski *et al.*, 2010) of a robust method for discriminating between the source being a nuclear explosion or a civilian reactor source. The method is based on isotopic activity ratios of radioxenon in the atmosphere. This paper builds on those findings and explores the impact of various fractionation scenarios on the radioxenon isotopic activity ratios. A companion paper explores the capability to use activity ratios of radioiodine isotopes for source characterization (Kalinowski *et al.* 2012).

In the first section of this paper, the simulations of various nuclear explosion types and fractionation scenarios are presented. The second section uses the empirical release data of underground nuclear tests at the Nevada Test Site to explore the degree of fractionation by comparing their isotopic activity ratios with those of the simulations. The third section shows the effect of fractionation on the radioxenon isotopic activity ratios for source discrimination.

2. *Simulated Radioxenon Isotopic Activities Under Different Fractionation Assumptions*

A nuclear explosion takes place in a very short time where the nuclear chain reaction and possibly fusion reactions occur on a time scale of a few millionth of a second and less. In this short instant, the initial feed of hundreds of fission products is formed and radioactive decay dictates the subsequent activity changes with time. Three different scenarios are studied here with regard to the combination of fissile material used and the neutron energy spectrum. Fission of ^{235}U and ^{239}Pu is simulated at fission neutron energies using the Bateman equations (Bateman, 1910) implemented in Matlab (Kalinowski and Pistner 2006; Liao, 2011). The independent fission yields and the decay branching fractions are taken from England and Rider (1994). These were adopted for ENDF/B-VI. Differences for figures in other libraries are most significant for the short lived pre-cursors. One day after the fission process ended, the uncertainty originating from the library values is negligible for the purpose of this paper.

The development over time for simulated radioxenon and radioiodine isotopic activity ratios with regard to different nuclear explosion scenarios has been published before (Kalinowski, 2011). In the totally unfractionated case, all fission products are staying together, perhaps in an underground cavity, such that in-growth from precursors is fully allowed.

Fractionation changes the isotopic activity ratios. The temporal development of radioiodine activities with various fractionation scenarios after a nuclear explosion is plotted in the companion paper (Kalinowski *et al.*, 2012). For radioxenon, two different fractionation types are considered in this paper. In the standard type, radioxenon is fully removed from all its precursors immediately, or at a later time, and from then on decays without further in-growth from the decay chain. The alternative fractionation type considered here assumes that radioxenon and radioiodine are fully removed from all other precursors together and at the same time. The different impacts of these two fractionation types on the development over time of radioxenon activities can be seen in Fig. 1 using ^{235}U with fission energy neutrons as example. The activity in TBq refers to a 1 kT explosion. The resulting curves for isotopic activity ratios are presented together with data from the Nevada Test Site in the following section.

As can be seen in Fig. 1, there is a strong impact of fractionation that is most pronounced for early separation of the volatiles from the non-volatile precursors. The unfractionated case is represented by the upper curve while the simulated fractionations start at various delay times (0.01, 0.1, 1, 10 and 100 h) from that line and stay always below it. The impact gets less with increased delay between time zero and the separation time and eventually vanishes at a delay time related to the relevant half-lives. The impact is of course strongest for radioxenon separating from the precursors alone and less pronounced for xenon and iodine separating together. This can be noted in Fig. 1b by the lower curves all bending upwards before going down due to decay in contrast to Fig. 1a. For ^{131m}Xe, the maximum activity can be reduced by almost five orders of magnitude if separated alone from all precursors and by more than three orders of magnitude if iodine and xenon are separated together.

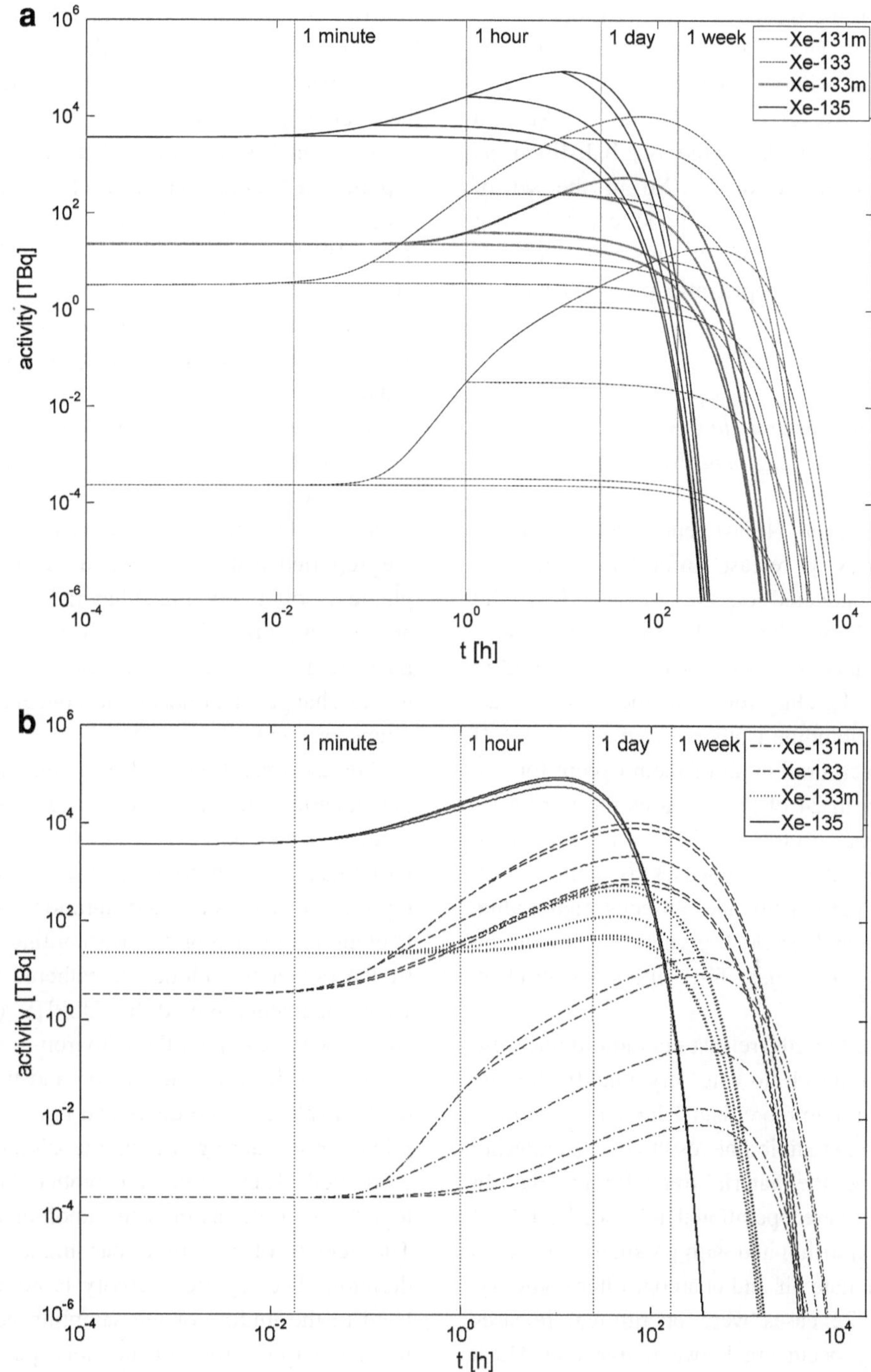

Figure 1
Different impacts of two fractionation types on the development over time of radioxenon activities using ^{235}U with fission energy neutrons as an example. **a** Radioxenon is removed from all precursors, **b** radioxenon and radioiodine are removed together. The activity in TBq refers to a 1 kT explosion

However, the fractionation of iodine and xenon together has no significant impact on the radioxenon activities, if it happens later than 1 h after the explosion. For xenon separating alone, the impact is still strong after 10 h and for ^{131m}Xe even after 100 h.

The parent of the isomer ^{131m}Xe, ^{131}I, has a half-life of 8.02 days. The other isotopes have shorter-lived precursors with a maximum of 2.19 days (^{133m}Xe) for ^{133}Xe, 20.8 h (^{133}I) for ^{133m}Xe and 6.61 h (^{135}I) for ^{135}Xe. Due to the long half-life of its precursor, the presence of ^{131m}Xe in the set of quantified isotopes provides a good capability for distinguishing different types of sources by separating them in a multi-isotopic-ratio plot (KALINOWSKI *et al.*, 2010).

3. Empirical Radioiodine Release Data of Underground Nuclear Tests

SCHOENGOLD *et al.* (1996) report detailed atmospheric radioactivity release information for 433 nuclear tests conducted on the Nevada Test Site (NTS) from 15 September 1961 through 23 September 1992. An analysis of these data can be found in KALINOWSKI (2011). Only for 45 of the 433 releases, two or more radioiodine isotopes were reported. ^{131}I and ^{133}I were measured in all of them except for one case each. Only in 24 of the 45 cases the third relevant radioiodine isotope (^{135}I) was measured. Information for all these releases is collected in Table 1. Both particulate and gaseous radioiodine were measured and Table 1 gives the sum of the activities collected on glass fiber filters and in charcoal cartridges.

Twelve of the reported releases of radioiodine were uncontrolled ventings following very quickly after the explosion but with no more than 1 h delay. Five incidents of measuring radioiodine resulted from cratering explosions that set the material free without delay. In six cases, these were operational releases.[1] At least 38 h after the explosion and with passing the releases through a particulate air and charcoal filter combination, a further 22 cases were operational releases without filtering occurring between five and 312 h after the explosion. The duration of uncontrolled releases is very short, up to half an hour. For operational releases with filtering, the releases are reported to last between 10 min and 114 h, without filtering between 6 and 130 h.

For some tests, more data are available than reported in SCHOENGOLD *et al.* (1996). For one of the prominent examples, the Sedan test conducted on 7 June 1962, PLACAK (1963) contains atmospheric concentrations observed at several stations. All 18 samples with two or three radioiodine isotopes measured in the charcoal cartridges are collected in Table 2.

Measurement uncertainties are reported neither in SCHOENGOLD *et al.* (1996) nor in PLACAK (1963) but all data is given with two significant digits. The largest uncertainty is related to the time for which the reported values are valid. The passage of a plume is often not matching the sampling periods and uncertainties in fixing a time to which the measurement is corrected bears uncertainties due to the change of atmospheric concentrations during plume passage.

For the operational releases, the sampling times are identical to the release periods reported in SCHOENGOLD *et al.* (1996). For the ventings and cratering tests, the sampling times are different from the reported release delay and duration. For these tests, beginning and ending of gas sampling at the sites hit by the radioactive cloud are gathered from the relevant documents issued by USPHS (U. S. Public Health Service) and EPA (Environmental Protection Agency). With this information and the known time of the explosion, the delay and duration of the sampling period during which the cloud passed were calculated. These times are plotted in Figs. 3, 4, 5 together with the associated radioiodine ratio values. The length of the time bar marks the sampling duration. The reported activity is normally extrapolated to the middle of the sampling period and in a few cases to the time of the cloud passage. Figure 2 demonstrates this for the example of the test called Pampas as part of Operation Nougat conducted on 1 March 1962. The 9.5 kt charge was detonated at 11:10 PST. The Gunderson's Ranch station began sampling at 12:45 and ended at 17:50 at the same day (USPHS, 1964). The delay of sampling start can be calculated to be 2 h 35 min (marked with A in the

[1] Operational releases result from purging of tunnels or sometimes shafts to minimize the exposure to personnel, from drill-back operations to recover samples for diagnostic purposes, from gas sampling or from sealing the drill hole with a plug and cementing it to the surface (SCHOENGOLD *et al.*, 1996).

Table 1

The table lists all historical releases of radioiodine from nuclear explosions at the Nevada Test Site between 1961 and 1992 as reported by SCHOENGOLD *et al. (1996)*

Days	Month	Years	Test name	Depth (m)	Delay of sampling start (h)	Sampling duration (h)	I-131 (Ci)	I-133 (Ci)	I-135 (Ci)	Type	Yield (kt)
15	9	1961	Antler	400	2.52	4.32	4.2	102	450	1	2.6
22	12	1961	Feather	250	7.79	0.79	–	27	94	1	0.15
1	3	1962	Pampas	360	4.125	2.54	0.012	0.29	0.47	1	9.5
5	3	1962	Danny Boy	34	2.542	2.46	73,000	290,000	470,000	4	0.43
14	4	1962	Platte	190	4.25	1.42	11.4	75	120	1	1.85
13	6	1962	Des Moines	200	2.25	2.92	33,000	38,0000	5,400,000	1	2.9
6	7	1962	Sedan	195	3.03	3.58	8,80,000	2,400,000	11,000,000	4	100
11	7	1962	Johnnie Boy	1	11.75	2	70,000	130,000	320,000	4	0.5
19	10	1962	Bandicoot	240	28.25	0.083	9,000	–	73,000	1	12.5
14	11	1963	Anchovy	260	24	6	2.5	11	350	3	Low
23	1	1964	Oconto	265	38	26	0.001	0.0005	–	2	10.5
13	3	1964	Pike	114	7.17	0	360	13,000	–	1	<20
19	8	1964	Alva	166	85	129.6	0.041	0.055	–	3	4.4
5	12	1964	Crepe	220	217.75	38.4	1	1	–	3	3.4
18	2	1965	Wishbone	179	20	216	1.3	15	26	3	<20
3	3	1965	Wagtail	750	168	6	0.03	0.02	–	3	20–200
20	3	1965	Suede	143	31.37	15.5	0.1	1.6	0.89	3	<20
26	3	1965	Cup	537	79.27	89	0.3	0.7	–	3	20–200
5	4	1965	Kestrel	447	112.58	16.5	0.029	0.09	0.00029	3	<20
14	4	1965	Palanquin	85	11.02	0	9,06,000	3,500,000	7100000	4	4.3
21	5	1965	Tweed	281	96.87	120	0.014	0.0068	–	3	<20
23	7	1965	Bronze	531	150.17	6	0.23	0.23	–	3	20–200
21	8	1965	Ticking	208	47.75	16.25	0.16	0.13	–	3	<20
27	8	1965	Centaur	172	45.5	1.5	0.0022	0.0065	–	3	<20
12	11	1965	Sepia	241	96	6	0.0011	0.0035	0.0001	3	<20
18	1	1966	Sienna	275	46.42	6	0.0016	0.0056	0.0002	3	<20
3	2	1966	Plaid II	270	52.5	74.4	0.019	0.076	–	3	<20
23	4	1966	Fenton	167	5.08	115.2	0.056	2.4	7.3	3	1.4
12	9	1966	Derringer	255	0.83	47.1	1.5	41	152	1	7.8
5	11	1966	Simms	198	48.33	47.7	0.009	0.018	–	3	<20
25	10	1967	Cognac	301	24	6	0.0049	0.027	0.003	3	<20
26	1	1968	Cabriolet	52	2.33	4.66	6,000	90,000	–	4	2.3
12	3	1968	Buggy a-e	41	1.55	5.13	22,000	200,000	970,000	1	5.4
27	8	1968	Diana Moon	242	5	13	0.1	2.1	3.6	1	<20
12	9	1969	Minute Steak	264	0.083	3.92	0.05	3.4	34	1	<20
18	12	1970	Baneberry	278	0.5	259	80,400	1,200,000	–	1	10
25	7	1972	Atarque	294	144	6	1.7E − 06	1.4E − 06	–	3	<20
9	8	1972	Cebolla	287	312	48	2.4E − 07	2.4E − 08	–	3	<20
25	4	1973	Velarde	277	48	6	0.027	0.14	0.016	3	<20
14	8	1974	Puye	430	168	6	2.1E − 06	8.6E − 07	2.5E − 09	2	<20
3	8	1983	Laban	326	48	0.8	0.000011	0.000025	–	2	<20
22	3	1986	Glencoe	610	120	6	8.9E − 06	9.6E − 06	–	3	29
21	5	1986	Panamint	480	55	0.0617	0.0001	0.0009	–	2	<20
13	5	1988	Schellbourne	463	56.9	113.5	0.000032	0.00011	–	2	20–150
10	3	1990	Metropolis	469	48	48	0.000088	0.00019	–	2	20–150

Only those tests are included for which at least two radioiodine isotopes were measured. The activities include both particulate and gaseous radioiodine. A dash (–) means that the concentration was below the detection limit. The types of the releases are noted with the following numbers: type 1 uncontrolled rapid venting, type 2 operational release with filtering, type 3 operational release without filtering, type 4 cratering

Table 2

The table lists all observations of radioiodine from the cratering nuclear explosion with a yield of 100 kt TNT equivalent that was called Sedan and conducted at a depth of 1 5 m on 6 July 1962

Delay of sampling start (h)	Sampling duration (h)	Time activities apply to (h)	I-131 ($\mu\mu Ci/m^3$)	I-133 ($\mu\mu Ci/m^3$)	I-135 ($\mu\mu Ci/m^3$)
0	6.62	3.03	260	13,000	60,000
6.75	3.52	8.367	28	780	–
0	21.83	3.033	7.2	240	500
21.83	23.167	33.42	11	170	–
45	24	57	7.1	34	–
0	21.5	9.5	0.27	4.4	–
21	24.5	33.25	19	860	–
0	10	10	5.2	250	–
10	12	10	70	2,400	–
0	11	3.83	0.22	4.2	–
0	26.25	12.17	6.5	220	510
0	21	9	–	5.0	6.8
3.25	24	5.5	49	200	–
1.083	15.67	2.25	75	3,900	9,300
23.5	7.167	27.083	10	260	–
0	20	4.3	12	79	–
20	24	32	6.9	190	–
0.67	24	12.67	0.23	3.3	

Only those measurements are included which are based on iodine collected on charcoal (gaseous fraction) and for which at least two radioiodine isotopes were quantified as reported by Placak (1963). A dash (–) means that the concentration was below the detection limit

figure) after the explosion and the sampling duration is 5 h 5 min (marked with B).[2]

4. Further Comparisons of Observed and Simulated Data

The simulated radioiodine isotopic activity ratios are compared to the values of reported releases from nuclear tests at the Nevada Test Site. Figure 3 applies for the ratio between ^{135}I and ^{133}I, Fig. 4 for ^{135}I and ^{131}I and Fig. 5 for ^{133}I and ^{131}I. The data are picked from Schoengold *et al.* (1996) to show all releases from Nevada tests in the three plots labeled a) and from Placak (1963) to show all data from the Sedan test only in the three plots labeled b). The simulations use ^{235}U (a) and ^{239}Pu (b) with fission energy neutrons and assume complete fractionation at various time steps.

In each figure, the unfractionated case is represented by the lower curve while the simulated fractionations start at various delay times (0.001, 0.01, 0.1, and 1 h) from that line and stay always above it, except for some cases at separation times in the order of minutes and shorter. The simulated curves range over many orders of magnitude and show the largest spread in the time frame between an hour and a day after the explosion. However, the dynamic range is dominated by early fractionations. Obviously, a separation between radioiodine and its precursors later than 1 h after the explosion has little impact on the isotopic activity ratio and remains close to the unfractionated case.

Most of these data in the plots of Figs. 3a, 4, 5a (Nevada releases as reported by Schoengold *et al.*, 1996) lie at the lower end of the simulated curves indicating either no fractionation of radioiodine from its precursors at all or a fractionation later than 0.1 h after the explosion. However, some data are found outside the range of the theoretical curves, most of them show a lower than expected isotopic activity ratio. In general, the agreement of low lying data points is better with the simulation for ^{239}Pu and this

[2] The data point used in Fig. 2 was deliberately selected to fit well with the simulation in order to demonstrate how delay and duration are graphically represented.

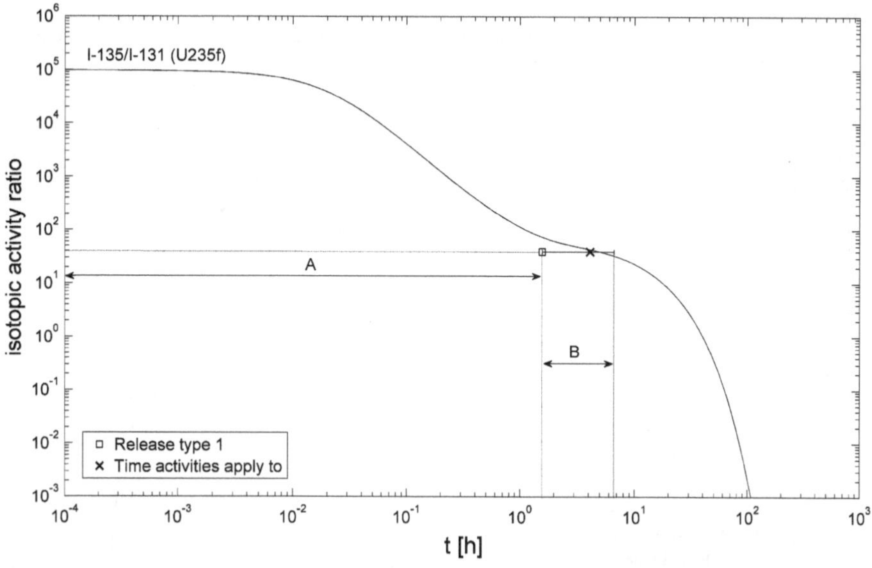

Figure 2
Model and experimental (x) isotopic activity ratio between ^{135}I and ^{131}I for the 1962 test "Pampas" in a shaft at Nevada Test Site. *A* is the delay between the explosion time and the start of the sampling period. *B* is the duration of the sampling period

is reasonable because most underground tests in Nevada were dominated by plutonium fission. The assumption of ^{235}U being the fission material catches a few data points that lie above the highest simulation line for ^{239}Pu. Only two single outliers on the high end of activity ratios cannot be understood by the simulations. For the test called "Anchovy" on 14 November 1963 at 8 o'clock local time, the ^{135}I to ^{133}I activity ratio is as high as the simulation is only within a few hours after the test explosion (see Fig. 3a, highest data point at the 1-day delay line). However, for the delay reported, it is one order of magnitude higher than the case of instantaneous fractionation immediately after the time of the explosion. Even though the time of the release is not reported by the hour, it definitely happened on the calendar day after the test, i.e., with a 16-h delay for the unrealistic assumption that the drill-back operation was conducted right at midnight. More likely, the delay was 24 h. For the Puye test on 14 August 1974, the activity ratio (see Fig. 3a, lowest data point at the 1-week delay line) could be understood, if the drill-back took place 5 days after the test instead of the reported delay of 7 days. Two further data have their x mark (for the time the decay corrected concentration is valid) to the right of the simulation curves while the respective sampling periods start left to the curves, i.e., earlier. A possible explanation might be a too large delay reported for the plume passage.

On the low end of activity ratios, the deviations of reported data from the simulation curves cannot be considered as spurious outliers and need to be explained. The picture is less serious for the ^{135}I to ^{133}I activity ratio (5 out of 27), a little worse for the ^{135}I to ^{131}I activity ratio (6 out of 26) and most pronounced for the ^{133}I to ^{131}I activity ratio (14 out of 43).

The likely explanation for isotopic activity ratios lying below the theoretical curves are inconsistencies in the measurements of the different isotopes. The activity of ^{131}I is often reported as the sum of the activity on the filter and in the charcoal. However, due to the measurement method (gross-beta counting over several days after allowing the short-lived background to decay) and because of their short half-lives, the activity of ^{135}I and ^{133}I on the filter is in most cases below the detection limit at count time. The activity on the charcoal is measured by gamma spectroscopy soon after the sampling ended. Accordingly, only the gaseous part of their activity is reported, and the atmospheric concentrations of ^{135}I and ^{133}I are underestimated. As result, their isotopic

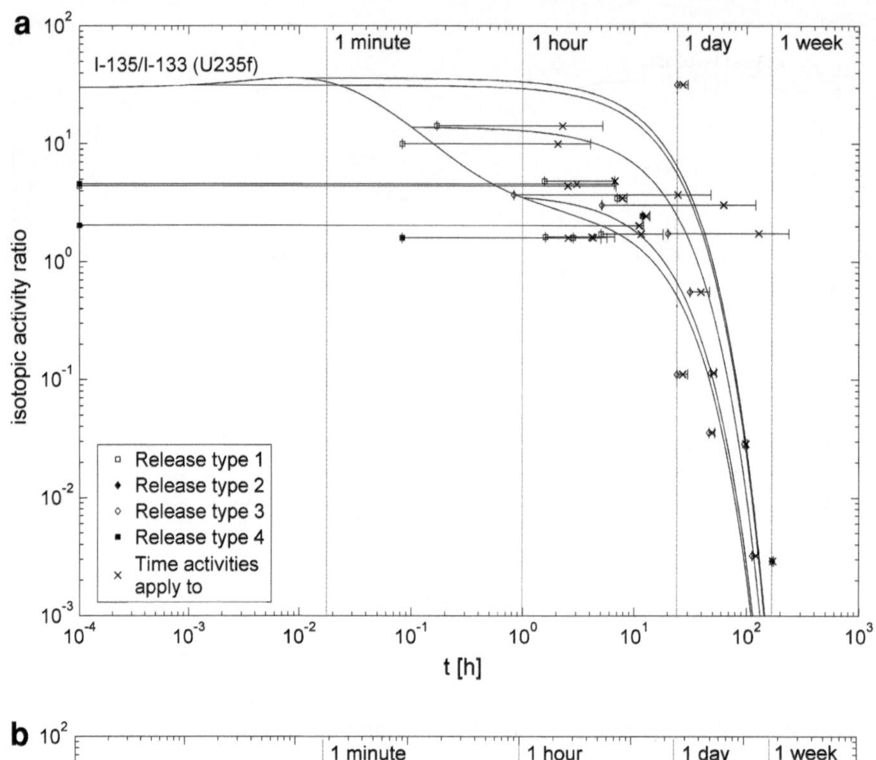

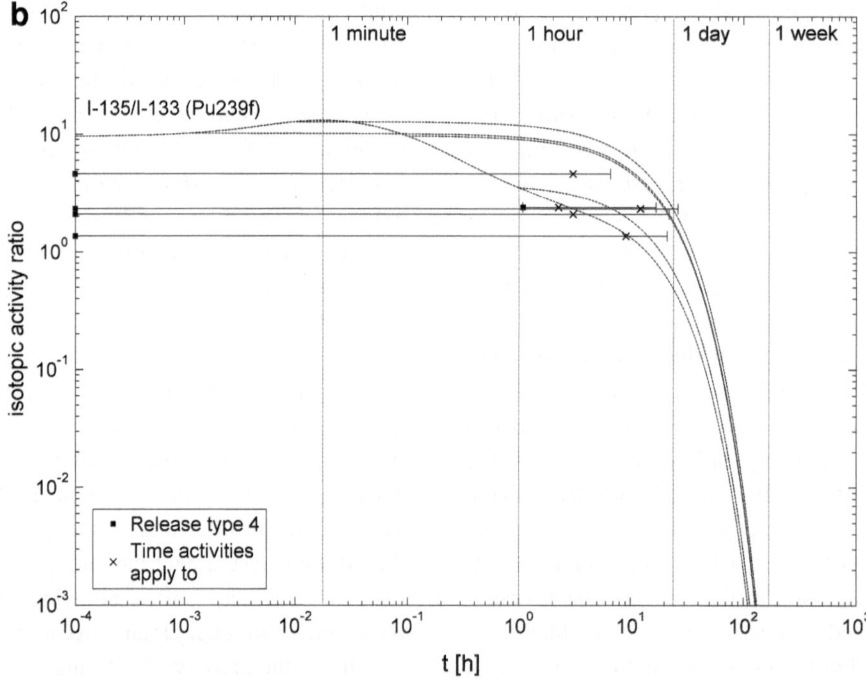

Figure 3

Comparison of simulated and reported radioiodine isotopic activity ratios between ^{135}I and ^{133}I released from nuclear tests at the Nevada Test Site. The plot **a** shows all releases as reported by Schoengold *et al.* (1996), the **b** part applies to all observations made in the wake of the Sedan test and reported by Placak (1963). The simulations use ^{235}U (**a**) and ^{239}Pu (**b**) with fission energy neutrons and assumes complete fractionation at various time steps. The **b** part (Sedan test) is based on gaseous radioiodine only. For venting and cratering tests, the *horizontal bar* indicates the sampling duration, for controlled releases it marks the reported duration of the release (see also Table 1). The X is put at the time to which the reported activities apply to. These are normally calculated for the mid-point of the collection time but apply to the cloud passage time, if that was determined

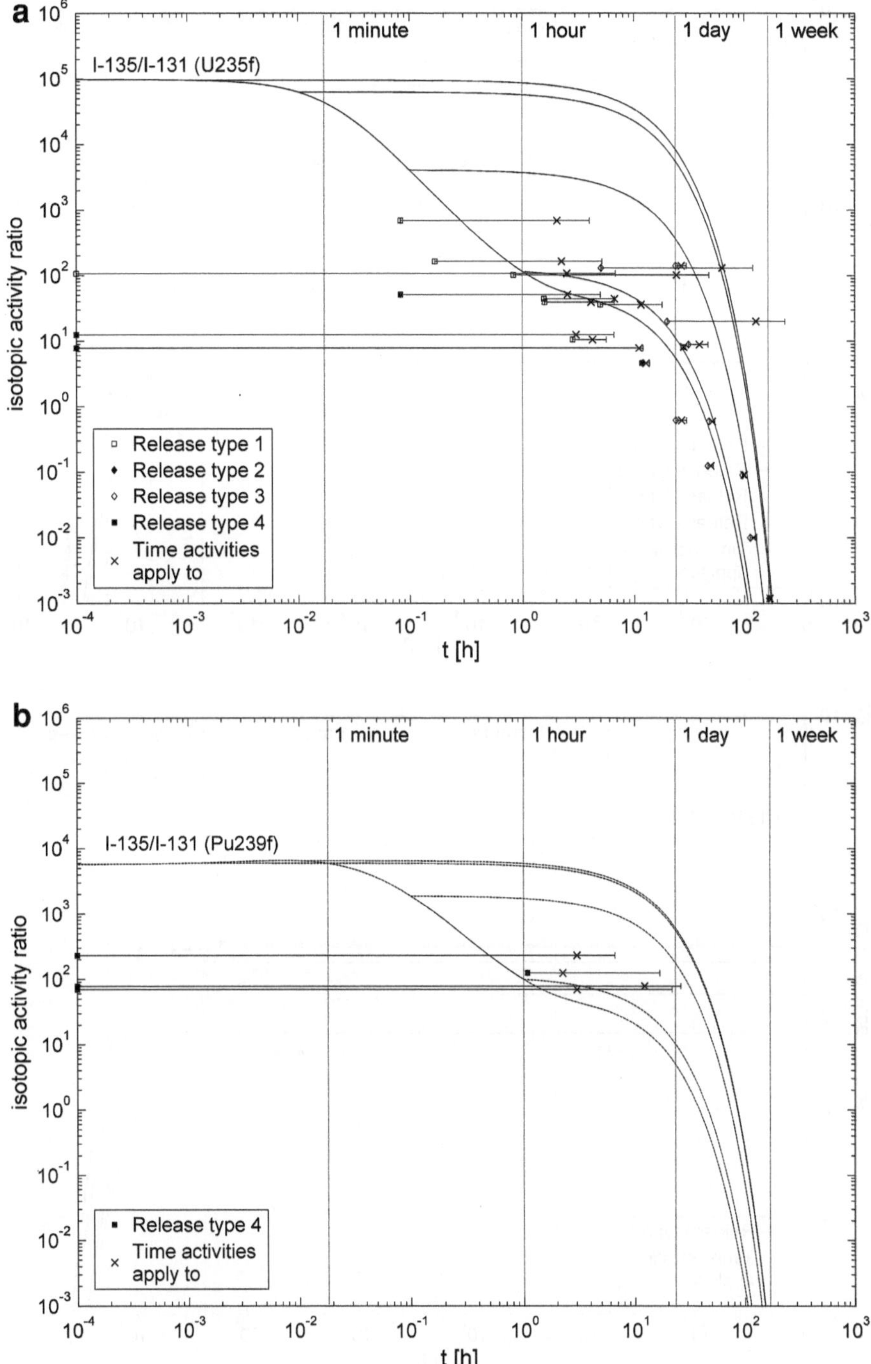

Figure 4
Same as in Fig. 3 but for the radioiodine isotopic activity ratios between ^{135}I and ^{131}I

activity ratios with respect to the ^{131}I are shifted towards too low values.

In order to test this hypothesis and investigate the isotopic activity ratios without any disturbance by inconsistent measurements, the plots in Figs. 3b to 5b display data for the gaseous fraction of radioiodine only that is trapped on charcoal. All are measurements taken in wake of the Sedan explosion on 6 July

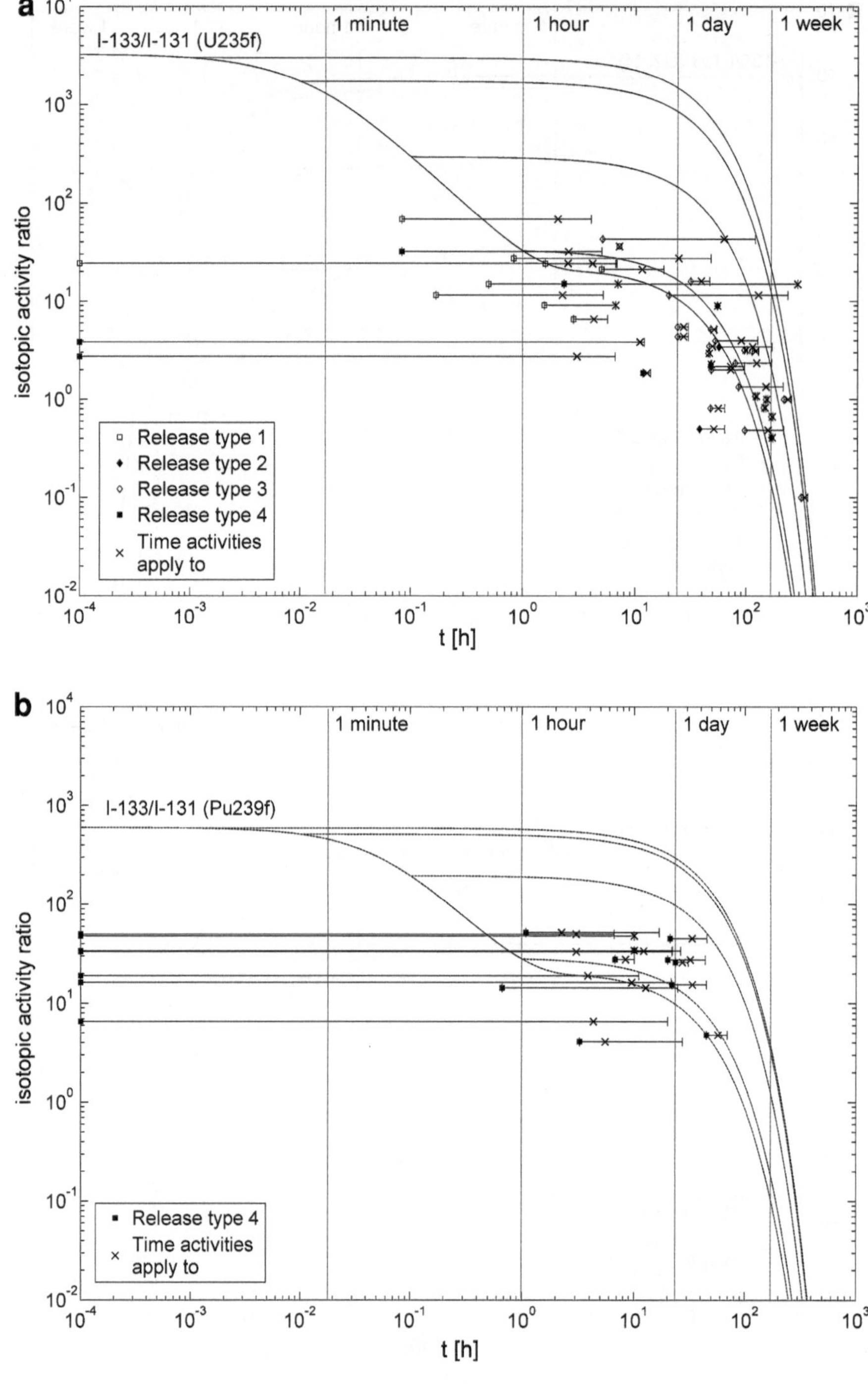

Figure 5
Same as in Fig. 3 but for the radioiodine isotopic activity ratios between ^{133}I and ^{131}I

1962 as reported by PLACAK (1963). Only the sample with the highest concentration is shown in the plot a) as well. This is the sample taken at Diablo, Nevada, including the plume cloud passage 3 h after the explosion. For this sample SCHOENGOLD *et al.* (1996) reported a ^{135}I/^{131}I activity ratio for total radioiodine of 12.5. This is entered in plot of Fig. 4a and lies below the ensemble of simulated lines. If for both isotopes only the gaseous fraction is counted as reported by Placak (1963), the activity ratio is 231 and the entry is found within the range of the theoretical curves as seen in the plot of Fig. 4b. For the $^{133}I/^{131}I$ the same correction would change the value from 2.7 (Schoengold *et al.*, 1996) to 50 according to Placak (1963). This would shift the low entry in Fig. 5 into the simulated range.

These examples confirm the hypothesis that inconsistent activity measurements for the different isotopes could explain why isotopic ratios are found too low in comparison to the simulated curves. In fact, the data in the plots of Figs. 3b and 4b are all found close to or within the range of theoretical lines. In the plot of Fig. 5b, the majority of data points lie within the simulation curves, but again two entries are below. What is more striking however is the fact that the isotopic ratios are spreading over one order of magnitude even for delay times differing by only 3 h. Even for samples taken at the same site and at the same time, the activity ratios differ by a factor of up to 3.3. It is unlikely that any physical or chemical process in the cloud causes these differences. Apparently, the activity measurements are not as reliable as indicated by a two digit reporting precision.

A further strange finding is that the plume has passed the most distant station (Ely, Nevada) 10 h after the explosion, but 7 out of 10 data points in Fig. 5b are determined for times with a delay beyond 10 h and up to 57 h after the explosion. This observation at Ely is a bit doubtful, because it was a cratering test that releases most of its activity instantaneously. Even stranger is a detection of radioiodine at Lockes, Nevada, determined for about 7 h before the explosion time. These shifted delays between explosion and detection hint at the possibility that the associations of activity observations to certain times bear some uncertainties and may in some instances be off by a few hours for early detections (before one day has passed) and by a day for later detections (with delays of several days).

Except for the explained deviations and discussed outliers, the overall agreement between the measurements at the Nevada Test Site and the model is reasonable. Most data lie within the ensemble of simulated curves and their major trends are also followed by the observational data. These agreements give confidence in further conclusions drawn from comparing the reported data and the theoretical calculations. The spread of radioiodine activity ratios of the observational data centers close to the two lowest simulation curves, the unfractionated case and the line that indicates a fractionation time of 1 h after the explosion. Almost all data are consistent with a fractionation no earlier than 0.1 h after the explosion. Accordingly, the radioiodine isotopes and their precursors remained basically non-fractionated before the releases took place.

5. Separation of Radioxenon and Radioiodine in Operational Releases at the Nevada Test Site

Figure 6 shows the change over time of four different isotopic activity ratios between radioxenon and its precursor radioiodine. All ratios are chosen for the isotopes and isomers that are relevant for nuclear explosion monitoring and are in the same decay chain even if no data are available for comparison. The reported Nevada test release data are shown together with the theoretical curves for three nuclear explosion scenarios (the different combinations of fissile isotope and neutron energy spectrum introduced in section 1) and no fractionation is assumed. Figure 7 shows the same as in Fig. 6 but only for ^{235}U and fission neutron energy with fractionation of xenon and radioiodine together from all other precursors at delays of 0.001, 0.01, 0.1 and 1 h after the explosion. Again, the effect of fractionation can be seen with the curve ending its further build-up through the decay chain and start decaying according to the effective half-life of the respective radioxenon to radioiodine activity ratio.

The theoretical curves in Figs. 6 and 7 do not exhibit a wide spread, in particular not in the timeframe for data of emissions at the Nevada Test Site that are available. All data are found below the simulation curves spreading over many orders of magnitude. This indicates that radioiodine is in most cases strongly depleted in the air samples compared to radioxenon. This depletion does not result from the

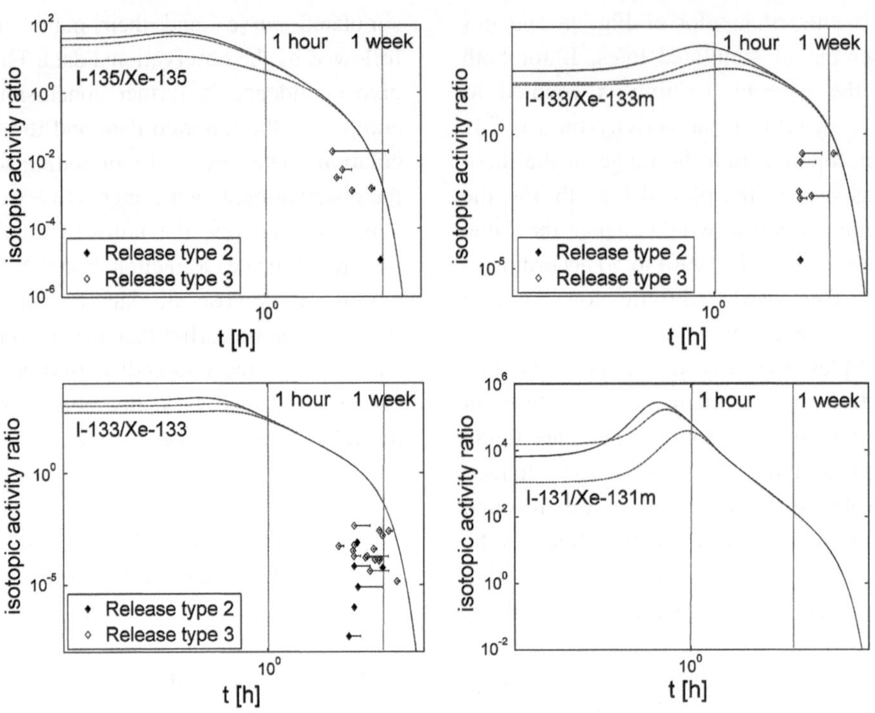

Figure 6
Change over time of four different isotopic activity ratios between radioxenon and its precursor radioiodine. The simulations with three nuclear explosion scenarios are compared to reported Nevada test releases. No fractionation is assumed

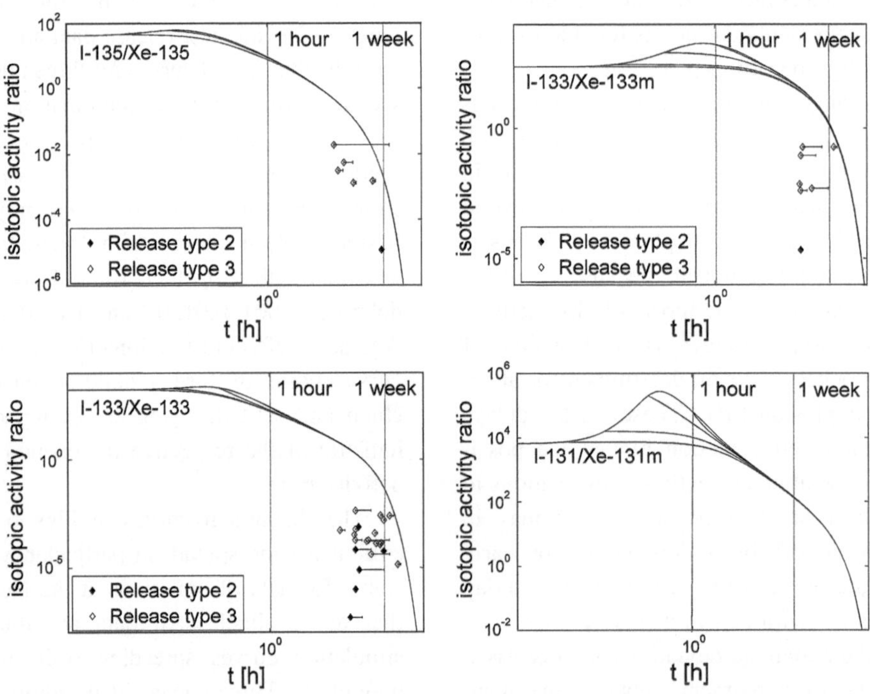

Figure 7
The plots show the same as in Fig. 6 but only for ^{235}U and with fractionation of radioxenon and radioiodine together from the other precursors

common fractionation of iodine and xenon. It is due to a fractionation between radioxenon and all its precursors including radioiodine. Partly this may take place on the way from the explosion cavern to the surface; partly it can be explained by the filtering of air prior to its operational release. The data points for these filtered releases (noted as type 3 in Table 1) are found to be at least two orders of magnitude lower than all other entries for the masses 135 and 133 with the metastable state of radioxenon. For the activity ratio of ^{133}I and ^{133}Xe, the operational releases with and without filtering are mixed but most of the filtered releases have a stronger depletion in radioiodine than most of the unfiltered releases. The strong efficiency of the filtering procedure is obvious with reduction of the radioiodine content by two to eight orders of magnitude compared to the theoretical values.

It remains to be understood why the unfiltered releases are depleted in radioiodine by one to four orders of magnitude. The previous section and Figs. 3, 4, 5 reveal that the radioiodine isotopes and their precursors remained basically non-fractionated before the releases took place. This is consistent with the result of Kalinowski (2011) according to which no significant fractionation between the radioxenon isotopes and the precursors occurs prior to release on any of the relevant pathways for operational releases. The depletion in radioiodine is most likely due to the passing of the release through a ventilation system with a filter without this being explicitly reported. This would imply that the distinction made here between filtered and unfiltered operational releases is not strictly applicable. The clear distinction visible in the figures and described above gives evidence that at least a difference in the extent of filtering exists. There are two less plausible explanations for the depletion in radioiodine which both cannot be verified from the available data and information. Either the released radioiodine is not fully accessible to the measurement or the gas stream that is released by some kind of operation at the test site is passing through a natural filtering material prior to the release and keeping part of the iodine back while fully releasing xenon.

Since xenon is a noble gas, its release from an underground explosion depends only on the availability of an unplugged path from the explosion cavity to the surface. For iodine, the mobility depends on its presence in the air stream at the moment a path opens up. It may be present either in gaseous form or attached to aerosol particles. Iodine that has condensed on or been adsorbed on non-mobile surfaces will remain underground. On its way out, aerosol particles may deposit on surfaces and gaseous iodine may be partially held back by adsorption on surfaces and by chemical reactions transforming it into a non-volatile compound. The former is particularly of relevance, if the air stream happens to be filtered through a soil column or is purposely pumped through a particulate filter and activated charcoal before being released to the atmosphere. According to Clément *et al.* (2007) adsorption of iodine has a significant impact on its volatility and gaseous iodine can react with radiolysis products of humid air to form non-volatile oxides or iodate ions.

The conclusion from the investigation of activity ratios of radioiodine and radioxenon of the same mass number is that practically for every operational release radioiodine was to some extent removed from the gas stream and less radioiodine was released than the theoretically predicted activity based on the radioxenon activity.

6. Effect of Fractionation on Source Discrimination Based on Radioxenon Isotopic Activity Ratios

A robust method for source discrimination has been demonstrated earlier based on the relationship of two different isotopic activity ratios (Kalinowski *et al.*, 2010). This requires three or four radioxenon isotopes to be quantified. In some cases the method works with the detection limit by substituting a missing concentration value, if certain isotopes are not detected. Hence, it is possible to use this method if only two isotopes are detected. A special advantage of this new method is its independence on the time periods elapsed between generation and release as well as between release and detection, i.e., it is invariant to dilution and decay.

In Fig. 8, this discrimination method is shown for the case that all four radioxenon isotopes are measured and made use of. Two isotopes are used on the

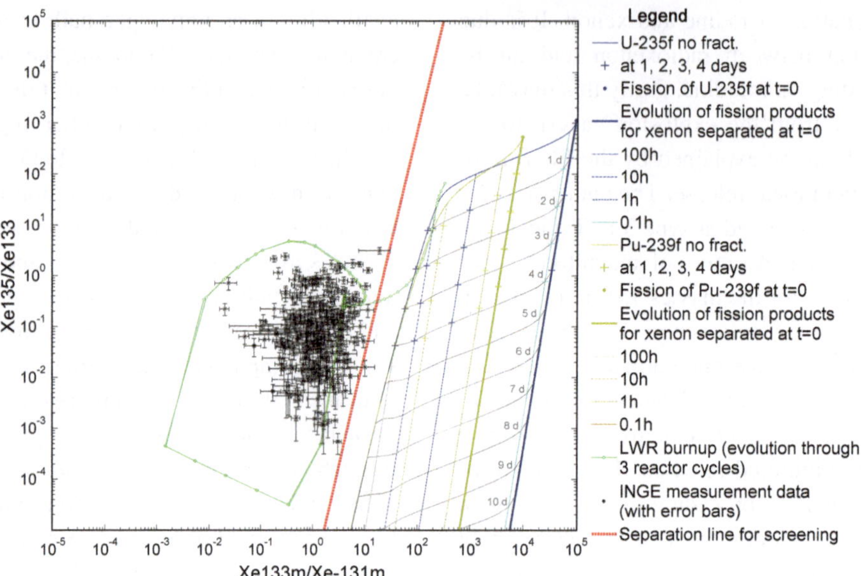

Figure 8
Time-invariant source discrimination based on radioxenon isotopic activity ratio relationship with reactor emission data for the case that all four isotopes are measured (Kalinowski *et al.*, 2010). The *dashed line* marks the time-invariant screening separation. All isotope ratio relations found above (i.e., *left* to) this *line* can be screened out, i.e., the related samples are definitely irrelevant for CTBT verification purposes, because they cannot be explained by a nuclear explosion. All samples that have ratio dependencies found below (i.e., *right* to) the *line* might be relevant for CTBT monitoring purposes. The exact location of the separation *line* is subject to further studies. The expected isotopic ratios present at $t = 0$ are denoted with the "*closed circle*" symbol, and different lines *styles* are used to show the change in the isotopic ratios for the different fractionation scenarios. Time steps of the first four full days are marked on these lines with "+" symbols. For ^{235}U, the daily time steps for 1–10 days of delay after the explosion are marked with *thin solid lines* to indicate were other fractionation scenarios would be at the daily time steps

abscissa; the other two are taken for the activity ratio on the ordinate. The simulation curves entered here separate the plot area into two distinctive domains. The trajectories for three operational cycles of a nuclear reactor follow a circular pattern in the left half of the plot (Kalinowski and Pistner, 2006). The simulation curves for nuclear explosion scenarios remain in the right half of the plane. In order to illustrate the power of this source discrimination method, the observational data of the International Noble Gas Experiment (INGE) are shown in the same plot as used in the similar plot by Kalinowski *et al.* (2010). This demonstrates that the method is robust against the typical measurement errors and a separation line can be defined that leaves most observations of ambient air in the reactor domain.

The change in the isotopic ratios is shown for complete fractionation involving only the independent yields (pure radioxenon curve) and without any fractionation allowing for in-growth (unfractionated decay chain curve). Depending on the amount of fractionation of the radioxenon isotopes from their parents, the actual ratios will fall between the two or precisely on one of the lines.

Figure 9 shows the same as Fig. 8 but with the assumption that radioxenon and radioiodine are together separated from the precursors.

In each plot, a separation line is drawn between the reactor and the explosion domain. Any isotopic activity ratio combination on the separation line moves down along that line with the radioactive decay. Any isotopic activity ratio combination off that line remains on the same side of the line forever. Radioactive decay changes the isotopic activity ratios along a line that is parallel to the separation line.

In Fig. 9, the left lines bounding possible isotopic activity ratios of the three nuclear explosion scenarios remain the same as in Fig. 8. These are valid for the case without any fractionation. Accounting for radioiodine separating from the non-volatile precursors together with radioxenon affects all other lines, in

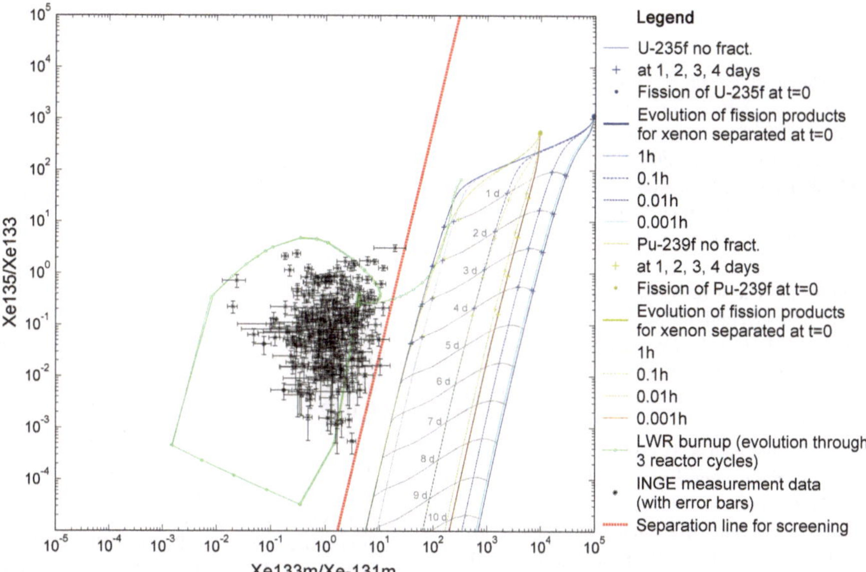

Figure 9
This plot shows the same as in Fig. 8 but with the assumption that radioxenon and radioiodine are together separated from the precursors

particular the position of the right boundary lines. They are bent towards the separation line in Fig. 8. The conclusion of this comparison is that no kind of fractionation compromises the clear separation of the nuclear explosive domain from emissions of nuclear reactors.

7. *Conclusions and Implications for Monitoring for Nuclear Explosions*

This paper studies the temporal development of radioxenon and radioiodine activities with two different assumptions on fractionation. In the first case only the noble gases are released, in the second case radioiodine is released as well while the precursors remain underground. For the second case, the simulated curves of activity ratios are compared to prompt and delayed atmospheric radioactivity releases from underground nuclear tests at Nevada as a function of the time of atmospheric air sampling for concentration measurements of ^{135}I, ^{133}I and ^{131}I. Most of the data from the Nevada Test Site lie at the lower end of the simulated curves indicating either no fractionation of radioiodine from its precursors at all or a fractionation not earlier than 0.1 h and more likely later than 1 h after the explosion. Accordingly, the radioiodine isotopes and their precursors remained basically non-fractionated before the releases took place.

The conclusion from the investigation of activity ratios of radioiodine and radioxenon of the same mass number is that practically for every operational release iodine was to some extend removed from the gas stream and less radioiodine was released than the theoretically predicted activity based on the radioxenon activity.

The final conclusion is that no kind of fractionation can compromise the clear separation of the nuclear explosive domain from emissions of nuclear reactors that can be used for source discrimination based on the relationship of two different radioxenon isotope activity ratios.

Acknowledgments

This work was funded by the German Foundation for Peace Research (DSF) and the University of Hamburg. The authors are grateful to the two reviewers for their questions and suggestions that helped to improve on the manuscript.

References

Bateman, H. (1910), *The solution of a system of differential equations in the theory of radio-active transformation.* Proc. Cambridge Phil. Soc., *16*, 423.

Clément, B., Cantrel, L., Ducros, G., Funke, F., Herranz, L., Rydl, A., Weber, G., Wren, C. (2007), *State of the art report on iodine chemistry, Nuclear Energy Agency, Committee on the Safety of Nuclear Installations.* Report NEA/CSNI/R(2007)1, February 2007.

De Geer, L.-E. (2001), *Comprehensive Nuclear-Test-Ban Treaty: Relevant radionuclides.* Kerntechnik *66*(3), 113–120.

England, T.R., and Rider, B.F. (1994), *Evaluation and Compilation of Fission Product Yields: 1993.* Los Alamos report LA-UR-94-3106 (ENDF-349), Appendix A (Set A Yields Evaluated and Compiled), October 1994.

Hoffmann, W., Kebeasy, R., and Firbas, P. (1999), *Introduction to the verification regime of the Comprehensive Nuclear-Test-Ban Treaty.* Phys. Earth Planet. Interiors, *113*, 5–9.

Kalinowski, M.B. (2006), *Comprehensive nuclear-test-ban treaty CTBT verification.* In: R. Avenhaus, N. Kyriakopoulos, M. Richard, G. Stein (Eds.), Verifying Treaty Compliance. Springer Berlin, Heidelberg 2006, pp. 135–152.

Kalinowski, M.B. (2011), *Characterisation of prompt and delayed atmospheric radioactivity releases from underground nuclear tests at Nevada as a function of release time.* J. Environ. Radioact. *102*(9), 824–836. doi:10.1016/j.jenvrad.2011.05.006.

Kalinowski, M.B., Liao, Y.-Y., and Pistner, C. (2012), *Discrimination of nuclear explosions against civilian sources based on atmospheric radioiodine isotopic activity ratios.* Submitted to Pageoph Topical Volume II on Recent Advances in Nuclear Explosion Monitoring.

Kalinowski, M.B., and Pistner, Ch. (2006), *Isotopic signature of atmospheric xenon released from light water reactors*, J. Environ. Radioact. *88*(3), 215–235.

Kalinowski, M.B., and Schulze, J. (2002), *Radionuclide monitoring for the comprehensive nuclear-test-ban treaty.* J. Nuclear Mater. Manag. *30*(4), Summer 2002, 57–67.

Kalinowski, M.B., Axelsson, A., Bean, M., Blanchard, X., Boeyer, T.W., Brachet, G., Hebel, S., McIntyre, J.I., Peters, J., Pistner, C., Raith, M., Ringbom, A., Saey, P., Schlosser, C., Stocki, T.J., Taffary, T., and Ungar, R.K. (2010), *Discrimination of nuclear explosions against civilian sources based on atmospheric xenon isotopic activity ratios.* In: B ecker, A., Schurr, B., Kalinowski, M.B., Koch, K., Brown, D. (Eds.), Recent Advances in Nuclear Explosion Monitoring. Pure and Applied Geophysics Topical, vol. 167/4-5, S.517–539. doi: 10.1007/s00024-009-0032-1.

Liao, Y.-Y. (2011), *Fraktionierung bei der Freisetzung von Leitnukliden für die Entdeckung von unterirdischen Kernwaffentests.* Diplomarbeit, Universität Hamburg, September 2011.

Placak, O.R. (1963), *Project Sedan, Final Off-Site Report, U.S. Public Health Service*, Report PNE-200F, 25 April 1963.

Schoengold, C.R., DeMarre, M.E., and Kirkwood, E.M. (1996), *Radiological effluents released from U.S. continental tests 1961 through 1992, United States Department of Energy - Nevada Operations Office.* DOE/NV-317 (Rev.1) UC-702, Las Vegas, August 1996.

USPHS (1964), *Final Report of Off-Site Surveillance for OPERATION NOUGAT September 15, 1961–June 30, 1962.* U. S. Public Health Service, Department of Health Education and Welfare, Off-Site Radiological Safety Program, Southwestern Radiological Health Laboratory, Las Vegas, Nevada, April 24, 1964. (SWRHL-1r).

(Received March 13, 2012, revised August 2, 2012, accepted August 12, 2012, Published online September 7, 2012)

Pure Appl. Geophys. 171 (2014), 693–697
© 2012 Springer Basel AG
DOI 10.1007/s00024-012-0577-2

Pure and Applied Geophysics

Potential of Spectrum Categorisation Concepts using Radionuclide Ratios for Comprehensive Nuclear-Test-Ban Treaty Verification

FREDERIK POSTELT[1]

Abstract—In order to develop further existing categorisation concepts for CTBT verification, xenon ratios have been used to help identify nuclear explosions. 25,726 noble gas spectra have been analysed and an additional state-of-health criterion has been introduced.

Key words: CTBT, xenon, xenon ratio, noble gas, noble gas spectrum, noble gas spectra, categorisation.

1. *Introduction*

Most of the fission products from a nuclear weapons explosion are radioactive and can be effectively detected using state of the art detection systems. Some fission products are noble gases and, therefore, chemically inert. They are most likely to escape even from underground nuclear explosions and remain in the atmosphere. The xenon isotopes Xe-131m, Xe-133m, Xe-133 and Xe-135 have the most suitable fission yields and half-lives for verification purposes: from 9.1 h to 11.8 days, long enough to enable reliable detection off-site and short enough to have a very low background in the atmosphere. Stable xenon has a concentration of 0.087 ppm by volume in the atmosphere (AUER *et al.,* 2010).

Radioactivity measurements are used to verify compliance with the (not yet entered into force) Comprehensive Nuclear-Test-Ban Treaty: Even 0.1 g of Xe-133 mixed equally in the earth's atmosphere can be detected (AUER, 2010). Radionuclides can be used as evidence of nuclear explosions and particularly the named radioxenon isotopes have unique properties for nuclear explosion monitoring (AUER *et al.,* 2010).

The International Monitoring System (IMS) of the Comprehensive Nuclear Test-Ban-Treaty Organisation (CTBTO) includes 80 radionuclide monitoring stations, of which 40 are planned as noble gas stations (WOTAWA *et al.,* 2010). Most of the collected data does obviously not indicate nuclear events and is therefore of little interest for CTBT verification. Computer algorithms are very helpful in supporting the analysts to handle all incoming data and focus on the most significant samples only.

In this work an existing categorisation concept using xenon ratios has been tested and further developed. All noble gas spectra acquired by the IMS at 21 different radionuclide stations between June 2007 and June 2010 were analysed. These 21 stations are distributed all over the world, using SAUNA, SPALAX and ARIX detectors and have low as well as medium and relatively high background xenon concentrations (AUER *et al.,* 2010). The total number of analysed noble gas spectra equals 25,726. An additional state-of-health criterion is introduced and tested.

2. *Xenon Ratios*

A major challenge for the IMS is to distinguish between potential nuclear explosions and other sources. Civil sources such as nuclear power plants and medical isotope production facilities release radionuclides, which can resemble releases from nuclear explosions. KALINOWSKI *et al.* (2010) have shown the possibility of discriminating between nuclear explosions and other sources by analysing the ratios of different radioxenon isotopes. Figure 1

[1] Carl Friedrich Von Weizsäcker-Centre for Science and Peace Research, University of Hamburg, Beim Schlump 83, 20144 Hamburg, Germany. E-mail: fPostelt@physik.uni-hamburg.de

shows such a plot of the four xenon ratios and how they can help to distinguish between civil sources and nuclear explosions. By plotting the ratios against each other, a separation line can be drawn, which divides the emissions expected from nuclear explosions and those measured by the International Noble Gas Experiment (INGE). Despite its potential, this technique has not been implemented yet as categorisation criteria, but is only supposed to be used as additional flag. In addition it shows the categorisation levels used in this paper and the ratio thresholds, which are defined in Eq. 1:

$$\frac{C^{-}_{\mathrm{Xe-133m}}}{C^{+}_{\mathrm{Xe-131m}}} < 2, \frac{C^{-}_{\mathrm{Xe-135}}}{C^{+}_{\mathrm{Xe-133}}} < 5, \frac{C^{-}_{\mathrm{Xe-133m}}}{C^{+}_{\mathrm{Xe-133}}} > 0.3. \quad (1)$$

One threshold is set at a Xe-135/Xe-133 ratio of 5 and has to be seen as quality check. As a matter of fact, it is very unlikely to occur, as Fig. 1 shows a detection has to be made within a few days (depending *inter alia* on whether fractionation takes place), while the sampling, decay time and acquisition alone take between 36 and 72 h. Kalinowski (2011) has shown that usually no fractionating is taking place.

The other threshold ratio is set at Xe-133m/Xe-131m and equals 2 and is a first approach, while a final definition should make use of another line, as already discussed by Kalinowski *et al.* (2010). The definition applied here is simpler than the one proposed by Kalinowski *et al.* (2010) and nuclear explosions might get screened out, especially if the signal reaches the station later than 1 to 3 days after the explosion, depending on the type of weapon Kalinowski *et al.* (2010).

The Xe-133m/Xe-131m ratio might also be moved to smaller values by a memory effect due to the relatively long half-live of Xe-131m, resulting in a high background and a small Xe-133m/Xe-131m ratio. This has been observed in the aftermath of the 2006 DPRK nuclear test by Ringbom *et al.* (2009). This issue is being alleviated here by using the Xe-133m/Xe-133 ratio as an additional flag, which can help to assess a spectrum.

3. State-of-Health Information

The proposed categorisation concept includes an additional state-of-health (SOH) flag. According to the sample qualities summarised in Table 1, every spectrum is flagged as GREEN, YELLOW or RED. This should help the human analyst to assess every sample. In addition, the SOH flag is used as precondition for the automatic categorisation. The algorithm depends on the data quality of the analysed samples, especially where samples are interconnected, as for the abnormal concentration. Therefore, the preconditions are essential for the automatic

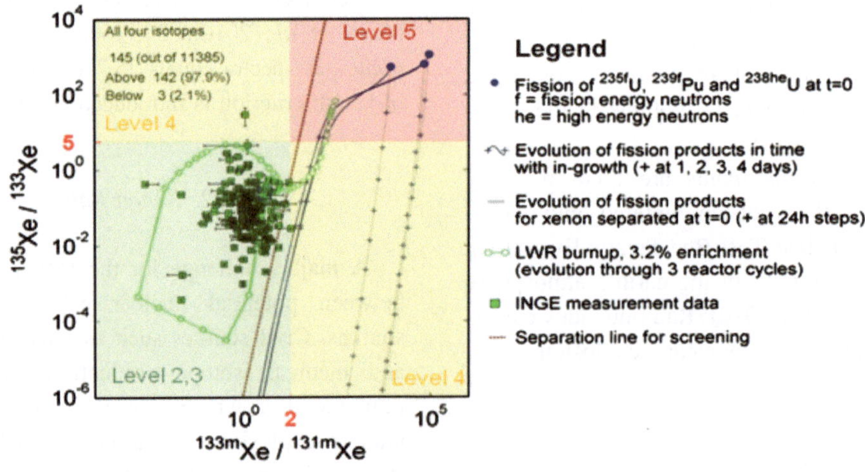

Figure 1
Plot of the xenon ratios that shows their potential for spectrum categorisation by Kalinowski *et al.* (2010). The four levels and thresholds were added by the author

Table 1

State-of-health criteria for spectra acquired at SPALAX stations as developed at the PTS plus an additional criterion, the minimum detectable concentration for Xenon 135

SoH-criteria for SPALAX stations						
Sampling Time		12 h	21.6 h	26.4 h	48 h	
Acquisition Time		12 h	21.6 h	26.4 h	48 h	
Xenon Volume		0.87ml				
Reporting Time	10 h				48 h	96 h
MDC-133	0.001 mBq/m³		1 mBq/m³		5 mBq/m³	
MDC-135	0.001 mBq/m³		1 mBq/m³			10 mBq/m³

Introducing the latter significantly facilitated the algorithm. The SOH criteria for SAUNAII and ARIXI stations vary only in numbers and not for the MDC's

analysis in order to achieve a reliable statistic. Only spectra flagged as GREEN or YELLOW are taken into account and categorised.

Table 1 shows the state-of-health criteria as developed at the IDC plus an additional criterion, a maximum and minimum threshold for the minimum detectable concentration (MDC) for xenon-135, which typically ranges from 0.5 to 1 mBq/m^3. However, in case of malfunction (for example low volume or a lot of Radon in the sample or when the quality control source is visible to the sample chamber), the MDC value for Xe-135 can be considerably higher.[1] Introducing a SOH criterion for the Xe-135 MDC significantly strengthened the algorithm, as it helps to eliminate an important number of spectra which the algorithm has difficulties in dealing with. Its introduction is one of the major findings of this study.

4. *Categorisation Concept*

The categorisation concept has been developed at the IDC and was presented in working group B, the policy making organ of the CTBTO Preparatory Commission by Nikkinen *et al.* (2011). Having passed the preconditions, every noble gas spectrum is searched for the presence of radioxenon. If no xenon isotope is detected, the spectrum is categorised as LEVEL 1, see Fig. 2. In case of detection it is checked whether this activity concentration is typical for this particular station or not. The abnormal concentration is defined in Eq. 2, by using all spectra taken at this particular station in the previous year:

$$C_{\mathrm{abn}} = \text{Median} + 3 \times \text{Spread}\,. \qquad (2)$$

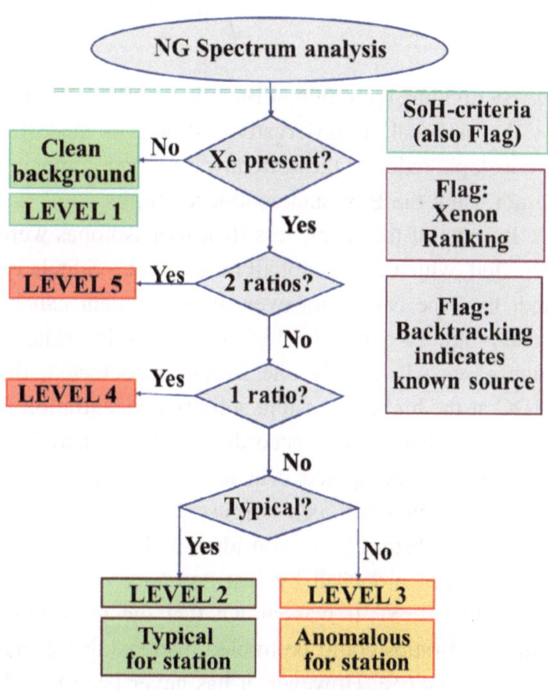

Figure 2
Categorisation concept as used in this work. A similar categorisation concept was presented by Nikkinen *et al.* (2011) in WGB 36, which is the policy making organ to the CTBTO PrepCom

[1] Personal communication with Mika Nikkinen, International Data Centre Division to the CTBTO PrepCom, Vienna.

If the activity concentrations of the three xenon isotopes Xe-133m, Xe-133 and Xe-135, are below this limit, the spectrum is categorised as LEVEL 2; if at least one concentration is above the threshold, as LEVEL 3. Xe-131m is not considered for a categorisation in cases of an abnormal concentration, because it is less significant than the other isotopes and carries a higher risk of a memory effect (RINGBOM *et al.*, 2009). Equation 1 defines the radioxenon ratios and according thresholds used for categorisation.

C^+ and C^- are the maximum and minimum Bayesian confidence limits of the corresponding isotope and have been defined by ZÄHRINGER *et al.* (2008), see Eq. 3:

$$C^- = C + S \cdot f^{-1}[1 - 0.975 \cdot f(C/S)],$$
$$C^+ = C + S \cdot f^{-1}[1 - 0.025 \cdot f(C/S)]. \qquad (3)$$

where C is the measured activity concentration, S the statistical error, calculated interactively from the station history and $f(x)$ the cumulative Gaussian distribution function, given in Eq. 4:

$$f(x) = (2\pi)^{-1/2} \int_{-\infty}^{x} e^{-\frac{z^2}{2}} dz. \qquad (4)$$

By dividing the minimal Bayesian confidence limit by the maximal, a conservative estimate is made for the isotopic ratios as these are always larger compared to not using the Bayesian confidence limits.

In most of the cases, less than four isotopes were detected, which means not all ratios can be calculated and, in some cases, not even an assessment can be done at all. To avoid this highly unsatisfying state in such cases where only one isotope is detected, the MDC at the highest possible activity concentration is used to calculate the according ratios. Therefore, every ratio is computed if at least one of the corresponding xenon isotopes is present. The use of the MDC as substitute for non-identified isotopes has already been proposed by KALINOWSKI *et al.* (2010, see Tables 4, 5). It was shown that the number of calculated ratios could be trebled, increasing the rate from 16 to 47 %. However, it has never been tested. This is done in this work for the first time.

Mixing the two statistical concepts of Bayesian confidence limits and the MDC might introduce some inconsistency. It is, however, justified through the high number of samples which can be additionally included in the automatic analysis compared to previous concepts. The MDC is only used in those cases, where an analysis would otherwise not be possible at all as the concentrations are unknown.

Only the first two thresholds of Eq. 1 are used for the categorisation whereas the Xe-133m/Xe-133 threshold serves as an additional flag. This choice has been made, as the four-isotope plot allows the most reliable distinction between nuclear explosions and other sources, see KALINOWSKI *et al.* (2010).

If only one of the first two thresholds is above the corresponding limit, the spectrum is categorised as LEVEL 4; if both values are above the limit the category is set to LEVEL 5. In addition to the third ratio a number of other flags is provided: besides the SOH-Flag, the Xenon ranking gives the rank number with regard to absolute activity concentrations (between 1 and 365) out of the up to 365 samples of the previous year. Backtracking known sources using atmospheric transport modelling may explain abnormal concentrations and serves as a further flag.

5. *Results*

The automatic categorisation of all 25,726 spectra acquired by the PTS between June 2007 and June 2010 is summarised in Table 2: 25,726 spectra have been analysed and only 60/1 been characterised as Level 4/5, respectively. By a quick human review, these numbers could be reduced to only 1/0, respectively.

Table 2

25,726 spectra have been analysed and only 60/1 been characterised as Level 4/5, respectively

Total samples: 25,726	Bad (SoH)	Level 1	Level 2	Level 3	Level 4	Level 5
Automatic						
Total	4,843	7,243	12,173	1,366	60	1
Percentage	–	34.80	58.40	6.60	0.29	0.01
Reviewed						
Total					1	0

By a quick human review this numbers could be reduced to only 1/0, respectively

By establishing the MDC for the xenon-135 isotope as an additional SOH-criterion, the screening could be significantly improved and 4,883 spectra identified, which the algorithm could not categorise automatically. Most of these spectra can be processed by human intervention, e.g., through recalibration by an analyst. The high number of not analysed spectra shows that the algorithm still needs to be improved. However, the presented work proves that the categorisation with xenon ratios can be used for fully automatic analysis of 81 % of the unreviewed raw data even if only two relevant isotopes have been detected. This number is much higher than in previous studies.

Out of the remaining 20,843 spectra only 0.3 % are categorised as Level 4 or 5 cases. This reduced number of samples can be easily and efficiently assessed by human analysts with higher priority. In a next step the other spectra can be reviewed. Automated processing can facilitate and speed up the analysis of noble gas spectra and, therefore, help to guarantee an effective verification of the Comprehensive Nuclear-Test-Ban Treaty.

This work will be pursued by using the developed algorithm to categorise historical nuclear underground tests whose xenon releases have been measured.

Acknowledgments

This work could not have been done without the valuable help and research of Matthias Zähringer, Mika Nikkinen, Marco Verpelli, Abdelhakim Gheddou and Martin B. Kalinowski.

References

M. Auer, T. Kumberg, H. Sartorius, B. Wernsperger, C. Schlosser, *Ten years of development of equipment for measurement of atmospheric radioactive xenon for the verification of the CTBT*, Pure and Applied Geophysics, Vol. *167*, 2010, Pages 471–486.

M. Auer, *Noble Gas Data and Processing – IMS Perspective*, presentation at the IDC, June 11th, 2010.

G. Wotawa, A. Becker, M. B. Kalinowski, P. J. R. Saey, M. Tuma, M. Zähringer, *Computation and Analysis of the Global Distribution of the Radioxenon Isotope 133-Xe based on Emissions from Nuclear Power Plants and Radioisotope Production Facilities and its Relevance for the Verification of the Nuclear-Test-Ban Treaty*, Pure and Applied Geophysics, Vol. *167*, 2010, Pages 541–557.

M. B. Kalinowski, A. Axelsson, M. Bean, X. Blanchard, T. W. Bowyer, G. Brachet, S. Hebel, J. I. McIntyre, J. Peters, C. Pistner, M. Raith, A. Ringbom, P. R. J. Saey, C. Schlosser, T. Stocki, T. Taffary, K. Ungar, *Discrimination of Nuclear Explosions against Civilian Sources Based on Atmospheric Xenon Isotopic Activity Ratios*, Recent Advances in Nuclear Explosion Monitoring, Pure and Applied Geophysics Topical Volume *167*/4–5, 2010, Pages 517–539.

M. B. Kalinowski, *Characterisation of prompt and delayed atmospheric radioactivity releases from underground nuclear tests at Nevada as a function of release time*, Journal of Environmental Radioactivity *102*, 2011 Pages 824–836.

A. Ringbom, K. Elmgren, K. Lindh, J. Peterson, T. Bowyer, J. Hayes, J. McIntyre, M. Panisko, R. Williams, *Measurements of radioxenon in ground level air in South Korea following the claimed nuclear test in North Korea on October 9, 2006*, Journal of Radioanalytical and Nuclear Chemistry, 2009, Volume *282*, Pages 773–779.

M. Nikkinen, U. Stoehlker, A. Gheddou, M. Verpelli, *Noble Gas Categorisation Scheme*, Scientific Methods, Software Application Section, International Data Centre Division to the Provisional Technical Secretariat of the CTBTO PrepCom, Vienna, Working Group B 36 Presentation, 2011.

M. Zähringer, G. Kirchner, *Nuclide ratios and source identification from high-resolution gamma-ray spectra with Bayesian decision methods*, Nuclear Instruments and Methods in Physics Research Section A: Accelerators, Spectrometers, Detectors and Associated Equipment, Volume *594*, Issue 3, 11 September 2008, Pages 400–406.

(Received December 6, 2011, revised August 1, 2012, accepted August 12, 2012, Published online September 7, 2012)

Pure Appl. Geophys. 171 (2014), 699–705

DOI 10.1007/s00024-012-0499-z

Pure and Applied Geophysics

Impact of Monthly Radioxenon Source Time-Resolution on Atmospheric Concentration Predictions

Michael Schöppner,[1,2] Martin Kalinowski,[3] Wolfango Plastino,[1,2] Antonio Budano,[2] Mario de Vincenzi,[1,2] Anders Ringbom,[4] Federico Ruggieri,[2] and Clemens Schlosser[5]

Abstract—The general characterisation of the global radioxenon background is of interest for the verification of the Comprehensive Nuclear-Test-Ban Treaty. Since the major background sources are only a few isotope production facilities, their source term has an emphasized influence on the worldwide monitoring process of radioxenon. In this work, two different datasets of source terms are applied through atmospheric transport modelling, to estimate the concentration at two radioxenon detection stations in Germany and Sweden. One dataset relies on estimated average annual emissions; the other includes monthly resolved measurements from an isotope production facility in Fleurus, Belgium. The quality of the estimations is then validated by comparing them to the radioxenon concentrations that have been sampled at two monitoring stations over the course of 1 year.

Key words: Medical isotope production facility, Radioxenon measurement, Environmental monitoring, Treaty verification, Comprehensive Nuclear-Test-Ban Treaty Organization (CTBTO).

1. Introduction and Background

Once entered into force, the Comprehensive Nuclear-Test-Ban Treaty (CTBT) prohibits its ratifying member states to conduct any kind of nuclear explosion within their control. The physical verification of the treaty obligations is based on three waveform technologies, i.e. seismic, hydroacoustic and infrasound, as well as radionuclide monitoring. For this purpose, the International Monitoring System (IMS) is built and already more than 80 % operational. The noble gas component of the radionuclide monitoring is based on the emission of radioxenon during a nuclear explosion, which is transported through the atmosphere, even after being released from an underground explosion. A worldwide network of monitoring stations is constantly measuring the radioxenon concentration with daily sampling (Zähringer *et al.*, 2009).

However, nuclear explosions are not the only sources of radioxenon. Nuclear power plants and isotope production facilities (IPF), e.g. for medical isotopes, have been recognized as the main background sources that could compromise the ability to detect nuclear weapon tests. When the location and emission strength of a source is known, the influence on the daily samples from the radionuclide monitoring stations can be simulated with atmospheric transport modelling (ATM). Therefore, the further characterization of the background source terms has been the aim of several works (Kalinowski and Tuma 2009; Saey *et al.* 2010; Wotawa *et al.*, 2010).

Isotope production facilities have been identified as the strongest emitters of radioxenon, with every IPF emitting about the same order of magnitude of radioxenon as all nuclear power plants together (Saey, 2009). Therefore, the radioxenon source terms of the IPFs are of high interest for the understanding of the background signal. However, it cannot be taken for granted that the producing companies would publish the released quantities of radioxenon in near real time. Thus, estimations about their average yearly emissions have been made (Kalinowski *et al.*, 2012).

[1] Department of Physics, University of Roma Tre, Via della Vasca Navale 84, 00146 Rome, Italy. E-mail: schoeppner@fis.uniroma3.it

[2] INFN, Section of Roma Tre, Via della Vasca Navale 84, 00146 Rome, Italy.

[3] Carl Friedrich von Weizsäcker Center for Science and Peace Research (ZNF), Beim Schlump 83, 20144 Hamburg, Germany.

[4] Swedish Defence Research Agency (FOI), Gullfossgatan 6, S-164 90 Stockholm, Sweden.

[5] Federal Office for Radiation Protection (BfS), Rosastraße 9, 79098 Freiburg, Germany.

In this work, emission data from the National Institute for Radioelements (IRE) in Fleurus, Belgium, with a monthly resolution, is used to simulate the concentrations at the IMS radionuclide stations. The estimations over one year are compared to the sampled concentration data of two IMS radionuclide stations. These results, based on the more accurate source terms, are then compared to the concentrations based on the previously estimated average yearly emissions.

2. Materials and Methods

2.1. Atmospheric Transport Modelling

When particles are emitted in a certain time interval from a given point on the globe, they can be transported through the atmosphere. Thus, they can arrive at a certain different location at a later point in time in a diluted concentration. For this matter, ATM can be used to estimate the time-dependent relation between two locations on a global (or regional) grid. These two points shall be called source and receptor; and the relation between them, source-receptor-sensitivity (SRS). If the emission of the source or the diluted concentration at the receptor is known, the other one can be estimated with the SRS value as calculated by ATM. Of course, the results strongly depend on the meteorological conditions of the regarded time period, as well as on local atmospheric patterns that are not resolved by the simulation, which can lead to altered signals (Plastino *et al.*, 2010). Particles, which have been emitted in one time interval, can contribute at different arrival times to the concentration at the receptor site, via their various trajectories through the atmosphere.

The relation between a source and a receptor can be described with a source-receptor sensitivity matrix. The concentration c (Bq/m^3) at any given receptor can be expressed as the product of a spatio-temporal source field S (Bq) and a corresponding source-receptor sensitivity field M (m^{-3}) at discrete locations (i, j) and time intervals n:

$$c = M_{ijn} \cdot S_{ijn}.$$

The field S is a multidimensional array of sources, which is transformed by the multidimensional array of multiplicators M into the concentration c that is measured at the receptor (Wotawa *et al.*, 2003). Here M presents the sensitivity between source and receptor and has the dimension of m^{-3}, whereas the inverse element of M can be depicted as a dilution volume. However, while the underlying calculations are naturally three-dimensional in space, the produced SRS matrix M is only two-dimensional, with time as a third dimension. General ATM software can simulate the transport of particles released from point, line, area or volume sources. The simulations can include long-range and mesoscale transport, diffusion, and deposition, as well as radioactive decay, into the calculations. In ATM, it is usually distinguished between forward and backward modelling, where both methods have advantages and disadvantages. Forward modelling is more efficient when the number of known sources is limited and the receptors are undefined. Backward modelling is more efficient when the number of receptors is limited and the sources are numerous or unknown. Therefore, in the case of IMS radionuclide sampling, the backward mode is preferred, as the location of the receptor site is well known.

2.2. Data Status

The concentration at IMS radionuclide stations can be estimated with ATM and the use of an emission database. In this case, such a database contains information about the location and strength of all contemplated sources with regard to ^{133}Xe, i.e. 200 nuclear power plant sites (Kalinowski, Tuma, 2009), and five IPF are considered. The emissions of nuclear power plants are estimated to be in the orders of 1E + 10 and 1E + 13 Bq per year. Usually, emissions from NPPs are batch-released and below the boundary of semi-continuous regime (Kalinowski, Tuma, 2009), but since the detailed, time-resolved emission data is generally unknown, only continuous emissions can be used in the simulation. The impact of batch emissions is lower for NPPs away from the detector than for NPPs closer to the detector. On the other hand, isotope production facilities are rarer, but they usually produce a higher output of radioxenon, i.e. orders of 1E + 13 and 1E + 16 Bq per year are estimated. Three of the five IPF are in the northern hemisphere and typically influence the radioxenon measurements in Europe, namely the ones in Fleurus (Belgium), Chalk River (Canada) and Petten (Netherlands); the latter usually emits radioxenon in the order of magnitude of average NPPs. Each known source is allocated to the nearest grid point as resolved by the ATM, i.e. in this case $1^\circ \times 1^\circ$. One database (a) handles all these sources as constant emitters of radioxenon, based on commonly accepted estimations of their average yearly emission. The second database (b) is similar with the exception that the IPF in Fleurus, Belgium, is handled as a varying source, based on reported emission data with monthly resolution from 2008. As these data are confidential, they are not published here, but are only described quantitatively. The year 2008 is characterized by the first five months with average emission strength, then a five months period during summer/autumn with emission particularly below the average, and towards the end of the year, two months with emissions clearly above average. The overall reported emissions are slightly lower than the usually applied estimation of 10^{15} Bq per year. Thus, the total radioxenon emission inventory for the year 2008 is lower in dataset (b) than in dataset (a).

To compare the effect of these two datasets on ATM-based radioxenon estimations, two IMS radionuclide stations have been selected; the German DEX33 at Schauinsland Mountain close to Freiburg, and the Swedish SEX63 close to Stockholm. With regard to ^{133}Xe, a total of 197 24 h-samples are available from the German station, resulting in a total time coverage of 54 % for the calendar year 2008. From the Swedish station, 581 12-h samples are available, resulting in a total time coverage of 79 % for the calendar year 2008. These samples also include non-detections of radioxenon, i.e. the concentrations are below the minimum detectable concentration (MDC), which has to be at 1 mBq/m^3 or lower, by design criteria (Saey, Schlosser *et al.*, 2010). However, after being reviewed from the International Data Centre (IDC) of the CTBTO the data is still subject to uncertainties due to the complex detection process (Auer *et al.*, 2010). For each of these 779 samples, a Flexpart simulation has been conducted to produce the accordant SRS fields.

The meteorological fields are provided from the European Centre for Medium-Range Weather Forecasts (ECMWF), and have spatial resolution of 1° in latitude/longitude and a time resolution of 3 h. Thus, the calculated SRS fields also have the same resolution in space and time.

In terms of estimated source terms, other relevant isotopes and isomers of xenon (namely ^{131m}Xe, ^{133m}Xe and ^{135}Xe) could be included, but the accordant data basis of IMS samples is not sufficient to cover a significant part of 2008. However, the emissions of these radionuclides are included in the simulation—particularly to simulate the feeding of ^{133}Xe through ^{133m}Xe.

3. *Analysis and Discussion*

The two radioxenon emission databases have been applied on the calculated SRS fields, to determine the estimated radioxenon concentrations at the two selected stations. The simulated concentration values for the German (Swedish) station are presented and compared to the measurements in Fig. 1 (Fig. 2); the 197 samples (581 samples) are chronologically ordered with regard to their date and time of sampling. This means that the horizontal axis does not fully reflect the calendar year 2008, but just the order in which the samples were taken. The horizontal axis crosses the vertical axis at 1e-03 Bq/m^3, which is the official MDC for IMS radioxenon sampling (Saey, Schlosser *et al.*, 2010).

As seen, for both stations, both simulations depict the characteristics of the time series rather well, including many maxima and minima; only for some time periods, the signal is off, or e.g. even predicts a minimum, where in reality there is a maximum (or vice versa). However, it has to be clearly stated that both kinds of simulation produce results of nearly the same quality. The statistical covariance and correlation have been calculated for each simulated time series, with regard to the experimental data. Table 1 shows that they are either similar for the simulations based on the different datasets, or even better for the simulation that is based on the all-constant dataset (a). The reason for this will become clear, when looking closer at the composition of the samples, cp. below. Nevertheless, due to the regional density of legitimate radioxenon emitters, each with emissions

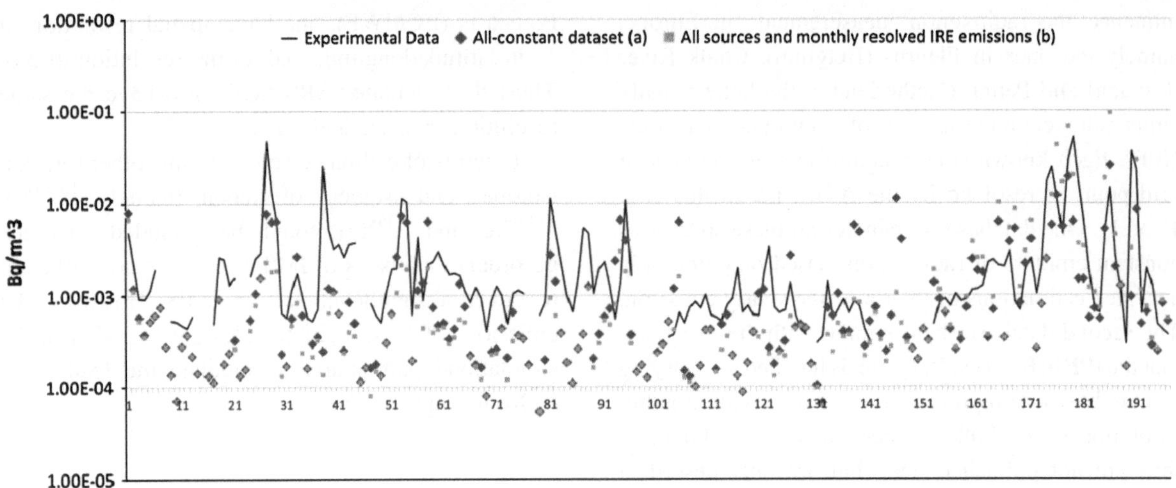

Figure 1
Comparison of the simulations based on two radioxenon emission datasets with the experimental data for the German IMS station DEX33

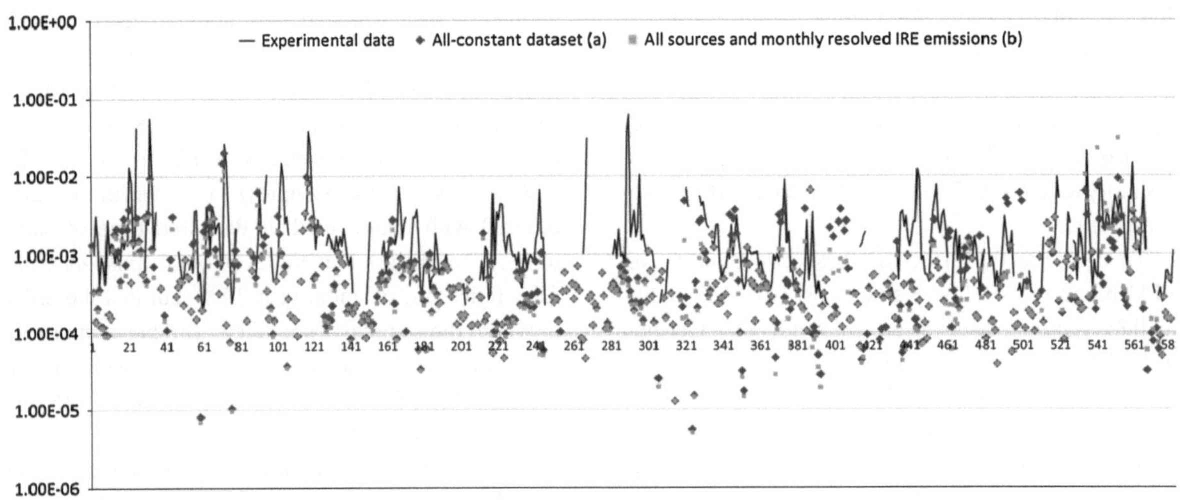

Figure 2
Comparison of the simulations based on two radioxenon emission datasets with the experimental data for the Swedish IMS station SEX63

Table 1

Statistical parameters of simulations for the German and the Swedish station based on datasets (a) and (b)

	Covariance σ		Correlation ρ	
	Dataset (a)	Dataset (b)	Dataset (a)	Dataset (b)
DEX33	1.13E−05	3.06E−05	0.50	0.42
SEX63	4.01E−06	3.08E−06	0.40	0.30

that can only be estimated, in none of the presented scenarios does the correlation come close to the ideal value of one.

3.1. Total Yearly Concentration

Before going into the statistical analysis of the data, it is obvious that the simulated concentration values are, over the course of one year, rather too low than too high. When all collected samples available from 2008 are summed, they accumulate to 0.58 Bq/m^3 for the German station, while the simulation produces only 0.30 Bq/m^3 with the all-constant sources dataset (a), and 0.45 Bq/m^3 with the dataset (b), including the IRE time resolved emissions. The situation is similar for the Swedish station; while the collected samples add up to a total of 1.12 Bq/m^3, the simulations only produce

0.47 Bq/m^3 and 0.42 Bq/m^3, respectively. This is most likely due to a generally underestimated global radioxenon emission inventory.

However, within the simulated radioxenon signals, the contributions from IRE, Belgium, to the total concentration of 2008, are as follows: At the German station for dataset (a) 0.22 Bq/m^3 (of the aforementioned total of 0.30 Bq/m^3) and for dataset (b) 0.38 Bq/m^3 (of 0.45 Bq/m^3) are accounted due to IRE. At the Swedish station for dataset (a) 0.23 Bq/m^3 (of 0.47 Bq/m^3) and for dataset (b) 0.19 Bq/m^3 (of 0.42 Bq/m^3) are accounted due to IRE.

Though the IPF in Fleurus, Belgium, is usually the strongest regional emitter of radioxenon in Europe, supra-regional sources also contribute to the signals at the German and the Swedish stations, and have to be taken into account. During the year 2008, the IPF in Chalk River, Canada, regularly contributed to the concentrations at both stations: The estimated constant source term of 10^{16} Bq/m^3 contributes with 0.04 Bq/m^3 to the German station and 0.14 Bq/m^3 to the Swedish station, over the total period of 2008. Compared to the total impact from IRE, this means that the IPF in Chalk River, Canada, plays usually only a minor role in the daily samples of the German station, but can significantly contribute to the daily samples of the Swedish station.

3.2. Quality of Simulations of Single Samples

In order to validate the quality of the simulations for single samples, two values have been calculated for each simulated sample: (1) The fraction of the IRE-contribution to the total sample concentration, and (2) the ratio of the simulated total sample concentration to the accordant experimental value. This means that (1) can vary between 0 and 1, depending on how much of the simulated concentration of one sample results from IRE emissions; and (2) can take any positive value, but ideally should be 1, since it is simply the ratio of simulation to experimental data.

Both values are then put in one diagram, and are sorted in ascending order with regard to the share of the IRE-contribution to the total simulated concentration (1). This means the samples with the lowest fractions of IRE-contributions are seen in the left end of the diagram, while the samples with the highest fractions are on the right end of the diagram; i.e. the series is monotonically increasing with regard to (1). Of course, the ratio of simulation to experimental data (2) stays attached to each sample.

The results for the German station are shown in Fig. 3 for the simulation based on dataset (a) and (b). As seen, there is a correlation between (1) and (2); for samples with a high share of IRE-contribution, the ratio between simulated and experimental data is wider spread, but on average closer to one. This signifies that, when, due to atmospheric conditions and/or a high emission, the share of the IRE-contribution to the simulated sample concentration is high, and the simulation of the total concentration is more likely to reflect the experimental data (i.e. on average closer to 1). The smaller spread on the left end results from the more steady contributions from nuclear power plants, which can act as an area source, while the larger spread on the right end results from an insufficient time resolution of the IRE source term.

For the Swedish station, the results are presented in the same manner in Fig. 4. Here, the correlation between (1) and (2) is not as noticeable as it is for the German station. A slight convergence of (2) towards 1 can be observed for high values of (1). However, the effect is more prominent for the simulation based on the all-constant sources dataset (a). The fact that the correlation between IRE-contribution and the quality of the simulation is weaker than for the German station is most likely due to the fact that for the Swedish station, the IPF in Chalk River, Canada, also plays a comparably important role. Therefore, samples from the Swedish station can have up to two major contributors, IRE and CLK, and are thus less dependent on a single one of them.

4. Conclusion and Outlook

It has been shown that, although the simulations can depict quite well the general characteristics of the experimental signal, the dataset based on time-resolved source terms did not improve the quality of the simulation. This is equally valid for both the German and the Swedish station. Compared to the previously used estimations of annual total emissions, the reported

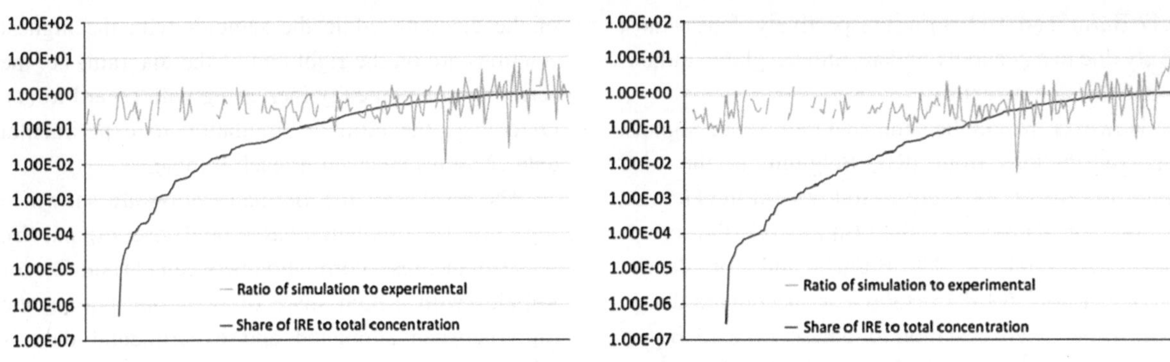

Figure 3
The effect of the share of IRE of the total concentration on the quality of the simulation for the German station; *left* the simulation for the all-constant dataset (a), *right* the simulation for the monthly resolved IRE dataset (b)

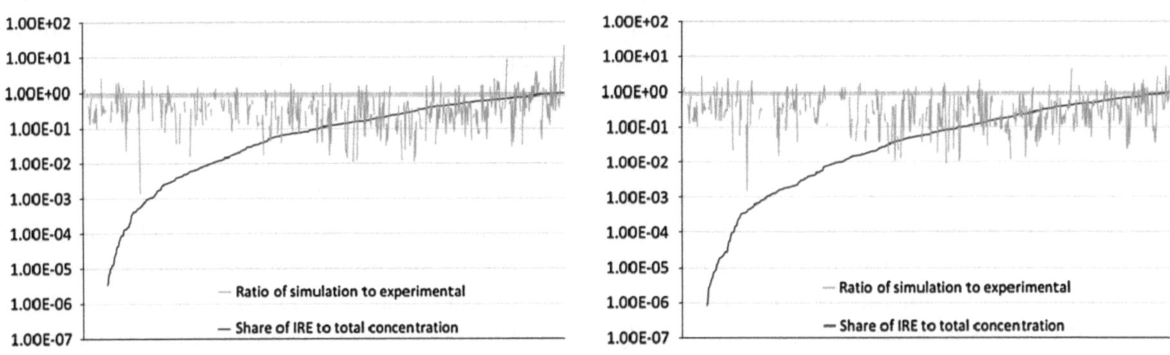

Figure 4
The effect of the share of IRE of the total concentration on the quality of the simulation for the Swedish station; *left* the simulation for the all-constant dataset (a), *right* the simulation for the monthly resolved IRE dataset (b)

monthly emission data can produce better results only in some cases. Generally, the previously used dataset (a) leads to a simulated time series of statistically higher quality. This is most likely due to the fact that the global radioxenon emission inventory is generally underestimated and dataset (b) is built on an even lower radioxenon emission inventory resulting from the IRE emission reports.

The reasons for a general underestimation of the radioxenon detections can be manifold, i.e. be found within the atmospheric transport model; additional/unknown sources of radioxenon exist; or the real source terms are higher than used in the simulation. The model used here, Flexpart, has been proved in the analysis of tracer experiments not to produce particular over-estimations or under-estimations, when emissions and detections are well known (Stohl *et al.*, 1998). Since simulated samples for the German station with high contributions from IRE are more likely to reflect the experimental data, it is believed that the underestimation refers to the collective of nuclear power plants found in Western Europe, rather than to IRE itself. This conclusion cannot be drawn from the simulations for the Swedish station, because here the situation is more complicated, involving IRE and CLK in equal measure.

However, the presented results suggest that a time resolution higher than monthly is necessary, e.g. daily emission values from major radioxenon emitters, in order to significantly improve the quality of the ATM-based simulation for radioxenon concentrations in the future.

Acknowledgments

The authors highly acknowledge the support by the National Scientific Committee Technology of INFN for the ERMES project, and the European Commission under the FP7 programme for the EUMEDGRID project (Grant RI-246589). The authors are grateful to IRE, Belgium, for the provision of data, specifically Benoît Deconninck, and to the Grid Lab of INFN and the Department of Physics of University of Roma Tre, specifically to Federico Bitelli.

References

Auer, M., Kumberg, T., Sartorius, H., Wernsperger, B., Schlosser, C. (2010), Ten years of development of equipment for measurement of atmospheric radioactive xenon for the verification of the CTBT, Pure Appl. Geophys. *167*, 471–486.

Kalinowski, M.B., and Tuma, M.P. (2009), Global radioxenon emission inventory based on nuclear power reactor reports, Journal of Environmental Radioactivity *100*, 58–70.

Kalinowski, M.B., Grosch, M., Hebel, S. (2012), Assessment of radioxenon emissions from medical isotope production based on worldwide technetium-99 m consumption, Pageoph Topical Volume II.

Plastino, W., Plenteda, R., Azzari, G., Becker, A., Saey, P.R.J., Wotawa, G. (2010), Radioxenon time series and meteorological pattern analysis for CTBT event categorisation. Pure and Applied Geophysics *167*, 559–573.

Saey, P. (2009), The influence of radiopharmaceutical isotope production on the global radioxenon background, Journal of Environmental Radioactivity *100*, 396–406.

Saey, P., Bowyer, T., Ringbom, A. (2010), Isotopic noble gas signatures released from medical isotope production facilities—Simulations and measurements, Applied Radiation and Isotopes *68*, 1846–1854.

Saey, P., Schlosser, C., *et al.* (2010), Environmental Radioxenon Levels in Europe: a Comprehensive Overview, Pure and Applied Geophysics. doi:10.1007/s00024-009-0034-z.

Stohl, A., Hittenberger, M., Wotawa, G. (1998), Validation of the Lagrangian particle dispersion model Flexpart against large-scale tracer experiment data, Atmospheric Environment Vol. *32* No. 24, pp. 4245–4264.

Wotawa, G., *et al.* (2003), Atmospheric transport modelling in support of CTBT verification—overview and basic concepts, Atmospheric Environment *37*, 2529–2537.

Wotawa, G., Becker, A., Kalinowski, M.B., Saey,P., Tuma, M., Zähringer, M. (2010), Computation and Analysis of the Global Distribution of the Radioxenon Isotope 133-Xe based on Emissions from Nuclear Power Plants and Radioisotope Production Facilities and its Relevance for the Verification of the Nuclear-Test-Ban Treaty, Pure Appl. Geophys. *167*, 541–557.

Zähringer, M., Becker, A., Nikkinen, M., Saey, P., Wotawa, G. (2009), CTBT radioxenon monitoring for verification: today's challenges, J Radioanal Nucl Chem *282*, 737–742.

(Received December 1, 2011, revised May 4, 2012, accepted May 5, 2012, Published online June 13, 2012)

Pure Appl. Geophys. 171 (2014), 707–716

DOI 10.1007/s00024-013-0687-5

Pure and Applied Geophysics

Global Xenon-133 Emission Inventory Caused by Medical Isotope Production and Derived from the Worldwide Technetium-99m Demand

Martin B. Kalinowski,[1] Martina Grosch,[2] and Simon Hebel[2]

Abstract—Emissions from medical isotope production are the most important source of background for atmospheric radioxenon measurements, which are an essential part of nuclear explosion monitoring. This article presents a new approach for estimating the global annual radioxenon emission inventory caused by medical isotope production using the amount of Tc-99m applications in hospitals as the basis. Tc-99m is the most commonly used isotope in radiology and dominates the medical isotope production. This paper presents the first estimate of the global production of Tc-99m. Depending on the production and transport scenario, global xenon emissions of 11–45 PBq/year can be derived from the global isotope demand. The lower end of this estimate is in good agreement with other estimations which are making use of reported releases and realistic process simulations. This proves the validity of the complementary assessment method proposed in this paper. It may be of relevance for future emission scenarios and for estimating the contribution to the global source term from countries and operators that do not make sufficient radioxenon release information available. It depends on sound data on medical treatments with radio-pharmaceuticals and on technical information on the production process of the supplier. This might help in understanding the apparent underestimation of the global emission inventory that has been found by atmospheric transport modelling.

Key words: Medical isotope production, radioxenon, xenon-133, emission inventory, CTBT.

1. Introduction

The Comprehensive Nuclear-Test-Ban Treaty bans all nuclear test explosions. To verify this treaty, a complex International Monitoring System is in place to detect nuclear explosions. Among other technologies, it relies on atmospheric radionuclide measurements, most importantly on the measurement of short-lived Xenon isotopes. To facilitate interpretation of the measurements, a profound knowledge of the atmospheric background is crucial. Radioxenon is emitted from nuclear facilities and operations like power reactors and medical isotope production Kalinowski and Tuma (2009). The latter have been recognized as causing the most significant impact on explosion monitoring Saey (2009).

The quantification of radioxenon emissions from medical isotope production facilities can be approached from two directions. The problem can be solved by simulating the radiation of the required HEU targets in the reactor in order to obtain the resulting radioxenon production and batch release. This approach can be problematic with respect to the available information regarding the irradiation procedure, including data such as radiation times, neutron flux and spectrum, and duration of target storage for cool down and decay after irradiation. Any attempt to derive xenon emissions from isotope production must make assumptions regarding these factors.

Also unpublished are the actual amounts of molybdenum-99 that are produced and shipped. Operators consider this confidential information. Some information hints at the fact that there exists a production overcapacity (Bonet *et al.*, 2005; Reistad *et al.*, 2007). The molybdenum-99 batches are produced on demand, making it difficult to estimate the production, even if detailed information about the irradiation procedure was available.

At the very least, it would be possible to calculate a ratio of xenon per molybdenum based on some

The views expressed in this publication are those of the authors and do not necessarily reflect the views of the CTBTO Preparatory Commission nor those of the University of Hamburg.

[1] Preparatory Commission for the Comprehensive Nuclear-Test-Ban Treaty Organization (CTBTO), Provisional Technical Secretariat, Vienna International Centre, P.O. Box 1200, 1400 Vienna, Austria. E-mail: martin.kalinowski@ctbto.org

[2] Carl Friedrich von Weizsäcker-Centre for Science and Peace Research, University of Hamburg, Beim Schlump 83, 20144 Hamburg, Germany.

basic irradiation parameters. It is estimated that radiation times vary from 2 to 10 days with a subsequent storage time of 2 days NRC (2009). Until the end of the last decade, typical targets were composed of HEU Vandegrift (2005). Today more producers are working with LEU and, further, are considering converting to LEU WOSMIP II (2011). Because the used data were collected before 2008, we assume an enrichment of 93 %, and an exposure to a thermal neutron spectrum.

The approach chosen here is to estimate the xenon production based on information from the consumer side. Technetium-99m consumption in hospitals is documented and, to some extent, is open source information. The Tc-99m consumption data has been collected and processed by Grosch (2008). Based on the consumption, it is possible to calculate the Mo-99 production required to meet the demand, and, using the calculated xenon-to-molybdenum ratios, derive the amounts of radioxenon that are produced and potentially emitted from the medical isotope production facilities.

Occasionally, hospitals emit Xe-133 and Xe-131m directly when isotopes are employed in a medical examination. These releases amount to approximately 1 MBq/day Heimbigner *et al.*, (2002).

2. *Previous Assessments of the Global Xe-133 Emission Inventory*

In order to determine the source strength of Xe-133 emitters, several different approaches are possible and the methods have different levels of accuracy.

- Measurements in the facility by the operator or a third party provide the highest accuracy that may be in the range of 1–10 % depending on the measurement technology and procedures.
- The second best option is the facility specific simulation of the production processes. This depends on technical details of the production process to be known. Simulations may in the ideal case achieve a similar accuracy as direct measurements, but the variability of processing parameters has to be taken into account.
- The third approach is to use ambient concentration observations and apply atmospheric transport modelling (ATM) to determine the size of the source term. With single observations at large distances, this can result in over- or underestimates by one order of magnitude. When tens of observations at multiple receptor sites are used in a linear regression analysis, the source strength best explaining the observations may still be off by a factor of two.
- Another approach complementary to the previous but with uncertainties of up to an order of magnitude is described for the first time in this paper. It assesses the demand of Tc-99m and derives the related Xe-133 releases from plausible production scenarios. It has the advantage to provide an independent approach that may help to evaluate and refine the complementary and more accurate methods. This may partially be the only option in case no sufficiently definite information is available for one or more production facilities. This may in particular be relevant for future scenarios or isotope productions with a lack of detailed or updated production information. However, the information whether Tc-99m is generated by fission or by another nuclear reaction is a minimum information requirement for this complementary approach to be applicable.

In order to compare results based on these different approaches, Table 1 provides an overview over various assessments of the major contributors to the global Xe-133 emission inventory.

All nuclear power plants are estimated to have a total annual release of 0.74 PBq/year Kalinowski and Tuma (2009). This assessment is based on quarterly and annual emission reports for 111 reactor sites and for the remaining 85 sites on scaling reactor-specific generic values calculated from the data available for the 111 sites. The regression analysis based on atmospheric transport simulations of multiple atmospheric observations by Wotawa *et al.* (2010) found a total annual release of 1.6 PBq, i.e., more than 100 % larger than estimated based on emission reports.

Hoffmann *et al.* (2009) investigated the isotope production facility with the highest Xe-133 emissions (Chalk River Laboratories, CRL). They apply ATM based on a few episodes of elevated concentrations at

Table 1

Assessments of the major contributors to the global Xe-133 emission inventory

Emittent (site and operator)	Assessment methodology	Annual emission	Daily emission, average—maximum[a]	Days per year	Period of available data	References
All nuclear power plants (NPPs), Earth	Atmosph. observations and ATM		Few GBq per reactor	365	25 Oct–16 Nov 2004 and Jan 2006	Achim *et al.* (2007)
	Reports by the operator	0.75 PBq	2.1 TBq 4.8 GBq/ reactor		1995–2005	Kalinowski and Tuma (2009)
	Atmosph. observations and ATM	1.6 PBq	*4.4 TBq*		July 2007–June 2008	Wotawa *et al.* (2009)
Fleurus, Belgium	Atmosph. observations and ATM		Few TBq–few tens TBq	220	25 Oct–16 Nov 2004 and Jan 2006	Achim *et al.* 2007
Institut des Radioéléments (IRE)	Production processes simulation	1 PBq	4.6 TBq		2005	Saey (2009)
	Atmosph. observations and ATM	2 PBq	*9.1 TBq*		July 2007–June 2008	Wotawa *et al.* (2009)
Chalk River, Canada	Atmosph. observations and ATM		Few tens TBq	315	25 Oct–16 Nov 2004 and Jan 2006	Achim *et al.* (2007)
Chalk River Laboratories (NRU)	Reports by the operator	*5.5 PBq*	17.3 ± 2.9–52 TBq		Aug–Nov 2008	Hoffman *et al.* (2009)
	Production processes simulation	6 PBq	16 TBq		2005	Saey (2009)
	Atmosph. observations and ATM	10–15 PBq	*35–50 TBq*		July 2007–June 2008	Wotawa *et al.* (2009)
Petten, Netherlands Nuclear Research and Consultancy Group (NRG)	Production processes simulation	0.7 TBq	*2.5* GBq	290	2005	Saey (2009)
Pelindaba, South Africa Nuclear Technology Products Radioisotopes Pty. (NTP)	Production processes simulation	4.1 PBq	13 TBq	315	2005	Saey (2009)
Lucas Heights, Australia Australian Nuclear Science and Technology Organisation (ANSTO)	Gamma measurement in the facility	*0.15 PBq*	0.49–5.3 TBq	300	Nov 2008–Feb 2009	Saey *et al.* (2010a)
Global Xe-133 emission inventory	Best estimate	18 ± 6 PBq				

[a] Numbers in italics are derived based on operational days per year

several sites in North America and Europe for the period between August and November 2008. They determined a wide spread of source term estimations ranging from 5 to 27,000 TBq/day and compared these with daily stack emissions as reported by the operator (17.3 ± 2.9 TBq/day). For a receptor on the same continent, the median source term estimate is in the same order of magnitude as the reported one.

However, for receptors in Europe the estimations are an order of magnitude higher, sometimes even more. This can be expected in view of the source term variations, interference with other strong sources in the region of the receptor and uncertainties in meteorological data and due to model limitations.

More accurate is the multiple linear regression analysis carried out by Wotawa *et al.* (2010) with the goal of optimizing the explanation of the global distribution as observed in time-series at 12 receptor locations. It assumes annual emissions of 3 PBq for each of the major three isotope production facilities in addition to the release from nuclear power plants. Confidence in the model was provided by the fact that the global distribution of sources is well understood. The study simulates the resulting atmospheric concentrations with ATM. For the isotope production facility at Fleurus (IRE) Wotawa *et al.* (2010) found 2 PBq/year and for the one at Chalk River either 9.6 or 16.2 PBq/year, depending on the selected methodology (95 % percentile values of the twelve stations vs. the median values). For IRE, this is 100 % higher than estimated by Saey (2009) with a production process simulation, though the studies relate to different years. For CRL, this is almost two to three times higher than the release figures reported by the operator, while the results of Saey (2009) compare well within 10 % with the reports (see Table 1).

The most comprehensive article on isotope production emissions so far, Saey (2009) estimates that the four largest medical isotope producers in the world together release about 11 PBq Xe-133/year. The best estimate for the global emission inventory of Xe-133 using all reported assessments and process simulations of the major contributors summarized in Table 1 is 12.5 PBq/year. When using the values derived by ATM based on observations and whenever this is not available using reported, respectively simulated emissions the best estimate is 21.0 PBq/year. In summary, by comparing the reported or simulated process releases with those found by ATM based on observational data the latter appear to indicate that the global emission inventory based on known sources may be significantly underestimated by about a factor of two.

The best estimate for the global emission inventory that treats reported, respectively simulated assessments on par with those using ATM based on observations would be 18 $\pm$ 6 PBq/year. The next sections of this paper will describe the new approach for assessing the global emission inventory of Xe-133 based on estimating the demand for Tc-99m and the results will be compared to the estimates of other methods as summarized above.

3. Medical Isotope Consumption

3.1. Role of Technetium-99m

Radioisotopes are used mostly for diagnostic examinations, but can also be used for therapy. Technetium-99m (Tc-99m) is the most widely used radioisotope in medicine [80 % according to IAEA (2010)] and it is produced when molybdenum-99 decays. The half-time of Tc-99m is 6 h (Brühl, 2006, p. 3).

Hospitals buy the Tc-99m required for examinations in the form of Tc-99m generators. The Mo-99 decays into the daughter nuclide Tc-99m with a half-life of 66 h, which then can be eluted, meaning that the Tc-99m produced is washed out of the generator for use. The Mo-99 remains bound to the column. The generator is also called a "technetium cow", as it can be "milked" several times as needed. The generators are acquired once per week and used for 1 week.

3.2. Worldwide Consumption of Technetium-99m

The goal is to estimate the amount of Tc-99m used worldwide. Each country is considered according to its health care level as defined by UNSCEAR (2000, p. 297). The four levels are defined by one physician for less than 1000 citizens (care level I), per 1,000–3,000 citizens (care level II), per 3,000–10,000 citizens (care level III), for more than 10,000 citizens (care level IV). The demand for Tc-99m is assessed for Germany, Sweden and the US and then used to extrapolate to the world consumption. Table 2 presents a comparison of these countries regarding their health care situation. For further refinement of this approach, specific data have to be acquired from more countries. Since the inhabitants of health care

Table 2

Factors for comparing health care in different countries and the fraction of the population getting examined using radionuclides (DESTATIS, 2007, pp. 698–700; BfS, 2006; SSI, 2008; Kahn and von Hippel, 2007)

Country	Inhabitants per doctor	Inhabitants per hospital bed	Population getting examination with radionuclides [%]
Germany	264	157	4.4
Sweden	328	341	1.2
United States	332	307	3.9
Mean			3.2

level I countries receive 90 % of all nuclear examinations worldwide (UNSCEAR 2008), the analysis of the available data should yield a representative result.

In Germany on average 3.6 million procedures using radionuclides were made per year in the years 1996–2004. On average 44 examinations were made per 1,000 inhabitants, which means that approximately 4.4 % of the population were examined per year (BfS, 2006, p. 242). The most commonly performed procedures using radionuclides, which together accumulate to over 90 %, are scintigraphies (Stamm-Meyer *et al.*, 2006, p. 5).

In Sweden the most commonly used radioisotopes are Tc-99m, I-131, In-111, I-123 and Cr-51 in that order, with Tc-99m being the most commonly used (Starck, 2008; SSI, 2008). In total 106,360 examinations using radionuclides were processed in the year 2007 and out of these 85,156 were performed using Tc-99m. The variation of examinations made with Tc-99m between the years 1999–2007 are 84,000–88,000 and no trend in the number of examinations is observed SSI, (2008). The population of Sweden was 9.2 million in the year 2007 (SCB, 2008). Hence, the percentage of the population that gets examined using radionuclides overall are 1.2 % and the number of the population getting examinations with Tc-99m is 0.95 %. The number of applications with Tc-99 m out of all radionuclides is exactly 80 % for Sweden, confirming what is often claimed in the literature (von Hippel and Kahn, 2006, p. 152; UNSCEAR, 2000, p. 316).

The United States is responsible for half of the global Mo-99 demand, around 12 million procedures per year (von Hippel and Kahn, 2006, p. 157; Kahn, 2008). The American population was around 304 million in 2007, with a percentage of 3.9 % of the population getting examined per year.

4. Production Procedure

4.1. Irradiation of Targets

There are two principal ways to produce Mo-99, which rely on the neutron irradiation of either Mo-98 or of U-235. The former method has a relatively low efficiency and is responsible for less than 5 % of the worldwide production (Lambrecht *et al.* 1999; IAEA 2010). The rest is produced by irradiation of uranium targets in a nuclear reactor. This procedure will be described in the following.

As target material either HEU or LEU is possible, although the production efficiency decreases if lower uranium enrichments are used. In 2008 95 % of the world's Mo-99 supply was produced by HEU targets. Efforts are underway to determine the extent of the fraction produced by LEU IAEA (2010) which could lead to a transition away from HEU WOSMIP II (2011).

The targets are irradiated with a thermal neutron flux of about 10^{13}–5×10^{14} n/(cm^{2}s) for 2–12 days Saey *et al.*, (2010b). The exact maximum irradiation time is determined by the by-production of unwanted nuclides and the state when Mo-99 formation rate equals the loss rate due to decay. After irradiation took place, the targets are cooled in water for at least several hours in order to reduce the level of heat and radiation generated by short-lived radioisotopes. The cooled targets are then brought to a processing facility. The total time between end of irradiation and radioisotope production is typically in the range of 1–2 days.

4.2. Extraction of Radioisotopes and Release of Noble Gases

The chemical extraction process for Mo-99 depends on the target material NRC, (2009): If uranium/aluminium alloy is present, sodium hydroxide (NaOH) is used to dissolve the entire target including the cladding. Further steps in this alkaline dissolution process lead to high purified Mo-99 with a recovery yield in the range of 85–95 %. If the target consists of uranium metal or oxide, an acid dissolution process is chosen. Here, the target and the cladding are first physically separated before the uranium is dissolved in nitric acid. The recovery yield and purification level are similar to the alkaline dissolution process.

Noble gases like krypton and xenon are released during the extraction process and are either recovered for other purposes (frozen out with liquid nitrogen, separated during warming and then absorbed on molecular sieves) or they are treated as waste. In the latter case, the gases first pass charcoal delay lines in order to reduce their activity before they are released to the atmosphere. The recovery or release of I-131 and its impact on the xenon isotopic ratios is not discussed in this paper since it focuses on the assessment of Xe-133 releases.

5. Deriving Xe Emission From Tc-99m Consumption

5.1. Xe-Mo-Ratios

During irradiation of the targets in the reactor, the induced fission of HEU yields, among others, Mo and Xe isotopes depending on the flux and energy of the incoming neutrons. Each isotope accumulates until irradiation ceases. Afterwards, during the storage of the target, the decay of predecessor isotopes continues to produce Xe and Mo. Upon separation from the target, Xe and Mo decay according to their decay constants without further in-growth from the predecessors in the decay chain. The irradiation and decay were simulated using a simple MATLAB model described in Hebel (2010) and verified against the results of Saey (2009). The required yields and cross sections are taken from England and Rider (1994) and ENSDF (2007).

Based on the supposition of a 93 % enriched uranium target and a thermal neutron flux of 10^{14} n/(cm^{2}s), the development of Xe-Mo-ratios over time has been calculated for irradiation periods of 2, 5 and 10 days, as shown in Fig. 1. Of interest is the ratio of each respective xenon isotope to molybdenum-99 at the time of separation, which is assumed here to take place 48 h after ceasing irradiation.

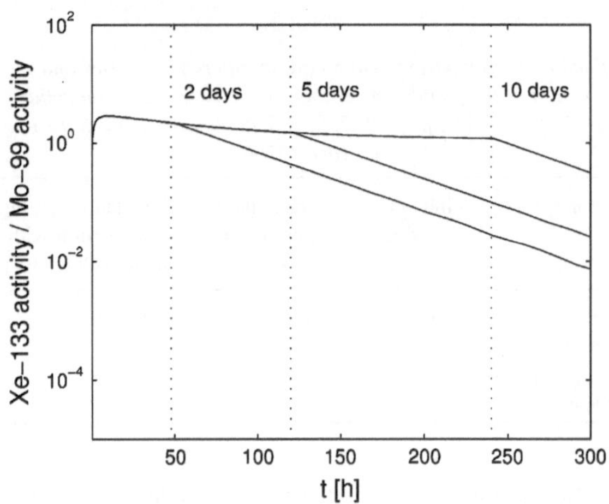

Figure 1
The activity ratio of Xe-133 and Mo-99 during and after various irradiation durations, before separation

The simulation is validated by comparison with similar calculations provided by Saey (2009) with HEU as target material after the same irradiation time and assuming the noble gases are released directly upon separation. For this purpose, the radioxenon isotope ratios of the emitted xenon batch, are plotted in Fig. 2 together with the ratios of a light-water reactor burn up and a nuclear test. The dashed line marks an arbitrary threshold to distinguish between the potential origins of the isotopes: explosion or nuclear power reactor.

The validity of the results shown in Fig. 2 is supported by comparison Saey (2009). The ratio of Xe-135/Xe-133 presented there is slightly higher than the one calculated here. The model used here ignores neutron capture of Xe-135, but the resulting small uncertainties are negligible: after a short cool down period, the model results match those of Saey (2009).

The irradiation time, as well as possible delay lines at the separation facility, have the largest influence on the isotopic ratios, whereas neutron flux, enrichment and cooling time are only of minor importance (also see Saey *et al.*, 2010b). Notably, the three examined scenarios display isotope ratios to the right of the separation line, well within the nuclear test range. Longer irradiation periods lead to an isotope ratio within the reactor domain defined by Kalinowski and Pistner (2006). However, longer irradiation times negatively affect the isotope production efficiency.

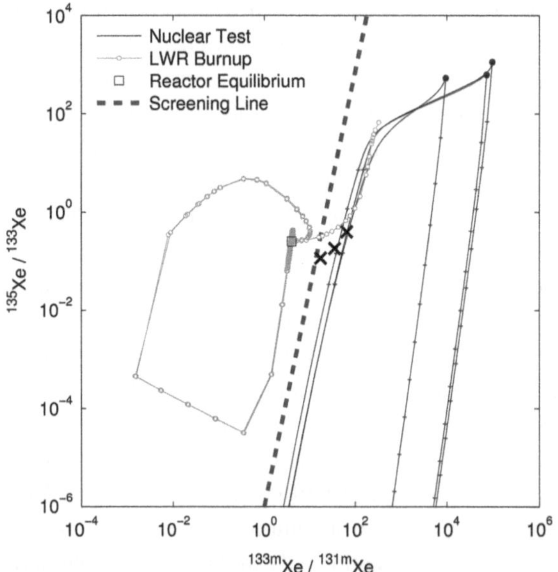

Figure 2
Isotopic ratios for xenon released after an irradiation period of 2 (*upper right black X*), 5 and 10 (*lower left black X*) days and 2 days of cool-down

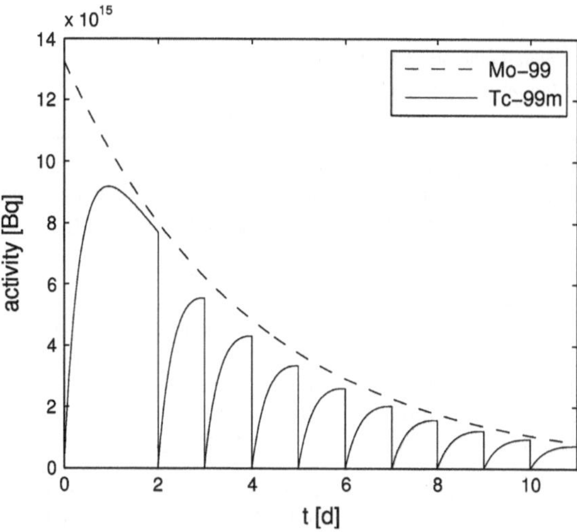

Figure 3
Consumption scenario in which Mo-99 is transported for 1 day and then extracted daily for 10 days

5.2. *Transport and Consumption*

After target removal, the molybdenum is separated and transported to the customer hospital within 2–6 days, which we assume based on oral communication with experts from the producers' side. During transport, the Mo-99 constantly decays into Tc-99m, which has a much shorter half-life resulting in a transient equilibrium.

In the hospital, the Tc-99m is removed before each treatment. Before removal, Tc-99m will almost reach transient equilibrium again. The development of the Mo-99 and Tc-99m amounts during transport and periodic removal is shown in Fig. 3.

The sum of Tc-99m removed for treatment is commonly reported by each hospital and can be used to determine how much Mo-99 must have been shipped by the supplier to meet the according demand. As described above, GROSCH (2008) has compiled information on the global Tc-99m consumption in hospitals. In the estimate below, the mean value taken from Table 2 is that 3.2 % of the population in health care level I countries annually get an examination using radionuclides and 80 % of these are done with Tc-99m. With the present data base, no variance can be given, but an indication for the uncertainty margin is provided by the low percentage in Sweden (1.2 %) and the high value for Germany (4.4 %).

These assumptions are extrapolated to the whole world population of six billion using the relative number of examinations between the different health care levels reported in UNSCEAR (2000) report. The intermediate data and the final result are presented in Table 3. For each region the number of people getting a Tc-99m examination is the product of the population base times the percentage of radio-medical treatment (3.2 % for level I countries and scaled for the other levels as described) times the fraction of Tc-99m (80 %). From this, the total activity is calculated for each region by multiplying the number of treated patients with the average administered amount of 0.6 GBq of Tc-99m per treatment.

The estimation presented in Table 3 results in an annual worldwide Tc-99m usage of about 29 PBq. In order to determine the uncertainty range, the Tc-99m usages is calculated with extreme assumptions using the highest and lowest percentage of the three countries. A 1.2 % assumption results in about 11 PBq and assuming 4.4 % leads to 40 PBq.

Table 3

Estimated amount of Tc-99m used worldwide, assuming that 3.2 % of the population in Health care level I get examined or treated with radionuclides

Health care levels[a]	Fraction of world population	Usage of Tc-99m [‰]	Number of people getting Tc-99m examination	Total activity administered [TBq]
I	0.26	25.6[b]	39,900,000	24,000[c]
II	0.53	3.1[d]	7,850,000	4,700
III	0.11	0.18	118,000	71
IV	0.10	0.009	5,400	3
Total (rounded)			47,900,000	28,800

[a] The health care levels are defined by UNSCEAR (2000, p. 297) as one physician for less than 1,000 citizens (care level I), per 1,000–3,000 citizens (care level II), per 3,000–10,000 citizens (care level III), for more than 10,000 citizens (care level IV)

[b] Assuming 3.2 % of the population get a radioisotope dependant procedure, 80 % of which utilise Tc-99m (UNSCEAR, 2000, p. 316)

[c] Assuming a total world population of six billion and an average administered amount of 0.6 GBq of Tc-99m

[d] Assuming that the same relationship exists between usage of Tc-99m in the different health care levels as the relationship for the different health care levels as for the number of examinations performed 1991–1996

5.3. Xenon Release and Comparison with Other Approaches

With an increasing world population and technical developments one might expect a growth in Tc-99m usage. Numbers from 1991 to 1996 of UNSCEAR (2008) estimated that 16 PBq was used annually, which would imply an increase and is within our uncertainties.

From the Mo-99 production, the according xenon emissions can be calculated using the xenon-molybdenum ratios established above. Table 4 shows the xenon release for three production scenarios that yield 29 PBq of Tc-99m, assuming 2 days storage of the targets after irradiation and 6 days between separation and first recovery.

Assuming optimal transport conditions and an arrival and recovery of the Mo-99 at the medical facility within 2 days after separation at the production site, only 13 PBq of Mo-99 would have to be produced (see Fig. 3) and the amount of emitted xenon would be considerably smaller (Table 5). The lower value is more likely as producers will try to minimise irradiation times by using highly enriched uranium. The enrichment's impact on isotope ratios is minimal.

The amount of required Mo-99 strongly depends on the cool down and transport times, as half of the produced Tc-99m is lost every 66 h.

Assuming extreme values for the annual Tc-99m usage calculated above and extreme assumptions for the transportation and irradiation time, we are able to estimate the range of uncertainty. Two days transportation time, 2 days irradiation time and an annual Tc-99m usage of 11 PBq produces 4 PBq of Xe-133 emissions. On the other side, 6 days transport time, 10 days radiation time and an annual Tc-99m usage of 40 PBq leads to 61 PBq Xe-133 emissions. Even with these extreme assumptions, the uncertainty of the Xe-133 emission is only in the range of one order of magnitude. A more precise and realistic result is provided by the scenarios described above.

In conclusion, this paper finds a total annual Xe-133 emission in the range of 11–45 PBq/year and more likely in the lower range. Adding the known emissions from nuclear power plants yields about 12–46 PBq/year for the total global Xe-133 emission inventory.

These values can be compared with those values found by previous publications as summarized in Sect. 3. The results of the new approach described here compare fairly well with the best estimate for the global emission inventory 18 $\pm$ 6 PBq/year. The slight overestimation of the results derived in this work might be explained by actual operations achieving a shorter delay between separation and clinical use. The results of the approach based on estimating the global Tc-99m demand are in close agreement with the results derived by ATM based on observations (21.0 PBq/year). The estimate using all reported assessments and process simulations of the major contributors (12.5 PBq/year) lies at the lower end of the range found in this paper, even though it can be considered the most accurate one. This seems to back the finding of the ATM approach based on observations that the known sources are underestimated or more likely that not all existing sources are accounted for. This finding highlights the benefit of using the independent approach for estimating the global Xe-133 inventory presented in this paper.

Table 4

Global annual xenon emission for different scenarios of isotope production with varying irradiation times a delay of 6 days before first clinical use

Rad. time [d]	Xe-131m [PBq]	Xe-133 [PBq]	Xe-133m [PBq]	Xe-135 [PBq]
2	0.02	29.83	1.45	11.72
5	0.04	37.13	1.49	6.68
10	0.08	45.57	1.45	5.20

Table 5

Global annual xenon emission for different scenarios of isotope production with varying irradiation times and a delay of 2 days before clinical use

Rad. time [d]	Xe-131m [PBq]	Xe-133 [PBq]	Xe-133m [PBq]	Xe-135 [PBq]
2	0.008	10.90	0.53	4.28
5	0.015	13.56	0.54	2.44
10	0.030	16.66	0.53	1.90

6. *Summary*

This work has shown that a rough but realistic estimate of xenon emissions from isotope production can be made independently from producer declarations and process simulations as well as independent from atmospheric transport simulations based on ambient air observations.

The new approach presented in this paper is based on declarations of the number of radioisotope doses administered in hospitals for estimating the global demand for Tc-99m and from this deriving the related Xe-133 emission inventory. This estimate does not vary greatly with production parameters like neutron flux, uranium target enrichment or overall production, but mainly depends on less flexible parameters like irradiation time, cool down and transport time. There is a strong financial incentive to keep these times short; therefore, it is unlikely that they will be significantly altered from the scenarios assumed in this paper.

The hospitals have no particular interest in keeping the number of Tc-99m applications confidential, and a more direct and focused enquiry may significantly improve on the rough estimates presented here. Monitoring Tc-99m consumption is, therefore, a viable method of validating xenon emission estimations and exploring otherwise unknown emission rates.

Acknowledgments

This work was carried out at the Carl Friedrich von Weizsäcker Center for Science and Peace Research, University of Hamburg, and revised after M.K. became staff member with the Provisional Technical Secretariat of the Preparatory Commission for the Comprehensive Nuclear-Test-Bean Treaty Organisation (CTBTO). This work was sponsored by the German Foundation for Peace Research (DSF).

References

ACHIM, P.; BELLIVIER, A.; ARMAND, P.; TAFFARY, T.; FONTAINE, J.P. & PIWOWARCZYK, J.C. (2007), Contributions of xenon releases in the atmosphere from radionuclide production facilities and nuclear nuclear power plants to the detection of 133Xe by SPALAX systems in western Europe. In: Proceedings of the 11th International Conference on Harmonisation Within Atmospheric Dispersion Modelling for Regulatory Purposes, Cambridge, UK, July 2-5, 2007, pp. 180-184.

BONET, H.; DAVID, B. & PONSARD, B. (2005), Production of Mo-99 in Europe: status and perspectives, in 'International Topical Meeting on Research Reactor Fuel Management'.

BRÜHL, K. & COMBA (supervisor), P. (2006), 'Technetium in der Radiodiagnostik', seminar presentation held at the University Clinic Eppendorf (Hamburg, Germany) during the Anorganic Chemistry advanced practical training.

CTBT/WGB/TL-2/40 (2000). Recommended Standard List of Relevant Radionuclides for IDC Event Screening. Report of Informal Workshop, Melbourne, Australia. 17–21 January 2000.

DESTATIS (2007), Statistical Yearbook for the Federal Republic of Germany, Federal Statistical Office, Wiesbaden.

ENSDF (2007), Evaluated Nuclear Structure Data File.

ENGLAND, T.R., and RIDER, B.F. (1994), Evaluation and Compilation of Fission Product Yields. Tech. Rep. ENDF-349, Los Alamos National Laboratory, http://t2.lanl.gov/publications/yields. LA-UR-94-3106.

German Federal Office for Radiation Protection (BfS) (2006), Strahlenexposition durch medizinische Maßnahmen (Radiation exposures from medical applications), Bundesministerium für Umwelt, Naturschutz und Reaktorsicherheit (BMU), chapter 4, pp. 239-248.

GROSCH, M. (2008), 'Complications of the Medical Radioisotope Production for the Non-proliferation Regime', Master's thesis, Institute for Peace Research and Security Policy at the University of Hamburg (Germany).

HEBEL, S. (2010), Genesis and Equilibrium of Natural Lithospheric Radioxenon and its Influence on Subsurface Noble Gas Samples for CTBT On-site Inspections, Pure and Applied Geophysics,

From the issue entitled "Special Issue: Recent Advances in Nuclear Explosion Monitoring", Volume *167*, Numbers 4-5, 463-470, DOI: 10.1007/s00024-009-0037-9.

Heimbigner, T.R.; McIntyre, J.I.; Bowyer, T.W.; Hayes. J.C.; Panisco, M.E. (2002), Environmental monitoring of radioxenon in support of the radionuclide measurement system of the international monitoring system. 24th Seismic Research Review – Nuclear Explosion Monitoring: Innovation and Integration. NNSA, Ponte Cedra Beach, Florida.

Hoffman, I.; Ungar, K.; Bean, M.; Yi, J.; Servranckx, R.; Zaganescu, C.; Ek, N.; Blanchard, X.; Petit, G.L.; Brachet, G.; Achim, P. & Taffary, T. (2009), *Changes in radioxenon observations in Canada and Europe during medical isotope production facility shut downs in 2008*. Journal of Radioanalytical and Nuclear Chemistry *282* (3), 767-772.

IAEA (2010), Nuclear Technology Review 2010. International Atomic Energy Agency, Vienna, http://www.iaea.org/Publications/Reports/ntr2010.pdf.

Kahn, L. H. (2008), 'The potential dangers in medical isotope production', Bulletin of the Atomic Scientists, Online edition, http://www.thebulletin.org/web-edition/columnists/laura-h-kahn/thepotential-dangers-medical-isotope-production.

Kahn, L. H. & von Hippel, F. (2007), *'How the radiologic and nuclear medical communities can improve nuclear security'*, Journal of the American College of Radiology *4*, 248–251.

Kalinowski, M. B. & Pistner, C. (2006), *'Isotopic signature of atmospheric xenon released from light water reactors'*, Journal of Environmental Radioactivity *88*, 215–235.

Kalinowski, M. & Tuma, M. (2009), *Global radioxenon emission inventory based on nuclear power reactor reports*, Journal of Environmental Radioactivity, Volume *100*, Issue 1, 58-70, ISSN 0265-931X, 10.1016/j.jenvrad.2008.10.015.

Lambrecht, R.M.; Sekine, T.; Ruiz, H.V. (1999), The Accelerator Production of Molybdenum-99, IAEA-TECDOC, 1065, pp. 75–85.

NRC (2009), Medical Isotope Production Without Highly Enriched Uranium, Committee on Medical Isotope Production Without Highly Enriched Uranium, National Research Council.

Reistad, O.; Bremer Maerli, M. & Hustveit, S. (2007), 'Toward elimination of HEU as a reactor fuel', Draft Version.

SCB (2008), 'website of the Statistiska centralbyrån (Statistics Sweden)', www.scb.se.

SSI (2008), 'website of the Statens Strålskydds Institut (Swedish Radiation Safety Authority)', http://www.ssi.se.

Saey, P. (2009), *The influence of radiopharmaceutical isotope production on the global radioxenon background*, Journal of Environmental Radioactivity *100*, 396 – 406.

Saey, P.; Auer, M.; Becker, A., Hoffmann, E.; Nikkinen, M.; Ringbom, A.; Tinker, R.; Schlosser, C. & Sonck, M. (2010a): *The influence on the radioxenon background during the temporary suspension of operations of three major medical isotope production facilities in the Northern Hemisphere and during the start-up of another facility in the Southern Hemisphere*, Journal of Environmental Radioactivity *101* (2010) 730-738.

Saey, P.; Bowyer, T.W.; Ringbom, A. (2010b): Isotopic noble gas signitures from medical isotope production facilities – Simulations and measurements, Applied Radiation and Isotopes 68, 1846-1854.

Stamm-Meyer, A.; Noßke, D.; Schnell-Inderst, P.; Hacker, M.; Hahn, K. & Brix, G. (2006), *'Diagnostic nuclear medicine procedures in germany between 1996 and 2002 – Application frequencies and collective effective doses'*, Nuklearmedizin *45*, 1–9.

Starck (2008), 'personal communication', Swedish Society of Nuclear Medicine.

UNSCEAR (2000), 'Sources and Effects of Ionizing Radiation', Report to the General Assembly with Scientific Annexes, United Nations Scientific Committee on the Effects of Atomic Radiation.

UNSCEAR (2008), 'Sources and Effects of Ionizing Radiation', Report to the General Assembly with Scientific Annexes, United Nations Scientific Committee on the Effects of Atomic Radiation.

Vandegrift, G. F. (2005), Facts and myths converning Mo-99 production with HEU and LEU targets, in 'Proceedings of the International Conference on Reduced Enrichment for Research and Test Reactors'.

von Hippel, F.; Kahn, L.H. (2006), 'Feasibility of Eliminating the Use of Highly Enriched Uranium in the Production of Medical Radioisotopes', Science and Global Security *14*, 151–162.

WOSMIP II (2011), 'Summary report', Workshop on Signatures of Medical and Industrial Isotope Production.

Wotawa, Gerhard; Becker, Andreas; Kalinowski, Martin; Saey, Paul; Tuma, Matthias; Zähringer, Matthias (2010), Computation and Analysis of the Global Distribution of the Radioxenon Isotope Xe-133 based on Emissions from Nuclear Power Plants and Radioisotope Production Facilities and its Relevance for the Verification of the Nuclear-Test-Ban Treaty, Pure and Applied Geophysics, Volume *167*, Issue 4-5, pp. 541-557.

(Received December 2, 2011, revised May 27, 2013, accepted May 29, 2013, Published online June 27, 2013)

Pure Appl. Geophys. 171 (2014), 717–734

DOI 10.1007/s00024-012-0563-8

Pure and Applied Geophysics

Detection of Noble Gas Radionuclides from an Underground Nuclear Explosion During a CTBT On-Site Inspection

CHARLES R. CARRIGAN[1] and YUNWEI SUN[1]

Abstract—The development of a technically sound approach to detecting the subsurface release of noble gas radionuclides is a critical component of the on-site inspection (OSI) protocol under the Comprehensive Nuclear Test Ban Treaty. In this context, we are investigating a variety of technical challenges that have a significant bearing on policy development and technical guidance regarding the detection of noble gases and the creation of a technically justifiable OSI concept of operation. The work focuses on optimizing the ability to capture radioactive noble gases subject to the constraints of possible OSI scenarios. This focus results from recognizing the difficulty of detecting gas releases in geologic environments—a lesson we learned previously from the non-proliferation experiment (NPE). Most of our evaluations of a sampling or transport issue necessarily involve computer simulations. This is partly due to the lack of OSI-relevant field data, such as that provided by the NPE, and partly a result of the ability of computer-based models to test a range of geologic and atmospheric scenarios far beyond what could ever be studied by field experiments, making this approach very highly cost effective. We review some highlights of the transport and sampling issues we have investigated and complete the discussion of these issues with a description of a preliminary design for subsurface sampling that addresses some of the sampling challenges discussed here.

Key words: Noble gas, soil gas transport, CTBT, underground nuclear explosion, on-site inspection, soil gas sampling, radionuclide background.

1. Introduction

Many characteristics of an underground nuclear explosion (UNE) are not definitive of its nuclear nature. However, the detection of noble-gas (NG) radionuclides such as Xe-131m, Xe-133[1] and Ar-37 (PURTSCHERT *et al.*, 2007; SAEY, 2007) significantly above any background levels beneath the surface of a suspect site during an on-site inspection (OSI) is generally considered to be an extremely strong indicator of the recent occurrence of an underground nuclear explosion (UNE). A possible approach to detection of these noble gases might involve regularly collecting, during a period of days to weeks or even months, many cubic-meter-sized soil gas samples from multiple sampling stations distributed about the surface of a likely underground detonation site. The gas samples may be compressed for transport in high-pressure gas bottles and returned to the base of operations (BOO) for pre-processing and analysis. At the BOO, Ar and Xe components are separated from the samples and counted using systems such as the Swedish Automatic Unit for Noble Gas Acquisition (SAUNA) or the Russian Analyzer of Xenon Radioisotopes (ARIX) to detect explosion-produced radioxenons and the Chinese Movable Argon-37 Rapid Detection System (MARDS) for detecting Ar-37 produced by interaction of fast neutrons and Ca-40 atoms in the soil surrounding the nuclear test (see SAEY (2007) for additional discussion of the systems).

The views expressed here do not necessarily reflect the opinion of the United States Government, the United States Department of Energy, or Lawrence Livermore National Laboratory.

[1] Lawrence Livermore National Laboratory, 7000 East Avenue, Livermore, CA 94550, USA. E-mail: carrigan1@LLNL.gov

[1] DE GEER (2001) and SAEY (2007) also identify Xe-133m and Xe-135 as CTBT-relevant isotopes that are not considered here either because their half lives are so short or their production is so small that the possibility of their being detected during an on-site inspection is significantly reduced although they could conceivably be detected by the International Monitoring System (IMS) immediately following a prompt release of gases from an underground nuclear test. If the xenon isotopes mentioned here are in fact present in atmospheric samples, KALINOWSKI *et al.*, (2010) make the highly compelling argument that ratios of these isotopes, under certain conditions, can be used to distinguish between atmospheric releases produced by nuclear tests and civilian nuclear activities which could prove to be very important for avoiding false alarms in the IMS regime.

The current version of the draft OSI Operational Manual (CTBT/WGB/TL-18/47, June 2012) contains guidelines for subsurface and atmospheric noble gas sampling that are generally consistent with the discussion presented here. However, at the level of technical detail that it is intended to address, the manual cannot be expected to include significant discussions about physical processes involved in noble gas transport and sampling that may have important implications for the detectability of a subsurface noble gas release from a UNE. In the material that follows, we attempt to consider, either for the first time or in more detail than previously, some issues of transport and sampling of noble gases that are likely to have an impact on the detectability of a UNE. They include (1) updated temporal detection "windows" for xenon and argon isotopes of interest based on the NPE barometric pumping results; (2) capturing gases produced by fractures that are overlain or obscured by soil layers; (3) the possibility of using radon and other radionuclides for locating producing subsurface pathways and as a sampling diagnostic; (4) large volume (~ 1 m^3) sampling and the potential for atmospheric infiltration and contamination; (5) estimating the degree of soil gas "imprinting" of atmospheric noble gas contamination; (6) generalization of the analytical model of Riedmann and Purtschert (2011) for Ar-37 soil gas background and implications for soil gas sampling and (7) the use of a barometer-based protocol for sampling to enhance the potential for detecting trace gases brought to the surface by barometric pumping. In conclusion, we present a possible approach to sampling during an OSI that both addresses some of the issues (e.g., buried producing fractures) and uses to advantage physical processes (e.g., barometric pumping) that will enhance the probability of capturing gases from a UNE.

2. *Possible OSI Scenarios*

A challenge that confronts any attempt to develop an OSI radionuclide monitoring protocol is the wide range of scenarios that might be considered with many variables (e.g., geology, hydrology, site engineering, etc.) contributing to both the range of containment as well as the range of detection possibilities. Fortunately, the inherent uncertainty in the ability of earth materials to fully contain a detonation translates into a significant probability that detonation gases will not be fully contained within the earth after a length of time following a UNE. For example, prompt release of gases from the subsurface may indeed be prevented upon detonation, which means that an event is technically contained, however, the underground test site might still be the source of detectable gas leakage weeks to months later as a result of the pervasive barometric pumping mechanism (Nilson *et al.*, 1991; Carrigan *et al.*, 1996; Auer *et al.*, 1996).

Because of the wide range of possible containment and geologic scenarios, the discussion that follows is more likely to have heuristic than quantitative value. Furthermore, some of our assumptions, especially those involving mass and energy transport, are most appropriate for low yield events. Figure 1 illustrates a range of possible UNE scenarios from the perspective of the amount of gas transported to the surface. Following an underground detonation, a cavity is formed within hundreds of milliseconds. In addition to the pressurized cavity, detonation gases may also be injected into the surrounding geologic regime. As some of the detonation products such as water vapor cool and condense, pressure in the cavity may be dramatically reduced. This loss of pressure in the cavity may induce the roof of the cavity to collapse giving rise to stoping or the process of fracturing of overlying material which then falls into the cavity to form an upward propagating void or chimney. A depression or crater is formed if the chimney propagates to the surface. Even if the chimney does not reach the surface, new pathways for gas transport connecting the chimney and cavity to the surface may be created as the chimney intersects the natural and explosion-created fracture system of the overburden. Dubasov (2010) provides some insight into the containment of gases from Soviet underground tests at sites having two different geologies. About 47 % of the 340 tests at the Semipalatinsk Test Site were not fully contained with 43 % or 145 explosions involving at least a weak release of noble gases. Of the 39 tests at the Novaya Zemlya Test Site, 23 or about 58 % experienced some release of gases to the surface.

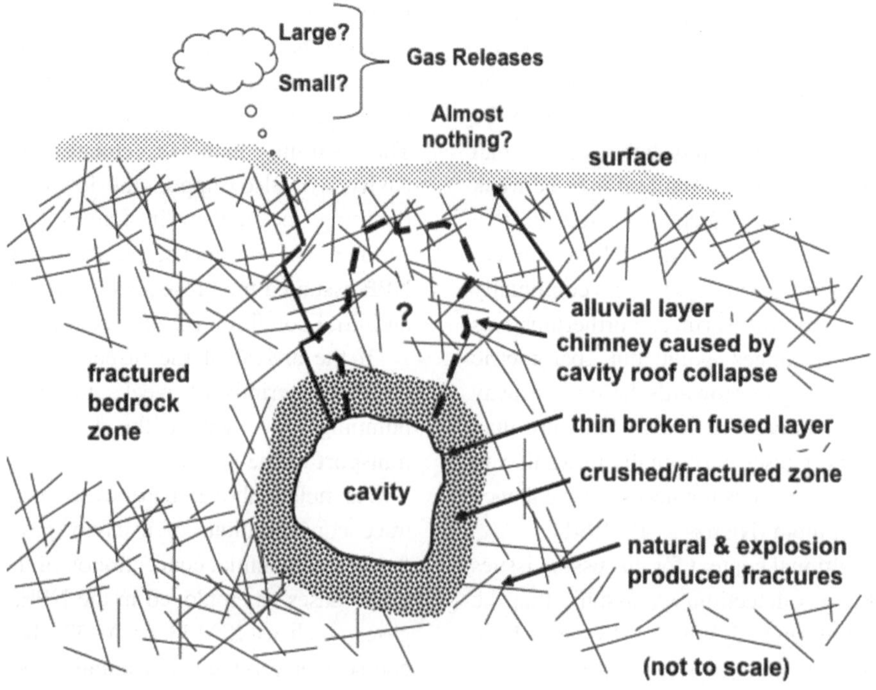

Figure 1
Cartoon showing relationship between detonation cavity, possible formation of chimney (*dashed line*) resulting from cavity-roof collapse and pathways (*solid line*) to the surface

Depending on how the timing for gas pathway formation coincides with the time scale for the presence of gas transport mechanisms, such as pressure drive from the cavity (early time), multiphase convection (intermediate time) or barometric pumping (late time), it is expected that gas transport to the surface may be manifested in different ways and to different degrees. An early time cavity-pressure driven leak following the detonation by only seconds to hours may produce over a period of hours a highly concentrated particulate and gas plume at the surface that is subject to atmospheric transport and subsequently detected at a distance of hundreds to thousands of kilometers by the International Monitoring System (IMS) (Bowyer *et al.*, 2002; Kalinowski *et al.*, 2008). At intermediate times, days to weeks, leaks driven by multiphase convection using cavity heat as a source of energy are expected to be highly effective at transporting cavity gases along fractures to the surface in concentrations that some models predict could be readily detectable on site (Sun and Carrigan, 2012). A longer term but weaker transport mechanism involves the effect of barometric pressure fluctuations on gas transport along the fracture network in the subsurface (Nilson *et al.*, 1991; Carrigan *et al.*, 1996; Auer *et al.*, 1996; Sun and Carrigan, 2012). The barometric pumping mechanism depends on both advection along a vertical fracture pathway as well as transverse diffusion of the gases into the porous walls of the fracture. During a period of low atmospheric pressure at the surface, the upward flowing gas species will also diffuse into the porous walls of a fracture. During the next period of high pressure at the surface, the direction of flow along the fracture pathway is reversed with gases flowing in the downward direction along the pathway back toward the cavity. However, all the gas advected upward during the previous period of low pressure does not participate in this reversed flow as some small fraction is retained or stored in the porous walls of the pathway. Then when flow in the fracture again changes to the upward direction, some fraction of the stored cavity gas in the matrix is transported back into the upward flow path and is carried yet further away from its source while also continuing to diffuse into the porous walls of the fracture. The net effect is for cavity gases to be gradually transported to the surface by this cyclic but not entirely reversible transport process.[2] Cavity gas transport by barometric pumping is far faster than chemical diffusion alone although the concentration of a gas reaching the surface may be several or more orders of magnitude below the original gas concentration in the cavity. While it may be weak compared to other advection mechanisms such as flow driven by a pressurized cavity or multiphase convection, barometric pumping is always present continually drawing gases towards the surface. Furthermore, as Nilson and his colleagues point out, barometric pumping always draws gases towards the surface away from a subsurface source while cavity pressure may just as easily transport gases horizontally into adjacent soil/rock regimes, which does not necessarily enhance the ability to detect them (Nilson *et al.*, 1991).

To provide additional context for discussing issues pertinent to noble gas detection, we assume that the International Monitoring System (IMS) has been triggered by seismicity and possibly wind-transported noble gases or particulates indicating a high likelihood that a UNE has occurred. We anticipate that an OSI might typically be initiated no earlier than about 10–14 days following a suspected UNE. By this time any cavity-pressure-induced seepage or venting of noble gas radionuclides may be minimal to nonexistent. While cavity pressure may be depleted within hours to a few days following a UNE, the intermediate time convective transport mechanism involving heat released by the explosion can set up subsurface convective circulations driving detonation gases along the fracture network towards the surface. Such circulations have not been previously explored in detail with application to transporting noble gases to the surface from an underground nuclear test. New simulations involving cavity heat released into partially water-saturated, fractured environments suggest that thermally driven multiphase circulations may temporarily be more effective than barometric pumping in moving tell-tale gases towards the surface (Sun and Carrigan, 2012). Thus, thermally driven circulations involving vaporization of pore and fracture moisture will tend to enhance the potential for detecting NG at the surface by producing a more or less continual flow of detonation gases into a fracture-filled subsurface regime. The resulting "halo" of detonation gases around the cavity will also support the longer-term barometric pumping mechanism in transporting gases to the surface. The barometric pumping simulations of the NPE assumed the presence of such a halo of gases as an initial condition (Carrigan *et al.*, 1996).

If the arrival of the inspection team follows the UNE by periods of weeks to months, barometric pumping may then be the dominant means of gas transport to the surface.

As noted above, barometric pumping transports trace concentrations of a cavity gas at a significant "cost" to the initial concentration of the inert chemical tracer gases as employed in the NPE. For radioactive gases, such as Xe-133 or Ar-37, the situation is, of course, worse as their concentrations additionally decay with time irrespective of transport. Simulations of Xe-133 and Ar-37 transport calibrated by the inert chemical tracer arrivals (i.e., sulfur hexafluoride and helium-3) following the 1993 NPE one-kiloton chemical explosion were used to suggest the detectability during an OSI of both gases at the surface by soil gas sampling methods. With the detection capabilities assumed at the time (Carrigan *et al.*, 1996: 1 Bq m^{-3}), both radionuclides were predicted to initially arrive at the surface with concentrations of almost one to two orders of magnitude above the assumed detection limit (Carrigan *et al.*, 1996). The model predicted that had the explosion been nuclear, creating both Ar-37 and Xe-133 in the explosion cavity, it was reasonable to expect the Xe-133 to arrive in concentrations at the surface above the minimum detectable concentration after about 50 days post-detonation, while the higher diffusivity Ar-37 might be expected to be detectable after 80 days. Both detections would fall within the 130-day OSI inspection period initiated days to weeks following the detonation.

Figure 2 illustrates the detection windows that follow from the published model for NPE barometric pumping. If the simulated arrival concentrations for Ar-37 and Xe-133 are also assumed to be the maximum concentrations of those gases at the surface, which is a conservative assumption, radioactive decay as determined by the half lives of the gases (e.g., Xe-133:

[2] In their excellent paper, Nilson *et al.*, (1991) provide a much more quantitative discussion of barometric transport than is presented here along with an exact analytical solution for the process assuming a sinusoidal pressure response in a fractured permeable regime.

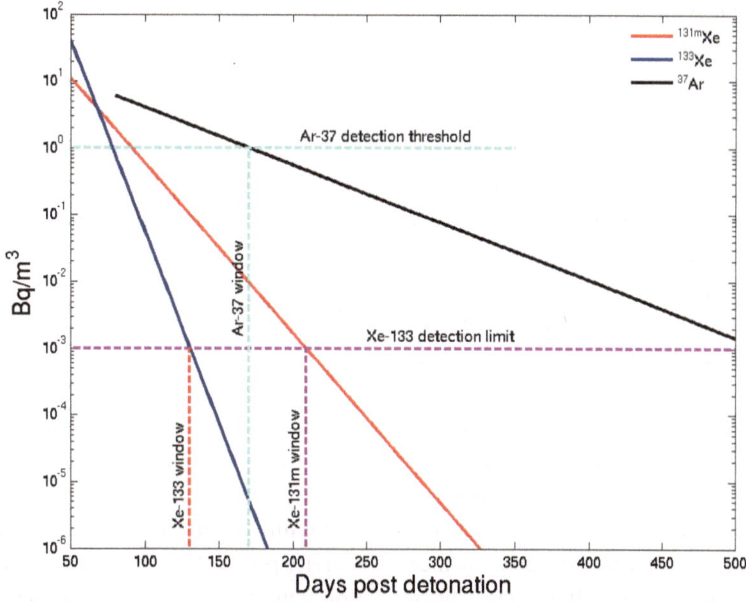

Figure 2

Detection windows for xenon and argon isotopes transported to surface by barometric pumping based on NPE tracer gas detections. The conservative detection threshold for Ar-37 is based upon estimates of soil gas background owing to natural production of Ar-37 by interaction of cosmic ray neutrons and calcium in shallow soil (see Fig. 11 with associated discussion). To obtain window widths, 80 days, corresponding to the time required to transport the gas to the surface in detectable amounts, must be subtracted from the time at which Ar-37 reaches the threshold level (~170 days post detonation) while 50 days must be subtracted from the radioxenon detection levels to obtain estimates of window widths

5.2 days, Ar-37: 34.8 days) governs the period or window of detection along with the minimum detectable concentration for each Xe isotope or the approximate limiting detection threshold for Ar-37 based on background soil gas concentrations. Using updated sensitivities for Xe-133 and a conservative detection threshold for Ar-37 that takes into account the existence of background soil gas concentrations, the Xe-133 window of detection is roughly 85 days in this calculation, while the Ar-37 window is roughly 90 days in width. Assuming an isotopic production of about 25 % of the Xe-133 (Saey, 2007), the window for Xe-131m was estimated to be 160 days wide. We emphasize that these estimates, while they are intended to be conservative in predicting Xe and Ar arrival concentrations and the widths of detection windows, are based on a deterministic model and are scenario and assumption dependent. The effect of considering different parameter combinations or scenarios is considered in more detail in a companion paper (Sun and Carrigan, 2012). The assumptions related to subsurface sampling during an OSI or capturing detonation gases and the potential for background levels of the gases of interest to be present is discussed in the following sections.

3. *Towards a Subsurface Noble Gas Sampling Concept of Operation*

To develop a concept for sampling gases in the subsurface, it seems reasonable to first decide what are the fundamental limitations of the sampling technique. Subsurface gas sampling is generally laborious. Holes must be augered and suitable packers with integrated sampling tubes installed in the holes. Alternatively, sampling tubes may be "pushed" into the soil with a hydraulic ram if the soil is accommodating and bentonite clay can be used in principle to seal the sampling hole instead of a packer. The site should be covered with a tarp or plastic sheeting to prevent or minimize infiltration, and other sampling quality control measures might be used (e.g., releasing a tracer such as sulfur hexafluoride, SF_6, beneath a tent or tarp to quantify infiltration levels as discussed later). We further assume that subsurface gas sampling is used primarily for the purpose of confirming an underground nuclear detonation, while atmospheric gas sampling may be used as a tool for both locating the site of a subsurface detonation as well as aiding in the confirmation that the detonation was nuclear in origin. We recognize that the subsurface technique involves a labor-intensive effort that may not be appropriate for use in a broad-area search scheme, especially where manpower limitations exist and/or restrictions on time in the field are in effect, as was assumed in the Comprehensive Test Ban Treaty Organization Noble Gas 2009 Field Test (Tanaka, 2011; Carrigan, 2010). Subsurface gas sampling should only be considered as a means of search area reduction for evaluating the most likely UNE sites in the inspection area if other better suited wide-area search methods (e.g., seismic after-shock monitoring, visual observations, surface gamma spectroscopy surveys, overflight, etc.) have failed to adequately locate the detonation site.

4. On-Site Inspection Concept of Operation Issues

4.1. Visible Bedrock Fractures Versus Alluvium Covered Pathways

A network of fractures or other pathways, natural and/or UNE produced, with some vertical trending elements is a necessary condition for transporting gases to the surface from an underground detonation point. The non-proliferation experiment (NPE) demonstrated that both visible and alluvium-covered fractures can produce detectable gases from an underground detonation. Three types of pathways were investigated in the NPE. Surface cracks and fractures were observed and successfully used for subsurface sampling, which involved only small (0.01 L) samples where infiltration would not be a problem. Geologic surveys near the NPE surface ground zero (SGZ) also located gas-producing areas where non-visible fracturing was present in the subsurface (e.g., buried fault zone). Finally, sample sites near SGZ, randomly selected without reference to any surface artifacts or geologic observations, were found to have a very low potential for producing gas from the explosion.

Both visible and alluvium-covered pathways present challenges for successful sampling. Because of their direct connection to the atmosphere, visible fractures or cracks are most susceptible to allowing infiltration and sample contamination during the sample acquisition process while alluvium-covered pathways are more difficult or impossible to find. We discuss these issues in greater detail as well as possible solutions in what follows.

4.2. Detection of NG Transport When Pathways are Not Observable

Simulations show that a 10 m-thick alluvium layer blanketing the fracture regime may still permit noble gas transport by barometric pumping along vertical trending fractures terminating at the base of the alluvium. If the mechanism driving gas transport is cavity pressure or convective flow, no limitation exists on alluvium thickness for detonation gases to be released into the base of the alluvium layer. However, as mentioned above, a major concern with the presence of alluvium is that it hides a producing fracture from visual observation. Furthermore, subsurface exploration methods such as ground penetrating radar or electromagnetic imaging as proposed for the continuation period of an OSI have insufficient spatial resolution to detect the existence of such fractures. Fortunately, NPE sampling results show that the presence of producing subsurface fracture networks can still be inferred from observable geologic features such as fault scarps or grabens. Thus, when surface fracturing is not visible as a guide for emplacing sampling stations, it is critical to understand, to the extent possible, the connectivity of the explosion-induced fracture regime and the natural fracture network associated with the local geology of the area.

4.3. Distributed Sample Point Pattern Versus Single Sample Point

One possible sampling-based solution for capturing noble gases from hidden fractures involves extracting soil gases over a broader region spanning the separation between producing fractures. Instead of sampling from a single tube in the alluvium at a sampling station, multiple sample tubes can be distributed over the surface (e.g., 5-spot pattern as found on dice) drawing uniformly subsurface gas from each tube. The scale of the pattern should approximately correspond to the anticipated average distance between fractures or rock joints. Although such scale information will not be directly known, statistical inferences may possibly be made from observations of locally visible outcroppings of bedrock (e.g., a road cut, construction site, mine tunnel or quarry). We have performed simulations of sampling with spatially distributed sampling tubes around an unknown fracture and find this approach to be superior to sampling only using one tube when the position of the underlying fracture is unknown. A numerical experiment was conducted with five randomly distributed sample points in a 4-m thick alluvium within 2 m of an underlying producing fracture. In this realization, only one of the five sample points happens to fall near the fracture. For a detailed model description, the reader is referred to SUN and CARRIGAN (2012). Xenon concentrations from the five tubes are averaged and then normalized by the actual concentration in the fracture at the time the samples are taken (Fig. 3). That the ratio sometimes approaches unity for three different values of permeability in the overlying soil suggests that spanning the region between fractures with multiple sampling tubes is a reasonable alternative to actually sampling at the fracture itself when the fracture location is not known. Another advantage of drawing gas samples from multiple tubes rather than one tube is that smaller volumes will be required from each tube, thus minimizing the possibility that infiltration of atmospheric gases will occur. In addition, a large-volume gas sample can be acquired more quickly from several sample tubes than from only one tube as subsurface sampling tends to be flow-rate limited. (Installation of five sampling tubes instead of one at each site is clearly more time consuming and represents a trade-off that must be considered). The 5-spot approach is discussed further in a later section from a concept of operation perspective.

4.4. *Using Radon and Other Radionuclides to Detect Producing Subsurface Fractures*

Field studies suggest that subsurface noble gas producing fractures may also be potentially located by detecting other natural and detonation-produced radioactive gases that emanate from them. Naturally occurring radon has been used by geophysicists to detect faults and fractures in the subsurface by traversing a zone and sampling periodically near the surface for peaks in the concentration of gases such as radon, carbon dioxide and helium (KING *et al.*, 1996; CIOTOLI *et al.*, 1999; YANG *et al.*, 2003). It appears that existing radon soil gas detection technology can serve as the basis for developing a rapid field survey capability using a near-surface probe sampling system to locate zones of anomalous radon concentration (FRONKA, 2010). In an area where a suspected UNE has occurred, detonation produced radionuclide gases may well be deposited at shallow depths above fractures. Thus, the detection of any significant peaks in subsurface radioactivity of any type may imply that a producing fracture network is nearby. Quick subsurface gas sampling with a probe using a walking survey approach (i.e., probing the subsurface at points on a laid-out grid) to locate hidden producing fractures remains experimental requiring further development before its value as an OSI technique can be determined.

4.5. *Large Volume Sampling*

The detection of tracer gases in the NPE required subsurface sample volumes of less than 0.01 L for analysis. This was often readily accomplished by hydraulically pushing a Geoprobe® tube with sampling screen into the soil of Rainier Mesa to a depth of up to 5 m. Following purging of the sample line, a small amount of gas was drawn into 0.25 m long, 0.006 m diameter copper tubes which were then crimped at both ends to hold the sample for subsequent analysis. The requirement for Xe-133 (SAUNA) and Ar-37 (MARDS) analyses is currently a minimum of one to two thousand liters (1–2 m^3, SAEY, 2007). This large volume requirement creates new sampling issues that did not exist during the NPE sampling process. These issues arise primarily from infiltration of atmospheric gas into the subsurface during soil gas sample extraction. At the least, mixing of atmospheric gas into a soil gas sample results in dilution of the sample and loss of Xe-133 detection sensitivity. At worst, atmospheric gas containing Xe-133 released from a nearby legitimate Xe-133 source (e.g., a medical isotope production facility) can potentially result in a false positive detection. Thus, it is important from an OSI

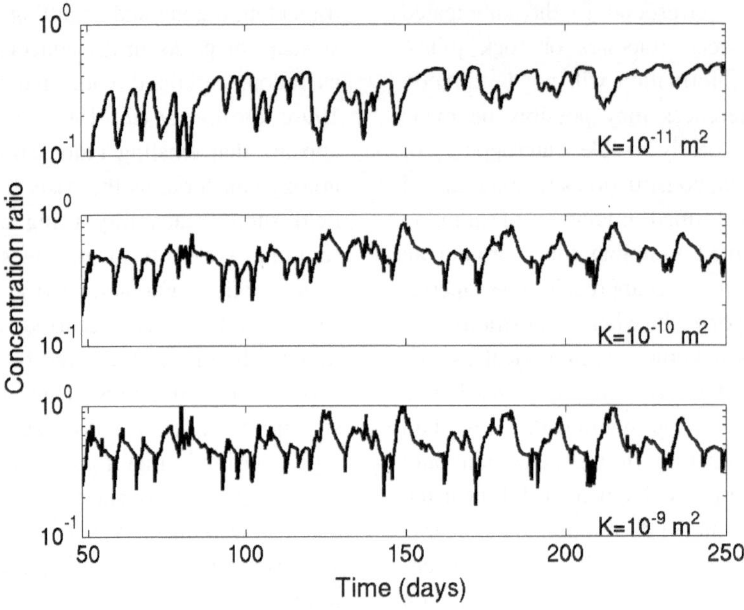

Figure 3
For three different permeabilities, the ratios of averaged xenon concentrations from five randomly selected locations in the alluvium overlying a producing fracture to xenon concentrations in the fracture itself are plotted as a function of time. In this realization only one of the sample points happens to be emplaced near the producing fracture. The range of permeabilities assumed corresponds to a soil layer consisting of a mixture of sand and gravel (most permeable) to a layer of very fine sand, loess or peat (least permeable)

protocol development perspective to both understand the problem as well as possible approaches to minimizing its impact on interpreting the results of Xe-133 analyses during an OSI.

We performed large-volume sampling experiments as part of the CTBTO Noble Gas 2009 Field Test held at the Turecky Vrch military site in Slovakia. The basic large-volume sampling station setup used in the experiment is illustrated in Fig. 4. A station was constructed by augering a 5 cm ($\sim$2 in.) diameter hole about 1.5 m deep in alluvium. A sampling tube mounted within an inflatable rubber packer was inserted into the hole above any water residing in the hole. The packer was inflated against the hole wall to eliminate air leakage along the hole. A 3 m $\times$ 3 m plastic sheet tent was placed over the sample tube and sealed around the tube to prevent air leakage. The edges of the tent were buried in the soil to prevent leakage directly from the atmosphere along the edges of the tent. The sample tube was connected to a small pump, which was in turn connected to a plastic sample bag. Turning on the electric pump would draw gas from the hole, transferring it to the sample bag. A small sample port on the outlet side of the pump allowed us to take periodic syringe samples for later analysis.

Prior to turning on the pump and starting the gas-sample extraction, charges of sulfur hexafluoride (SF_6) tracer gas were released under the tent. Figure 5 summarizes the sampling results obtained at two hydrologically different locations. The Quarry experiment (Fig. 5a) involved sampling through a thin layer of alluvium into an underlying zone of fractured bedrock. The sampling hole was augured using a water spray to cool the auger. At this site it was found that water drained from the hole unexpectedly quickly, thus indicating that the soil and rock below the sample point were highly permeable. The sampling pump extracted soil gas at a rate of approximately 6 L/min filling the sampling bag over a period of several hours.

During that period, small samples of gas flowing through the pump were periodically withdrawn from the sample port while noting the cumulative volume that had been withdrawn at the time of sampling. The syringe samples were capped and later analyzed. In many respects, the hydrologic characteristics of this site appear ideal for subsurface sampling as it

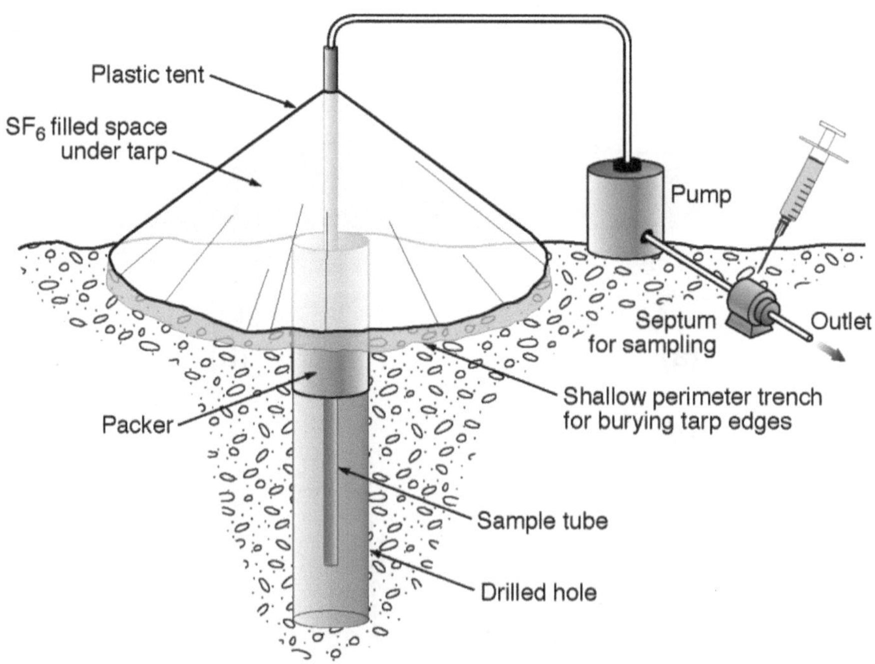

Figure 4
Experimental arrangement to monitor for infiltration and leakage of atmospheric gas into a soil-gas sample. A plastic tent is erected over the sample point and filled with SF_6 tracer. During the period of obtaining a 1–2 m^3 soil gas sample, syringe samples of the soil gas were periodically withdrawn using 60 mL syringes which were later analyzed for SF_6 concentration. If gas is drawn past the packer or through the surrounding alluvium during soil-gas extraction, a non-zero tracer concentration will be measurable during sampling

involves a thin cover of alluvium, which reduces infiltration, over what seems to be highly fractured rock, which can provide pathways for gases in the deep subsurface to reach the surface, which is in marked contrast with the hydrologic performance of the Turkish Hill site. Soil gas extraction was performed at the Quarry site only during the day and then restarted on the morning of the following day. Syringe samples were taken periodically during the sampling and the Quarry results are shown in Fig. 5c. Except for one anomalous reading, probably due to sample contamination, extraction of the first cubic meter at the Quarry site during the first day shows no detectable SF_6 tracer from the tent. The sampling site was then left overnight with sampling resuming the next morning. Surprisingly, the first syringe sample yielded a very high concentration of SF_6 tracer. It was also noticed that the inflatable rubber packer used to seal the sampling hole had lost significant pressure over night. The sudden very high concentration of tracer along with the packer pressure loss indicates that the main pathway for leakage of surface gas to the sampling point was directly through the hole past the packer and not by infiltration through the porous medium. The packer did not leak but plastic deformation of the hole resulting from the force of the packer on the hole wall apparently compromised the seal between the hole and the packer which resulted in leakage of tracer into the sample during extraction. The Turkish Hill site (Fig. 5b) turned out to be a far less ideal site for capturing gas samples from the deep subsurface. The site had been selected because of the underlying stratigraphy and surface grading. It was found that residual water was slow to drain from the hole and the soil was largely saturated from a recent storm, causing a significant decrease in gas permeability. Unlike the Quarry site, the inlet of the sample tube was not in proximity to any apparent underlying fractures. Immediately after the initiation of sampling, real-time measurements of the extracted soil gas (Fronka, 2010) yielded radon levels that were unexpectedly low which was also supported by the tracer results obtained from subsequent laboratory SF_6 analyses as shown in Fig. 5c that indicate approximately 30 % of the sample was mixed with

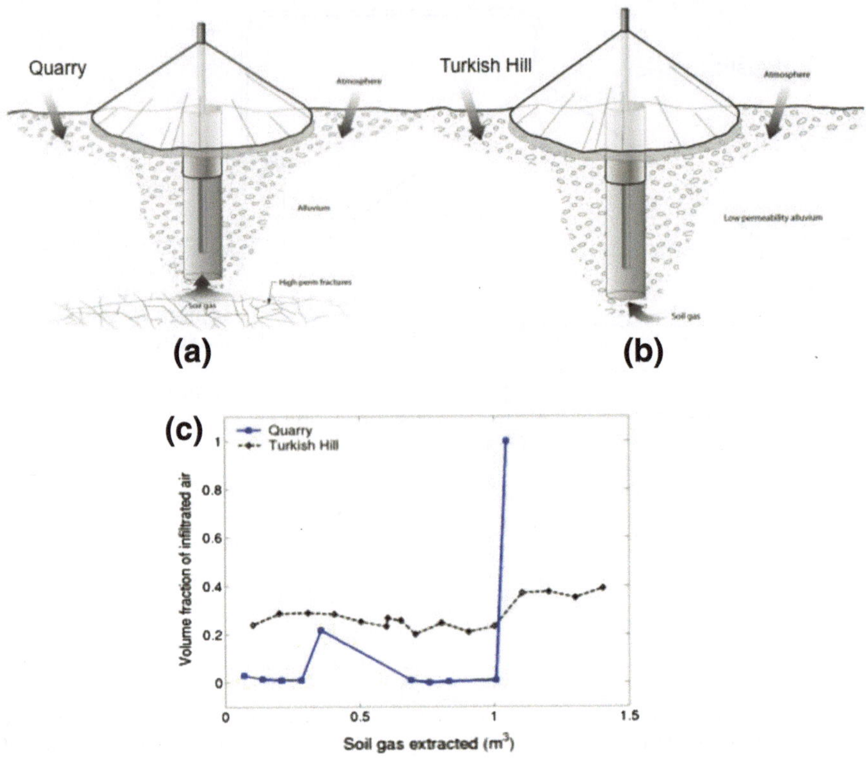

Figure 5
Sampling was performed at two different sites. **a** The high-permeability Quarry site consisted of a fractured layer underlying a thin soil layer. **b** The much lower permeability Turkish Hill site consisted of nearly saturated soil with low to vanishing gas permeability. **c** The Quarry site (*blue line*) readily provided gas from the subsurface with negligible infiltration (except for one point resulting evidently from sample contamination). Only on the second day when pumping resumed did atmospheric gas leak past the packer. The Turkish Hill site failed almost immediately evidently due to leakage past the packer

gases from the tent. Again, it is likely that plastic deformation of the soil in response to the pressurized packer sealing against alluvium resulted in leakage of tracer gas past the packer in response to the vacuum created by attempting to withdraw soil gas from a porous regime characterized by very low gas permeability.

In both the hydrologically ideal Quarry and far less ideal Turkish Hill sampling regimes, that the extraction process introduced tracer gas into the sample provides a lesson in the need to always maintain the packer well sealed against the wall of the hole. Because the hole wall is somewhat plastic and moves in its response to the inflating packer, it may be necessary to continually offset the resulting decrease in packer pressure using a pressurized gas cylinder and a pressure regulator to maintain a constant packer pressure. For long-term sampling, use of wet bentonite clay for sealing the sampling hole above the packer should also be considered. A bentonite seal was applied recently in another SF_6 tracer experiment and no leakage was found to occur during the course of the sampling in which extraction occurred continuously for more than 16 h.

In the two cases presented above, the appearance of tracer in a gas sample is most likely due to failure of the seal between the packer and sampling hole wall. However, large volume soil gas extraction is also subject to infiltration of atmospheric gas through the overlying porous medium. As gas is extracted from the inlet end of the sample tube, pore gas will be drawn both from above and below the sample point to replace the gas that is removed. Because only soil gas is desired in the sample, significant flows from above that eventually include atmospheric gas are to be avoided or at least minimized. The very small samples, as required by the NPE, did not result in any atmospheric gas being drawn down into the sample tube through the porous medium. On the other hand, the massive volumes of gas required during an OSI (as demanded primarily by the Xe analysis requirements) risk the possibility that atmospheric gas will be eventually drawn down into the buried sample tube unless it is emplaced very deeply. Unfortunately, sufficiently deep emplacement for the volumes of gas involved may not be possible owing to inadequate alluvial layer thickness or the sensitivity of the Inspected State Party (ISP) to the perceived degree of invasiveness of the subsurface sampling.

We have performed several simulations with the NUFT code (NITAO, 1998) in an attempt to better understand how infiltration is influenced by the parameters that define the relationship between the permeable subsurface and atmosphere. Figure 6 Illustrates a general model of a permeable alluvium layer (4 m thick) overlaying a permeable fractured regime (400 m thick) as may be common to locations of interest as possible underground testing sites. In the model used, the transport in the fractured rock is conceptualized using a dual-permeability approach (SUN *et al.*, 2010) with first-boundary condition on the water table at the bottom of the fractured rock and the barometric-pressure boundary condition on the ground surface.

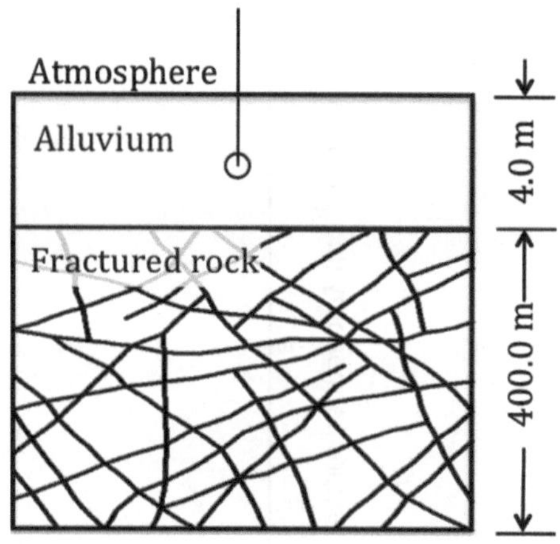

Figure 6
Conceptual model intended to demonstrate the relationship between atmospheric infiltration into alluvium and the flow of gas from underlying fractured bedrock during the extraction of a gas sample from the sampling point indicated by a *circle*. Alluvium is 4 m thick with a sampling point at a depth of 2 m and the underlying fractured zone is 400 m thick. (The alluvium thickness is exaggerated in the figure to emphasize emplacement of the sample tube)

In the model, the sampling point is located at a depth of 2 m corresponding to the middle of the alluvium layer. This seems to be a reasonable sampling depth based on augering and direct push attempts from a recent field test carried out at the Turecky Vrch site although any sampling model is highly scenario dependent. Figure 7 shows how the ratio of the amount of gas extracted from the fracture zone normalized by the amount of gas extracted from the atmosphere varies with both the soil and fracture permeability. Thus, for a fracture zone permeability of 8.0×10^{-11} m^2 and a soil layer permeability of 1.0×10^{-10} m^2, the contribution of gas in a sample volume produced by fractures is only 10 % of the contribution of gas from the atmosphere. However, increasing the fracture permeability by an order of magnitude effectively reverses the situation so that the contribution of gases from the fracture is now ten times greater than the contribution from the atmosphere.

For the particular ranges of fracture and alluvium permeability assumed here, the ratio is far more sensitive to changes in fracture permeability than it is to changes in alluvium permeability. A change of three orders of magnitude in the fracture permeability can change the ratio of the fracture to atmospheric contribution by up to seven orders of magnitude while the same change in the soil permeability changes the ratio by just over three orders of magnitude. We learn from this that atmospheric infiltration is not only determined by the hydrologic character of the alluvium layer but also by the characteristics of the underlying fracture layer, if present. The Quarry site appeared as if the underlying fracture layer was of high permeability while the Turkish Hill site appeared as if it had an equivalent low-permeability fractured layer. In either case atmospheric infiltration is expected and some attempt to mitigate infiltration seems prudent.

One means of reducing the impact of infiltration is to use a tarp or plastic sheeting to cover as much surface area around the sample point as is practical. Simulations (Figs. 8 and 9) show the effect of using a 3-m tarp in a sampling arrangement similar to that assumed in Fig. 6. The concentration of atmospheric gas drawn downward from above compared to the

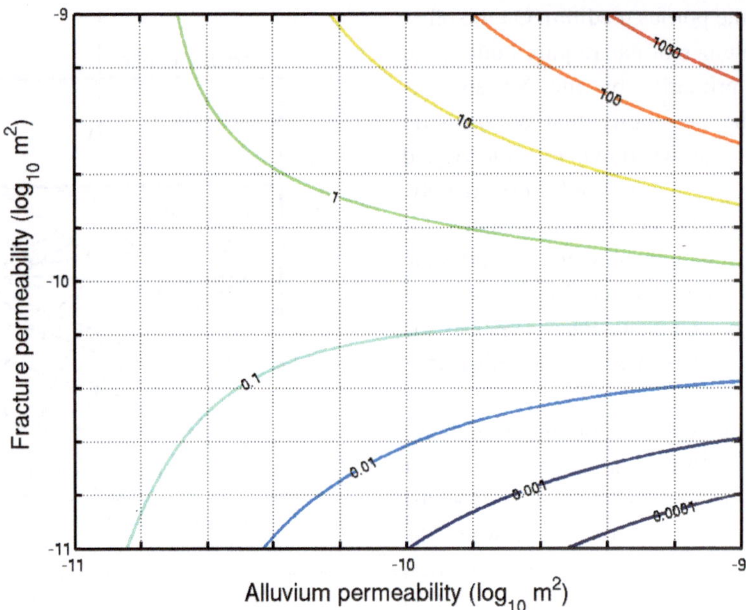

Figure 7
Ratio of gas flow from fractured rock to that from atmosphere captured in a sample after flow in the alluvium layer had reached steady state as a function of the alluvium and underlying fracture permeability

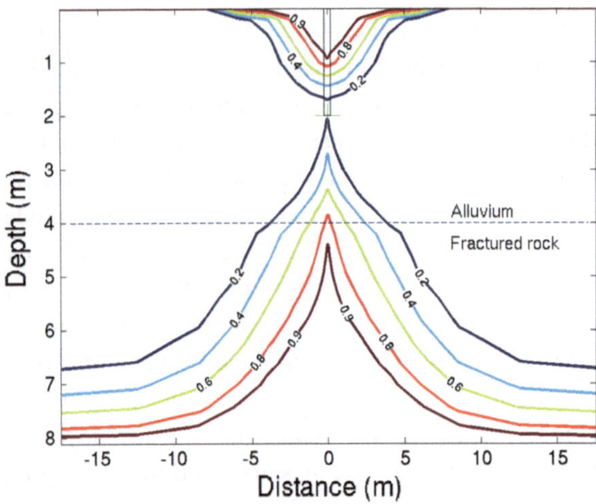

Figure 8
Concentration profiles for atmospheric gases drawn downward to the sample point and soil gases drawn upward to the sample point from the fractured rock for the case with no tarp to attenuate infiltration. The model assumes large volume sampling

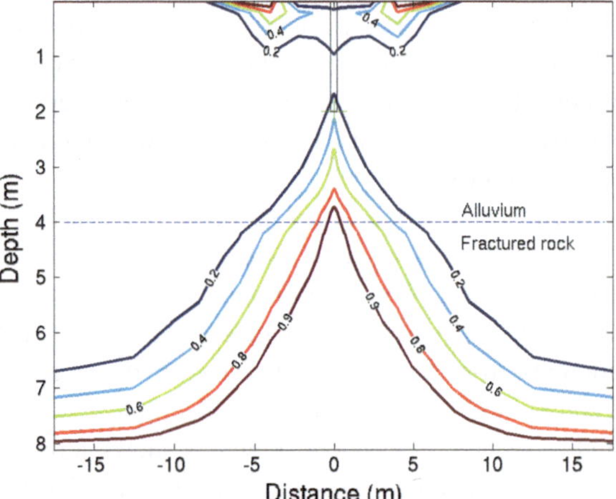

Figure 9
Concentration profiles for atmospheric gases drawn downward to the sample point and soil gases drawn upward to the sample point from the fractured rock for the case with tarp to attenuate infiltration. The model assumes large volume sampling

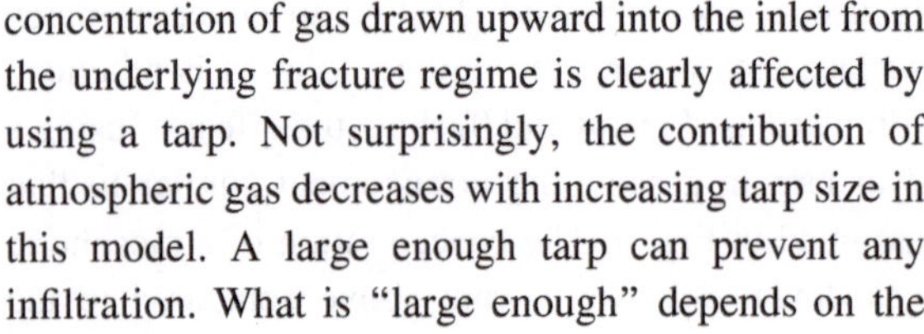

concentration of gas drawn upward into the inlet from the underlying fracture regime is clearly affected by using a tarp. Not surprisingly, the contribution of atmospheric gas decreases with increasing tarp size in this model. A large enough tarp can prevent any infiltration. What is "large enough" depends on the properties of the alluvial layer in which the sampling is performed, sample tube depth and the volume of the sample required. Furthermore, the presence of a sufficiently broad cover on the surface can also inhibit the ability of barometric pumping to transport gases to the surface locally. We are currently exploring how these parameters contribute to determining the optimal size of a surface cover or tarp.

4.6. Xenon-133 Background in Soil

We have only considered the effect of atmospheric infiltration in diluting a soil-gas sample while it is being extracted. In addition, contamination of a soil gas sample by radioxenon in the air can also occur as a result of drawing atmospheric gases directly into the sample. Even with plastic sheeting barriers to prevent atmospheric infiltration during sampling operations, the prevention of atmospheric contamination of a sample cannot be guaranteed without further precautions. Natural gas exchange between the atmosphere and soil caused by barometric fluctuations can also introduce contaminant gases from the atmosphere. During periods when the atmospheric pressure exceeds the pressure of gases in the soil, the atmospheric gas composition can be "impressed" into the shallow soil regime creating a "memory" of the composition of the gases passing over the surface. While HEBEL (2010) has argued for low values of a background in Xe-133 resulting from the decay of naturally occurring uranium in the soil, atmospheric sources of Xe-133 could conceivably result in the transport of low concentrations of this isotope into the shallow subsurface creating the very low probability of a false Xe-133 detection during an OSI. Besides normal releases from nuclear power reactors or larger accidental releases (>40 Bq m^{-3}) as occurred during the Fukushima nuclear reactor accident (BOWYER *et al.*, 2011), other possible common sources of radioactive Xenon isotopes are byproducts of the production of radiopharmaceutical isotopes (SAEY, 2009).

We have performed simulations in which an atmospheric concentration of radioxenon amounting to only 1 Bq m^{-3} is released into the atmospheric domain of our NPE soil model several days following the modeled arrival of Xe-133 at the surface and that atmospheric concentration is assumed to exist for only a day. From Fig. 10 it is seen that a detectable level of xenon is impressed into the shallow surface although the signal is far less than that anticipated for the arrival of xenon as predicted by the NPE calculations. However, this observation is tempered somewhat by comparing the averaged Xe concentration (i.e., from five randomly placed sample tubes near an unknown fracture) resulting from the memory effect with the averaged concentration associated with the NPE on the 60th day as it is found that the concentration averages for both sources of Xe in the soil are actually comparable. Furthermore, concentration levels of 100 Bq m^{-3} have been detected 10–20 km downwind from radiopharmaceutical production facilities (BOWYER, personal communication, 2011) and could result in very much larger "impressed" signals that might interfere with the detection of detonation-produced isotopes in soil gases. In addition, atmospheric concentrations of 1 Bq m^{-3}, the value of the Xe-133 concentration assumed here, have been observed at least 80 km downwind from an isotope production facility (RINGBOM *et al.*, 2008). The fact that significant atmospheric Xe-133 concentrations can occur at large distances from production facilities, and that imprinting of this contamination in the soil can also occur subject to certain atmospheric and soil conditions, strongly suggests that atmospheric xenon monitoring be performed in the vicinity of sites where soil gases are being collected. In addition, general soil gas background monitoring of suspect regions should

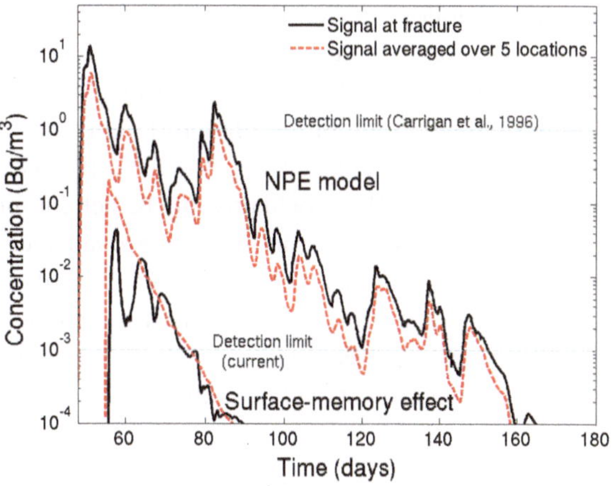

Figure 10
The signal due to the surface-memory effect is compared against the simulated signal from the NPE. It is assumed that a 1 Bq m^{-3} Xe-133 concentration passes over the ground surface on day 55 following the NPE event. Barometric pressure fluctuations "impress" some of the atmospheric gas into the near-surface soil-gas pore space. The averaged signals produced by five randomly situated sampling tubes in the soil over a fracture (*dashed line*) for the surface-memory component of gas is comparable at about 60 days to the averaged signal from the NPE model

be initiated following the arrival of the inspection team at locations of interest to establish a history of possible radio-isotopic backgrounds pertinent to confirming the occurrence of a violation. Evaluation of the ratios of detected xenon isotopes in the soil caused by this "memory" effect could possibly be helpful for evaluating the source of soil-gas background measurements (e.g., civilian versus weapon related as considered by KALINOWSKI *et al.*, 2010) as fractionation from atmospheric isotopic values caused by transport into the soil should be minimal.

4.7. Argon-37 Background in Soil

For UNE confirmation purposes, Ar-37 is highly attractive as a short-lived (half life 35.0 days) noble gas isotope because it is very rare and significant amounts of it appearing above any local background level are indicative of a recent subsurface nuclear event. While natural background levels are extremely low, coincidence-counting methods are so sensitive that even such exceedingly low background levels may be detectable in practice. Ar-37 is produced through a spallation interaction with naturally occurring Ca-40 and fast neutrons in the subsurface resulting from nuclear explosions. Similarly, the natural background in the near-surface regime results from interaction of Ca-40 in the soil with cosmic-ray neutrons. RIEDMANN and PURTSCHERT (2011) have argued that the naturally occurring equilibrium Ar-37 concentration in shallow soil gas is a function of an exponentially decreasing production rate from cosmic ray neutrons with increasing soil depth, diffusive transport in the soil air, and radioactive decay. They also show that the highest activities of 100 mBq m^{-3} air are two orders of magnitude larger than in the atmosphere peaking in the 1.0–2.0 m depth range and rapidly decrease with greater depth. RIEDMANN and PURTSCHERT (2011) find good agreement between their predicted profile of Ar-37 activity in the soil and measurements obtained from several sampling sites in Europe.

To better understand the range of applicability of their model for the production of Ar-37 in the shallow subsurface, we have included the effect of gas migration in the upper few meters of the soil resulting from daily fluctuations in the barometric pressure. We use actual pressure records to simulate the pressure changes occurring at the soil surface with soil properties and bedrock stratigraphy characteristic on average of what might be encountered in a high desert OSI site with an arid climate. As a result of the variability inherent in the parameters determining the Ar-37 activity, we used a Monte Carlo approach performing more than 100,000 emulations of geologic scenarios spanning the ranges of thirteen variables that define the problem (Table 1). Figure 11 summarizes the effect of including this barometric pressure dependence of the Ar-37 activity profile. We find that the temporally constant vertical profile of activity obtained by RIEDMANN and PURTSCHERT (2011) becomes an upper bound on possible profiles resulting from the inclusion of the fluctuating barometric transport.

The message that can be taken from studies of noble gas backgrounds (PURTSCHERT *et al.*, 2007; RIEDMANN, 2011) and the effect of barometric fluctuations on near-surface soil gas migration (SUN and CARRIGAN, 2012) is that deeper sampling is better as we are more likely to obtain samples from deep sources of noble gases, such as UNEs, that are uncontaminated by near-surface or atmospheric sources of these gases. Again, the limit on how deep we can emplace sample sites depends on the depth of any overlying alluvial layer, the available equipment for auguring holes in the context of OSI limitations

Table 1

Uncertain parameters and their ranges assumed during emulation of Ar-37 activity profiles

Uncertain parameters	Minimum value	Maximum value
Sand volume fraction (–)	0.35	0.65
Correlation length in x (m)	280	320
Correlation length in z (m)	1.50	2.50
Sand porosity (–)	0.25	0.50
Clay porosity (–)	0.33	0.60
Calcium content in sand (–)	0.05	0.20
Calcium content in clay (–)	0.30	0.60
van Genuchten m in sand (–)	0.5236	0.7920
van Genuchten m in clay (–)	0.0624	0.3208
van Genuchten α in sand ($\log_{10}$ Pa^{-1})	−4.6938	−3.8069
van Genuchten α in clay ($\log_{10}$ Pa^{-1})	−5.4958	−4.1358
Permeability in sand ($\log_{10}$ m^2)	−12.0	−9.0
Permeability in clay ($\log_{10}$ m^2)	−18.0	−15.0

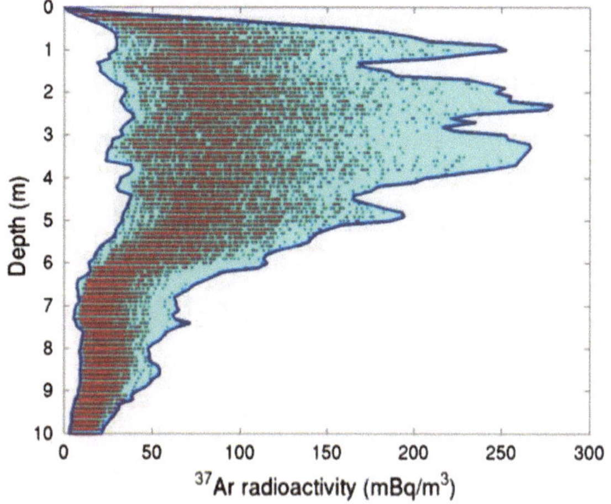

Figure 11
Ar-37 activity profile as a function of depth according to a stochastic model that includes the effects of variable clay/sand soil distributions, variable natural calcium content, variable porosity, permeability etc. Owing to geologic uncertainty, thirteen different parameter ranges (see Table 1) were assumed to obtain this result requiring the emulation of more than 100,000 different sampling scenarios spanning this parameter space. The figure summarizes the ranges of activity expected within this parameter space. Note that *blue lines* denote the minimum and maximum activities and *red dots* represent actual concentration profiles

and the perception by the ISP of the invasiveness of the sampling procedure.

4.8. Barometric Triggering of Sampling Versus Rapid Bulk Extraction

Another issue is the value of performing barometrically triggered gas extraction during the acquisition of a sample. Both NPE field experiments, simulations, and previous work (Nilson *et al.*, 1991) have shown that shallow detection of gases emanating from a deep source is tied to extracting gas samples at a site during a falling barometer and before reaching the minimum of a barometric low. In the case of the NPE, this was usually accomplished by obtaining field samples before the arrival of a storm on Rainer Mesa at NTS. This conclusion assumes that the samples are small as obtained in the NPE experiment. When the sample volumes are thousands of times larger, as required for Xe-133 and Ar-37 analyses, acquiring a sample during a single barometric low of an hour or two may not be possible; sample acquisition in such cases may be over the time periods of several barometric lows during possibly 1 or 2 days. Preliminary evaluation shows that taking bulk samples at one time can be effective if the sampling point is deeper in the soil. How optimal large ($\sim$1 m^3) samples are obtained during an OSI is of critical importance and we are continuing to investigate this issue.

4.9. A Possible Approach to Sampling During an OSI

Taking into account the concerns and issues already mentioned (e.g., hidden producing fractures, infiltration, flow-rate limiting of sample acquisition, etc.) we propose a preliminary, generalized model of subsurface noble gas sampling. Sample tubes for a given station are emplaced in a pattern, such as a 5-spot (Fig. 12), at the greatest depth that is reasonably attainable given field conditions, manpower and equipment availability and any Inspected State Party concerns about degree of invasion. The tubes may be installed in augered holes using inflatable packers. To further guard against infiltration, gas-tight plugs are formed above each packer by filling the remaining hole to the surface with a water-saturated, radionuclide-free bentonite clay mixture. The 5-spot pattern is then covered with plastic sheeting. If it is deemed necessary, SF_6 infiltration monitoring gas may be injected under the tarp to provide a measure of quality control monitoring during the sampling process. Each of the sampling tubes is connected by rigid vacuum tubing to individual rate-controllable electric pumps. All equipment used in the sampling station is constructed of materials having no memory to noble gases or any tracer gases (e.g., SF_6) used for monitoring atmospheric infiltration.

More sophisticated realizations of this design may include computerized control of the electric pumps allowing decreases in local barometric pressure to trigger the sampling of the site (e.g., a "smart" sampling system). The results of the NPE and other experiments along with computer simulations indicate that especially for shallow sampling ($\sim$1–2 m) barometric triggering of soil gas extraction may produce a greater likelihood of detection. In addition, monitoring pressure changes at the buried end of the sample tube will permit pumps to be turned off before seals are broken and infiltration occurs as a result of

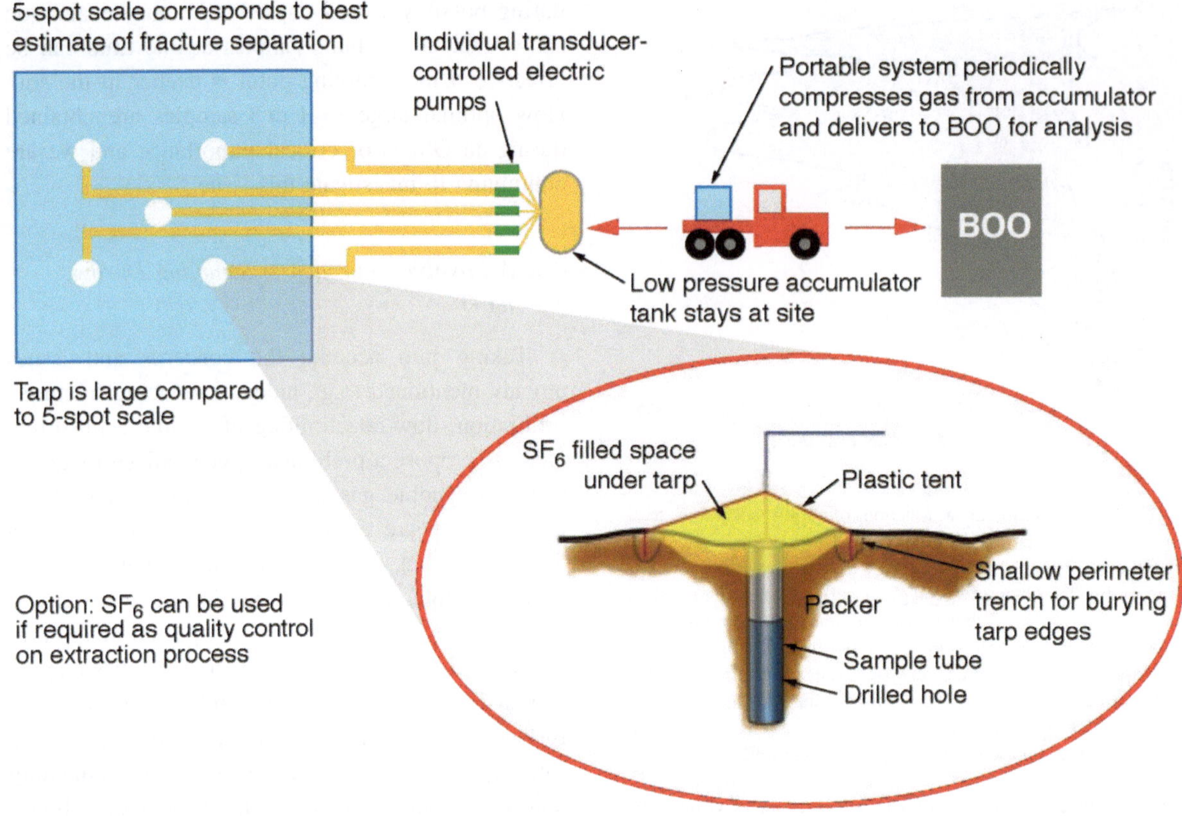

Figure 12
Possible approach to capturing gases from unknown subsurface fractures and minimizing potential for infiltration from each sample point. The scale of this particular pattern spans the characteristic or average fracture spacing as estimated from visual observations of joint and fracture distributions in a particular geologic zone. This suggested pattern is applicable when underlying fractures cannot be observed. *BOO* refers to the base of operation

leakage past a packer. Each of the electric pumps draws gas from its respective sample tube (or alternatively one pump drawing gas through a manifold connected to the five sampling lines) releasing the gas into a common low-pressure tank or container (e.g., high quality, puncture resistant inflatable sample bags or bladders) that remains at the site. The contents of this low-pressure container can be periodically transferred via a portable compressor to a pressure-rated tank, which is then transported to the base of operations (BOO) for analysis.

5. *Conclusions*

In the process of developing a sound concept of operation for noble gas sampling during an OSI, it is necessary to confront a variety of technical issues. Some of these issues are sufficiently specific to the OSI regime that they have not been well studied in other applications of soil gas sampling. In this paper we have described highlights of the issues and concerns that we have either addressed or are currently being considered with no attempt having been made to develop a sampling concept of operations. We have also suggested solutions to sampling challenges, based on field and modeling results, which are intended to optimize the probability of detecting UNE-produced noble gases. We believe that we are now in a position to begin formulating a noble gas concept of operation while recognizing that it will be a dynamic product that should benefit from continued research and improvements in understanding of the subsurface regime. While computer simulations are helpful during this formulation process, we also recognize that additional field experiments and trials will be crucial for its continued successful development.

Acknowledgments

This work was funded by the Office of Nuclear Verification (NA-243) with additional support provided by the Office of Proliferation Detection (NA-221), US Department of Energy. C. R. CARRIGAN also thanks the US Fulbright Program for support while on a research sabbatical at Cambridge University during which some of the problems considered here were initially formulated. Some of our thoughts have benefited from recent work at the Nevada Nuclear Security Site with support from NSTec staff. Finally, we thank Jerry Sweeney (LLNL) and Guy Brachet (CEA, France) and two anonymous reviewers for their insightful and supportive comments. This work was performed under the auspices of the US Department of Energy by Lawrence Livermore National Laboratory under Contract No. DE-AC52-07NA27344.

REFERENCES

AUER, L.H., ROSENBERG, N.D., BIRDSELL, K.H., WHITNEY, E.M.: *The effect of barometric pumping on contaminant transport*, J. Contam. Hydrol., *24*, 145–166, 1996.

BOWYER, T.W., SCHLOSSER, C., ABEL, K.H., HAYES, J.C., HEIMBIGNER, T.R., MCINTYRE, J.I., PANISKO, M.E., REEDER, P.L., SATORIUS, H., SCHULZE, J., and WEISS, W.: *Detection and analysis of xenon isotopes for the comprehensive nuclear-test-ban treaty international monitoring system*, J. Environ. Radioactiv., *59*, 139–151, 2002.

BOWYER, T.W., BIEGALSKI, S.R., COOPER, M., ESLINGER, P.W., HAAS, D., HAYES, J.C., MILEY, H.S., STROM, D.J., and WOODS, A.: *Elevated radioxenon detected remotely following the Fukushima nuclear accident*, J. Environ. Radioactiv., *102*, 681–687, 2011.

CARRIGAN, C.R., HEINLE, R.A., HUDSON, G.B., NITAO, J.J., ZUCCA, J.J.: *Trace gas emissions on geological faults as indicators of underground nuclear testing*, Nature, *382*(6591), 528–531, 1996.

CARRIGAN, C.R.: 2009 *noble gas field operations test: Towards detecting "the smoking gun" during an on-site inspection*, CTBTO Spectrum, *15*(1), 2010.

CIOTOLI, G., ETIOPE, G., GUERRA, M. and LOMBARDI, S., *The detection of concealed faults in the Ofanto Basin using the correlation between soil-gas fracture surveys*, Tectonophysics, *301*, 321–332, 1999.

DE GEER, L.-E.: *Comprehensive Nuclear-Test-Ban Treaty: relevant radionuclides*, Kerntechnik, *66*(3), 113–120, 2001.

DUBASOV, Y.: *Underground nuclear explosions and release of radioactive noble gases*, Pure Appl. Geophys., *167*, 455–461, 2010.

FRONKA, A.: personal communication, SURO, CZ, 2010.

HEBEL, S.: *Genesis and equilibrium of natural lithospheric radioxenon and its influence on surface noble gas samples for CTBT on-site inspections*, Pure Appl. Geophys., *167*, 463–470, 2010.

KALINOWSKI, M.B., BECKER, A., SAEY, P.R.J., TUMA, M.P., and WOTAWA, G.: *The complexity of CTBT verification, Taking noble gas monitoring as an example,* Complexity, *14*, 89–99, 2008.

KALINOWSKI, M.B., AXELSSON, A., BEAN, M., BLANCHARD, X., BOWYER, T.W., BRACHET, G., HEBEL, S., MCINTYRE, J.I., PETERS, J., PISTNER, C., RAITH, M., RINGBOM, A., SAEY, P., SCHLOSSER, C., STOCKI, T., TAFFERY, T., and UNGAR, K.R.: *Discrimination of nuclear explosions against civilian sources based on atmospheric xenon isotopic activity ratios*, Pure Appl. Geophys., *167*, 517–539, 2010.

KING, C.-Y., KING, B.-S., EVANS, W.C., and ZHANG, W.: *Spatial radon anomalies on active faults in California*, Appl. Geochem., *11*, 497–510, 1996.

NILSON, R.H., PETERSON, E.W., LIE, K.H., BURKHARD, N.R., and HEARST, J.R.: *Atmospheric pumping: a mechanism causing vertical transport of contaminant gases through fractures permeable media,* J. Geophys. Res., *96*(B13), 21933–21948, 1991.

NITAO, J.J.: User's manual for the USNT Module of the NUFT code, V. 2 (NP-Phase, NC-Component, Thermal), Lawrence Livermore National Laboratory, UCRL-MA-130653, 1998.

PURTSCHERT, R., RIEDMANN, R. and LOOSLI, H.H: Evaluation of Argon-37 as a means for identifying clandestine subsurface nuclear tests, Proceedings of 4th Mini Conference on Noble Gases in the Hydrosphere and in Natural Gas Reservoirs, GFZ Potsdam, Germany, 2007.

RIEDMANN, R., and PURTSCHERT, R.: *Natural ^{37}Ar concentrations in soil air: Implications for monitoring underground nuclear explosions*, Environ. Sci. Technol., *45*(20), 8656–8664, 2011.

RIEDMANN, R.: Separation of Argon from atmospheric air and measurements of Argon-37 for CTBTO purposes, Ph.D. dissertation, University of Bern, 2011.

RINGBOM, A., ANDERSSON, P., BAN, S., DE GEER, L.-E., ELMGREN, K., LINDH, K., PETERSON, J., SÖDERSTRÖM, C. and TOOLOUTALAIE, N.: Measurements of atmospheric radioxenon in Belgium in June and July 2008 – final report, EU/JA field campaign (CTBTO ref: 2008-0009/DELI), FOI Memo 2516, August 2008.

SAEY, P.R.J: *Ultra-low-level measurements of argon, krypton and radioxenon for treaty verification purposes*, ESARDA Bulletin, *36*, 42, 2007.

SAEY, P.R.J.: *The influence of radiopharmaceutical isotope production on the global radioxenon background*, J. Environ. Radioactiv., *100*, 396–406, 2009.

SUN, Y., BUSCHECK, T.A., LEE, K.H., HAO, Y., and JAMES, S.C.: *Modeling thermal-hydrological processes for a heated fractured rock system: impact of a capillary-pressure maximum*, Transport Porous Med., *83*, 501–523, 2010.

SUN, Y. and CARRIGAN, C.R: *Modeling noble gas transport and detection for the Comprehensive Nuclear -Test-Ban Treaty*, Pure Appl. Geophys., 2012. doi:10.1007/s00024-012-0514-4

TANAKA, J.: Technical Report: 2009 On-Site Inspection Noble Gas Field Operation Tests (NG09), CTBT/PTS/TR/2011-1, 2011.

YANG, T.F., CHOU,,C.Y., CHEN, C.-H., CHYI, H.H., and JIANG, J.H: *Exhalation of radon and its carrier gases in SW Taiwan*, Radiat. Meas., *36*, 425–429, 2003.

(Received November 29, 2011, revised July 7, 2012, accepted July 18, 2012, Published online August 10, 2012)

Pure Appl. Geophys. 171 (2014), 735–750

DOI 10.1007/s00024-012-0514-4

Pure and Applied Geophysics

Modeling Noble Gas Transport and Detection for The Comprehensive Nuclear-Test-Ban Treaty

YUNWEI SUN[1] and CHARLES R. CARRIGAN[1]

Abstract—Detonation gases released by an underground nuclear test include trace amounts of ^{133}Xe and ^{37}Ar. In the context of the Comprehensive Nuclear Test Ban Treaty, On Site Inspection Protocol, such gases released from or sampled at the soil surface could be used to indicate the occurrence of an explosion in violation of the treaty. To better estimate the levels of detectability from an underground nuclear test (UNE), we developed mathematical models to evaluate the processes of ^{133}Xe and ^{37}Ar transport in fractured rock. Two models are developed respectively for representing thermal and isothermal transport. When the thermal process becomes minor under the condition of low temperature and low liquid saturation, the subsurface system is described using an isothermal and single-gas-phase transport model and barometric pumping becomes the major driving force to deliver ^{133}Xe and ^{37}Ar to the ground surface. A thermal test is simulated using a nonisothermal and two-phase transport model. In the model, steam production and bubble expansion are the major processes driving noble gas components to ground surface. After the temperature in the chimney drops below boiling, barometric pumping takes over the role as the major transport process.

Key words: Noble gas, transport, CTBT, detection, modeling.

1. Introduction

Noble gas isotopes, such as ^{133}Xe and ^{37}Ar, have been studied for detection of nuclear explosions for the Comprehensive Nuclear-Test-Ban Treaty (CTBT, BOWYER *et al.*, 2002; AALSETH *et al.*, 2011). Because of their short half-lives, the ambient background concentrations for those gas components are extremely low (DRESEL and WAICHLER, 2004) and the International Monitoring System (IMS) has developed the measurement sensitivity for ^{133}Xe exceeding 1.0 mBq m^{-3}. In addition to nuclear explosions, radioxenon isotopes may be produced and released by civil events such as medical isotope production (KALINOWSKI *et al.*, 2010). ^{37}Ar is also produced in the subsurface due to cosmic-neutron activation of calcium by ^{40}Ca(n, α)^{37}Ar reaction. To better understand under what hydrogeologic conditions the Xe and Ar signals will reach the surface in detectable amounts and to better estimate the value of subsurface noble gas sampling during an on site inspection, the transport mechanisms from an underground nuclear test to ground surface have to be studied.

Gas-phase transport has been studied for various purposes (WEBB, 1996; WU *et al.*, 1998; TIDWELL, 2006; PRUESS, 2006). The barometric-pumping phenomenon has been identified as a driving force for gas-phase transport in the subsurface system (NILSON *et al.*, 1991; LINDSTROM *et al.*, 1994; WYATT *et al.*, 1995; CARRIGAN *et al.*, 1996; AUER *et al.*, 1996; MARTINEZ and NILSON, 1999; NEEPER, 2001, 2002; ROSSABI, 2006). The gas-phase advection in response to the pressure oscillation at ground surface behaves as cyclical "breathing" or push–pull gas flow through the permeable alluvium or fractures. Gradual upward movement of the gas phase in vertical trending fractures from the subsurface detonation point is in response to barometric pressure oscillations at the surface. Because of diffusive exchange of gases between the resulting oscillatory fracture flow and porous rock matrix, a net upward migration or "ratcheting" of gases from the point of origin occurs. A number of variables, such as gas diffusivity and fracture aperture determine the rate of migration, which may be of the order of a few meters per day.

In addition to these breathing processes that have been studied, we developed (1) a single-gas and isothermal model, and (2) two-phase and thermal model to study the Non-Proliferation Experiment

IM release number: LLNL-JRNL-501357.

[1] Lawrence Livermore National Laboratory, Livermore, CA, 94550, USA. E-mail: sun4@llnl.gov

involving a subsurface chemical explosion and the 1957 Rainier underground nuclear test, respectively. In the single-gas and isothermal model, we quantitatively studied the detectability under an isothermal condition and how the overburden alluvium affects the concentration signal at the ground surface. In the thermal model, we demonstrated how the phase change and heat-pipe phenomena affect the gas-phase transport and how to capture the concentration signal at or near ground surface.

2. *Physical Systems*

The physical processes that control noble gas migration from the working point of a nuclear detonation to ground surface are complex and interdependent. In an underground nuclear explosion, vaporization and compression of geologic media will create and expand an open, approximately spherical cavity surrounding the working point. The cavity reaches its full size within a fraction of a second after detonation. The cavity radius is a function of the energy of the detonation, its depth of burial, and the strength of overlying geologic units (Carle *et al.*, 2008). A steam bubble, together with radionuclides including noble gas isotopes, is formed and initially well contained within the cavity because of the highly compressed and impermeable cavity wall. Afterwards, steam-driven fractures (hydro-fractures) may be propagated from the cavity into the surrounding area and significantly reduce the cavity pressure. When the pressure within the cavity drops below the pressure resulting from the overburden rock, the overlying rock may collapse into the cavity creating a rubblized chimney, where radioactive gases are redistributed over the void volume of the chimney. Under the conditions of thermally-driven convection from the working point and of barometric pressure oscillation at the ground surface, noble gas nuclides migrate upward toward the ground surface through the rubblized chimney into overlying fractured layers and make detection possible.

2.1. *Definition of Detectability*

The success of noble gas detection from an underground nuclear test depends on the gas transport in the geological system. The geological structure, as well as many physical and chemical properties, controls the concentration profile of gas nuclides at and near ground surface. Taking the geological model of Carrigan *et al.*, (1996) as an example, if the detection sensitivity (threshold) for ^{133}Xe is 1.0 mBq m^{-3} (Auer *et al.*, 2004, 2010; Saey *et al.*, 2009), as shown in Fig. 1, the detectability can be characterized by (1) arrival time; (2) detection window width; and (3) detection window height (above the threshold). The arrival time, which is mainly determined by diffusivity of gas nuclides in the rock matrix and permeability in fractures, is defined as the time when the concentration of the noble gas component of the soil gas reaches its detection threshold. The detection window width is the time duration between the arrival time and the time when the concentration drops below the threshold. The window height is measured as the order of magnitude of the peak concentration exceeding the threshold. For a given radionuclide with a specific half-life (e.g., 5.24 days of ^{133}Xe), monitoring is expected to be conducted within its detection window.

2.2. *Non-Proliferation Experiment and The Rainier Test*

The Non-Proliferation Experiment (NPE) detonated on September 22 in 1993, was a low-yield (1 kt) chemical explosion (Carrigan *et al.*, 1996) conducted in the south-central N-tunnel below the surface of Rainier Mesa. Gas components, ^{3}He and sulfur hexafluoride (SF_6) were released at the source (433.4 m depth) and monitored over 517 days. The experimental data suggested that ^{133}Xe and ^{37}Ar would have been detectable after the detonation if they were released at the source (Carrigan *et al.*, 1996). A discrete fracture model was developed to describe the transport of ^{3}He and SF_6. In this study, we replace stable gas tracer ^{3}He with the ^{133}Xe radionuclide in the numerical model and investigate its detectability.

The Rainier test, which was conducted on September 19 in 1957 at depth of 274.3 m, is considered as the first fully contained underground test (Tompson *et al.*, 2011). In the years following the test, extensive re-entry mining and drilling operations

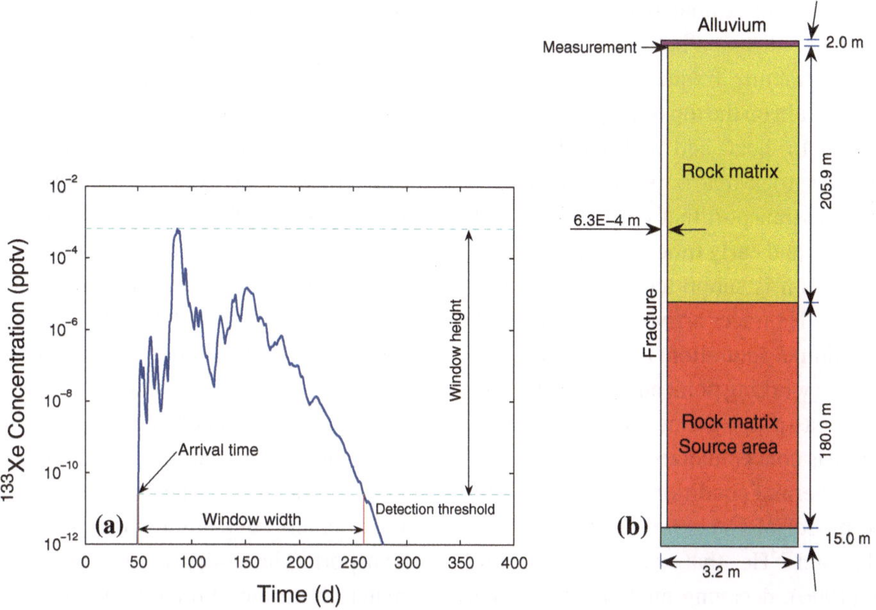

Figure 1
Study domain of single-gas-phase transport for the Non-Proliferation Experiment. **a** An example of ^{133}Xe concentration history at the outlet of the fracture. **b** Geometry of study domain. Note that the horizontal dimension is exaggerated

were undertaken to characterize the chimney and altered environment.

The Non-Proliferation Experiment, which was detonated near the Rainier test, was primarily a seismic study that also included the two-tracer-gas release experiment to study transport of detonation gas in fractured rock. Although both the Rainier and the NPE tests were conducted in the same bedded tuff system, the gas migration is conceptualized as either thermal or isothermal in terms of the pertinent transport mechanisms. In the thermal case, the residual heat produced by a nuclear detonation becomes the dominant driving force for gas-phase transport. In the absence of the thermal drive, the barometric pumping at the ground surface is the only driving force for the NPE chemical test. A full-scale Rainier-test model is developed to understand (1) the physical processes that affect the gas-phase radionuclide redistribution under transient pressure and temperature conditions; (2) the effects of variably saturated conditions on gas-phase radionuclide transport under thermal conditions; (3) the role of fracture-matrix interactions and matrix diffusion; and (4) the role of barometric pumping on noble gas detection.

3. Conceptual Models

Two conceptual models are presented in this section. The Non-Proliferation Experiment for ^{133}Xe transport is described as an isothermal and single-gas-phase transport in rock matrix with a discrete fracture (Fig. 1b). Because of the much greater gas production of a chemical explosion, heat will tend to be transported from the cavity and into the surrounding rock mass much more effectively than by a nuclear explosion. This results in a lower cavity temperature for the chemical event. Therefore, early time two-phase and cavity-pressure driven flow in the fractured wall-rock around the NPE detonation point has been implicitly lumped and included in the form of a "halo" of tracer surrounding the cavity. The Rainier test is described using a two-phase and four-component (water, air, ^{133}Xe, and ^{37}Ar) transport model under thermal conditions in a dual-permeability domain.

3.1. Conceptual Model of NPE

As shown in Fig. 1b, a single vertical fracture penetrates the rock matrix. The fracture aperture size is based on the mean value of estimated aperture size and matrix block size (i.e., distance between fractures) is estimated using the fracture frequency. An alluvium layer of 2.0 m is overlain on the top of fractured rock. It is assumed that both heat and radionuclides are distributed over the rock-matrix source area in a "halo" by early time transport that may involve both cavity-pressure drive and early time multi-phase flow. This model assumption is supported by the observations of the SF_6 and ^3He tracer arrival during the NPE experiment. The liquid saturation is assumed to be constant between the working point and ground surface and does not affect gas-phase permeability. Therefore, the physical system is conceptualized as a single-phase transport under isothermal condition and is essentially the same as the model of Carrigan *et al.* (1996). In addition to SF_6 and ^3He that were modeled by Carrigan *et al.* (1996), decaying nuclides, ^{133}Xe and ^{37}Ar are added in this model. Detailed description about rock properties, initial and boundary conditions, and model grids is referred to Carrigan *et al.* (1996). Three sensitive and uncertain parameters, aperture size of fracture, permeability in alluvium, and gas diffusivity in rock matrix are considered to characterize the detectability indices, arrival time, and detection window size (width and height).

3.2. Conceptual Model of Rainier Test

The shallower and more energetic Rainier test is simulated using three-stage sequential models that represent multiple-phase flow and multiple-species transport under thermal conditions in the fractured rock environment surrounding the test cavity and chimney. The models are developed in a two-dimensional and radial symmetric domain extending vertically from the ground surface to the water table and radially from the vertical axis passing through the working point to the outer boundary at a significant distance (1,000.0 m), where the assumption of the ambient thermal and hydrologic condition is justified.

The use of a sequence of model stages for the Rainier test allows important aspects of the thermo-hydrologic dynamics and evolution to be addressed over periods (1) before detonation, (2) after detonation and before chimney collapse, and (3) after chimney collapse. Transition between the stages involves the changing material domains and the remapping of initial conditions. The pre-test model in stage 1 is used to simulate the ambient thermal and hydrologic conditions, under which internally consistent and steady-state saturation, pressure, and temperature conditions are generated within an unaltered system before detonation. The pre-test model is considered as the conversion from a boundary-value problem to a steady-state condition. The test model in stage 2 is developed to present the processes and physical conditions after the detonation and before the chimney collapse. A spherical cavity, which holds explosively produced thermal energy, pressure, and radionuclides including ^{133}Xe and ^{37}Ar, is embedded in the model grids. The post-collapse model in stage 3 is designed to examine the longer-term gas-phase transport. The collapse causes reassignment of rock materials in the chimney domain and physical property alteration in the surrounding zone.

Differently from the NPE model, all models of Rainier test are based upon a nonisothermal, two-phase flow formulation involving water and air as principal components, and specifically considering ^{133}Xe and ^{37}Ar as minor components in determining their migration and redistribution for quantifying the detectability. It is assumed that sufficient calcium is available in the rock surrounding the explosion to produce ^{37}Ar by $^{40}Ca(n, \alpha)^{37}Ar$ reaction. The fractured rock system is conceptualized as a fracture-matrix dual permeability medium (DKM, Ho, 1997; Sun *et al.*, 2010) and simulations were conducted using the NUFT code (Nitao, 1998; Hao *et al.*, 2010). The NPE model can be considered as a simplified version of the Rainier test model in the post-collapse period under an isothermal condition.

3.3. Study Domain and Grids of Rainier Test Model

The study domain is composed of 17 geologic units, covers the entire unsaturated zone from ground surface to the water table, and includes the detailed geometry of the cavity, chimney, and altered zone, which were produced by the Rainier detonation. Grid-block sizes range from a few tens of centimeters around the chimney to tens of meters in the far field. As shown in Fig. 2, the physical system is discretized in a radially-symmetric (cylindrical) coordinate

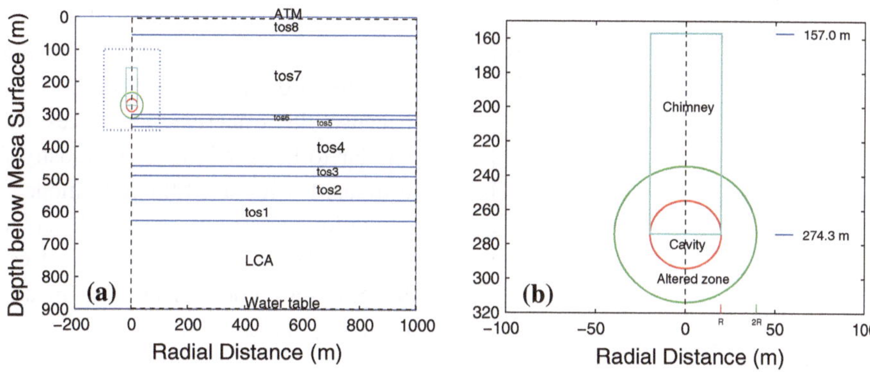

Figure 2
Study domain of the Rainier test model. **a** Vertical cross-section of the Rainier test model. "tos1", "tos2", $\cdots$, and "LCA" are geologic units where tos7 is subdivided into nine subunits. "ATM" denotes atmosphere layer. The *black dashed line* defines the study domain. **b** Local geometry of the cavity, chimney, and altered zone. The detonation location is at (0.0, 274.3), the cavity radius is 19.8 m, and the chimney height is 117.3 m between the depth of 157.0 m and the working point. "R" stands for the cavity radius

system. The conceptual model is based on a dual-permeability presentation of overlapping fracture and matrix continua (Sun *et al.*, 2010). The local geometry of chimney, cavity, and altered zone as shown in Fig. 2b is referred to Wadman and Richards (1961) and Tompson *et al.* (2011).

3.4. Physical Properties

Physical properties include component, phase, and rock properties. Rock properties include permeability, porosity, thermal conductivity, specific heat, and the van Genuchten parameters α, m, and residual saturation, to specify saturation-dependent gas and liquid permeability and capillary pressure (Carle *et al.*, 2008; Sun *et al.*, 2010; Tompson *et al.*, 2011).

3.4.1 Phase and Component Properties

The phase-equilibrium partitioning coefficient, K_{eq}, for a dissolved component in adjoining gas and liquid phases is defined as the ratio of its mole fractions in the gas phase, y, to its mole fraction in the liquid phase, x. A Clever gas solubility model (Clever, 1979, 1980) is applied to the gas phase-equilibrium relations for ^{133}Xe and ^{37}Ar such that

$$K_{eq} = \frac{y}{x} = \frac{p_g^r}{S \times p_g} \tag{1}$$

where S is the mole fraction (solubility) of component ^{133}Xe and ^{37}Ar dissolved in the liquid phase measured at the reference gas pressure, p_g^r, and p_g is the gas-phase pressure. The solubility is determined using experimental data as a function of temperature (Ekwurzel, 2004)

$$\ln S = A + \frac{B}{T_n} + C \ln T_n + F T_n \tag{2}$$

where A, B, C, and F are model parameters specific to xenon and argon, and T_n is the temperature in Kelvin divided by 100. The dependence of K_{eq} on temperature for xenon and argon is shown in Fig. 3.

The free molecular diffusion coefficients in gas and liquid phases, D_g and D_l, can be modeled as a function of temperature, T_k (Kelvin), and gas-phase pressure, p_g (Pa)

$$D_g = D_g^0 \frac{p_g^r}{p_g} \left(\frac{T_k}{273.15} \right)^{n_g}, \quad D_l = D_l^0 \left(\frac{T_k}{273.15} \right)^{n_l}. \tag{3}$$

The diffusion model parameters, D_g^0, D_l^0, n_g, and n_l should be fitted to experimental data. Because of the lack of experimental data for xenon and argon diffusion in the gas phase, we estimate the D_g^0 accordingly to molecular mass of gas-phase components and assume n_g to be zero. This assumption may result in an underestimation of the gas-phase transport in the high temperature zone. The diffusivity model in liquid phase was fitted using the experimental data of Holocher *et al.* (2002). Equilibrium partitioning and free molecular diffusivity model parameters for xenon and argon are given in Table 1.

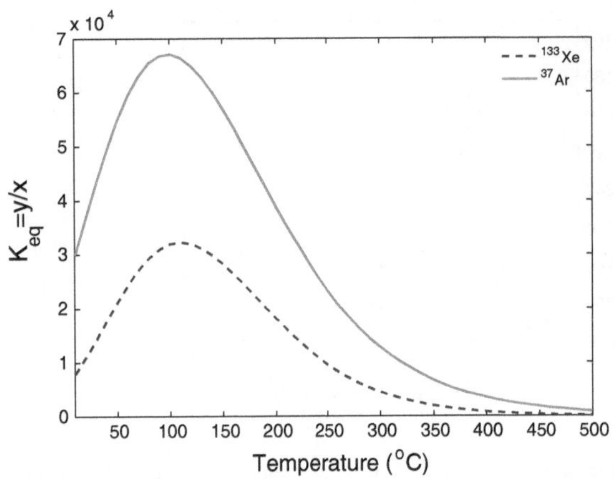

Figure 3
Phase-equilibrium coefficient of ^{133}Xe and ^{37}Ar as a function of temperature

3.4.2 Rock Properties

DKM rock properties are further categorized as medium-dependent hydrologic and thermal properties and partitioned physical properties (Buscheck *et al.*, 2002; Sun *et al.*, 2010). The hydrologic properties include relative permeability, porosity, van Genuchten α and m, residual saturation, and transition saturation (Sun *et al.*, 2010). The maximum saturation of all materials is assumed to be 1.0. For the fracture medium, three additional parameters, fracture connectivity, fracture frequency, and specific interaction area between fracture and matrix media, are included. Single-continuum materials, such as alluvium, use the same properties in fracture and matrix.

Appropriate values for the tortuosity factor are selected for the matrix and fracture continuum on the basis of the parameter range given by de Marsily (1986), which ranges from a value of 0.1 for clay to 0.7 for sand. A value of 0.2 is estimated for the matrix continuum because the pore sizes for the matrix are closer to that of clays than to that of sands. A value of 0.7 is assumed for the fracture continuum because the effective pore sizes for fractures are similar to those of sands. Tortuosity factors are also estimated with a Millington model using residual saturation from the van Genuchten parameters (Nitao, 1998). The effective diffusion coefficient for gas-phase components is a linear function of the tortuosity factor,

$$D_{\mathrm{e}} = \tau\phi(1 - S_{\mathrm{l}})D_{\mathrm{g}} \tag{4}$$

where τ is the tortuosity factor, ϕ is the porosity, and S_{l} is the liquid saturation.

3.4.3 Boundary/Initial Conditions

The ground surface and water table are considered to be the upper and lower boundaries of the Rainier test models. At the ground surface boundary, gas-phase conditions are specified by the air composition and the liquid-phase saturation is fixed to be zero. At the water table boundary, liquid-phase conditions are specified based on ground water compositions with a full liquid saturation. Initial conditions between ground surface and water table are obtained by running the pre-test model using the specified boundary conditions, physical properties, and stratigraphy. Fig. 4 shows a time series of the measured pressure and temperature at Rainier Mesa over 9,130-day period between January 1, 1983 and December 31, 2007. The data show the diurnal, annual, and daily variations. Pre-calibration exercise indicates that the variation of temperature does not significantly affect gas-phase flow while the daily variation of surface pressure significantly drives the push-pull processes between fractures and rock matrix. Therefore, the mean value of temperature over the NPE

Table 1

Equilibrium partitioning and free molecular diffusivity model parameters

Component	A	B	C	F	D_{g}^0 ($10^{-5}\mathrm{m}^2\mathrm{s}^{-1}$)	n_{g}	D_{l}^0 ($10^{-9}\mathrm{m}^2\mathrm{s}^{-1}$)	n_{l}
Ar	−57.67	74.76	20.14	0.0	2.00	0.0	1.44	7.22
Xe	−74.74	105.21	27.47	0.0	1.24	0.0	0.66	9.25

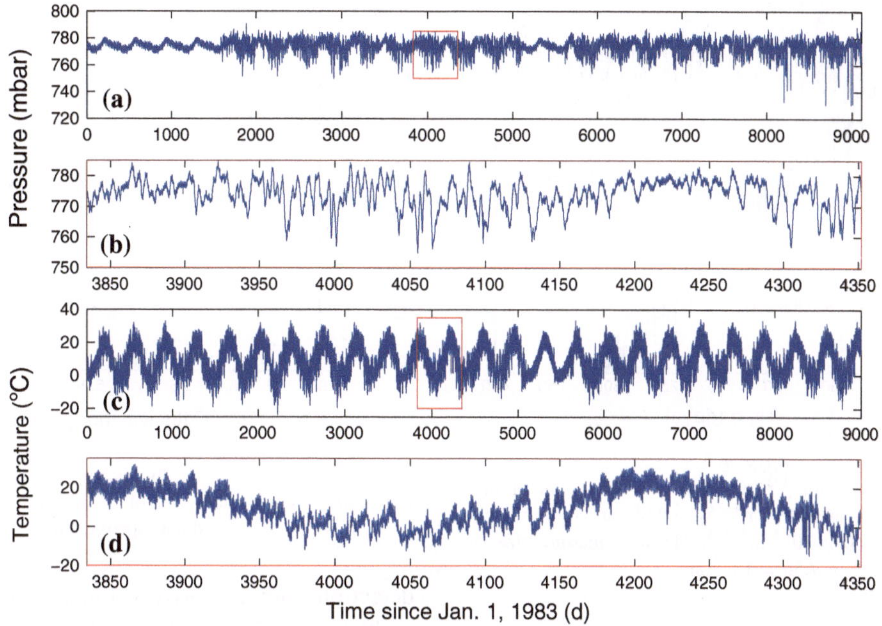

Figure 4
Barometric and temperature boundary conditions at ground surface. **a** Hourly pressure data from January 1, 1983 to December 31, 2007 (TOMPSON *et al.*, 2011). **b** Hourly pressure data used in the NPE experiment (CARRIGAN *et al.*, 1996). **c** Hourly temperature data from January 1, 1983 to December 31, 2007 (TOMPSON *et al.*, 2011). **d** Hourly temperature data during the NPE experiment

monitoring period (517 days), 9.98 °C is assumed as a constant temperature boundary at the surface.

Two sets of temperature data are available for estimating thermal conductivities of 17 stratigraphic layers at the locations underneath Rainier Mesa. At shallow depth, temperature was measured from ground surface to the working point (274.3 m depth) in well UCRL-3 before the detonation (WARNER and VIOLET, 1959). A more current measurement of 29.86 °C was obtained at the water table in nearby well ER-12-3 (TOMPSON *et al.*, 2011). Taking 9.98 °C and 29.86 °C as boundary temperatures at the surface and water table, respectively, and assuming a constant vertical heat flux of 3.30×10^{-2} (W m^{-2}) (WARNER and VIOLET, 1959), the thermal conductivities of those 17 layers are identified by fitting the measured temperature data as shown Fig. 5.

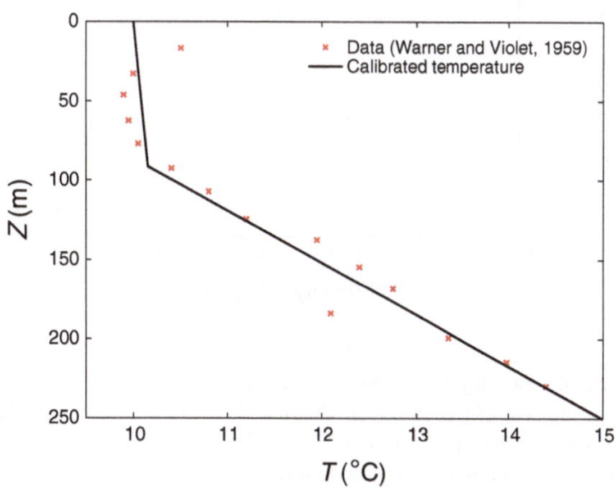

Figure 5
Pre-test temperature profile measured in well UCRL-3 on August 13, 1957 (WARNER and VIOLET, 1959)

In the source area (cavity and chimney), the cumulative production rates, 7.7×10^{-3} mol (6.7×10^{15} Bq) for ^{133}Xe and 7.0×10^{-5} mol (9.7×10^{12} Bq) for ^{37}Ar per kiloton yield (CARRIGAN *et al.*, 1996), are treated as the initial concentration conditions. The qualitative data of noble-gas concentrations in the region surrounding the Nevada test site can be referred to MULLEN and BARSH (1999). The initial temperature distribution over the entire study domain in the post-test period is estimated with a Hugoniot phenomenologic function of radial distance from the working point, yield, and other physic properties, as described by (BUTKOVICH, 1971; CARLE *et al.*, 2008).

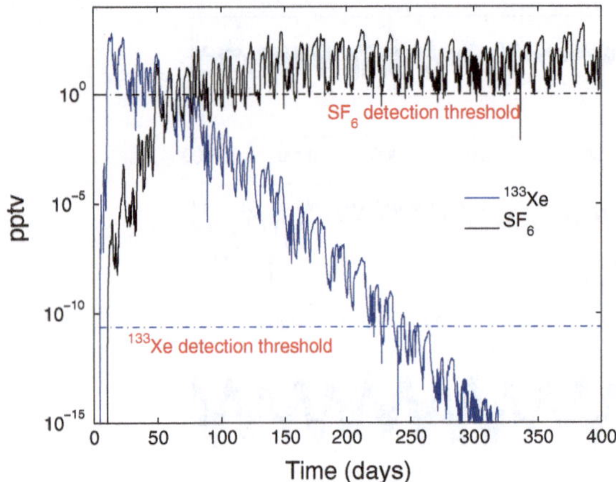

Figure 6
Concentration histories for ^{133}Xe and SF_6 at the outlet of the fracture (Fig. 1b). Diffusivities for ^{133}Xe and SF_6 are, respectively, 3.16×10^{-7} and 7.60×10^{-6} $m^2 s^{-1}$. The half aperture size is 7.50×10^{-4} m

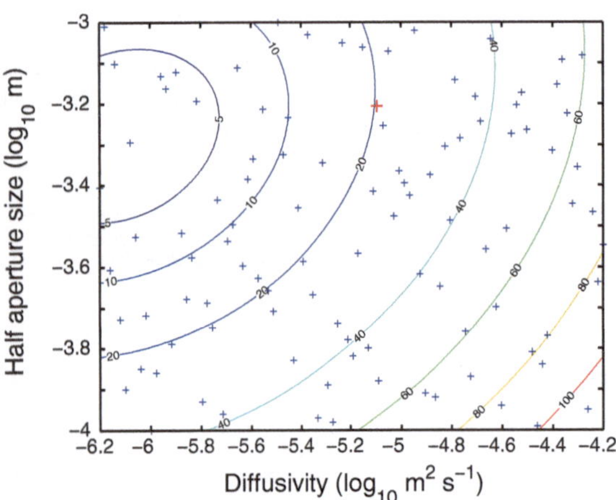

Figure 7
Arrival time (days) of ^{133}Xe as a function of diffusivity and half-aperture size of fracture. Note that the *blue* "+" indicates sample locations in the space of diffusivity and half-aperture size and the *red* "+" represents the deterministic model of CARRIGAN *et al.* (1996)

4. Modeling Results and Analyses

4.1. Model Results of Non-Proliferation Experiment

In the NPE model, the barometric pumping is the only driving force for gas-phase convection in fractures to transport halo gases to the surface. Among many other parameters, component diffusivity, fracture aperture size, and permeability and thickness of alluvium (if there is any) dominate the

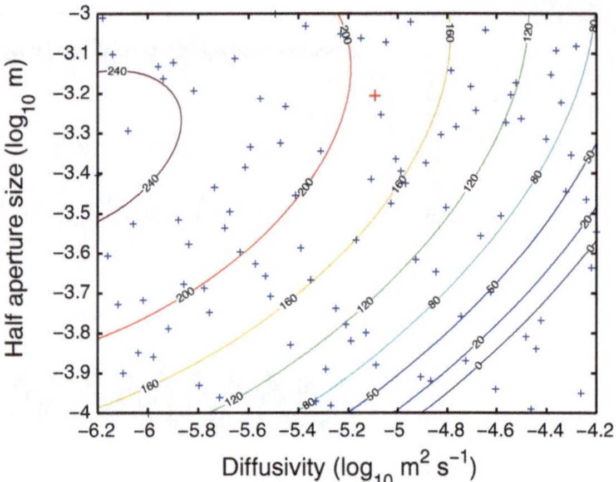

Figure 8
Detection window width (days) of ^{133}Xe as a function of diffusivity and half-aperture size

detection indices (arrival time, detection window width and height). The model of CARRIGAN *et al.* (1996) is modified using the USNT module (NITAO, 1998) with and without alluvium coverage. Two sets of 100 UNST models are developed for (1) possible combinations of the diffusivity and aperture size without the alluvium coverage, and (2) possible combinations of the diffusivity and alluvium permeability with the coverage. Model samples are produced using the LP-τ method in the PSUADE code (TONG, 2005).

4.1.1 Effect of Diffusivity and Fracture Aperture Size

In the isothermal case, the barometric pumping of surface pressure is the only driving force for gas component release through ground surface. Figure 6 shows the concentration histories for ^{133}Xe and SF_6 at the mouth of the fracture. Pure diffusion without considering the barometric pumping never delivers ^{133}Xe to ground surface at the detection level.

Taking 1.0 mBq m^{-3} (equivalent to 2.48×10^{-11} ppvt in air) as the measurement threshold for ^{133}Xe, the response surface of the arrival time is constructed using the PSUADE code (TONG, 2005). The first arrival time of the threshold concentration is proportional to the diffusivity in rock matrix and inversely proportional to the fracture aperture size when the permeability in alluvium takes the logarithmic mean value. As shown in Fig. 7, diffusivity appears more sensitive than aperture size to the arrival time. The red "+" in Fig. 7 is from the deterministic model of CARRIGAN *et al.* (1996). The arrival time (about 20 days) calculated based on the threshold of 1.0 mBq m^{-3} at the point is much shorter than 50 days (CARRIGAN *et al.*, 1996), which was simulated by using the threshold value of 1.0 Bq m^{-3}.

The detection window width is also examined as a function of the diffusivity and half-aperture size (Fig. 8). These two Figs. 7 and 8 can be used for detecting other gas components with specific diffusivity ranges.

4.1.2 *Effect of Diffusivity and Alluvium Permeability*

When the fracture aperture size is fixed with the logarithmic mean value, the detectability is evaluated in terms of matrix diffusivity and alluvium permeability. Alluvium layers or low-permeability rock on the top of the study domain reduce the effect of barometric pumping on gas-phase advection, dampen the magnitude of concentration signal, and cause a delay (phase shift) of the cyclic concentration history. Using the PSUADE code (TONG, 2005), a sensitivity analysis of detection characteristics is conducted in terms of the diffusivity and permeability.

Figures 9 and 10, show the arrival time and detection window width, respectively, as functions of ^{133}Xe diffusivity and the permeability in alluvium.

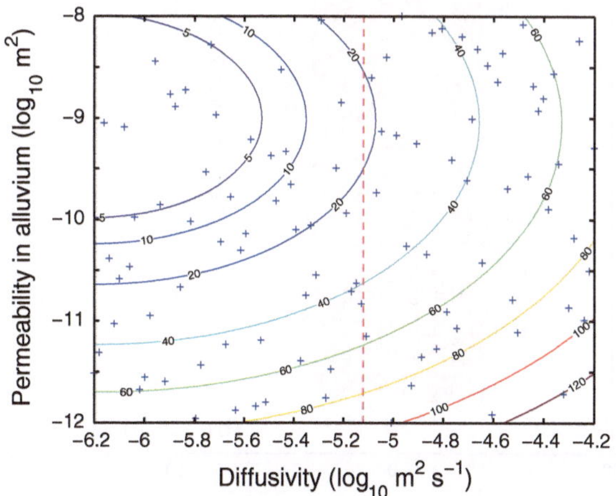

Figure 9
Arrival time (days) as a function of diffusivity and permeability in alluvium. The *dashed line* represents the deterministic model of CARRIGAN *et al.* (1996)

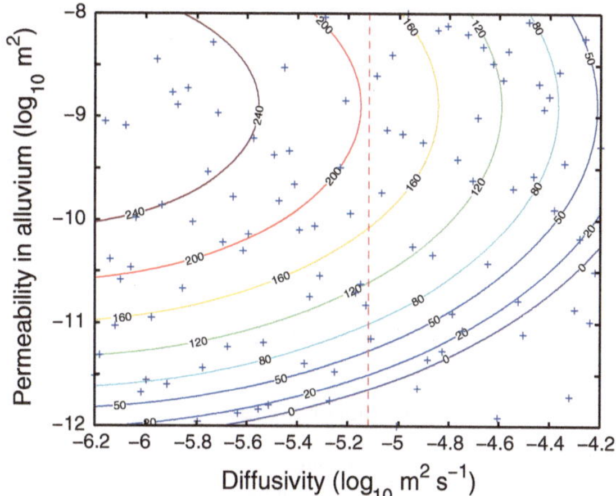

Figure 10
Detection window width (days) as a function of diffusivity and permeability in alluvium

The lower right corner defined by the zero-value contour in Fig. 10 indicates the nondetectable zone. The concentration history of ^{133}Xe on ground surface may never reach the detection threshold for high diffusivity in rock matrix and low permeability in alluvium. Although the alluvium thickness is fixed (2.0 m) in the model, it is expected that the nondetectable area grows monotonically with the thickness.

4.2. *Model Results of Rainier Test*

In the Rainier test model, we consider two-phase, four-component transport in a dual-permeability medium under thermal conditions. The main transport mechanisms include gas- and liquid-phase advection, dispersion and diffusion, and vapor-liquid partitioning. Test residual heat maintains the working point and surrounding area above boiling point for about 2 years (GOODALE *et al.*, 1958; WARNER and VIOLET, 1959).

The countercurrent flow of steam and liquid water is formed between the boiling and condensation zones when the vapor and liquid fluxes are of sufficient magnitude. Because of the high efficiency of latent heat transport, the temperature gradient in the heat-pipe zone is minimal, resulting in nearly isothermal conditions (SUN *et al.*, 2010). Capillary pressure plays an important role in determining gas-phase flux from the hot end to the cool end of the heat pipe, and liquid-phase flux back to the hot side (BIRKHOLZER, 2006).

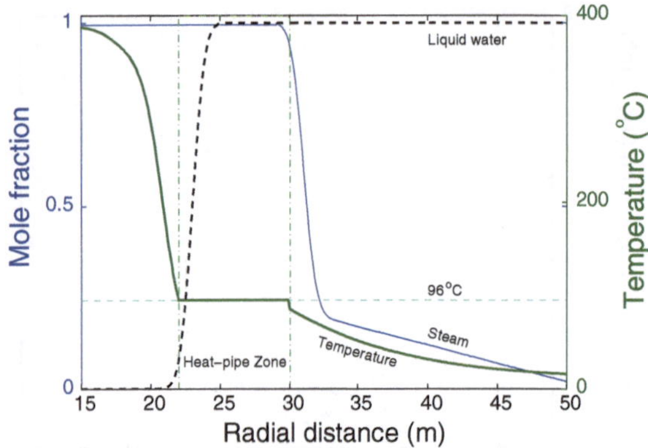

Figure 11
Temperature profile for z = 274.3 m and t = 60 days. The *dashed green box* represents the heat-pipe zone, in which gas phase flows from *left* to *right* in fractures and liquid phase flows from *right* to *left* in matrix by capillary pressure. The temperature in the heat-pipe zone remains constant

4.2.1 Steam Expansion and Heat-Pipe Phenomenon

Prior to the chimney collapse, the total water mass is estimated using the initial saturation, porosity, and melted volume of rock within the cavity (equivalent to the puddle glass volume, about 10 % of cavity volume, Tompson *et al.*, 2011). The steam bubble is formed by the test heat and well contained in the cavity by the compressed and impermeable cavity wall. Immediately after cavity collapse, the steam bubble expands, under thermally induced pressure gradient, to the chimney and altered zones (within two cavity radii). Temperature profile drops rapidly with the radial distance from the cavity. Therefore, most water outside the cavity after detonation will initially be in the liquid phase. The puddle zone temperature, however, will be much higher than the boiling point.

As shown in Fig. 11, a geologic heat pipe is formed by the countercurrent flow of steam in fractures and liquid water in the matrix between the boiling and condensation zones. The 8 m wide plateau on the temperature profile results from the gas-phase flux from the hot end to the cool end of the heat pipe and the liquid-phase flux toward to the working point. The steam bubble grows from the working point and shrinks after reaching the maximum size. As liquid water is supplied to the chimney zone, the steam production results in high pressure and drives xenon and argon away from the chimney.

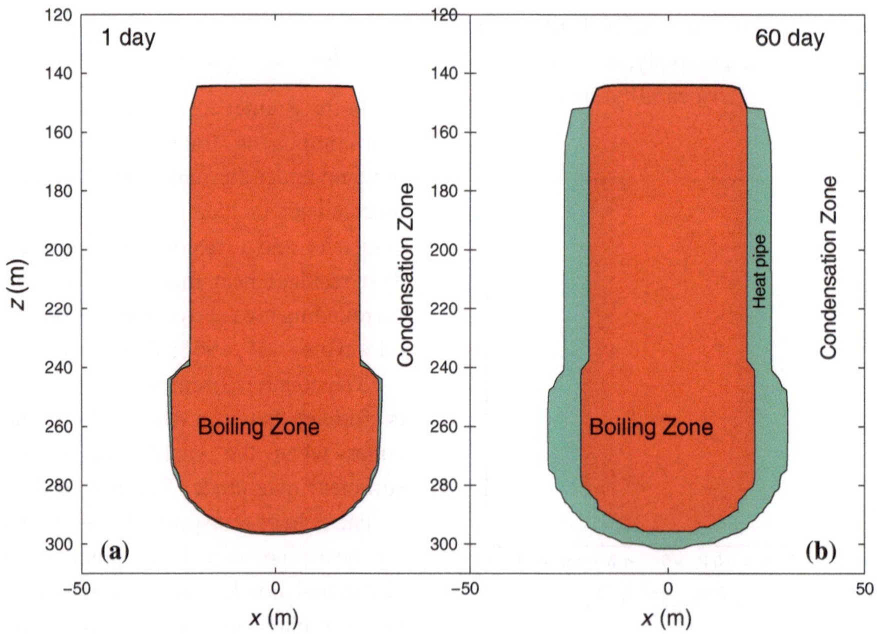

Figure 12
Heat-pipe zone at 1 day (**a**) and 60 days (**b**). The heat-pipe zone is defined between the boiling front (*red* zone) and condensation front (*blue* zone)

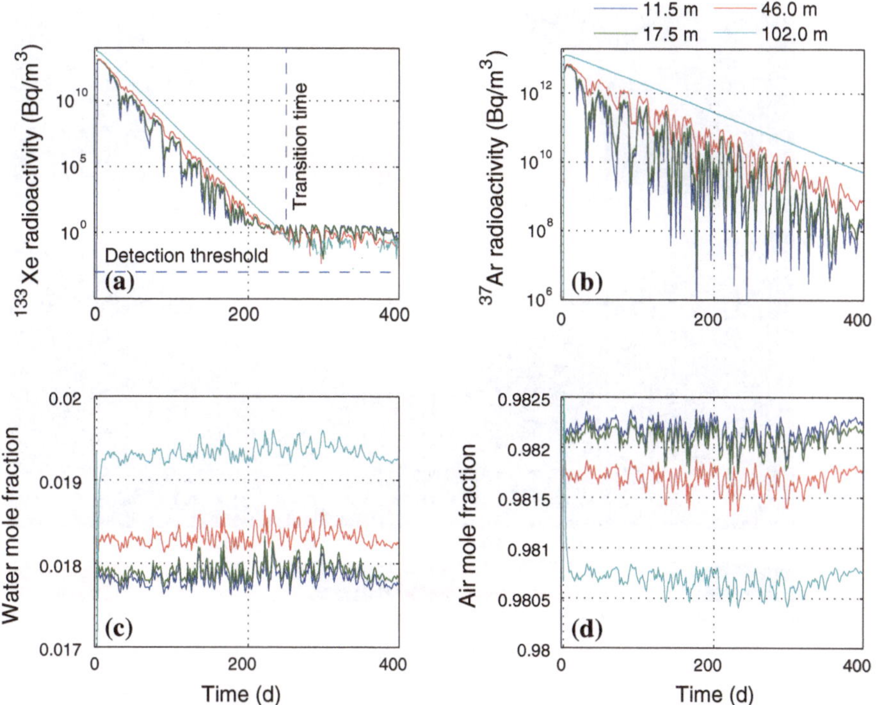

Figure 13
Concentration histories of ^{133}Xe (**a**), ^{37}Ar (**b**), water (**c**), and air (**d**) at depths of 11.5, 17.5, 46.0, and 102.0 m

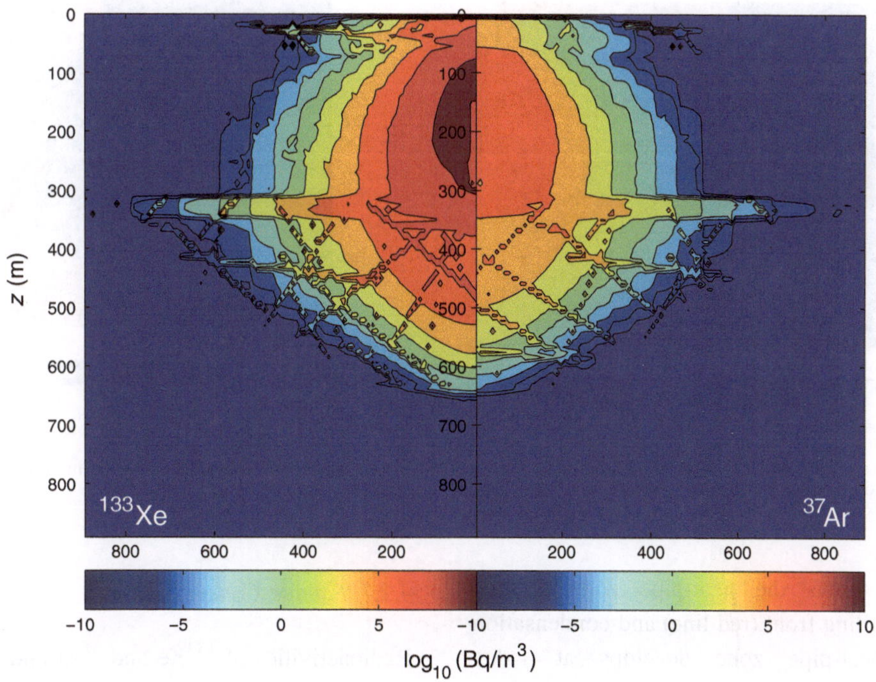

Figure 14
Contour plots of ^{133}Xe and ^{37}Ar radioactivities (Bq/m^3) at 1 day. Note that the half-symmetry domain of ^{133}Xe plume is flipped to the left for comparison purposes. Fracture aperture size is exaggerated for the purpose of visualization

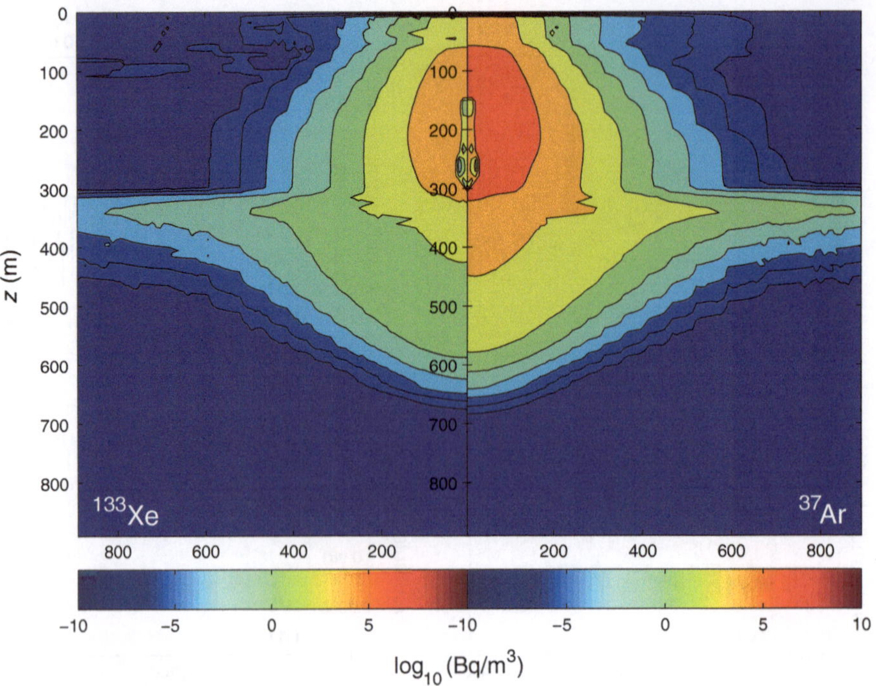

Figure 15
Contour plots of ^{133}Xe and ^{37}Ar radioactivities (Bq/m^3) at 100 days

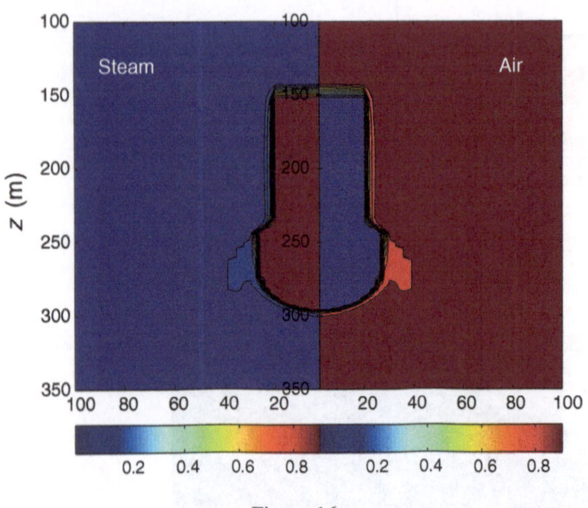

Figure 16
Mole-fraction distribution of steam and air at 1 day

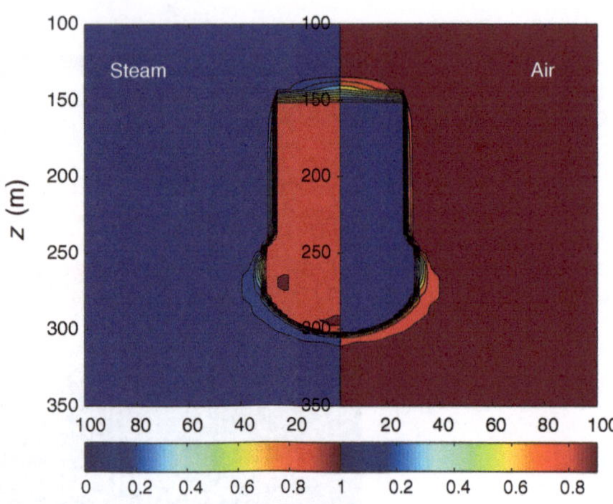

Figure 17
Mole-fraction distribution of steam and air at 100 days

Figure 12 shows the heat-pipe zone defined between the boiling front (red line) and condensation front. The heat-pipe zone develops at 1 day (Fig. 12a) and reaches 8 m wide (at $z = 274.3$ m) at 60 days. The inner line bounds the steam zone while the outer line holds the liquid water.

4.2.2 Sample Concentrations

Radioactivities of ^{133}Xe and ^{37}Ar and mole fractions of water and air are measured in the fracture medium at four selected depths (11.5, 17.5, 46, and 102 m) on the vertical central line and above the chimney

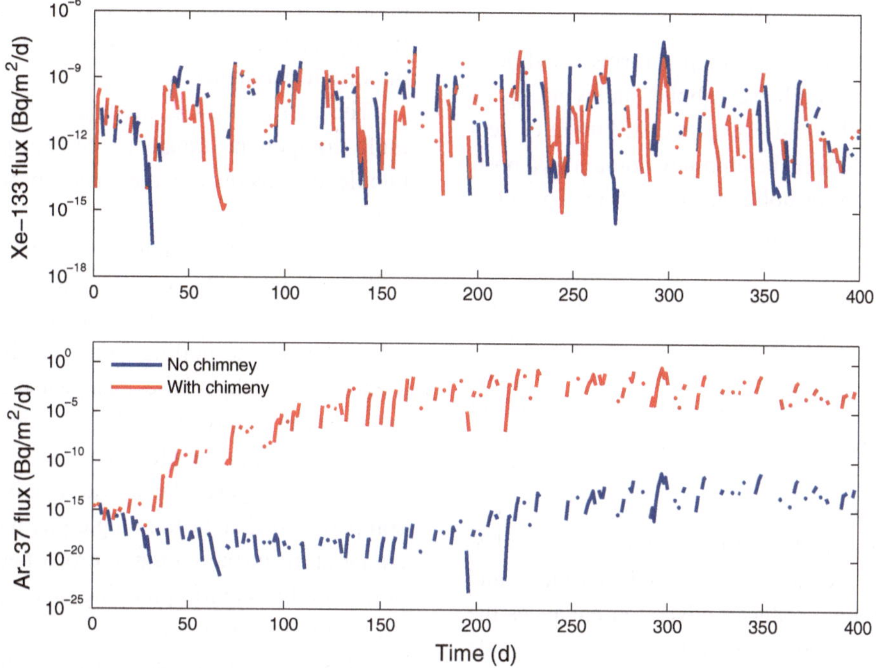

Figure 18
Upward fluxes of ^{133}Xe and ^{37}Ar at $x = 0.0$ m and $z = 8.0$ m. Note that the downward fluxes are not plotted

($x = 0$, Fig. 1). As shown in Fig. 13, both ^{133}Xe and ^{37}Ar reach their peak concentrations within a few hours and then decay accordingly to their half-lives. The detection window width for ^{133}Xe is about 260 days. The concentration history of ^{133}Xe at those points (at depths of 11.5, 17.5, 46.0, and 102.0 m) is dominated by the barometric boundary condition at low concentration magnitudes. Figure 13 shows the different mechanisms before and after 260 days. The thermally driven advection and the first-order decay mainly control ^{133}Xe concentration profile in the first 260 days. When the concentration drops to a certain magnitude near the threshold level, the effect of advection driven by barometric pumping becomes more apparent. For this reason, ^{133}Xe concentration after 260 days fluctuates with the boundary pressure. However, because of the longer half-life of ^{37}Ar, its concentration magnitude is much higher than the boundary concentration and the surface-boundary effect does not dominate the concentration history of ^{37}Ar with the simulation time (500 days).

4.2.3 Spatial Distributions of Noble Gas Components

Test-related ^{133}Xe and ^{37}Ar radioactivities (Bq/m^3) at various times reflect the transport processes in the fractured rock. Figures 14 and 15 show the ^{133}Xe and ^{37}Ar concentrations at 1 day and 100 days. The ^{37}Ar plume is widely spread because of its higher diffusivity and longer half-life. Discrete fractures are generated accordingly to fracture frequency and orientation in each geological layer for visualizing the concentration difference between fractures and matrix. Figure 14 tells that gas components migrate significantly faster in fractures than in matrix in early time.

Figures 16 and 17 show the distribution of the mole fraction of steam and air at 1 day and 100 days. The steam bubble is expanded from the working point in the chimney and mainly contained within the boiling front. Both boiling and condensation fronts expand and drive the test-related noble gas away from the chimney.

4.3. *Effect of Chimney Collapse*

Test-induced chimney collapse forms a cylindrical "chimney" of collapse rubbles and void space (Carle *et al.*, 2008) and causes redistribution of heat and radionuclides over the chimney domain as a post-collapse condition for noble-gas migration. To understand the effect of chimney collapse on noble-gas concentrations at or close to ground surface, we conducted numerical experiments of the Rainier-test system with and without the chimney (see Fig. 1). As shown in Fig. 18, the 117-m chimney makes ^{37}Ar (half life $t_{1/2} = 35.1$ days) upward flux much higher after 20 days. However, the chimney does not make a significant difference in the upward flux of short half-life nuclides, such as ^{133}Xe ($t_{1/2} = 5.24$ days). The result indicates that the chimney elevates the radionuclide source closer to ground surface and barometric-pumping induced advection enhances the gas migration towards to ground surface for stable isotopes and long half-life nuclides. In the reactive transport, which is dominated by fast reactions (short half-life decay), the effect of chimney collapse is limited.

5. *Conclusions*

The detectability of noble gas ^{133}Xe and ^{37}Ar is quantitatively characterized using (1) arrival time, (2) detection window width, and (3) detection window height. The transport of ^{133}Xe and ^{37}Ar is described by a single-gas-phase model and a two-phase multi-component model. In this study, we demonstrated the processes of ^{133}Xe and ^{37}Ar transport in fractured rock.

When the temperature is below the boiling point and the subsurface system is sufficiently dry, the system can be modeled as an isothermal and single-gas-phase transport as in the case of the NPE. Then, the barometric pumping becomes the major driving force to deliver ^{133}Xe and ^{37}Ar to ground surface.

If the fractured rock is not covered by alluvium, the diffusivity in rock matrix and fracture aperture size are sensitive parameters contributing the detectability of noble gases at ground surface. When the ground surface is covered by alluvium, the permeability in the overburden layer replaces the fracture aperture size to become a sensitive and more important parameter.

The thermal test is modeled as a two-phase and multi-component transport. Steam production and bubble expansion are the major processes to drive noble gas components to ground surface. Geologic heat pipe is kept constant (at boiling point) from its boiling front to the condensation front. Capillary-pressure in the rock matrix drives liquid water to move toward the boiling front and the pressure gradient elevated by phase change (steam production) drives steam away from the chimney through fractures. The steam bubble becomes a major driver for noble gas transport. The arrival time is usually shorter compared to that in the single-gas model. After the temperature in the chimney drops below the boiling point, the barometric pumping takes over the role as the major driving force.

This study is limited to two specific (isothermal and thermal) models. Although the isothermal and single-phase transport model is computationally less expensive, the thermal and two-phase transport model is capable of addressing noble-gas migration by thermal expansion. Cumulative yields of ^{133}Xe and ^{37}Ar are considered as the initial concentrations in cavity and chimney. The effect of the parent-daughter decay nature of source terms on noble-gas migration deserves further investigation. Geo-statistical modeling is also necessary to establish the transport behavior of these gases over a wide range of explosive yields and geological conditions.

Acknowledgments

The authors wish to thank anonymous reviewers and Chuanhe Lu and Jerry J. Sweeney at Lawrence Livermore National Laboratory for their careful review and helpful comments that led to an improved manuscript. This research was funded by Office of Nuclear Verification (NA-243), US Department of Energy and performed under the auspices of the US Department of Energy by Lawrence Livermore National Laboratory under Contract No. DE-AC52-07NA27344.

References

AALSETH, C.E., DAY, A.R., HAAS, D.A., HOPPE, E.W., HYRONIMUS, B.J., KEILLOR, M.E., MACE, E.K., ORRELL, J.L., SEIFERT, A., WOODS, V. T.: *Measurement of ^{37}Ar to support technology for On-Site Inspection under the Comprehensive Nuclear-Test-Ban-Treaty*, Nucl. Instrum. Meth. A, *652*(1), 58–61, 2011.

AUER, L.H., ROSENBERG, N.D., BIRDSELL, K.H., and WHITNEY, E. M.: *The effect of barometric pumping on contaminant transport*, J. Contam. Hydrol., *24*, 145–166, 1996.

AUER, M., AXELSSON, A., BLANCHARD, X., BOWYER, B.W., BRACHET, G., BULOWSKI, Y., ELMGREN, K., FONTAINE, J.P., HARMS, W., HAYES, J.C., HEIMBIGNER, T.R., MCINTYER, J.I., PANISKO, M.E., POPOV, Y., RINGBOM, A., SARTORIUS, H., SCHMID, S., SCHULZE, J., SCHLOSSER, C., TAFFARY, T., WEISS W., and WERNSPERGER, B.: *Intercomparison experiments of systems for the measurement of xenon radionuclides in the atmosphere*, Appl. Rad. Isotopes, *60*, 863–877, 2004.

AUER, M., KUMBERG, T., SARTORIUS, H., WERNSPERGER, B., and SCHLOSSER, C.: *Ten years of development of equipment for measurement of atmospheric radioactive xenon for the verification of the CTBT*, Pure Appl. Geophys., doi:10.1007/s00024-009-0027-y, 2010.

BIRKHOLZER, J.T.: *Estimating liquid fluxes in thermally perturbed fractured rock using measured temperature profiles* J. Hydrol., *327*, 496–515, 2006.

BOWYER, T.W., SCHLOSSER, C., ABEL, K.H., AUER, M., HAYERS, J. C., HEIMBIGNER, T.R., MCINTYRE, J.I., PANISKO, M.C., REEDER, P.L., ATORIUS, H., SCHULZE, J., and WEISS, W.: *Detection and analysis of xenon isotopes for the comprehensive nuclear-test-ban treaty International Monitoring System*, J. Environ. Radioactiv., *59*, 139–151, 2002.

BUSCHECK, T.A., ROSENBERG, N.D., GANSEMER, J., and SUN, Y.: *Thermohydrologic behavior at an underground nuclear waste repository*, Water Resour. Res., *38*(3), 1431–1447, 2002.

BUTKOVICH, T.R., *Influence of water in rocks on effects of underground nuclear explosions*, J. Goephy. Res., *76*(8), 1993–2011, 1971.

CARLE, S.F., ZAVARIN, M., SUN, Y., and PAWLOSKI, G.A.: *Evaluation of hydrologic source term processes for underground nuclear tests in Yucca Flat, Nevada Test Site: Carbonate Tests*, LLNL-TR-403485, 2008.

CARRIGAN, C.R., HEINLE, R.A., HUDSON, G.B., NITAO, J.J., and ZUCCA, J.J.: *Trace gas emissions on geological faults as indicators of underground nuclear testing*, Nature, *382*(6591), 1996.

CLEVER, H.L.: Ed., *Helium and Neon, IUPAC Solubility Data Series*, Vol. *1*, Pergamon Press, Oxford, 1979.

CLEVER, H.L.: Ed., *Argon, IUPAC Solubility Data Series*, Vol. *4*, Pergamon Press, Oxford, 1980.

DE MARSILY, G.: *Quantitative Hydrogeology: Groundwater Hydrology For Engineers*, Academic Press, 1986.

DRESEL, P.E. and WAICHLER, S.R.: *Evaluation of xenon gas detection as a means for identifying buried transuranic waste at the radioactive waste management complex*, Idaho National Environmental and Enginerring Laboratory, Pacific Northwest National Laboratory, PNNL-14617, 2004.

EKWURZEL, B.: *LLNL Isotope Laboratories Data Manual*, Lawrence Livermore National Laboratory, UCRL-TR-203316, 2004.

GOODALE, T.C., RAGENT, B., and SAMUEL, A.H.: *Temperatures from underground detonation, Shot Rainier*, University of California Radiation Laboratory, Mercury, Nevada, WT-1527, 1958.

HAO, Y., SUN, Y., and NITAO, J.J.: *Overview of NUFT—a versatile numerical model for simulating flow and reactive transport in porous media*, Lawrence Livermore National Laboratory, LLNL-BOOK-42714-DRAFT, 2010.

HO, C.K.: *Models of fracture-matrix interaction during multiphase heat and mass flow in unsaturated fractured porous media*, Sixth Symposium on Multiphase Transport in Porous Media, Dallas, TX, Nov. 16–21, 1997.

HOLOCHER, J., PEETERS, F., AESCHBACH-HERTIG, W., HOFER, M., BRENNWALD . M. S., KINZELBACH, W., and KIPFER, R.: *Experimental investigation on the formation of excess air in quasi-saturated porous media*, Geochim. Cosmochim. Ac., *66*(23), 4103–4117, 2002.

KALINOWSKI, M.B., AXELSSON, A., BEAN, M., BLANCHARD, X., BOWYER, T.W., BRACHET, G., HEBEL, S., MCINTYRE, J.I., PETERS, J., PISTNER, C., RAITH, M., RINGBOM, A., SAEY, P.R.J., SCHLOSSER, C., STOCKI, T.J., TAFFARY, T., and UNGAR, R.K.: *Discrimination of nuclear explosions against civilian sources based on atmospheric xenon isotopic activity ratios*, Pure Appl. Geophys., *167*, 517–539, 2010.

LINDSTROM, F.T., CAWLFIELD, D.E., and BAKER, L.E.: *Sensitivity analysis of the noble gas transport and fate model*: CASCADR9, DOE/NV/11432-129, 1994.

MARTINEZ, M.J., and NILSON, R.H.: *Estimates of barometric pumping of moisture through unsaturated fractured rock*, Transport Porous Med., *36*, 85–119, 1999.

MULLEN, A.A. and BARTH, J.: *Noble gas and tritium-in-air offsite environmental monitoring program summary from 1970–1995*, EPA-402-R-99-005, USEPA, Las Vegas, 1999.

NEEPER, D.A.: *A model of oscillatory transport in granular soils, with application to barometric pumping and earth tides*, J. Contam. Hydrol., *48*, 237–252, 2001.

NEEPER, D.A.: *Investigation of the vadose zone using barometric pressure cycles*, J. Contam. Hydrol., *54*, 59–80, 2002.

NILSON, R.H., PETERSON, E.W., LIE, K.H., BURKHARD, N.R., and HEARST, J.R.: *Atmospheric pumping: a mechanism causing vertical transport of contaminant gases through fractures permeable media*, J. Geophys. Res., *96*(B13), 21933–21948, 1991.

NITAO, J.J.: *User's manual for the USNT module of the NUFT code, version 2 (NP-phase, NC-component, thermal)*, Lawrence Livermore National Laboratory, UCRL-MA-130653, 1998.

PRUESS, K.: Numerical codes for continuum modeling of gas transport in porous media, Ho, C. and Webb, S. (eds.), *Gas transport in porous media*, 213–220, Springer, 2006.

ROSSABI, J.: Analyzing barometric pumping to characterize subsurface permeability, HO, C. and WEBB, S. (eds.), *Gas transport in porous media*, 279–290, Springer, 2006.

SAEY, P.R.J.: *The influence of radiopharmaceutical isotope production on the global radioxenon background*, J. Environ. Radioactivity, *100*, 396–406, 2009.

SUN, Y., BUSCHECK, T.A., LEE, K.H., HAO, Y., and JAMES, S.C.: *Modeling thermal-hydrological processes for a heated fractured rock system: impact of a capillary-pressure maximum*, Transport Porous Med., *83*, 501–523, 2010.

TIDWELL, V.: Scaling issues in porous and fractured media, HO, C. and WEBB, S. (eds.), *Gas transport in porous media* , 201–212, Springer, 2006.

TOMPSON, A.F.B., ZAVARIN, M., MCNAB, W.W., CARLE, S.F., SHUMAKER, D.E., LU, C., SUN, Y., PAWLOSKI, G.A., HU, Q., and ROBERTS, S.K.: *Hydrologic source term processes and models for underground nuclear tests at Rainier Mesa and Shoshone Mountain, Nevada National Security Site*, Lawrence Livermore National Laboratory, LLNL-TR-483852-DRAFT, Livermore, California, 2011.

TONG, C.: *PSUADE User's Manual*, Lawrence Livermore National Laboratory, LLNL-SM-407882, 2005.

WADMAN R.E. and RICHARDS, W.D.: *Postshot geologic studies of excavations below Rainier ground zero*, Lawrence Livermore National Laboratory, UCRL-6586, 31p, Livermore, California, 1961.

WARNER, S.E. and VIOLET, C.E.: *Properties of the environment of underground nuclear detonations at the Nevada Test Site, UCRL-5542*, Lawrence Radiation Laboratory, Livermore, California, 1959.

WEBB, S.W.: *Gas-phase diffusion in porous media*, SAND96-1197, UC-403, 1996.

WU, Y., Pruess, K., and Persoff, P.: *Gas flow in porous media with Klinkenberg effects*, Transport Porous Med., *32*, 117–137, 1998.

WYATT, D.E., RICHERS, D.M., and PIRKLE, R.J.: *Barometric pumping effects on soil gas studies for geological and environmental characterization*, Environ. Geol., *25*(4), 243–250, 1995.

(Received November 8, 2011, revised March 4, 2012, accepted May 30, 2012, Published online July 7, 2012)

Pure Appl. Geophys. 171 (2014), 751–761

DOI 10.1007/s00024-012-0639-5

Pure and Applied Geophysics

Application of Geophysical Techniques in Identifying UNE Signatures at Semipalatinsk Test Site (for OSI Purposes)

A. Belyashov,[1] V. Shaitorov,[1] and M. Yefremov[1]

Abstract—This article describes geological and geophysical studies of an underground nuclear explosion area in one of the boreholes at the Semipalatinsk test site in Kazakhstan. During these studies, the typical elements of mechanical impact of the underground explosion on the host medium—fracturing of rock, spall zones, faults, cracks, etc., were observed. This information supplements to the database of underground nuclear explosion phenomenology and can be applied in fulfilling on-site inspection tasks under the Comprehensive Nuclear-Test-Ban Treaty.

Key words: Underground nuclear explosions, semipalatinsk test site, comprehensive nuclear-test-ban treaty, on-site inspection, post-explosive disintegrated zones, geological and geophysical methods, seismic methods, atmogeochemical methods, thermometry.

1. Introduction

To fulfill on-site inspection (OSI) tasks under the Comprehensive Nuclear-Test-Ban Treaty verification regime, particularly to detect and identify underground nuclear explosion (UNE) consequences on the inspected area, one should know how UNE signatures appear in geophysical fields. A full-fledged study of UNE phenomenology is complicated by a lack of sufficient international information and data on studies of consequences from real underground nuclear tests.

UNEs are divided into two main groups—borehole explosions and tunnel explosions, which differ considerably in their mechanical impact on host rock. This paper will describe UNE phenomenology aspects revealed within geophysical fields evident from real borehole explosions conducted in the former Semipalatinsk test site (STS).

According to Adushkin and Spivak (2004), the following disintegration zones are formed as a result of destructive effects of borehole UNEs on a host medium:

- Camouflet cavity with relative radius 10–13.6 m/kton$^{1/3}$;
- Zone of rock contortion with relative thickness up to 3–4 m/kton$^{1/3}$;
- Damaged zone to relative distances up to 24–34 m/kton$^{1/3}$;
- High-density fissure zone to relative distances 50–55 m/kton$^{1/3}$;
- Block fractured zone with radius to 65–70 m/kton$^{1/3}$;
- The zone of local irreversible changes to the distances significantly greater than 100 m/kton$^{1/3}$.

Figure 1 schematically presents a typical configuration of the borehole UNE's central zone (Adushkin and Spivak, 2004).

General sense of tunnel UNE phenomenology can be acquired from descriptions of results of studies of UNE impacts on a host medium from explosions conducted in Nevada Test Site (Johnson *et al.*, 1959).

As it comes from the above characteristics of a typical UNE phenomenology, absolute dimensions and configurations of disintegrated host rocks resulting from borehole UNEs depend, in the first place, on the charge yield (kton); in the general case, a directly proportional relation between the charge yield and disintegrated zone size is observed (Adushkin and Spivak, 2004; Arhipov *et al.*, 2001). In this context, zones of post-explosive disintegrated rocks against explosion yield of tens kilotons may

[1] Institute of Geophysical Research, National Nuclear Center, Kurchatov, Kazakhstan. E-mail: abelyashov@igr.kz

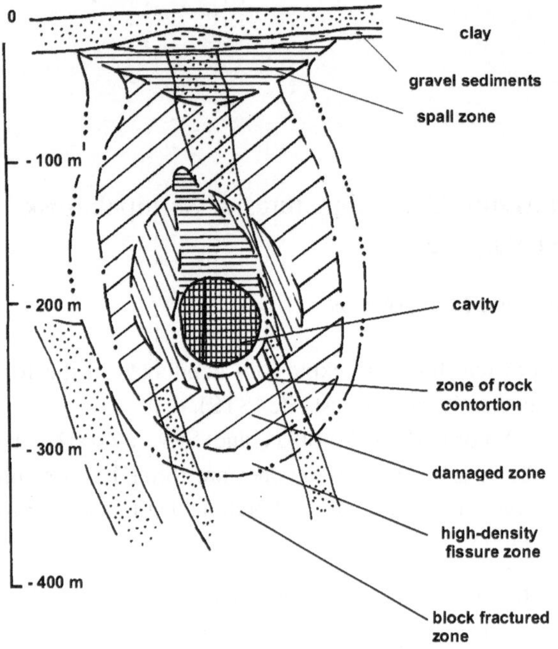

Figure 1
Typical configuration of the borehole UNE's central zone

reach several hundred meters in size (ADUSHKIN and SPIVAK, 1993; CHADWICK *et al.*, 1964).

Other factors influencing the size and form of zones of disintegration are host geology itself and the depth of charge laying. With the increase of the concentrated (point) charge location to values not <7–10 m/ton$^{1/3}$ (a situation of camouflet explosion) lithostatic pressure and density of rocks grow at the depth of an explosion source that, generally, results in reduction of the sizes of zones of deformation of the medium which can be considered in spherically symmetric representation (ADUSHKIN and SPIVAK, 1993). The small depths of a charge, when a free surface (contact "soil–air") near the explosion source significantly complicates mechanisms of formation of post-explosive zones and, as a rule, results in increase in their sizes (ADUSHKIN and SPIVAK, 1993; ARHIPOV *et al.*, 2001).

Considering a huge variety of geological types of rocks and the general heterogeneity of the geophysical environment, to estimate influence of petrophysical properties on formation process of zones of post-explosive disintegration and their sizes is rather difficult. All the variety of rock units may for clarity be subdivided into three types of petrophysical environments (STAROSTIN *et al.*, 1995): viscous and solid (basalts, diabases, porphyritic rocks, gabbro and other mafic rocks), plastic and soft (terrigenous-carbonates, tuff siltstones, slates and others) and brittle rocks (felsic rocks: granitoids, lava and diorite—rhyolitic subvolcanic rocks, tuffites). The above listed types differently affected by destructive impact of explosion. In particular, brittle rocks are very susceptible to mechanical destruction by explosion while plastic and soft rocks are less susceptible to explosive effects (ORLENKO *et al.*, 2002).

In the course of time and under the influence of various external factors such as lithostatic gravitational pressure, watering of host geology, natural and artificial seismic effects, the initial structure and sizes of post-explosive disintegration zones would have to be modified to become denser and to consolidate and a current structure, consequently, should differ from that one presented in Fig. 1. Nevertheless, zones of residual post-explosive disintegration existing currently allow studying UNE phenomenology to accomplish OSI tasks.

Various geological and geophysical technologies are applied in the studies of borehole UNE areas on the territory of the STS. The main goal of this research is study of the UNE's phenomenology and the monitoring of geodynamical processes proceeding in near-focal areas. The paper presents the main results of geological and geophysical observations in studying consequences from the underground nuclear explosion in borehole #104, site Sary-Uzen (one of the test sites in the central part of the STS).

The underground explosion in borehole #104 was conducted on 21 July 1970. The charge yield was from 0.001 to 20 kton (according to some sources, 7 kton) and the charge depth was 225 m. This UNE was fully camouflet (i.e., without products release to the daylight surface and to the atmosphere). The host geology is built up with metamorphogene sedimentary strata of the Middle Devonian, covered by Neogene clays and Quaternary alluvium and sliderocks. A lack of detailed prior geological information about UNE site makes interpretation difficult, on the one hand, but on the other hand this makes the situation more realistic in terms of OSI.

Since the studied underground explosion was conducted more than 40 years ago, and many short-life artifacts (for example, active aftershocks) no longer exist, seismic technique of remote sounding was applied to discover long-life consequences of the explosion. Apart from underground changing geology, revealed by indirect remote techniques, UNEs are also featured by surface consequences in form of various geochemical and temperature anomalies.

Geological and geophysical observations at borehole #104 were conducted in 2009.

2. *Seismic Techniques*

The main objects of seismic study at borehole #104 were rock disintegration areas and related fracture zones, displacements, micro-faults and other environment discontinuity elements generated by the effect from explosion shock wave on host rocks (Adushkin and Spivak, 1993).

2.1. *Refracted Wave Method/Diving Wave Tomography*

The refracted wave method/diving wave tomography (RWM/DWT) is applied to detect velocity heterogeneities in the near-surface section confined to the fracture zones due to the UNE. The observations were conducted along two profiles, west-east and south-north strikes of 1,200 m length. The distance between seismic energy injection points was set at 100 m and the distance between recording points was 10 m. Nonaccelerated drop weight was used as the seismic source ($\approx$15 kJ).

Figure 2a displays diagrams of boundary velocity, and Fig. 2b shows a tomographic velocity sections along the profiles line running through the UNE's epicenter. The boundary velocity means a velocity of the head wave, travelling along the border between the upper low-velocity layer and the top of the bedrock (at the depth of 40 m), and its value was obtained using Hagedorn plus-minus method (Hagedoorn, 1959). The error margin of the boundary velocity value determination is not more than 10 %. Tomographic sections were built by standard iteration procedures using "X-Tomo" software (www.xgeo.ru).

The specific feature of post-explosion velocity distribution of elastic waves is a readily diagnosable low-velocity area in the bedrock near the UNE epicenter. At the top of the bedrock minimal boundary velocity values (3.0–3.5 km/s) are measured in the radius of 100 m from the explosion epicenter. A peculiar feature of elastic waves velocity distribution in the profiles is the area of 250 m in radius with low-velocity values that can be traced through the whole depth by diving wave sounding.

Taking into consideration spatial coherence of the identified area with the UNE epicenter and its locality, it was interpreted as post-explosion fractured zone. In this regard such fracturing to a greater degree is detected within subsidence crater formed in interlocked natural and man-made region characterized by increased rocks disturbance adjoining focal area of UNE.

2.2. *Diffracted—Scattered Wave Method*

Physical underlay of the diffracted—scattered wave method (DSWM) is increased scattering and diffraction of seismic waves through acoustic discontinuities especially fractured zones (Shapiro and Faizullin, 1986).

Taking into account that scattered (diffracted) waves are specified by reduced (by 10–100) intensity, the use of which as an informative parameter shall require observation systems with multiple coverage to be available with further summing of the obtained wave fields for a common diffraction (scattering) point. This summation is made by using the focusing conversion algorithm suggested by Starobinets (1978) to enhance adequacy of tectonically disturbed zones mapping by means of seismic refracted wave method. According to this algorithm a studied geological block is scanned with the set resolution vertically and horizontally at points for which scattered (diffracted) wave energies are computed and the areas with maximal values of this parameter are further interpreted as disturbed structures.

A similar approach is done in applying scattered waves in seismic technology of a side-look location (Faizullin and Chirkin, 1998). Its high performance is acknowledged through detection and delineation of hydrocarbon reservoirs, in studying morphology of

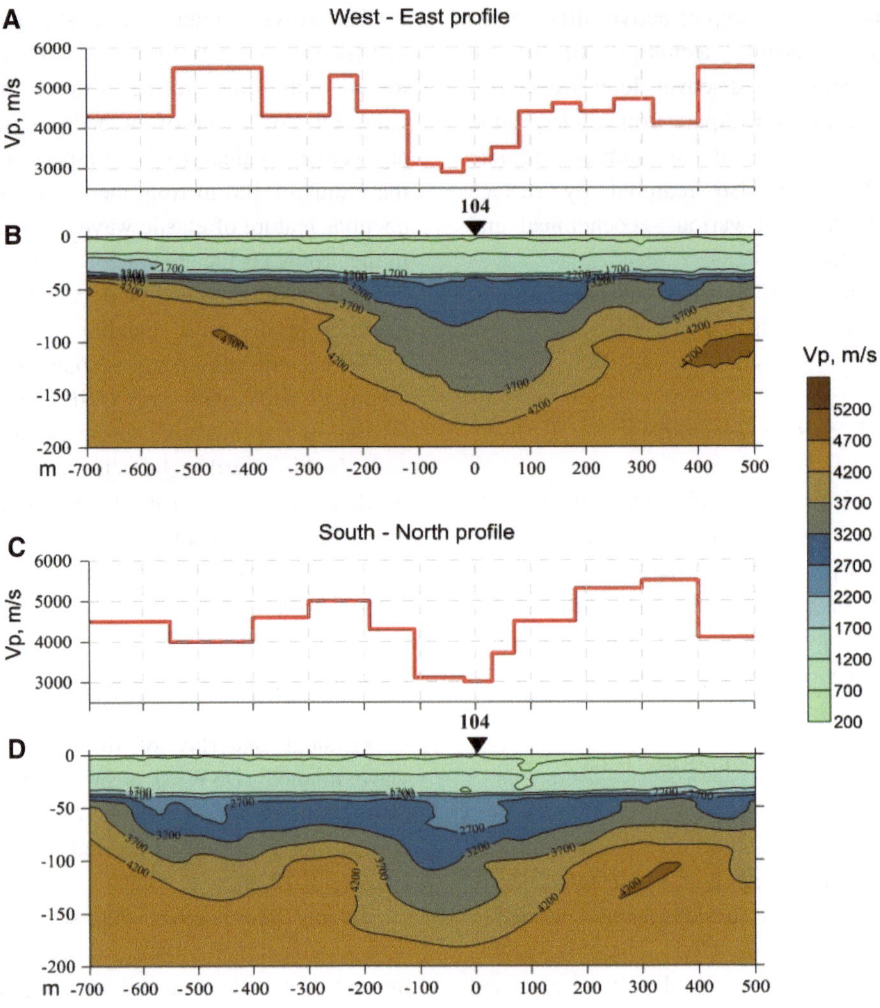

Figure 2
Seismic observation results using RWM/DWT method in the vicinity of borehole #104 (**a**, **c** diagrams of P-wave boundary velocities at the *top* of the bedrock; **b**, **d** tomographic P-wave vertical velocity sections)

ruptured structures and similar objects (DNISTRYANSKIY *et al.*, 2007; KUZNETSOV *et al.*, 2000).

This method was applied to reveal in the hypocentral area of the UNE anomalous high energies of diffracted waves which are related to active post-explosion destruction zones. Configuration of DSWM observations is presented in Fig. 3. Again, nonaccelerated drop weight was used as the seismic source ($\approx$15 kJ).

The results of field data processing were applied in building vertical sections of diffracted–scattered wave energy fields in the area of borehole #104 for two orthogonal profiles. The energy of diffracted–scattered waves is an accumulating amplitude of the signals at the specified time in the seismic traces, obtained using a focusing conversion algorithm. Generally, this method is described as Kirchhoff diffraction-stack migration (SHERIFF and GELDART, 1984). This energy is measured in conventional units as a relative quantitative parameter that defines a value of signal amplitude deviation at the certain parts of the section from the average amplitude value on the section (background). The areas with the significant amplitude deviation from the background values are considered post-explosive zones of rock disintegration. The vertical sections for the west-east and south-north profiles are given in Fig. 4.

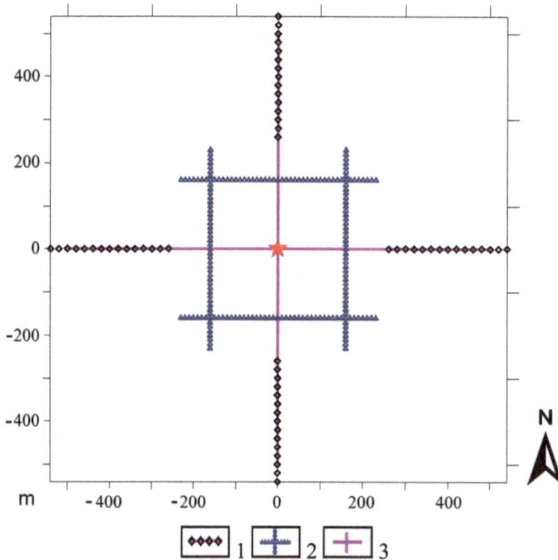

1 – seismic source of elastic oscillations; 2 – recording points;
3 – profiles used to plot vertical sections of energy fields.

Figure 3
DSWM method observation system configuration

According to DSWM data the zone of maximum fracture and fragmentation in the vicinity of the UNE hypocenter is a vertically stretched heterogeneity with dimensions 100 × 180 m, interlaying between 200 and 380 m. This zone differs in higher energy values of diffracted waves, exceeding the background level up to 150,000 conventional units. The high fracture zone is 300 × 450 m in size, interlaying in 150–450 m, and is shifted eastwards for west-east section from what was reflected by the boundary velocity distribution (Fig. 2).

2.3. Earthquake Converted Wave Method

Earthquake converted wave method (ECWM) is based on the reaction of elastic waves to the objects in the upper section and confined to UNE—e.g., fractures, micro-faults, spall zones. Spall zones are typical for almost all underground explosions. They are result of broken off ground parts above the focus in response to tensile radial stresses occurred due to reflection of shock wave from free (daylight) surface (Adushkin and Spivak, 1993). ECWM uses travel-time and dynamic attributes of seismic waves and it doesn't depend on precise information of focal point of an event (earthquake or explosion), origin time and absolute arrival times.

All the above-mentioned elements of post-explosion destruction, subject to their petrophysical

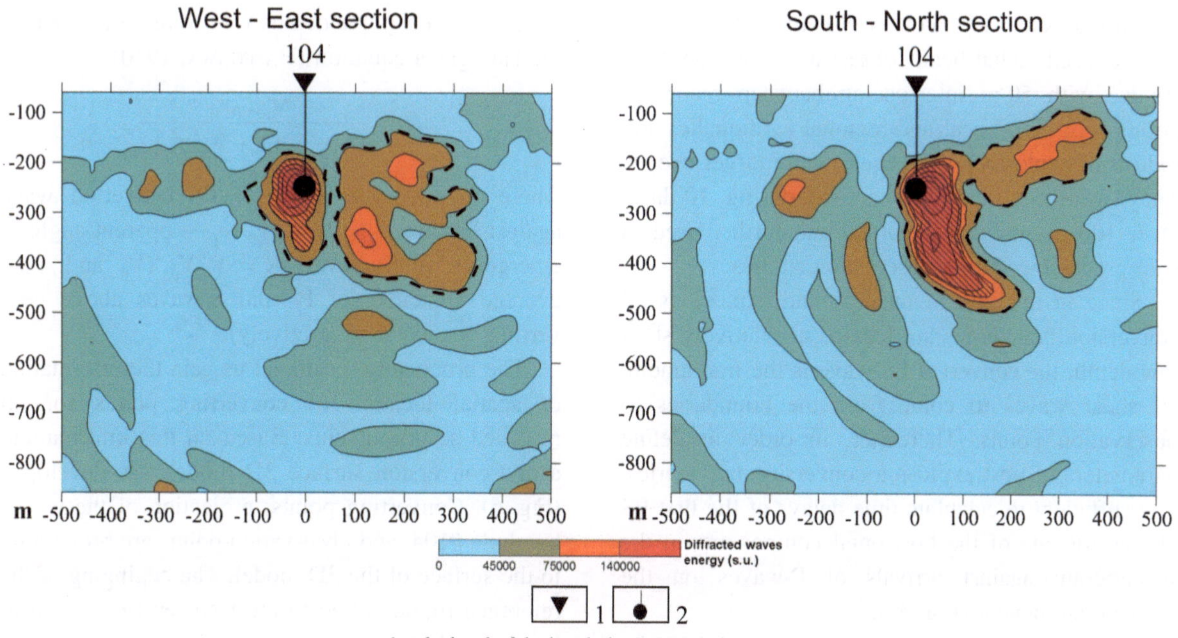

1 – the head of the borehole, 2- UNE's hypocenter

Figure 4
The vertical sections of diffracted-scattered waves energy fields

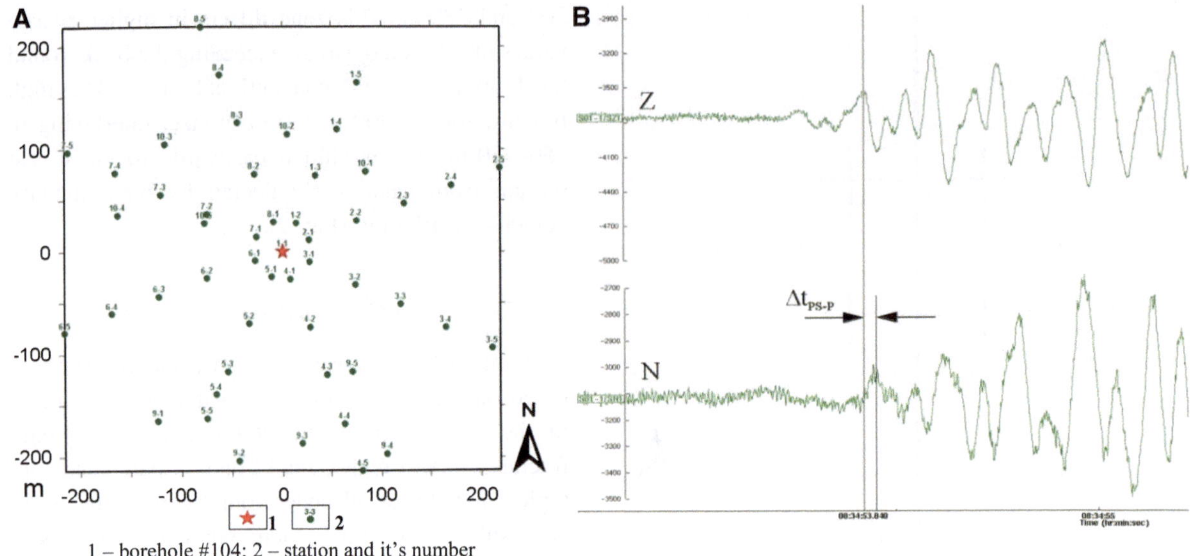

Figure 5
Seismic investigations using ECWM at the borehole #104 (**a** observational system configuration; **b** example of ΔtPS-P time delay for the distant earthquake)

properties, are conversion-forming boundaries—when upcoming seismic waves are refracted at them, a part of P-wave energy is transformed to converted PS phase. Parameters of conversion-forming boundary occurrences are estimated by arrival time differences at recording stations for P- and PS-waves (Pomerantzeva and Moszhenko, 1977; Puzyryov *et al.*, 1985).

We used radial-beam observations in the field studies with 50-m interspacing between recording points (Fig. 5a). Local and regional earthquakes and industrial explosions were used as the target events. They were continuously recorded during 10 days. Over 100 events were recorded and further used in data processing during observation studies.

Since in the case of underground explosions all conversion boundaries are located at relatively shallow depth, the converted PS-wave is the first among all shear waves to come from the boundaries to observation points. Thereafter, in order to define parameters of post-explosion conversion boundaries, it is sufficient to calculate time delays of the first PS phases for one of the horizontal components in the seismogram against arrivals of P-waves on the vertical component (Fig. 5b).

The arrival azimuth of seismic wave to the observation area and apparent velocities of P-waves (V_p^*) were estimated by areal travel-time diagram. Average P-wave and S-wave velocities above converting boundaries (V_p and V_s) were obtained from the results of active seismic observations and constituted at borehole #104 2.0 and 1.1 km/s, respectively. The angle of P-wave departure was defined from the ratio $\cos e = V_p/V_p^*$.

Conversion point depths *(H)* were calculated by the Hasegawa equation (Hasegawa, 1930):

$$H = \frac{\Delta t_{ps-p} v_p}{(K-1)\left(1+\frac{K}{2}\sin^2\frac{i_p}{2}\right)}$$

where Δt_{PS-P}—time delay of PS converted wave against P-wave, $i_p = 90°—e_p$ (e_p—apparent angle of emergence of P-wave), $K = V_s/V_p$ (V_p and V_s—average velocities of P- and S-waves above converting boundary, respectively).

The processing results let us gain the information of spatial location of converting points for all recorded seismic events. Based on this information, a first-conversion surface 3D model was developed (Fig. 6). Converting points, projection of the top of borehole #104, and observation points are brought up to the surface of the 3D model. The beginning of the reference frame is tied to the UNE epicenter (top of borehole #104).

The structure of the model consists of several general blocks. The first to note is the central part of

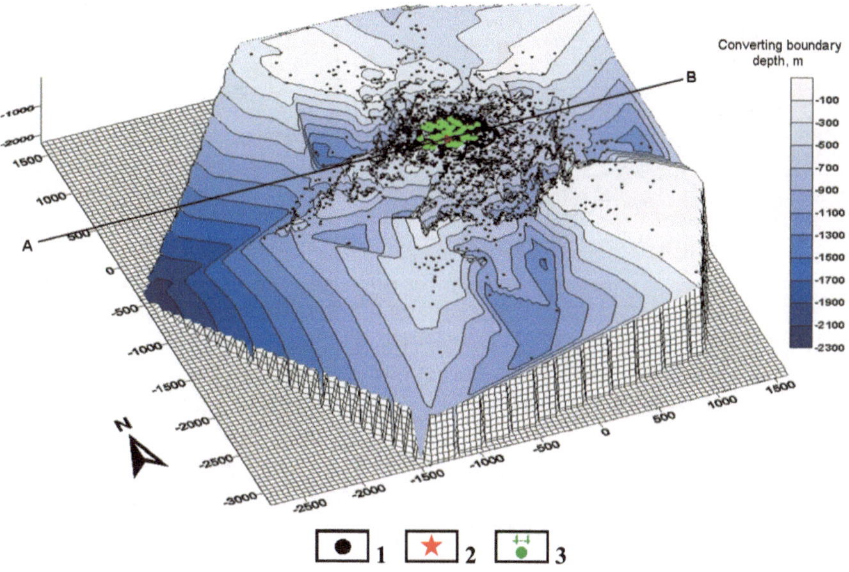

Figure 6
Conversion surface model developed by using ECWM method around borehole #104 (*1* converting points; projections to conversion surface: *2 top* of borehole #104; *3* observation points)

the model of isotropic-radial form with dimensions 2,200 × 2,200 m, reproducing a highly destructed block of rocks. This whole conventional boundary, accounting for the highly fractured pattern due to the UNE explosion, is laying in the interval between −20 and −150 m. The areas of large depth of converting boundary bedding (up to −2,300 m) concern to the geological media without interior boundaries and any other natural or post-explosive considerable converting objects and are interpreted as undisturbed by UNE zones.

These relatively large zones of active destruction near borehole #104 can be explained by technical conditions of the UNE—the charge depth was 225 m. For the relatively small depth of the charge, with a low geostatic pressure the energy impact of pulse explosion source on the host rocks is not isotropic-radial, and most energy is transferred to the shock wave propagating to the daylight surface, increasing the size of the zone of general irreversible destruction.

Besides the central part, there are also four large monolithic blocks in the model. These blocks are interlaying in intervals from 30 to 60 m and separated from each other by collapsed zones with depths up to 1,500 m. Such complex block structure can be explained by the impact of the consequences of other UNEs which were conducted in that region—there were five nuclear explosions conducted within 2 km eastward, south-eastward, and north-westward from borehole #104.

In general, according to the conversion surface model, the impact of UNE on the host rocks is characterized by the following features:

- formation of dome-like spall surface, interlaying in the intervals from −20 m (central part) to −150 m (in circumference) above the focal point of the nuclear explosion. In the horizontal plane this surface is ~500 × 500 m^2 in size.
- within the given model is a complex block structure (~2,200 × 2,200 m^2) with noncontinuous first conversion surface and large number of various vertical and horizontal discontinuity boundaries and separate conversion sites. This fact testifies high extent of rock destruction in this area.

For quantitative determination of post-explosion destruction of geological medium, we used integrated relative density of converting points parameter. This parameter was obtained as follows: along the given

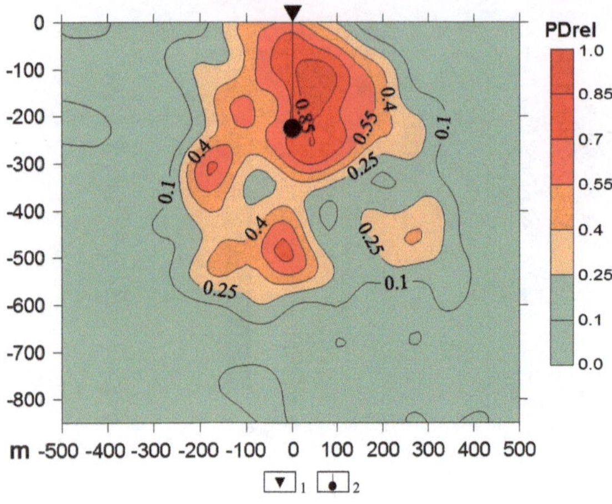

Figure 7
Vertical section of relative density of converting points in the area of borehole #104 (*1* the *top* of the borehole, *2* UNE's hypocenter)

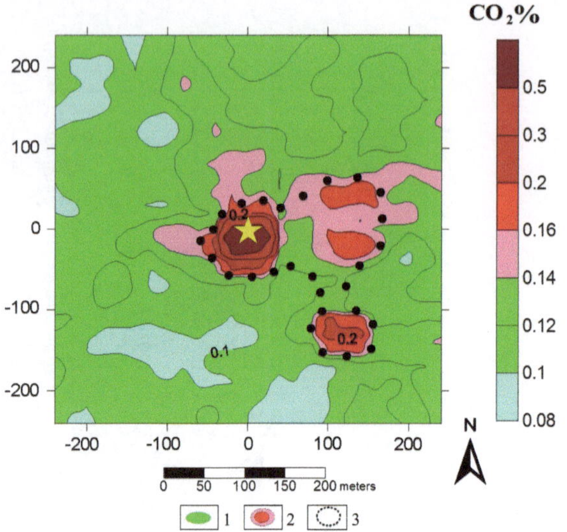

Figure 8
Determined zones in blanket sediments with anomalous concentrations of CO_2 in the subsurface air at borehole #104 (*yellow star*—the wellhead #104; *1* CO_2 background concentrations; *2* zone of high concentrations CO_2; *3* contours of geodynamic active area)

line A–B (Fig. 6) in the range of 50 m wide conversion points were combined in vertical plane. Density of conversion points (number of points per area unit) at the obtained section varied depending on degree of disturbance—in the areas with maximal rocks disintegration with the increased value of converting elements (fractures, splits, etc.) density of points was higher. Values of density are given in the form of integral parameter of density of conversion points—(PDrel).

Vertical section with distribution of density converting points parameter values is presented at Fig. 7. The use of this parameter helped us to obtain a more detailed picture of the area adjoining to the hypocenter. This area is featured with a high density of converting points, which distinguishes it as the seismoacoustic discontinuity with the most contrast in converting wave boundaries. According to ECWM data, the horizontal thickness of this discontinuity is 500 m, spreading to 550 m deep. Below this level PDrel values do not exceed standard deviations (± 0.125) from the average (0.1), taken as the background, with no regularities in distribution of converting points.

3. *Atmogeochemical Methods (CO_2 Measurements)*

Carbon dioxide surface measurements were conducted near borehole #104 to detect permeable structures and faults (fractures, microfaults) in the epicentral area which were generated by explosive impact on the host medium.

The observations were conducted in the surface area with 480 × 480 m coverage and with even spacing 20 × 20 m. A gas analyzer X-am-7000 was used as a measuring instrument with a mechanical air sampling device. The analyzed soil air was sampled at 0.5 m deep. To prevent CO_2 samples from stuffing with atmospheric air, the sampling procedure was carried out as follows: a hole of 22–25 mm in diameter and 0.7–0.8 m in depth was drilled. The sampler of a conic form was inserted into the hole with effort, soil around the sampler was tamped using a hammer and then gas samples were taken.

Field measurements resulted in plotted contour map of carbon dioxide concentration in the subsurface air (Fig. 8). The general regularity of CO_2 distribution is presence of the zone with maximal CO_2 content (stable in time) in space adjacent to the UNE epicenter at the distance not more than 100 m. This area was interpreted as contingent to post-explosion disturbance and gas-permeability of the near-borehole space, from where gas mainly escaped in the moment of the explosion.

The particular feature of CO_2 distribution in the subsurface air across this area is occurrence of two zones with anomalously high concentrations at 200–250 m eastward and south-eastward from the epicenter. Taking into consideration ECWM survey data indicating post-explosion fracturing in this part of the section, these zones can be explained by gas-permeable geodynamic active zones in blanket formations.

4. *Thermometry*

Thermometry was applied in studies at borehole #104 to detect local thermal flows in explosion-after permeable structures in the epicentral area of the UNE.

The survey was conducted in the meshed area 480 × 480 m in size and with a mesh step of 40 m. During measurements, plastic tubes (1 m long and 20 mm in diameter) filled with dielectric oil and driven full-length into the ground were used as observation points. Measurements were taken at 1-m depth using a thermoresistor as a temperature sensor. The measuring device was a digital multimeter to measure resistance and was followed by further conversion of the measured resistance values to temperature values according to the specifications reference table of the thermal probe. Vertical temperature gradient ΔT (differential temperature measured at 0.5 m and 1.0 m depths) was calculated for each observation point.

Figure 9 shows distribution of the vertical temperature gradient over borehole #104 area, anomalous values of which were interpreted as contingent to local thermal flows.

According to theoretical calculations, at a temperature on a daylight surface about 20 °C, and at a depth of explosion about 7–8 °C, existence of deep rising local thermal flows through permeable structures has to be shown in the form of zones with increased values of a negative temperature gradient. At the same time, influence on the daylight surface of various temperature factors (first of all, solar radiation) results in formation of descending thermal flows. In this case, zones of the lowered negative temperature gradient are formed in the area of

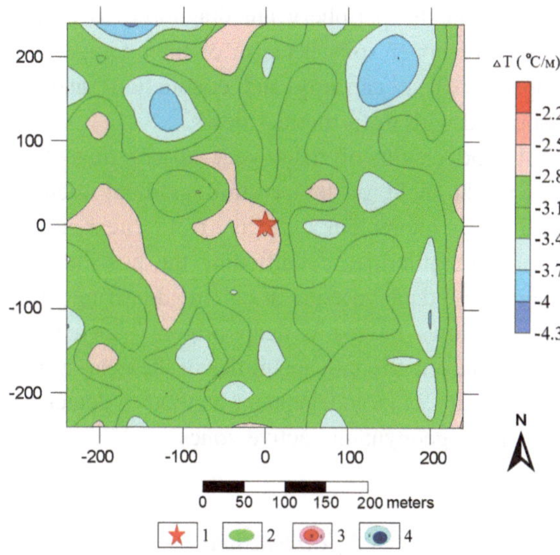

Figure 9
The determined geodynamic active zones in the blanket sediments with anomalous vertical temperature gradients at borehole #104 (*1* the wellhead; *2* background vertical temperature gradient; *3* high—level vertical temperature gradient; *4* low-level vertical temperature gradient)

outcropping permeable structures. To solve the OSI tasks, when searching post-explosive temperature artifacts a main factor will be spatial stability of alternating temperature field gradients revealed within the same place and measured at various extents of heat of upper soils.

During measurements in the vicinity of the borehole #104 head, the area of lowered temperature values of a negative gradient about −2.2 °C/m against background level of −3.1 °C/m was revealed. This fact testifies to heating of a daylight surface at the time of measurements that resulted in prevalence of a descending thermal flow (from solar radiation) over a deep thermal flow. It should be noted that at repeated temperature measurements at the same place increased gradient values (up to −4.5 °C/m) were discovered.

Also in peripheral parts of the site of investigations zones of abnormal (both increased, and lowered) gradient values were identified. In view of the fact that according to the results of atmogeochemical surveys concentration of carbon dioxide in the soil air on these sites doesn't exceed background values, it may be concluded that the revealed local lowering and increasing of vertical temperature gradient on the periphery are diagnosed as heterogeneity of upper soils because of their temperature properties (heat capacity and conduction) and not associated with deep thermal flows.

As the results of many-year observations have shown, the field of temperature gradient includes significant anomalous effects from near-surface thermal irregularities conditioning the possibility of ambiguous data interpretation in detecting local thermal flows having a deep origin. This case is limiting informational capabilities of the near-surface thermometry as an independent method to detect and to monitor geodynamic active zones.

5. *Conclusion*

As a result of geology and geophysical observations within charge borehole #104 at Semipalatinsk test site the following artifacts caused by underground nuclear explosion both in geological environment and on the surface were revealed:

- anomaly of reduced values for P-waves above the hypocenter of the explosion testify significant mechanical disturbances of bedrock roofing (result of seismic refracted waves method/diving waves tomography);
- areas with increased energy values of diffracted–scattered waves with horizontal dimensions up to 400 m in hypocentral part of UNE (result of diffracted–scattered wave method);
- areas peculiar for post-explosive rocks disintegration in the form of spall zones, block structures, increased density of conversion points (result of earthquake converted wave method);
- subsurface CO_2 local anomalies with 100 m radius within epicenter of UNE allow us revealing post-explosive permeable structures and therefore they serve as additional criterion while detecting epicenters of UNEs at one of the last OSI search stage (result of atmogeochemical method);
- subsurface temperature anomalies (result of thermometry).

Taking into account the age of this UNE (more than 40 years) and mechanical changes of geological environment occurred during this period under impact of various external factors, all above mentioned anomalies can be referred to long-term artefacts peculiar for borehole tests. Absence of detailed prior information on geology-tectonic parameters of UNE area approached task to reveal post-explosive artifacts close to real OSI. We should mention that the detected anomalies are of artificial nature and they do not correspond to natural behavior of rocks at Semipalatinsk test site.

In compliance with the above, the first main conclusion for the OSI purposes is that the typical UNEs are characterized by some artifacts that could be revealed not exact after the testing, but a long time after.

Additional investigation issue was to assess applicability of the applied geology and geophysical techniques to reveal consequences of UNEs and therefore solve OSI tasks. Based on the results of work done we can draw a conclusion on sufficient quality of the applied techniques.

The information of UNE phenomenology gained as the result of the described above studies can be used to supplement to the information data base in studying UNE traces and consequences. The knowledge of UNE display features in different geological and technical conditions can be useful in prompt and effective conducting of the on-site inspection under the Comprehensive Nuclear-Test-Ban Treaty verification regime.

References

Adushkin, V. V. and Spivak, A. A. (1993), *Geomechanics of large-scale explosions,* "Nedra" Publisher, Moscow, 319 (in Russian).

Adushkin, V. V. and Spivak, A. A. (2004), *Changes in Properties of Rock Massifs Due to Underground Nuclear Explosions*, Physics and Astronomy, Combustion, Explosion, and Shock Waves, Vol. *40*, No. 6, 624–634, doi:10.1023/b:cesw.0000048263.34894.58

Arhipov, V. N. *et al.* (2001), *Mechanical impact of the nuclear explosion*, scientific edition, "Physics and Mathematics Literature" Publisher, Moscow, 381 (in Russian).

Chadwick, P., Cox, A. D. and Hopkins, H. G. (1964), *Mechanics of Deep Underground Explosions*, Philosophical Transactions of the Royal Society of London. Series A, Mathematical and Physical Sciences Vol. *256*, No. 1070, 235-300.

Dnistryanskiy, V. I., Poberezhskiy, S. M. and Gorelikov, V. I. (2007), *A problems of gas deposits prospecting at the big depths in the difficult geological factor and probable ways of its solution (on the example of the Pre-Ural foredeep works)*, "The territory of Oil and Gas" magazine, No. 4, 26-31 (in Russian).

FAIZULLIN, I. S. and CHIRKIN, I. A. (1998), *Seismoacoustic methods of the rock fracturing study*, "Geoinformatics" magazine, No. 3, 24-27 (in Russian).

HAGEDOORN, J. G. (1959), *The Plus-Minus method of interpreting seismic refraction sections*, Geophysical Prospecting No. 7, 158-182.

HASEGAWA, M. (1930), *Die Wirkung der obersten Erdschicht auf die Anfangsbewegung einer Erdbeben*, "Zeitschrift für Geophysik", Heft 2, 78-98.

JOHNSON, G. W., HIGGINS, G. H. and VIOLET, C. E. (1959), *Underground Nuclear Detonations,* Journal of Geophysical Research, 64, No. 10, 1457-1470.

KUZNETSOV, O. L., KURYANOV, YU. A., MUSLIMOV, R. H., FAIZULLIN, I. S., HISAMOV, R. S. and CHIRKIN, I. A. (2000), *Space-time changes of the geomedia jointing on the results of 4D SLBO method measurements*, "Geoinformatika" magazine, No. 3, 32-39 (in Russian).

ORLENKO, L. P. *et al.* (2002), *Physics of the explosion,* Vol. *1*, scientific edition, "Physics and Mathematics Literature" Publisher, Moscow, 823 (in Russian).

POMERANTZEVA, I. V. and MOSZHENKO, A. N. (1977), *Seismic research with "Earth" equipment*, "Nedra" Publisher, Moscow, 256 (in Russian).

PUZYRYOV, N. N., TRIGUBOVA, A. V. and BORODOV, L. YU. (1985), *Seismic exploration by shear-wave and converted wave methods,* "Nedra" Publisher, Moscow, 277 (in Russian).

SHAPIRO, S. A. and FAIZULLIN, I. S. (1986), *Features of seismic wave attenuation in rocks as discrete diffusing media*, "Physics of the Earth" magazine, Moscow, No. 9, 56-63 (in Russian).

SHERIFF, R. E. and GELDART, L. P. (1984), *Exploration Seismology, Vol. 2: Data-Processing and Interpretation*, Cambridge University Press, 240, ISBN-10:0521250641, ISBN-13:978-0521250641.

STAROBINETS, A. E. (1978), *Use of diffracted waves in seismic exploration*, "Geophysics" magazine rev., Moscow, 96 (in Russian).

STAROSTIN, V. I., VELICHKIN, V. I., PETROV, V. A., VOLKOV, A. B. and KOCHKIN, B. T. (1995), *Structure-petrophysical and geodynamical aspects of the crystalline rock massifs selection in connection with the problem of radioactive waste disposal,* "Geo-ecology" magazine, No. 6, 17-26 (in Russian).

(Received September 20, 2011, revised December 26, 2012, accepted December 29, 2012, Published online January 30, 2013)

Pure Appl. Geophys. 171 (2014), 763–777

DOI 10.1007/s00024-012-0574-5

Pure and Applied Geophysics

Overhead Detection of Underground Nuclear Explosions by Multi-Spectral and Infrared Imaging

John R. Henderson,[1] Milton O. Smith,[1] and Michael E. Zelinski[1]

Abstract—The Comprehensive Nuclear Test Ban Treaty allows for Multi-Spectral and Infrared Imaging from an aircraft and on the ground to help reduce the search area for an underground nuclear explosion from the initial 1,000 km^2. Satellite data, primarily from Landsat, have been used as a surrogate for aircraft data to investigate whether there are any multi-spectral features associated with the nuclear tests in Pakistan, India or North Korea. It is shown that there are multi-spectral observables on the ground that can be associated with the nominal surface ground zero for at least some of these explosions, and that these are likely to be found by measurements allowed by the treaty.

Key words: Comprehensive Nuclear Test Ban Treaty, CTBT, remote sensing, multi-spectral imaging, infrared imaging, underground nuclear explosion, on-site inspection.

1. Introduction

The Comprehensive Nuclear Test Ban Treaty (CTBT) permits Multi-Spectral and InfraRed Imaging (MSIR) to be performed as part of an On-Site Inspection (OSI) for the purpose of reducing the search area for the location of a possible underground nuclear explosion (UNE). Dedicated airborne MSIR measurements were not made in conjunction with historical or recent UNE's, so satellite data has been used to determine if there are MSIR observables associated with recent UNE's. In this work, MSIR data from commercial satellites has been used to show that there are detectable surface observables which can be used to greatly reduce the search area for the location of the UNE. This has been demonstrated using Landsat data of the Indian, Pakistani, and North Korean UNE's in the last 12 years, and with GeoEye-1 data for the North Korean tests. The techniques used typically identify a region of interest less than 1 km^2 in size (compared to the nominal 1,000 km^2 search area), and the few false positives have been resolvable as such by using visible imagery.

The results of this study show that MSIR data from satellites can be used to help prepare the inspection team for an OSI. The Landsat data used here were chosen for their ready availability, the expectation that the satellite spatial resolution and spectral bands would be useful, and the 16-day site revisit time of the satellite. Data from other satellites may have greater utility—for one site, GeoEye data with 3 m spatial resolution were used to find regions that were not detected in the 30 m spatial resolution Landsat data. The analysis techniques used here are fairly straightforward change detection or surface categorization techniques and a more sophisticated spatial/spectral algorithm that uses both properties to find anomalous regions in a data set. Improvements in sensitivity and reduction of false alarms are expected with the development of more sophisticated techniques.

[1] LLNL, Livermore, CA, USA. E-mail: henderson9@llnl.gov

The GeoEye data show that more sensitive detection of regions of interest is possible with higher spatial resolution, although that data suffers from a minimal set (4) of spectral channels. Additionally, this work has shown that automated algorithms are particularly useful in sifting through large data volumes and detecting specific types of anomalies that are not readily apparent to visual inspection of the data.

The results here provide confidence that airborne MSIR data will be useful for an On-Site Inspection under the CTBT.

2. Background and Purpose

The Comprehensive Nuclear Test Ban Treaty (CTBT) allows for Multi-Spectral and Infrared Imaging (MSIR) measurements as part of an On-Site Inspection (OSI) (CTBT 1996). The objective of MSIR measurements is to help narrow the search area for the location of a potential nuclear explosion that violates the CTBT. The application and utility of MSIR measurements for OSI require further study: MSIR measurements have only recently been made (September 2011 and May 2012) at field exercises conducted by the CTBT Organization; there are no vetted requirements for the specification of equipment that might be used for MSIR; and the current level of experience is insufficient to generate a Concept of Operations (CONOPS) or analysis procedures for the use of MSIR to support an OSI.

In general, the UNE observables that might be detected by MSIR fall into five categories (Henderson 2010):

1. disturbed earth at the surface (due to the shock wave from the explosion),
2. plant stress in the vicinity of the UNE,
3. artifacts of human activity,
4. thermal effects, and
5. novel materials at the surface.

The first has not been explicitly measured for a UNE, but visible observations of earth movement, surface fissures, and measurements of surface upheaval from prior UNE's indicate that disturbed earth is possible for a UNE of sufficient size (Eisler and Chilton 1964). Further, measurements for other purposes have shown that disturbed earth can be detected with MSIR imaging (Rockett 1999).

Airborne MSIR measurements of plant stress were made during the Non-Proliferation Experiment (NPE), so this observable has been demonstrated under relevant conditions. The amount of plant stress was observed to peak one to two days after the explosion, and then fade back to pre-explosion levels after about a week (Pickles 1995).

Observables associated with activities in support of a UNE are expected to vary with the specific conditions of executing the UNE, but MSIR measurements are often able to detect activities such as recent traffic on dirt and gravel roads, as well as thermal indications of activity in buildings. Temperature related observables lie in the long-wave infrared, while road traffic and disturbed earth can have observables in the visible and infrared spectral regions (Rockett 1999).

Thermal effects at the surface might be due to either hot gases from the UNE blast escaping to the surface, or underground water flows being redirected to the surface by changed sub-surface geology. Novel materials could be brought to the surface through either venting of materials from the UNE, or by migration of native sub-surface material to the surface because of the escape of hot gases or water (Hall *et al.* 1997).

Maturing the use of MSIR for an OSI requires that the MSIR observables of an UNE be well characterized so the MSIR equipment, deployment CONOPS, and data analysis techniques can be specified. Given the absence of relevant airborne data (except for plant stress), satellite data was used to determine whether MSIR observables are reliably associated with a UNE, and to characterize those observables. Since MSIR instruments on satellites have improved significantly over time, this work focused on UNE's conducted in the last 14 years.

There is an extensive body of prior work on using satellite measurements for treaty monitoring. Much of this work focuses on using satellite imagery to detect activities prior to a nuclear test (Jasani 1995; Jasani 2000), or the use of satellite imagery and data to localize a nuclear test after the detection of a seismic signal (Canty and Schlittenhardt 2001b).

Additionally, there is a growing body of work on the use of satellite SAR data to detect surface movement induced by the underground explosion (Cong *et al.* 2007; Vincent *et al.* 2003, 2011).

A 2010 paper by Schlittenhardt *et al.* (2010) provides an excellent description of the use of change detection algorithms with satellite data, as well as a good summary of satellites of interest. There are a number of additional papers on analysis methods and change detection techniques for this application (Jasani and Canty 2001; Canty *et al.* 2001a; Canty and Nielsen 2006; Nussbaum *et al.* 2005).

The purpose of this paper is to demonstrate that (1) there are MSIR observables associated with UNE's, and (2) those observables have unique spectral/spatial properties which can be exploited with a basic set of analysis tools to greatly reduce the search area for purposes of an OSI. The methodology in this paper distinguishes it from previous work because we have used the satellite data as surrogate data for the expected aerial MSIR data that might be acquired during an OSI, and have chosen to use a subset of the satellite data to represent the 1,000 km^2 inspection area roughly centered on the seismic center, as prescribed by the CTBT. The results here are expected to be directly applicable to what inspectors in an OSI would obtain, and the emphasis is on the use of the multi-spectral capabilities, since visual observations are a mature technology for use in an OSI. A key aspect of this paper, which we have not seen specifically addressed elsewhere, is the question of false positives (a positive finding at a location other than the site of the UNE). Since any information which reduces the search area down from 1,000 km^2 is useful, techniques which generate false positives are acceptable, as long as they find the area of interest and do not generate too many false positives.

3. Selection of Data

The six UNE's in the last 12 years were chosen as the focus of this study because they represent the size of UNE (approximately one kiloton of yield) of concern to the CTBT, and because they are recent enough that a variety of potentially relevant satellite measurements exist. While Landsat 5 (whose Thematic Mapper data was used here) was launched in 1984 (well before the cessation of nuclear testing in 1992), satellites with higher spatial resolution and better spectral capabilities have launched more recently. The Landsat data was used because the data is readily downloadable; it has relevant spectral bins, including a standard plant stress data product; the spatial resolution is acceptable (30 m versus an expected feature size of over 150 m for disturbed earth or plant stress), the data volumes are manageable, and the frequency of observation is acceptable for the disturbed earth measurements. The Landsat data also allow us to extract a 900 km^2 region (30 km × 30 km sub-image) from a single data set, which avoids having to stitch together two satellite data sets for our desired processing. Schlittenhardt *et al.* (2010) needed to merge two ASTER data sets to cover the region of interest since the ASTER swath width is only 60 km and the center did not pass the center of the region of interest. Similar concerns would apply to SPOT imagery (60 km swath width). Google Earth imagery, and higher spatial resolution data from additional satellites were used to understand the context of results obtained from the Landsat data.

The six UNE's are, in order of ease of detection of MSIR observables, (not chronological order):

1. 28 May 1998, Pakistan (seismic magnitude $m_b = 4.8$) at 28.90°N 64.89°E
2. 30 May 1998, Pakistan ($m_b = 4.6$) at 28.49°N 63.78°E
3. 11 May 1998, India ($m_b = 5.2$) at 27.07°N 71.76°E
4. 13 May 1998, India (no seismic signal detected)
5. 9 October 2006, North Korea ($m_b = 4.1$) at 41.31°N 129.02°E
6. 25 May 2009, North Korea ($m_b = 4.5$) at 41.29°N 129.04°E

The Pakistani and Indian test magnitudes are from Walter *et al.* (1998) and the test locations from Barker *et al.* (1998). The North Korean test magnitudes and locations are from the CTBT Organization (CTBTO 2009; Bras *et al.* 2007).

4. UNE Detection Algorithms

The techniques used here for detecting observables associated with a UNE fall into three categories. The first is change detection, the second is surface categorization, and the third uses the spatial and spectral properties of the data to search for anomalies. In change detection, reference data is used from before (or after) the seismic event to establish the level and variability in each spectral channel for each pixel in the scene. The post-event data is then compared to the pre-event baseline to determine if any statistically significant changes have occurred for any pixels. These regions are flagged (by their spatial location and statistical significance) for consideration as candidate locations for the UNE.

The second technique is to use the spectral properties of a scene to group pixels into a few categories, where the pixels in each category have a similar spectral shape. Categories typically correspond to scene features such as soil, rock, vegetation or shadows. The pre-event and post-event categories are compared spatially, and any differences evaluated. For example, a region that showed as vegetative pre-event and then shows as soil post-event would be a candidate for the UNE location since surface disruption might overturn vegetation and expose bare ground. An advantage of this technique is that it does not require the use of pre-event data, although it is probably more sensitive when pre-event data is available. For example, a region that shows up as soil might be a candidate for proximity to the UNE if it can be distinguished from roads or farms by its size, shape, location or spectral difference from adjoining regions. The first and second techniques are based on known techniques (Malpica and Alonso 2008; Reed and Yu 1990).

The third technique is similar to the scene categorization algorithm, except that the spatial properties of the scene are incorporated. For example, regions of a scene that show as bare earth might be either dirt roads or farms. Each of those has distinct spatial properties. A region of disturbed earth may appear to have the spectral shape of bare earth, but will have different spatial characteristics and can therefore be identified as anomalous. This is useful because it can identify potential regions of interest in a complex scene that might be missed by visual inspection of the scene categories map (from the second method). This work draws on prior spectral–spatial algorithms (Mitchley *et al.* 2009; Myint 2001; Rivard *et al.* 2008; Srinivas and Wu 2005; Subotic *et al.* 1997; Thomas 2008; Thyagarajan and Patterson 2010; Van der Werff and Lucieer 2006; Zhang *et al.* 2005).

It should be noted that application of these detection algorithms to specific sites is presently an iterative process. Detection thresholds need to be adjusted for the properties of the location being imaged to achieve consistent and robust results, and comparison of the results from the different algorithms was used to provide insight into interpretation of the results and refinement of the algorithms for this application. That new understanding is the basis for future work to develop more sensitive detection algorithms.

5. Analysis Process

For each UNE, published estimates of the location of the UNE were evaluated for a best estimate of the location of the UNE. The Landsat data nearest that location was downloaded for several data sets before and after the event, subject to data availability and the data being sufficiently cloud-free. A 30 km $\times$ 30 km region around the best guess location was extracted from the data for primary analysis, intended to be similar to the 1,000 km^2 search region allowed by the CTBT.

6. Results

6.1. First Pakistani Test, 28 May 1998

The 28 May 1998 Pakistani nuclear test was chosen as a first case because there is an obvious surface change in the visible imagery, which is probably due to a rock slide proximate to the location of the UNE. In this case, since the visible imagery clearly indicates significant changes in the surface material, spectral analysis should produce similar results.

Landsat data were acquired for two dates before the event (26 April 1998 and 28 May 1998), and 13

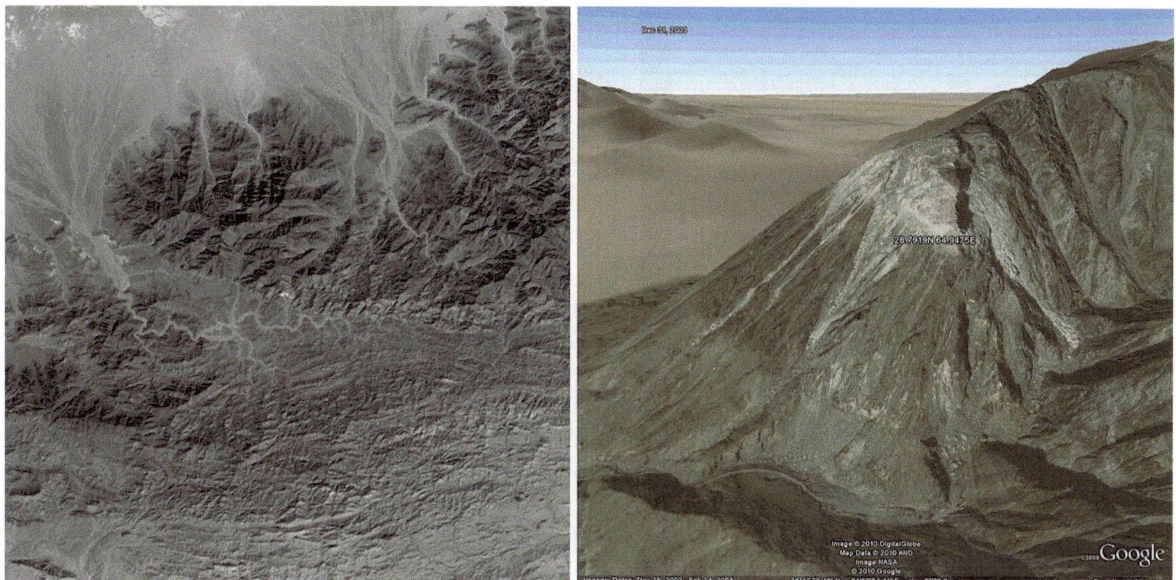

Figure 1
Left landsat blue band data for a 30 km × 30 km (1,000 pixel × 1,000 pixel) region centered on the suspect UNE location (*white spot in center*). The data is from 29 June 1998, 1 month after the event. *Right* imagery of that location from 2003, with apparent rock slides down the slopes

dates after the event (13 June 1998 to 7 January 1999). Figure 1 shows Landsat data 1 month after the event which illustrates that the event location is obvious in visible imagery, although this single image is not sufficient to unambiguously identify those pixels as proximate to the UNE. Comparison to pre-event data, and inspection of detailed visible imagery (right image in Fig. 1) show that the bright region in the image was not present pre-event, and that the bright pixels are due to rock slides.

A subset of the data sets were used for change detection—one date before the event (28 May 1998), and 7 dates after the event (13 June 1998–19 October 1998). Some of the data were excluded because they included very thin clouds which would increase the apparent pixel variability and reduce the algorithm sensitivity. The most recent data sets were excluded because of concerns that weathering of the exposed material and human activities would also add unwanted changes to the reference data. (Spectral changes due to material weathering were observed using the material categorization algorithm, confirming the concerns here and illustrating the benefit of using and comparing several algorithms).

The change detection algorithm determined the statistical significance of change for each pixel (in a multivariate sense) between the one data set taken before the UNE and the larger set taken after the UNE. The spectra of the statistically significant pixels are shown in the left graph in Fig. 2 (the black spectra are from the UNE, all other colors are from other regions showing significant change), and the location of those pixels is shown in the right plot in Fig. 2. Note that the graph of the spectra shows two distinct behaviors, which correspond to two physically separate regions on the ground. Inspection of the visible imagery suggests that the changed pixels in the upper left are due to farm activities, possibly tilling the ground or crop growth. The remaining region corresponds to the suspect region in the visible imagery.

The spectral categorization techniques were also applied to this data. Figure 3 shows the results of that analysis. Again, the exposed material from the UNE is readily apparent, and the "false positive" for the same material along the stream is readily distinguished as not relevant. Note that there are no other significant false positives. These results were obtained for each of the 13 post-event images that extend for 6 months after the event, indicating the

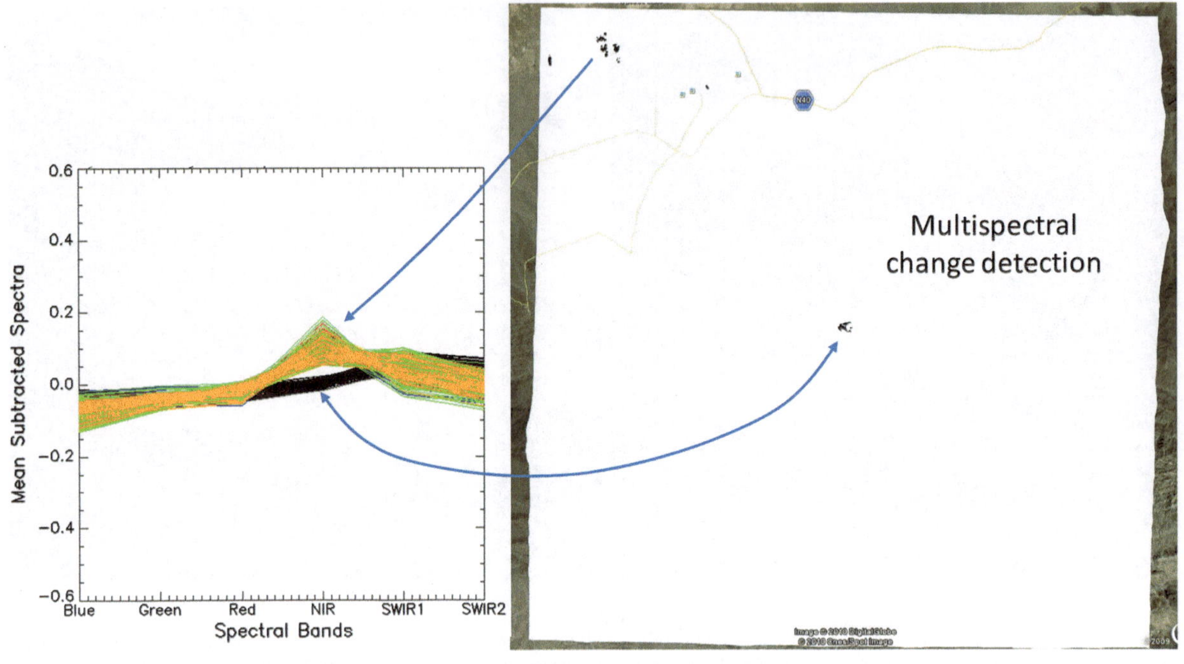

Figure 2

Left spectral difference *graph* shows spectra of pixels identified as having changed significantly after the event. Note that there are two types of spectral behavior observed. *Right* the location of pixels with significant change (*black dots*) shows them to be in two spatially distinct regions. The spatial extent of the *right* image corresponds to the *left* image in Fig. 1

spectral persistence of the observable and that no pre-event data is needed to locate the observable. The ability to use only post-event data is directly relevant to an OSI where any MSIR aircraft data will only be collected after the event.

Here, we also looked at whether the spectrum of the different surface materials changed with time and found that they did for the exposed rock material over the 7-month period after the event. This indicates either that continued rock slides were exposing material with slightly different spectral properties than the original material, or that the exposed material was weathering and its spectral properties were changing over time. In either case, the variation with time of the spectral shape of this region also indicates a region of interest.

While these results are not surprising, given the obvious visible observable, they are important for three reasons. First, it demonstrates that the multispectral data and analysis techniques can find observables associated with a UNE. Second, the only "false positives" identified by the techniques were in fact regions one would want to investigate further since they represent disturbed earth (and were readily identified as false positives from the context of the visible imagery). Third, these techniques provide an automated means of sifting through millions of pixels, which is important if higher spatial resolution data is used, and is likely to be essential when rapid and objective data analysis is needed for an OSI.

6.2. *Second Pakistani Test, 30 May 1998*

The 30 May 1998 Pakistani nuclear test does not have an obvious feature in any visible imagery. Figure 4 shows the 30 km × 30 km regions selected for analysis, along with two estimates of the UNE location, based on seismic analysis with differing assumptions. Two candidate locations for the UNE derived from analysis of the Landsat multispectral data are also shown in the figure for comparison. Thirteen sets of Landsat data were acquired and used in the analyses here, three pre-event (27 April 1990–19 May 1998) and 10 post-event (4 June 1998–8 May 2000). Again, only some of the Landsat data were used for the change detection algorithm (all

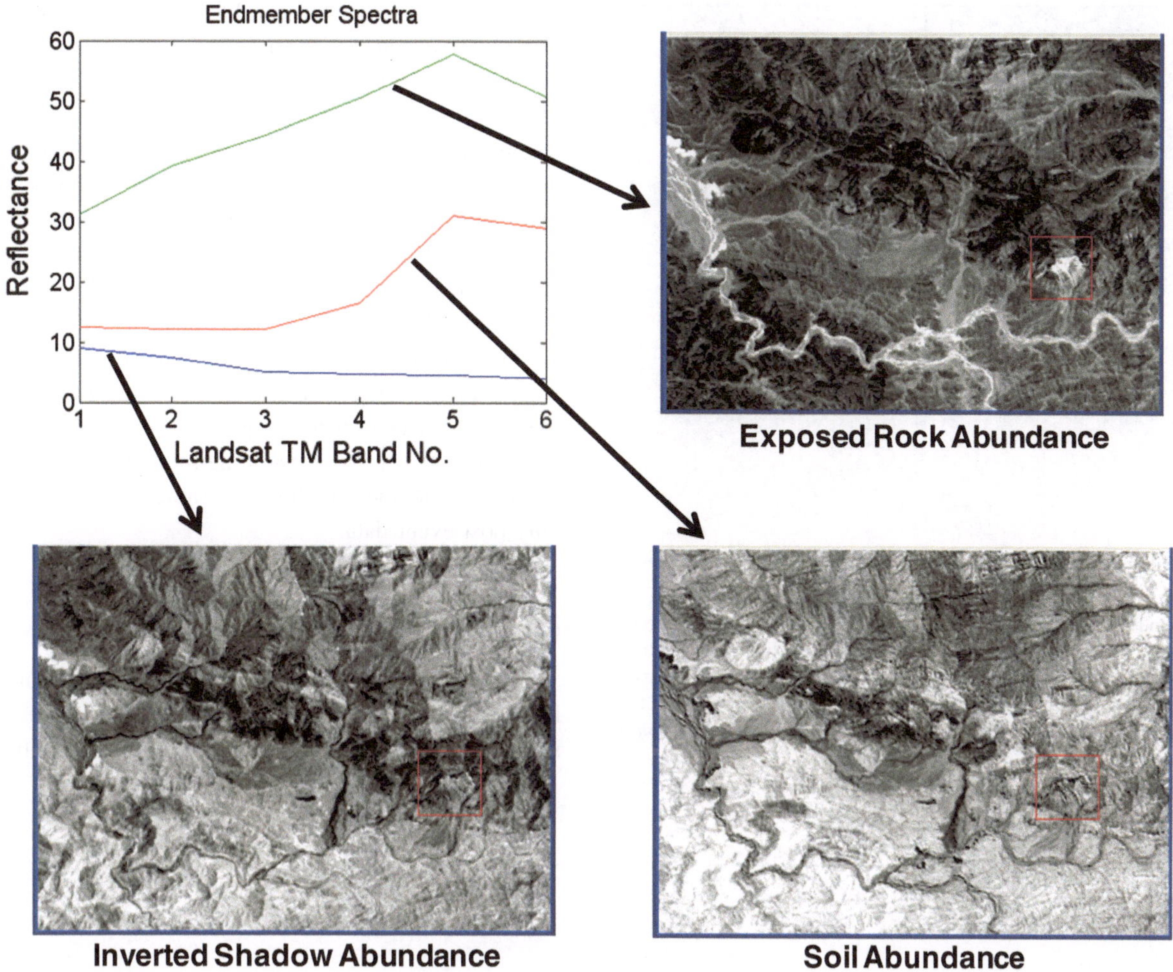

Figure 3
Results of using six spectral channels to classify the scene into three types of surface material: rock, soil and shadow. *Upper left* spectra of the three types of surface material. The other images are abundance maps for a 12 km × 8.8 km central portion of the 30 km × 30 km image, showing how similar each pixel on the ground is to one of the three material types. Note that the exposed *rock map* shows both the region exposed by rock slides from the UNE (*red box*) and material exposed by erosion and roads along the stream

three pre-event data sets and only the 4 June 1998 post-event data set).

Figure 5 shows the results of the change detection algorithm. Note that the two candidate UNE locations are in a different location than either of the two locations suggested by the seismic data. This is a good demonstration of the limits of accuracy of the ability to geo-locate the UNE seismically for a low yield nuclear test, and of the ability of the multi-spectral data to reduce the search area. The two suspect areas are the regions in the image with the strongest change signal. The other areas in the image showing some significant amount of change are associated with geological features and are likely due to seasonal variations in solar illumination of the scene.

Figure 6 shows the results of using the categorization algorithm to distinguish surface materials by their spectral properties. Here the algorithm has been enhanced to include a spatial filter to identify regions where their spatial scale distinguishes them from other parts of the scene with similar spectral properties. This enhances sensitivity and reduces the false positive rate. The results of this analysis are a single high-contrast spectral/spatial region detected for all data on and after 19 May 1998. This is before the test

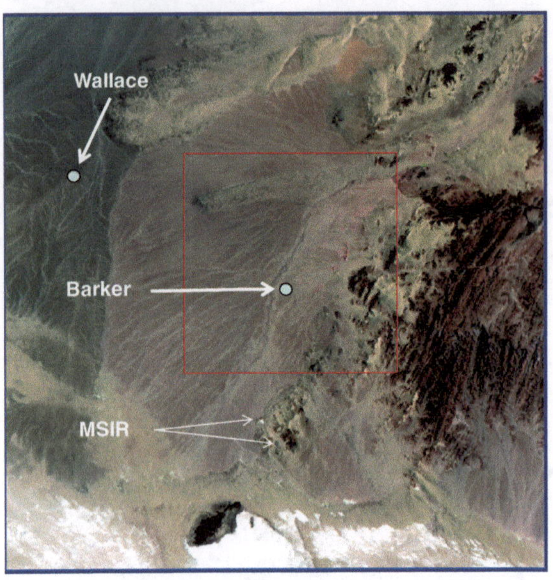

Figure 4
Landsat false color data for a 30 km × 30 km region around the 30 May 1998 Pakistani nuclear test. The test location inferred from seismic data is shown for estimates with two different methods (WALLACE 1998; BARKER *et al.* 1998). Two candidate locations for the test location derived from the MSIR data are shown by the *thin arrows*

was executed, so the observable is probably due to human activity at that location. Note that this region matches the upper left (North East) suspect region identified from change detection, but that only a single data set of post-event data was used in its detection. The spatial/spectral analysis shows a marginally significant detection for some of the post-event data at the location of the lower right suspect area from the change detection analysis. It is interesting that two very different analysis techniques for the MSIR data give similar results for candidate areas for the nuclear test. It is also not surprising that the change detection technique appears to be more sensitive, but it is very encouraging that the spatial/ spectral technique can find the same regions using only post-event data.

6.3. *First Indian Test, 11 May 1998*

The 11 May 1998 Indian nuclear test does not have an obvious feature in visible imagery. Sixteen sets of Landsat data were acquired, three before the

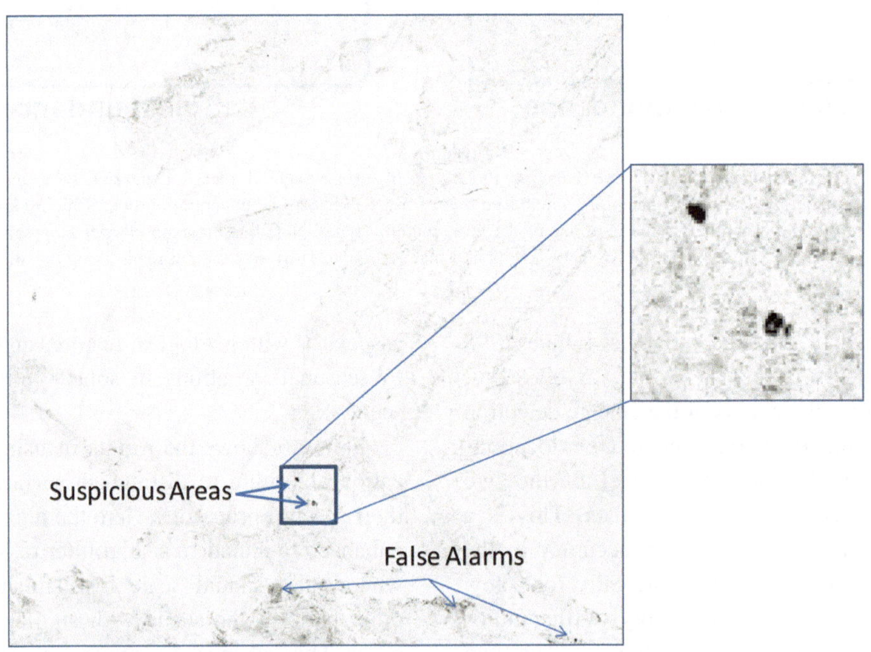

Figure 5
Change detection image for the sets of Landsat data for the 30 May 1998 Pakistani test. The degree of statistically significant change is shown in shades of *gray*, *black* indicating the most significant change. The main image corresponds to the 30 km × 30 km image in Fig. 4

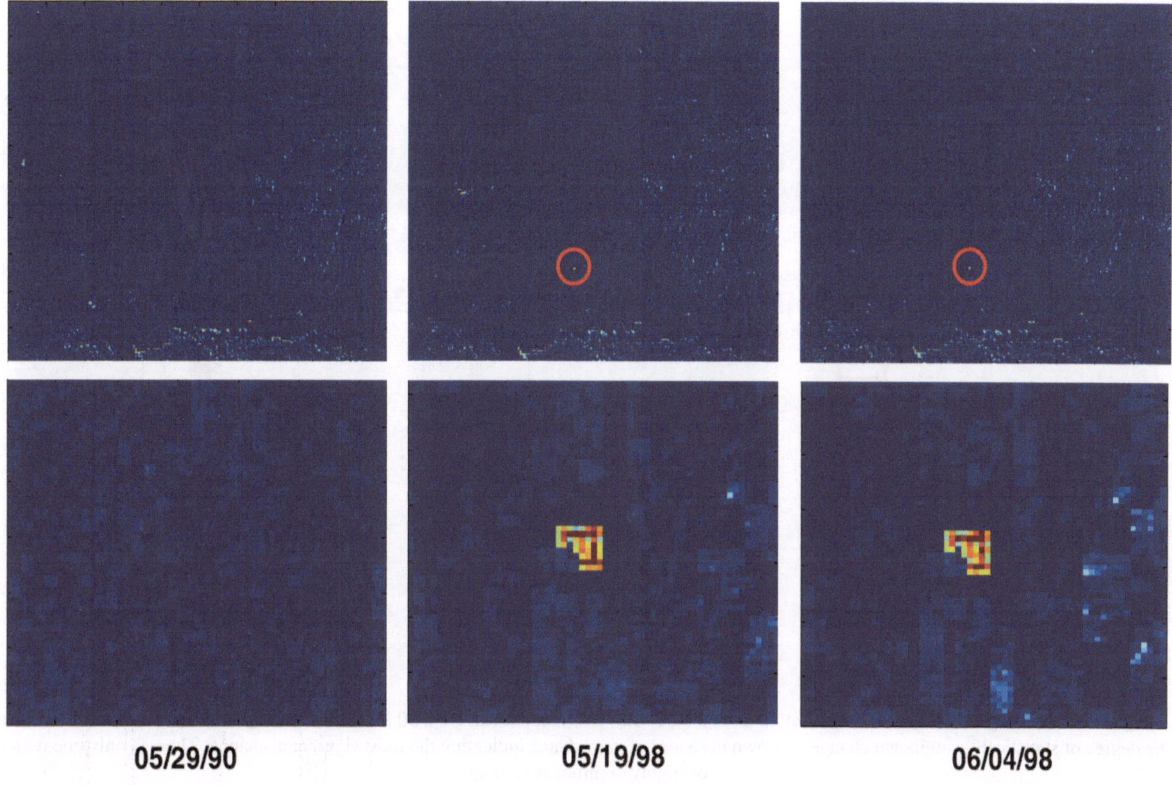

Figure 6
Anomaly detection images (*top row*) of the 30 km × 30 km area in Fig. 4 using both spectral and spatial information on single data sets for the dates indicated. The *bottom row* shows the pixel-by-pixel detail for the *circled* region in the *top row*. Each pixel is 30 m × 30 m in size. The results show no anomalies in the data 8 years before the test, and show an anomaly for all data sets on and after 19 May 1998. The anomaly here is in the same location as the *upper left* suspicious area in Fig. 5

test (9 February 1998–14 April 1998) and 13 after the test (16 May 1998–21 May 2000). Four of the data sets were used for change detection, three before the test (9 February 1998–14 April 1998) and one after the test (16 May 1998). Figure 7 shows the change detection image for the 30 km × 30 km regions selected for analysis based on the best published information for the likely location. Figure 8 shows the region indicated by the change detection analysis superimposed on a current visible image. There is nothing to indicate any surface disturbance in the visible image, but there is about 10 years between the acquisition of the MSIR data and of the visible data. It is interesting to note that the indicated region lies adjacent to an area of obvious human activity. An earlier analysis (Canty *et al.* 2001a) found the same hammer-like region using change detection on the 29 March 1998 and 16 May 1998 Landsat data, and identified the hammer-like region as the result of a brush fire by using commercial satellite imagery. While not necessarily related to a UNE, the technique was successful in identifying a region of interest through change detection.

Spectral image categorization was performed on this data and indicated the same region as a region of interest.

6.4. *Second Indian Test, 13 May 1998*

The 13 May 1998 Indian nuclear test was of lower yield than the 11 May test, and public reporting by India indicated it was about 10 km away from the 11 May test. Here we use the same imagery as before, and lower the threshold for change detection and inspect the area about 10 km distant from the first test for candidate locations. Figure 9 shows that there are two

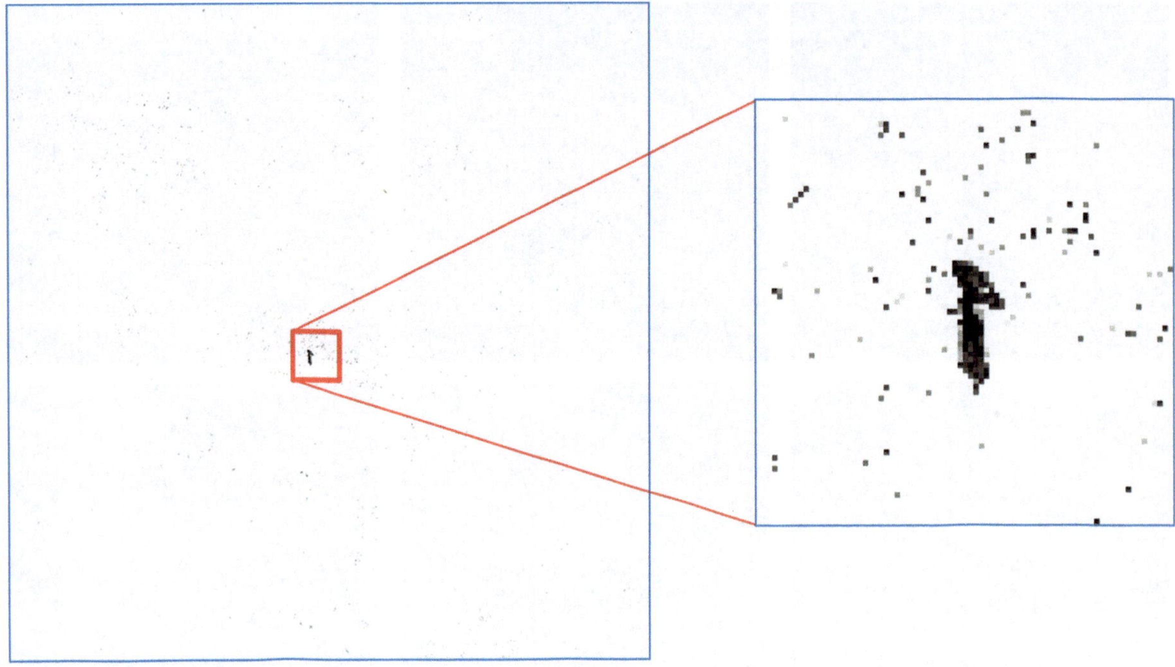

Figure 7
Change detection image for the sets of Landsat data for the 11 May 1998 Indian test for a 30 km × 30 km region around the seismic center. The degree of statistically significant change is shown in shades of *gray*, *black* indicating the most significant change. There is only one region of highly significant change

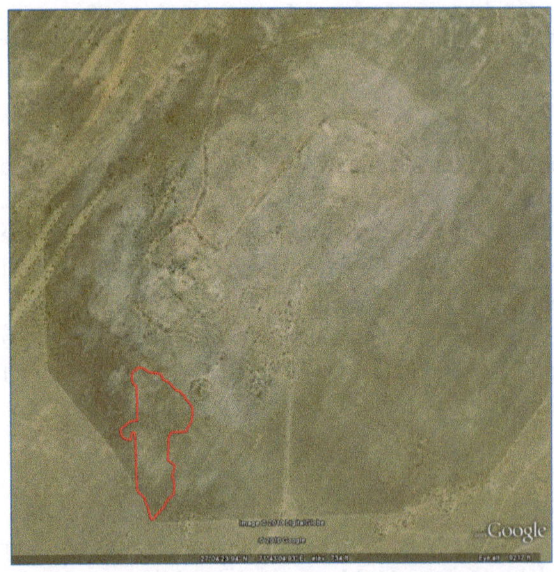

Figure 8
Visible image showing the location (*red outline*) of the feature determined from the change detection analysis. There is no obvious visible feature which corresponds to the change region. It is close to an area of human activity

candidate regions for inspection that fit the constraints. Since there is no corroborating information for these regions (no visible data, and the image categorization algorithm has not yet been run for these regions), they should be considered as they would for an actual OSI—regions of high interest for early inspection, but not necessarily a definitive location.

6.5. *North Korean Tests, 9 October 2006 and 25 May 2009*

6.5.1 *Landsat Data for North Korean Tests*

The North Korean nuclear tests illustrate some of the limitations of using satellite data. There are frequently clouds in this region, so the Landsat data is often unavailable. For both of these tests, there were only seven Landsat data sets readily available, but there is a nearly 2-year gap with no data around 2008 due to weather and technical problems with the satellite. The data sets are from:

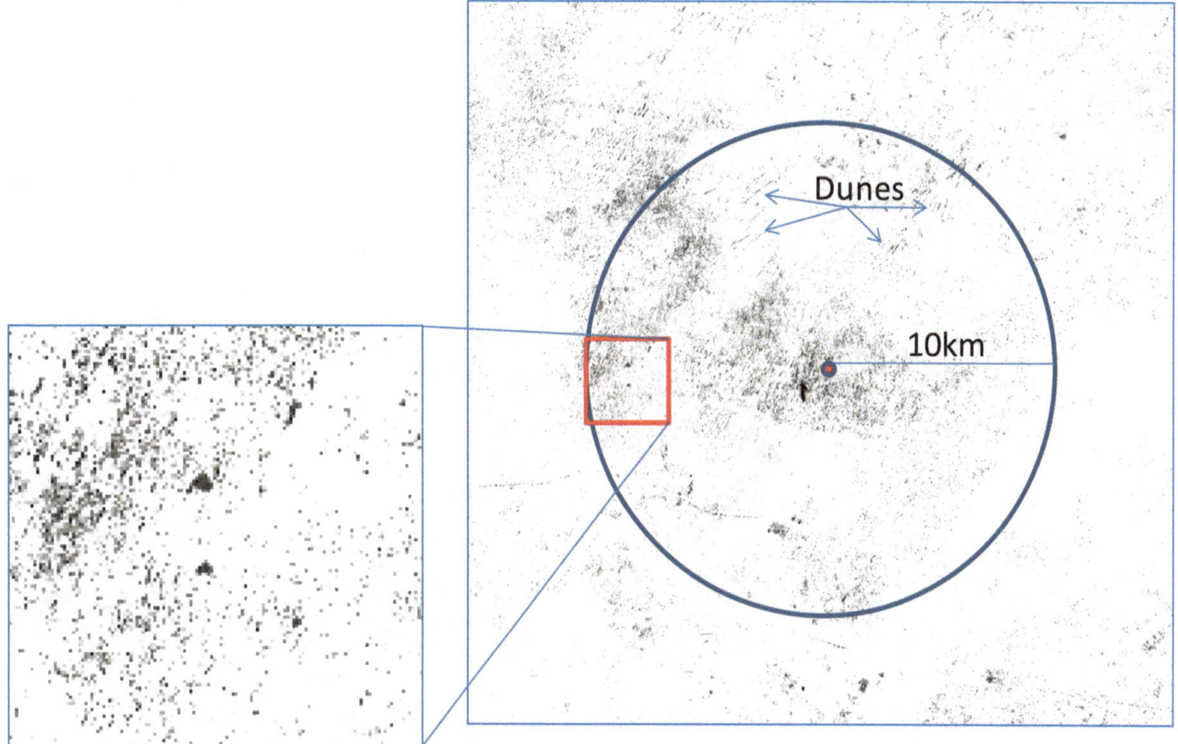

Figure 9
Change detection image for the 30 km × 30 km sets of Landsat data for the 13 May 1998 Indian test (*right*). The detection threshold has been turned down to attempt to detect features from the lower yield test. The higher yield 11 May test feature is near the center of the data shown by the *red dot*. The *circle* indicates the publicly announced approximate separation between the 11 May test and the 13 May test. The two features in the *red box* are candidate locations for inspection for the 13 May test since they show significant change, they are within the stated range of the first test, and they do not correspond to false positives from geographical features

- 22 September 2006 (before 1st test),
- 24 October 2006 (after 1st test, before 2nd test),
- 24 August 2007 (after 1st test, before 2nd test),
- 25 September 2007 (after 1st test, before 2nd test),
- 29 August 2009 (after both tests),
- 29 September 2009 (after both tests),
- 30 September 2009 (after both tests).

Some partially clouded Landsat data may also be available, but adapting the algorithms to account for clouds in some of the data was beyond the scope of this initial effort. It is important to note that the presence of clouds can introduce false positives (detections unrelated to the UNE) for the change detection algorithm, and false negatives (missed detections) for the spectral/spatial algorithm. Also, Landsat 7 data is available for some of this time, but there were problems with the scan mirror on that satellite (the data used here is from Landsat 5), which would have required additional work to attempt to correct for the pointing problems and would likely have resulted in artifacts in the change detection algorithms due to imperfections in pixel co-registration.

Because of the geography and significant plant cover in the region, the North Korean data has larger seasonal variability than the other data analyzed (see Fig. 10). This is manifested as varying amounts of snow on the ground, seasonal variations in the vegetation (due to natural variations, logging, and farming), and variations in the shadowing in the imagery due to the mountainous terrain combined with seasonal changes in solar illumination, which impacts pixel signal levels and surface material categorization in the images.

There are two paths that might prove fruitful for demonstrating MSIR utility for the North Korean tests. First, algorithms might be developed that can use partially cloudy data and account for the seasonal

Figure 10
Comparison of two sets of color imagery (710 m × 840 m) of the North Korean test area, before (*left*) and after (*right*) the two tests. The *left* image was taken in February in the middle of winter with the trees and ground bare, and the *right* in the Fall when the trees and the ground were still vegetated. The *white material* in the *right* image appears to be mining tailings. The *right* image is a false color image generated from the *red*, *green*, and *blue* channels of the 4-channel spectral data from GeoEye-1

effects observed. Second, other satellite data sources might be used, which would mitigate the data availability problem, and might provide inherently more useful data depending on the spatial and spectral resolution. We have started investigating this second path, as described in the next section.

6.5.2 GeoEye-1 North Korean Data

We acquired GeoEye-1 multi-spectral imagery of the North Korean test area with an acquisition date of 12 October 2009, about 5 months after the second North Korean nuclear test. This is relatively high spatial resolution (approximately 3 m) with a nearly 30 km swath width. (These are about twice the nominal satellite values of 1.6 m and 16 km swath width since this imagery was apparently acquired at a significant slant angle.) GeoEye-1 provides panchromatic imagery and 4-band multispectral imagery (blue, green, red, and near-infrared) with four times the pixel size as the panchromatic imagery. We ran the spatial/spectral algorithm to detect anomalies and identified several facilities of potential interest. While these anomalies are apparent from visual inspection of a color presentation of the imagery (subset shown in Fig. 10), the difficulty is that visual inspection results in hundreds of features of comparable apparent significance, often small buildings or clearings in the trees. The spatial/spectral algorithm reduces this number by a factor of 100 or more.

Figure 10 shows Google imagery from February 2005 and GeoEye-1 multispectral imagery from October 2009. The first thing to note is how starkly different the two images are, even though they are both nominally true color images. The difference is that the February 2005 data is in winter with bare ground and bare trees. The October 2009 data still has vegetation present. These large seasonal differences make change detection very difficult with this location. The bright material is probably mining tailings, and is the spatial/spectral observable that caused this portion of the 30 km × 30 km image to be flagged as a region of interest.

It is important to note that this analysis of the GeoEye-1 data used only a single post-event data set to find anomalies, and that the spatial resolution is approaching that expected of airborne MSIR measurements, where sub-meter spatial resolution is readily achievable. Further, the spatial/spectral algorithm highlighted regions of interest automatically, so it was not necessary to evaluate the entire data set, only the regions-of-interest identified.

7. Discussion and Path Forward

These early results of the analysis of Landsat Thematic Mapper data show that MSIR observables can be detected from overhead measurements, and that they can be correlated with the likely location of an underground nuclear test. In some cases this is due to disturbed earth, and in other cases it appears human activities generated the observable. It is also very encouraging that two of the early algorithms used here do not require pre-event data, and hence are relevant to the OSI problem where aircraft data will only be acquired post-event. Also relevant is that the regions of interest are few in number and roughly 1 km^2 in size or smaller, which is a significant reduction from the nominal 1,000 km^2 initial search area.

The Landsat data were only likely to find disturbed earth observables, and large-scale human activity observables because the 30 m pixel size is only likely to detect features on that scale or larger. While plant stress might be observed at spatial scales comparable to the disturbed earth observables, it may also require 0.5 m spatial resolution to separate the vegetation from the underlying ground materials. Similarly, many human artifacts would not be resolved with 30 m data. The GeoEye-1 data show that higher spatial resolution is useful for detecting human activities.

Despite these limitations, this work has shown that MSIR can find useful observables with several detection algorithms and a low false positive rate, and that the resulting regions of interest can be used in conjunction with high-resolution visible imagery to provide further discrimination between regions warranting further investigation and those unlikely to be relevant, such as farmlands. The fact that the change detection and the spatial/spectral algorithms identify the same regions using different techniques and different subsets of the data provides further confidence in these results.

The GeoEye-1 data show that higher spatial resolution is useful for detecting human activities. However, this data had only four spectral channels, and more spectral channels with higher spatial resolution may provide additional benefit for identifying regions of interest and eliminating false positives. We have started to look at sets of GeoEye-1 and other comparable satellite data (such as that from QuickBird) with the hope that the time series of data can be used to identify those features which are of highest interest, and that information can in turn be used to refine the search parameters so the high-interest regions can be more reliably identified from a single data set.

Finally, it is important to note that an important concern for the maturation of MSIR is to develop the requirements and deployment CONOPS for the MSIR instrumentation that might be flown as part of an OSI. The improvement in ability to identify regions of interest in going from 30 to 3 m spatial resolution is dramatic, despite the reduction of spectral channels from 6 to 4. This shows that it is likely that aircraft data will have significant utility for reducing the search area in an OSI, and that use of other data with higher spatial resolution and more and different spectral channels might go a long way toward developing the MSIR instrument requirements for an OSI. In the absence of a nuclear test where a post-event aircraft overflight with an MSIR instrument is allowed, a combination of satellite data on UNE's and other events (e.g., mining blasts or earthquakes) along with aircraft data on surrogate events or activities will have to be used to develop the MSIR instrument requirements.

Acknowledgments

This work was performed under the auspices of the U.S. Department of Energy by Lawrence Livermore National Laboratory under Contract DE-AC52-07NA27344.

References

Barker, B., *et al.*, "Monitoring Nuclear Tests", Science, Vol. *281*, 25 September 1998, pp. 1967–68.

Bras, R. L., Hampton, T., Coyne, J., Bobrov, D. and Zerbo, L., "CTBTO seismic processing and the announced DPRK nuclear test of October 9, 2006", Geophysical Research Abstracts, Vol. *9*, 07286, 2007.

Canty, M. J., Jasani, B., and Schlittenhardt, J., "Wide Area Change Detection with Satellite Imagery for Locating Underground Nuclear Testing", Symposium on International Safeguards Verification and Nuclear Material Security 2001. IAEA publication IAEA-SM-367/16/02.

Canty 2001b Canty, M. J., Schlittenhardt, J., "Satellite data used to locate site of 1998 Indian nuclear test", Eos Trans. AGU, *82*(3), 25–29 (2001).

Canty, M. J., and Nielsen, A. A., "Visualization and unsupervised classification of changes in multispectral satellite imagery", International Journal of Remote Sensing *27*(18), 3961–3975 (2006).

Cong, X., Schlittenhardt, J., Gutjahr, K., Soergel, U., Canty, M., and Nielsen, A., "Using differential SAR interferometry for the measurement of surface displacement caused by underground nuclear explosions and comparison with optical change detection results", Global Monitoring for Security and Stability (GMOSS) Integrated Scientific and Technological Research Supporting Security Aspects of the European Union (edited by G. Zeug & M. Pesaresi), European Commission - Joint Research Centre, 282–293 (2007).

CTBT 1996. Text of the Comprehensive Nuclear Test Ban Treaty, Protocol Paragraphs 69b and 80.

CTBTO 25 May 2009 press release. (http://www.ctbto.org/press-centre/press-releases/2009/ctbtos-initial-findings-on-the-dprks-2009-announced-nuclear-test/).

Eisler, J. D. and Chilton, F., "Spalling of the Earth's Surface by Underground Nuclear Explosions," Journal of Geophysical Research, Vol *69*, No. 24, p5285, 1964.

Hall, G. E. M., Vaive, J. E., and Button, P., "Detection of past underground nuclear events by geochemical signatures in soils," Journal of Geochemical Exploration, 1529 (1997).

Henderson, J. R., Primer on use of Multi-Spectral and Infra Red Imaging for On-Site Inspections, LLNL report LLNL-TR-463081, October 2010.

Jasani, B, "Could civil satellites monitor nuclear tests?", Space Policy, February 1995, 11(1), pp 31–40.

Jasani, B, "Contribution of Remote Sensing Satellite to CTBT Verifiability", Report of the Independent Committee on the Verifiability of the CTBT, Nov 2000. (http://www.ctbtcommission.org/jasanipaper.htm).

Jasani, B., and Canty, M., "Change detection methods applied to observation of nuclear tests by commercial remote sensing satellites", 3rd Workshop on Science and modern technology for safeguards, 13-16 November 2000, Tokyo, Japan, Proceedings, European Communities, Report EUR 19943 EN, pp 189–97 (2001).

Malpica, J. A. and Alonso, M. C., "A method for change detection with multi-temporal satellite images using the RX algorithm", The International Archives of the Photogrammetry, Remote Sensing and Spatial Information Sciences. Vol. XXXVII. Part B7. Beijing 2008.

Mitchley, M., Sears, M., and Damelin, S., "Target detection in hyperspectral mineral data using wavelet analysis", IGARSS(4): 881–884 (2009).

Myint, S., "A robust texture analysis and classification approach for urban land-use and land-cover feature discrimination", Geocarto International, 16(4):27–38 (2001).

Nussbaum, S., Niemeyer, I., and Canter, M. J., "Feature recognition in the context of automated object-oriented analysis of remote sensing data monitoring Iranian nuclear sites," Proc SPIE 5988 598805 (2005).

Pickles, W. L., Observations of Temporary Plant Stress Induced by the Surface Shock of a 1-kt Underground Chemical Explosion, LLNL, UCRL-ID-122557, December 1995.

Reed, I. S., and Yu, X., "Adaptive multiple-band CFAR detection of an optical pattern with unknown spectral distribution", IEEE Transactions on Acoustics, Speech and Signal Processing, *38*, pp. 1760–1770 (1990).

Rivard, B., Feng, J., Gallie, A., Sanchez-Azofeifa, "Continuous wavelets for the improved use of spectral libraries and hyperspectral data", *Remote Sensing of the Environment*, *112*: 2850–2862 (2008).

Rockett, P., "Multi-spectral Imaging in a CTBT OSI," On-Site Inspection Workshop-5, November 1999.

Schlittenhardt, J., Canty, M., Grünberg, I. "Satellite Earth observations support CTBT monitoring: a case study of the nuclear test in North Korea of Oct. 9, 2006 and comparison with seismic results", Pure Appl. Geophys., *167*, 601–618, doi: 10.1007/s00024-009-0036-x (2010). http://www.springerlink.com/content/0t333072860x7521/.

Srinivas, A. and Wu, Y., "Multiresolution histograms for SVM-based texture classification", *Computer Science*, *3656*:754–761 (2005).

Subotic, N., Thelen, B., Gorman, J. and Reiley, M., "Multiresolution detection of coherent radar targets", IEEE *Trans. Image Processing* Vol. *6*, no.1, pp. 21–35 (1997).

Thomas, A., "Extending the RX anomaly detection algorithm to continuous spectral and spatial domains", Proc. of IEEE Southeast Con, pp. 557–562 (2008).

Thyagarajan, K. and Patterson, R., "Systems and methods of using spatial/spectral/temporal imaging for hidden or buried explosive detection", U.S. Patent Appl. No. 12/613,430 U.S. Pub. No. 2010/0166330 (2010).

Van der Werff, H. and Lucieer, A., "A contextual algorithm for detection of mineral alteration halos with hyperspectral remote sensing", Remote Sensing Image Analysis: Including the spatial domain, Chapter 11, 201–210 (2006).

Vincent, P., Larsen, S., Galloway, D., Laczniak, R. J., Walter, W. R., Foxall, W., and Zucca, J. J., "New signatures of underground nuclear tests revealed by satellite radar interferometry", Geophysical Research Letters, Vol. *30*, NO. 22, 2141, doi:10.1029/2003GL018179, 2003.

Vincent, P., Buckley, S. M., Yang, D., and Carle, S. F., "Anomalous transient uplift observed at the Lop Nor,China nuclear test site using satellite radar interferometry time-series analysis", Geophysical Research Letters, Vol. *38*, L23306, doi: 10.1029/2011GL049302, 2011.

Wallace, T. C., 1998. "The May 1998 India and Pakistan Nuclear Tests", Seismological Research Letters, September 1998.

Walter, W. R., Rodgers, A. J., Mayeda, K., Meyers, S. C., Pasyanos, M., and Denny, M., "Preliminary Regional Seismic Analysis of Nuclear Explosions and Earthquakes in Southwest Asia", LLNL, UCRL-JC-130745, July 1998.

Zhang, X., Younan, N., and O'Hara, C., "Wavelet domain statistical hyperspectral soil texture classification", IEEE *Trans. Geosci. Remote Sens*. Vol. *43*, no. 3, pp. 615–618 (2005).

(Received November 22, 2011, revised June 8, 2012, accepted August 12, 2012, Published online September 22, 2012)